STEEL
BUILDINGS
ANALYSIS AND DESIGN

4th Edition

STANLEY W. CRAWLEY, M.ARCH, P.E.
Architect and Structural Engineer,
Professor of Architecture,
University of Utah

ROBERT M. DILLON, M.A.ARCH.; AIA, M.ASCE, AIC,
Architect and Professional Member, AISC

The Evans & Sutherland Building, Salt Lake City, Utah

Architects: Erlich-Rominger Architects AIA, Los Altos, CA
Structural Engineers: Reaveley Engineers & Associates, Inc.
Salt Lake City, Utah

Photo Courtesy of VULCRAFT

STEEL
BUILDINGS
ANALYSIS AND DESIGN

John Wiley & Sons, Inc.
New York   Chichester   Brisbane   Toronto   Singapore

In recognition of the importance of preserving what has been
written, it is a policy of John Wiley & Sons, Inc., to have books
of enduring value published in the United States printed on
acid-free paper, and we exert our best efforts to that end.

This publication is designed to provide accurate and
authoritative information in regard to the subject
matter covered. It is sold with the understanding that
the publisher is not engaged in rendering legal, accounting,
or other professional services. If legal advice or other
expert assistance is required, the services of a competent
professional person should be sought. *From a Declaration
of Principles jointly adopted by a Committee of the
American Bar Association and a Committee of Publishers.*

Library of Congress Cataloging in Publication Data:
Crawley, Stanley W.
     Steel buildings: analysis and design/Stanley W. Crawley, Robert
M. Dillon.—4th ed.
          p.   cm.
     Includes index.
     ISBN 0-471-84298-2
     1. Building, Iron and steel.   2. Structural engineering.
I. Dillon, Robert M. (Robert Morton), 1923–   . II. Title.
TH1611.C73   1993
624.1′821—dc20   92-27769
                                                              CIP

Printed in the United States of America

10 9 8 7

Preface

The purpose of this text is to present the general principles of structural analysis and their application to the design of the more common types of low- and intermediate-height building frames. This fourth edition retains the general scope and method of presentation of the first three editions; however, the material has been updated and, where necessary, extended.

A knowledge of the elementary principles of statics and strength of materials continues to be assumed, although much of this material is reviewed throughout the text as a prelude to design. Again, no attempt has been made to include complete tables of the properties of structural shapes or to give the text of pertinent codes, standards, and specifications. Sufficient information and data of this type are included, however, to enable the reader to follow the presentation without difficulty. This is not the case with many of the problems. Apart from the impracticality of supplying additional information and data of this kind, it is important that the reader become familiar as early as possible with the various reference materials used in practice.

The American Institute of Steel Construction (AISC) *Manual of Steel Construction: Allowable Stress Design* is recommended for use in connection with this text. Most of the discussion and illustrative examples, as well as answers to problems, have been keyed to the 1989, (ninth) edition of the *Manual* and to the June 1, 1989 AISC Specification, which it contains. Reference also is made to the major model building codes, to the ASCE, ASTM, and BSSC (American Society of Civil Engineers, American Society for Testing Materials, and Building Seismic Safety Council, respectively), to the Steel Joist Institute, and to the AISI (American Iron and Steel Institute), where their recommended re-

quirements or procedures seem appropriate. Even though effort has been made to include the most recent data, it can be expected that code, standard, and specification changes will continue to be made as these organizations endeavor to keep abreast of newly acquired knowledge and experience. These newer references should, of course, be sought out, evaluated, and used. It is our intention to update this text again when new information becomes available to an extent that major change is necessary, and when further or more rigorous analytical and design techniques are in need of assimilation, e.g., to extend the treatment of load- and resistance-factor design as it comes into more general usage.

As noted, the material and method of presentation in this new edition have remained basically unchanged. As in the first three editions, our intent is to bridge the gap between academic work and professional practice, i.e., to carry analysis and design beyond that applied to individual members and components and into total structural frames.

Professor Winfred O. Carter, Ph.D., Utah State University, Logan, Utah, is again recognized for his considerable contribution to the original development of the presentation in Chapter 12 and Appendices D and E. The computer programs in Appendices D and E were initially created by Dr. Carter. Since the earlier editions, however, these programs have undergone significant change. Credit for these changes is due Dr. Wayne Rossberg, Adjunct Assistant Professor of Architecture at the University of Utah. Dr. Rossberg modified these programs to make them conform more closely to the FORTRAN 77 standard, and to improve their user interface. These two programs—i.e., the programs presented in Appendices D and E—have broad application and have been provided on disk in the inside back cover of this text. Also included on each disk is an example problem using that program— truss and frame. These programs can be run on any PC using MS-DOS 2.0 or higher.

We express our appreciation to the users of the third edition who have provided helpful suggestions.

Finally, appreciation is expressed to the American Society of Civil Engineers (which has now supplanted the American National Standards Institute as the promulga- tor of the accepted national standard on minimum design loads for buildings), and once again to the American Institute of Steel Construction, the American Iron and Steel Institute, the Steel Joist Institute, the NIBS Building Seismic Safety Council, U.S. Naval Facilities Engineering Com- mand, and the model-building-code orga- nizations, for their cooperation in making inclusion of reference materials possible.

Stanley W. Crawley
Robert M. Dillon

Contents

STEEL
BUILDINGS
ANALYSIS AND DESIGN

1

GENERAL CONSIDERATIONS

1.1 Introduction

Modern buildings are constructed in many different ways. However, when speaking of buildings in a structural sense, the majority can be classified as wall-bearing, skeleton-frame, or a combination of both. With wall-bearing construction, floors and roof are supported by load-bearing walls; the thicknesses of the walls are determined largely by the number of stories and the magnitude of the loads thus brought to them for support. With skeleton-frame construction, walls as well as floors and roof are supported by a structural framework of beams, girders, columns, and similar members.

Generally speaking, wall-bearing construction has ceased to be the method of choice in the United States except in the case of certain low-rise building types. Indeed, it was the practical and economical limitations of such construction when applied to medium-rise and high-rise buildings that led to the development of the skeleton-frame. It is true, however, that with the emergence of higher-strength masonry units and, particularly, high-bond mortars, over the past 20 years or so there has been some resurgence of interest in wall-bearing high-rise construction. There also has been interest in a variety of other construction concepts that do not rely, or rely only in part, on a skeleton-frame for support—e.g., various load-bearing, panelized, and volumetric module solutions for low-rise buildings and even for buildings of considerable height. In addition, it can be anticipated that continued attention will be given in years ahead to construction concepts based on dimensionally and functionally coordinated *subsystems*—i.e., exterior wall, floor-ceiling, partitioning, and similar building elements that can be assembled to create buildings of a wide variety of sizes and types. Structural integration will continue to be extremely important, and the skeleton-frame can be expected to play an important role. The skeleton-frame, therefore, will probably continue to be as significant in the future as it is in today's construction. Because the exterior walls in skeleton-frame construction are relieved of a load-carrying function, they serve primarily as an enclosing environmental control envelope—i.e., to control light, temperature, sound, moisture, etc. Freeing the exterior walls of a primary structural function permits a much wider choice of materials and methods of fabrication. The skeleton-frame by also freeing interior walls and partitions of a primary structural role, allows much greater flexibility in architectural planning, including the planning for changes in interior space. The skeleton-frame offers advantages in terms of ease of fabrication and transport, speed of erection, and coordination of building trades as well. All of these characteristics are invaluable in meeting today's complex building requirements.

Although the structural frame for any given building will be developed from the standpoints of structural adequacy and economy, column locations and spacing will usually be determined by architectural considerations arising from anticipated as well as immediate occupancy requirements. It is important to recognize that, just as there is generally more than one architectural scheme that will satisfy occupancy requirements, so also will there be several satisfactory structural solutions. The architectural and structural schemes should be compatible, i.e., they should "build" well without resorting to unduly complex and extravagant structural arrangements. (See also Art. 13.1.)

In multistory buildings of skeleton-frame construction, the most economical center-to-center column spacing for average loads is on the order of 22 to 28 ft. Spacings less than 20 ft are seldom economical from a structural standpoint, and those 30 ft and more, even in one direction, can usually be justified only when occupancy considerations call for the greater unobstructed floor area that such spans provide. These guidelines do not necessarily apply to high-rise (tall) buildings—such buildings frequently pose unique structural problems. Although as mentioned in the Preface, the principles developed in this text are generally applicable to all types of steel frame buildings, structural design emphasis is placed upon the more common types such as commercial and industrial buildings, apartment houses, schools, hospitals, and similar structures.

1.2
Overall Design Procedure

To reiterate, it is of utmost importance that the structural design of a building be coordinated with both the architectural scheme and the mechanical-electrical-electronic requirements from the inception of the project. The general arrangement of floor framing, and especially the placement of columns, should be borne in mind during development of the architectural scheme. Preliminary framing plans should be made and column dimensions approximated before a final scheme is adopted. This is necessary because the size of columns and the clearances required, especially in the lower stories, may materially affect the architectural layout. As soon as the floor framing arrangement has been determined, beams and girders can be designed, followed by the final design of columns and foundations. Throughout development of the structural design, the architectural, structural, and mechanical-electrical-electronic plans must constantly be checked against one another to ensure accuracy and overall efficiency of design and construction, and ultimate building operation.

Although for small buildings the structural framing is sometimes shown directly on the architectural plans, this practice is not recommended. For projects of any appreciable size, a separate set of framing plans is essential if the location, size, and joining of structural members are to be recorded and made readable without the confusion of other information. On all but the smallest projects, the modest amount of time and effort required to provide separate framing plans is well worth the effort. The general character of framing plans and their relationship to the architectural drawings will be readily apparent from a brief study of the set of working drawings for the building presented in Chapter 13.

1.3
Design Loads

The loads for which a building is designed are classified as dead, live, and environmental. In the past, environmental loads were frequently considered part of the live load—e.g., those due to wind, snow, rain, earthquake forces, and soil and hydrostatic pressures (the latter two acting horizontally on walls below grade). However, current practice tends to recognize these as three distinct classifications.

Dead Loads / These are the loads due to the weight of the permanent parts of the building such as floors, beams, girders, walls, roof, columns, stairways, fixed partitions, etc., and include fixed service equipment such as mechanical, electrical, and electronic system components, water tanks, and similar items supported by the structure. The weights of different building materials to be used in determining the dead loads are specified in local building codes.

Where such data are incomplete, or in localities where no code is operative, the reader is referred to the comprehensive lists given in the model building codes or the American Society of Civil Engineer's *Minimum Design Loads for Buildings and Other Structures*, ASCE 7-88[1] (formerly ANSI A58.1). For example, a list of cubic-foot weights of representative basic building materials, most of which are as recommended in that publication, is given in Table 1.1. The net effect of any prestressing of members also is classified as a part of the dead load.

Live Loads / These represent the probable loads on the building structure due to occupancy and use, and generally are considered to be uniformly distributed over the floor area. The value assessed, expressed in pounds per square foot (psf), is enough to cover the effect of ordinary concentrations that may occur. Except for the dead load, all vertical loads are included, e.g., the weight of occupants, furniture, other than fixed equipment, and stored materials. On roof surfaces, it includes an allowance for maintenance workers and equipment, and other movable objects and people. Buildings that will contain heavy machinery or similar large concentrations of live load must, of course, be designed specifically for such concentrations.

There continues to be some lack of uniformity among different building codes as to proper live-load allowances for various types of occupancy; however, there are continuing efforts by a number of organizations to bring about an ever-higher degree of uniformity.[2] Presented in Table 1.2

---

[1]Available from ASCE, 345 E. 47th St., New York, N.Y. 10017-2398.

[2]Some of these are: the Council of American Building Officials (CABO), including the International

Table 1.1
Weights of Typical Building Materials

Materials	Pounds per Cubic Foot	Materials	Pounds per Cubic Foot
Cast-stone masonry		Masonry, rubble stone:	
(cement, stone, sand)	144	Granite	153
Cinder fill	57	Limestone, crystalline	147
Concrete, plain:		Limestone, oolitic	138
Cinder	108	Marble	156
Expanded-slag aggregate	100	Sandstone	137
Haydite (burned-clay aggregate)	90	Metals:	
Slag	132	Aluminum	165
Stone (including gravel)	144	Brass	526
Vermiculite and perlite		Bronze	552
aggregate, non-load-		Lead	710
bearing	25–50	Steel, cold-drawn	492
Other light aggregate,		Mortar, cement, or lime	130
load-bearing	70–105	Terra cotta, architectural:	
Concrete, reinforced:		Voids filled	120
Cinder	111	Voids unfilled	72
Slag	138	Timber, seasoned:	
Stone (including gravel)	150	Ash, commercial white	41
Earth	63–120	Cypress, southern	34
Masonry, ashlar:		Fir, Douglas, Coast	
Granite	165	region	34
Limestone, crystalline	165	Oak, commercial reds and	
Limestone, oolitic	135	whites	47
Marble	173	Pine, southern yellow	37
Sandstone	144	Redwood	28
Masonry, brick:		Spruce, red, white, and	
Hard (low absorption)	130	Sitka	29
Medium (medium absorption)	115	Western hemlock	32
Soft (high absorption)	100		

Conference of Building Officials (ICBO), Building Officials and Code Administrators, International (BOCA), and Southern Building Code Congress, International (SBCC); and the American Society of Civil Engineers (ASCE). In addition, under the auspices of the National Institute of Building Sciences, and with financial support by the Federal Emergency Management Agency, the Building Seismic Safety Council has developed recommended building seismic safety standards as part of the National Earthquake Hazard Reduction Program (NEHRP), the latest version of which (1992) is referenced herein. Currently, the ICBO, headquartered in California, has achieved state and local governmental adoptions of its code principally in the Far West; the SBCC, principally in the South and Southwest; and the BOCA, principally in the North Central and Northeast. However, there is a considerable overlap.

are typical recommended live-load allowances for various occupancies abstracted from several well-known model codes and standards. A reduction in these live loads generally is permitted for large areas.

Some codes also require that provision be made in the live loads of office and loft buildings for the effect of partitions, which may be either movable or not fixed as to location until after the building is erected. When this is the case, an allowance of 20 psf of floor area is often used. In addition, codes usually stipulate minimum concentrated live loads.

Table 1.2
Typical Minimum Live-Load Requirements

Classes of Occupancy	Minimum Live Loads per Square Foot (psf) of Floor Area			
	Std. Bldg. Code 1989[a]	Basic Bldg. Code 1989[b]	Uniform Bldg. Code 1991[c]	Amer. Soc. of Civil Engrs. Design Loads 1988[d]
Assembly				
Fixed seats	50	50	50	60
Movable seats	100	100	100	100
Dwellings	40[e]	40	40	
First floor	—	—	—	40[i]
Habitable second floor	—	30	—	30[i]
Garages				
Passenger car	120[f]	50	50[f]	50
Other	—[h]	—[f]	100[f]	—[f]
Hospitals				
Private rooms	40	40	—	40
Operating rooms	60	60	—[f]	60
X-ray rooms	—[g]	—	—[f]	—
Hotels				
Guest rooms	40	40	40	40
Manufacturing				
Light	100	125	75[f]	125
Heavy (factories)	150	250	125[f]	250
Office space				
Typical rooms	50	50	50	50
Schools				
Class rooms	40	40	40	40
Sidewalks				
Over areaways, etc.	200[f]	250	250[f]	250
Stores				
Retail	75	75–100	75[f]	75–100
Wholesale	100	100	100[f]	125
Warehouses	125–250	125–250	125–250	125–250

[a]Southern Building Code Congress, International Inc. (SBCC).

[b]Building Officials and Code Administrators, International (BOCA).

[c]International Conference of Building Officials (ICBO).

[d]American Society of Civil Engineers (ASCE); *Minimum Design Loads for Buildings and Other Structures*, ASCE 7-88 (formerly ANSI A58.1).

[e]Except sleeping rooms and attics with storage, where the specified load is 30 psf.

[f]Requirements for maximum wheel loads, and/or concentrated or special loads.

[g]To be approved by the building official.

[h]Use AASHTO lane loads.

[i]Habitable attics and sleeping areas, 30 psf; all other areas, 40 psf.

Note: In most codes there are additional classes and/or subclasses of occupancy; also, in many cases, there are additional qualifiers that could be shown. The actual documents should be referred to.

Special provision also must be made for impact loads such as those caused by the operation of elevators. It is common practice, for example, to provide for this impact effect in the design of beams, girders, and the first tier of columns supporting elevator machinery, by increasing the actual loads a specified percentage—e.g., 100 per cent. Also, it usually is required that accessible roof-supporting members be designed to support a concentrated load of some 2000 lb that may be suspended from them.

Environmental Loads / Included under this load classification are those previously mentioned—i.e., wind, snow, rain, earthquake forces, soil and hydrostatic pressures, and self-restraining forces such as those caused by temperature and moisture.

Building codes generally specify minimum wind pressures which a building must be designed to withstand. The design of buildings for wind is introduced in Chapter 8.

The snow load is an essential element of the load on roofs in many geographical areas. Maps indicating ground snow loads in pounds force per square foot for various parts of the United States are shown in ASCE 7-88. Snow loads and their computation are discussed in greater detail in Chapter 9. Rain loads, ponding of water on roofs, and the effects of rain on snow are also treated in Chapter 9.

In many localities, earthquake resistance is a critical factor, and building codes in these areas generally require earthquake-resistant design and/or special details. Such attention to earthquake effects is well known in the western United States. Loads resulting from earthquakes are discussed in some detail in Chapter 8.[3]

[3]The previously cited Building Seismic Safety Council (BSSC) was created in 1979 under the auspices of

1.4 Working Stresses

The term *working stress* is a carry-over from early design procedures and can be somewhat loosely defined as the unit stress to which the steel will be subjected in actual use (based on elastic performance of the structure). The *allowable* working stress is considerably less than the breaking strength of the steel. The allowable working stresses recommended by the American Institute of Steel Construction (AISC) are widely accepted throughout the United States. These stress levels are listed in the 1989 AISC "Specification for Structural Steel Buildings: Allowable Stress Design and Plastic Design (ASD).[4]

In the 1989 AISC Specification, the allowable working stress is listed as a percentage[5] of the *yield* stress, which varies with the type and grade of steel. Since the yield stress is closely associated with that point where permanent deformation takes place, it is a basic and important physical property of the material and a guide to its design strength (Appendix A). Furthermore, the allowable working stress de-

the National Institute of Building Sciences (NIBS) to bring together the many U.S. building-community organizations to foster the development and appropriate applications of improved seismic safety provisions in building design and regulation throughout the United States. Evidence of this work and that of the National Bureau of Standards, which is involved in the BSSC work and provided the Secretariat for the former ANSI A58.1, is already apparent in the content of the ASCE 7-88 standard, and the earthquake provisions of the BOCA and SBCC model codes. The provisions of the UBC also are becoming more consistent with those recommended in the NEHRP by the BSSC.

[4]Contained in the *Manual of Steel Construction: Allowable Stress Design*, American Institute of Steel Construction, Ninth Edition, and available separately from AISC, 1 East Wacker Dr., Suite 3100, Chicago, Illinois 60601.

[5]Under certain circumstances, the working stress is listed as a function of the modulus of elasticity.

pends on the type of stress under consideration, i.e., axial tension, bending, shear, etc. For example, except in the case of plates over 8 in. in thickness, a steel bearing the designation A36 (Art. 1.6) has a yield point of 36,000 lb per sq in. (psi) and, under certain circumstances, would have an allowable working stress in bending that is 60 percent of 36,000, or 21,600 psi, which is rounded off to 22,000 psi. It is of interest to note that the tensile breaking strength of this same steel is between 58,000 and 80,000 psi.

Although the design of light-gage steel members is not specifically discussed in this text (except for joists, covered in Chapter 5), the attention of the reader is directed to the following applicable specifications:

Standard Specifications for Open Web Steel Joists, issued by the Steel Joist Institute.

Specification for the Design of Light Gage Cold-Formed Steel Structural Members and *Specification for the Design of Light Gage Cold-Formed Stainless Steel Structural Members*, both issued by the American Iron and Steel Institute.

The 1989 AISC specification is still quite recent; therefore, the designer may still encounter local building-code jurisdictions that do not permit its stress values. The local building code should always be consulted in actual design work.

1.5
Factor of Safety

No structural member of a building frame is ever designed to carry a load that will develop its full ultimate strength under normal service conditions. There are too many elements of uncertainty, both as to loading and as to uniformity in quality of materials and construction, to permit such

a degree of precision in structural design. Consequently, some margin of safety must be provided, and in allowable-stress design this is accomplished by setting allowable working stresses at values well below the ultimate strength. The ratio between ultimate strength and working stress has been defined as the *factor of safety*. However, this definition is not wholly satisfactory, since failure of a structural member in a building actually begins when the stress exceeds the yield point (or more precisely the elastic limit). This is due to the fact that deformations produced by stresses above this value are *permanent* and thus change the shape of the structure, even though there may be no danger of collapse. However, even though there is no general agreement on an exact definition of *factor of safety*, the above discussion will serve to indicate the concept. The relationships among ultimate strength, elastic limit, yield point, and deformation under stress are reviewed briefly in Appendix A and in Chapter 11, LRFD and Plastic Design.

1.6
Structural Design Procedures

The AISC now supports two structural design procedures for buildings, as distinct from methods of analysis. These are known as ASD (allowable-stress design) and LRFD (load- and resistance-factor design).[6]

Beginning with the first AISC manual in 1923, and until 1989, there was only one version of the AISC manual and accompanying specification, the title of the manual being simply *Manual of Steel Construction*. Beginning in 1986, however, the AISC published the first edition of a new manual and specification based on load- and resistance-factor design procedures. This new

[6]Or simply "stress design" and "strength design."

version was titled *Manual of Steel Construction*: *Load and Resistance Factor Design*. Then in 1989, when the ninth edition of the standard AISC manual and accompanying specification were published, its title was changed to *Manual of Steel Construction*: *Allowable Stress Design*. ASD is the design approach that has been used in all previous editions of this text and is the one thus far referred to in this chapter. Because ASD is still the approach most in use by the design profession, it is the one that will principally be used in this, the fourth edition. However, LRFD is being recognized increasingly in standards and codes as well as design offices. Indeed, many who have used this approach feel that it is easier, and in many respects more reliable, in the way both loads and resistances are factored. For this reason, some discussion of the nature of LRFD, and the differences between LRFD and ASD procedures, and between elastic and plastic analysis, is in order. Therefore, LRFD, which in Chapter 8 will only be referred to in relation to contemporary design for earthquake forces, will be introduced as an alternative design procedure in Chapter 11.

Regardless of the methods of analysis and design employed, however, one must never lose sight of the fact that although most types of design calculations are carried out with considerable precision, the assumptions—i.e., loads and the responses they engender—on which these calculations are based often are, to say the least, less than precise. This is why it is said over and over that *judgment* is an essential ingredient in any design endeavor.

1.7
Structural Steels

There is a wide variety of structural steels available to designers today. Major im-provements have been made in strength, ductility, and corrosive resistance. As noted in Art. 1.4, the yield stress becomes the index for structural strength, and for many years, the yield stress for structural steel was effectively limited to 33,000 psi. Today, structural steel is readily available in a range of yield stresses from 32,000 to 130,000 psi, and in the future it is expected to be available in yield stresses reaching 160,000 to 200,000 psi. Ductility is the ability of a material to flow with a constant (or nearly constant) stress and still maintain its strength. The greater the ductility, the more a structure can adjust to peak stress resultants, thus providing more reserve strength. This characteristic forms the basis for plastic design, which is presented in Chapter 11. It should be pointed out, however, that except for certain earthquake specifications, at present there is no required minimum ductility for steel structures.

Corrosion-resistant steels and new techniques for fireproofing steel members have significantly altered the appearance of steel buildings in recent years. Steels that are not corrosion-resistant cannot be left exposed to the weather without painting and maintenance; not only will they rust and look unsightly, but also the steel will lose strength as a result of corrosive action. However, when the new corrosion-resistant or "weathering" steels are used, a natural oxide coating quickly forms on exposed surfaces, protecting the steel much the same as painting. Care must be taken with architectural detailing, however, to be sure that moisture on such steel surfaces will be properly drained away so as not to streak windows or stain concrete and other surfaces.

In addition, it must always be remembered that steel will begin to lose its load-carrying ability when its temperature reaches 600°F and will have little strength left at 1200°F. Therefore, structural steel that could be exposed to such temperatures in fire situations must be protected. In the past, providing such protection has always implied insulating steel members by shielding or encasement with such materials as concrete (as shown in the examples in this text) and gypsum in the form of blocks, sheet, or plaster. Both life safety and property protection must be considered, and the direction of such efforts probably will be toward design approaches that deal with such variables as fire loads, ignition sources, flame and smoke generation and spread, fire suppression, and human as well as structural and vital-equipment isolation and shielding. The structural integrity of buildings obviously must be preserved until there is no danger to life either within or outside; and, because the repair and replacement of fire-damaged structural members can be both difficult and costly, structural-member protection must be commensurate with the risk of such damage. For the near future, however, code-stipulated encasement and shielding requirements based on fire experience, rather than design procedures, will undoubtedly continue to govern fire protection of structural steel members. New techniques are emerging, such as isolating steel members outside buildings when there is no danger of damaging fire exposure from adjacent structures and circulating liquid coolants within structural elements exposed to fire inside buildings; however, much more needs to be done on the total issue of building fire safety and fire protection.

It is customary practice to assign names to the various types of steel in accordance with the corresponding designation given by the American Society for Testing Materials (ASTM).[7] For example, all steels covered by the ASTM A36 "Specification for Structural Carbon Steel" are called A36

---

[7]Or simply ASTM.

Steel. Currently, the steel most frequently used for structural purposes is A36 Steel; therefore, it can be considered to be the standard. As noted in Art. 1.4, A36 Steel, in all structural shapes and in plates up to 8 in. thick, has a yield stress of 36,000 psi (36 ksi). Six types of steel most frequently used in buildings are listed in Table 1.3, together with their corresponding yield stress and other pertinent data. It will be noted from Table 1.3 that A242 and A588 Steels provide inherent resistance to corrosion. A514 Steel is the strongest and currently is available only in plate form; it also can be obtained in grades that are corrosion resistant. Another feature common to all steels except A36 is that the higher-stress grades apply to the smallest thicknesses. For example, A441 and A242 Steels have a 50-ksi yield stress only if the thickest part of the structural piece is not more than $\frac{3}{4}$ in.; the yield stress is reduced to 42 ksi when the thickness is over $1\frac{1}{2}$ in. but not more than 4 in.

The higher-strength steels understandably cost more; however, this does not mean that the overall cost of a structure will necessarily be higher if such steels are used. Steel prices are established on a per-pound basis, and when higher-strength steels are used, the resultant poundage of steel required can well be less. Following are selected steels and their approximate mill price in Salt Lake City (1992):

A36	25¢ per lb, rolled shapes
A572 (GR 65)	31¢ per lb, rolled shapes
A588 (GR 50)	29.5¢ per lb, rolled shapes

It is apparent, however, that some of the higher-strength steels also are more costly on a per-pound basis; therefore, it is not only the reduction in weight of steel used

Table 1.3
Structural Steels for Buildings

	ASTM Designation[a]	*Strength* *Minimum Yield Stress (ksi)*	*Strength* *Tensile Stress*[b] *(ksi)*	*Available forms (ASTM rolled shape groups*[c] *and plates*[d]*)*	
	A36	36	58–80[e]	All shapes, plates, bars	
	A441	40	60	No shapes	Plates over 4 to 8
	A441	42	63	Shape groups 4, 5	Plates over $1\frac{1}{2}$ to 4
	A441	46	67	Shape group 3	Plates over $\frac{3}{4}$ to $1\frac{1}{2}$
	A441	50	70	Shape groups 1, 2	Plates to $\frac{3}{4}$
Corrosion-resistant steel	A242	42	63	Shape groups 4, 5	Plates over $1\frac{1}{2}$ to 4
Corrosion-resistant steel	A242	46	67	Shape group 3	Plates over $\frac{3}{4}$ to $1\frac{1}{2}$
Corrosion-resistant steel	A242	50	70	Shape groups 1, 2	Plates to $\frac{3}{4}$
Corrosion-resistant steel	A588	42	63	No shapes	Plates over 5 to 8
Corrosion-resistant steel	A588	46	67	No shapes	Plates over 4 to 5
Corrosion-resistant steel	A588	50	70	All shapes	Plates to 4
	A572[f]	42	60	All shapes	Plates over $\frac{1}{2}$ to 6
	A572[f]	50	65	All shapes	Plates to 2
	A572[f]	60	75	Shape groups 1, 2	Plates to $1\frac{1}{4}$
	A514	90	100–130	No shapes	Plates over $2\frac{1}{2}$ to 6
	A514	100	110–130	No shapes	Plates to $2\frac{1}{2}$

[a]It should be noted that A440 Steel, which was not recommended for welding, is no longer produced.
[b]Minimum unless range is given.
[c]Shape groups are in accordance with ASTM A6 (Art. 3.6).
[d]All thicknesses are in inches.
[e]Minimum 58 ksi for shapes over 426 lb per ft.
[f]Also available in stress grade.

that must be considered in any decision to use the higher-strength steels. If the higher-strength steels could be used in such a manner that each would be allowed to develop the same percentage of its yield stress, then, for example, less A572 Steel would be needed than A36 Steel. The actual amount would depend on the type of critical stress involved. In all fairness, it must be pointed out that only in special cases will the more restrictive specifications applied to the higher-strength steels allow development of the same percentage of yield stress as with A36 Steel.

The average cost of the corrosion-resistant steels is somewhat greater for equal-strength steels and more than for A36 Steel. Therefore, if these steel qualities are being sought, the cost-benefit analysis will need to include other factors such as life-cycle maintenance costs. In any event, the basic price of the steel can never be the sole basis for choice decisions; the least expensive part of a steel structure is the material itself. To these basic prices must be added the freight costs to the fabricator, the fabrication itself, transport to the building site, and, finally, the erection. Often overlooked is the fact that some higher-strength steels are more costly to fabricate than A36 Steel.

A more useful figure in estimating ultimate costs would be the per-pound (or ton) price of the erected steel. Fabrication can easily double the initial price for the material from the steel producer, increasing the in-place cost even more. A more accurate measure might be the per-square-foot cost of the erected steel. The amount of steel in structures of average height and size[8] ranges from 2 to 12 psf, making the in-place cost per square foot, a rather significant factor. Finally, such other factors as the availability and quantity of material used, location of the site, simplicity and repetition of details, costs of

labor and equipment, and programmed time of construction can have a pervasive influence on ultimate costs.

For the type and height of buildings dealt with in this text, A36 Steel generally will be the economical choice. However, there may well be cases were a lesser depth of section is desirable—e.g., in columns or in beams to maintain height clearances—in which case, A572 Steel would be preferable. In the case of columns, this would be particularly true where there were unusually high transmitted loads; in the case of beams the same could be said so long as any added deflection would not pose a problem. And, of course, there may be aesthetic or other reasons for selecting the corrosion-resistant steels. However, as a general rule the higher-strength steels will have greater applicability in the taller buildings and in specialized applications such as prefabricated steel buildings.

In the final analysis, each must be evaluated on its own merits.

---

[8]Tall buildings are excluded, e.g., the Sears Tower in Chicago has a steel frame weighing 33 psf.

2

REACTIONS, SHEAR, AND BENDING MOMENT

2.1
Introduction

A beam is a structural member that is subjected to transverse loads and distributes these loads in one or two directions. As such, it is considered to be part of a linear system.[1] Beams generally are positioned horizontally, and loads are applied to them vertically. To produce beam action as traditionally defined, the span length must be several times the depth of the member (Fig. 2.1a). Little beam action would result under conditions shown in Fig. 2.1b, because the proportions are such that the loads would be transferred to supports by an arching action. Furthermore, for pure bending, the width of the beam cross section relative to its depth must be such as to provide sufficient rigidity to prevent failure due to buckling or twisting, and the loads causing bending must be in a plane containing the central longitudinal axis of the beam (Fig. 2.1c). These prerequisites for pure and simple bending are not always met in an exact sense. This does not mean that a member not meeting these exacting criteria cannot be classified as a beam; however, in their absence, it is the responsibility of the designer to make such corrections or adjustments in the design process as are necessary. Twisting and buckling are discussed in greater detail in Art. 5.5; in Arts. 5.15 and 5.19, methods of design are presented for the situation in which the loading plane does not coincide with the central longitudinal axis.

Beams in turn may be classified according to the manner in which they are sup-

---

[1]Nonlinear structural elements containing bending but distributing transverse loads in three or more directions are usually referred to as plates or grids.

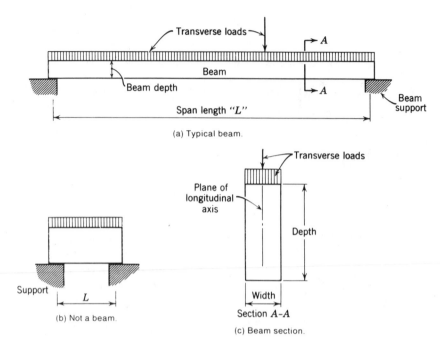

Figure 2.1/Requirements for a beam.

ported. If a beam rests on two supports (Fig. 2.2a), it is called a *simple beam*. If one or both ends of a beam so supported project beyond a support (Fig. 2.2b), it is termed an *overhanging* beam. When it projects from a single support, as would be the case when built into a wall (Fig. 2.3a) or when otherwise rigidly anchored at one end (Fig. 2.3b), it is called a *cantilever beam*. The cantilever beam, however, must be fastened to its support in such a manner that the fixed end cannot rotate or tilt,

regardless of the tendency to do so induced by the forces acting on it. If a beam has two supports, each being fixed (Fig. 2.4a), it is said to be a *restrained beam*. And a beam having more than two supports (Fig. 2.4b) is called a *continuous beam*. Continuous and restrained beams cannot be analyzed by the usual method of statics and are therefore referred to as "statically indeterminate." Methods of indeterminate beam analysis are presented in Chapter 10.

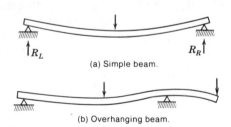

Figure 2.2/Beams on simple support.

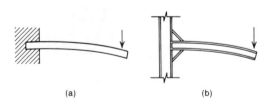

Figure 2.3/Cantilever beams.

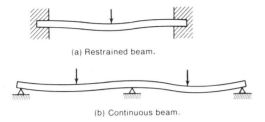

(a) Restrained beam.

(b) Continuous beam.

Figure 2.4/Statically indeterminate beams.

2.2 Loading

Beam loads are classified as either concentrated or distributed. A concentrated load is one that extends over so small an area that it may be assumed to act at a point. For example, Fig. 2.2a shows a simply supported beam subjected to a concentrated load at midspan. A girder in a building is another example; it receives concentrated loads at the points where the floor beams frame into it.

A distributed load is one that extends over a significant portion or the entire length of a beam. It may be uniformly distributed (Fig. 2.5a) or uniformly varying (Fig. 2.5b). Uniformly varying loads are often referred to as triangular or trapezoidal. The weight of the beam itself is a good example of a uniformly distributed load, assuming that the beam is of constant cross section. Another example is parallel beams in a build-

ing that support a floor slab; the weight of the slab will produce a uniformly distributed load over the full beam length. If the beams are not parallel, the distributed load may be uniformly varying.

Concentrated loads are usually expressed in pounds, kips, or tons. The term kip or kilopound denotes 1000 pounds. Distributed loads, on the other hand, are expressed in terms of weight per unit of beam length, usually pounds or kips per foot.

2.3 Reactions

The loads discussed in the preceding article are referred to as external forces acting on the beam. In order that a beam may remain in static equilibrium, balancing external forces must also be present. The supports which develop these balancing forces are called *reactions*.

In Fig. 2.2a, the left reaction has been designated as R_L, and the right, as R_R. Reactions are usually treated as concentrated forces. This, of course, is not always true, as can be seen from the case of a beam resting on a masonry wall (Fig. 2.6). The beam extends a given distance past the face of the wall; the reaction is thus distributed over the area of contact between the beam and the wall. This area is usually sufficiently small, however, so that no appreciable error is introduced by considering the reaction to act at the center of the supporting area. The *span* of a simple beam is the distance between its reactions.

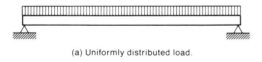

(a) Uniformly distributed load.

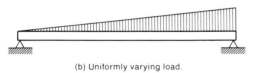

(b) Uniformly varying load.

Figure 2.5/Beam loading.

Span

Figure 2.6/Beam span.

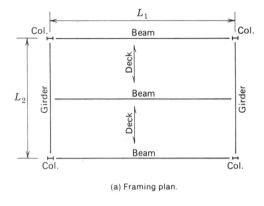

(a) Framing plan.

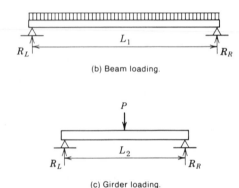

(b) Beam loading.

(c) Girder loading.

Figure 2.7/Beam and girder loads.

In a typical skeleton-frame composed of beams, girders, and columns, the span is generally taken conservatively to be the centerline distance between horizontal members. Figure 2.7a shows a rectangular floor framing plan (see also Art. 5.17). The member spans are designated L_1 (beam) and L_2 (girder), and are taken to be the centerline distances between columns. If it is assumed that the flooring is composed of a one-way floor deck spanning between beams (Fig. 2.7a), then that deck will bring a uniform load to the beams (Fig. 2.7b). The two beams spanning between the columns will carry their load directly to the columns. The beam spanning between the girders will carry its load to each girder in the form of a concentrated load (Fig. 2.7c), and the girders, in turn, will carry

their load to the columns. Note that each center-beam reaction (Fig. 2.7b) becomes the concentrated girder load (Fig. 2.7c).[2]

The reaction conditions at the fixed end of a cantilever beam (Fig. 2.3) and at the ends of a restrained beam (Fig. 2.4a) are more complicated than those of the simply supported types shown in Fig. 2.2 (a) and (b). This is due to the fact that forces are developed within the support that prevent free rotation of the beam ends. This will be discussed in more detail in subsequent chapters.

2.4
Determination of Reactions

After the nature and magnitude of loading have been determined, and the span and type of support have been established, the next step in the design of a beam is the calculation of the reactions. This is accomplished by applying the conditions for static equilibrium which require that (1) the sum of the components in any direction of all forces be zero, and (2) the sum of the moments of all forces about any axis or center of rotation be zero.

For the purpose of structural analysis, the components of the forces are generally referenced to the horizontal and vertical axes of Cartesian coordinates. A restatement of the conditions of equilibrium then may be made as follows: (1) The algebraic sum of the horizontal components of all forces acting on a body is zero; (2) the algebraic sum of the vertical components of all forces acting on a body is zero; and (3) the algebraic sum of the moments of all forces acting on the body, using any point in the plane of the forces as a center of moments, is zero. These conditions may be

[2]The load distribution shown in Fig. 2.7 is based on flexible connections as discussed in Chapter 7.

expressed as follows:[3]

$$\Sigma H = 0, \qquad \Sigma V = 0, \qquad \Sigma M = 0$$

It is evident that with horizontal beam supporting vertical loads only, there will be no horizontal components and hence no horizontal components of the reactions. The reactions will therefore be vertical, and it is necessary to consider only $\Sigma V = 0$ and $\Sigma M = 0$.

Three equations may be written for a simple beam: (1) the summation of moments with the left reaction as the center of moments; (2) the summation of moments with the right reaction as the center of moments; and (3) the summation of vertical loads and reactions. In order to solve for the reactions, at least one moment equation must be used. This will give the value of one reaction; the other reaction may be found from the summation of the vertical forces, or by taking moments about the then known reaction. Until one becomes accustomed to handling these equations, it is best to solve for the reactions using the two moment equations and then check by $\Sigma V = 0$.

Example 1

Compute the reactions of the beam shown in Fig. 2.8. Neglect the weight of the beam.

Solution

Taking moments about the left reaction ($\Sigma M_L = 0$) and considering counterclockwise moments as positive,

$$12R_R - 2400(4) = 0$$
$$12R_R = 9600$$
$$R_R = 800 \text{ lb}$$

Using $\Sigma M_R = 0$ (taking moments about the

[3]The sign Σ (large) Greek letter (sigma) indicates a summation, i.e., an algebraic addition of all similar terms involved in the problem.

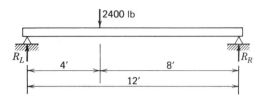

Figure 2.8

right reaction and clockwise as positive),

$$12R_L - 2400(8) = 0$$
$$12R_L = 19,200$$
$$R_L = 1600 \text{ lb}$$

Using $\Sigma V = 0$ as a check, the sum of the loads must equal the sum of the reactions:

$$R_L + R_R = 1600 + 800 = 2400 \text{ lb}$$

Example 2

Compute the reactions of the beam shown in Fig. 2.9. The beam weighs 40 lb per ft.

Note. When computing reactions, a distributed load may be considered as acting as its center of gravity.

Solution

$\Sigma M_B = 0$ (taking B as the center of moments and clockwise as positive):

$$16R_A - [5000(20)] - [4000(2)]$$
$$- [100(14)(7+2)] - [40(20)10] = 0$$
$$16R_A = 100,000 + 8000 + 12,600 + 8000$$
$$16R_A = 128,600$$
$$R_A = 8037.5 \text{ lb}$$

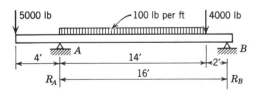

Figure 2.9

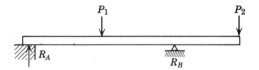

Figure 2.10/Overhanging beam.

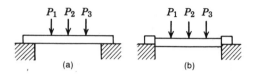

Figure 2.11/Shear failure.

$\Sigma M_A = 0$ (taking A as the center of moments and counterclockwise as positive):

$$16R_B + [5000(4)] + [40(4)2]$$
$$- [4000(14)] - [100(14)7]$$
$$- [40(16)8] = 0$$
$$16R_B = -20{,}000 - 320 + 56{,}000$$
$$+ 9800 + 5120$$
$$16R_B = -20{,}320 + 70{,}920 = 50{,}600$$
$$R_B = 3162.5 \text{ lb}$$

Note. In writing the ΣM_A equation, the moments produced by the loads to the left of R_A were given the same sign as the moment produced by R_B because they tend to rotate the beam in the same direction.

Using $\Sigma V = 0$ as a check, the sum of the loads must equal the sum of the reactions:[4]

Loads:

$$5000 + 4000 + 1400 + [40(20)] = 11{,}200 \text{ lb}$$

Reactions:

$$R_A + R_B = 8037.5 + 3162.5 = 11{,}200 \text{ lb}$$

In the analysis of an overhanging beam such as that shown in Fig. 2.10, it is often impossible to determine the direction of R_A by inspection. If the load P_2 is a great

[4]It is neither necessary nor desirable to carry the values of the reactions to the nearest pound or decimal thereof. Three significant figures will produce a degree of accuracy which is consistent with the assumptions of structural theory and conditions met in practice. On this basis, R_A and R_B of Example 2 would have values of 8040 lb and 3160 lb, respectively. The figures have been extended in this text, as in this case, only where such extensions of the numerical work seemed to clarify the explanation of the problem.

deal larger than P_1 it may be that R_A will act downward. A satisfactory method of solution is to assume that R_A acts upward and solve the problem in the usual manner. If the value for R_A comes out positive (+), the assumption was correct. If R_A comes out negative (−), the assumption was wrong and the reaction acts in the opposite direction. The numerical value, however, will be correct.

2.5
Shear

Figure 2.11a represents a beam supporting a system of concentrated loads. It is evident from the sketch that the beam might fail by simply dropping between the supporting walls as shown in Fig. 2.11b. This type of failure is called a shearing failure. It is probable that the beam would fail in some other manner before being sheared off vertically, but the tendency to fail in this way would nevertheless be present. The force that measures this tendency is called the vertical shear, or simply the shear, and is designated by the symbol V.

As indicated in Fig. 2.12, beams may be considered to be composed of an infinite number of vertical segments, though for practical reasons a finite number is shown. The magnitude of the shearing force V at any right section[5] is the amount necessary to produce static equilibrium when the portion to the right or left of the section is

[5]A right section is a section perpendicular to the longitudinal axis of the beam.

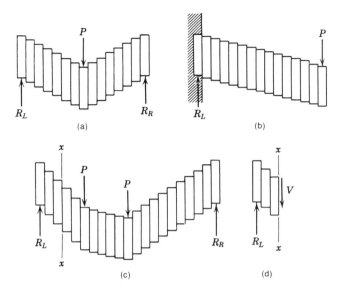

Figure 2.12/Shear variation.

isolated and treated as a free body.[6] This shearing force is determined from the general equation $\Sigma V = 0$.

The portion of the beam to the left of the section x-x in Fig. 2.12c is shown as a free body in Fig. 2.12d. It must be in equilibrium if the entire beam is to be in equilibrium. Therefore, there must also be a force acting at the cut face; this force is the internal shear. In Fig. 2.12d, it is denoted by V and is equal in magnitude, but opposite in direction, to R_L. From the above explanation the following definition may be formulated:

The shear at any right section of a beam is the algebraic sum of all the transverse forces on one side of the section.

The loads on a beam, therefore, tend either to cause the portion to the right of

the section to descend and the portion to the left to ascend (Fig. 2.13), or vice versa (Fig. 2.14). The former is arbitrarily designated positive (+) shear, and the latter, negative (−) shear.

In addition, due to the fact that the shearing force may vary along the length of a

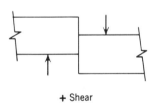

+ Shear

Figure 2.13/Positive shear.

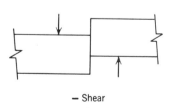

− Shear

Figure 2.14/Negative shear.

[6]A free body means the entire structure or some part of it, considered separately from its surrounding parts. When the entire structure is in equilibrium, every part of it (every free body) also must be in equilibrium. All forces acting on a free body must be considered in setting up the equilibrium condition.

beam, it is usually necessary for purposes of design to know how it will vary, and to locate and calculate its maximum value.

2.6
Shear Diagrams

The shear diagram is a convenient device for graphically representing variations in shear that occur along the beam length.

A horizontal line at any convenient scale equal in length to the span of the beam is first drawn. This line is one axis of Cartesian coordinates, and may be pictured as the longitudinal axis of the loaded beam. The magnitude of positive shear is plotted above the line, and negative, below the line. The resulting diagram may be thought of as a graphical representation of the shear equation.

2.7
Shear Diagrams—Concentrated Loads

Figure 2.15 represents a beam with equal concentrated loads at the quarter points. The weight of the beam has been neglected. From the symmetry of the loading it is evident that the reactions are equal

and have a value of $3P/2$. At some convenient distance below the space diagram, a base line representing zero shear has been laid off.

If a section of the beam just to the right of the left reaction is considered, the sum of all the forces acting upward on the left of the section (+ shear) is $3P/2$. Hence, from the definition of shear, the shear at the support equals $+3P/2$. This is termed the end shear and in this case is equal to the reaction. This value is now plotted to some scale along a vertical line representing the line of action of the left reaction.

If another section is taken through the beam at A-A, from the definition of shear, the sum of all the forces acting upward on the left of the section (+ shear) is still $3P/2$. Therefore, the shear at section A-A is also $+3P/2$. (It should be kept in mind that in this example the weight of the beam has been neglected.) Section A-A might have been any section between the left reaction and the load P_1. It is thus evident that the shear is constant between these two loads and the shear line is horizontal.

If another section is taken just to the right of the load P_1, the shear at that section will be $+3P/2 - P = P/2$. This value is plotted on the line extending down from

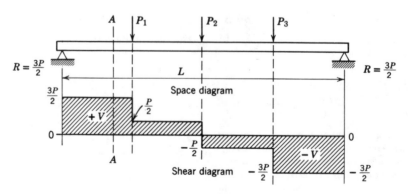

Figure 2.15/Shear diagram, concentrated loads.

P_1. The shear between P_1 and P_2 is constant, and therefore the shear line between these loads is also horizontal. This process is continued across the beam. The fact that the numerical value of the shear at the right end of the beam is equal to the right reaction will serve as a check.

2.8
Shear Diagrams—Distributed Loads

Figure 2.16 is the shear diagram for a simple beam supporting a uniformly distributed load over its entire span. The total load on the beam is wL, and each reaction is $wL/2$.

As in the preceding illustration, the end shear is equal to the reaction. This is plotted to scale as before. If the next section is taken at the center of the span, the shear will be $wL/2 - wL/2 = 0$. If another section such as y-y is taken at the quarter point of the span, the shear will be $wL/2 - wL/4 = + wL/4$. At a point three-quarters of the span length from the left reaction, the shear is $+ wL/2 - 3wL/4 = - wL/4$. It should be noted that under a uniform load, the shear varies as a straight inclined line. This is also evident from the equation expressing the magnitude of the

shear in terms of a variable of the span. Letting x be any distance from the left reaction, from Fig. 2.16 it is seen that the value of the shear at x ft to the right of the left reaction is

$$V = \frac{wL}{2} - wx$$

Since the only variable in the equation is x, and it is to the first power, the equation of a straight line results. The point of zero shear may be calculated by setting the above equation equal to zero:

$$\frac{wL}{2} - wx = 0$$

$$x = \frac{L}{2}$$

It is evident from the foregoing discussion of downward-acting loads that the maximum shear for a simple beam or cantilever occurs adjacent to the reaction. However, from Example 3 it will be seen that it is not always equal to the reaction in the case of overhanging beams.

Example 1

Draw the shear diagram for the beam shown in Fig. 2.17. Neglect the weight of the beam.

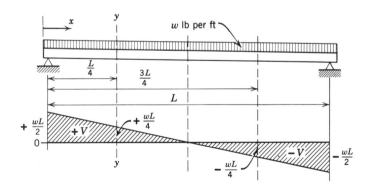

Figure 2.16/Shear diagram, distributed loads.

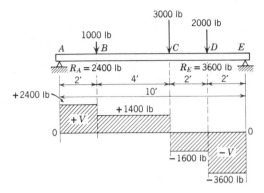

Figure 2.17

Solution

Solving for reactions, $\Sigma M_E = 0$ (taking E as the center of moments)

$$10R_A - [1000(8)] - [3000(4)]$$
$$- [2000(2)] = 0$$
$$10R_A = 8000 + 12,000 + 4000$$
$$= 24,000$$
$$R_A = 2400 \text{ lb}$$

and writing $\Sigma M_A = 0$ (taking A as the center of moments),

$$10R_E - [2000(8)] - [3000(6)]$$
$$- [1000(2)] = 0$$
$$10R_E = 16,000 + 18,000 + 2000$$
$$= 36,000$$
$$R_E = 3600 \text{ lb}$$

Using $\Sigma V = 0$ as a check, the sum of the loads must equal the sum of the reactions.

Loads: $1000 + 3000 + 2000 = 6000 \text{ lb}$

Reactions: $2400 + 3600 = 6000 \text{ lb}$

These reactions have been recorded on the diagram.

The end shear at A is $+2400$ lb. There are no loads between A and B; therefore, at any section between these points, the only force on the left of the section is 2400 lb acting upward. Thus, the shear for all sections between A and B is 2400 lb

and is positive. The shear for all sections between B and C is $2400 - 1000 = 1400$ lb; for all sections between C and D, $2400 - 1000 - 3000 = -1600$ lb; and for all sections between D and E, $2400 - 6000 = -3600$ lb. With these values determined, the shear diagram may be drawn as shown.

Example 2

Draw the shear diagram for the beam shown in Fig. 2.18, neglecting the weight of the beam.

Solution

The reactions, and thus the end shears, each equal 8000 lb.

The shear at the center is

$$8000 - [1000(8)] = 0$$

The shear 4 ft from the left reaction is

$$8000 - [1000(4)] = 4000 \text{ lb}$$

With these values, the shear diagram may be constructed. It should be noted that for beams loaded as in Fig. 2.18 the shear line may be drawn by simply connecting the ordinates of the end shears. The shear line should cross the base line at the center of the span.

The analysis may also be made as follows: Since the intensity of the load does not change over the span length, and there are

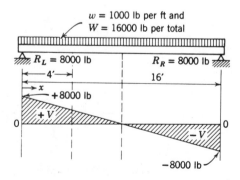

Figure 2.18

no concentrated loads, only one equation is necessary for the entire beam. Using the left reaction as the origin for measuring x, the equation for shear at any point becomes

$$V = 8000 - 1000x$$

when $x = 0$,

$$V = +8000 \text{ lb}$$

when $x = 16$,

$$V = -8000 \text{ lb}$$

Again, the equation is that of a straight line, two points of which have been determined, and may now be shown on the shear diagram.

Example 3

Construct the shear diagram for the beam shown in Fig. 2.19. Neglect the weight of the beam.

Solution

Solving for the reactions,

$$\sum M_D = 0$$

(taking D as the center of moments),

$$16R_B - [2000(20)] - [4000(4)]$$
$$- [1200(6+4)] = 0$$
$$16R_B = 40,000 + 16,000 + 12,000 = 68,000$$
$$R_B = 4250 \text{ lb}$$

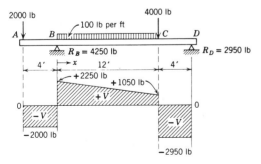

Figure 2.19

The sum of the loads is $2000 + 4000 + 1200 = 7200$ lb. Therefore $R_D = 7200 - 4250 = 2950$ lb. This reaction should be checked, using B as the center of moments.

The end shear at A is -2000 lb (see Fig. 2.19). As there are no loads between A and B, the shear for any section in this panel is -2000 lb. At the section just to the right of B, the shear is $-2000 + 4250 = +2250$ lb. At the section just to the left of C, the shear is $-2000 + 4250 - 1200 = +1050$ lb. The load between B and C is a uniform load; thus the shear line is a straight line sloping downward from left to right. At a section just to the right of C, the shear is $-2000 + 4250 - 1200 - 4000 = -2950$ lb. This is also the shear for any section between C and D. Therefore, the shear at $D = -2950$ lb. This value checks with the reaction.

The shear equation for the portion of the beam directly underneath the distributed load is

$$V = -2000 + 4250 - 100x$$
$$V = 2250 - 100x$$

when the origin of x is at the left reaction.

Example 4

A cantilever beam projects 14 ft from the face of a wall. It supports a uniform load of 200 lb/ft, extending from the fixed end to within 2 ft of the free end, and a concentrated load of 1000 lb applied at the free end. Draw the shear diagram. Neglect the weight of the beam.

The fixed end may be assumed to be at the left as shown in Fig. 2.20a, or at the right as shown in Fig. 2.20b.

Solution 1

Assume the fixed end to be at the left as in Fig. 2.20a. The shear for all sections between A and B will be 1000 lb. If a

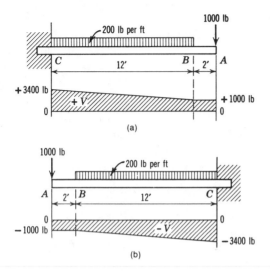

(a)

(b)

Figure 2.20

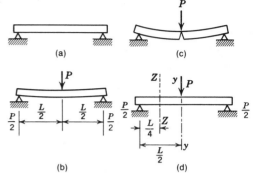

Figure 2.21/Bending-moment failure.

section is taken just to the left of A, the right segment of the beam will tend to move downward relative to the left segment. Therefore, the shear in panel AB is positive (Fig. 2.13). Under the distributed load, the shear will increase uniformly, making the shear at C equal to $1000 + [200(12)] = 3400$ lb. The diagram is constructed as shown.

Solution 2

Assume the fixed end at the right as in Fig. 2.20b. The shear in panel AB will be 1000 lb. If a section is taken just to the right of A, the left segment will tend to move downward relative to the right segment. Therefore, the shear in the panel is negative (Fig. 2.14). Under the distributed load, the shear increases uniformly, making the shear at C equal to -3400 lb. The diagram is constructed as shown.

Although the sign of shear at any section of the beam is $(+)$ for Solution 1 and $(-)$ for Solution 2, the numerical value is the same. Since in the design of beams it is the numerical value of the shear that is used, this ambiguity of signs presents no difficulty.

2.9
Bending Moment

The external forces acting on a beam, that is, the loads and reactions, deform the beam by bending it. Each such external force acts to produce a specific moment at any given point along the beam length.

For example, if the beam shown in Fig. 2.21a, has a concentrated load P applied at the center, it will deflect as shown in Fig. 2.21b.[7] If the load were heavy enough it is conceivable that the bending would continue until the beam failed as indicated in Fig. 2.21c.

The tendency of the beam to fail in this manner is measured by the moment of the reaction about section y-y (Fig. 2.21d), through the point of application of the load. From Fig. 2.21b, this moment is $(P/2)(L/2) = PL/4$ and is called the *bending moment* at section y-y.

The tendency of the beam to fail by bending is present at every point along its length; therefore, for any section that may be taken through the beam, there exists a definite value for the bending moment. For example, consider any right section in

---

[7]In most steel or concrete beams this deflection may not be visible to the eye, but it is nevertheless present and could be observed if sufficiently sensitive instruments were used.

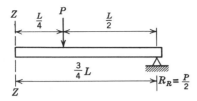

Figure 2.22/Bending moment at Z.

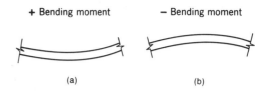

Figure 2.23/Bending-moment sign convention.

the above beam, e.g., section Z-Z. Using the left segment, the bending moment at Z-Z is $(P/2)(L/4) = PL/8$. Figure 2.22 shows the right segment of the same beam. The product of each external force and its distance to the cut section (moment arm) causes a moment either clockwise or counterclockwise. Thus, from Fig. 2.22, the bending moment at section Z-Z is

$$M = \frac{P}{2}\left(\frac{3L}{4}\right) - P\left(\frac{L}{4}\right)$$

$$= \frac{3PL}{8} - \frac{PL}{4} = \frac{PL}{8}$$

where M = bending moment; in the first line, $(P/2)(3L/4)$ is the product of the right reaction R_R multiplied by its distance from the section Z-Z, and $P(L/4)$ is the product of the load P times its distance from Z-Z. The term $P(L/4)$ is given the opposite sign from the preceding term because the moment caused by it tends to counteract the effect of the right reaction.

When the left segment of the beam was used, the only force acting was the left reaction, and its moment arm is $L/4$. Thus, the previously calculated bending moment,

$$\frac{P}{2}\left(\frac{L}{4}\right) = \frac{PL}{8}$$

is the same as that determined when using the right segment. This leads to the generalization that the value for the bending moment at any given section is the same

regardless of whether it is computed using the left or right segment of the beam. In order to evaluate the moment at any section, the following definition may be used:

The bending moment at any right section of a beam is the algebraic sum of the moments of all the forces on one side of this section.

This definition assumes that the beam is horizontal and that all the forces lie in a vertical plane.

A sign convention must be adopted to further define the type of bending moment occurring at any section. If the beam tends to become concave upward at the section under consideration (Fig. 2.23a), the bending moment is said to be positive ($+$). If the beam tends to become concave downward at the section (Fig. 2.23b), the bending moment is negative ($-$).

One method which will give the correct sign for the bending moment is to consider the moments of all upward forces as positive ($+$) and the moments of all the downward forces as negative ($-$). This system enables one to work from either end of the beam.

2.10
Bending-Moment Diagrams

The variation of the bending moment from section to section of a beam throughout its length may be represented by means of diagrams similar to those used for shear.

2.11
Bending-Moment Diagrams— Concentrated Loads

Figure 2.24 represents a beam with equal concentrated loads at the quarter points. The weight of the beam has been neglected. From the symmetry of the loading it is evident that the reactions are equal and have a value of $3P/2$. At some convenient distance below the loading diagram a base line representing zero bending moment is laid off.

If a section is taken at A, the algebraic sum of the moments of all the forces to the left of the section is zero, since there are no forces to the left. Therefore, the bending moment at A equals zero.

If a section is taken at B, directly under the load, the algebraic sum of the moments of all the forces to the left of the section is $+(3P/2)(L/4)$, or $3PL/8$. This value is plotted to some convenient scale on the line representing the section at B.

If the bending moment at any other section between A and B is plotted, it will be found to lie on a straight line connecting A' and B'. Thus, for a section halfway between A and B, the bending moment is $+(3P/2)(L/8) = 3PL/16$. This value is just half the value at B. It may be noted,

then, that for a system of concentrated loads, the bending moment between loads varies as a straight line. This line may be inclined.

Referring again to its definition, the bending moment at C is $+[(3P/2)(L/2)] - [P(L/4)] = 3PL/4 - PL/4 = 2PL/4$, or $PL/2$. This value is plotted on the line representing the section at C'.

In a similar manner, the bending moments at D and E may be computed and plotted. The completed diagram is shown in the figure.

Equations also may be written to determining the bending moment at *any* point in the beam. The presence of the three concentrated loads shown in Fig. 2.24 dictates the necessity for four distinct equations, one for each of the four segments. For every equation, consider x to be zero at the left reaction. Then, for any point between A and B, the moment will be

$$M_{A-B} = \frac{3P}{2}x$$

Similarly, between B and C

$$M_{B-C} = \frac{3P}{2}x - P\left(x - \frac{L}{4}\right) = \frac{Px}{2} + \frac{PL}{4}$$

Each is the equation of a straight line;

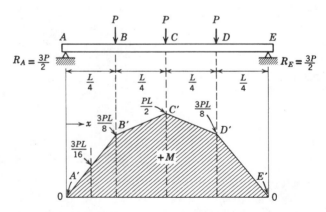

Figure 2.24/Bending-moment diagram, concentrated load.

therefore, the value of the bending moment between A and B, or B and C, may be obtained by substituting the appropriate value for x in the proper equation.

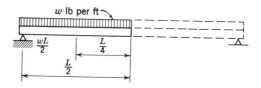

Figure 2.26/Bending moment at midspan.

2.12
Bending-Moment Diagrams—Distributed Loads

Figure 2.25 represents a beam carrying a uniformly distributed load over its full span. The total load on the beam is wL, and each reaction is $wL/2$.

If a section is taken at the center of the span, the algebraic sum of the moments of all the forces to the left of the section is

$$M = +\frac{wL}{2}\left(\frac{L}{2}\right) - w\left(\frac{L}{2}\right)\frac{L}{4}$$

$$= \frac{wL^2}{4} - \frac{wL^2}{8} = \frac{wL^2}{8}$$

where M is the bending moment, $(wL/2)(L/2)$ is the left reaction times its distance from the section under consideration, and $w(L/2)L/4$ is the load per foot, times the length of beam covered, times the distance from the center of gravity of the load to the section. This is indicated more clearly in Fig. 2.26, where only the left segment is shown.

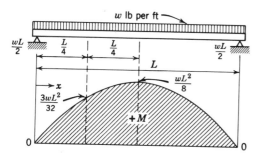

Figure 2.25/Bending-moment diagram, distributed loads.

If the distances are measured in feet and the loads are given in pounds, the bending moment is expressed in foot-pounds.[8]

In a similar manner, the bending moment at any other section could be found. If a section were taken through the beam at a distance of one-quarter of the span length from the left reaction (Fig. 2.27), the equation would be

$$M = +\frac{wL}{2}\left(\frac{L}{4}\right) - w\left(\frac{L}{4}\right)\frac{L}{8}$$

$$M = \frac{wL^2}{8} - \frac{wL^2}{32} = \frac{4wL^2}{32} - \frac{wL^2}{32}$$

$$M = \frac{3wL^2}{32}$$

If the bending moments at additional points were computed, they would be found to lie on a parabolic curve, the vertex of which is at the center of the span. Therefore, it may be seen that under a uniform load, the bending-moment curve is a parabola. In order to plot the bending-moment diagram for a beam supporting a uniform load, it is necessary to compute the moment at even intervals of, say, 1 or 2 ft along the span length, or to construct the parabola by graphical methods after determining the maximum ordinate.

[8]In many textbooks on mechanics, this unit of measure is designated as pound-feet. Although pound-feet is a more precise term from the standpoint of theoretical mechanics, foot-pounds has been used throughout this text because of its wide acceptance among practicing architects and engineers.

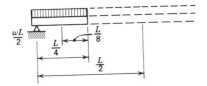

Figure 2.27/Bending moment at quarter-span.

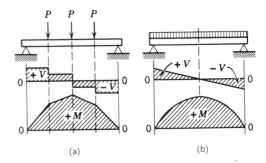

Figure 2.28/Relation between shear and bending moment.

An alternative method for establishing the bending-moment diagram is to develop an equation representing the bending moment at any point. Only one equation is necessary in this case, as the load is uniformly distributed over its entire length and no concentrated loads occur (other than the reactions). Consider x to be zero at the left reaction; then the bending moment at any distance x, measured to the right from that reaction, will be

$$M = \frac{wL}{2}x - wx\left(\frac{x}{2}\right) = \frac{wL}{2}x - \frac{wx^2}{2}$$

This is the equation of a parabola, and the curve of this equation plotted graphically is the bending-moment diagram for the beam.

2.13
Relation between Shear-Force Diagram and Bending-Moment Diagram

In Art 2.7 the shear diagram for a beam supporting equal concentrated loads at the quarter-points was constructed, and in Art 2.11 the bending-moment diagram for the same beam was drawn. These diagrams have been combined in Fig. 2.28a.

Figure 2.28b combines the shear diagram in Fig. 2.16 and the bending-moment diagram in Fig. 2.25, both of which were for a beam supporting a uniform load.

From Fig. 2.28 it is evident that:

1. On any portion of a beam where there are no loads, the shear may be repre-

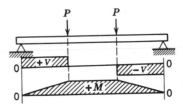

Figure 2.29/Relationship between shear and moment.

sented by a straight horizontal line and the bending moment by a straight line (uniformly varying).

2. On any portion of a beam supporting a uniformly distributed load, the shear may be represented by a straight inclined line and the moment by a parabolic curve.

3. At any point where the line representing the shear passes through zero, there is a maximum ordinate of the bending-moment diagram.

A fourth relationship, brought out by Fig. 2.29, is that if the shear is zero for any distance, the bending moment is constant for that distance and hence is represented by a straight, horizontal line.

These four statements relate to specific conditions as illustrated. In Art 2.15, a more general treatment of beam diagrams[9] is presented, and it is shown mathemati-

[9]*Beam diagrams* is a general term meaning load, shear, and bending-moment diagrams.

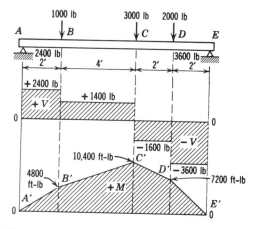

Figure 2.30/Example 1.

Solution

See Example 1 following Art. 2.8 for the computation of reactions and construction of the shear diagram.

Working from the left, the bending moments are

$$M_A = 0$$
$$M_B = +2400(2) = 4800 \text{ ft-lb}$$
$$M_C = +[2400(6)] - [1000(4)]$$
$$= 14,400 - 4000$$
$$= 10,400 \text{ ft-lb}$$
$$M_D = +[2400(8)] - [1000(6)] - [3000(2)]$$
$$= 7200 \text{ ft-lb}$$
$$M_E = 0$$

cally that the above relationships hold for any beam under any system of vertical loads. Article 2.15 also demonstrates use of the equations representing the diagrams.

Beam diagrams and formulas for the more common types of loading are shown in Art. 2.14 and Appendix B, and in the AISC Manual.[10] The reader should check a representative number of these formulas by deriving them.

In the examples which follow, the loads and spans are the same as those used for the examples with corresponding numbers following Art. 2.8. The shear diagrams have been reproduced to show the relation between shear and bending moment and to aid in the construction of the bending-moment diagrams.

Example 2

Draw the shear and bending-moment diagrams for the beam shown in Fig. 2.31. Neglect the beam weight.

Solution

See Example 2 following Art. 2.8 for the computation of reactions and construction of the shear diagram.

The maximum bending moment occurs at the center and is

$$M = +[8000(8)] - [1000(8)4]$$
$$= 64,000 - 32,000 = 32,000 \text{ ft-lb}$$

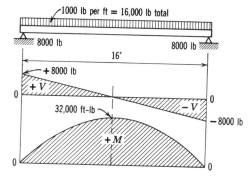

Figure 2.31/Example 2.

Example 1

Draw the shear and bending-moment diagrams for the beam shown in Fig. 2.30. Neglect the beam weight.

[10]See *Manual of Steel Construction* (ASD), Ninth Edition, American Institute of Steel Construction, Chicago, 1989.

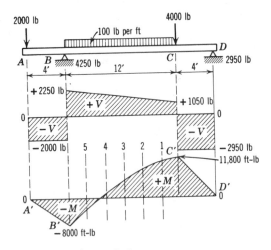

Figure 2.32/Example 3.

The moments at 1- or 2-ft intervals along the span are computed in a similar manner, and a curve drawn through the plotted points.

Example 3

Draw the shear and bending-moment diagrams for the beam shown in Fig. 2.32. Neglect the beam weight.

Solution

See Example 3 following Art. 2.8 for the computation of reactions and construction of the shear diagram.

Working from the left, the bending moments are

$$M_A = 0$$
$$M_B = -[2000(4)] = -8000 \text{ ft-lb}$$

See Fig. 2.33 (a) and (b) for determination of sign.

Working from the right,

$$M_C = +[2950(4)] = +11,800 \text{ ft-lb}$$

Solving for the bending moment at 2-ft intervals between C and B,

$$M_1 = +[2950(6)] - [4000(2)] - [100(2)1]$$
$$= 9500 \text{ ft-lb}$$
$$M_2 = +[2950(8)] - [4000(4)] - [100(4)2]$$
$$= 6800 \text{ ft-lb}$$
$$M_3 = +[2950(10)] - [4000(6)] - [100(6)3]$$
$$= 3700 \text{ ft-lb}$$
$$M_4 = +[2950(12)] - [4000(8)] - [100(8)4]$$
$$= 200 \text{ ft-lb}$$
$$M_5 = +[2950(14)] - [4000(10)] - [100(10)5]$$
$$= -3700 \text{ ft-lb}$$

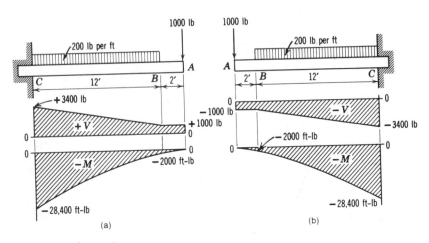

(a)

(b)

Figure 2.33/Example 4.

Example 4

A cantilever beam projects 14 ft from the face of a wall. It supports a uniform load of 200 lb per ft extending from the fixed end to within 2 ft of the free end, and a concentrated load of 1000 lb applied at the free end. Draw the shear and bending-moment diagrams. Neglect the beam weight.

Solution 1

The beam is fixed at its left end as in Fig. 2.33a. For construction of the shear dia-

gram, see Example 4 following Art. 2.8. Working from the free end, the bending moments are

$$M_A = 0$$
$$M_B = -[1000(2)] = -2000 \text{ ft-lb}$$

(see Fig. 2.23(a) and (b) for the determination of the sign), and

$$M_C = -[1000(14)] - [200(12)6]$$
$$= -28,400 \text{ ft-lb}$$

The moment at a point halfway between C and B is

$$M = -[1000(8)] - [200(6)3]$$
$$= -11,600 \text{ ft-lb}$$

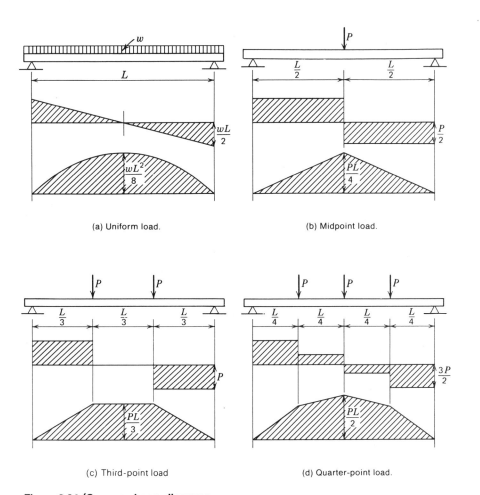

(a) Uniform load.

(b) Midpoint load.

(c) Third-point load

(d) Quarter-point load.

Figure 2.34/Common beam diagrams.

The moments at other points between C and B are computed in a similar manner.

Solution 2

The beam is fixed at its right end as in Fig. 2.33b.

The moment curve will be as shown. The proof of this is left to the reader. Referring to Fig. 2.33 (a) and (b), it will be seen that the sign of the bending moment in a cantilever beam is the same whether the beam is fixed at the right or the left.

2.14
Common Beam Diagrams

Economies can be realized through simplicity and repetition in structural framing patterns and details. When the framing pattern selected is both simple and repetitive, one such economy that is available to the designer is the opportunity to use readily recognizable member-loading diagrams and the formulas that accompany them to arrive at maximum values for design. Four such diagrams are shown in Fig. 2.34 along with formulas for maximum shear and moment. As was noted in Art.

2.13, it would be well for the reader to derive such values—in this case, those shown in Fig. 2.34 (c) and (d).

As long as the beam material stays within its elastic limit (Appendix A), the rules of superposition can be applied to beam diagrams i.e., shear or moment values derived from a given loading at any point along the length of a beam can be added directly to the corresponding shear or moment values derived from another loading at that same point. This permits the determination of shear and moment values due to combined loadings by dealing with each separately and then adding values at a given point to arrive at aggregate values. An example of this is illustrated in Fig. 2.35. A simple beam having both a uniform load (Fig. 2.35a) and a midpoint concentrated load (Fig. 2.35b) would result in a combined diagram as shown in Fig. 2.35c.

2.15
Analysis of Beam Diagrams

Presented in this article is more generalized analysis of the relation between shear and bending moment than that developed in Art. 2.13. Although some of the steps

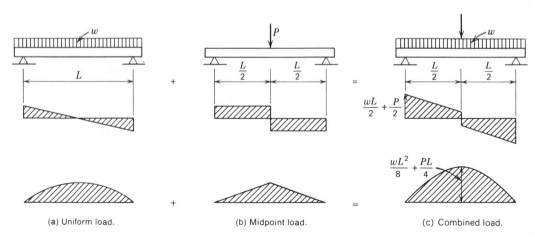

(a) Uniform load.   (b) Midpoint load.   (c) Combined load.

Figure 2.35/Superimposing diagrams.

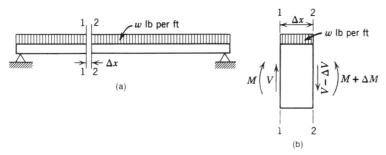

Figure 2.36/Analysis of diagrams.

involved may not be clear to those unfamiliar with calculus, the conclusions reached and the explanatory diagrams should be studied carefully.

Figure 2.36a indicates a beam with the portion between two right sections (1-1 and 2-2) removed. This is shown as a free-body diagram at larger scale in Fig. 2.36b. The two sections are assumed to be very close together, the distance between them being designated Δx. The quantities ΔM and ΔV represent, respectively, the increments of change in bending moment and shear between the two sections. Since the free body is in equilibrium, the sum of moments at section 1-1 is zero. Therefore,

$$M - (M + \Delta M) + (V - \Delta V)\,\Delta x$$
$$+ w(\Delta x)\frac{\Delta x}{2} = 0$$

and

$$\Delta M = V\Delta x - \Delta V\Delta x + w\left(\frac{(\Delta x)^2}{2}\right)$$

Dividing both sides of the equation by Δx,

$$\frac{\Delta M}{\Delta x} = V - \Delta V + w\left(\frac{\Delta x}{2}\right)$$

Now, the limit of $\Delta M / \Delta x$, as Δx approaches zero, is the derivative dM / dx, and at the limit, the quantities ΔV and $w(\Delta x / 2)$ become zero. Consequently,

$$\frac{dM}{dx} = V$$

and, since the derivative measures the rate of change of a function (the slope), it may be concluded that the shear at any section through a beam measures the rate of change of the bending moment at that section (Fig. 2.37).

Referring again to Fig. 2.36b and applying the condition that $\Sigma V = 0$, it may similarly be shown that

$$\frac{dV}{dx} = w$$

(downward w is considered negative). This means that the intensity of the loading at any point on a beam governs the slope of the shear diagram at that point.

The first basic rule relating to beam diagram relationships may now be stated as follows:

The slope of the bending-moment diagram at any point is equal to the ordinate of the corresponding shear diagram at the same point; and the slope of the shear diagram at any point is equal to the ordinate of the corresponding load diagram[11] at the same point.

It will be seen from the foregoing discussion that a shear curve or diagram may be derived from the corresponding bending-

[11]A load diagram is simply a graphical representation of the variation in the intensity of the load (unit load) on the beam. Loads acting downward take a negative sign.

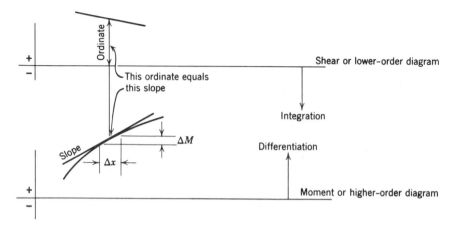

Figure 2.37/Ordinate-slope relationship.

moment diagram by the process of differentiation. This operation proceeds from a "higher-order" diagram to a "lower-order" diagram, i.e., of the diagrams for bending moment, shear, and loading the bending-moment diagram is the highest-order of the three, and the loading diagram, the lowest-order (Fig. 2.37).

The mathematical construction of the next higher-order diagram from a given diagram involves the process of integration, i.e., to arrive at the bending-moment diagram from the shear diagram it is necessary to integrate the shear curve. The construction of shear and bending-moment curves by integration is illustrated in the example given at the end of this article.

The second basic rule relating to beam-diagram relationships follows from the above discussion and may be stated as follows:

The change in bending moment between any two points is equal to the area of the shear diagram between the same two points; and the change in shear between any two points is equal to the area of the load diagram between the same two points.

Figure 2.38 illustrates this relationship for shear and bending moment.

One of the conclusions stated in Art. 2.13 was that at any point where the shear line passes through zero, there is a maximum ordinate of the bending-moment curve. That this is the case will also be evident from a study of the first basic rule given earlier in this article, since the tangent to

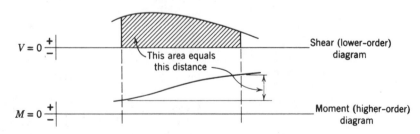

Figure 2.38/Area-ordinate change relationship.

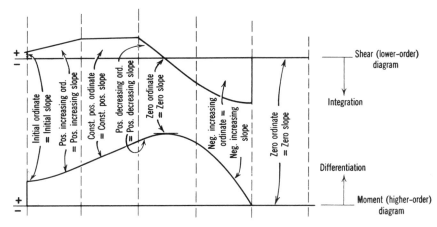

Figure 2.39/Relationship between higher and lower diagrams.

the bending-moment curve can be horizontal (zero slope) only where the value of the shear (dM/dx) is zero. Figure 2.39 illustrates this point, as well as other slope considerations.

Example

A simply supported beam with a span length of 20 ft sustains a uniformly distributed load of 2 kips per ft over its entire length, and a concentrated load of 4 kips applied 5 ft from the left support (Fig. 2.40a). Construct the load, shear, and bending-moment diagrams.

Solution

(1) Computing the reactions, it will be found that $R_L = 23$ kips and $R_R = 21$ kips. These values are noted on the "load diagram" shown in Fig. 2.40b. Both the distributed load and the concentrated load are plotted below the base line and given negative signs (see footnote 11 above). Since the intensity of the distributed load (2 kips per ft) is constant for the entire length of the beam, it may be represented

by the equation

$$w = -2$$

and this will be its value at *any* point a distance x to the right of R_L.

(2) The equation for the shear at any point between R_L and 5 ft to the right (the point of application of the concentrated load) is the integral of the load equation between these same limits, i.e.

$$V_{0\text{-}5} = \int (-2)\,dx = -2x + C$$

The constant of integration (C) is determined by observing that the value of the shear at one point, between the limits indicated, is known, i.e., at R_L. Therefore, when $x = 0$, $V = 23$ kips. Substituting in the above equation,

$$V = -2x + C$$
$$23 = -2(0) + C$$

or

$$C = 23$$

and the final equation of the shear diagram between the limits $x = 0$ and $x = 5$ ft is

$$V_{0\text{-}5} = 23 - 2x$$

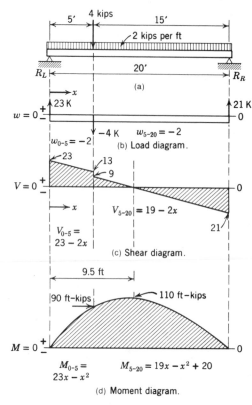

(a)

(b) Load diagram.

(c) Shear diagram.

(d) Moment diagram.

Figure 2.40/Example: analysis of beam diagrams.

Therefore, the shear just to the left of the concentrated load at $x = 5$ is

$$V_{5L} = 23 - 2(5) = 13 \text{ kips}$$

and the shear just to the right of the concentrated load is

$$V_{5R} = 13 - 4 = 9 \text{ kips}$$

These values have been plotted on the shear diagram (Fig. 2.40c).

(3) For the portion of the beam between the concentrated load and the right reaction (R_R), the shear equation for the distributed load has the same general expression, since the intensity of the loading is the same over the entire length of the

beam. Therefore,

$$V_{5\text{-}20} = \int(-2)\,dx = -2x + C$$

However, when $x = 5$, $V = 9$, so the constant of integration is given by

$$9 = -2(5) + C$$

or

$$C = 19$$

Therefore, the shear at *any* point between the concentrated load and R_R is given by the expression

$$V_{5\text{-}20} = 19 - 2x$$

where x is again measured to the right from R_L. Since the shear just to the left of the right reaction must be numerically equal to R_R, the equation may be checked as follows:

$$V_{(x=20)} = 19 - 2x = 19 - 2(10) = -21 \text{ kips}$$

(4) The point of zero shear, which occurs between $x = 5$ and $x = 20$, may be found algebraically by setting the applicable shear equation equal to zero and solving for x, i.e.

$$V = 19 - 2x = 0$$
$$x = 9.5 \text{ ft}$$

(5) The equation for the bending-moment curve, at any point between R_L and the concentrated load, may be found by integrating the corresponding shear equation. Thus,

$$V_{0\text{-}5} = 23 - 2x$$

and

$$M_{0\text{-}5} = \int(23 - 2x)\,dx = 23x - x^2 + C$$

Evaluating the constant of integration, it will be observed that $M = 0$ when $x = 0$ and consequently the value of C is also zero. Therefore,

$$M_{0\text{-}5} = 23x - x^2$$

From this equation, the moment at 5 ft

from R_L is

$$M_{(x=5)} = 23(5) - 5^2$$
$$M_{(x=5)} = 115 - 25 = 90 \text{ ft-kips}$$

(6) For the portion of the beam between the concentrated load and R_R,

$$V_{5\text{-}20} = 19 - 2x$$

and

$$M_{5\text{-}20} = \int (19 - 2x)\, dx = 19x - x^2 + C$$

Evaluating C, it will be noted that when $x = 5$, $M = 90$. Therefore,

$$90 = 19(5) - 5^2 + C$$

from which

$$C = 20$$

Therefore,

$$M_{5\text{-}20} = 19x - x^2 + 20$$

The maximum moment occurs at $x = 9.5$ ft and is

$$M_{(x=9.5)} = 19(9.5) - (9.5)^2 + 20$$
$$= 110 \text{ ft-kips}$$

(7) The above values of the shears and bending moments have been plotted in Fig. 2.40 (c) and (d), respectively. As an illustration of the second basic rule of beam-diagram relationships, it will be observed from the figure that the area of the shear diagram between $x = 5$ ft and $x = 9.5$ ft is $9(4.5)/2 \approx 20$, which is equal to the change in the value of the bending moment between these same two points, i.e., $110 - 90 = 20$ ft-kips.

2.16
Bending-Moment Equations

Expressions for the bending moment at any point along a beam can, of course, be established by procedures other than integration of the shear curve. However, a

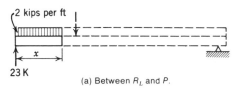

(a) Between R_L and P.

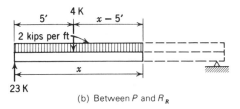

(b) Between P and R_R

Figure 2.41 / Bending-moment equations.

graph of the principles and relationships involved in this generalized method will be found helpful in understanding beam deflections as discussed in Chapter 4. A more direct procedure for establishing the same bending-moment equations developed in the example of Art. 2.15 is demonstrated below.

Using the beam shown in Fig. 2.40a, pass a right section through it at any distance x between the left reaction and the concentrated load (see Fig. 2.41a). Taking moments about the cut section,

$$M_{0\text{-}5} = 23x - 2x\left(\frac{x}{2}\right)$$
$$M_{0\text{-}5} = 23x - x^2$$

It will be observed that this is the same equation as that developed in step (5) of the preceding example.

Repeating this procedure for a right section taken at *any* point along the beam between the concentrated load and R_R (Fig. 2.41b),

$$M_{5\text{-}20} = 23x - 2x\left(\frac{x}{2}\right) - 4(x-5)$$
$$M_{5\text{-}20} = 19x - x^2 + 20$$

which is the same as the equation derived in step (6) of the example in Art. 2.15.

PROBLEMS

1–7. Determine the reactions for the beams shown in Figs. 2.42 through 2.48, neglecting the beam weight. (Answers to Problems 1 through 3 given in Appendix G.)

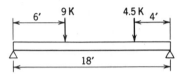

Figure 2.42

8. Draw the shear and bending-moment diagrams for the beam shown in Fig. 2.42, using the reactions determined in Problem 1. Use a length scale L of $\frac{1}{8}$ in. = 1 ft, a shear scale V of 1 in. = 20 kips, and a bending-moment scale M of 1 in. = 60 ft-kips. (Answer for shear and bending-moment values given in Appendix G.)

9. Draw the shear and bending-moment diagrams for the beam shown in Fig. 2.43, using the reactions determined in Problem 2. Scales: L, $\frac{1}{8}$ in. = 1 ft; V, 1 in. = 20 kips; M, 1 in. = 60 ft-kips. (Answers for shear and bending-moment values given in Appendix G.)

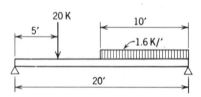

Figure 2.43

10. Draw the shear and bending-moment diagrams for the beam shown in Fig. 2.44, using the reactions determined in Problem 3. Scales: L, $\frac{1}{8}$ in. = 1 ft; V, 1 in. = 10 kips; M, 1 in. = 60

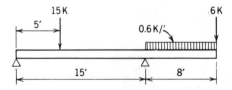

Figure 2.44

ft-kips. (Answers for shear and bending-moment values given in Appendix G.)

11. Draw the shear and bending-moment diagrams for the beam shown in Fig. 2.45, using the reactions determined in Problem 4. Scales: L, $\frac{1}{8}$ in. = 1 ft; V, 1 in. = 10 kips; M, 1 in. = 50 ft = kips.

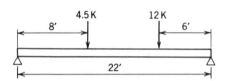

Figure 2.45

12. Draw the shear and bending-moment diagrams for the beam shown in Fig. 2.46, using the reactions determined in Problem 5. Scales: L, $\frac{1}{8}$ in. = 1 ft; V, 1 in. = 20 kips; M, 1 in. = 50 ft-kips.

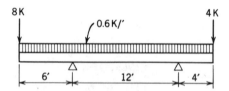

Figure 2.46

13. Draw the shear and bending-moment diagrams for the beam shown in Fig. 2.47, using the reactions determined in Problem 6. Scales: L, $\frac{1}{8}$ in. = 1 ft; V, 1 in. = 20 kips; M, 1 in. = 60 ft-kips.

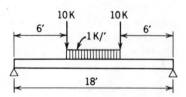

Figure 2.47

14. Draw the shear and bending-moment diagrams for the beam shown in Fig. 2.48, using the reactions determined in Problem 7. Scales: L, $\frac{1}{4}$ in. = 1 ft; V, 1 in. = 20 kips; M, 1 in. = 60 ft-kips.

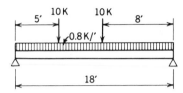

Figure 2.48

15. Compute the reactions, draw the shear and bending-moment diagrams, and write their corresponding equations for the beam shown in Fig. 2.49. Scales; L, $\frac{1}{4}$ in. = 1 ft; V, 1 in. = 10 kips; M, 1 in. = 10 ft-kips.

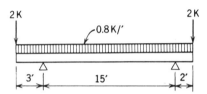

Figure 2.49

16. Draw the shear and bending-moment diagrams for the cantilever beam shown in Fig. 2.50 and write their corresponding equations. Scales: L, $\frac{1}{4}$ in. = 1 ft; V, 1 in. = 10 kips; M, 1 in. = 60 ft-kips.

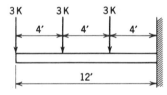

Figure 2.50

17. Draw the shear and bending-moment diagrams, and write their corresponding equations, for the beam shown in Fig. 2.51. Scales: L, $\frac{1}{4}$ in. = 1 ft; V, 1 in. = 10 kips; M, 1 in. = 30 ft-kips.

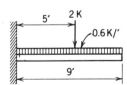

Figure 2.51

18. A simple beam 14 ft long carries a uniformly distributed load of 800 lb per ft over its entire length, and a concentrated load of 4 kips located 6 ft from the left end. Draw the shear and bending-moment diagrams and write their corresponding equations. Scales: L, $\frac{1}{4}$ in. = 1 ft; V, 1 in. = 5 kips; M, 1 in. = 20 ft-kips.

19. Solve Problem 13 by parts i.e., by drawing the shear and bending-moment diagrams for the uniform load and the concentrated loads separately and then combining them.

20. A simple beam 18 ft long carries a uniformly distributed load of 800 lb per ft over its entire length, and a concentrated load of 3 kips located 4 ft from the left end. Start with the load diagram and its equations, and, through subsequent integration, construct the shear and bending-moment diagrams and write their corresponding equations. For all equations, use $x = 0$ at the left reaction. Determine, algebraically, the location and magnitude of the maximum bending moment. Scales: length, $\frac{1}{4}$ in. = 1 ft; load, 1 in. = 1 kip per ft; V, 1 in. = 10 kips; M, 1 in. = 20 ft-kips. (Answers given in Appendix G.)

21. A simple beam 20 ft long carries a uniformly distributed load of 500 lb for 12 ft of its length starting from the left end, and another uniformly distributed load of 1000 lb per ft for the remaining 8 ft of length. Start with the load diagram and its equations, and, through subsequent integration, construct the shear and bending-moment diagrams and write their corresponding equations. Determine, algebraically, the location and magnitude of the maximum bending moment. Scale: length, $\frac{1}{4}$ in. = 1 ft; load, 1 in. = 1 kip per ft; V, 1 in. = 6 kips; M, 1 in. = 20 ft-kips.

22. A simple beam 14 ft long carries a uniformly varying load. The load varies from 500 lb per ft at the left end to 1200 lb per ft at the right end. Start with the load diagram and its equations, and, through subsequent integration, construct the shear and bending-moment diagrams and write their corresponding equations. Determine the point of maximum bending moment. Scales: length, $\frac{1}{4}$ in. = 1 ft; load, 1 in. = 1 kip per ft; V, 1 in. = 6 kips; M, 1 in. = 20 ft-kips. Compare this solution with one using an average uniformly distributed load of 850 lb per ft acting over the entire length. (Partial answers given in Appendix G.)

3

BEAMS—Bending and Shear

3.1
Resisting Moment

In the preceding chapter, the external forces acting on a beam were discussed, and a method for measuring their effect in terms of shear and bending moment was developed. It is now necessary to consider those forces acting within a beam that resist the tendency of the external forces to cause failure. The first of these is the action within the beam that resists bending; this is called the *resisting moment*.

Before proceeding with the derivation of an expression for the resisting moment, it will be well to recall the principle that for every action there is an equal and opposite reaction. Referring to Fig. 2.25, consider that the uniform load is being increased slowly. Under these conditions the maximum value of the bending moment at the center of the span will increase with the load, as will the bending moment at all other points along the span. In order to maintain equilibrium, an increasing resisting moment must be built up within the beam by the internal stresses developed. The maximum value of the resisting moment will, of course, occur at the point of maximum bending moment, but the bending moment at *every* section along the span will be opposed by a resisting moment which is equal in magnitude to the bending moment and opposite in sign.

3.2
Theory of Bending

The derivation of the expression for resisting moment is based on two fundamental assumptions. The first of these is:

A plane section of a beam before bending remains a plane section after bending.

41

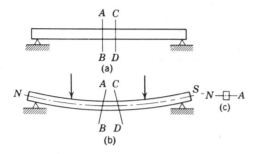

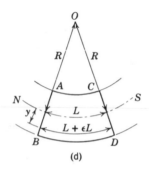

Figure 3.1/Theory of bending.

The parallel plane sections AB and CD in Fig. 3.1a remain plane sections after the beam is bent as shown in Fig. 3.1b. They are, however, no longer parallel.

Figure 3.1d shows the segment of the beam between the planes in an exaggerated curve. From this figure it is clear that the portion of the beam between the two sections was rectangular before bending and became trapezoidal after bending. Therefore, the fibers[1] in the upper portion of the beam must have become shorter than their original length, and those in the lower portion, longer. Furthermore, it is evident from Fig. 3.1d that the maximum shortening of the fibers occurred at the top surface and the amount of shortening decreased toward the center of the beam. Likewise, the maximum lengthening decreased toward the center. From this it follows that there must be some surface

---

[1]Steel, of course, is not a fibrous material in the sense that wood is, but the concept of infinitely small fibers is very useful in studying stress relationships within any structural material.

between the top and bottom of the beam where no change in length occurred. This surface is called the *neutral surface* and is designated in the figure by the line *NS*. The line in which this neutral surface cuts a right section of the beam is called the neutral axis *NA* (Fig. 3.1c).

The second assumption used in the derivation of the resisting moment is:

Within the elastic limit of the material, the stress varies directly as the deformation and, therefore, as the distance from the neutral axis.

In other words, those fibers having greater than average changes in length carry more stress. This means that the maximum compressive stress in the beam shown in Fig. 3.1b occurs at the extreme top fiber and decreases uniformly to zero at the neutral axis. The maximum tensile stress occurs at the extreme bottom fiber and decreases uniformly to zero at the neutral axis.

The proposition that the value of the fiber stress at any point in a beam cross section is proportional to the distance of that fiber from the neutral axis can be demonstrated very simply. Referring to Fig. 3.1d, assume that planes AB and CD are taken very close together so that when extended they meet at point O.

Then, let

R = radius of curvature of neutral surface (NS),

L = original length of the fibers before bending (the length at the neutral surface does not change),

y = distance from NS to *any* fiber of the beam in the plane,

ϵ = unit deformation (strain) at a fiber y distance from the neutral surface.

By geometry,

$$\frac{R}{L} = \frac{R + y}{L + \epsilon L}$$

and

$$\epsilon L = \frac{L}{R}(R + y) - L$$

Simplifying $\epsilon = y / R$. Now, referring to Appendix A, it is found that $\epsilon = f / E$, i.e., strain = unit stress divided by modulus of elasticity. The strains, therefore, may be equated and, solving for the stress,

$$\frac{f}{E} = \frac{y}{R}$$

$$f = \frac{E}{R}y$$

Since the modulus of elasticity is constant for any given material, and the radius of curvature is constant at any given point for any loading, it is apparent that the fiber stress is proportional to the distance from the neutral axis. Also, since most common structural materials have the same modulus of elasticity in tension as in compression, the extreme fiber stress in tension and compression, for beams of the same material and symmetrical cross section, is equal.

3.3
Position of Neutral Axis

In symmetrical sections such as rectangular beams and I-shaped beams, the neutral axis is midway between the top and bottom surfaces. Stated more generally, the neutral axis passes through the center of gravity (or centroid) of the section. This is true regardless of the shape of the beam cross section, although a rectangular section has been used for simplicity in the demonstration below (Fig. 3.2).

It should be borne in mind that the varying stresses indicated in Fig. 3.2b act perpendicular to the outline of the beam cross section shown in Fig. 3.2a. Also, it is assumed here that the section resists the action of a positive bending moment, i.e., stresses above the neutral axis are com-

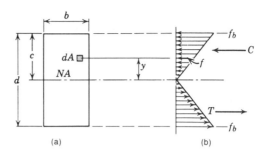

Figure 3.2/Bending stress distribution.

pressive, and those below, tensile. The area denoted $\overline{dA}$ represents a very small area (differential area) of the section at a distance y from the neutral axis. The force acting on area $\overline{dA}$ is equal to the value of the fiber stress at y times $\overline{dA}$. However, it was shown above (Art. 3.2) that the fiber stress at *any* distance y from the neutral axis is equal to $(E / R)y$. Therefore, the force on area $\overline{dA}$ becomes $(E/R)y\,\overline{dA}$.

Since $\overline{dA}$ is *any* differential area *above* the neutral axis, the sum of all such terms as $(E/R)y\,\overline{dA}$ will equal the total compression on the section—shown in Fig. 3.2b as C. Using differential areas *below* the neutral axis, the total tension T can be obtained by the same procedure. Since the only horizontal forces present are C and T, their algebraic sum must be zero. Consequently, the sum of the product of all the differential areas and their respective stresses for the entire section must be zero. Expressed in equation form,

$$T - C = 0 = \sum \frac{E}{R}y\,\overline{dA}$$

But the ratio E/R, being a constant, is not equal to zero. Therefore,

$$\sum y\,\overline{dA} = 0$$

From this expression, it may be stated that the product of each differential area and the distance from its center of gravity to the neutral axis of the section, when summed over the entire section, is equal to zero. Therefore, $\sum y\,\overline{dA}$ is the moment of the total area about the neutral axis. The moment is zero only if the neutral axis passes through the centroid of the total area. Consequently, the neutral axis must pass through the centroid.

3.4
The Beam Bending Formula

The expression for resisting moment that will now be developed on the basis of the foregoing discussion is known as the *beam formula* or *flexure formula*. From Fig. 3.2 it is apparent that the resisting moment of the section is equal to the couple formed by the forces C and T. However, rather than evaluating this couple, a simpler approach to derivation of the beam formula is through the summation of differential areas.

It was shown in Art. 3.3 that the force on the differential area is $(E/R)y\,\overline{dA}$. The moment of this force about the neutral axis may be denoted dM and expressed as follows:

$$dM = \frac{E}{R}y\,\overline{dA}\,y$$

$$dM = \frac{E}{R}y^2\,\overline{dA}$$

The total moment at the section would be the sum of all the differential moments, i.e.

$$\sum dM = \sum \frac{E}{R}y^2\,\overline{dA}$$

or

$$M = \frac{E}{R}\sum y^2\,\overline{dA}$$

However, since

$$\frac{E}{R} = \frac{f}{y}$$

this equation becomes

$$M = \frac{f}{y}\sum y^2\,\overline{dA}$$

The quantity $\sum y^2\,\overline{dA}$ is called the *moment of inertia* of the section with respect to the neutral axis and is designated by the symbol I. In a mathematical sense, the moment of inertia of an area may be described as the sum of the products obtained by multiplying each elementary area (composing the given area) by the square of its distance from an axis (in this case the neutral axis). In a physical sense, the moment of inertia may be thought of as a factor which influences the resisting moment by relating it to the size and shape of the beam cross section. The units for I will be distance (usually inches) raised to the fourth power (inches^4). Substituting I for $\sum y^2\,\overline{dA}$, the *general* expression of the beam formula becomes

$$M = \frac{f}{y}I$$

or, as commonly written,

$$M = \frac{fI}{y}$$

In this general expression, f is the unit stress at any point in a section located at a distance y from the neutral axis. It may be tension or compression. In design, the most critical stress usually is the maximum stress, which occurs on the outermost fibers, or where y is a maximum. Since the letter c is usually used to designate the distance of the extreme fiber from the neutral axis, the beam formula may be written

$$M = \frac{fI}{c}$$

With any given material (such as steel), the maximum allowable unit stress in bending will be established by specification;[2] therefore, the maximum *allowable* resisting moment of a particular section will be

$$M_R = \frac{F_b I}{c}$$

3.5
Section Modulus

The quantity I/c is called the section modulus and is dependent upon the size and shape of the section. It is defined as the moment of inertia divided by the distance from the extreme outer fiber to the neutral axis. Its value is in inches cubed, and it is commonly represented by the letter S. The *AISC Manual* and other structural handbooks give the section moduli of rolled shapes as well as their moments of inertia. These are usually listed in "Dimensions and Properties" or some similarly titled tables. For example, when the familiar I-shaped sections[3] are used as beams to support vertical loads, and are placed in the customary manner with webs vertical, the values of S and I to be used in the beam formula are those for the x-x axis shown in the properties tables.

3.6
Beam Sections

It is evident, from the beam formula, that the resisting moment of a beam section varies with its moment of inertia, i.e., the

[2] In the general nomenclature of the 1989 AISC Specification, a distinction is made between permitted or allowable bending stress (F_b) and computed or actual bending stress (f_b); therefore, these symbols will be used in design discussions, rather than the general bending-stress symbol (f) used in theoretical discussions.

[3] I-shaped sections, as used in this text, refer to those sections designated in the Ninth Edition of the AISC Manual as **S**, **W**, and **M** (see Art. 3.6 and Fig. 3.3).

larger the moment of inertia, the larger the resisting moment. Therefore, the most economical section would be one that provided the largest moment of inertia with the least area. Also, since the maximum value of the fiber stress occurs at the extreme top and bottom of the beam cross section, it is desirable to place as much as possible of the material near the top and bottom faces of the section.

The familiar I structural-steel beam shapes resulted from the effort to satisfy these criteria.

Figure 3.3 (a), (b), (c), (d), (e), and (f) show various sections, both symmetrical and unsymmetrical, produced at steel rolling mills. Tee sections (Fig. 3.3e) are manufactured in small sizes; however, the large sizes used for structural purposes generally are made by splitting **W**, **M**, or **S** sections. Plates and bars (Fig. 3.3f) also are stocked by fabricators, so that as the need arises they can be used along with standard rolled shapes to produce built-up sections (Fig. 3.3g). Many such combinations of plates and shapes are possible. Built-up sections are discussed in greater detail in Chapter 5.

There are times—e.g., when a member is to be exposed—that the designer may prefer to use steel pipe or structural tubing (Fig. 3.3h). Because of the different manufacturing process involved, however, structural tubing is not produced in the same ASTM designations as other rolled shapes of the type shown in Fig. 3.3. Currently, steel pipe is available in ASTM A53, Grade B (F_y = 35 ksi), and on a limited basis in ASTM A501 (F_y = 36). Steel tubing is available in ASTM A500, Grade B (F_y = 46 ksi).

The x-x axis is shown on each of the Fig. 3.3 beam sections. This axis passes through the centroid of the section and becomes the neutral axis when bending occurs about that axis. In other words, the plane of loading must be perpendicular to the neu-

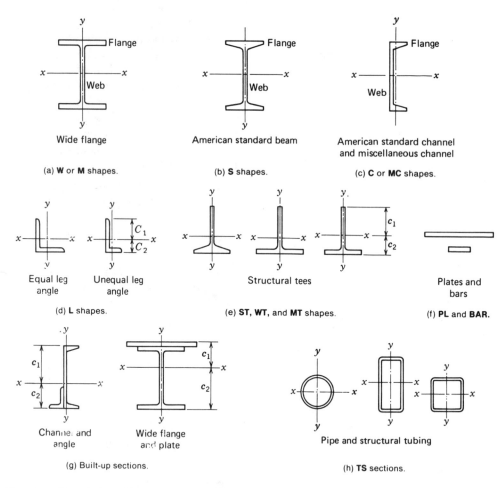

Figure 3.3/Manufactured beam shapes and sections.

tral axis to utilize the quantities I, c, and S computed with reference to that axis. On rare occasions, beam sections are placed so that bending occurs about the y-y axis. When such is the case, y-y becomes NA, and the beam properties for designing must be referenced to the y-y axis. Part I of the AISC Manual lists the various dimensions and properties for designing, for most available beam shapes.

When the beam is symmetrical about NA, the value for c is half the depth of the beam. Consequently, there is only one value for the section modulus, and the maximum fiber stresses developed in tension and compression are exactly the same, occurring on opposite outermost fibers. When the beam is not symmetrical about NA there will be two values for c (Fig. 3.3d, e, and g). As a result, the stresses in the extreme top and bottom will not be the same. It is evident that there are two possible values of S, i.e., $S = I/c_1$, or I/c_2. If the allowable stresses in tension and compression are the same, the smallest S ($S = I/c$), should be used in computing the allowable resisting moment, and is the only one listed in the AISC Manual. (The reader should prove that this is a valid conclusion.)

The steel shapes shown in Fig. 3.3 (a), (b), (c), (d), and (e) are designated by a shape or group symbol, followed by a nominal depth and the actual weight per foot of length. The designation **W** 10×49 indicates a wide flange shape, approximately 10 in. deep, and weighing 49 lb per linear foot. The shape or group symbols, their names, and principal characteristics are as follows:

W   Wide-flange shapes. Parallel inner and outer flange surfaces (available in the largest range of sizes and thus the most frequently used).[4]

M   Miscellaneous shapes. Special lightweight shapes having a profile similar to **W** shapes (available from a limited number of producers or is infrequently rolled).

S   American standard beams. Webs thicker and flanges narrower than the **W** shapes; also, tapered inner flange surface.

C   American standard channels. C-shaped, with tapered inner surface on flanges.

MC   Miscellaneous channels. Special-purpose shapes, having the same profile as **C** shapes (available from a limited number of producers or is infrequently rolled).

L   Angles. Angle legs, either equal or unequal in length, set at a right angle to each other.

ST, WT, MT   Structural tees. **S**, **W**, and **M** shapes, split at mid-depth.

PL and **BAR**   Plates and bars.

TS   Steel pipe and structural tubing, circular, rectangular, or square in shape.

The designation for steel pipe and structural tubing is different from other shapes. The shape group is given first, followed by the depth and width dimensions (or the diameter in the case of pipe), and the outside wall thickness—e.g., **TS** $6 \times 4 \times$ .375, or, for circular sections, **TS** 6 OD × .280.

The actual depth of **W** and **M** sections usually varies somewhat from the nominal depth, since the heavier sections of any one nominal depth are produced by adding thickness to the outer face of the flanges and to the web. Some of the heavier 14-in. **W** sections, for example, have actual depths significantly larger than the nominal size given.

As has been noted, not all rolled beam sections are available in all the types and grades of steel. The various types of steel were discussed in Art. 1.6, and Table 1.3 placed them in stress grades along with their availability according to an ASTM A6 grouping. In general, those heavier beams with large flange thicknesses are available only in the lower-stress grades. A notable exception to this is A36 and most of A588. Before specifying any beam size and grade of steel, it is always good practice to check its availability both locally and nationally. Table 2 in the AISC Manual summarizes the A6 structural-shape size grouping, which then can be cross-checked against Table 1.3.

---

[4]**HP** bearing pile shapes, having essentially parallel flanges and equal web and flange thicknesses, also are available but not generally used for above-ground structural building frames.

3.7
Use of the Beam Formula

The beam formula may be presented in three different ways:

(1) $M = \dfrac{F_b I}{c}$; (2) $f_b = \dfrac{Mc}{I}$; (3) $\dfrac{I}{c} = \dfrac{M}{F_b}$;

or, letting $I/c = S$,

(1) $M = F_b S$; (2) $f_b = \dfrac{M}{S}$; (3) $S = \dfrac{M}{F_b}$.

With form (1), when the dimensions of the beam are known (represented by I/c) and F_b is the maximum allowable fiber stress for the material, solving for M gives the maximum bending moment that the beam can resist, i.e., the maximum resisting moment.

With form (2), when the maximum bending moment due to the loading is known, as well as the dimensions of the beam, solving for f_b gives the developed bending stress in the extreme fiber.

With form (3), when the bending moment and allowable extreme fiber stress are known, solving for I/c gives the section modulus required. This is the form used in design. When the required section modulus has been computed, a beam section is selected that has an I/c equal to or greater than the one required.

Frequently, the allowable bending stress is difficult to establish and cannot be precisely determined until the section itself has been selected. Under these conditions, the value of F_b can be approximated, and later in the design process it can be refined after trial sections have been selected. The basic allowable bending stress (F_b) is taken as 60 per cent of F_y. Under certain conditions, this value can be increased to 66 per cent of F_y, and yet under other conditions it could be reduced to 20 or 30 per cent of F_y. Chapter 5 treats this topic in greater detail. For the remain-der of this chapter, the basic value (60 per cent of F_y) will be used.

It should be noted that F_b and f_b are expressed in pounds per square inch, c in inches, S in inches^3, and I in inches^4. M, therefore, must be in inch-pounds. M, as usually computed from the reactions and loads, is expressed in foot-pounds and must be converted to inch-pounds by multiplying its value by 12 before it is used in the formula.

Later, in Chapter 5 and in subsequent chapters, there will be ample opportunity to show correlations between developed and handbook solutions. These correlations will not always be in exact agreement, because of a difference in number roundoff procedures. Some examples in this text round off the allowable stress $F_b = 0.60(36) = 21.6$ to 22 ksi, and $F_b = 0.66(36) = 23.8$ to 24 ksi. The tables and charts in the AISC Manual are inconsistent in that the charts are based upon this roundoff procedure and the tables use the more exact number.

Example 1

(a) Find the maximum resisting moment of a **W** 16×36 if its allowable bending stress F_b is $0.6F_y$ and the steel A36.

(b) Would the section in (a) be considered safe in bending for a simply supported beam if the span were 28 ft and it carried a superimposed load of 1000 lb per ft?

(c) What would be the maximum actual fiber stress (f_b) for this loading condition?

Solution 1

(1) Referring to Part 1 of the AISC Manual, "**W** Shapes—Dimensions and Properties," the elastic section modulus for the **W** 16×36 is found to be 56.5 in.^{3}. Also,

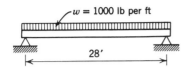

Figure 3.4/Example 1

$F_b = 0.6(36,000) \simeq 22,000$ psi. Then

$M = F_bS = 22,000(56.5) = 1,240,000$ in.-lb

$M = \dfrac{1,240,000}{12} = 103,000$ ft-lb

Solution 2

(1) Make a sketch of the beam showing the loads (Fig. 3.4). The beam must support its own weight in addition to the superimposed load; therefore, the unit load is $1000 + 36 = 1036$ lb per ft.

(2) From Fig. 2.34 or the AISC Manual ("Beam Diagrams and Formulas"), it is seen that for this loading condition the maximum moment occurs at midspan and is equal to

$M = \dfrac{wL^2}{8} = \dfrac{1036(28)^2}{8} = 102,000$ ft-lb

The section is safe as long as the allowable resisting moment (103,000 ft-lb determined above) is equal to or greater than the developed bending moment (102,000 ft-lb).

Solution 3

(1) The maximum fiber stress will be at the outermost fibers of the beam at the point where the largest bending moment occurs (midspan), or

$f_b = \dfrac{M}{S} = \dfrac{102,000(12)}{56.5} = 21,700$ psi

(2) This also proves as a check on the safety of the beam, as the actual fiber

stress f_b (21,700 psi) is less than the allowable, F_b (22,000 psi).

Example 2

An **S** 10×35 has a span of 10 ft. How great a concentrated load at the center will the beam support in bending if the steel is type A242 and the allowable bending stress is $0.6F_y$?

Solution

(1) Make a sketch of the beam showing the loads (Fig. 3.5).

(2) Referring to Table 1.3 for A242 steel and AISC Manual, Table 2, it is observed that the **S** 10×35 falls into group 1. Consequently, $F_y = 50,000$ psi and

$F_b = 0.6(50,000) = 30,000$ psi

(3) Referring to the AISC Manual, **S** Shapes—Properties, the section modulus is found to be 29.4 in.[3]. The maximum resisting moment of the beam is

$M = F_bS = 30,000(29.4) = 882,000$ in.-lb

$M = \dfrac{882,000}{12} = 73,500$ ft-lb

(4) From Fig. 2.34, it is seen that the maximum moment resulting from the concentrated load is at midspan and is equal to

$$M = \dfrac{PL}{4}$$

The moment resulting from the beam

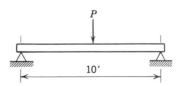

Figure 3.5

weight occurs at the same point and is equal to $M = wL^2/8$. The numerical sum of these moments represents the total maximum moment caused by the external forces:

$$M = \frac{PL}{4} + \frac{wL^2}{8}$$

$$M = \frac{P10}{4} + \frac{35(10)^2}{8} = 73,500 \text{ ft-lb}$$

$$P = 29,200 \text{ lb}$$

Example 3

Design for resistance to bending the most economical beam that will carry a superimposed uniformly distributed load of 1900 lb per ft, if the beam is simply supported over a span of 24 ft. The maximum allowable stress in bending (F_b) is $0.6F_y$, and the steel, A36.

Solution

(1) Make a sketch of the beam showing the loads (Fig. 3.6).

(2) Determine the maximum moment resulting from the superimposed load.

$$M = \frac{wL^2}{8} = \frac{1900(24)^2}{8} = 137,000 \text{ ft-lb}$$

(3) The allowable bending stress is

$$F_b = 0.6(36,000) \simeq 22,000 \text{ psi}$$

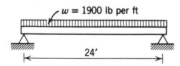

Figure 3.6

(4) Determine the required section modulus for this moment:

$$S = \frac{M}{F_b} = \frac{137,000(12)}{22,000} = 74.7 \text{ in.}^3$$

(5) Allowance must be made for the beam weight. The problem now is to select a beam having a section modulus slightly greater than 74.7 in.^{3}. To aid in this, Part 2 of the AISC Manual provides a table entitled "Allowable Stress Design Selection Table," where shapes are listed in order of the magnitude of their section modulus. Referring to this table, a **W** 21 × 44 is selected having an S of 81.6 in.^{3}. All other beams listed within this group, having a smaller section modulus, are greater in weight.

(6) The **W** 21 × 44 should be checked, taking into consideration the dead load of the beam. The moment at the center due to the weight of the beam is

$$M = \frac{wL^2}{8} = \frac{44(24)^2}{8} = 3200 \text{ ft-lb}$$

The total bending moment at the center is

$$M = 137,000 + 3,200 = 140,200 \text{ ft-lb}$$

The section modulus required for this moment is

$$S = \frac{M}{F_b} = \frac{140,200(12)}{22,000} = 76.5 \text{ in.}^3$$

Inasmuch as this required value is less than the section modulus of the selected beam (81.6 in.^{3}), the **W** 21 × 44 is adequate.

Example 4

Design for resistance to bending the lightest-weight beam that will carry a superimposed uniformly distributed load of 2 kips per ft and a concentrated load of 8 kips

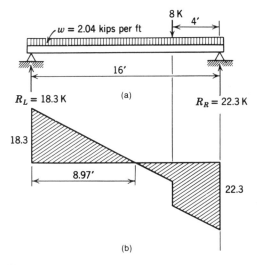

Figure 3.7

acting 4 ft from the right reaction. The beam is simply supported and has a span of 16 ft. The allowable stress in bending (F_b) is $0.6F_y$, and the steel, A36.

Solution

(1) Make a sketch of the beam showing the loads (Fig. 3.7a), and compute the reactions. It may be observed from the loading that the maximum moment does not occur at midspan but at some other point. From Chapter 2 it is known that the maximum moment will occur at the point where the shear is equal to zero.

(2) Draw the shear diagram, and calculate the point where the shear is equal to zero as indicated in Fig. 3.7b. (When the point of maximum moment is unknown, the effect of the beam weight is more easily accounted for by estimating its magnitude and including it at the beginning of the computations.) Assume a beam weight of 40 lb per ft.

(3) Calculate the maximum moment. It is not necessary to draw the bending-moment diagram, as only the maximum moment is used in the solution. Using a free body to the left of the point of zero shear,

$$M = 18.3(8.97) - 2.04(8.97)\frac{8.97}{2}$$

$$= 82.1 \text{ ft-kips}$$

This value also may be found by determining the area of the shear diagram to the left of the point of zero shear, as follows:

$$M = \frac{18.3(8.97)}{2} = 82.1 \text{ ft-kips}$$

(4) Determine the required section modulus:

$$S = \frac{M}{F_b} = \frac{82.1(12)}{22} = 44.8 \text{ in.}^3$$

(5) Referring to the "Allowable Stress Design Selection Table" (AISC Manual), select a **W** 16×31 having an $S = 47.2$ in.^{3}. Inasmuch as the assumed weight of the beam was 40 lb per ft, no further refinement is necessary.

3.8
Design Procedure

The last example illustrates steps necessary for the design of a simple beam to resist bending stresses. For beams with more complicated systems of loading, additional operations are often required. The following procedure gives the usual steps necessary for a complete design to resist bending. In Chapter 5, more extensive use will be made of the AISC Manual in design.

1. Make a sketch of the beam showing the loads. This may or may not include an estimated weight for the beam, depending upon the designer's experience and the complexity of the problem. (See Art. 3.9.)

2. Compute the reactions.

3. Find the points of maximum bending moment. These will occur where the shear line passes through zero. For the more complicated cases it may be necessary to plot the shear diagram.

4. Compute the maximum bending moment. For beams having overhangs there are maximum positive and maximum negative bending moments. It is the greatest numerical value, regardless of sign, that is generally used in designing.[5] For complicated loadings it may be advantageous to plot the bending-moment diagram.

5. Select an allowable unit stress (F_b) based on the structural steel to be used, and find the required section modulus using the beam formula

$$S = \frac{M}{F_b}$$

6. Select a beam with a section modulus somewhat *larger* than that determined necessary in step 5 in order to allow for the additional bending moment caused by the beam weight. (When the estimated weight of the beam has been included in step 1 with reasonable accuracy, this excess is not needed and the solution is complete without performing step 7.)

7. Recompute the maximum bending moment due to the combined effect of the superimposed load and the beam weight,

[5]It is evident from a comparison of Fig. 2.23a and the discussion in Art. 3.2 that positive bending moment induces tensile stress in the bottom fibers of a beam and compressive stress in the top fibers. A study of Fig. 2.23b will show that this condition is reversed in the case of negative bending moment, the top fibers being in tension and the bottom fibers in compression. Since, in this chapter, the allowable stresses for structural steel in tension and compression are assumed equal, this consideration is not important with symmetrical sections such as **W** or **S** sections except where overhanging or continuous beams are to be spliced. However, other conditions, as presented in Chapter 5, will require more detailed consideration of positive and negative bending moments and their influence on design.

and from this again determine the required section modulus. This value should not be greater than the section modulus of the beam previously assumed. If it is, a new beam must be assumed and this step repeated.

When there are no other governing conditions, the lightest-weight section that will carry the load should be used. In many instances, however, the required clear story height or other architectural considerations limit the depth, thereby necessitating the use of a heavier beam of equal strength but less depth. The above procedure may be abbreviated as one becomes proficient in design.

3.9
Effect of Beam Weight

In cases where the weight of the beam is very small compared with the applied load, it may be neglected. However, in fireproof construction, where the steel beams may be surrounded by concrete, the weight of the beam plus the fireproofing is often an appreciable part of the load. Furthermore, the maximum bending moment, with the beam weight included, may not occur at exactly the same location as the maximum bending moment due to the superimposed loads alone (Example 4). In cases where the weight of the beam is large compared to the superimposed load, the position and amount of the maximum bending moment should be checked after the weight of the beam is known.

PROBLEMS

The following problems call for design for resistance to bending stresses only. A handbook giving properties of structural shapes is required for their solution. It is recommended that the Ninth Edition of the AISC Manual be used.

1. A **W** 12×40 is simply supported on a 16-ft span. Find the total uniform load it will support in addition to its own weight. The steel is A36, and the allowable bending stress is $0.6F_y$. (Answer given in Appendix G.)

2. An **S** 18×54.7 is simply supported on a 21-ft span. It supports a uniformly distributed load of 2,000 lb per ft (including its own weight) plus a concentrated load of 3 kips acting 3 ft from the left reaction. Determine the magnitude and position of the maximum computed bending stress (f_b).

3. A beam spans 24 ft and carries a uniformly distributed load of 2 kips per ft plus concentrated loads of 10 kips placed at its third-points. Find the lightest **S** shape and lightest **W** shape that will sustain this loading if the steel is A36 and the allowable bending stress is $0.6F_y$. (Answer given in Appendix G.)

4. A beam spans 16 ft and carries concentrated loads of 2 kips at its quarter points. Find the lightest nominal 10-in. beam required if the steel is A36 and the allowable bending stress is $0.6F_y$. (Answer given in Appendix G.)

5. A **TS** $6 \times 3 \times \frac{1}{4}$ spans 8 ft. Determine the largest concentrated load that may be placed at its midspan. The steel is A501, and the allowable bending stress is $0.6F_y$.

6. A **W** 14×30 spans 21 ft. Neglect beam weight. Equal concentrated loads are applied at the third-points. Determine the maximum value of these loads if the steel is A36 and the allowable stress is $0.66F_y$.

7. An **S** 5×10 acts as a cantilever beam that projects 6 ft from a wall. The steel is A36, and the allowable stress is $0.66F_y$. Determine the maximum concentrated load that can be placed 2 ft from the free end.

8. A **W** 21×83 spans 20 ft. It has a uniform load of 500 lb per ft over its entire length, a concentrated load of 32 kips at 6 ft from the left reaction, and a concentrated load of 21 kips at 8 ft from the right reaction. Determine the location and magnitude of the maximum computed bending stress (f_b).

9. A **W** 14×30 rests on two supports 18 ft apart. The beam extends 6 ft beyond the left support and 4 ft beyond the right support. A concentrated load of 12 kips is applied midway between the two supports, another concen-

trated load of 6 kips is applied at the left end of the beam, and a third concentrated load of 10 kips is applied at the right end of the beam. In addition, a uniform load of 400 lb per ft, including the beam weight, extends over the entire length of the beam. Find the value of the computed bending stress (f_b):

(a) over the left support;

(b) under the 12 kip load;

(c) over the right support.

In each case, indicate the position of the maximum computed bending stress in tension and compression, i.e., top or bottom of beam.

10. Select an American Standard beam (**S** shape) to carry a single concentrated load of 20 kips at the center of an 18-ft span. The steel is A36, and the allowable stress is $0.66F_y$. The beam will be fully encased in concrete (2-in. covering on all sides). Assume weight of concrete to be 144 lb per cu ft.

11. Select a wide-flange section (**W** shape) 24 ft long to support concentrated loads of 10 kips located at each third-point of the span. The steel is A572, Grade 50, and the allowable stress is $0.6F_y$.

12. Select an S-shape beam to carry a superimposed load of 1500 lb per ft over a span of 22 ft. The steel is A588, Grade 50, and the allowable bending stress is $0.6F_y$.

13. Select a **W** shape for the following conditions: The span is 18 ft; 6 ft from the right reaction a concentrated load of 6 kips is applied, and a uniform load of 500 lb per ft extends from the left reaction to the concentrated load. The steel is A36, and the allowable stress is $0.66F_y$.

14. Select a **W** shape for the following conditions: The beam span is 24 ft, three equal concentrated loads of 4 kips each are applied at the quarter-points, and the nominal depth of beam is limited to 10 in. The steel is A36, and the allowable bending stress is $0.6F_y$.

15. Design a beam for the following conditions: The overall length is 22 ft; the beam overhangs the right support by 6 ft; a concentrated load of 6 kips is applied 4 ft from the left reaction and another of equal magnitude at 8 ft from the left reaction; a third concentrated load of 8 kips is applied at the extreme end of the overhang;

and a uniform load of 1200 lb per ft extends over the length of the beam. The steel is A36, and the allowable bending stress is $0.66F_y$.

16. A C 8×18.75 lies on its web with the flanges pointing up. It has a span length of 8 ft and carries a uniform load of 200 lb per ft. Determine the maximum computed bending stress (f_b) in tension and in compression.

17. Select a structural tee section to meet the following conditions: the flange is on the bottom and must have a width of approximately 8 in.; the stem is vertical, and the span is 12 ft and must carry a uniform load of 350 lb per ft. Use A36 steel; allowable stress $= 0.6F_y$. For the section that is selected, determine the maximum computed bending stress in tension and in compression.

18. Select an angle to carry a load of 300 lb per ft over a $6\frac{1}{2}$-ft simple span. The bottom leg of the angle is to be $3\frac{1}{2}$ in., placed horizontally. The steel is A36, and the allowable bending stress is $0.6F_y$.

3.10
Shearing Resistance

After a beam has been designed for bending, its resistance to shear should be investigated. In most cases this investigation will not result in an increase in size. However, where the span is short, and/or the beam is heavily loaded or subject to large concentrated loads near the ends, shear failure is a possibility.

In Arts. 2.5, 2.6, 2.7, and 2.8 means were presented for determining the magnitude of the vertical shearing force at any section along the beam span in terms of the external forces. In order to maintain equilibrium there must be an equal and opposite internal resisting force developed by the beam at every section. This is called the *resisting shear force*. The capacity of a beam to resist bending moment was measured by the resisting moment in terms of fiber or bending stress. In a similar manner the magnitude of the resisting shear force must be measured in terms of a shearing stress. This shearing stress is the *intensity* of the

Figure 3.8/Horizontal shear.

force at any given point, and consequently must consist of units of force per unit of area (usually lb per sq in.). As with bending stresses, it is necessary to determine how the shear stress varies throughout the section and how to calculate its maximum value.

Figure 2.12 shows the tendency of all vertical planes to slide past one another due to vertical shearing forces. Figure 3.8 shows that there is an additional shearing action present, called horizontal shear.

Figure 3.8a represents a beam which has deflected due to a load P. Figure 3.8b represents a beam of the same dimensions as that shown in part a, but made up of three independent strips; the deflection in this case is more than that of the solid beam. Furthermore, it is evident that slipping has occurred along the surfaces of contact of the three independent pieces. This tendency for one layer of the beam to slip past another is also present in the solid beam of Fig. 3.8a but is prevented by the resistance of the beam to horizontal shear. There is a horizontal shearing force acting along every horizontal plane in the beam. Consequently, there must exist a horizontal shear stress at all horizontal planes.

3.11
Relation between Horizontal and Vertical Shear

Figure 3.9a represents a portion of the beam shown in Fig. 3.8a. The vertical lines represent two sections through the beam taken so close together (differential distance dx) that the shear on the sections

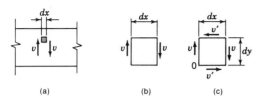

Figure 3.9/Vertical and horizontal shear.

top face is $v'(1)\,dx$ and that on the right side face is $v(1)\,dy$. Therefore, taking moments about O,

$$v'\,dx\,dy = v\,dx\,dy$$

or

$$v' = v$$

That is, the intensity of horizontal shear at any point in a beam is equal to the intensity of the vertical shear at that point. Vertical shear, then, produces another shearing force at right angles to itself, of equal intensity, acting lengthwise in a beam. The symbol v is used to indicate the shearing stress, either horizontal or vertical. In later discussions, the symbol f_v will be used to denote the developed shear stress in steel beams.

may be assumed equal. Consider that a small particle of the beam between these sections (having dimensions of dx and dy) has been removed. This particle is shown at a larger scale in Fig. 3.9b with the vertical forces (v) acting on it. It is evident from this figure that the particle is not in equilibrium and would rotate in a clockwise direction if no other forces were present. We know, however, that the particle is in equilibrium, and, therefore, some other forces such as those represented by v' in Fig. 3.9c must exist.

Since the particle is in equilibrium, moments taken about any point, such as the corner O, must equal zero. If the depth of the particle perpendicular to the paper is taken as unity, then the total stress on the

3.12
Intensity of Shear Stress

In order to determine how the shear stress varies throughout the cross section of a beam, it is necessary to develop a general expression representing the magnitude of

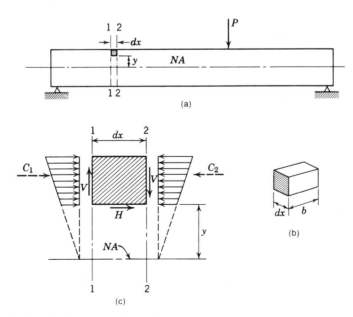

Figure 3.10/Derivation of shear-stress formula.

the stress at any point. To accomplish this it is convenient to isolate a small element of a beam as a free body and apply the laws of statics. This element is taken from the beam shown in Fig. 3.10a. For the sake of clarity, the beam is shown straight rather than in its deflected form.

The element to be considered falls between the two vertical planes 1-1 and 2-2. These planes are assumed to be very close together, only a differential distance dx separating them. Another plane, parallel to the neutral axis and a distance y from it, completely isolates the small rectangular block shown in Fig. 3.10b. Note that the block or element has a width b, equal to the beam width.

The bending moments at planes 1-1 and 2-2 will be designated M_1 and M_2, respectively. These moments will produce flexural stresses on the sides of the element as shown in Fig. 3.10c, and if a summation of these flexural stresses is made over their respective areas, they will have the resultants C_1 and C_2. If M_2 is greater than M_1, it follows that C_2 is greater than C_1, and the element will tend to move to the left. To maintain equilibrium, an additional stress H is required, acting over the area $b\,dx$. Consequently,

$$C_2 - C_1 = H = v(b)\,dx$$

The values of the stress resultants, C_1 and C_2, may be found by summing the stresses on the areas of the side faces of the block ($\Sigma f\,dA$). However, in Art. 3.4 it was shown that the fiber stress at any point was

$$f = \frac{My}{I}$$

Therefore,

$$C_1 = \Sigma \frac{M_1 y}{I}\,dA$$

or

$$C_1 = \frac{M_1}{I} \Sigma y\,dA$$

Also,

$$C_2 = \frac{M_2}{I} \Sigma y\,dA$$

and

$$C_2 - C_1 = \frac{M_2 - M_1}{I} \Sigma y\,dA$$

Since the two planes 1-1 and 2-2 are only a differential distance dx apart, the difference between the moments M_2 and M_1 is only a differential moment dM. Summarizing,

$$C_2 - C_1 = v(b)\,dx = \frac{dM}{I} \Sigma y\,dA$$

$$v = \frac{dM}{dx}\left(\frac{1}{Ib}\right) \Sigma y\,dA$$

Further analyzing the above equation, it has been shown that $dM/dx = V$ (see Art. 2.15). Also, $\Sigma y\,dA$ is the static moment, denoted by the symbol Q. It is the moment, about the neutral axis, of the area included between the plane of the desired stress and top edge of the section. Hence, the general expression for shear stress is

$$v = \frac{VQ}{Ib}$$

3.13
Distribution of Shear Stresses

Further examination of the shear formula shows that the magnitude of the shear varies inversely as the width of the section, so that the smaller the width, the larger the shearing stress. If the width remains constant for any given section, the only variable is Q, which takes a maximum value when the plane of stress is at the neutral axis.

Consider the rectangular beam section shown in Fig. 3.11a. The width b remains constant (2 in.) and the static moment Q is determined below with reference to the

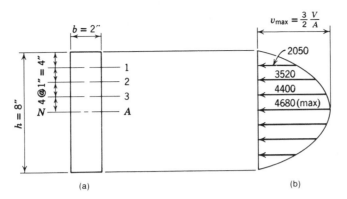

Figure 3.11/Shear distribution: rectangular shape.

four planes 1, 2, 3, and NA:

$$Q_1 = 2(1)3.5 = 7 \text{ in.}^3$$

$$Q_2 = 2(2)3 = 12 \text{ in.}^3$$

$$Q_3 = 2(3)2.5 = 15 \text{ in.}^3$$

$$Q_{NA} = 2(4)2 = 16 \text{ in.}^3$$

The moment of inertia with reference to the neutral axis $(bh^3/12)$ is 85.5 in.^{4}.

Substituting these values in the equation $v = VQ/Ib$, an external shearing force V of 50 kips would produce the stresses shown in Fig. 3.11b. These stresses are shown graphically to illustrate their distribution throughout the section. Although they are plotted horizontally, it should be remembered that there is also an equal stress at each point acting in the vertical direction.

A designer is primarily interested in the maximum stress. This will occur at the neutral axis. Using the rectangular section shown in Fig. 3.11a, the maximum shear stress at the neutral axis is

$$v = \frac{VQ}{bI} = \frac{V(b)(h/2)(h/4)}{b(bh^3/12)}$$

Letting $A = bh$ and substituting in the

above,

$$v = \frac{3}{2}\left(\frac{V}{A}\right)$$

Thus, for rectangular sections, the maximum shearing stress is $\frac{3}{2}$ times the average shearing stress.

In Art. 3.6 it was shown that the most efficient shape of beam to resist bending moment is a flange-type section. However, the flanges have very little influence on the resistance to shear. In order to illustrate this, the cross section of a **W** 12×40 is shown in Fig. 3.12a with the position of six horizontal planes indicated for use in computing the static moment. They are labeled NA, B, C, D, E, and F. Another plane is located at point G. In calculating the shear stress for the six planes in the web, the value of b in the formula VQ/Ib is the thickness of the web. For calculating the shear stress at point G, the value of b is the thickness of the flange. If the total vertical shear on the section is assumed to be 40 kips, then carrying out an analysis similar to the one above will result in unit stresses as recorded in Fig. 3.12b. It will be noted that although the flange areas are relatively large, they are not particularly effective. Approximately 90 per cent of the shearing resistance is carried by the beam web.

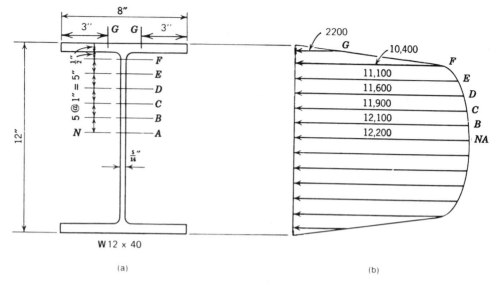

Figure 3.12/Shear distribution: flanged beams.

3.14
Designing for Shear

For routine design, the method illustrated in the preceding article for determining the maximum shear stress at the neutral axis of flanged beams is too cumbersome. Furthermore, there is some question as to the precise value that should be used for the allowable unit shear stress (F_v).[6]

Therefore, an approximate method is used in practice, one in which it is assumed that the web of the steel beam resists all the shear. An allowable value for the average unit shearing stress is used, which is low enough so that the actual maximum unit shear developed in the beam is within the limits of safety. For example, if the specified value for the allowable *average* unit shearing stress (F_v) was 14,500 psi when figured on this basis, the actual maximum stress developed might be in the neighborhood of 15,500 psi. In practically all build-

[6]In the General Nomenclature of the 1989 AISC Manual, F_v is the "allowable shear stress," f_v is the "computed (actual) shear stress," and V is the "maximum permissible web shear." These symbols will be used in this text in all design discussions.

ing codes, the value specified for allowable unit shearing stress in steel beams is based on this approximation. In conformance with this general practice, the term f_v as used in this text will denote *average unit shearing stress*, rather than unit shearing stress, and the symbol A_w will indicate area of web (actual depth of beam times web thickness). The shear formula, therefore, becomes

$$f_v = \frac{V}{A_w}$$

For the beam shown in Fig. 3.12, assumed to be sustaining a total vertical shear of 40 kips, the average unit shearing stress, using the above formula, is 11,400 psi, compared to the actual unit shear of 12,200 psi. The AISC Specification states that the allowable shear stress F_v on the webs of beams shall be not more than $0.4F_y$. For A36 Steel, $0.4F_y = 14.4$ ksi but is always rounded off to 14.5 ksi. Also, for 42-ksi-yield-stress steels, the allowable shearing stress is rounded off to 17.0 ksi. This assumes that the web resists all the shear as computed in the above formula. In some rolled sections made from grades higher than A36 Steel, the allowable shear stress must be reduced from $0.4F_y$ to account for diagonal buckling as discussed in Art. 5.8.

3.15
Use of the Formula
for Unit Shearing Stress

The formula given in Art. 3.14 is used in two ways. First, when a beam has been designed to resist bending, the shear in the web is checked by solving for f_v. This value should not exceed that set forth in the specifications. The unit shearing stress is, of course, greatest at the section of maximum total vertical shear. For simple beams with no overhangs this is adjacent to the support; V then becomes the shear at the support, and A_w is the product of the depth of the beam and the web thickness. Should the value of f_v come out greater than the one specified, a beam with a thicker web or greater depth must be selected and investigated.

When it is desired to find the maximum permissible total shear that a beam will carry, the specified unit shearing stress is multiplied by the area of the web,

$$V = F_v A_w$$

Depth and web thicknesses of **S**, **W**, and other shapes will be found in the AISC Manual. It should be noted that in many instances the actual depths of **W** shapes vary considerably from the nominal depths. The web area should always be computed by using the actual depth of the section.

Comparison of the web thicknesses of **S** and **W** shapes will reveal that for beams of comparable nominal depths and weights, the web thicknesses of **S** shapes are greater. This variation results from the placement of more of the metal in the flanges of the **W** shapes and, consequently, less in the web. This is done in order to produce a more efficient beam shape from the standpoint of bending. Where heavy shears are involved, the greater thickness of the web of the **S** shapes frequently makes them advantageous. However, the majority of steel beams used in present-day practice are **W** shapes because bending resistance controls the design of steel beams much more frequently than does shear. As previously stated, shear is most likely to become a controlling factor when heavy loads are carried on short spans or when large concentrated loads occur near the ends of a span.

It should be noted too, that, because of the physical properties of the material, structural steel beams subjected to excessive shearing stress do not fail by actual shearing of the metal along a plane section such as x-x in Fig. 2.12 (Art. 2.5) or by the horizontal splitting characteristic of wood beams, as indicated in Fig. 3.8b. When excessive shearing stresses produce failure, it occurs by diagonal buckling of the web, a phenomenon caused by the combined action of vertical and horizontal shear and the bending stresses. However, the web thicknesses and depths of rolled structural steel beam sections have been so proportioned that buckling will not develop if the usual allowable shearing and bending stresses are not exceeded. (See Art. 5.8.)

Example 1

If an **S** 12×40.8 is used to carry the loads shown in Fig. 3.13, what is the maximum average unit shearing stress f_v developed

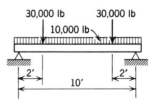

Figure 3.13/Example 1.

within the beam? (The 10,000-lb uniform load includes an allowance for the beam weight.)

Solution

(1) Compute the reactions. Since the loading is symmetrical, each reaction will be equal to half of the total load, or

$$R_L = R_R = \frac{10,000 + 2(30,000)}{2} = 35,000 \text{ lb}$$

(2) The maximum vertical shear occurs adjacent to, and is equal to, the reaction. Its value, therefore, is 35,000 lb.

(3) Referring to the AISC Manual for the dimensions of the **S** 12×40.8, the web area of the section is found to be $12(0.462) = 5.54$ sq in. Therefore,

$$f_v = \frac{V}{A_w} = \frac{35,000}{5.54} = 6318 \text{ psi}$$

Example 2

What is the maximum permissible total shear on an **S** 10×25.4, if the allowable unit shearing stress (F_v) is 14,500 psi (A36 Steel)?

Solution

$$V = F_v A_w$$

and

$$F_v = 14,500 \text{ psi}$$

Referring to the AISC Manual, the area of the web of an **S** 10×25.4 is $10(0.311) = 3.11$ in.^{2}. Therefore,

$$V = 14,500(3.11) = 45,100 \text{ lb}$$

PROBLEMS

The problem numbers cited below refer to the problems immediately following Art. 3.9.

1. Determine the maximum (average unit) shearing stress for Problem 1. Determine the total uniform load if shear is the controlling factor and $F_v = 14,500$ psi (A36 Steel). (Answer given in Appendix G.)

2. Determine the maximum shearing stress in the beams for Problems 2, 3, and 4. (Answers given in Appendix G.)

3. Determine whether the beams in Problems 7, 8, and 9 are safe for shear if $F_v = 14,500$ psi (A36 Steel).

4

BEAMS—
DEFLECTION

4.1
General

A beam may be designed to resist failure adequately, either in bending or in shear, and yet may deflect to such an extent as to be unsuitable for its intended use. The most common results of excessive beam deflection are damage to adjacent materials, a curvature which is noticeable and thus psychologically disturbing, or functional inadequacies such as an out-of-level or excessively springy floor. In order to avoid such conditions it is common practice to design beams so that, when loaded, they will not deflect more than a specified amount, established as a small fraction of the span length. It has been a long-standing practice in structural design to limit the computed live load deflection of floor and roof framing to a maximum value of $\frac{1}{360}$ of the span when supporting plaster. Obviously, however, all structures do not support plaster. Therefore, there is need to establish criteria based on conditions of occupancy, type of materials involved, and similar related factors.

A floor system can meet all requirements with regard to strength and yet be unsatisfactory with regard to stiffness. Such is the case when people, moving about, create a floor vibration that can be felt by the occupants. One way to avoid this is to set the upper limits for elastic deflection to a low value, thereby increasing the required stiffness. Another method is to investigate the natural frequency of the elements that make up the system in order to avoid those that experience has shown to be objectionable. This method, described under the topic of "serviceability" is discussed more fully in Chapter 11.

Deflection limitations are discussed in more detail in Chapter 5; this chapter deals only with determining actual deflections of statically determinate beams.

A study of deflection phenomena is also important from the theoretical standpoint as an aid in the analysis of statically indeterminate beams and frames. The design of continuous beams (three or more supports), rigid frames, and other indeterminate structures involves the relative stiffness of members, which is related directly to the deformation pattern. Consequently, the study of deflection and of the principles derived therefrom provides tools for the analysis of statically indeterminate structures. The application of these principles to the design of continuous beams is treated in Chapter 10.

The treatment presented here concerns only that deflection resulting from the bending moment. Small amounts of deflection occur in beams due to the shearing stresses present; however, except in very deep and short beams this is such a small percentage of the total deflection that it may be neglected. Since it is the bending moment that produces the deflection, any expression for deflection will contain about the same factors that appear in the expression for bending strength. As will be demonstrated in subsequent articles, the general form of the equations for deflection is

$$\Delta = K \frac{WL^3}{EI}$$

This expression shows that the magnitude of the deflection depends on the amount of the load W, the span length L, the stiffness of the beam material as measured by the modulus of elasticity E, and the size and shape of the cross section as indicated by the moment of inertia I. The coefficient K is a constant, the value of which depends on the distribution of the load and support conditions. The above expression also indicates (as would be expected from the physical factors involved) that deflec-

tion increases with an increase in load or span length, and decreases with an increase in modulus of elasticity or moment of inertia. It is of particular importance to observe that deflection varies as the third power of the span length. These relationships may be fixed firmly in mind by studying Fig. 4.15, which presents deflection formulas for several loading conditions frequently encountered in building structures.

The derivation of expressions for deflection under various conditions of loading may be accomplished by several methods. Two are presented in this chapter: The area-moment method and the elastic-curve method. The latter requires the use of calculus, since it is based on the solution of the differential equation for the elastic curve by double integration. The area-moment method, although based on the equation of the elastic curve, may be applied without involving calculus, and will therefore be developed first.

AREA-MOMENT METHOD

4.2
Area-Moment Principles— Derivation

The area-moment principles and the method which results from their application provide one of the more useful sets of tools both for determining simple beam deflections and for analyzing indeterminate structures.

In Chapter 3 it was shown that the neutral surface of a homogeneous beam is a continuous plane which passes through the centroid of each right section. When the beam deflects, the neutral surface follows a continuous curve, which is called the elastic curve. Deflections are measured to the elastic curve from the original position of the neutral surface. The derivation of the area-moment equations is shown in semigraphic form in Fig. 4.1, as an aid to those not familiar with calculus.

First Area-Moment Principle / The curved line *CD* in Fig. 4.1 represents the neutral surface of a beam after bending. It is the elastic curve of a deflected beam of indefinite length. *A* and *B* are points on this elastic curve and are separated by a very small distance *dx*. When two lines are constructed perpendicular to the elastic curve at points *A* and *B*, their extensions will intersect at the center of curvature *O*, a distance *R* from the curve, and will form the small angle *dθ*. The curvature is actually very slight; therefore, *AB* may be considered to be the horizontal length *dx*.

Then from geometry,

$$dx = R\,d\theta$$

or

$$d\theta = \frac{1}{R}\,dx$$

In Arts. 3.2 and 4.6 it is shown that

$$\frac{1}{R} = \frac{M}{EI}$$

Therefore,

$$d\theta = \frac{M}{EI}\,dx$$

Tangents to the elastic curve are next drawn at points *A* and *B*, perpendicular to the radius of curvature at *A* and *B*,

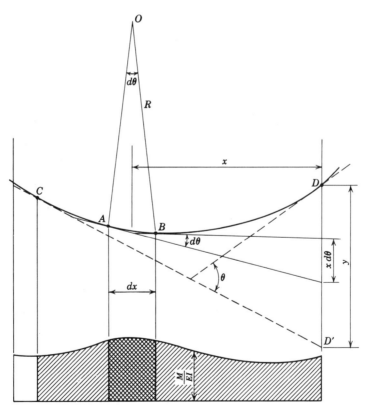

Figure 4.1/First area-moment principle.

respectively. Then, from geometry, the angle between the two is also equal to $d\theta$.

The corresponding M/EI diagram is shown directly below the elastic curve. The above-stated relationship between $d\theta$ and

the M/EI diagram may be observed from Fig. 4.1, as follows:

The elemental angle, $d\theta$, formed by two tangents drawn to the elastic curve at points A and B, equals the elemental area

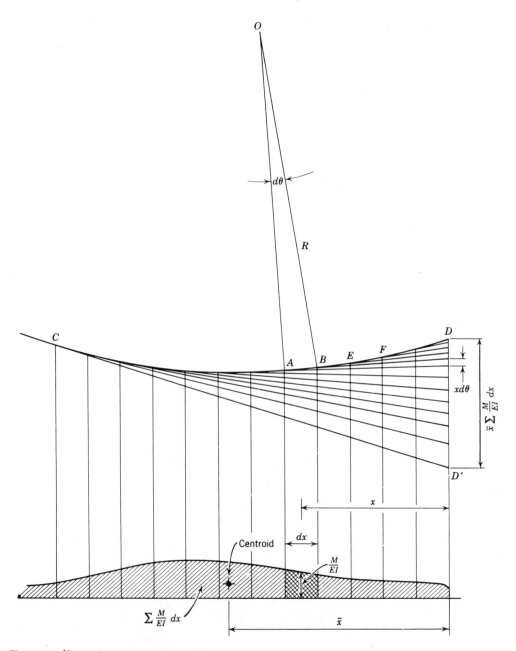

Figure 4.2/Second area-moment principle.

$(M/EI)\,dx$ (shown by the double-cross-hatched area). Thus, any angle such as θ, formed by any two tangents to the elastic curve (such as the ones at C and D), can be determined by a summation of all elemental areas of the M/EI diagram between the two tangent points (the entire cross-hatched area shown). Presented in equation form,

$$\theta = \sum \frac{M}{EI}\,dx$$

This is the first area-moment principle. It may be stated as follows:

The angle in radians, or the change in slope, between two tangents drawn to the elastic curve is equal to the area of the bending-moment diagram between the two tangent points divided by the flexural rigidity (EI) of the beam.

Second Area-Moment Principle / It is now desirable to establish a means of measuring the distance y shown in Fig. 4.1. This is the distance between a second point on a continuous elastic curve (D), measured in a direction perpendicular to the original straight axis of the beam (DD'), and the tangent to the elastic curve at a first point (C). This distance y is frequently referred to as the tangent deviation. It is seen that the elemental section AB contributes $x\,d\theta$ to this total distance. Therefore, since

$$d\theta = \frac{M}{EI}\,dx$$

it follows that

$$x\,d\theta = x\,\frac{M}{EI}\,dx$$

This is the first moment of the area $(M/EI)\,dx$ about the line DD'.

Consequently, the total distance y must be the sum of the moments of all elemental areas between C and D about DD'. This

is clearly shown in Fig. 4.2, where each of several elemental sections AB, BE, EF, etc., contributes $x\,d\theta$ to the overall distance DD'. Thus, the sum of all the elemental areas is equal to the total area of the M/EI diagram between C and D, and $\bar{x}$ is the moment arm or horizontal distance to the centroid of the M/EI area between C and D. Presented in equation form,

$$y = \sum x\frac{M}{EI}\,dx = \bar{x}\sum\frac{M}{EI}\,dx$$

The second area-moment principle can now be stated as follows:

The vertical distance between a second point on a continuous elastic curve and the tangent at a first point on the elastic curve is equal to the moment of the area of the M/EI diagram between these two points about the second point.

In some cases the tangent deviation is equal to the desired deflection (Example 1). In other cases it becomes a critical theoretical dimension from which the desired deflection can be subsequently determined. This is shown in Example 2 in Article 4.5.

4.3 Application of Area-Moment Method

In Figs. 4.1 and 4.2, the bending-moment diagram was shown irregular so as to represent any possible shape. However, most bending-moment diagrams encountered in structural design problems may be broken down into one or more of the basic geometric shapes shown in Fig. 4.3. An area formula and a centroid location are given for each shape shown in the figure.

To achieve simplified shapes from which to work it may be necessary to draw the

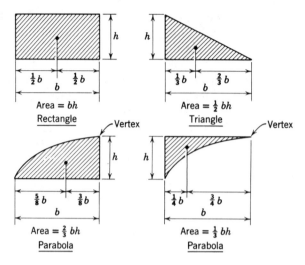

Figure 4.3/Basic geometric shapes.

bending-moment diagram in parts. This is accomplished by treating each load or reaction independently, as illustrated in Fig. 4.4a which shows a simply supported beam having a concentrated load at midspan and a uniformly distributed load over its entire length.

Taking the right reaction as the center of moments, and considering only the left reaction, the diagram of Fig. 4.4b will result. The left reaction by itself tends to bend the beam concave upward, producing a positive moment for its entire length. The bending-moment diagram for the distributed load, again using R_R as the center of moments, is shown in Fig. 4.4c, and is negative for its entire length. The moment diagram shown in Fig. 4.4d is that for the concentrated load at midspan. If all the parts shown in Fig. 4.4(b), (c), and (d) are added algebraically, the net diagram will be as shown in Fig. 4.4e.

Example 1

Determine the maximum slope and deflection of a cantilever beam having a total length L and a concentrated load P at the free end, using the area-moment method. Neglect the weight of the beam. E and I remain constant.

Solution

(1) Make a sketch of the beam showing the loads (Fig. 4.5a).

(2) Compute reactions, and sketch the bending-moment diagram. Indicate the value of the bending moment at the support (Fig. 4.5b).

(3) Sketch the assumed deflected shape of the beam (Fig. 4.5c), and study the elastic curve to determine the most effective way to apply the area-moment principles to yield the desired solution. In this case, it is known that the fixed end of the beam cannot rotate. Therefore, a tangent constructed at this point will be horizontal. Another tangent constructed at C' on the elastic curve will form the angle θ at B with the tangent drawn at A. This angle is also the slope of the elastic curve at the free end.

(4) Construct the tangents as discussed in step (3). Then, applying the first area-moment principle, the slope at the free

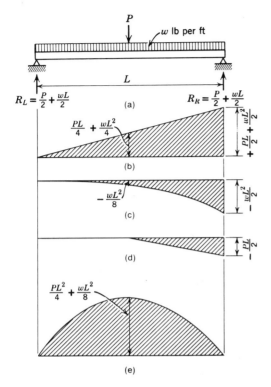

Figure 4.4/Moment diagram by parts.

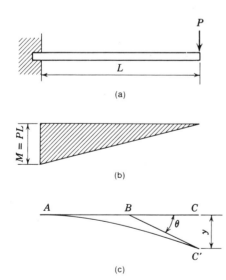

Figure 4.5/Example 1: area moment.

end is the shaded area times $1/EI$:

$$\theta = \frac{1}{2}PL(L)\frac{1}{EI}$$

$$\theta = \frac{PL^2}{2EI}$$

(5) As the tangent at the fixed end is horizontal, all distances from this tangent to the elastic curve are actual deflections. Thus it is seen that the tangent deviation is the deflection for that point. The maximum deflection at the free end will be measured by the vertical distance between the tangent at A and the elastic curve of the beam at C'.

Applying the second area-moment principle, the deflection at the free end is

$$y = \frac{1}{2}PL(L)\frac{2}{3}L\left(\frac{1}{EI}\right)$$

$$y = \frac{PL^3}{3EI}$$

Example 2

Derive the formula for the maximum deflection of a cantilever beam having a uniformly distributed load w over its entire span length L. E and I remain constant. Neglect the weight of the beam. (Whenever distributed loads are encountered, the bending-moment diagram will be a curved line in the shape of a parabola. If the load is *uniformly* distributed, the parabola will be of the 2nd degree and the data shown in Fig. 4.3 are applicable.)

Solution

Steps (1) through (3) are similar to those for the preceding example, and the data have been recorded in Fig. 4.6.

(4) Applying the second area-moment principle, the maximum deflection at the

68 / BEAMS—DEFLECTION

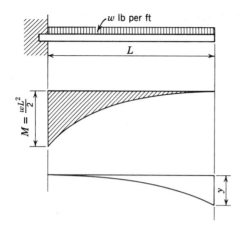

Figure 4.6/Example 2: area moment.

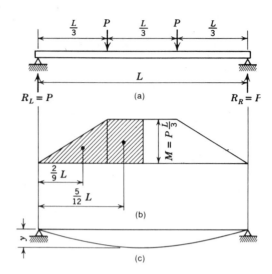

Figure 4.7/Example 1: symmetry.

free end is

$$y = \frac{1}{3}\left(\frac{wL^2}{2}\right)L\left(\frac{3L}{4}\right)\frac{1}{EI}$$

$$y = \frac{wL^4}{8EI}$$

4.4
Area-Moment: Symmetry

When simply supported beams without overhangs are symmetrically loaded, the area-moment method is greatly simplified. This is due to the fact that the slope of the elastic curve at midspan is zero, indicating maximum deflection. Also, the tangent is a horizontal line, and a vertical distance to the elastic curve at any point (commonly referred to as "tangent deviation") is simple to calculate, and this deviation at the support is equal to the deflection at midspan (see Fig. 4.7). The following two examples illustrate the area-moment method as applied to simply supported beams, symmetrically loaded.

Example 1

Determine the maximum deflection of a simply supported beam (no overhangs),

carrying equal concentrated loads at its third-points. Neglect the beam weight. E and I remain constant.

Solution

(1) Make a sketch of the beam showing loads and reactions (Fig. 4.7a).

(2) Construct the bending-moment diagram (Fig. 4.7b). Because the net bending-moment diagram is composed of simple shapes with centroids easily located, there is nothing to be gained by drawing it in parts.

(3) Sketch the deflected shape of the neutral axis. By observation it is apparent that the maximum deflection will occur at midspan. Construct a tangent to the neutral axis at midspan. This is a horizontal line; thus the vertical distance y from this tangent to the neutral axis at the support (which does not move) is equal to the deflection at midspan.

(4) Applying the second area-moment principle, calculate the distance y. This is equal to the moment of the cross-hatched area divided by EI, about a vertical line

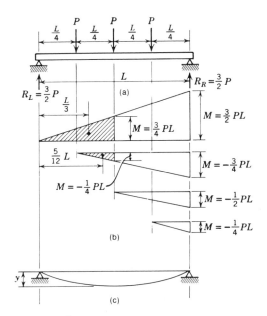

Figure 4.8/Example 2: symmetry.

shown in parts. Centroidal distances are located as shown.

(3) Sketch the approximate shape of the elastic curve (Fig. 4.8c) and, observing that the maximum deflection occurs at midspan, construct a tangent at this point. The distance y (as shown) is the maximum deflection.

(4) Applying the second area-moment principle, calculate the distance y:

$$y = \frac{1}{EI} \left\{ \left[\frac{1}{2} \left(\frac{3}{4} PL \right) \frac{L}{2} \left(\frac{L}{3} \right) \right] \right.$$

$$\left. - \left[\frac{1}{2} \left(\frac{1}{4} PL \right) \frac{L}{4} \left(\frac{5L}{12} \right) \right] \right\}$$

$$y = \frac{19}{384} \left(\frac{PL^3}{EI} \right)$$

drawn through the left support:

$$y = \frac{1}{EI} \left\{ \left[\frac{1}{2} \left(\frac{PL}{3} \right) \frac{L}{3} \left(\frac{2L}{9} \right) \right] \right.$$

$$\left. + \left[\frac{PL}{3} \left(\frac{L}{6} \right) \frac{5L}{12} \right] \right\}$$

$$y = \frac{23}{648} \left(\frac{PL^3}{EI} \right)$$

4.5
Area-Moment:
Unsymmetrical Loading

The analysis of a statically determinate beam becomes more complex when the point of maximum deflection cannot be determined by observation. Such is the case, for example, when the loading is not symmetrical as in Figs. 4.9 and 4.10. Here, the maximum deflection will not occur under the load, nor at the center, but rather at some point between. When confronted with such a problem, it is possible that another analytical method would lead to a more rapid solution.[1] However, this is not to say that the area-moment method cannot be effectively used, as will be illustrated by the following two examples.

Example 2

Determine the maximum deflection of a simply supported beam (no overhangs) carrying equal concentrated loads at its quarter points. Neglect the beam weight. E and I remain constant.

Solution

(1) Make a sketch of the beam showing loads and reactions (Fig. 4.8a).

(2) Construct the bending-moment diagram (Fig. 4.8b). For the sake of illustration, the bending-moment diagram is

Example 1

Determine the deflection under the load P of the simply supported beam shown in

---

[1] For example, the differential equation for the elastic curve (as discussed later in this chapter) or the theory of elastic energy could be used.

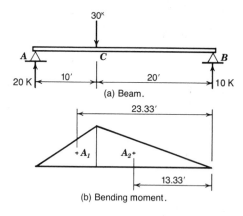

(a) Beam.

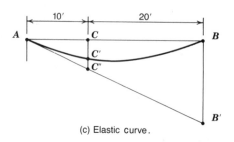

(b) Bending moment.

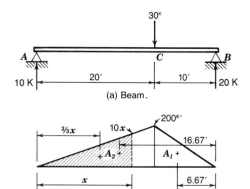

(a) Beam.

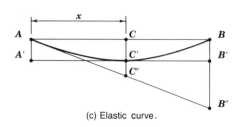

(b) Bending moment.

(c) Elastic curve.

Figure 4.9/Example 1: unsymmetrical.

(c) Elastic curve.

Figure 4.10/Example 2: unsymmetrical.

Fig. 4.9. Neglect the beam weight. E and I remain constant.

Solution

Since E and I remain constant, it becomes convenient to calculate y in terms of EI. Therefore, place EI on the other side of the equals sign and in the numerator. Observe the answer to Example 2 in the previous article, which was

$$y = \frac{19}{384}\left(\frac{PL^3}{EI}\right)$$

This is the same as

$$EIy = \frac{19}{384}(PL^3)$$

Proceeding with the solution:

(1) Make a sketch of the beam showing loads and reactions (Fig. 4.9a).

(2) Construct the bending-moment diagram (Fig. 4.9b). Calculate the two trian-

gular areas in the moment diagram, and locate their centroids from point B:

$$A_1 = \tfrac{1}{2}(200)(10) = 1000 \text{ kip-ft}^2$$

$$A_2 = \tfrac{1}{2}(200)(20) = 2000 \text{ kip-ft}^2$$

(3) Sketch the deflected shape of the neutral axis (Fig. 4.9c). This is the elastic curve. Construct a tangent to the neutral axis at point A, and, applying the second area-moment principle, calculate the distance BB'. This is equal to the moment of the entire area of the bending-moment diagram about BB':

$$EIBB' = 1000(23.33) + 2000(13.33)$$

$$= 50,000 \text{ kip-ft}^3$$

(4) From similar triangles in the geometry, calculate the distance CC'':

$$EICC'' = \frac{10}{30}(50,000) = 16,667 \text{ kip-ft}^3$$

(5) Applying the second area-moment principle again, calculate the tangent deviation at point C. This is seen to be the distance $C'C''$ and is equal to the area A_1 times its moment arm from point C:

$$EIC'C'' = 1000(3.33) = 3333 \text{ kip-ft}^3$$

(6) The actual deflection under the load is seen to be $CC'' - C'C''$:

$$EICC' = 16{,}667 - 3333 = 13{,}333 \text{ kip-ft}^3$$

(7) Assuming the steel beam to have a moment of inertia of 600 in.^{4}, the deflection under the load is calculated to be

$$y = \frac{13{,}333(12)^3}{29{,}000(600)} = 1.32 \text{ in.}$$

Example 2

Determine the maximum deflection of the simply supported beam shown in Fig. 4.10. Neglect the beam weight. E and I are constant.

Solution

Steps (1) and (2) are the same as Example 1.

(3) The maximum deflection CC' will occur somewhere between the center of the span and the load P, an unknown distance x from point A. Construct a tangent to the elastic curve at C'. Because this is the point of maximum deflection, the slope is zero and $AA' = CC' = BB'$. Also, the bending moment at C is equal to $10x$. Applying the second area-moment principle, AA' is equal to the moment of the moment diagram area between A and C (the cross-

hatched area) about AA':

$$EIAA' = \frac{10x}{2}(x)\left(\frac{2}{3}x\right) = 3.33x^3$$

Observe that this value is equal to the distance CC', which is also equal to $CC'' - C'C''$.

(4) Construct a tangent to the elastic curve at A, and find the distance BB'':

$$EIBB'' = 2000(16.67) + 1000(6.67)$$
$$= 40{,}000 \text{ kip-ft}^3$$

(5) From similar triangles calculate the distance CC'':

$$EICC'' = \frac{40{,}000}{30}x = 1333x \text{ kip-ft}^3$$

(6) Find the distance $C'C''$. This is the tangent deviation at point C and is equal to the first moment of the cross-hatched area about point C:

$$EIC'C'' = \frac{10x}{2}(x)\frac{x}{3} = 1.67x^3 \text{ kip-ft}^3$$

(7) The following summation and algebra solves for the unknown distance x:

$$EICC' + EIC'C'' = EICC''$$
$$3.33x^3 + 1.67x^3 = 1333x$$
$$5x^2 = 1333$$
$$x = \pm 16.33 \text{ ft}$$

(8) Substituting the positive value of x above in the equation of step (3) gives the maximum deflection:

$$EIy = 3.33(16.33)^3 = 14{,}500 \text{ kip-ft}^3$$

(9) Assuming the steel beam to have a moment of inertia of 600 in.^{4}, the deflection under the load is calculated to be

$$y = \frac{14{,}500(12)^3}{29{,}000(600)} = 1.44 \text{ in.}$$

ELASTIC-CURVE METHOD

4.6
Equation of the Elastic Curve—Derivation

In Art. 3.2 it was shown that the magnitude of the bending stresses varies directly with the distance from the neutral axis, or

$$\frac{f}{y} = \frac{E}{R}$$

where R is the radius of curvature of the elastic curve. Taking the general flexure expression $f = My/I$ and substituting this value of f in the above equation,

$$\frac{1}{R} = \frac{M}{EI}$$

This may be thought of as the companion expression to the flexure formula. It is the basis for calculations of deflection due to bending. The product EI is often called the *flexural rigidity*, since it is a direct measure of the resistance to bending deformation.

In order to determine an equivalent expression for $1/R$, assume point p on the elastic curve of a deformed beam as shown in Fig. 4.11a. If the origin of coordinates for the elastic curve is taken at the left end of the beam, then p has coordinates x and y, and y is the deflection of p. R is the radius of curvature of the elastic curve at p. Then, from calculus, the reciprocal of the radius of curvature of any curve is

$$\frac{1}{R} = \frac{\dfrac{d^2y}{dx^2}}{\left[1 + \left(\dfrac{dy}{dx}\right)^2\right]^{3/2}}$$

However, the curves involved in beam deformations are extremely flat, and the slope of the curve at $p(dy/dx)$ is very small relative to unity. Consequently, it may be assumed that the powers of dy/dx may be neglected without introducing appreciable error. The preceding equation then becomes

$$\frac{1}{R} = \frac{d^2y}{dx^2}$$

Equating the two values for $1/R$,

$$\frac{d^2y}{dx^2} = \frac{M}{EI}$$

This is the differential equation of the elastic curve.

4.7
Application of Elastic Curve—Double Integration

Starting with the equation of the elastic curve and integrating once gives an expression for the slope θ of the elastic curve at any point x:

$$\theta = \frac{dy}{dx} = \int \frac{M}{EI}\, dx$$

Integrating a second time yields an expression for the deflection (y coordinate) at any point x on the curve:

$$y = \int\int \frac{M}{EI}\, dx^2$$

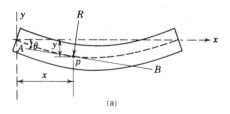

(a)

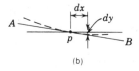

(b)

Figure 4.11/Elastic-curve derivation.

Also, since as shown in Art. 2.15

$$M = \int \int w\,dx^2$$

it follows that

$$y = \int \int \frac{\int \int w\,dx^2}{EI}\,dx^2$$

In Chapter 2 it was shown that the bending-moment equation may be written in terms of the variable x. If proper values of E and I are used, a single integration will give the value of the slope of the elastic curve in terms of x, and a second integration will yield the value of the deflection also in terms of x. Consequently, the slope or the deflection at some specific point, measured along the beam, may be determined by substituting this distance for x in the appropriate equation.

The point of maximum deflection frequently cannot be located by observation. However, from the principles established in Art. 2.15, "Analysis of Beam Diagrams," it is known that the maximum deflection will occur at the point where the slope of the elastic curve is zero (except in cantilever beams). Consequently, equating the expression for slope to zero, solving for x, and then substituting this value of x in the expression for deflection, the maximum deflection may be obtained. This is best illustrated by examples.

Example 1

Determine the maximum slope and deflection of a cantilever beam having a total length L and a concentrated load P at the free end. Neglect weight of the beam. E and I remain constant.

Solution

(1) Make a sketch of the beam showing the loads (Fig. 4.12a).

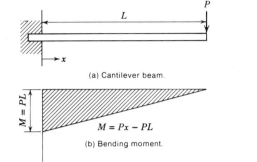

(a) Cantilever beam.

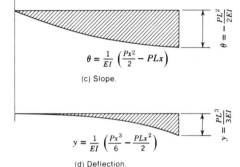

(b) Bending moment.

$$\theta = \frac{1}{EI}\left(\frac{Px^2}{2} - PLx\right)$$

(c) Slope.

$$y = \frac{1}{EI}\left(\frac{Px^3}{6} - \frac{PLx^2}{2}\right)$$

(d) Deflection.

Figure 4.12/Example 1: cantilever beam.

(2) Establish the equation for the bending moment, measuring x from the fixed end, and draw the bending-moment diagram (Fig. 4.12b):

$$M = -P(L - x)$$
$$M = Px - PL$$

(3) Divide the bending-moment equation by EI and integrate once to obtain the equation for the slope of the elastic curve:

$$\theta = \int \frac{M}{EI}\,dx = \int \frac{Px - PL}{EI}\,dx$$
$$= \frac{1}{EI}\left(\frac{Px^2}{2} - PLx + C\right)$$

The constant of integration may be evaluated from the known condition in a cantilever beam, i.e., the slope at the fixed end is zero. Thus, when $x = 0$, $\theta = 0$; consequently, $C = 0$, or

$$\theta = \frac{1}{EI}\left(\frac{Px^2}{2} - PLx\right)$$

The curve for the slope is plotted in Fig. 4.12c.

(4) Calculate the value of the maximum slope. The maximum slope occurs at the free end, or when $x = L$:

$$\theta = \frac{1}{EI}\left(\frac{PL^2}{2} - PL^2\right)$$

$$\theta = -\frac{PL^2}{2EI}$$

(5) Integrating the expression for the slope of the elastic curve results in the equation for deflection

$$y = \int \theta\, dx = \frac{1}{EI}\left(\frac{Px^3}{6} - \frac{PLx^2}{2} + C\right)$$

The constant of integration may again be established by the given condition of the problem, i.e., there is no deflection at the fixed end. Therefore, when $x = 0$, $y = 0$; consequently, $C = 0$, and

$$y = \frac{1}{EI}\left(\frac{Px^3}{6} - \frac{PLx^2}{2}\right)$$

This curve, when plotted, represents the exact shape of the curve of the neutral axis resulting from the load P (Fig. 4.12d).

(6) Calculate the value of the maximum deflection. It is obvious that the maximum deflection will occur at the free end, or when $x = L$. Therefore,

$$y = \frac{1}{EI}\left(\frac{PL^3}{6} - \frac{PL^3}{2}\right)$$

$$y = -\frac{PL^3}{3EI}$$

An excellent exercise would be to rework this problem by measuring x from the free end. Although intermediate equations would take a different form, the final expression for maximum θ and y would remain the same.

4.8
Double Integration—Use of Symmetry

The double-integration method becomes quite laborious where the constants of integration are difficult to determine. However, as has been previously pointed out, when simply supported beams are symmetrically loaded, the solution by any method becomes much simpler. This is true because the maximum deflection occurs at midspan, which also means that the slope of the elastic curve is zero at midspan. The simplicity of the loading generally eliminates the necessity of sketching the complete beam diagrams to execute the analysis. The two examples which follow illustrate the double-integration procedure as applied to simple beams symmetrically loaded.

Example 1 illustrates the complete integration process, while Example 2 shows only that work necessary to provide the desired answer.

Example 1

Prepare the complete set of beam diagrams $(L, V, M, \theta, \Delta)$ and equations for a simply supported beam of 20 ft and a single concentrated load of 20 kips at midspan. Neglect the weight of the beam. E and I remain constant. Measure x from the left reaction.

Solution

(1) Make a sketch of the beam showing the load and reactions (Fig. 4.13a).

(2) Prepare the load diagram (Fig. 4.13b). Since the weight of the beam is neglected, there is no distributed load and the equation is a constant zero.

(3) Prepare the shear diagram (Fig. 4.13c). In this example, the shear diagram is a

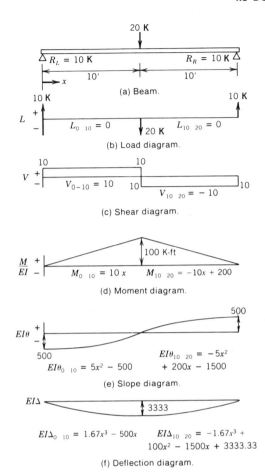

(b) Load diagram.

(c) Shear diagram.

(d) Moment diagram.

(e) Slope diagram.

(f) Deflection diagram.

Figure 4.13/Example 1: double integration.

constant horizontal line, positive for the left half and negative for the right half. (For more complex loadings having distributed loads, the equation for shear may be derived by integrating the load equations.)

(4) Construct the moment diagram, and develop the equations for the moment (Fig. 4.13d). For the left half, integrating the constant 10 gives

$$M_{0\text{-}10} = 10x + C$$

The constant of integration is determined by observing that where $x = 0$, M is also 0; consequently, the constant is zero. For the right half, integrating the constant -10

gives

$$M_{10\text{-}20} = -10x + C$$

The constant of integration is determined by observing that where $x = 10$, the moment is 100. Substituting this in the above equation,

$$100 = -10(10) + C$$
$$C = 200$$

In this example, E and I are assumed constant; therefore, the *shape* of the M/EI diagram will remain the same as the M diagram. Only the values of the ordinates will change, permitting EI to be treated as an outside constant in subsequent operations.

(5) Construct the $EI\theta$ diagram, and develop the equations for its shape (Fig. 4.13e). For the left half, integrating the moment equation gives

$$EI\theta_{0\text{-}10} = 10\frac{x^2}{2} + C$$

The constant of integration is determined by observing that the slope is zero at midspan. Substituting $x = 10$ and $EI\theta = 0$ in the above equation,

$$0 = 10\frac{(10)^2}{2} + C$$
$$C = -500$$

For the right half, integrating the moment equation gives

$$EI\theta_{10\text{-}20} = -10\frac{x^2}{2} + 200x + C$$

The constant of integration is determined by observing that the slope is zero at midspan; then, using the above equation,

$$0 = -10\frac{(10)^2}{2} + 200(10) + C$$
$$C = -1500$$

(6) Construct the $EI\Delta$ diagram and develop the equations (Fig. 4.13f). For the

left half, integrating the $EI\theta$ equation

$$EI\Delta_{0\text{-}10} = 5\frac{x^3}{3} - 500x + C$$

The constant of integration is determined by observing that the deflection at the left reaction is zero; then, using the above equation

$$0 = 5\frac{(0)^3}{3} - 500(0) + C$$
$$C = 0$$

For the right half, integrating the $EI\theta$ equation

$$EI\Delta_{10\text{-}20} = -5\frac{x^3}{3} + 200\frac{x^2}{2} - 1500x + C$$

The constant of integration is determined by observing that the deflection at the right reaction is zero; then, using the above equation,

$$0 = -5\frac{(20)^3}{3} + 200\frac{(20)^2}{2} - 1500(20) + C$$
$$C = 3333.33$$

(7) It should be apparent that the value for $EI\Delta$ at midspan (when $x = 10$) should be the same using either equation. Using the equation for the left half,

$$EI\Delta = 1.67(10)^3 - 500(10)$$
$$EI\Delta = -3330 \text{ kip-ft}^3$$

Using the equation for the right half,

$$EI\Delta = -1.67(10)^3 + 100(10)^2$$
$$- 1500(10) + 3333$$
$$EI\Delta = -3337 \text{ kip-ft}^3$$

The slight differences in values are, of course, due to rounding off numbers.

Finally, the maximum deflection for any beam can be obtained by using its E and I values and adjusting for units. If, for example, a **W** 12×40 steel beam were

used, its maximum deflection would be

$$\Delta = \frac{3330(12)^3}{310(29,000)} = 0.64 \text{ in.}$$

the units being

$$\Delta = \frac{\text{kip-ft}^3(12 \text{ in./ft})^3}{\text{in.}^4(\text{kip/in.}^2)} = \text{in.}$$

Example 2

Determine the maximum deflection of a simply supported beam having a span length L and subjected to a uniformly distributed load of w lb per ft. E and I remain constant.

Solution

(1) Make a sketch of the beam showing the load, reactions, and the origin for x (Fig. 4.14).

(2) Write the equation of the bending moment curve and divide by EI:

$$\frac{M}{EI} = \frac{1}{EI}\left(\frac{wLx}{2} - \frac{wx^2}{2}\right)$$

(3) Determine the equation of the slope diagram by integrating the M/EI equation:

$$EI\theta = \frac{wLx^2}{4} - \frac{wx^3}{6} + C$$

Determine the constant of integration C as follows: When $x = L/2$, $EI\theta = 0$.

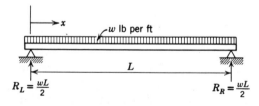

Figure 4.14/Example 2: double integration.

Therefore,

$$EI\theta = \left[\frac{wL}{4}\left(\frac{L}{2}\right)^2\right] - \left[\frac{w}{6}\left(\frac{L}{2}\right)^3\right] + C = 0$$

$$C = -\frac{wL^3}{24}$$

and the complete equation for the slope curve is

$$EI\theta = \frac{wLx^2}{4} - \frac{wx^3}{6} - \frac{wL^3}{24}$$

(4) Determine the equation of the elastic curve by integrating the $EI\theta$ equation:

$$EI\Delta = \frac{wLx^3}{12} - \frac{wx^4}{24} - \frac{wL^3}{24}(x) + C$$

When $x = 0$,

$$EI\Delta = 0$$

and so

$$C = 0$$

(5) Calculate the maximum deflection. This occurs at midspan, or when $x = L/2$:

$$EI\Delta = \left[\frac{wL}{12}\left(\frac{L}{2}\right)^3\right] - \left[\frac{w}{24}\left(\frac{L}{2}\right)^4\right]$$
$$- \left[\frac{wL^3}{24}\left(\frac{L}{2}\right)\right]$$

$$\Delta = \frac{5wL^4}{384EI}$$

4.9
Deflection Formulas

The actual calculation of maximum deflection for purposes of design is usually accomplished by the simple application of a formula. It is virtually impossible to list formulas for every conceivable beam support condition and system of loads; this, of course, is the reason for the general treat-

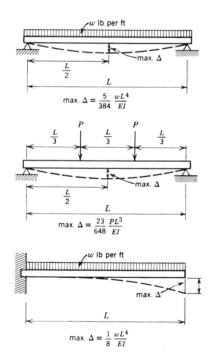

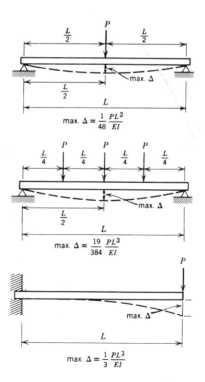

Figure 4.15/Deflection formulas.

ment in the preceding articles. Also, there are situations where calculation of deflections can be quite complex; in such situations, computer programs are often developed and used.

Figure 4.15, however, shows deflection formulas for the beams and loadings most frequently encountered in building structures. Additional formulas are presented in the AISC Manual.

To use these formulas it is necessary to select the one which corresponds to the actual design conditions and to know the properties of beam sections.[2]

It should be noted that if a beam is subjected to distributed and concentrated loads, the total deflection at any one point is the sum of the deflections figured separately for the same point.

PROBLEMS

1. A simply supported beam has a span length of 30 ft. There is a 10-kip concentrated load 10 ft from the left reaction. The distributed load from the weight of the beam is negligible.

(a) Construct the complete beam diagrams and determine the equation for each curve. Use the left reaction as the origin for measuring x in every equation. In constructing the slope and deflection diagrams, use the differential equation of the elastic curve.

(b) Determine the point of maximum deflection.

(c) Calculate the maximum deflection if a **W** 14×34 is used.

(Answers given in Appendix G.)

2. A beam has a total length of 20 ft. The left reaction is at the left end, and the right reaction is 5 ft to the left of the right end. The load (including the beam weight) is 10 kips per ft over the entire length.

---

[2]The moment of inertia (I) may be selected from a steel handbook, and the modulus of elasticity (E) for structural steel is assumed to be 29,000,000 psi in all examples in this text. (See Appendix A.)

(a) Construct the complete beam diagrams, and determine the equation for each curve. Use the left reaction as the origin in measuring x for the equations between the reactions, and use the right reaction as the origin in measuring x for the equations for the cantilevered portion.

(b) Determine the point of maximum deflection.

(c) Calculate the maximum deflection and the deflection at the free end of the cantilever if a **W** 14×90 is used.

3. A **W** 8×31 cantilevers 12 ft from a wall. It carries a uniformly distributed load of 400 lb per ft from the wall to within 4 ft of the free end. There are also two 1-kip concentrated loads, one located at the free end and the other 4 ft from the free end. Using the area-moment method, determine:

(a) The maximum deflection.

(b) The deflection of the beam at 4 ft from the free end.

Note. Draw the moment diagram in parts. (Answers given in Appendix G.)

4. A **W** 12×40 is simply supported on a 24-ft span. It carries a uniformly distributed load of 800 lb per ft over the entire length, and two 6-kip concentrated loads, one located 6 ft from each support. Using the area-moment method, determine the maximum deflection. (Answer given in Appendix G.)

5. An **S** 12×35 spans 20 ft. It carries a uniformly distributed load of 1200 lb per ft over its entire length. Determine the maximum deflection.

6. A beam spans 24 ft and supports a concentrated load of 10 kips at its midspan. Neglecting the beam weight, select the lightest-weight 10-in. **W** section that will not deflect more than $\frac{1}{2}$ in. (Answer given in Appendix G.)

7. Determine the maximum deflection of a **W** 12×19 that cantilevers 11 ft from a wall, and carries a distributed load of 200 lb per ft over its entire length and a concentrated load of 2 kips located at its free end.

8. Determine the maximum deflection of a **W** 12×26 that spans 24 ft and carries a distributed load of 500 lb per ft and a 4-kip concentrated load acting at each third-point.

5

BEAMS— DESIGN PROCEDURES

5.1
General

The complete design of beams requires consideration of (1) bending and shear resistance, (2) deflection limitation, (3) lateral buckling, and (4) local buckling.

Bending and shear resistance using assumed values of F_b were discussed in Chapter 3, and in Chapter 4 methods for calculating beam deflections were developed. In this chapter, items (2), (3), and (4) are treated in detail, and a general beam design procedure is established.

Such a beam design procedure must include examination of the consequences of excessive deflection, as well as of buckling associated with sections not having adequate lateral support, because both could affect the value of F_b to be used in the design. The general beam design procedure and its use are described, first without resorting to the extensive design aids furnished by the AISC Manual, and then, later in the chapter, with some of these design aids. Consideration also is given to special beam types and loading conditions. However, only statically determine beams are discussed; statically indeterminate (continuous) beams are considered in Chapter 10.

5.2
Deflection Limitations

It is difficult to establish deflection limitations, because the *safety* of a beam is seldom governed by the amount it deflects under load.

As a general rule, building codes offer little guidance, because they are principally an instrument for ensuring the safety of structures. The one notable exception to this is treatment of *roof ponding*, which will be discussed later. The issue of limitation on deflection, therefore, is one that generally must be resolved by the designer's own judgment.

Maximum permissible deflection is generally expressed either as an absolute dimension (e.g., inches or fraction of an inch) or as a proportion of the span length (e.g., $L/360$). An absolute dimensional limitation may result from a need to provide a specified amount of construction clearance —i.e., it is known that if a beam deflects at some one or more points more than a given amount, it will cause damage to adjacent materials or items of equipment, such as breaking windows, cracking partitions, or causing doors to stick. The most frequently encountered proportional limitation (discussed at the outset of Chapter 4) specifies that the maximum live-load deflection shall not exceed $\frac{1}{360}$ of the span for beams and girders supporting plastered ceilings. Presumably, a greater deflection, occurring after plaster has been applied, will cause cracking. This limitation is explicit in the AISC Specification, under the heading of "Serviceability, Design Considerations." However, it is frequently used as a guide whether plaster ceilings are used or not. It is important to emphasize that this limitation is for live load only.

Even if none of the above considerations is important, it still may be desirable to impose a limitation for visual or psychological reasons. This limit is frequently taken as $\frac{1}{300}$ or $\frac{1}{240}$ of the span under total load.

When the uses to which a building will be put throughout its life are unknown, it is wise to use the more restrictive $\frac{1}{360}$ limitation. However, when future use is obvious and such restrictions are unnecessary, the designer will want to take advantage of the economy provided by a greater permissible

deflection. Also, when it is known that specific limits even more restrictive than $\frac{1}{360}$ of the span are needed, they, of course, should be used.

It should be apparent from Chapter 4 that the flexural rigidity (EI) is the property of a beam that controls deflection; i.e., the larger the value of E and I, the smaller will be the deflection. It must be reemphasized, however, that the value of E for all types and grades of steel is 29,000 ksi (see Appendix A). Therefore, if deflection is controlling the size of a beam, the use of a higher grade of steel will not be beneficial and, in fact, could be considered wasteful. In truth, then, it is the moment of inertia (I) that controls deflection in steel buildings.

For a given cross-sectional area, the value of I will vary with the square of the depth. Consequently, establishing the depth of the beam is an important step in the design procedure and is frequently the first step. One helpful guide to establishing this depth with simply supported beams is what is known as the "$L/24$ ratio." This ratio is often stated as "the beam depth in inches equals one-half the span in feet"—e.g., a 24-ft simple beam span would require a section having a 12-in. minimum depth.

The AISC Manual furnishes similar guides —e.g., for large open floor areas, free of partitions that can dampen the effects of repeated live loadings as would be produced by foot traffic, use a minimum beam depth equal to $\frac{1}{20}$ of the span. This limitation will minimize the feeling of floor movement under foot traffic.

Another depth guide provided in the AISC Manual is that established for fully stressed beams and girders used for floors. For such members, the ratio given is $F_y/800$;[1] however, this ratio can be increased to

---

[1] F_y and all other stress values given in the AISC Specification are expressed in kips per square inch.

$F_y/1000$ for sloping roof purlins.[2] For example, an A36 Steel floor beam over a simple span of 28 ft and using $F_b = 24$ ksi, the minimum depth should be

$$\frac{36}{800}(28)12 \approx 15 \text{ in.}$$

If lesser depths are to be used, it is desirable to effect a directly proportional reduction in the allowable stress—e.g., if a 12-in. beam depth is selected, the allowable bending stress should be reduced to

$$F_b = \frac{12}{15}(24) = 19.2 \text{ ksi}$$

These guidelines can be tempered by other conditions, such as overhangs, restrained ends, built-in camber,[3] etc., all of which reduce the deflection.

The one case of excessive deflection that could affect the safety of a building is that which could result in the ponding of water on a roof. For this reason, the AISC Specification is explicit concerning such deflections. Most roofs are sloped or pitched so that they can shed water to perimeter gutters or to internal drains. However, when flat roofs are used, there is the possibility that water will accumulate, causing deflection of the roof and thus further accumulations of water until there is sufficient weight to cause collapse of the roof. Adequate stiffness in the roof system will prevent such an occurrence. A roof system is composed of the roof deck, the beams supporting the deck, and the girders supporting the beams. There is an AISC Specification for achieving adequate roof-sys-

tem stiffness; if this specification is not satisfied, then some other investigation must be undertaken to assure safety from the effects of ponding. Example 5, Art. 5.6, illustrates application of the AISC ponding criteria.

5.3 Use of Deflection Formulas in Design

Procedures for direct calculation of beam deflections were shown in Chapter 4, as were procedures for developing and using standard deflection formulas for those support and loading conditions most frequently encountered in practice. There are, of course, situations where determining the location and value of the maximum deflection is quite complicated and may influence the approach that must be taken in design of the beam; however, the usual design procedure is as follows:

1. Construct the beam loading (or space) diagram.

2. Assume a beam depth and a beam weight.

3. Construct shear and moment diagrams to determine maximum values of M and V (or use appropriate formulas).

4. Solve for S(required) $= M$(actual)$/F_b$(allowable), and select a trial beam section from the AISC "Allowable Stress Design Selection Table" or "Dimensions and Properties" tables. The value of F_b must be assumed to be some percentage of F_y and either verified or refined according to the trial section selected. (As noted earlier, both deflection and flange buckling may affect the value of F_b.)

5. Compare the actual beam weight with that assumed. If the assumed weight is significantly more or less than the actual,

---

[2]A purlin is a small roof beam that directly supports the roof deck.

[3]Camber in beams is the reverse curvature, usually built in during fabrication, used to counter the actual deflection due to loading. Frequently, a camber equal to the deflection caused by the dead load is recommended. Under such conditions, the beam will be level, or flat, until it is subjected to live loads.

repeat steps 3 through 5 until there is reasonable agreement.

6. Check for shear resistance:[4]

$$f_v(\text{actual}) = \frac{V(\text{actual})}{A_w(\text{actual})} \leq F_v(\text{allowable})$$

If f_v is greater than the allowable, return to step 4 and select another trial section having a greater web area (A_w). Then, repeat steps 3 through 6 until the trial section meets requirements for both bending and shear. (With standard rolled sections, resistance to shear will seldom be a problem.)

7. Determine the desired allowable deflection (Δ) in inches, and calculate the maximum deflection, using a standard formula if available; e.g., for a concentrated load at the center of a simple span,

$$\Delta(\text{actual}) = \frac{PL^3}{48EI} \leq \Delta(\text{allowable})$$

If no standard formula fits the design situation, compute the actual deflection in accordance with the general procedure presented in Chapter 4. If the Δ(actual) exceeds the Δ(allowable), return to step 4 and select another trial section having a larger moment of inertia (I). Then, repeat steps 3 through 7 until the trial section[5] meets requirements for bending, shear, and deflection.

It is apparent that this procedure can become time-consuming unless one is fortunate enough to select the proper trial section in step 4 on the first attempt. However, as mentioned in step 6, shear will seldom be a problem with rolled sections,

and deflection usually will be critical only where light loads are carried over long spans, or where clear story height or other architectural considerations require use of a beam of less depth than would otherwise be selected to meet bending requirements. Furthermore, as one becomes more proficient in design, it is possible to abbreviate the above procedure in ways such as that shown in footnote 5 or by use of handbook design aids as discussed in Art. 5.9.

5.4
Application of Design Procedure

The following examples will serve to explain the application of the method for treating deflection developed in Art. 5.3, as well as the total design procedure for bending, shear, and deflection, but without consideration of buckling. (The numbering of steps will differ from that of the general procedure described in Art. 5.3.)

Example 1

Design for the lightest-weight section that will support a uniform load of 750 lb per ft over its entire simple span of 24 ft (Fig. 5.1). Use A36 Steel with $F_b = 0.66F_y$ and

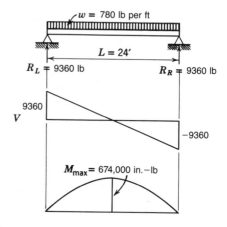

Figure 5.1

---

[4]In the AISC nomenclature, A_w is defined as "area of girder web," i.e., web thickness times depth of section.

[5]When available, deflection formulas also may be solved for I(required); e.g. $I(\text{required}) = PL^3/48E\Delta$. A beam section having an I equal to or greater than that required may then be selected.

$E = 29,000$ ksi. Limit maximum total load deflection to $\frac{1}{300}$ of the span.

Solution

(1) Assuming a beam weight of 30 lb per ft, $w = 780$ lb per ft. If $L/24$ is used as a guide for the depth, $d_1 = 12$ in. If $F_y/800$ is used as a guide for the depth,

$$d_2 = \frac{36}{800}(24)12 = 13 \text{ in.}$$

This information will be useful in selecting a trial section (step 4):

$$R_L = R_R = V = \frac{780(24)}{2} = 9360 \text{ lb}$$
$$V_{max} = 9360 \text{ lb}$$

(2) The maximum bending moment at midspan is

$$M = \frac{wL^2}{8}$$
$$= \frac{780(24)^2}{8}$$
$$= 56,200 \text{ ft-lb}$$
$$M = 56,200(12)$$
$$= 674,000 \text{ in.-lb}$$

(3) Compute the required section modulus, assuming the depth will be adequate to use

$$F_b = 0.66(36) = 24 \text{ ksi}$$
$$S = \frac{M}{F_b} = \frac{674,000}{24,000} = 28.1 \text{ in.}^3$$

(4) Select the lightest-weight beam furnishing the required section modulus. Referring to the AISC "Allowable Stress Design Selection" table, select a **W** 14×22:

$$S = 29.0 \text{ in.}^3 > 28.1 \text{ in.}^3 \qquad \text{as required}$$

The 14-in. depth $> 13 > 12$ as determined in step (1); therefore $F_b = 24$ ksi (verified).

(5) Check the **W** 14×22 for shear, where $F_v = 14,500$ psi, $A_w = t$ (thickness of web) times d (depth of section):

$$f_v = \frac{V}{td} = \frac{9360}{0.230(13.74)} = 2960 < 14,500 \text{ psi}$$

(6) Determine the maximum deflection at midspan:

$$\Delta(\text{allowable}) = \frac{\text{span}}{300} = \frac{24(12)}{300} = 0.96 \text{ in.}$$
$$\Delta = \frac{5wL^4}{384EI} = \frac{5(780)(24)^4(12)^3}{384(29,000,000)199}$$
$$= 1.01 > 0.96 \text{ in.}$$

Because the actual deflection exceeds the allowable, a beam having a greater I must be selected and checked for adequacy in bending (S) and shear (f_v).

(7) From the AISC "Elastic Section Modulus" table, the next largest and most economical section is a **W** 14×26. The previously assumed beam weight (30 lb per ft) is adequate. Therefore,

$$S = 35.3 \text{ in.}^3 > 28.1 \text{ in.}^3 \qquad \text{as required}$$

and

$$f_v = \frac{V}{td} = \frac{9360}{0.255(13.91)}$$
$$= 2640 < 14,500 \text{ psi}$$
$$\Delta = \frac{5wL^4}{384EI}$$
$$= \frac{5(780)(24)^4(12)^3}{384(29,000,000)245}$$
$$= 0.82 < 0.96 \text{ in.}$$

Therefore, the **W** 14×26 satisfies all three requirements (bending moment, shear, and deflection) and is adequate. The assumed beam weight of 30 lb per ft is slightly greater than the actual, but not enough greater to justify a recalculation in the hope of being able to use a lighter section. This can be checked by computing Δ based upon a 776-lb-per-ft load. Had the assumed weight been considerably less than the actual, a like decision would have had to be made, i.e., whether the difference would be apt to have sufficient effect to justify a recalculation to see if a heavier beam is required. Experience will help in making such decisions.

Example 2

The problem of Example 1 also may be handled beyond step (6) by solving the deflection equation for the moment of inertia required as indicated in footnote 5, Art. 5.3.

Solution

Steps (1) through (6) are the same.

(7) Determine required moment of inertia to limit deflection to the 0.96 in. specified:

$$I = \frac{5wL^4}{384E\Delta} = \frac{5wL^4}{384E\,0.96}$$

$$= \frac{5(780)(24)^4(1728)}{384(29{,}000{,}000)0.96} = 209 \text{ in.}^4$$

(8) Select the lightest-weight section furnishing an $I \geq 209$ in.^{4} and an $S \geq 28.1$ in.^{3}. Referring to the elastic-section-modulus table (AISC "Allowable Stress Design Selection" table), make a trial selection based on section modulus only (**W** 14×22). Next refer to the "Moment of Inertia Selection" tables. From the table, it will be seen that the lightest-weight section having $I \geq 209$ is the **W** 14×26. This section has an $S > 28.1$; therefore, it is selected.

As seen from step (5), shear is not critical, and the actual weight is close enough to the assumed weight to consider the selection adequate.

Example 3

Redesign the beam in Example 1 using a 12-in. nominal beam depth and the $F_y/800$ depth criteria, everything else remaining the same as in Example 1.

Solution

Steps (1) and (2) are the same.

(3) Compute the required section modulus. If fully stressed, the allowable bending stress is

$$F_{b1} = 0.66(36) \approx 24 \text{ ksi}$$

and the depth must be

$$d = \frac{36}{800}(24)12 \approx 13 \text{ in.}$$

When using the 12-in.-deep beam, the stress must be reduced to

$$F_{b2} = \frac{12}{13}(24) = 22.2 \text{ ksi}$$

and so

$$S = \frac{674}{22.2} = 30.4 \text{ in.}^3$$

(4) Compute the required moment of inertia.

$$I = 209 \text{ in.}^4$$

[see Example 2, step (7)].

(5) Referring to the AISC Manual "**W** Shapes: Dimensions and Properties" table, select a **W** 12×30 from the 12-in.-deep group ($I = 238$ in.^{4} and $S = 38.6$ in.^{3}). It is observed that the value of I controls the selection. Again, the actual weight is close enough to that assumed, and shear is not critical.

PROBLEMS

(Answers to all problems are given in Appendix G.)

1. A beam is simply supported on a 28-ft span. There is a single concentrated load of 15 kips placed at its midspan. Neglecting the weight of the beam, select the most economical section that can be safely used. Limit the maximum total-load deflection to $L/240$. Use A36 Steel, $F_{b1} = 0.66F_y$, and the $F_y/800$ depth-span guide.

2. A beam is simply supported on a 36-ft span. There are two 10-kip concentrated loads, one at each third-point. Neglecting the weight of the beam, select the most economical section that can be safely used. Limit the maximum total-load deflection to $L/300$. Use A36 Steel, $F_{b1} = 0.66F_y$, and the $L/24$ beam-depth guide.

3. A beam is simply supported on a 32-ft span. There are three 5-kip concentrated loads symmetrically placed at the quarter-points. The maximum permissible depth of beam is 14 in. Neglecting the weight of the beam, select the most economical section that can be safely used. Limit the maximum total-load deflection to $L/300$. Use A36 Steel, $F_{b1} = 0.66F_y$, and the $F_y/800$ depth-span guide.

4. A beam is simply supported on a 26-ft span. There is a concentrated load of 2 kips at the center of the span. In addition, there is a 400-lb-per-ft uniform load (including the beam weight) over the entire span. Select the most economical section that can be safely used. Limit the maximum total-load deflection to $L/240$. Use A572, Grade 50 Steel, $F_{b1} = 0.66F_y$, and the $F_y/800$ depth-span guide. Redesign the beam using A36 Steel, all other requirements remaining the same.

5. A beam cantilevers 12 ft from a wall. There is a 2-kip concentrated load at the free end and a 4-kip concentrated load 6 ft from the free end. Consider these concentrated loads to be live loads. The dead load, including the weight of the beam and its surrounding construction, is 100 lb per ft. The live-load deflection at the free end must be limited to $\frac{1}{2}$ in. Select the most economical section that can be safely used. Use A36 Steel and $F_b = 0.66F_y$.

6. A simply supported beam, 32 ft long, has an 8-ft overhang at one end, making the distance between supports 24 ft. Select the most economical section that can be safely used if there is a uniformly distributed load of 1200 lb per ft (including the beam weight) over the entire span. Limit the total-load deflection between supports to 0.80 in. and that at the free end of the overhang to 0.40 in. (up or down). Use A36 Steel, $F_b = 0.66F_y$, and no depth-span guide because of the overhang.

5.5 Lateral Buckling

A beam subjected to bending is in tension at one flange and in compression at the other. It is the compression flange which tends to buckle, and this tendency increases as the stress increases. If buckling occurs, it will be in the direction of least resistance. As the web firmly holds both flanges vertically, this direction is sideways, or laterally, and is nearly always accompanied by some twisting. Figure 5.2 illustrates the hypothetical buckled position of a simply supported beam. The vertical displacement shown is normal deflection and is not due to buckling.

The shape of a beam also affects its tendency to buckle. An economical beam usually has a large moment of inertia about its

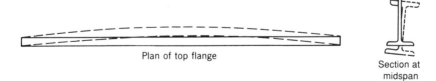

Plan of top flange

Section at midspan

Figure 5.2/Lateral buckling.

x-x axis and a relatively small moment of inertia about its *y-y* axis. The tendency to buckle increases as the ratio $I_{x\text{-}x}/I_{y\text{-}y}$ increases. Therefore, the deeper and narrower the section, the more susceptible it is to buckling.

Obviously, no beam could buckle if the surrounding constructing prevented it. When a beam is thus prevented from lateral movement, it is classified as laterally supported. The task of deciding when a beam is laterally supported, however, is

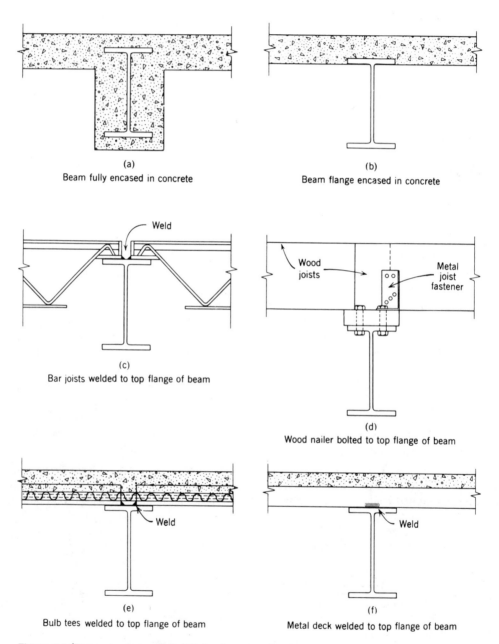

(a)
Beam fully encased in concrete

(b)
Beam flange encased in concrete

Weld

(c)
Bar joists welded to top flange of beam

Wood joists

Metal joist fastener

(d)
Wood nailer bolted to top flange of beam

Weld

(e)
Bulb tees welded to top flange of beam

Weld

(f)
Metal deck welded to top flange of beam

Figure 5.3/Beams having complete lateral support for the top flange.

not always a simple one. When in doubt, the beam should be designed as laterally unsupported, as this is the more critical assumption. Figure 5.3 shows several types of construction that are generally recognized as adequate means of providing complete lateral support of the top flange. The bar joists shown in Fig. 5.3c must be spaced fairly close together (as discussed later in Art. 5.18). Precast floor systems held in place by clips, for example, are *not* generally considered sufficiently rigid to provide lateral support. Nor is a beam embedded in a conventional masonry wall necessarily provided lateral support by the wall alone.

Special or regular intermediate members also may be introduced at predetermined points along the length of the beam to provide needed lateral support. However, details of connections must be given careful attention. Figure 5.4(a) and (b) show two ways of connecting beams to provide lateral support. In Fig. 5.4a only the top flange is supported. Figure 5.4c shows roof beams, or purlins, inclined to the direction of gravity loads, which are laterally supported by tie rods placed as shown.

Even beams which receive lateral support from other members can buckle between the points of support; therefore, the shorter the laterally unsupported length, the greater the resistance to buckling. The beams shown in Fig. 5.3 have nearly continuous support, thus effectively preventing buckling.

It should be noted, too, that because it is the compression flange which may buckle, this will be the bottom, rather than the top flange, over portions of any beam having negative moment, such as at overhangs.

It is necessary to establish early in the design process whether a beam should be classified as one having adequate lateral support or as one that is laterally unsupported. The proper classification depends on the proportions of the beam, and as a

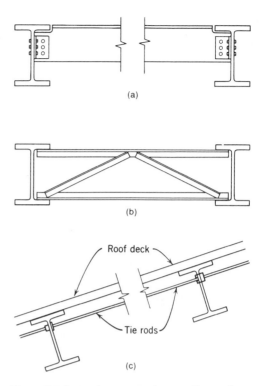

Figure 5.4/Lateral support at intermediate points.

general rule, these proportions cannot be determined until the design process has begun and a trial selection has been made. When this is the case, the designer must use a successive approximation procedure, i.e., he must assume a condition to exist, then modify it as subsequent calculations indicate the need.

The two articles that follow discuss the design process first for beams with adequate lateral support and then for laterally unsupported beams. In each process, the principal task is to establish the allowable bending stress F_b. The flowchart shown in Fig. 5.5(a) and (b) should be helpful in understanding those conditions that govern the selection of a particular formula for F_b. This flowchart is for flanged beams (**W, M, S, C, MC**) only. For other shapes— i.e., box, tubular, angles—refer to the AISC Specification.

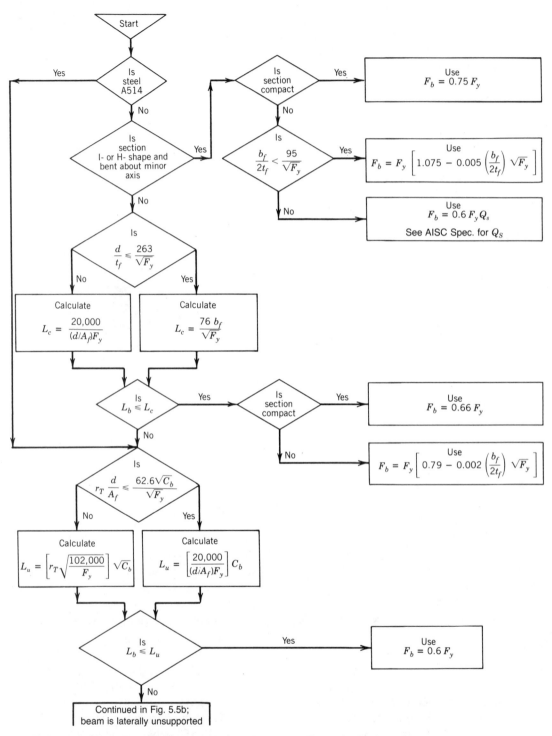

Figure 5.5a/Flow diagram for allowable bending stress F_b except for hybrid and box beams.

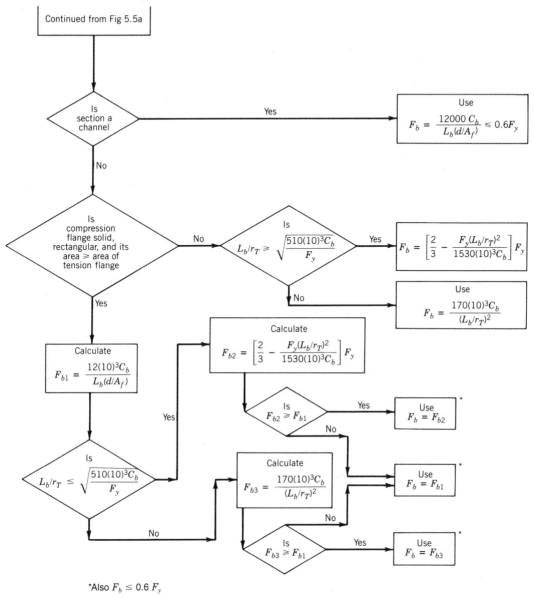

Figure 5.5b/Flow diagram for allowable bending stress F_b except for hybrid and box beams, continued.

5.6
Beams: Laterally Supported

Two principal factors must be considered —conditions that (1) constitute adequate lateral support and those that (2) dictate the allowable stress (F_b) for beams having adequate lateral support. Constructions like those shown in Fig. 5.3 provide continuous lateral support for the top flange. Under such conditions, any beam, regardless of its proportions, can automatically be classified as having adequate lateral

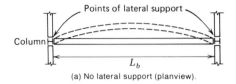

(a) No lateral support (planview).

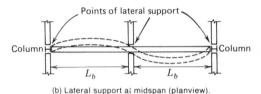

(b) Lateral support at midspan (planview).

Figure 5.6/Lateral-support buckling length (L_b).

support, provided the beam is subjected to positive bending only (i.e., compression on the top flange). Other constructions do not provide continuous lateral support but do furnish support at specific points along the length of the beam (Fig. 5.4). If the distance from one point of lateral support to another is excessive, then the compression flange could buckle between these points (Fig. 5.6). This would also be the case if the bar joists in Fig. 5.3c were spaced far apart.

When beams have adequate lateral support, the allowable bending stress varies between $0.6F_y$ and $0.66F_y$. The higher value ($0.66F_y$) can be used only if the beam is classified as compact and its unbraced length (L_b) does not exceed a special L_c length for each flanged beam. Most rolled beam shapes are compact (particularly if made from A36 Steel). Compactness is described more thoroughly later in this article.

Since each flanged beam will have an identifiable beam width, depth, and flange area, each beam for a specific grade of steel will have a unique length for the limits of its bracing points if it is to be classified as laterally supported. There are two formulas for calculating the L_c length of a given beam. According to the AISC Specification, both conditions must be met in order

to use the higher allowable stress, $0.66F_y$. The first formula, based upon lateral buckling, is given as

$$L_c = \frac{76b_f}{\sqrt{F_y}}$$

and controls when

$$\frac{d}{t_f} \leq \frac{263}{\sqrt{F_y}}$$

The second formula, based upon torsional twisting, is given as

$$L_c = \frac{20,000}{(d/A_f)F_y}$$

and controls when

$$\frac{d}{t_f} > \frac{263}{\sqrt{F_y}}$$

Both values for L_c can be calculated and then the smaller of the two used for comparing with L_b; or the limting formula can be established first by observing its proportions (d/t_f) and comparing this value with $263/\sqrt{F_y}$. The factor d/A_f in the twisting formula is the depth of the section divided by the area of the compression flange.[6] All the other quantities in the formulas are identified in Fig 5.8. The controlling value of L_c for each rolled section is listed in the *AISC Manual*, along with other design values, in both the "Allowable Stress Design Selection" table and the "Allowable Loads on Beams" table.

When the actual buckling length L_b exceeds L_c, $0.66F_y$ cannot be used for the allowable bending stress. Under these conditions, the stress drops abruptly to $0.6F_y$ or less. There is a unique length L_u for each flanged beam associated with the allowable stress $0.6F_y$. When the actual

[6]The value of d/A_f for each rolled beam is listed along with its other properties in the *AISC Manual*.

buckling length L_b exceeds the L_c of a beam but does not exceed the L_u of the beam, it is still classified as having adequate lateral support, but the allowable stress is equal to $0.6F_y$. When L_u is exceeded, the beam is considered laterally unsupported.

The complete explanation of L_u includes two additional features. An important property of a shape in resisting buckling is its radius of gyration. Its symbol is r, and it can be calculated from the formula $r = \sqrt{I/A}$. This property is described in more detail in the next chapter in reference to column buckling. It also is discussed in the next article under laterally unsupported beams. Of concern here are the proportions of the compression flange when referenced to the y axis of the beam (Fig. 5.13), which includes one-third of the area of the compression part of the web. This specific radius of gyration is referred to as r_T, and its value is listed in the AISC Manual for every rolled section, together with its other dimensions and properties. The use of r_T in determining L_u is shown in subsequent formulas.

The other feature influencing the buckling length is the so-called moment gradient multiplier. Its symbol is C_b, and its value ranges from 1 to 2.3. It can be applied only if the beam is loaded in the plane of its web, it is bent about its major axis, and the rectangular compression flange has an area not less than that of the tension flange.

The value of C_b is frequently taken as unity (1.0), which represents a conservative assumption; i.e., when $C_b = 1.0$, the assumption is that the compressive bending stress remains constant between braced points or that its maximum value is located somewhere between the braced points and not at the ends. If less critical conditions are known to exist, there will be a lesser tendency to buckle, and the designer may wish to take advantage of this fact. The value of C_b depends on the variation in bending moment throughout the length of the beam in relation to the location of the bracing points. This is another reason why it is good practice to sketch the bending-moment diagram.

The 1989 AISC Specification provides the following formula for obtaining C_b:

$$C_b = 1.75 + 1.05\frac{M_1}{M_2} + 0.3\left(\frac{M_1}{M_2}\right)^2 \leq 2.3$$

where M_1 is the smaller and M_2 the larger bending moment at the ends of the unbraced length, taken about the strong axis of the member, and where the ratio of the end moments (M_1/M_2) is positive for reverse-curvature bending and negative for single-curvature bending. When the bending moment at any point within an unbraced length is larger than that at both ends of this length, the value of C_b is taken as unity.

For further explanation of C_b, refer to the beam shown in Fig. 5.7. Bracing points are located at the reactions and at each applied concentrated load (denoted by the symbol $\times$).

The cantilever length is

$$C_b = 1.75 + 1.05\left(-\frac{0}{16}\right) + 0.3\left(-\frac{0}{16}\right)^2$$
$$= 1.75$$

The length between the right reaction and

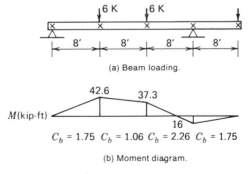

(a) Beam loading.

(b) Moment diagram.

Figure 5.7/Moment-gradient example.

the first 6-kip load is

$$C_b = 1.75 + 1.05\left(+\frac{16}{37.3}\right) + 0.3\left(+\frac{16}{37.3}\right)^2$$
$$= 1.75 + 0.45 + 0.06$$
$$= 2.26$$

The length between the two concentrated loads is

$$C_b = 1.75 + 1.05\left(-\frac{37.3}{42.6}\right) + 0.3\left(-\frac{37.3}{42.6}\right)^2$$
$$= 1.75 - 0.92 + 0.23$$
$$= 1.06$$

Two formulas are used to establish L_u, the length that cannot be exceeded if $0.6F_y$ is to be used as the allowable bending stress. The formula based upon twisting is

$$L_u = \left[\frac{20,000}{(d/A_f)F_y}\right]C_b$$

and controls when

$$r_T\frac{d}{A_f} \leq \frac{62.62\sqrt{C_b}}{\sqrt{F_y}}$$

The formula based upon buckling is

$$L_u = \left[r_T\sqrt{\frac{102,000}{F_y}}\right]\sqrt{C_b}$$

and controls when

$$r_T\frac{d}{A_f} > \frac{62.62\sqrt{C_b}}{\sqrt{F_y}}$$

Both values of L_u can be calculated and then the larger of the two used for comparing with L_b, or the governing formula can be established first by observing the product of r_T and d/A_f and comparing with $62.62\sqrt{C_b}/\sqrt{F_y}$.

When $C_b = 1$, the twisting formula for L_u is seen to be the same as for L_c. The controlling value of L_u for each rolled section when $C_b = 1$ is listed in the AISC Manual, together with other design values,

in both the "Allowable Stress Design Selection" table and the "Allowable Loads on Beams" table.

Although not explicitly described in the AISC Specification, it should be apparent from the equations that the listed value of L_u, if based upon the twisting formula, can be multiplied by C_b and the resulting product used for comparison with L_b. However, when the buckling formula controls, the listed value should be multiplied by $\sqrt{C_b}$. Expressed in equation form, a beam has adequate lateral support for a stress of $0.6F_y$ if

$$L_b \leq L_u C_b \qquad \text{when twisting controls}$$

or

$$L_b \leq L_u\sqrt{C_b} \qquad \text{when buckling controls}$$

The Ninth Edition of the AISC Manual lists over 300 shapes (**W**, **M**, and **S**) commonly used as beams. Of these, only 19 have properties such that twisting controls the L_u value listed in the Manual. Identifying these 19 sections could prove useful for future design problems if C_b is not taken as 1.

For an example of adequate lateral support, consider a **W** 10×33 beam of A36 Steel. It has a depth-to-flange thickness ratio of $9.73/0.435 = 22.37$, which is less than $263/\sqrt{36} = 43.83$, so buckling controls the L_c length. Consequently,

$$L_c = \frac{76(7.96)}{\sqrt{36}\,(12)} = 8.4 \text{ ft}$$

Checking the product of r_T and d/A_f, $2.14(2.81) = 6.01$, which is less than $62.62/\sqrt{36} = 10.44$. Consequently, the twisting formula controls the L_u value. Taking $C_b = 1$, L_u is calculated.

$$L_u = \frac{20,000}{2.81(12)36} = 16.5 \text{ ft}$$

For the same beam section but of A572, Grade-50 Steel, the same controlling formulas apply, but the allowable length must be reduced to $L_c = 7.1$ ft and $L_u = 11.8$ ft. The length can be calculated as shown here, or it may be obtained directly from the AISC Manual if the beam is a regular rolled section.

In the above example for A36 Steel, the beam is considered as having adequate lateral support if the actual unbraced length L_b is equal to or less than 16.5 ft. However, should the moment gradient C_b be equal to 1.25, the unbraced length could be increased to $1.25(16.5) = 20.6$ ft.

Once it is established that a beam has adequate lateral support, the allowable bending stress (F_b) will range between $0.6F_y$ and $0.66F_y$,[7] the exact value depending on L_c and the degree of compactness of the particular section. To explain: in an effort to use the amount of steel in a given beam section most effectively, there is a wish to distribute the metal (Fig. 5.8) so as to increase flange widths (b_f) and beam depth (d), and to decrease flange thickness (t_f) and web thickness (t_w). However, if such distribution is carried to excess, the section will become flimsy and have little ability to develop compression stress. To control this tendency, the AISC Specification contains "width-thickness ratio" limitations. The upper limits of the ratio for rolled sections are

$$\frac{b_f/2}{t_f} \leq \frac{95}{\sqrt{F_y}}$$

$$\frac{h}{t_w} \leq \frac{253}{\sqrt{F_y}}$$

Most beam sections do more than simply meet this minimum requirement. Those

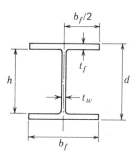

Figure 5.8/Beam proportions.

rare sections not meeting these requirements are classified as *slender elements*.

It was stated earlier that the basic allowable bending stress (F_b) is $0.6F_y$. This basic value should be used only when there is adequate lateral support for the beam and the limitations cited above for width-thickness ratios are met (which is the case with all rolled sections listed in the AISC Manual).

Compact beams are defined as those sections meeting the following more restrictive width thickness ratios for the flange:

$$\frac{b_f/2}{t_f} \leq \frac{65}{\sqrt{F_y}}$$

When bending occurs alone in a beam, as discussed in this chapter, there is no limiting width-thickness criterion for a compact web.

Compact sections that are braced so that $L_b \leq L_c$ are permitted the higher allowable bending stress, i.e., $F_b = 0.66F_y$. Most rolled beam sections meet the requirement for compactness when A36 Steel is used. The *AISC Manual* "Dimensions and Properties" tables list, for each section, values of F_y', which is the theoretical yield stress at which the section becomes noncompact.

Beams that are braced so that $L_b \leq L_c$, but are not fully compact, have values of F_b ranging between $0.6F_y$ and $0.66F_y$ and are established by the so-called blending

---

[7]Exceptions to this are beams of A514 Steel, some built-up box beams, and built-up beams of two different grades of steel (hybrid).

formula

$$F_b = F_y \left[0.79 - 0.002 \left(\frac{b_f}{2t_f} \right) \sqrt{F_y} \right]$$

Observe in the above formula that when $b_f / 2t_f$ becomes $95/\sqrt{F_y}$, the quantity within the bracket becomes 0.6.

Following is a summary of the allowable bending-stress formulas for beams that are laterally supported.

When $L_b \leq L_c$, $L_b \leq L_u$, and the beam is compact,

$$F_b = 0.66 F_y$$

When $L_b \leq L_c$, $L_b \leq L_u$, and the beam is not compact,

$$F_b = F_y \left[0.79 - 0.002 \left(\frac{b_f}{2t_f} \right) \sqrt{F_y} \right]$$

When $L_b > L_c,[8]$ but $L_b \leq L_u$,

$$F_b = 0.6 F_y$$

(compactness not a consideration).

When $L_b > L_c$ and $L_u,[8]$ the beam is not adequately laterally supported and

$$F_b < 0.6 F_y$$

Examples 1, 2, 3, and 4, which follow, illustrate use of the above formulas. Exam-

[8]If C_b is not equal to 1.0, these formulas should be

$$L_b > L_c \quad \text{but} \quad L_b \leq C_b L_u$$
$$L_b > L_c \text{ and } C_b L_u$$

when twisting controls;

$$L_b > L_c \quad \text{but} \quad L_b \leq \sqrt{C_b} L_u$$
$$L_b > L_c \text{ and } \sqrt{C_b} L_u$$

when buckling controls.

ple 5 uses a general beam design procedure for laterally supported beams.

Example 1

A **W** 14×34 of A36 Steel has a total simple span of 28 ft and is loaded as shown in Fig. 5.9 (the distributed load includes an allowance for the weight of the beam). The beam has complete lateral support, and there is no depth criterion or deflection limitation. Determine the maximum value of the concentrated loads (P).

Solution

(1) Since there is complete lateral support, $L_b = 0$. Determine if the section is compact. The limit of the width-thickness ratio for the flange is

$$\frac{65}{\sqrt{36}} = 10.8$$

Because there is no axial stress (f_a), web proportions are not a criterion for compactness. From the AISC Manual the value of $b_f / 2t_f$ for a **W** 14×34 is seen to be $7.4 < 10.8$:

$$\frac{b_f}{2t_f} = \frac{6.745}{2(0.455)} = 7.41 < 10.8 \qquad \textbf{OK}$$

The section is compact. (This can be verified by observing from the AISC Manual "Allowable Stress Design Selection" table

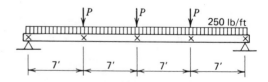

Figure 5.9/Examples 1, 2, and 3.

or from the "Dimensions and Properties" table that F_y' has no value listed, which means that it is larger than 60 ksi and the section is compact.)

(2) Determine the allowable bending stress. Since $L_b = 0$ and the section is compact,

$$F_b = 0.66(36) \simeq 24 \text{ ksi}$$

(3) Calculate the maximum resisting moment. From the AISC Manual, $S = 48.6$ in.^{3} for a **W** 14×34; therefore,

$$M_R = SF_b = 48.6(24) = 1166 \text{ in.-kips}$$

$$M_R = \frac{1166}{12} = 97.2 \text{ ft-kips}$$

(4) Calculate the maximum moment caused by the load, and equate to the resisting moment to determine P. From Fig. 2.34

$$M_{\max} = \frac{0.25(28)^2}{8} + \frac{P(28)}{2}$$

$$= 24.5 + 14P$$

$$24.5 + 14P = 97.2$$

$$14P = 72.7$$

$$P = 5.2 \text{ kips}$$

Example 2

Determine the maximum value of the loads (P) for the beam used in Example 1 if instead of continuous lateral support there is point lateral support located at the concentrated loads and at the reactions. Use $C_b = 1$.

Solution

(1) Calculate L_c and L_u for the beam. From the AISC Manual, obtain the values of d, t_f, b_f, r_T, and d/A_f for a **W** 14×34.

Then

$$\frac{d}{t_f} = \frac{13.98}{0.455} = 30.73 < \frac{263}{\sqrt{36}} = 43.83$$

Therefore,

$$L_c = \frac{76(6.745)}{\sqrt{36}} = 85.4 \text{ in.} = 7.1 \text{ ft}$$

Also, since

$$(r_T)\frac{d}{A_f} = (1.76)(4.56)$$

$$= 8.03 < \frac{62.62}{\sqrt{36}} = 10.44$$

The value of L_u is calculated as

$$L_u = \frac{20,000}{4.56(36)} = 122 \text{ in.} = 10.2 \text{ ft}$$

These values can be verified from the AISC "Allowable Stress Design Selection" table, or the "Allowable Loads on Beams" table.

(2) Determine if the section is compact. This is step (1) in Example 1 (section is compact).

(3) Determine the allowable bending stress. Since $L_b = 7.0 < 7.1 < 10.2$ and the section is compact,

$$F_b = 0.66(36) \simeq 24 \text{ ksi}$$

(4) Calculate the maximum resisting moment and allowable load (P). This is the same as steps (3) and (4) in Example 1:

$$P = 5.2 \text{ kips}$$

Example 3

Determine the maximum value of the loads (P) for Example 1 if the beam is a **W** 16×40, made from A572, Grade-50 Steel, and has point lateral support at the ends and at concentrated loads.

Solution

(1) Determine if the beam has adequate lateral support. Calculate L_c and L_u for the **W** 16×40:

$$\frac{d}{t_f} = \frac{16.01}{0.505} = 31.7 < \frac{263}{\sqrt{50}} = 37.19$$

$$L_c = \frac{(76)6.995}{\sqrt{50}} = 75.2 \text{ in.} = 6.3 \text{ ft}$$

$$r_T\frac{d}{A_f} = (1.82)(4.53) = 8.24 < \frac{62.62}{\sqrt{50}} = 8.86$$

$$L_u = \frac{20,000}{(4.53)(50)} = 88.3 \text{ in.} = 7.4 \text{ ft}$$

These values can be verified by reference to the AISC Manual. Since $L_b = 7.0$ ft > 6.3 ft, there is no need to determine if the section is compact. Since $L_b = 7.0$ ft < 7.4 ft, there is no need to calculate C_b.

(2) Determine the allowable bending stress. Since $L_c < L_b < L_u$, $F_b = 0.6F_y$:

$$F_b = 0.6(50) = 30 \text{ ksi}$$

(3) Calculate the maximum resisting moment and the maximum allowable load P:

$$M_R = SF_b = 64.7(30) = 1941 \text{ in.-kips}$$

$$M_R = \frac{1941}{12} = 162 \text{ ft-kips}$$

$$M_{max} = \frac{0.25(28)^2}{8} + \frac{P(28)}{2}$$
$$= 24.5 + 14P$$
$$24.5 + 14P = 162$$
$$P = 9.82 \text{ kips}$$

Example 4

A **W** 10×33 beam of A572, Grade-60 Steel is simply supported and spans 22 ft. It

carries a uniformly distributed load and has complete lateral support. There are no depth criteria or deflection limitations. Determine the value of the unit load it can carry, including its own weight.

Solution

(1) Determine if the section is compact. The width-thickness ratio limit for the flange is

$$\frac{65}{\sqrt{60}} = 8.4$$

Determine values of b_f and t_f from the AISC Manual:

$$\frac{b_f}{2t_f} = \frac{7.960}{2(0.435)} = 9.15$$

Since $9.15 > 8.4$, the section is not compact.

(2) Determine if the blending formula can be used:

$$\frac{95}{\sqrt{60}} = 12.3$$

Since $9.15 < 12.3$, it can be used.

(3) Calculate the allowable bending stress and the maximum resisting moment:

$$F_b = 60\left[0.79 - 0.002(9.15)\sqrt{60}\right]$$
$$= 38.8$$
$$M_R = SF_b = 35(38.8) = 1358 \text{ in.-kips}$$

$$M_R = \frac{1358}{12} = 113 \text{ ft-kips}$$

(4) Find the maximum moment developed by the loads, and calculate w. From Fig. 2.34,

$$M_{max} = \frac{wL^2}{8} = \frac{w(22)^2}{8} = 60.5w$$

$$60.5w = 113$$
$$w = 1.87 \text{ kips per ft}$$

Example 5/Design of a Roof System

A plan indicating the location of beams and girders of a typical 24- by 28-ft interior bay is shown in Fig. 5.10. The beams, girders, and columns are to be attached to one another with flexible connections (see Art. 7.3, Type 2 construction); therefore, only simple reactions are generated.

The deck will consist of 20-ga. metal decking, covered with 2-in. reinforced-stone concrete; i.e., it is a one-way deck that will transfer all its load directly to the beams. The top flange of the beams will be 1 in. above the top flange of the girders; therefore, the deck will not come in contact with the girders. The deck is to be spot-welded (with welding washers) to the beam, providing the beam with complete lateral support.

One inch of rigid insulation and a 5-ply built-up roof will be applied over the deck, and a metal-lath and gypsum-plaster ceiling will be hung from the deck. Using a combined live and environmental load of 30 psf, design the beam and girder using A36 Steel. Check the system for roof ponding.

Solution

(1) Itemize and total the loads:[9]

5-ply built-up roof	6.0 psf
1-in. rigid insulation	1.5
2-in. reinforced-stone concrete deck	25.0
20-ga. metal decking	2.5
Metal-lath and gypsum-plaster hung ceiling	10.0
Total dead load	45.0 psf

(2) Sketch the beam loading, and calculate loads, maximum moment, and shear (Fig.

[9]See building code or manufacturer's specification.

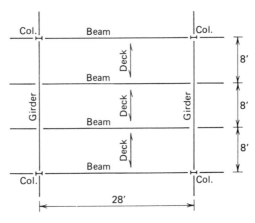

Figure 5.10/Example 5: typical interior bay framing plan.

5.11):

Live and environmental load	$30(8) = 240$ lb per ft
Dead load	$45(8) = 360$
Assumed beam weight	$= 30$
Total	$= 630$ lb per ft

From Fig. 2.34,

$$M_{max} = \frac{0.63(28)^2}{8} = 61.7 \text{ ft-kips}$$

$$V_{max} = R = \frac{0.63(28)}{2} = 8.82 \text{ kips}$$

(3) Estimate the beam depth and allowable bending stress, using a span-depth guide of $\frac{1}{24}$ [actual deflection check will be made in step (4)]:

$$d_{min} = \frac{28(12)}{24} = 14 \text{ in.}$$

Since the beam's compression flange (top)

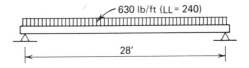

Figure 5.11/Beam for Example 5.

has complete lateral support, and probably the selected beam will be compact, the stress can be assumed as

$$F_b = 0.66(36) \simeq 24 \text{ ksi}$$

(4) Determine required moment of inertia. From step (2), the combined live and environmental load $w = 240$ lb per ft. The maximum live-load deflection for a plastered ceiling is

$$\Delta_{max} = \frac{28(12)}{360} = 0.93 \text{ in.}$$

From Fig. 4.15

$$I_{req} = \left(\frac{5}{384}\right)\frac{0.24(28)^4(12)^3}{29,000(0.93)} = 123 \text{ in.}^4$$

(5) Calculate required section modulus and select a trial section:

$$S_{req} = \frac{61.7(12)}{24} = 30.9 \text{ in.}^3$$

Referring to the AISC Manual "Dimensions and Properties" table and the 14-in.-deep beam group, select the lightest-weight section having an $S \geq 30.9$ in.^{3} and an $I \geq 123$ in.^{4}. Select **W** 14×26.

(6) Verify assumptions and modify the selection if required; check to see if the **W** 14×26 is compact:

Flange maximum width-thickness ratio,

$$\frac{65}{\sqrt{36}} = 10.8$$

Actual ratio,

$$\frac{b_f}{2t_f} = 6.0 \qquad \text{OK}$$

The section is compact and $F_b = 24$ ksi (verified); the assumed beam weight (30 lb/ft) is satisfactory. No revisions are necessary.

(7) Check shear:

$$f_v = \frac{V}{A_{web}} = \frac{8.82}{13.91(0.255)} = 2.49 \text{ ksi}$$

and

$$2.49 < 14.5 \qquad \text{OK}$$

(8) Calculate and sketch girder loading, and determine maximum moment and shear. The girder supports an area of $28(24) = 672$ sq ft. Most codes permit a live-load reduction for floors. In some instances, a similar reduction is allowed for roofs, but this depends greatly on the geographic location and source of live load on the roof. In this example, no reduction will be made:

Girder point live load

$$0.24(28) = 6.7 \text{ kips}$$

(beam frames in from each side);
Girder point dead load

$$0.39(28) = 10.9$$

Total point load, 17.6 kips

Assume a girder weight of 45 lb per ft. See Fig. 5.12 for girder loading. From Fig. 2.34,

$$M_{max} = \frac{17.6(24)}{3} + \frac{0.045(24)^2}{8}$$
$$= 141 + 3 = 144 \text{ ft-kips}$$

$$V_{max} = 17.6 + 0.045\left(\frac{24}{2}\right) = 18.1 \text{ kips}$$

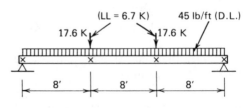

Figure 5.12/Girder for Example 5.

(9) Estimate the girder depth and allowable bending stress. Use a depth-span ratio of $\frac{1}{24}$:

$$\text{min. depth} = \frac{24(12)}{24} = 12 \text{ in.}$$

Beams provide point lateral support at 8-ft intervals; therefore, $L_b = 8$ ft. If the girder selected has an $L_c \geq 8$ ft and is compact, $F_b = 0.66F_y$. If the girder selected has $L_c < 8$ ft, but $L_u > 8$ ft, then $F_b = 0.6F_y$. Assume $F_b = 0.66(36) = 24$ ksi.

(10) Determine the required moment of inertia. The maximum live- and environmental-load deflection is $24(12)/360 = 0.8$ in. From Fig. 4.15

$$I = \left(\frac{23}{648} \right) \frac{6.7(24)^3(12)^3}{29,000(0.8)} = 245 \text{ in.}^4$$

(11) Calculate the required section modulus, and select a trial section:

$$S_{\text{req}} = \frac{144(12)}{24} = 72 \text{ in.}^3$$

Referring to the *AISC Manual* "Allowable Stress Design Selection" table and "Moment of Inertia Selection" table, select as a trial section the **W** 21×44 having an $S = 81.6$ in.^{3} and $I = 843$ in.^{4}.

(12) Verify assumptions and modify the selection if required. It is observed from the "Allowable Stress Selection" table that for A36 Steel, the **W** 21×44 has L_c and L_u values (6.6 ft and 7 ft, respectively) that are less than $L_b = 8$ ft. Also, note that $C_b = 1.0$, because the middle 8 ft has a maximum moment between braced points. Therefore, this first trial section does not have adequate lateral support, and $F_b < 0.6F_y$, which does not verify the assumptions made in step (9). (It must be emphasized here that this section might still be found to be satisfactory if it were examined as a beam without adequate lateral support, and, in fact, it is satisfactory. However, such examination is covered in

the next article; therefore, the solution will proceed using the criteria established to this point.) A **W** 18×46 is selected next for investigation (2 lb more in weight but 2.6 in. less in depth). From the AISC Manual for the **W** 18×46, $I = 712$ in.^{4}, $S = 78.8$ in.^{3}, $L_c = 6.4$ ft, and $L_u = 9.4$ ft. Since $L_b = 8$ ft > 6.4 ft but < 9.4 ft,

$$F_b \neq 0.66F_y \text{ (assumed)}$$

but

$$F_b = 0.6F_y$$

Therefore, $F_b = 0.6(36) \simeq 22$ ksi and

$$f_b = \frac{144(12)}{78.8} = 21.9 \text{ ksi}$$

Since $21.9 < 22$, the bending stress is acceptable.

(13) Check shear:

$$f_v = \frac{18.1}{18.06(0.360)} = 2.78 \text{ ksi} \qquad \text{OK}$$

(14) Check roof system for ponding (AISC Specification K2):

$$C_p = \frac{32L_sL_p^4}{10^7I_p} = \frac{32(28)(24)^4}{10^7(712)} = 0.0418$$

$$C_s = \frac{32SL_s^4}{10^7I_s} = \frac{32(8)(28)^4}{10^7(245)} = 0.0642$$

$$0.0418 + 0.9(0.0642) = 0.106 < 0.25 \qquad \text{OK}$$

The I_d does not have to be checked, since the structural deck is of reinforced concrete. The system is safe from ponding.

5.7
Beams: Laterally Unsupported

A laterally unsupported beam is one having lateral support of its compression flange at points that are a greater distance apart (L_b) than the unique L_u value for that beam. In determining the value of L_u, as well as the accompanying value of F_b, the coefficient C_b can either be conservatively assumed to be 1.0 or evaluated more accurately, as defined and discussed in the previous article. When $L_b > C_b L_u$ (or $\sqrt{C_b} L_u$ when buckling controls), the value of the allowable bending stress F_b will always be less than $0.6F_y$, and because this is so, the question of compactness is irrelevant. The procedure used to determine the value of F_b for laterally unsupported beams is different than for beams having adequate lateral support.

The 1989 AISC Specification provides two separate and distinct formulas for calculating F_b based on the lateral buckling of the compression flange. The susceptibility to buckling is expressed as the ratio of the buckling length to the "effective radius of gyration," or L_b / r_T.[10]

The radius of gyration (r) is measured in inches and can be determined for any given shape from the formula $r = \sqrt{I/A}$, where r is always referenced to the same axis as I. The AISC defines the effective section for r_T as the compression flange plus one-third of the compression-web area, with I and r_T referenced to the y-y axis (Fig. 5.13). For convenience, r_T values for all rolled sections are listed along with other properties in the AISC Manual "Dimensions and Properties" tables.

The first step in determining the value of F_b is to calculate the ratio L_b / r_T, making sure that L_b is measured in inches if r_T is taken from the AISC Manual (where it is listed in inches). If the ratio is small, i.e., less than

$$\sqrt{\frac{102(10)^3 C_b}{F_y}}$$

then the susceptibility to buckling is negligible, the beam is classified as having adequate lateral support, and consequently

---

[10] Buckling and radius of gyration are defined more extensively in Chapter 6, Columns and Struts.

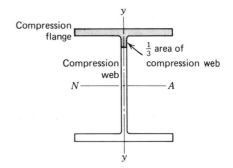

Figure 5.13/Area for calculating r_T.

$F_b = 0.6F_y$. If L_b / r_T is greater than this ratio, buckling could become critical and another limit for L_b / r_T must be calculated. The AISC Specification gives this second limit as

$$\sqrt{\frac{510(10)^3 C_b}{F_y}}$$

Once it has been established that L_b / r_T is equal to or greater than the smaller radical and also equal to or less than the larger radical, a valid formula for considering F_b is

$$F_{b2} = \left[\frac{2}{3} - \frac{F_y (L_b / r_T)^2}{1530(10)^3 C_b} \right] F_y$$

If L_b / r_T is greater than the larger radical, a valid formula for considering F_b is

$$F_{b3} = \frac{170(10)^3 C_b}{(L_b / r_T)^2}.$$

The formulas for both F_{b2} and F_{b3} are based on the lateral bending stiffness of the compression flange and a small part of the web, and do not account for the elastic torsional buckling strength. When F_{b2} and/or F_{b3} are calculated as shown above, the values may be somewhat conservative; therefore, the AISC Specification allows for further refinement if the compression flange is solid, is approximately rectangular in cross section, and has an area not less than that of the tension flange. Under these circumstances, the AISC Specification provides a simplified twisting formula for determining F_b, i.e.,

$$F_{b1} = \frac{12(10)^3 C_b}{L_b(d/A_f)}$$

This is a familiar formula, discussed earlier as a limiting condition for adequate lateral support. Summarizing, for laterally unsupported beams:

If F_{b1} is not a valid consideration, then the allowable bending stress will be either F_{b2} or F_{b3}, depending on which formula the ratio L_b/r_T indicates should be used.

If F_{b1} is a valid consideration, then it is compared with either F_{b2} or F_{b3} (only one of which is valid), and the largest value is taken as the allowable bending stress.

When determining the allowable bending stress (F_b) of a *given* section for a laterally unsupported condition, the procedure is direct and does not require assumptions (Examples 1 and 2, which follow). The flow chart (Fig. 5.5a and b) can be used to locate the proper allowable stress, However, when *designing* beams that are known to be laterally unsupported, the procedure is indirect and may require numerous assumptions, leading to many trial sections that fail to verify the assumptions and thus have to be rejected as unsatisfactory. This is particularly true if design aids, such as the charts and tables furnished in the AISC Manual, are not available. If the designer wishes to select the lightest-weight section that is structurally adequate, he may have to make even more trial selections (Example 3).

Example 1

A **W** 14×34 of A36 Steel has a simple span of 26 ft, no overhangs, and no lateral support except at the reactions. Calculate the largest uniform load it can carry, including its own weight.

Solution

(1) Determine the lower and upper limits of the allowable L_b/r_T ratio to indicate which formula for F_b is valid:

$$\sqrt{\frac{102(10)^3 C_b}{F_y}} = \sqrt{\frac{102(10)^3 1.0}{36}} = 53.2$$

$$\sqrt{\frac{510(10)^3 C_b}{F_y}} = \sqrt{\frac{510(10)^3 1.0}{36}} = 119.0$$

Note that in this example $C_b = 1.0$, and that the above calculated values could have been taken directly from the *AISC Manual* (Table 5 in "Numerical Values" at the end of the AISC Specification).

(2) Calculate the actual L_b/r_T for the **W** 14×34. The AISC Manual "Dimensions and Properties" tables list r_T as 1.76 in.; therefore

$$\frac{L_b}{r_T} = \frac{26(12)}{1.76} = 177$$

Since this ratio is larger than the upper limit for F_{b2} (177 > 119), one effective equation is

$$F_{b3} = \frac{170,000}{(L_b/r_T)^2}$$

(3) Determine the allowable bending stress:

$$F_{b3} = \frac{170,000}{(177)^2} = 5.43 \text{ ksi}$$

Since the beam is a rolled section, the twisting formula is also valid:

$$F_{b1} = \frac{12,000}{26(12)4.56} = 8.43 \text{ ksi}$$

Since $8.43 > 5.43$, the allowable bending stress is 8.43 ksi.

(4) Calculate the maximum resisting moment and the allowable uniform load:

$$M_R = F_b S = \frac{8.43(48.6)}{12} = 34.1 \text{ ft-kips}$$

and, from Fig. 2.34,

$$w = \frac{8(34.1)}{(26)^2} = 0.40 \text{ kips per ft}$$

Example 2

A **W** 21×50 of A588 Steel has a simple span of 26 ft, no overhangs, and point, lateral support at reactions and at midspan (Fig. 5.14). The weight of the beam is negligible. Calculate the maximum concentrated load that could be placed at midspan if:

(a) C_b is assumed as 1.0.

(b) C_b is calculated.

Solution (a)

(1) Since A588 Steel is used and not all sections are manufactured in A588, verify the use of a **W** 21×50 and its yield stress. Refer to the *AISC Manual*, Tables 1 and 2. The **W** 21×50 is included in Group 1, and Group 1 in A588 Steel has a yield stress of 50 ksi.

(2) Determine lower and upper limits of the allowable L_b/r_T ratio to determine

which formula for F_b is valid:

$$\sqrt{\frac{102,000}{50}} = 45.2$$

$$\sqrt{\frac{510,000}{50}} = 101.0$$

$$\frac{L_b}{r_T} = \frac{13(12)}{1.6} = 97.5$$

Since $45.2 < 97.5 < 101$, use

$$F_{b2} = \left[\frac{2}{3} - \frac{F_y(L_b/r_T)^2}{1530(10)^3 C_b}\right] F_y$$

(3) Calculate the allowable bending stress:

$$F_{b2} = \left[\frac{2}{3} - \frac{50(97.5)^2}{1,530,000(1)}\right]50$$

$$= 0.356(50) = 17.8 \text{ ksi}$$

$$F_{b1} = \frac{12,000}{13(12)5.96} = 12.9 \text{ ksi}$$

The allowable bending stress is 17.8 ksi.

(4) Determine the maximum resisting moment and calculate the midspan load (P):

$$M_R = F_b S = 17.8\left(\frac{94.5}{12}\right) = 140 \text{ ft-kips}$$

From Fig. 2.34,

$$P = \frac{4(140)}{26} = 21.5 \text{ kips}$$

Solution (b)

(1) From step (1) of Solution (a), $F_y = 50$ ksi.

(2) Calculate C_b. From the AISC Specification,

$$C_b = 1.75 + 1.05\left(\frac{0}{M_2}\right) + 0.3\left(\frac{0}{M_2}\right)^2 = 1.75$$

(3) Determine if the beam has adequate lateral support. From the AISC Manual, a **W** 21×50 has $L_u = 6.0$ ft. Check whether

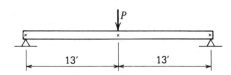

Figure 5.14/Example 2.

twisting or buckling controls F_b:

$$(r_T)\frac{d}{A_f} = 1.6(5.96) = 9.54 > \frac{62.62}{\sqrt{50}} = 8.86$$

Therefore, buckling controls and $\sqrt{C_b}\,L_u = \sqrt{1.75}\,(6.0) = 7.94$ ft. Since $7.94 < 13$ ft, the beam does not have adequate lateral support and $F_b < 0.6F_y$.

(4) Determine lower and upper limits for L_b/r_T:

$$\sqrt{\frac{102,000(1.75)}{50}} = 59.6$$

$$\sqrt{\frac{510,000(1.75)}{50}} = 134$$

$$\frac{L_b}{r_T} = \frac{13(12)}{1.60} = 97.5$$

Since $59.6 < 97.5 < 134$, use

$$F_{b2} = \left[\frac{2}{3} - \frac{F_y(L_b/r_T)^2}{1530(10)^3 C_b}\right]F_y$$

(5) Calculate the allowable bending stress:

$$F_{b2} = \left[\frac{2}{3} - \frac{50(97.5)^2}{1,530,000(1.75)}\right]50$$

$$= (0.489)(50) = 24.5 \text{ ksi}$$

$$F_{b1} = \frac{12,000(1.75)}{13(12)5.96} = 22.6 \text{ ksi}$$

The allowable bending stress is 24.5 ksi.

(6) Determine the maximum resisting moment and calculate the midspan load (P):

$$M_R = F_b S = 24.5\left(\frac{94.5}{12}\right) = 193 \text{ ft-kips}$$

$$P = \frac{4(193)}{26} = 29.7 \text{ kips}$$

Example 3

Design a beam to carry a uniform load of 400 lb per ft (including its own weight)

over a simple span of 26 ft. There is no lateral support except at reactions. The steel is A36. For deflection consideration, use the minimum depth-span ratio of $F_y/800$, and limit total load deflection to $\frac{1}{300}$ of the span.

Solution

(1) Calculate the minimum depth of the beam if the final design requires that it be fully stressed:

$$d = \frac{36}{800}(12)26 = 14 \text{ in.}$$

(2) A laterally unsupported length of 26 ft is so long that a beam of lesser depth and wider flange widths probably will be required. In anticipation of investigating sections of lesser depth than that recommended when fully stressed, tabulate reduced values of F_b for 12-in.- and 10-in.-deep sections. For 12-in. beams,

$$F_b = \frac{12}{14}(22) = 18.9 \text{ ksi}$$

For 10-in. beams,

$$F_b = \frac{10}{14}(22) = 15.7 \text{ ksi}$$

(3) Calculate the required moment of inertia for the deflection criteria:

$$\Delta_{max} = \frac{26(12)}{300} = 1.04 \text{ in.}$$

From Fig. 4.15

$$I_{req} = \frac{5(0.4)(26)^4(12)^3}{384(29,000)1.04} = 136 \text{ in.}^4$$

(4) Assume an allowable stress, calculate the required section modulus, and make some trial selections. Since $L_b = 26$ ft, which is rather long, the deeper lighter-weight beams will have a much reduced stress. Try $F_b = 12$ ksi. From Fig. 2.34

$$M_{max} = \frac{0.4(26)^2}{8} = 33.8 \text{ ft-kips}$$

$$S_{req} = \frac{33.8(12)}{12} = 33.8 \text{ in.}^3$$

Referring to the AISC Manual "Allowable Stress Design Selection" table and "Dimensions and Properties" table, a **W** 14×26 appears to be the best choice for a trial selection. Also, choose the next heavier 14-in. beam, which is the **W** 14×30. Both sections have adequate depth and moment of inertia.

(5) Check the trial beam sections for adequate bending strength, and revise selection as indicated by the calculations. Observe that $C_b = 1.0$. Calculate lower and upper limits of L_b/r_T:

$$\sqrt{\frac{102,000}{36}} = 53.2$$

$$\sqrt{\frac{510,000}{36}} = 119$$

For the **W** 14×26,

$$f_b = \frac{33.8(12)}{35.3} = 11.5 \text{ ksi}$$

$$\frac{L_b}{r_T} = \frac{26(12)}{1.28} = 244$$

$$F_{b3} = \frac{170,000}{(244)^2} = 2.86 \text{ ksi}$$

$$F_{b1} = \frac{12,000}{26(12)6.59} = 5.84 \text{ ksi}$$

NO GOOD

For the **W** 14×30,

$$f_b = \frac{33.8(12)}{42} = 9.66 \text{ ksi}$$

$$\frac{L_b}{r_T} = \frac{26(12)}{1.74} = 179$$

$$F_{b3} = \frac{170,000}{(179)^2} = 5.31 \text{ ksi}$$

$$F_{b1} = \frac{12,000}{26(12)5.34} = 7.20 \text{ ksi}$$

NO GOOD

This trial section appears to be close to adequate; therefore, in all probability, the next higher-weight 14-in. beam (**W** 14×34) would suffice and could be checked. However, there also could be a lighter-weight beam of lesser depth that worked. Try both a **W** 12×30 and a **W** 10×33. Both have an adequate moment of inertia. For the **W** 10×33,

$$f_b = \frac{33.8(12)}{35} = 11.6 \text{ ksi}$$

$$11.6 < 15.7, \quad \text{OK (Step 3)}$$

$$\frac{L_b}{r_T} = \frac{26(12)}{2.14} = 146$$

$$F_{b3} = \frac{170,000}{(146)^2} = 7.98 \text{ ksi}$$

$$F_{b1} = \frac{12,000}{26(12)2.81} = 13.7 \text{ ksi}$$

The **W** 12×33 section is adequate. For the **W** 12×30,

$$f_b = \frac{33.8(12)}{38.6} = 10.5 \text{ ksi}$$

$$10.5 < 18.9, \quad \text{OK (Step 3)}$$

$$\frac{L_b}{r_T} = \frac{26(12)}{1.73} = 180$$

$$F_{b3} = \frac{170,000}{(180)^2} = 5.25 \text{ ksi}$$

$$F_{b1} = \frac{12,000}{26(12)4.30} = 9.94 \text{ ksi}$$

NO GOOD

Final selection, **W** 10×33.

PROBLEMS

1. A **W** 12×26 beam of A36 Steel is simply supported on a 24-ft span and carries, in addition to its own weight, a concentrated load of 10 kips at midspan. There is no depth-span criterion or deflection limitation.

(a) Determine whether the beam is safe if it has complete lateral support.

(b) Determine whether the beam is safe if lateral support is provided only by the member bringing the 10-kip load to the beam.

(Answers given in Appendix G.)

2. A **W** 16×36 beam of A572, Grade-50 Steel is simply supported on a 26-ft span. There is no depth-span criterion or deflection limitation.

(a) Determine the maximum distributed load the beam could carry (including its own weight) if it had complete lateral support.

(b) Determine the maximum distributed load the beam could carry (including its own weight) if it had no lateral support.

(Answers given in Appendix G.)

3. A beam of A36 Steel is simply supported on a span of 24 ft and carries a uniformly distributed load of 590 lb per ft, including its own weight. There is no depth-span criterion or deflection limitation.

(a) Design the beam assuming that complete lateral support is provided.

(b) Redesign the beam assuming that lateral support is provided only at midspan.

4. A simply supported A36 Steel beam of 30-ft span carries two concentrated 6-kip loads, one located at each third-point of the span. In addition, there is a 200-lb-per-ft uniform load (including the beam weight) over the entire span. Lateral support is provided only at the reaction and points of concentrated loads. Use the minimum depth-span criterion for floors, i.e., $F_y/800$. Design the beam.

5. Design an A36 Steel beam which cantilevers 10 ft from a wall and must carry a uniform load of 200 lb per ft over its full length and a 2-kip concentrated load at its free end. There is no deflection limitation criterion. The construction is such that it does not provide lateral support,

but the support does provide bracing against twist. (Answer is given in Appendix G.)

6. An A36 Steel beam is 28 ft long. It is simply supported but has an 8-ft overhang on one end. It carries a 4-kip concentrated load at the end of the overhang and a 10-kip concentrated load midway between the supports. Lateral support is provided at the reactions and at the points where the concentrated loads are applied. There is no deflection limitation, and the weight of the beam can be neglected. Design the beam.

7. Design an A36 Steel beam which cantilevers 14 ft from a wall and is loaded as follows: a 10-kip concentrated load at the free end and a 1-kip concentrated load 4 ft from the wall. Lateral support is provided at the wall and at the point where the 1-kip load is applied. The wall provides resistance to twist. There is no deflection limitation. (Answer given in Appendix G.)

8. A typical interior floor framing system is shown in plan view in Fig. 5.15. A572, Grade-50 Steel is specified for the beams and girders. All connections are of the flexible type, generating simple spans. A one-way metal deck is used to span between the beams and, when welded to the beams, provides lateral support. The only point lateral support for the girders is that provided by the beams where they frame to the girders. There is a lightweight concrete fill

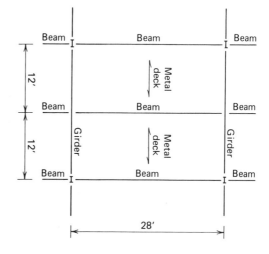

Figure 5.15/Typical interior floor framing plan (Problem 8).

placed on the metal deck, a carpet above, and a hung ceiling below, totaling 32 lb per sq ft, not including the weight of beams and girders. Use the floor $F_y/800$ depth-span criterion, and limit the live-load deflection to $\frac{1}{360}$ of the span. Do not use a live-load reduction. Design the beams and girders for a live load of 40 psf.

5.8
Local Buckling

Attention to local buckling is too often neglected in beam design. Whenever thin plates, such as webs and flanges, are subjected to a large compressive force, there is a tendency to buckle similar to that in columns (Chapter 6). When local buckling does occur, there are two major effects possible: (1) loss of load-carrying capacity in the buckled plate itself; and (2) more important, shortly thereafter (sometimes even before a noticeable buckle develops), redistribution of stress to other parts of the beam which, by subjecting these parts to forces for which they were not designed, could lead to subsequent failure.

When any structural failure occurs, the initial cause is frequently difficult to detect. This is particularly true when there is a more observable failure of a structural element that would not have occurred if that element had not suddenly been subjected to forces resulting from an initial failure elsewhere. More often than not, local buckling is the cause of initial failure.

Compression-flange buckling between points of lateral support was discussed in the previous article. This article deals with web buckling. The principal functions of the web of a beam are to separate the flanges as much as possible to gain maximum benefit from beam depth and to resist nearly all of the generally small shear stresses. As a consequence, there is a tendency to make webs as thin as possible

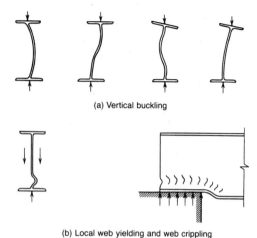

(a) Vertical buckling

(b) Local web yielding and web crippling

Figure 5.16/Local web buckling.

and the distance (depth) between flanges as large as possible, with the result that the web plate may be weak in the vicinity of concentrated loads (Fig. 5.16).

In Fig. 5.16a, four types of vertical web buckling are shown. Which one will occur depends on the nature of the support the top flange receives from the surrounding construction. The possibility of beam failure due to vertical buckling exists only in the vicinity of large concentrated loads, i.e., at reactions where the beam rests on its support or at points where columns or other beams rest on the top flange of the beam. At the same points, there is also the possibility of two other types of web failure, known as *local web yielding* and *web crippling*. Generally, when the web is safe from yielding and crippling it is also safe from vertical buckling; therefore it is the yielding and crippling which are investigated. However, in plate girder and other built-up members all local buckling should be investigated.

Local Web Yielding / It is assumed for purposes of design (AISC Specification) that the web acts as a column, with cross-sectional dimensions equal to $N + 2.5k$ times the thickness t_w of the web at reactions (Fig. 5.17a), and $N + 5k$ times t_w at interior loads applied at a distance from the member end that is greater than the depth of the member (Fig. 5.17d). The distance N is equal to the length of bearing. The AISC procedure, permitting a value slightly greater than N, i.e., $2.5k$ or $5k$, is based on the fact that the portion of the web directly over or under N does not act as a free column. Instead, because of its continuity with the rest of the web and the flanges, it distributes a part of the load to the adjacent web material. The distance k is equal to the distance from the outside face of the flange to the web toe of the fillet (Fig. 5.17b). Values of k for the various rolled sections are given in the AISC Manual tables, under "Dimensions and Properties."

At end supports, the actual unit compressive stress developed on the cross section of the assumed column is equal to the beam reaction R divided by the effective column area and is limited to

$$0.66F_y = \frac{R}{t_w(N + 2.5k)}$$

Therefore, the required minimum length of bearing for any given reaction may be found by solving the above equation for N:

$$N_{min} = \frac{R}{0.66F_y t_w} - 2.5k$$

At an interior point of application of a heavy concentrated load P to the top flange, the required minimum length of bearing is determined as follows (Fig. 5.17d):

$$N_{min} = \frac{R}{0.66F_y t_w} - 5k$$

Where the required length of bearing cannot be obtained, a beam with a thicker web must be used, or the thinner web must be reinforced by vertical stiffener plates, which are usually placed on either side as indicated in Fig. 5.17(c) and (e).

Web Crippling / Vertical web bearing stiffeners may also be required to prevent web crippling. An interior condition for R is defined as one at a distance more than $d/2$ from the end of the member; other-

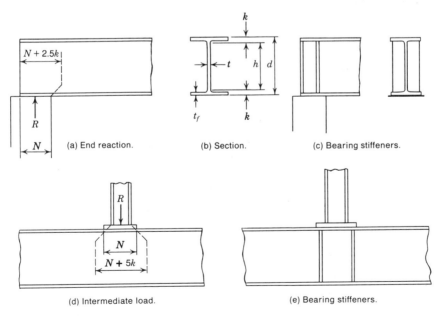

(a) End reaction. (b) Section. (c) Bearing stiffeners.

(d) Intermediate load. (e) Bearing stiffeners.

Figure 5.17/Design for local buckling.

wise it is an end condition. Recent research has shown that both the overall depth d of the section and the thickness t_f of the flange, along with the yield stress F_{yw} of the web, are influencing factors for web crippling. The 1989 AISC Specification requires use of bearing stiffeners if the force R exceeds that given by the following formulas:

For the end condition

$$R = 34(t_w)^2\left[1+3\left(\frac{N}{d}\right)\left(\frac{t_w}{t_f}\right)^{1.5}\right]\sqrt{\frac{F_{yw}t_f}{t_w}}$$

For the interior condition

$$R = 67.5(t_w)^2\left[1+3\left(\frac{N}{d}\right)\left(\frac{t_w}{t_f}\right)^{1.5}\right]\sqrt{\frac{F_{yw}t_f}{t_w}}$$

These are empirical formulas developed after extensive research and testing.

Example 1

The end reaction of a **W** 16×26 is 28 kips.

(a) Determine the minimum required length of bearing at the support to prevent local web yielding for both A36 Steel and A588 Steel.

(b) Assuming 3 in. of actual bearing, would stiffeners be required by the web-crippling criteria?

Solution (a)

(1) From the AISC Manual, $d = 15.69$ in.; $k = 1\frac{1}{16}$ in.; $t_w = 0.25$ in.; $t_f = 0.345$ in.

(2) Solving for N for the A36 Steel,

$$N = \frac{28}{0.66(36)0.25} - 2.5(1.06) = 2.06 \text{ in.}$$

(3) Solving for N for the A588 Steel ($F_y = 50$ ksi),

$$N = 0.74 \text{ in.}$$

Solution (b)

(1) Same as for solution (a).

(2) Solving for R for the A36 Steel,

$$R = 34(0.25)^2\left[1+3\left(\frac{3}{15.69}\right)\left(\frac{0.23}{0.345}\right)^{1.5}\right]$$
$$\times\sqrt{36\left(\frac{0.345}{0.23}\right)}$$
$$= 19.57 \text{ kips} < 28$$

Bearing stiffeners would be required, or the length N could be increased to 8 in. Solving for R for $F_y = 50$ ksi,

$$R = 23 \text{ kips} < 28$$

Bearing stiffeners would be required, or the length N could be increased to 5 in.

Web Diagonal Buckling / Figure 5.18a illustrates diagonal buckling, which is still another way the web of a steel beam can fail. Such failure is due to a combination of stresses resulting from shear and flexure, frequently referred to as principal

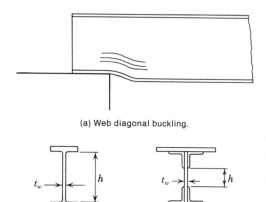

(a) Web diagonal buckling.

(b) Section.   (c) Section.

Figure 5.18/Local diagonal buckling.

stress. The deeper the section and the thinner the web, the more susceptible the beam is to diagonal buckling.

The precise analysis of principal stress and its tendency to produce diagonal buckling is beyond the scope of this text and certainly is unwarranted for the majority of the cases encountered in practice.

The AISC Specification sidesteps the issue of principal stress and prevents diagonal buckling by limiting the h/t_w ratio and/or by reducing the allowable shear stress below $0.4F_y$, the value of which depends upon the use and spacing of transverse stiffeners as shown in Fig. 5.31. These transverse stiffeners are similar to the bearing stiffeners shown in Fig. 5.17 except that they are spaced throughout the length of the beam rather than at points of concentrated loads.

The question of web buckling and reduced values of F_v must be examined when

$$\frac{h}{t_w} > \frac{380}{\sqrt{F_y}}$$

For rolled beams listed in the AISC Manual only a few **M** sections have h/t_w values exceeding this limit when $F_y = 36$ is used. All **W** and **S** shapes are adequate, and $F_v = 0.45F_y$ for A36 Steel. Higher-grade steels need further investigation.

Furthermore, the specification states that the actual or developed shear stress must be based upon the thickness of the web and the clear distance between the flanges:

$$f_v = \frac{V}{t_w h}$$

This is a departure from the method previously discussed, where the full depth d was used. Under these circumstances the allowable shear stress is calculated as follows:

$$F_v = \frac{F_y}{2.89} C_v \le 0.4F_y$$

The constant in the formula (C_v) depends upon whether transverse stiffeners are used, their spacing a, and the actual h/t_w of the section being investigated. As with the investigation for web crippling, the formulas for design are empirically developed from research and laboratory tests:

$$C_v = \begin{cases} \dfrac{45,000 k_v}{F_y(h/t_w)^2} & \text{when} \quad C_v < 0.8 \\[3mm] \dfrac{190}{h/t_w} \sqrt{\dfrac{k_v}{F_y}} & \text{when} \quad C_v > 0.8 \end{cases}$$

$$k_v = \begin{cases} 4.00 + \dfrac{5.34}{(a/h)^2} & \text{when} \quad \dfrac{a}{h} < 1.0 \\[3mm] 5.34 + \dfrac{4.00}{(a/h)^2} & \text{when} \quad \dfrac{a}{h} > 1.0 \end{cases}$$

Observe that when no stiffeners are used, the ratio a/h is very large and $k = 5.34$.

Example 2

Investigate further the condition described in Example 1 for A588 Steel to determine if transverse stiffeners are necessary.

Solution

(1) From the AISC Manual $d = 15.69$ in., $t_f = 0.345$ in., $t_w = 0.25$ in.

(2) Calculate the h/t_w and compare with the allowable:

$$h = 15.69 - 2(0.345) = 15.00 \text{ in.}$$

$$\frac{h}{t_w} = \frac{15.0}{0.25} = 60.0$$

$$\frac{380}{\sqrt{50}} = 53.7$$

Since $60.0 > 53.7$, $F_v < 0.4F_y$ and further investigation is necessary.

(3) Determine the developed shear stress:

$$f_v = \frac{28}{15.0(0.25)} = 7.45 \text{ ksi}$$

(3) Determine the allowable shear stress if no transverse stiffeners are used. Since a/h is very large, $k_v = 5.34$. Assume $C_v > 0.8$ (to be verified). Then

$$C_v = \frac{190}{60}\left(\frac{\sqrt{5.34}}{50}\right) = 1.03 \quad \text{(verified)}$$

$$F_v = \frac{50}{2.89}(1.03) = 17.8 \text{ ksi}$$

$$17.8 \text{ ksi} < 0.4(50) = 20.0$$

Since $7.47 < 17.8$, no stiffeners are required.

5.9
Beam Design Tables

The AISC Manual provides tables of "Allowable uniform load in kips for beams laterally supported." These tables are valid only if the beams are simply supported and are without overhangs. Such tables are an invaluable aid to the designer, provided their advantages and limitations are understood.

Table 5.1 illustrates a typical allowable-load table abstracted from the Ninth Edition of the AISC Manual. The manual provides tables for all rolled **W, S, M, C,** and **MC** shapes, both for A36 Steel and for steel having $F_y = 50$ ksi. Observe that Table 5.1 is based upon $F_y = 36$ ksi. Each beam shape has a resisting moment $M_R = SF_b$, which can be equated to $WL/8$. Solving for W

$$W = \frac{8SF_b}{L}$$

For example, a **W** 12×50, having a section modulus of 64.7 in.^{3}, can safely support a total uniform load of 51 kips for a span of 20 ft. This is calculated as follows:

$$W = \frac{8(64.70)(0.66)(36)}{(12)(20)} = 51 \text{ kips}$$

This 51 kips is the total load the beam can carry, which includes its own weight.

It is seen that the allowable bending stress is taken as $0.66F_y$, which is valid only if the beam has adequate lateral support ($L_b \leq L_c$) and the beam is compact (see Art. 5.8). If the beam has adequate lateral support but is not compact, the tabulated load is based upon the allowable stress using the blending formula (Art. 5.8).

These tables also list the lengths L_c and L_u for each beam, calculated according to the procedures described in Art. 5.8. The loads listed in the tables are valid only if $L_b \leq L_c$. If the unbraced length of the beam is larger than L_c but less than L_u, the listed load must be reduced (because F_b is reduced to $0.6F_y$). For compact beams the reduction is $0.6/0.66 = 0.91$.

For noncompact shapes the reduction is the ratio of $0.6F_y$ to the allowable stress used to compute its capacity. For channels the question of compactness does not arise ($F_b = 0.6F_y$) and only L_u lengths are listed.

These safe-load tables also list the maximum deflection in inches for each span length for the listed capacity. From Table 5.1 it is seen that the **W** 12×50, with a span of 20 ft, has a maximum deflection of 0.82 in. when carrying a total uniform load of 51 kips. The listed deflection is not always a precise value; it is based upon nominal depth of the beam. This approximation permits the listing of one constant deflection for all beams of the same depth for their listed load capacity. The formula for this approximate deflection is derived as follows.

The maximum deflection for the simply supported, uniformly loaded beam of Fig.

Table 5.1
Typical Beam Design Table*

F_y = 36 ksi	BEAMS W Shapes Allowable uniform loads in kips for beams laterally supported For beams laterally unsupported, see page 2-146										W 12
Designation	W 12			W 12			W 12				Deflection In.
Wt./ft	50	45	40	35	30	26	22	19	16	14	
Flange Width	8⅛	8	8	6½	6½	6½	4	4	4	4	
L_c	8.50	8.50	8.40	6.90	6.90	6.90	4.30	4.20	4.10	3.50	
L_u	19.6	17.7	16.0	12.6	10.8	9.40	6.40	5.30	4.30	4.20	

Span in Feet (F_y = 36 ksi)

Span	50	45	40	35	30	26	22	19	16	14	Deflection
3									76	69	.02
4							92	82	68	59	.03
5							80	67	54	47	.05
6				108	92	81	67	56	45	39	.07
7	130	116		103	87	76	57	48	39	34	.10
8	128	115	101	90	76	66	50	42	34	30	.13
9	114	102	91	80	68	59	45	37	30	26	.17
10	102	92	82	72	61	53	40	34	27	24	.20
11	93	84	75	66	56	48	37	31	25	21	.25
12	85	77	69	60	51	44	34	28	23	20	.29
13	79	71	63	56	47	41	31	26	21	18	.35
14	73	66	59	52	44	38	29	24	19	17	.40
15	68	61	55	48	41	35	27	22	18	16	.46
16	64	58	51	45	38	33	25	21	17	15	.52
17	60	54	48	42	36	31	24	20	16	14	.59
18	57	51	46	40	34	29	22	19	15	13	.66
19	54	48	43	38	32	28	21	18	14	12	.74
20	51	46	41	36	31	26	20	17	14	12	.82
21	49	44	39	34	29	25	19	16	13	11	.90
22	47	42	37	33	28	24	18	15	12	11	.99
23	45	40	36	31	27	23	17	15	12	10	1.08
24	43	38	34	30	25	22	17	14	11	10	1.18
25	41	37	33	29	24	21	16	13	11	9	1.28
26	39	35	32	28	24	20	15	13	10	9	1.38
28	37	33	29	26	22	19	14	12	10	8	1.61
30	34	31	27	24	20	18	13	11	9	8	1.84

Properties and Reaction Values

	50	45	40	35	30	26	22	19	16	14	
S_x in.³	64.7	58.1	51.9	45.6	38.6	33.4	25.4	21.3	17.1	14.9	For explanation of deflection, see page 2-32
V kips	65	58	51	54	46	40	46	41	38	34	
R_1 kips	30.2	24.9	21.9	17.8	14.5	12.0	13.5	11.3	9.80	8.17	
R_2 kips/in.	8.79	7.96	7.01	7.13	6.18	5.46	6.18	5.58	5.23	4.75	
R_3 kips	36.7	30.0	23.5	24.2	17.9	13.9	17.6	13.7	10.8	8.65	
R_4 kips/in.	3.97	3.32	2.56	2.54	1.98	1.60	2.06	1.87	2.05	1.83	
R kips	51	42	32	33	25	20	25	20	18	15	

Load above heavy line is limited by maximum allowable web shear.

*Courtesy of the American Institute of Steel Construction. "Manual of Steel Construction: Allowable Stress Design," Ninth Edition, p. 2-68.

111

4.15 is

$$\Delta = \frac{5}{384}\frac{wL^4}{EI}$$

Substituting W for wL,

$$\Delta = \frac{5}{384}\frac{WL^3}{EI}$$

Then, factoring for convenience and ease of observation,

$$\Delta = \frac{5}{48}\left(\frac{WL}{8}\right)\frac{L^2}{EI}$$

The factor in parentheses is seen to be the maximum moment for the beam, which can be equated to the resisting moment i.e.,

$$\frac{WL}{8} = F_bS = F_b\left(\frac{I}{d/2}\right)$$

Substituting this expression in the deflection formula,

$$\Delta = \frac{5}{48}\left[F_b\left(\frac{I}{d/2}\right)\right]\frac{L^2}{EI}$$

and simplifying and adjusting the units so that L will be in feet,

$$\Delta = \frac{5}{48}\left[F_b\left(\frac{I}{d/2}\right)\right]\frac{L^2(12)^2}{EI} = \frac{30F_bL^2}{Ed}$$

It has been established previously that E for all steels is 29,000 ksi; therefore,

$$\Delta = \left(\frac{30}{29}\frac{F_b}{d}\right)\frac{L^2}{1000}$$

This formula can be further simplified for

A36 Steel by substituting $0.66(36) = 23.8$ for F_b:

$$\Delta = \frac{30}{29}\left(\frac{23.8}{d}\right)\frac{L^2}{1000} = \frac{0.0246L^2}{d}$$

In the example of the **W** 12×50 carrying the 51-kip load, the listed deflection is calculated as

$$\Delta = 0.0246(20)^2/12 = 0.82 \text{ in.}$$

whereas the more precise value is calculated as

$$\Delta = \frac{5}{384}\left[\frac{51(20)^3(12)^3}{29,000(394)}\right] = 0.80 \text{ in.}$$

If the uniform load the beam is carrying is less than the listed capacity, the maximum deflection is reduced by the ratio of the actual load to the listed load capacity. The use of these tables for uniform loads is better understood through the use of examples.

Example 1

A simple floor beam of A36 Steel and span of 17 ft has a superimposed uniform load of 2000 lb per ft. It has complete lateral support. Design the beam using data contained in Table 5.1.

Solution

(1) Determine the total superimposed load to be supported:

$$W = (2)(17) = 34 \text{ kips}$$

(2) Select a trial beam section. Referring to Table 5.1 and entering with $L = 17$ ft, the lightest-weight section having a capacity slightly larger than 34 kips is seen to be a **W** 12×30 with 36-kip capacity.

(3) Check the selection. The weight of the beam is $30(17) = 510$ lb, $\simeq 0.5$ kips. The total uniform load including beam weight is

$$W = 34 + 0.5 = 34.5 < 36 \qquad \text{OK}$$

(4) Check shear. Referring to Table 5.1, the maximum uniform load the beam can carry is 92 kips (the listed value above the heavy line). Since $34.5 < 92$, the shear is adequate.

(5) Investigate the deflection. Since $b/2t_f = 6.56/2(0.52) = 6.31 < 65$, the section is compact and the listed load capacity of 36 kips is based upon $F_b = 0.66F_y = 23.8$ ksi. Consequently, the 36 kips creates 0.59 in. (from Table 5.1). Therefore, a 34.5-kip load would create

$$\Delta = \frac{34.5}{36}(0.59) = 0.57 \text{ in.}$$

Also, if the ratio of live load to total load was, for example, 0.75, the live-load deflection would be $0.75(0.59) = 0.44$ in. If this deflection is not considered excessive (the actual is less than the limit $L/360 = 0.60$ in.), and there are no objections to a 12-in.-deep beam, the **W** 12×30 should be adopted.

The most economical beam can be found by seeking progressively lighter beams that are also adequate. By referring to the beam design tables in the AISC Manual (Ninth Edition) for **W** 16 sections, the reader should verify that a **W** 16×26 will also satisfy the stated requirements of this example. If there were no restriction on depth because of ceiling height or other

mandatory clearances, the **W** 16×26 would be used.

Example 2

Using Table 5.1, determine the total safe uniform load, including the beam weight, that a **W** 12×26 will sustain on a span of 17 ft if lateral bracing is provided only at the center of the span.

Solution

(1) The maximum unbraced length for this condition is equal to half of the 17-ft span, or $8\frac{1}{2}$ ft.

(2) Referring to Table 5.1, it will be seen that L_c for this section is 6.9 ft. However, the $8\frac{1}{2}$-ft unbraced length is less than the L_u value of 9.4 ft. Therefore, the total load capacity must be based upon $F_b = 0.6F_y = 21.6$ ksi.

Since $b/2t_f = 6.49/2(0.38) = 8.5 < 65$, the section is compact, and the listed load (31 kips) is based upon $F_b = 23.8$ ksi.

Consequently, the safe load capacity is calculated as

$$W = \frac{21.6}{23.8}(31) = 28.1 \text{ kips}$$

Concentrated Load Equivalents / Although design tables are prepared for beams carrying uniformly distributed loads, they may also be used for other loading conditions provided an equivalent uniform load may be found. For example, a laterally supported beam will carry a single concentrated load at the center of its span equal to one-half the value for a uniform load. This relationship is derived by equating the formula for the maximum bending moment caused by a concentrated load P at the center of a span, to that for a uniform load W over the entire span:

$$M(\text{max, concentrated load}) = \frac{PL}{4}$$

$$M(\text{max, uniform load}) = \frac{WL}{8}$$

$$\frac{PL}{4} = \frac{WL}{8}$$

from which

$$P = \frac{4WL}{8L} = \frac{W}{2}$$

This may be written as

$$W = 2P$$

in which the number 2 is referred to as the coefficient for obtaining the equivalent uniform load. Consequently, when designing for this condition, it is necessary to multiply the concentrated load by this coefficient.

An equivalent deflection coefficient can be developed in a similar manner. The maximum deflection for a simple span with a concentrated load at the center is seen from Fig. 4.14 to be

$$\Delta = \frac{1}{48} \left(\frac{PL^3}{EI} \right)$$

For convenience and observation, it may be factored to read

$$\Delta = \frac{1}{12} \left(\frac{PL}{4} \right) \frac{L^2}{EI}$$

The middle factor in parentheses represents the maximum moment at the center and may be replaced by its equivalent, i.e.,

$$M = \frac{F_b I}{d/2}$$

Simplifying and adjusting units so that L will be in feet,

$$\Delta = \frac{1}{12} \left(\frac{F_b I}{d/2} \right) \frac{L^2(12)^2}{EI} = \frac{24}{29} \left(\frac{F_b}{d} \right) \frac{L^2}{1000}$$

Comparing this equation for maximum deflection for the concentrated load with that for the distributed load derived earlier, i.e.,

$$\Delta = \left[\frac{30}{29} \left(\frac{F_b}{d} \right) \right] \frac{L^2}{1000}$$

it is seen that the only difference is the value of the numerator in the first term, namely, 24 versus 30. Consequently, the deflection coefficient becomes

$$\frac{24}{30} = 0.80$$

When calculating deflections using the design table for uniform loads, it becomes necessary to multiply the equivalent uniform load by this coefficient if the load is concentrated at midspan. Figure 5.19 gives load and deflection coefficients for three

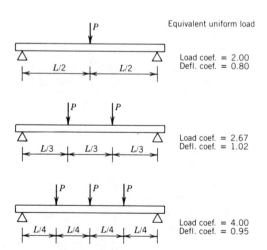

Figure 5.19/Concentrated local equivalents.

different concentrated load conditions that are frequently used in framing solutions. Similar coefficients may be developed for even more complex loadings; however, the solution of such problems using coefficients may actually entail much more work than the direct application of the basic beam design procedure. The AISC Manual contains a more complete "Table of Concentrated Load Equivalents." Coefficients for shear also could be developed, but since shear seldom governs in design, their use is not recommended. Nevertheless, shear should always be checked before accepting a beam as safe. The use of load coefficients in conjunction with beam design tables is illustrated by the example that follows.

Example 3

Using the AISC beam design tables, select a **W** section that will support concentrated loads of 7 kips applied at the third-points and a uniform load of 19 kips (including an allowance for beam weight) over the entire span of an 18-ft laterally supported floor beam. Check deflection.

Solution

(1) Sketch the beam loading diagram (Fig. 5.20).

(2) Refer to Fig. 5.19 for the load and deflection coefficients for equal concentrated loads at the third-points (2.67 and 1.02, respectively).

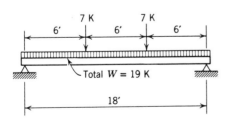

Figure 5.20

(3) Calculate the total equivalent uniform load on the beam. This is seen to be the actual uniform load plus the equivalent of the two concentrated loads:

$$W = 19 + 2.67(7) = 37.7 \text{ kips}$$

(4) Select the beam from Table 5.1 (the beam weight has already been accounted for). Select the **W** 12×35 with a capacity of 40 kips.

(5) Check shear. From Fig. 5.20 it is seen that the maximum reaction is $19/2 + 7 = 16.5$ kips, which is less than the listed value of V (54 kips) for the selected beam; therefore, the shear resistance is adequate.

(6) Check deflection. Since $b/2t_f = 6.56/2(0.52) = 6.31 < 65$, the load capacity and deflection are based upon $F_b = 23.8$ ksi. Consequently, the maximum deflection caused by the actual uniform load is

$$\Delta = \frac{19}{40}(0.66) = 0.31 \text{ in.}$$

The maximum deflection caused by the equivalent uniform load from the concentrated loads is

$$\Delta = 1.02\left(\frac{2.67(7)}{40}\right)0.66 = 0.31 \text{ in.}$$

The total deflection is

$$\Delta = 0.31 + 0.31 = 0.62 \text{ in.}$$

5.10 Beam Charts

The AISC Manual also provides charts for the selection of **W** and **M** shapes having different unbraced lengths. Two separate sets of charts are used: one for A36 Steel and the other for $F_y = 50$ ksi steels. A typical chart for A36 Steel is shown in Fig. 5.21. It can be seen that the allowable resisting moment for most **W** and **M** rolled shapes is plotted vertically for corresponding values of L_b, plotted horizontally.

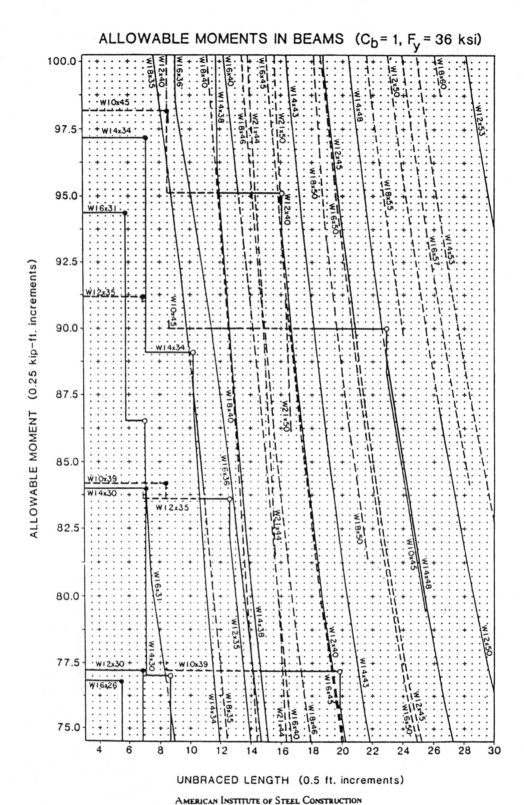

Figure 5.21/Typical beam design chart (laterally unsupported beams). Courtesy of American Institute of Steel Construction; *Manual of Steel Construction: Allowable Stress Design*, Ninth Edition, p. 2-172.

Some heavy **W** 14 shapes are not shown, because their principal use is for columns.

These charts are important design aids, because they show the resisting-moment profile for many shapes over long ranges of unsupported lengths and enable the designer to observe many suitable sections at one time. The unbraced lengths in feet are for the compression flange. In all cases, C_b has the conservative value of 1.0. If there is reason to believe a better selection can be made by calculating precise values of C_b, the general procedures of Arts. 5.6 and 5.7 can be followed or an adjustment can be made in the unbraced length criteria.

Article 5.6 described the affect C_b has on the L_u of a section and its dependence on whether the section has proportions such that buckling or twisting controls. Buckling controls most rolled sections; therefore, those few sections that are controlled by twisting should be identified. In Fig. 5.21 there are only two, the **W** 16×26 and **W** 21×44. Consequently, for all of the remaining sections, the actual unbraced length can be divided by $\sqrt{C_b}$ when entering the length along the bottom of the chart. Otherwise, the unbraced length can be divided by C_b before using the chart. These adjustments are only applicable for those portions of the curve where the stress is equal to or less than $0.6F_y$.

Much can be learned by tracing the moment profile of a single shape for increasing values of L_b. This will be done here for the **W** 14×34 shown in Fig. 5.21. It is a compact shape and, if adequately laterally supported, has $F_b = 23.8$ ksi, which for these charts is rounded off to 24 ksi. Therefore

$$M_R = SF_b = \frac{48.6}{12}(24) = 97.2 \text{ ft-kips}$$

This resisting moment will remain constant at 97.2 ft-kips for values of L_b from 0 to L_c, which is seen to be 7.1 ft; beyond this, there is a sudden drop in the allowable bending stress to 21.6 ksi, which for these charts is rounded off to 22 ksi. Therefore,

$$M_R = SF_b = \frac{48.6}{12}(22) = 89.1 \text{ ft-kips}$$

This resisting moment will remain constant at 89.1 ft-kips for values of L_b over 7.1 ft, until L_b reaches L_u (or $\sqrt{C_b}L_u$), which is observed to be 10.2 ft.

For unbraced lengths larger than 10.2 ft, F_b declines at a rapid rate. Its precise value can be calculated from one of the three formulas described in Art. 5.7, namely F_{b1}, F_{b2}, or F_{b3}. Since the basis for beam selection from the charts is bending moment, they are valid for any loading. Adjustment in the design may be required, however, for deflection limitations; i.e., if deflection is critical for the beam selected, another selection, possibly a deeper beam, will be necessary. This can only be determined by checking the deflection of the section chosen and returning to the chart for another selection if the first is not satisfactory.

After computing the maximum bending moment (M_b) in ft-kips, and determining the maximum unbraced length of compression flange, the chart is examined and the intersection of coordinates noted. Any beam curve above and to the right of the intersection is satisfactory. The nearest curve which is solid is the most economical in terms of beam weight.

Example 1

On the basis of bending moment, select (using the chart, Fig. 5.21) the lightest-weight beam for a simple span of 30 ft, carrying a uniform load of 400 lb per ft, and third-point concentrated loads of 3 kips each. The beam has lateral support at points of concentrated loads only. The steel is A36. Check the selected section using the $F_y/800$ depth-span criterion.

Solution

(1) Assuming a beam weight of 40 lb per ft, determine the maximum bending moment:

$$M_{max} = \frac{wL^2}{8} + \frac{PL}{3}$$

$$= \frac{0.44(30)^2}{8} + \frac{3(30)}{3}$$

$$= 49.5 + 30 = 79.5 \text{ ft-kips}$$

(2) Enter the chart (Fig. 5.21) with $L_b =$ 10 ft and $M = 79.5$, and note the intersection of the coordinates. The nearest solid beam curve, upward and to the right from the intersection is a **W** 14×34 and is selected.

(3) Calculate the minimum beam depth for a fully stressed beam:

$$d = \frac{F_y L}{800} = \frac{36(30)12}{800} = 16.2 \text{ in.}$$

(4) Since a 14-in.-deep beam has been selected, determine its reduced value of allowable bending stress:

$$F_b = \frac{14}{16.2}(22) = 19.0 \text{ ksi}$$

(5) Calculate the actual bending stress and compare with the allowable:

$$f_b = \frac{M}{S} = \frac{79.5(12)}{48.6} = 19.6 \text{ ksi}$$

f_b is seen to be slightly larger than the allowable, i.e., by 3 per cent, and judgment would be required as to the importance of the depth-span criterion before the next heavier section is selected.

PROBLEMS

1. An A36 Steel beam is 28 ft long. It is simply supported but has an 8-ft overhang on one end. It carries a 4-kip concentrated load at the end of the overhang and a 10-kip concentrated load midway between the supports. The beam selected is a **W** 10×19. (This is Problem 6 following Art. 5.7.) In order to prevent web yielding, determine the required length of bearing at the support having the largest reaction. Would web stiffeners be required if $N = 2\frac{1}{2}$ in.?

2. A structural member brings a 50-kip vertical load to rest on the top flange of a **W** 12×19 of A242 Steel. Determine the required length of bearing on the top flange to prevent web yielding. (Answer given in Appendix G.)

3. Using the AISC Manual "Allowable Uniform Load" table, design an A36 Steel floor beam to span a distance of 27 ft (simple supports). It carries a superimposed dead load of 500 lb per ft and a live load of 1 kip per ft over its entire length. The beam has complete lateral support. Limit the maximum live-load deflection to $L/360$. (Answer given in Appendix G.)

4. Using the AISC Manual "Allowable Uniform Load" table, design an A36 Steel beam to span a distance of 28 ft (simple supports). It carries a uniform load of 500 lb per ft (including the beam weight) and three concentrated loads of 2 kips each, located at the quarter-points. Lateral support is provided at reactions and the quarter-points. Use the $F_y/800$ depth-span criterion. (Answer given in Appendix G.)

5. Using the AISC Manual "Allowable Uniform Load" table, design an A588 Steel floor beam to span a distance of 26 ft (simple supports). It carries a uniform dead load of 600 lb per ft (including its own weight) and a concentrated live load of 10 kips at its midspan. The beam has complete lateral support. Also, limit the maximum live-load deflection to $L/360$.

6. Using the AISC Manual beam design charts, design an A36 Steel beam to span a distance of 20 ft (simple supports). It carries a uniform load of 500 lb per ft (including the beam weight) and a concentrated load of 10 kips located 8 ft from the left reaction. Lateral support is provided at reactions and at the point of concen-

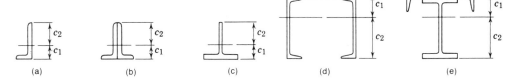

Figure 5.22/Unsymmetrical sections.

trated load only. (Answer given in Appendix G.)

7. Using the AISC Manual beam design charts, design an A36 Steel beam 27 ft long, having a simple span of 18 ft, and a 9-ft overhang at the right reaction. The beam carries a uniform load of 1.8 kips per ft (including its own weight). Lateral support is provided only at the reactions and the end of the overhang.

8. Using the AISC Manual beam design charts, design an A588 Steel floor beam that is to be simply supported over a 27-ft span and carries a dead load of 500 lb per ft (including its own weight) and two equal concentrated live loads of 10 kips each located at the third-points. Lateral support is provided only at the reactions and 10-kip loads. The $F_y/800$ depth-span ratio is recommended, and the maximum live-load deflection is limited to $L/360$. (Answer given in Appendix G.)

5.11
Unsymmetrical Sections

Several of the standard rolled sections are unsymmetrical about either one or both axes, e.g., the angle, tee, and channel. And when a beam section is built up using rolled sections and/or plates, the resulting gross shape can be unsymmetrical. For certain applications, such sections may be both necessary and desirable.

For example, the tee is well adapted to roof construction where the deck material is in the form of precast planks or slabs. Tees with their webs up may be used for the sub-purlins; the precast deck units can then span between the purlins, resting on the flanges and being separated by the

standing web of the tee. Angles, either singly or in pairs, are frequently used as lintels to support masonry over openings.

A beam which is unsymmetrical about its axis of bending is not economical in the sense that this term has been used thus far. Because the neutral axis of such a section is not at mid-depth, the top and bottom fibers will not be equally stressed (Fig. 5.22). It will be remembered that the section modulus is equal to the moment of inertia divided by the distance from the neutral axis to the extreme outside fiber. For the sections of Fig. 522, this is I/c_2. If c_2 is twice as great as c_1, it is evident that the *extreme* fiber stress of any given section will be twice that at the other face; e.g., when the extreme fiber of an A36 Steel section is stressed to an allowable limit of 24,000 psi, the fibers of the other face will be stressed only to 12,000 psi.

The design of unsymmetrical sections will be treated in greater detail in subsequent articles.

5.12
Lintels

Wherever an opening occurs in a unit masonry wall, some means must be provided for carrying the masonry over the opening. Beams used for this purpose are called lintels and are most commonly used over doors and windows. Figure 5.23 shows several types. Those shown in (a) and (b), made up of angles, are for relatively small openings in 8- and 12-in. brick walls. Sometimes the angles, which are placed

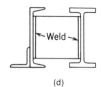

(a)   (b)   (c)   (d)

Figure 5.23/Typical lintels.

back to back, are welded or bolted to-gether, but often they are independent. The **W** beam and plate shown in (c) is for longer spans.

In the majority of cases, lintels are not designed with reference to twisting or tor-sion; therefore, special effort should be made to center the load on, or as near as possible to, the centroid of the lintel sec-tion.

Just how much of the masonry immedi-ately over an opening is borne by the lintel is uncertain. When there is a height of masonry above the opening at least equal to the span of the lintel and when there is also a substantial pier between openings, it can be assumed that there will be an arch action in the masonry which transmits some part of the load to the adjacent unbroken wall. Experience indicates that in such instances only a small triangular section of the wall is generally carried by the lintel (Fig. 5.24). The height of this triangle is assumed equal to about half the span of the lintel; this makes an isosceles triangle with base angles of 45° (Fig. 5.26).

In situations such as that illustrated in Fig. 5.25a, where there is insufficient height of masonry for the triangular loading to de-velop, the masonry to be supported is that shown by the shaded area. For cases simi-lar to that shown in Fig. 5.25b, the open-ings above would also destroy any arch action in the masonry. It should also be borne in mind that a lintel may receive some load from the floor above and this weight must be added to that of the ma-sonry to be supported.

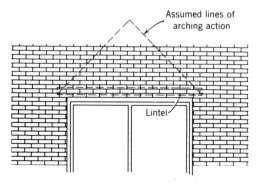

Figure 5.24/Arching over window in brick wall.

Lintels are frequently designed much larger than required for resistance to bending stresses in order to provide suffi-cient bearing surface for the masonry. For instance, the outstanding legs of the angles shown in Fig. 5.23a would normally be at least $3\frac{1}{2}$ in. The height of the vertical legs, of course, will vary with the span. Lintels over openings in brick walls should extend at least 4 in. beyond the face of the open-ing on each side. For wide openings, a greater bearing length is usually necessary and can be calculated (Art. 5.16).

There is always some question as to whether the wall furnishes adequate lat-eral support for the lintel. There is cer-tainly little lateral support until the mortar has cured, and even after curing, the abil-ity of the masonry to furnish lateral sup-port to the lintel will depend upon its thickness and proximity to floors, roofs, pilasters, etc.

When lateral support does not exist or there is doubt as to its adequacy, a flanged

Figure 5.25/Openings in brick walls.

beam with soffit plate to engage the masonry, as shown in Fig. 5.23c, may be used and designed in the manner illustrated in Art. 5.7. On the other hand, angles having no compression flange are entirely different. The allowable loads given in the AISC Manual are computed on the basis of an F_b not to exceed $0.60F_y$, and must be reduced when the angles are laterally unsupported. Such lintels require special investigation.

Example 1

Design a lintel of two A36 Steel angles to carry an 8-in. brick wall over a 68-in. opening. Conditions are such that a triangular loading may be assumed, but the angles are not attached to each other. The weight of the brickwork is assumed to be 130 lb per cu ft. Assume that lateral support is provided. The lintel is to have 4-in. bearing at the sidewalls (Fig. 5.26).

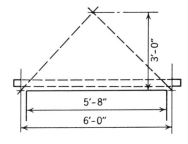

Figure 5.26/Example 1.

Solution

Note. In the triangle of load, the distances are based on the lintel span, i.e., center to center of bearing.

(1) Determine the span and area of brickwork to be supported. The length of bearing is 4 in.; therefore, the reactions will occur at the center of bearing, and the span is $68 + 2 + 2 = 72$ in., or 6 ft. The triangle has a base of 6 ft and an altitude of 3 ft. Therefore, the wall area carried is

$$\frac{6(3)}{2} = 9.0 \text{ sq ft}$$

(2) If the brickwork weighs 130 lb per cu ft, the weight of a 1 sq ft section is $(8/12)(130)$, or 86.7 lb. And the total weight is $9.0(86.7) = 780$ lb.

(3) The maximum bending moment will be at the center of the span. Each reaction is 390 lb. Taking moments about the center of the span,

$$M = \left[390(3)\right] - \left[390\left(\tfrac{1}{3}\right)3\right]$$
$$= 780 \text{ ft-lb} = 9360 \text{ in.-lb}$$

(4) Calculate the required section modulus and select the angle size:

$$F_b = 0.60(F_y) \simeq 22{,}000 \text{ psi}$$
$$S = \frac{9360}{22{,}000} = 0.43 \text{ in.}^3$$

Because two angles are called for, the

section modulus of each must be

$$\frac{0.43}{2} \simeq 0.22 \text{ in.}^3$$

To provide sufficient bearing for the brick, consider only those angles having an outstanding leg of $3\frac{1}{2}$ in. or greater. Referring to the AISC Manual, the smallest angle listed meeting these requirements has dimensions of $3\frac{1}{2} \times 2\frac{1}{2} \times \frac{1}{4}$ in. and a section modulus of 0.412 in.^{3}. This angle is selected.

Many designers have a standard minimum angle which they use for lintels, such as

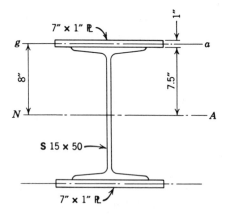

Figure 5.27/Symmetrical built-up section.

$3\frac{1}{3} \times 3 \times \frac{5}{16}$ in., or $3\frac{1}{2} \times 3 \times \frac{3}{8}$ in., placed with the short legs back to back. In such cases, the design is carried out to determine whether the standard lintel has the needed strength.

5.13
Built-Up Sections

It is frequently more economical to reinforce **W** or **S** sections with cover plates in regions of maximum bending moment than to supply a heavier section throughout the entire span length. When such plates are used, a symmetrical built-up section results, such as that shown in Fig. 5.27. Unsymmetrical built-up sections are shown in Fig. 5.28(a) and (b).

In order to investigate the strength of built-up sections, it is first necessary to determine the moment of inertia of the section about the axis of bending. Methods for computing the I of plane areas are given in standard textbooks on mechanics, and formulas for several geometric shapes are tabulated in the AISC Manual. The manual, of course, also gives the I of rolled shapes in the properties tables.

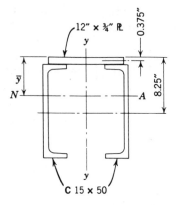

(a) Data for centroid computation.

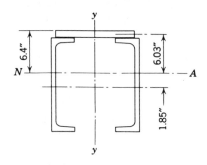

(b) Data for I computation.

Figure 5.28/Unsymmetrical built-up sections.

Symmetrical Sections / The simple built-up section shown in Fig. 5.27 is composed of a **S** 15×50 with two 7-in. by 1-in. plates welded to its flanges. The moment of inertia of the built-up section about its neutral axis is equal to the sum of the moments of inertia of all the members about the same axis. The first step in determining the I of the built-up section is to find the moment of inertia of the **S** 15×50, since its axis coincides with the neutral axis of the built-up section. From the AISC Manual "Dimensions and Properties" tables, this is found to be 486 in.[4]. The I of one of the plates about its own centroidal axis parallel to the longer side (*ga* in Fig. 5.27) is found from the formula for rectangular sections:

$$I = \frac{bt^3}{12} = \frac{7(1)^3}{12} = 0.583 \text{ in.}^4$$

It is now necessary to find the moment of inertia of the plate about the *NA* of the built-up section. This is accomplished by means of the *transfer equation*,[11] which may be expressed as follows:

The moment of inertia of a cross section about any axis parallel to the axis through its own centroid is equal to the moment of inertia of the cross section about its own centroid, plus its area times the square of the distance between the two axes—or, expressed mathematically,

$$I = I_0 + Az^2$$

where:

I = moment of inertia of the cross section about the required axis.

I_0 = moment of inertia of the cross section about a parallel axis through the centroid.

[11]For the derivation of the transfer equation, see any standard textbook on mechanics of materials.

A = area of the cross section.

z = distance between the two parallel axes.

Applying this equation to the *two* plates of the built-up section shown in Fig. 5.27,

$$I = 2\left[I_0 + Az^2\right]$$
$$= 2\left[0.583 + 7(8)^2\right]$$
$$= 2[0.583 + 448] = 2(448.583)$$
$$= 897.166, \text{ or } 897 \text{ in.}^4$$

The moment of inertia of the entire built-up section about its neutral axis *NA* is therefore $486 + 897 = 1383$ in.[4].

It is evident from this discussion that the moment of inertia (I_0) of a thin plate about an axis through its c.g. may be neglected when that axis is parallel to the longer side of the plate. When the members considered are angles or other shapes, however, the value of I_0 is appreciable and should not be omitted.

Unsymmetrical Sections / In order to determine the moment of inertia of an unsymmetrical section about its axis of bending (*NA*), it is first necessary to locate the centroid of the overall built-up shape, since this cannot be established by inspection as for symmetrical sections. It will be recalled from mechanics that the centroid of an area is that point about which the static moment of all component areas is equal to zero, and that it may be found by equating the moment of the area of the entire built-up section about a given reference axis to the sum of the moments of all its component areas about the same axis. Since the section shown in Fig. 5.28 is symmetrical about the vertical axis, the centroid will fall somewhere along the line *y-y*. If the two channels are each **C** 15×50 and the plate $12\times\frac{3}{4}$ in., the various moment arms about the top edge of the plate as an axis of moments will be as indicated in

Fig. 5.28a, where 8.25 is the distance from the gravity axis of the channels to the top edge of the plate, 0.375 is the corresponding distance from the gravity axis of the plate, and $\bar{y}$ the distance from the top of the plate to the gravity axis (NA) of the entire section.

From the AISC "Dimensions and Properties" tables, the area of a **C** 15×50 is 14.7 in.^{2}, and the area of the plate is 9 in.^{2}. Therefore, the total area is

$$A_t = 2(14.7) + 9 = 38.4 \text{ in.}^2$$

Equating moments of areas,

$$38.4\bar{y} = 2[14.7(8.25)] + 9(0.375)$$
$$\bar{y} = \frac{243 + 3.38}{38.4} = \frac{246.4}{38.4} \approx 6.4 \text{ in.}$$

With the center of gravity (and consequently the neutral axis) located, the moment of inertia of the built-up section about NA may now be calculated using the transfer equation discussed earlier in this article. The z distances for the plate and channels, required in the transfer equation, are determined as follows and recorded in Fig. 5.28b:

$$z(\text{plate}) = 6.40 - 0.375 = 6.03 \text{ in.}$$
$$z(\text{channels}) = 8.25 - 6.40 = 1.85 \text{ in.}$$

Then, neglecting I_0 of the plate,

$$I(\text{plate}) = I_0 + Az^2$$
$$= 0 + 9(6.03)^2 = 327 \text{ in.}^4$$

From the same tables, the moment of inertia of one **C** 15×50 about its own gravity axis is 404 in.^{4}. Therefore,

$$I(2 \text{ channels}) = 2\left[404 + 14.7(1.85)^2\right]$$
$$= 2[404 + 50] = 908 \text{ in.}^4$$

and consequently the moment of inertia of the entire built-up section about NA is $327 + 908 = 1235$ in.^{4}.

Section Modulus / The section modulus for built-up shapes is determined from the relationship $S = I/c$, where c is the distance from the neutral axis to the extreme fiber in bending. For the symmetrical section shown in Fig. 5.27,

$$S = \frac{I}{c} = \frac{1383}{8.5} = 163 \text{ in.}^3$$

For the unsymmetrical section of Fig. 5.28, it will be observed that the overall depth of the channels plus the plate is 15.75 in. Since the distance from NA to the top fiber was found to be 6.4 in., the corresponding distance to the bottom fiber is $15.75 - 6.4 = 9.35$ in. This is the larger of the two values, thereby corresponding to the distance c_2 in Fig. 5.22d. Consequently, S for this section is

$$S = \frac{I}{c} = \frac{1235}{9.35} = 132 \text{ in.}^3$$

5.14
Plate Girders

Although the heavier **W** beams are adequate for most long spans encountered in building construction, plate girders are nevertheless frequently used where unusually heavy loads occur over long spans. Such a condition is shown in Fig. 5.29 where two interior columns in the first and second stories have been omitted in order to achieve larger unobstructed ground-floor space. Figure 5.30 illustrates typical cross sections of both bolted and welded plate girders, and Fig. 5.31 shows the use of web stiffeners to prevent buckling of the relatively thin web plates. However, because of modern fabrication techniques, bolted plate girders are seldom used.

Procedures for detailed design of bolted and welded plate girders are not included in this text; however, the discussion of built-up sections in the preceding article indicates the basis for proportioning girders to resist bending, using the gross-moment-of-inertia method.

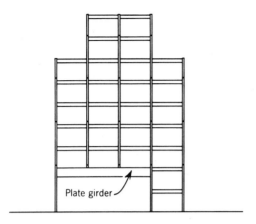

Figure 5.29/Building cross section.

was assumed to act at the centroid of the flange cross section, made up of the flange angles and cover plates. The flange-area method is no longer considered sufficiently accurate to obtain a precise and economical design, but it is useful in obtaining a trial section when designing bolted-plate girders, which may then be checked by the moment-of-inertia method.

The reader interested in pursuing the design of plate girders further should consult the AISC Manual. The Ninth Edition of the manual contains an extensive treatment of this topic, including illustrative examples.

Figure 5.32(a) and (b) show an assumed stress distribution used for two design procedures.

The earlier method of design known as the *flange-area* or *chord* method was based on the proposition that, since the major portion of the cross-sectional area of the plate girder is located in the flanges, and since the highest bending stresses occur in the flanges, a reasonable approximate stress distribution is that shown in Fig. 5.32b. Here, the resultant of the flange stresses

5.15
Bending about Two Axes

Occasionally a beam may be required to support loads inclined to its axes, or applied perpendicular to both axes. The former may occur when the beam is tilted and the loads are vertical as in Fig. 5.33a, and the latter, when there are two or more loads applied from different directions at the same time (Fig. 533b). With reference

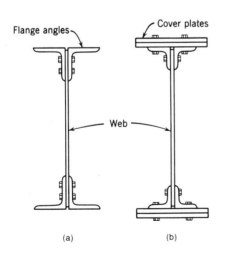

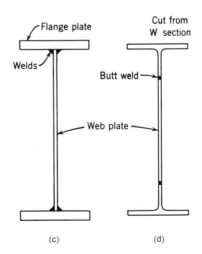

Figure 5.30/Typical plate girder sections.

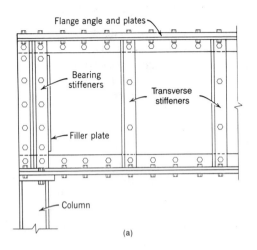

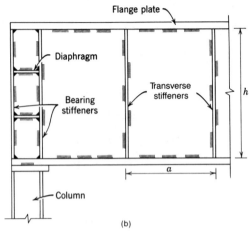

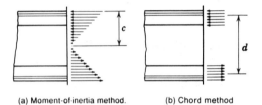

Figure 5.31/Typical plate girder elevations.

(a) Moment-of-inertia method.   (b) Chord method

Figure 5.32/Stress distribution.

to Fig. 5.33b, a girt is a secondary structural member used principally on the outside perimeter of structures to support materials which form the enclosing envelope. They span from column to column and are generally either channels or angles.

All of the theoretical formulas used thus far have been based on the assumption that the plane of loading passes through the centroid of the beam section. This is true for both beams shown in Fig. 5.33, and as a result their design is relatively simple. In the case of Fig. 5.33a, it is necessary only to resolve the load into components parallel to the x-x and y-y axes, producing a loading similar to that of Fig. 5.33b. For any point along the beam's length, the fiber stress is then the algebraic sum of the stresses resulting from bending

about each axis when computed independently. Expressed in equation form,

$$f = \frac{M_x c_x}{I_x} \pm \frac{M_y c_y}{I_y}$$

where:

M_x = moment resulting from bending about the x-x axis.

I_x = moment of inertia with respect to the x-x axis.

c_x = distance from the x-x axis to the fiber f, measured perpendicular to the x-x axis.

M_y = moment resulting from bending about the y-y axis.

I_y = moment of inertia with respect to the y-y axis.

c_y = distance from the y-y axis to the fiber f, measured perpendicular to the y-y axis.

A tilted beam, loaded through its center of gravity as shown in Fig. 5.33a, is seldom encountered in practice. The loads are almost always applied at the center of the top flange. In such cases, the plane of loading does not pass through the c.g. of the beam (Fig. 5.34). As a result, twisting,

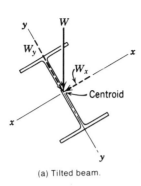

(a) Tilted beam.

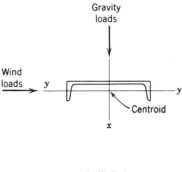

(b) Wall girt.

Figure 5.33/Biaxial bending.

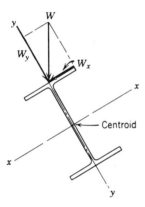

Figure 5.34/Biaxial bending, approximate method.

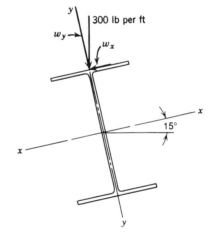

Figure 5.35

or *torsion*, occurs in addition to bending. A complete analysis of torsion is beyond the scope of this text. However, a satisfactory approximate method for dealing with a condition such as shown in Fig. 5.35 is given below.

The load component W_y passes through the c.g. and the resulting stresses are calculated in the usual manner. The load component W_x does not pass through the c.g., and nearly all resistance to bending resulting from this component is provided by the top flange. Therefore, the moment of inertia of the top flange alone may be used when computing flexural stresses resulting from W_x. As the flanges are equal

and the web contributes a negligible amount to the moment of inertia about the y-y axis, one-half the total I_y should be used.

Consequently, the formula for stress should now be

$$f = \frac{M_x c_x}{I_x} \pm \frac{M_y c_y}{I_y/2}$$

If the beam section is symmetrical about both axes, the section modulus may be used in lieu of I/c. The trial-and-error method of design is appropriate when designing beams having bending about two axes, keeping in mind the fact that most beams offer little resistance to bending about the *y-y* axis.

Example 1

A **W** 8×18 spans 14 ft between the two sloping roof girders. The load, including the weight of the beam, is 300 lb per ft over the entire span. The girders are inclined 15° to the horizontal. Check the **W** 8×18 to determine if it is safe in bending; $F_b = 23,800$ psi.

Solution

(1) Make a sketch of the cross section of the beam showing the gravity loading (Fig. 5.35).

(2) Determine the respective load components that act perpendicular to the *x-x* and *y-y* axes:

$$w_y = 300\cos 15° = 290 \text{ lb per ft}$$

$$w_x = 300\sin 15° = 77.6 \text{ lb per ft}$$

(3) Analyze the beam for bending both for the *x-x* axis and the *y-y* axis. As the only means of beam support are the inclined girders, it will be assumed the beam rests on the girders and, in addition, has its top flange clipped to the girder (Fig. 5.36). This means that, for bending, the beam is simply supported for 14 ft about both axes. For bending about the *x-x* axis, the maximum moment at midspan will be

$$M_x = \frac{w_y L^2}{8} = \frac{290(14)^2}{8} = 7110 \text{ ft-lb}$$

For bending about the *y-y* axis, the maximum moment also will be at midspan, or

$$M_y = \frac{w_x L^2}{8} = \frac{77.6(14)^2}{8} = 1900 \text{ ft-lb}$$

Because both maximum moments occur at midspan, the maximum stress will also occur at midspan:

$$f_b = \frac{M_x}{S_x} + \frac{M_y}{S_y/2}$$

$$= \frac{7110(12)}{15.2} + \frac{1900(12)}{1.52}$$

$$= 5613 + 15,000$$

$$= 20,613 \text{ psi (compression)}$$

$$< 23,800 \text{ psi (allowable)}$$

The stress as computed above would occur at the upper edge of the top flange (point *A* in Fig. 5.36).

5.16
Beam Bearing Plates

When the end of a beam rests on a masonry wall, it is sometimes necessary to provide a steel bearing plate in order to distribute the reaction over an area sufficient to keep the average pressure on the masonry within allowable limits.

The required area of the plate is found by dividing the end reaction of the beam by the allowable unit bearing pressure on the masonry. In the absence of specific building-code requirements, the values for safe bearing pressures on various kinds of masonry and concrete walls given in Table 5.2 should prove helpful.[12]

[12] Many codes specify that the allowable bearing stress be limited to 25 per cent of the ultimate compressive strength of the masonry unit or assembly.

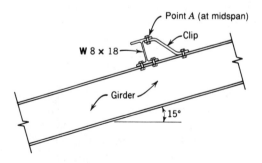

Figure 5.36/Purlin support detail.

Table 5.2
Safe Bearing Pressure on Masonry and Concrete Walls[a]

Type of Wall	Pressure (psi)
Brick:	
Soft	150
Medium	225
Hard	500
Concrete blocks:	
Hollow units	150
Solid units	300
Cinder blocks:	
Hollow units	120
Solid units	260
Cut and bedded stone:	
Bluestone	400
Sandstone	300
Limestone	500
Rubble stone	140
Poured concrete walls:	
3000-psi concrete	900
4000-psi concrete	1200

[a]Using average cement mortar and net areas.

Referring to Fig. 5.37a, the dimension C of the plate (parallel to the beam length) is frequently limited by the thickness of the wall. Even on a 13-in. brick wall, C would be limited to 8 in. so as not to interfere with the outside course of the brickwork. The dimension B of the plate (parallel to the face of the wall) is determined by dividing the required area by the value established for C.

The thickness of the plate is governed by bending considerations, which may be thought of as the tendency of the uniform bearing pressure on the bottom of the plate to curl the plate upward about the beam flange, as illustrated with great exaggeration in Fig. 5.37d. If the bearing plate is not thick enough to prevent distortion of this nature, then the beam reaction will not be uniformly distributed over the area

of contact between the plate and the masonry, and there will be greater pressure directly under the beam than at the edges of the plate.

It may be observed from Fig. 5.37(a) and (d) that the projection of the plate on either side of the beam acts as an inverted cantilever with a uniformly distributed load. There is some uncertainty, however, as to just where the maximum bending moment will occur. If the beam flange is comparatively thick and therefore, stiff, it might be assumed that the flange would not tend to curl as shown in Fig. 5.37d, but instead would remain flat. Under these conditions, the maximum bending moment in the plate would occur at the edge of the flange, and the cantilever projection n would have the value indicated in Fig. 5.37b. If the beam flange does not remain flat, however, the value to be used for the cantilever projection n will obviously have to be larger than indicated in Fig. 5.37b. One widely used procedure for determining the value of n is shown in Fig. 5.37c. Values of k (equivalent to the distance from bottom of beam to web toe of fillet) for rolled beam sections are given in the *AISC Manual* "Dimensions and Properties" tables. The AISC recommends this procedure, using an allowable fiber stress in the bearing plate of $F_b = 0.75F_y$. The AISC procedure also relates this process to the dimension N (Fig. 5.17), which replaces C (Fig. 5.37). The design of bearing plates is illustrated by the example which follows.

Example 1

Design a beam plate for a **W** 10×22, both of A36 Steel, that transmits a reaction of 24,000 lb to a wall built of medium brick laid in cement mortar that provides an 8-in. length of bearing. The allowable unit bearing pressure on this wall is $F_p = 225$ psi.

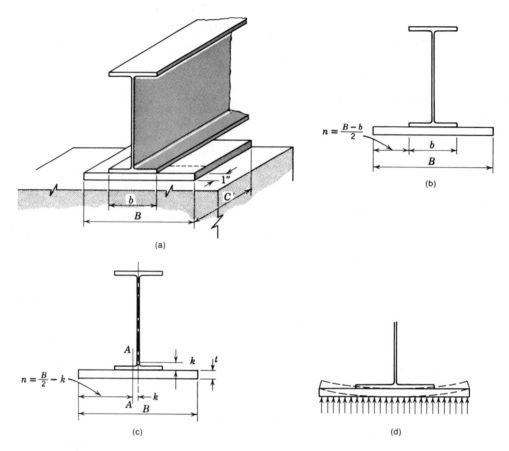

$$n = \frac{B-b}{2}$$

(b)

$$n = \frac{B}{2} - k$$

(c)

(d)

Figure 5.37/Bearing-plate design data.

Solution

(1) The required plate area is found by dividing the beam reaction R by the allowable unit pressure F_p:

$$A = \frac{R}{F_p} = \frac{24,000}{225} = 107 \text{ in.}^2$$

(2) This area requires an 8- by 13.5-in. plate. Referring to Fig. 5.37a, $C = 8$ in., $B = 13.5$ in., and the actual bearing pressure $f_p = 24,000/8(13.5) = 222$ psi.

(3) From tables in the AISC Manual, the value of k is found to be $\frac{3}{4}$, or 0.75 in. Referring to Fig. 5.37c, the cantilever pro-

jection is

$$n = \frac{B}{2} - k = \frac{13.5}{2} - 0.75 = 6.0 \text{ in.}$$

(4) Considering a strip of the plate 1 in. wide (Fig. 5.37a), the maximum bending moment at section A-A (Fig. 5.37c) is

$$M = \frac{f_b n^2}{2} = \frac{222(6.0)^2}{2} = 3996 \text{ in.-lb}$$

(5) Determine the allowable bending stress in the bearing plate, and calculate the required section modulus:

$$F_b = 0.75(36) = 27 \text{ ksi}$$

$$S = \frac{M}{F_b} = \frac{4.0}{27} = 0.148 \text{ in.}^3$$

(6) The required thickness $t = \sqrt{6S}$. This expression is derived from the definition of section modulus ($S = I/c$). The value of I for a rectangular section in the familiar form is $bh^3/12$, where b equals the width of the section, and h, the depth (or, in this case, the thickness of the plate). Substituting these expressions,

$$S = \frac{I}{c} = \frac{bt^3}{12c}$$

But c equals half the depth of the section —in this case, the thickness $t/2$. Substituting this value,

$$S = \frac{2bt^3}{12t} = \frac{bt^2}{6}$$

The width of the section in this case has been taken as 1 in. Therefore,

$$S = \frac{t^2}{6} \quad \text{and} \quad t = \sqrt{6S}$$

Substituting the value of S found in step (5),

$$t = \sqrt{6S} = \sqrt{6(0.148)} = \sqrt{0.89} = 0.94 \text{ in.}$$

An $8 \times 13.5 \times 1$-in. plate is adopted.

The above procedure may be abbreviated in practice by using the AISC 1989 formula for thickness, which does not involve direct computation of bending moment and section modulus. Applying the AISC formula,

$$t = \sqrt{\frac{3f_b n^2}{F_b}} = \sqrt{\frac{3(222)6.0^2}{27,000}} = 0.94 \text{ in.}$$

The reader can satisfy himself that

$$\sqrt{6S} = \sqrt{\frac{3f_b n^2}{F_b}}$$

reduces to an identity by substituting Mc/I for F_b, and then $f_b n^2/2$ for M on the right side of the equation.

When heavy beams rest on relatively thin walls, the dimension C shown in Fig. 5.37a is often so limited that B becomes very large in proportion. This results in a large projection from the edge of the flange and a corresponding increase in thickness. Figure 5.38a shows one method of providing bearing for such a condition. The load is assumed to be equally divided between the beams, and they are designed as inverted cantilevers. The shearing and bearing stresses in the webs of the supporting beams must also be investigated.

Beams bearing on masonry walls are usually provided with anchors as a means of tying the structure together. Figure 5.38(b) and (c) illustrate two common types of

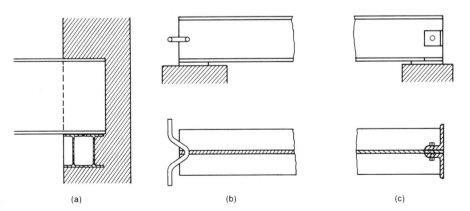

(a)   (b)   (c)

Figure 5.38/Bearing details and wall anchors.

anchorage. The bent rod, called a government anchor, shown at (b), is usually made $\frac{3}{4}$ in. in diameter, and the angles at (c), $\frac{3}{8}$ in. thick. Some steel companies have their own standard anchors, which are listed in their handbooks. In design for earthquake resistance, the anchors are of special significance.

5.17
Floor Framing

Figure 5.39 illustrates two methods of framing commonly employed in ordinary office building construction. Third-point concentration is shown at (a), and center

concentration at (b). When the span of a girder is much over 16 ft, third-point concentration is desirable. The area of floor supported by one beam is found by multiplying the span length by the sum of half the distances to the adjacent beams. Span lengths are generally figured from center to center of supporting members, although this distance is sometimes reduced where beams frame against the flanges of large columns. The arrangement and design of connections between various members is treated in Chapter 7.

Typically, flexible connections that generate simple reactions are used. This is the assumed case in Figs. 5.39, 5.40, and 5.41.

The arrangement of the framing is governed, to a large extent, by the type of floor system employed. For example, in steel-frame office buildings, apartment houses, and similar structures, where concrete slabs reinforced with wire fabric are often used, 6 to 10 ft is a desirable distance between beams.

Other factors influencing framing are floor openings (i.e., stairs and elevators) and

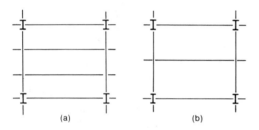

(a) (b)

Figure 5.39/Typical interior floor framing.

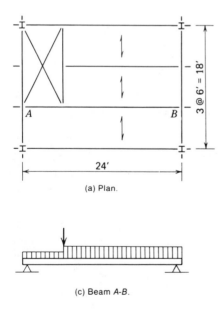

(a) Plan.

(c) Beam A-B.

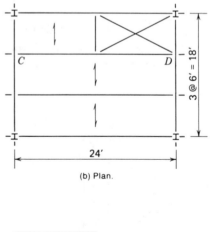

(b) Plan.

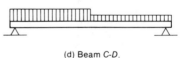

(d) Beam C-D.

Figure 5.40/Framing around openings.

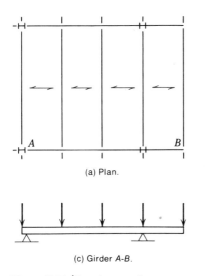

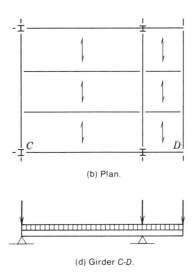

(a) Plan.

(b) Plan.

(c) Girder *A-B*.

(d) Girder *C-D*.

Figure 5.41/Framing overhangs.

overhangs on exterior bays. Figure 5.40(a) and (b) show two typical ways of framing around openings. The x-shaped portions denote openings and, in the examples shown, have dimensions of 6 by 12 feet. Each framing arrangement produces a unique type of loading on the framing members. For example, Fig. 5.40c illustrates the resulting load on member *A-B*, and Fig. 5.40d shows the load on member *C-D*, assuming no load from the opening.

Figure 5.41(a) and (b) show two ways of framing with an overhang on one side, and Fig. 5.41(c) and (d) show the resulting loads on girders *A-B* and *C-D*.

5.18
Open-Web Steel Joists

In addition to the standard types of reinforced-concrete slab construction and cellular steel flooring, several forms of steel joists are manufactured for supporting floors and roofs. Cold-formed steel members for the support of light loads also are popular.

Figure 5.42 illustrates typical light-weight joists, and Fig. 5.43 shows two types of light-gage metal deck that span the distance between joists. The deck shown is only $1\frac{1}{2}$ in. deep, and, when used alone, the joists must be fairly close together. The metal deck could be much deeper, or it could serve as a form for concrete, either of which would permit greater spacing between joists.

The joists in Fig. 5.42a are made from cold-formed, light-gage metal and generally range from 6 to 14 in. in depth. They are used most often in "pre-engineered buildings," which usually are all-metal buildings having components designed and made by one manufacturer.

Bar joists of the type shown in Fig. 5.42b are made by a large number of manufacturers. The steels used in these, as well as the larger "long-span joists" (Fig. 5.42c), usually are both A36 Steel and another steel having $F_y = 50$ ksi. The higher-strength steel is used for the top and bottom chords; these joists are referred to as the "K Series." Bar joists are very economical and consequently very popular for both floor and roof framing. They may be

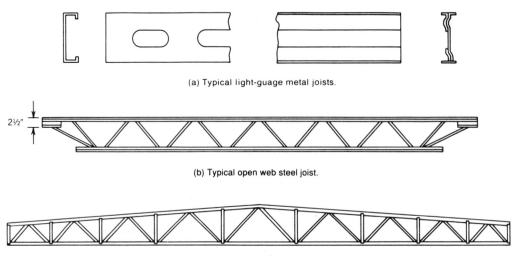

(a) Typical light-guage metal joists.

2½″

(b) Typical open web steel joist.

(c) Typical long-span steel joist.

Figure 5.42/Types of steel joists.

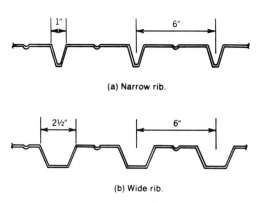

1″   6″

(a) Narrow rib.

2½″   6″

(b) Wide rib.

Figure 5.43/Light-gage metal deck.

supported on either bearing walls or steel girders in a skeleton-frame. Usually, the lighter the load and the longer the span, the more economical this system will be. Heavy, concentrated loads should be avoided or treated with special care.

The spacing of joists generally ranges from 18 in. to 8 ft, depending upon the type and strength of the deck that spans between the joists, and the capacity of the joist to carry the deck load to the supporting walls or girders. The joists are designed for uniform loads applied to their top edge. On occasion, the long-span joist is used as a girder having concentrated loads at intervals that coincide with vertical and diagonal intersections at the top of the joist. The Steel Joist Institute calls these *joist girders*. This system is discussed in more detail in Chapter 9.

Regular H-Series joists range in depth from 8 to 30 in. (in 2-in. increments) and are used for spans up to 60 ft. For longer spans (up to 96 ft), the long-span type called LH Series is available in depths from 18 in. to 4 ft. The largest type, called deep long-span joists or DLH, has depths of up to 6 ft and can span up to 144 ft, but is recommended for roofs only.

Open-web steel joists are designed essentially as simply supported, uniformly loaded trussed beams. When not otherwise laterally restrained, bridging is required in the form of continuous steel members attached to the top and bottom chords to position the joists and prevent twisting and buckling (placed at intervals depending on the clear span), and also diagonally, in the form of steel members attached to the chords between horizontal lines of bridging.

Properties of joist sections (as well as cellular steel flooring), together with safe-load tables and construction details, are given in the catalogs of the various manufactures.

Also, safe-load tables for the K, LH, and DLH Series are given in Appendix F. The loads listed in the table are the total uniform load in lb/ft along the top edge of simply supported joists without overhangs. The first number listed in the table for a given joist size and span (the largest) is the maximum load the joist can carry including its own weight. The second number, for the same joist size and spacing, is that load that will produce a maximum deflection at midspan of $L/360$. By multiplying this load by 1.5 (i.e., $360/240$), the load that will cause a deflection of $L/240$ can be obtained.

Example 1

Using the typical interior bay shown in Fig. 5.10 and the general design conditions shown in Example 5, Art. 5.6, select open-web joists to replace the beams.

The joists will span the 28-ft dimension and bear upon the steel girders. A 2-in. minimum cast-in-place concrete slab reinforced with ribbed metal lath will be used in place of that specified. Use K-Series open-web joists.

Solution

(1) Total imposed dead load = (43.0 psf plus an allowance of 2 psf for bridging) = 45.0 psf.

(2) Total live and environmental load = 30.0 psf.

(3) Select a joist spacing. A 2-in. slab can span 3 ft but not 4 ft; therefore, 24 in. will be selected for convenience.

(4) Determine the design load in pounds per linear foot (plf) of joist:

$$2(45+30) = 150 \text{ plf}$$

(5) Referring to Appendix F, *Standard Load Table for Open Web Steel Joists*, select a 14K3, with a safe load of 180 plf. The weight of the joist is 6 plf; therefore, its imposed load-carrying capacity is

$$180-6 = 174 > 150 \text{ plf}$$

(6) Check deflection. The combined uniform live and environmental load on the joist is $30(2) = 60$ plf, which is less than the listed value of 88; therefore, it can be concluded that the deflection for this load is less than $L/360$.

(7) Check end bearing. Assuming that the girder is the **W** 18×46 selected in Example 5, Art. 5.6, the 6-in. flange will provide ample length of bearing ($2\frac{1}{2}$ in.) for both abutting joists, even though it is not necessary in such cases. Two $\frac{1}{8}$-in. fillet welds (Fig. 5.3c and Chapter 7), 1 in. long, will be used to fasten the ends of joists to the girder and provide lateral support.

(8) Select bridging. Referring again to the specifications for K-Series joists, select two rows of horizontal bridging, placed at the third-points of the span (i.e., 9 ft 4 in. apart) on both the top and bottom chords and welded at points of contact. Diagonal bridging is to consist of cross-bracing between chords. (Horizontal bridging must be able to resist a horizontal force of 700 lb and have an $L/r < 300$; diagonal bracing must have an $L/r < 200$—see Art. 6.11 for design of struts.)

Example 2

Design a long-span floor joist for a live load of 30 psf and a span of 85 ft. The joist spacing is 4 ft and supports the following construction:

Roofing	$6\frac{1}{2}$ psf
Insulation	$1\frac{1}{2}$
Metal deck	
and infill	12
Ceiling	6
	26 psf

Solution

(1) Calculate the load on the joist:

$$w = 4(26+30) = 224 \text{ plf}$$

(2) Select a trial LH joist from Appendix F. Select a 44LH11 having a capacity of 247 plf. Observe that the weight of this joist is 22 plf, giving a total load of $224 + 22 = 246$ plf.

(3) Check deflection. The live load is $30(4) = 120$ plf, which is less than 123: OK.

(4) Check shear. The total load is $249(85) = 21,165$ lb which is less than 23,900 lb: OK.

(5) Design bridging. The maximum spacing between lines of bridging for this size joist is 16 ft (check manufacturer's specification or the specifications of the Steel Joist Institute). Use six lines of bridging.

5.19
Torsion

Occasionally, steel beams are subjected to torsional loading which generates torsional shear stresses. Whenever this occurs, the combined shear stress caused by bending and torsion should be investigated to assure that the maximum value remains within the limits prescribed by the AISC.

This article deals only with the simplified applied design aspect of torsion. For a more thorough theoretical discussion, refer to any standard text on strength of materials. For a more complete design dis-

cussion, refer to *Torsional Analysis of Rolled Steel Sections* by C. P. Heins, Jr. and P. A. Seaburg, a steel design file, available from Bethlehem Steel Corporation.

Torsional loading occurs when the line of action of the load does not pass through the shear center of the beam. The shear center for I-shaped beams, as well as box-, tubular-, and circular-shaped beams, coincides with the center of gravity. However, it does not coincide with the center of gravity of channel-shaped and other nonsymmetrical open section beams. The twisting moment, or *torque*, induced by torsional loading is similar to a bending moment in that its value is the product of the load and its perpendicular distance to the shear center. In Fig. 5.44 it is seen to be $T = Pe$ and remains constant for the full length of the beam.

When torsional loading cannot be avoided by appropriate detailing or if it appears to be a major part of the loading system, the use of a solid circular-, a box-, or a tubular-shaped beam is recommended. Of these, the most efficient shape to resist torsional shear stress is the circular shape, which may be either solid or tubular. The nature of shearing stresses in solid circular sections and tubular sections is somewhat different, as shown in Fig. 5.45.

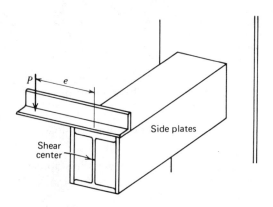

Figure 5.44/Torsion loading.

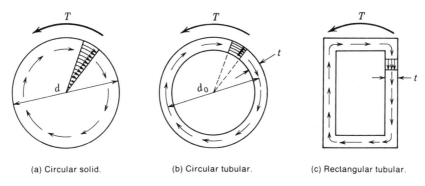

(a) Circular solid.   (b) Circular tubular.   (c) Rectangular tubular.

Figure 5.45/Torsional shear stress.

Circular shapes do not warp, and torsional stresses consist of only shear stresses. The magnitude of these shear stresses is proportional to the distance from the shear center and varies linearly from zero at the shear center to a maximum at the outermost fiber. The value of torsional shear stress for circular sections is given by the formula

$$f_v = \frac{Tc}{J}$$

where

T = torque in inch-pounds or inch-kips.

c = radial distance from the shear center to the point under consideration.

J = polar moment of inertia in inches^4.

Solid circular shapes have a J value of $\pi d^4/32$, while the hollow circular shape has a J value of $\pi[d^4 - (d - t)^4]/32$ (see Fig. 5.45). An alternative formula for calculating the J value for a circular tubular section is

$$J = \frac{\pi d_0^3 t}{4}$$

where d_0 is the mean diameter of the section (Fig. 5.45b).

When a noncircular thin-walled section (Fig. 5.45c) is subjected to torsion, warping will occur at each section along the length of the member. However, if the ends of the member are not restrained, the warping takes place in such a manner that no

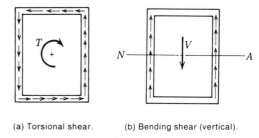

(a) Torsional shear.   (b) Bending shear (vertical).

Figure 5.46/Direction of shear stresses.

warping stresses occur. Only shear stresses are present, and they remain constant throughout the thickness of the tube. The magnitude of these shear stresses can be determined from Bredt's formula[13]

$$f_v = \frac{T}{2 t A_0}$$

where A_0 is the area of the tube enclosed by the wall median line.

Shearing stresses calculated by the formulas presented above must be added to other stresses caused by beam action. Consequently, the direction of independently calculated stresses becomes an important consideration.

Figure 5.46 shows that on one side of the vertical web the shearing stresses due to flexure and the shearing stresses due to torsion are directly additive. It should be

[13] W. Flugge, Editor, *Handbook of Engineering Mechanics*, First Edition, McGraw-Hill Book Co., New York, 1962, Section 36, page 13.

recalled that the value of the shearing stress due to flexure is calculated from the formula

$$f_{vb} = \frac{VQ}{Ib}$$

and is a maximum at the neutral axis. Consequently, it becomes necessary to add the torsional shearing stress (f_{vt}) to the bending shear (f_{vb}) at the neutral axis in order to obtain a maximum value.

Whenever an open shape, such as the customary I- and channel-shaped beams, is used where torsion is considered critical, the design process is considerably more involved, and the torsional-analysis steel design file referred to earlier should be consulted. The effect of the open shape causes a significant warping action, which, in turn, may generate stresses of larger magnitude than the torsional shear stress discussed earlier.

The torsion on open sections causes three distinct types of stresses: torsional shear stress, warping shear stress, and warping normal stress, all of which must be algebraically added to the stresses caused by the normal bending of the beam.

The distribution of torsional shear stress in a solid rectangular shape and in the typical I-shaped section is shown in Fig. 5.47, and its magnitude can be calculated from the formula

$$f_{vt} = \frac{Tt}{J}$$

The torsional constant J for the rectangular shape is approximately $\frac{1}{3}bt^3$; for the I-shaped section, it can be approximated by adding the three quantities $\frac{1}{3}bt^3$ for the individual rectangles making up the shape. Also, the AISC Manual lists J values for beams in its "Torsion Properties" tables. This shear distribution is called St. Venant torsion.

When warping shear stress and warping normal stress occur, their magnitudes may

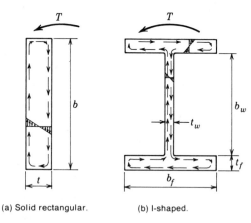

(a) Solid rectangular. (b) I-shaped.

Figure 5.47/Torsional shear stress (without warping).

be determined by solving the differential equation for nonuniform torsion,[14]

$$\frac{T}{EC_w} = \frac{1}{a^2}\frac{d\phi}{dx} - \frac{d^3\phi}{dx^3}$$

in which C_w is the *warping constant* of the cross section (a property listed in the "Torsion Properties" table of the AISC Manual), and $a^2 = (EC_w)/(GJ)$, where G is the modulus of elasticity in shear and J is the torsional constant previously defined. This differential equation relates derivatives of the angle of twist ϕ to the torque T applied to the beam and may be used for any beam having a thin-walled open cross section.

A more practical approach to the determination of warping stresses involves the use of charts from which the angle of twist and its derivatives can be determined for various combinations of torsional loadings and end restraints. Such charts, along with a complete explanation of their use, may be found in the *Steel Design File* by Heins and Seaburg, which was previously cited.

[14]For a complete derivation of this equation, see S. Timoshenko, "Theory of Bending, Torsion and Buckling of Thin-Walled Members of Open Cross Section," *Journal of the Franklin Institute,* Philadelphia, March-April-May, 1945.

PROBLEMS

1. Design a lintel composed of two angles that can carry an 8-in. masonry wall over an opening of 10 ft. The angles will be placed back to back and adequately secured together to act as a unit. The weight of the masonry is 140 lb per cu ft. Provide a 6-in. bearing at each end of the lintel. Conditions are such that the triangular loading of Fig. 5.24 may be assumed. There is adequate lateral support, and the steel is A36.

2. A built-up beam is composed of a **W** 10×22 with its web vertical and a **C** 8×11.5 with its web horizontal in the general arrangement indicated by Fig. 5.22e. The channel is welded to the top flange of the **W**-section. Determine the maximum uniform load this beam could carry on a laterally supported span of 20 ft, if A36 Steel is used. What will be the maximum deflection under such a load? (Answer given in Appendix G.)

3. Determine the total uniform load, including its own weight, that the symmetrical built-up section discussed in connection with Fig. 5.27 can support on a span of 28 ft. Assume A242 Steel and full lateral support.

4. Determine the total uniform load, including its own weight, that the unsymmetrical built-up section discussed in connection with Fig. 5.28 can carry on a span of 22 ft. Assume A36 Steel and no lateral support. What are the values of the fiber stress at top and bottom of the beam at the point of maximum bending moment under this load?

5. A channel girt of A242 Steel is to span between columns spaced 16 ft apart. The flanges of the channel will point downward. The horizontal wind load is 240 lb per ft, and the vertical load resulting from the siding and girt is 60 lb per ft. Assume that the vertical load acts at the center of the web. Design the girt using a maximum allowable bending stress of 30 ksi for tension and 15 ksi for compression.

6. A **W** 10×33 of A36 Steel spans 20 ft between two sloping girders that are inclined 12° to the horizontal. There is no lateral support. The uniform load of 550 lb per ft (including the weight of the beam) is vertical and is assumed to act at the center of the top flange. Deter-

mine the maximum critical extreme fiber stress. Is the beam safe? (Answer given in Appendix G.)

7. Design a steel bearing plate for an A36-Steel **S** 10×35 that transmits a reaction of 13 kips to a wall built of soft brick. The length of beam bearing on the wall is 8 in. (Answer given in Appendix G.)

8. A **W** 14×30 of A36 Steel transmits 34 kips to a cinder-block wall (grouted). Construction is such that only 6 in. of bearing is available, requiring the use of a grillage beam. Design the grillage beam of A36 Steel. (Answer given in Appendix G.)

9. In Fig. 5.39a, assume the longer dimension of the typical bay shown to be 24 ft, and the shorter, 18 ft. The beams span the 24-ft length. The total load to be supported on the floor (including live load, weight of the concrete floor construction, and an allowance for the weight of the beams and girders) is 150 psf. The total-load deflection is limited in all cases to $\frac{1}{240}$ of the span. The relation of concrete floor construction to the steel framing is as indicated in Fig. 5.3b. The slab is so reinforced that it transfers all its load to the supporting beams (none to the girders). Adjacent bays are identical with the one under consideration; therefore, corresponding beams will carry the same loads and have the same reactions. All steel connections are of the flexible type, creating simple reactions (nonrestraining). Use A36 Steel, and:

(a) Design the members for the arrangement shown, using the lightest-weight **W**-sections that will carry the loads.

(b) Redesign the members, placing the girders on the 24-ft span with the beams on the 18-ft span; girders are to receive the beams at third-points of the span, as previously.

(c) Compare the weight of steel required by each design, reducing figures to pounds of steel per square foot of floor area.

10. Design beam *A-B* in Fig. 5.40a (A36 Steel) for a total uniform load of 100 psf (including allowances for the weight of all steel members). Assume complete lateral support, and use the $F_y / 800$ depth-span criterion.

6

COLUMNS AND STRUTS

6.1
Introduction

The term *column* is usually applied to relatively heavy vertical members, while the lighter vertical and inclined members, such as braces and the compression members of roof trusses, are called struts. By definition, columns and struts are lineal compression members having a length substantially greater than their least lateral dimension.

The action of compression members can best be understood by beginning with a short compression block. The ultimate load that can be carried by such a block (Fig. 6.1a) is determined experimentally by increasing the applied load until failure by yielding occurs. The unit compressive stress in the block at the time of failure is the ultimate load divided by the area of the block, or $f_a = P/A$.

If a piece of the same material and cross section, but with a length substantially greater than its lateral dimensions, is subjected to a like test (Fig. 6.1b), it will fail before the applied load reaches the value which caused failure of the short block. In this case, failure is said to be due to buckling. (Buckling is a sudden yielding "knuckle-like" distortion due to compression.) Yielding would still be the mode of failure, even for this longer piece, if it were perfectly straight and of a homogeneous material, had no initial residual stresses, and sustained no loads which were not applied exactly along its longitudinal axis. But such conditions are theoretically ideal and not attainable in practice. Therefore, the distribution of stress over the cross section will not be uniform, and the resulting eccentricity, however slight, will cause a bending moment. This bending moment produces bending stresses,

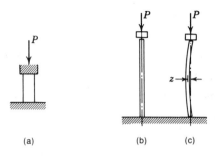

Figure 6.1/Compression.

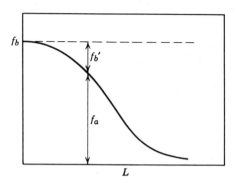

Figure 6.2/Buckling stresses.

which are referred to as buckling stresses simply to prevent confusion with bending stresses produced by eccentrically applied loads. That is, buckling and bending stresses are the same, except that the former, by definition, result from axial loads, and the latter from applied eccentric loads and/or residual stress. It should be emphasized that buckling stresses are in addition to direct compression stress from applied loads.

Returning to Fig. 6.1b, assume P to be a light load. As the load is increased, the column begins to buckle (Fig. 6.1c). If the load which first causes buckling is not increased, the column will continue to support the load, and the deflection z will remain constant. The column is, however, in a state of unstable equilibrium, and any slight increase in the load, or even the application of a horizontal force due to accidental jarring, may destroy this condition. The deflection z would then become greater, increasing the buckling stresses due to the moment Pz, which in turn would cause z to increase still more. The buckling stresses would continue to increase in this manner until the column failed.

In general, the buckling tendency of a column varies with the ratio of length to least lateral dimension. For tall, slender columns, this ratio is large; therefore, if failure occurs it will be due almost entirely to buckling. On the other hand, for short columns where this ratio is small, if failure

occurs it will be due principally to yielding. Between these extremes are the so-called intermediate columns with which failure, if it occurs, will be due to a combination of buckling and crushing. Most columns used in building construction are in the intermediate range.

Through Art. 6.14, discussion will be limited to columns having axial loads, i.e., loads applied parallel to the long axis of the column in such a manner as to be *concentric* with the centroid of the column cross section.

6.2
Buckling Stresses

The value of the actual maximum unit column stress is difficult to determine, but it is evident that the average unit stress, $f_a = P/A$, will be less than the crushing strength of the material by an amount dependent upon the tendency to buckle.

If these buckling stresses could be measured, they would be $f_b' = Mc/I = Pzc/I$. Figure 6.2 illustrates this conclusion for a given-size column. The ordinate f_b is the maximum permissible average stress and is assumed to remain constant[1] for any length

---

[1]This is an oversimplification which will be modified later. Actually, the AISC Specification requires a factor of safety which changes with the length, as discussed in Art. 6.5.

(*L*) of a given column. However, as the length increases, f_a must be decreased to allow for the presence of f_b'. The "reduction formulas" by which this is accomplished are based largely on the results of experimental test data rather than on a direct mathematical derivation. These will be treated in Art. 6.5.

6.3
Column Shapes

From the foregoing discussion it is evident that the strength of any given column (because of its tendency to buckle) will depend upon the area and shape of the cross section as well as the grade of steel from which it is made.[2]

A useful property of a section resulting entirely from its shape is the moment of inertia. Unbraced columns tend to buckle in a direction perpendicular to the axis about which the moment of inertia is least. Therefore, the ideal cross section is one having the same moment of inertia about any axis through its center of gravity. One example obviously is a circle.

As material near the center of gravity of a section contributes little to the moment of inertia, the most efficient column is one having as small an amount of material as possible placed near the axis. A hollow circular section (a pipe) approaches this ideal; however, pipe columns are used only to a limited extent in buildings and rarely as principal members in the frames of multistory structures. One difficulty encountered when using pipe columns is that of making effective beam connections.

A rolled **W**-shape column is shown in Fig. 6.3a, and the same section, reinforced with

[2]The grade of steel is not a factor if the column is very long and slender, placing it in the category of "elastic buckling." See Art. 6.5.

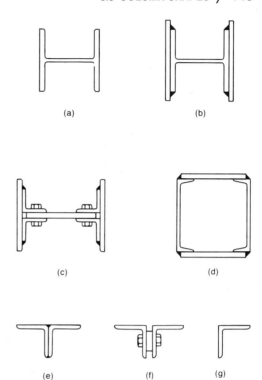

Figure 6.3/Typical column and strut sections.

flange plates, in (b). Columns built up of angles, plates, and channels are shown in (c) and (d), and they may be welded, bolted, or riveted. Angle sections used as struts are shown in (e), (f), and (g).

Exposed tubular shapes are increasing in popularity. Structural tubing (**TS**) is available in square, round, or rectangular shapes, and in a variety of wall thicknesses and overall dimensions. The AISC Manual includes data on the more frequently used square and rectangular shapes, and literature is available from producers if more complete listings are desired. Most of the available round steel pipe is of ASTM A501 Steel with $F_y = 36$ ksi and ASTM A53 Grade B Steel with $F_y = 35$ ksi, but may be designed at stresses allowed for $F_y = 36$ ksi. Most of the available rectangular steel tubing is of ASTM A500 grade B with $F_y = 46$ ksi.

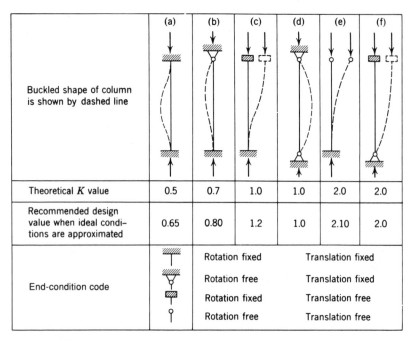

	(a)	(b)	(c)	(d)	(e)	(f)
Buckled shape of column is shown by dashed line						
Theoretical K value	0.5	0.7	1.0	1.0	2.0	2.0
Recommended design value when ideal conditions are approximated	0.65	0.80	1.2	1.0	2.10	2.0
End-condition code	Rotation fixed Translation fixed					
	Rotation free Translation fixed					
	Rotation fixed Translation free					
	Rotation free Translation free					

Figure 6.4/Effect of end conditions. Courtesy of American Institute of Steel Construction and Column Research Council; AISC *Manual of Steel Construction*, Ninth Edition, Allowable Stress Design, p. 5-135.

6.4
Radius of Gyration and Slenderness Ratio

In the design of beams (Art. 3.5), the section modulus was found to be an index of the strength of a member in bending. From its definition ($S = I/c$), it is evident that the value of the section modulus depends upon the size and shape of the section. In column design, the analogous quantity is radius of gyration. It too is dependent upon the size and shape of the section and is one measure of effectiveness in resisting buckling. The radius of gyration is expressed in inches by the formula $r = \sqrt{I/A}$, in which I is the moment of inertia of the section about the given axis and A is the area.[3] In Art. 6.1 it was stated that the tendency of a column to buckle varies in general with the ratio of

its unbraced length to its least lateral dimension. For structural shapes such as those shown in Fig. 6.3, the least lateral dimension is not an accurate criterion. Consequently, the radius of gyration, which relates more precisely to the stiffness of column sections in general, is used in column design formulas.

There are two other factors which affect the buckling tendency of columns—actual length and end conditions. These combine to produce what is termed "effective buckling length," i.e., the distance between points of contraflexure (between points where there is a change in curvature). For the pin-ended column (Fig. 6.4d), these points occur at the column ends where lateral movement (translation) is prevented. As may be seen from Fig. 6.4, the effective length may be quite different for various end conditions.

In design, end conditions are accounted for through use of an effective length fac-

[3]For a more extended treatment of radius of gyration, refer to any standard textbook on mechanics.

tor, i.e., a dimensionless number or *K value*, which when multiplied by the actual length of the column gives the effective length for buckling. The ratio of effective length to least radius of gyration (KL/r) is called the slenderness ratio. It is the absolute measurement of the column's tendency to buckle. *L* and *r* are both expressed in inches.

It is clearly stated in the 1989 AISC Specification that a great deal of engineering judgment must be used in selecting the appropriate *K* value. Figure 6.4 illustrates the six most common theoretical end conditions and the recommended *K* values for each. The difference between theoretical and recommended values is largely due to the fact that joint fixity or truly pinned ends are seldom realized in actual construction. Case (d), previously referred to, is the condition most frequently encountered.

Actually, the proper value of *K* is derived from an analysis of the degree of restraint imparted to the column end by other structural members framing into that end. The Commentary on the AISC Specification furnishes an *alignment chart* to aid in determining the proper *K* values. This alignment chart is reproduced and used in the Building Design Project, Chapter 13, Art. 13.24 and Fig. 13.30.

For a more thorough discussion of this topic, the reader is referred to Art. 13.24 or to the Commentary on the AISC Specification in Part 5 of the 1989 AISC Manual. A method for making additional refinements in *K* values once preliminary sizes of structural members have been established is also presented in the manual.

6.5
Column Formulas

From Art. 6.2, it is seen that the average unit stress in a column at the time of

failure is less than the yielding strength of the material by an amount dependent upon the buckling tendency of the column. It follows that the allowable average stress for use in design must also be influenced by this factor and will therefore depend upon the slenderness ratio as well as the compressive strength of the material. Using an average factor of safety of 1.67 and A36 Steel, the maximum permissible column design stress is

$$\frac{36,000}{1.67} = 21,600 \text{ psi}$$

It is this quantity that must be reduced to allow for buckling stresses.

Experimental tests of columns with high slenderness ratios (over 150) have produced rather consistent results. Such scatter of plotted data as does exist is attributed almost entirely to variations in end conditions. On the other hand, test data from columns with smaller slenderness ratios, i.e., in the intermediate range (50 to 130), show a wide range of scatter. Here, their erratic performance is attributed principally to the presence of residual stresses (produced when the column was made) and to the initial out-of-straightness of test columns. This scatter pattern persists until the slenderness ratio becomes so small that failure results from crushing rather than buckling, and any variation in test results may then be attributed entirely to material strength.

No matter how slight, these unpredictable conditions greatly affect column performance. Consequently, all columns are now classified in one of two categories: those that would fail due to *elastic buckling*, i.e., those having very large slenderness ratios, and those that would fail due to *inelastic buckling*.

Interestingly, test results continue to compare very well with the loads predicted by Leonhard Euler in 1757. The basic Euler formula[4] for the ultimate load of a long, slender, pinned-end column is

$$P_u = \frac{\pi^2 EI}{L^2}$$

It will be easier to establish the relationship between this formula and the 1989 AISC Specification, if both sides of the equality are divided by the column area A, i.e.,

$$\frac{P_u}{A} = \frac{\pi^2 EI}{L^2 A}$$

P_u / A is defined as the average unit stress at failure, or F_a. Substituting this and the relationship $I / A = r^2$ in the above formula,

$$F_a = \frac{\pi^2 E}{(L/r)^2}$$

The end-condition factor K has been omitted, since it was initially defined as unity (Fig. 6.4d).

It is reasonably estimated that the upper limit of elastic buckling failure will occur when the average column stress is equal to one-half the yield stress; therefore,

$$\frac{F_y}{2} = \frac{\pi^2 E}{(L/r)^2}$$

and, solving for L/r,

$$\frac{L}{r} = \sqrt{\frac{2\pi^2 E}{F_y}}$$

It is this formula which establishes the slenderness ratio that distinguishes elastic from inelastic buckling, in its dependence on the yield point of the steel. The symbol

[4] For derivation of this formula, refer to any standard textbook on mechanics.

for this value of L/r is C_c. Therefore,

$$C_c = \sqrt{\frac{2\pi^2 E}{F_y}}$$

For the various steels, this value is also given in the "Numerical Values" Table 4 of the AISC Specification (Part 5 of the AISC Manual). For example, the value for A36 Steel is 126.1, and for steels with a 50-ksi yield point, 107.0.

In conclusion, if the actual slenderness ratio $KL/r \geq C_c$, elastic buckling will be the assumed mode of failure. If, on the other hand, $KL/r < C_c$, inelastic buckling is to be assumed. The AISC Specification provides a special reduction formula for F_a in each case.

Inelastic Buckling / The maximum average unit stress should not exceed

$$F_a = \frac{\left[1 - \dfrac{(KL/r)^2}{2C_c^2}\right] F_y}{\text{F.S.}}$$

where F.S. is the factor of safety.

This factor of safety reflects the inconsistencies in column performance under tests previously mentioned. The recommended formula for the factor of safety derived from these considerations is

$$\text{F.S.} = \frac{5}{3} + \frac{3(KL/r)}{8C_c} - \frac{(KL/r)^3}{8C_c^3}$$

Note that when $KL/r = 0$, F.S. $= 1.67$, and when $KL/r = C_c$, F.S. $= \frac{23}{12} = 1.92$.

Elastic Buckling / The maximum average unit stress should not exceed that resulting from the Euler formula when a constant factor of safety of $\frac{23}{12}$ is used. Therefore,

$$F_a = \frac{\pi^2 E}{(KL/r)^2 \text{F.S.}} = \frac{12\pi^2 E}{23(KL/r)^2}$$

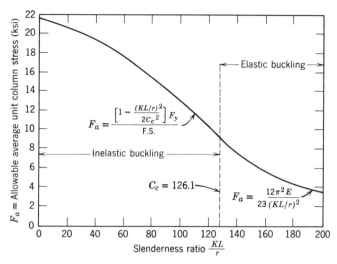

Figure 6.5/Allowable compression stress for ASTM A36 steel. Original source: Column Research Council.

It should be pointed out that this formula does not include a factor representing the stress grade of the steel. This is due to the fact that E does not vary appreciably with F_y, and E controls buckling. Consequently, the same limiting stress will be the result for all grades of steel.

The two limiting stress formulas for A36 Steel are plotted in Fig. 6.5. Note that the upper limit of the slenderness ratio is 200, which is the maximum permitted by the AISC Specification.

6.6
Investigation of Columns

To determine the safe axial load that a column can carry according to any particular specification, it is usually convenient to first compute the slenderness ratio and then compare it with C_c for the specified steel grade. In so doing, the column is placed in one of the two categories, i.e., inelastic buckling if $KL/r < C_c$, or elastic buckling if $KL/r \geq C_c$. The appropriate column formula is then selected, and the slenderness ratio is substituted in that for-

mula to determine the allowable average stress F_a. The safe load on the column will be equal to the allowable average stress times the area of the column section.

Example 1

Determine the total safe axial load on a W 12×72 column with an unbraced height of 14 ft. Column ends are assumed pinned and braced against translation. Use A36 Steel.

Solution

(1) From the AISC Manual "Dimensions and Properties" tables, the area and radii of gyration of this section are found to be

$$A = 21.1 \text{ in.}^2$$
$$r_x = 5.31 \text{ in.}$$
$$r_y = 3.04 \text{ in.}$$

(2) A 36 Steel has a yield point $F_y = 36,000$ psi; therefore,

$$C_c = \sqrt{\frac{2\pi^2 E}{F_y}} = \sqrt{\frac{2\pi^2 (29,000)}{36}} = 126.1$$

(3) The maximum slenderness ratio is

$$\frac{KL}{r} = \frac{1(14)12}{3.04} = 55.3$$

(4) Since the actual slenderness ratio is less than 126.1, the column is in the class that would fail due to inelastic buckling. From the AISC Specification, the factor of safety is

$$\text{F.S.} = \frac{5}{3} + \frac{3(KL/r)}{8(C_c)} - \frac{(KL/r)^3}{8(C_c)^3}$$

$$= \frac{5}{3} + \frac{3(55.3)}{8(126.1)} - \frac{(55.3)^3}{8(126.1)^3} = 1.82$$

and the average allowable stress is

$$F_a = \frac{\left[1 - \frac{(KL/r)^2}{2C_c^2}\right]F_y}{\text{F.S.}}$$

$$= \frac{\left[1 - \frac{(55.3)^2}{2(126.1)^2}\right]36,000}{1.82} \simeq 17,900 \text{ psi}$$

(5) The allowable axial load on the column, including its own weight, is equal to the allowable average stress times the area, or

$$P = F_a A = 17,900(21.1)$$
$$= 377,000 \text{ lb} = 377 \text{ kips}$$

Example 2

Determine the total safe axial load on a W 6×20 with an unbraced height of 18 ft. Column ends are assumed pinned and braced against translation. Use A36 Steel.

Solution

(1) From the AISC Manual "Dimensions and Properties" tables, the area and radii of gyration of the section are

$$A = 5.87 \text{ in.}^2$$
$$r_x = 2.66 \text{ in.}$$
$$r_y = 1.50 \text{ in.}$$

(2) A36 Steel has a yield-point stress $F_y = 36,000$ psi. Then,

$$C_c = \sqrt{\frac{2\pi^2 E}{F_y}} = \sqrt{\frac{2\pi^2(29,000)}{36}} = 126.1$$

(3) The actual slenderness ratio is

$$\frac{KL}{r} = \frac{1(18)12}{1.50} = 144$$

(4) Since the actual ratio is greater than 126.1, the column is in the class that would fail due to elastic buckling. From the AISC Specification, the average allowable stress is

$$F_a = \frac{12\pi^2 E}{23(KL/r)^2}$$

$$= \frac{12(3.14)^2 29,000}{23(144)^2}$$

$$= 7.2 \text{ ksi}$$

(5) The allowable axial load on the column, including its own weight, is

$$P = F_a A = 7.2(5.87)$$
$$= 42.3 \text{ kips}$$

Tables are available which will expedite column analysis. Based on the particular column formula and specified grade of steel, these tables provide the allowable stress for different values of the slenderness ratio. They may be found in Part 3 of the AISC Manual. In the Ninth Edition of the AISC Manual, there are two tables, one for A36 Steel and one for steels having an $F_y = 50$ ksi. The calculated values of F_a for each formula (elastic and inelastic cases) are tabulated for every value of KL/r from 1 to 200. With this table as a ready reference, F_a may be found without resorting to the calculations of steps (2) and (4) of Examples 1 and 2 above.

6.7
Built-Up Sections

In Example 1, Art. 6.6, the column used was a rolled **W** or **M** section such as that shown in Fig. 6.3a. Sections of this type are by far the most frequently used in steel building construction. However, at times it is necessary to reinforce ordinary rolled sections with plates as shown in Fig. 6.3(b) and (d), and it may even be necessary or desirable to fabricate a column using plates and angles or other standard shapes. The strength of built-up columns is investigated in the same manner as that outlined in Art. 6.6. Properties of the most widely used built-up sections, such as area and radii of gyration, may now be found in steel handbooks as readily as for rolled shapes. When such tables are not available, the moment of inertia of a built-up section must be determined in the same manner as for built-up beams (Art. 5.13). It is usually necessary to find the moment of inertia about both principal axes of the section and then to compute the least radius of gyration by substituting the least I

in the equation $r = \sqrt{I / A}$. With the least radius of gyration and the area determined, the methods of Art. 6.6 apply.

When designing built-up sections, care should be taken to conform to the limits of the width-thickness ratios set forth in Section B5 of the 1989 AISC Specification. Some of these limits are inversely proportional to the square root of the yield point stress F_y. Two categories are defined: unstiffened elements (projecting) and stiffened elements (supported along two edges).

Figure 6.6a illustrates three typical unstiffened elements and shows how to determine the dimensions b and t. Figure 6.6b illustrates three typical stiffened elements. The dimensions b, t_w, d, and h are determined as shown. The limits of these width-thickness ratios for these and other conditions are listed in Table B5.1 in the Ninth Edition of the AISC Manual.

6.8
Unbraced Height

In multistory steel-frame buildings, the unbraced height of a column is usually taken as the floor-to-floor distance. There are

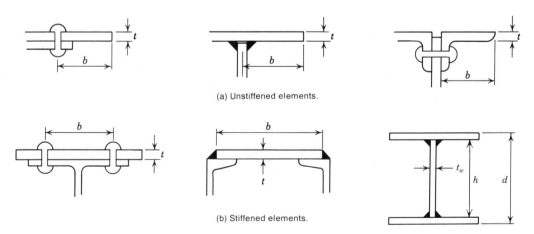

(a) Unstiffened elements.

(b) Stiffened elements.

Figure 6.6/AISC width-thickness ratios.

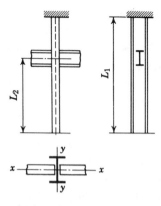

Figure 6.7/Unbraced heights.

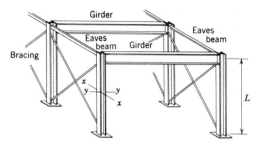

Figure 6.8/Frame: end conditions.

occasions, however, when there is need to brace a column against buckling at closer intervals in at least one direction. In Art. 6.4 it was stated that the slenderness ratio KL/r is determined by using the least radius of gyration. This is valid only when the column has equal unbraced heights for both axes, and end conditions are the same for both axes. When the unbraced height is different for each axis, the column buckling tendency is greatest for that axis having the largest slenderness ratio. In Fig. 6.7, the proper slenderness ratio for design is either KL_1/r_x or KL_2/r_y, whichever is greater. It is obvious that whenever intermediate supports are possible, it would be desirable to orient the column so that the axis having the smallest radius of gyration would be the one braced.

The effect of the end condition on the buckling tendency is further illustrated in Fig. 6.8. In this example, the girder is rigidly attached to the column, thus restricting end rotation about the x-x axis at the top of the column. However, due to the absence of bracing in this direction, end translation is not restricted. The eaves beam is attached to the column in such a manner as not to restrict column rotation about the y-y axis, but end translation is prevented by the bracing. The bottom of the column would be classified as a simple

pinned-end condition, restricted against translation in both directions. Therefore, the K value for the x-x axis is 2.0, and for the y-y axis, 1.0 (Art. 6.4 and Fig. 6.4).

There are additional procedures for calculating a more precise K value for a column. These procedures take into account the fact that the standard column base does not function as a true pin and that the top of the column is not fully fixed against rotation. Also, the K values and deflected shapes shown in Fig. 6.4 are for elastic conditions; should a column have a failure mode in the inelastic range, a further adjustment is permitted. The Commentary on the AISC Specification contained in the AISC Manual should be referred to; however, the building design project in Chapter 13 also includes these procedures.

To determine the maximum slenderness ratio for design, a calculation must be made for both axes, i.e.,

$$\frac{KL_x}{r_x}, \quad \text{where} \quad K = 2.0$$

and

$$\frac{KL_y}{r_y}, \quad \text{where} \quad K = 1.0$$

It is also possible to encounter conditions under which the three terms K, L, and r will all be different. The maximum value of KL/r always governs the tendency of a column to buckle and should be used to establish F_a.

Example 1

A column is built up from an **S** 6×12.5 with $\frac{1}{2} \times 6$-in. plates welded to each flange as shown in Fig. 6.9. The unbraced height for the *x-x* axis is 16 ft, and for the *y-y* axis, 8 ft. The *K* value for each axis is 1.0. Using A36 Steel, determine the maximum axial load permitted by the 1989 AISC Specification.

Solution

(1) Referring to the AISC Manual, the area, flange width, moments of inertia, and depth of the **S** 6×12.5 are

$$A = 3.67 \text{ in.}^2, \qquad I_x = 22.1 \text{ in.}^4$$

$$b_f = 3.332 \text{ in.}, \qquad I_y = 1.82 \text{ in.}^4$$

$$d = 6.00 \text{ in.}$$

(2) Check the width-thickness ratio as limited by Section B5 of the 1989 AISC Specification. (Only the cover plates need be considered, since all rolled **W** and **S** shapes meet the requirements of this section of

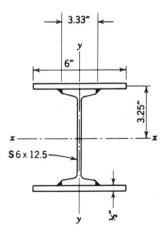

Figure 6.9/Example 1.

the Specification.) For projecting elements, from Fig. 6.6,

$$\frac{h}{t} = \frac{6.0 - (2)(0.359)}{0.5} = 10.6 < 70$$

Therefore $k_c = 1.0$ and

$$(b/t)_{max} = \frac{95}{\sqrt{F_y}} = \frac{95}{\sqrt{36}} = 15.8$$

$$(b/t)_{actual} = \frac{(6 - 3.33)/2}{0.5} = 2.67$$

For elements supported along two edges,

$$(b/t)_{max} = \frac{253}{\sqrt{F_y}} = \frac{253}{\sqrt{36}} = 42.1$$

$$(b/t)_{actual} = \frac{3.33}{0.5} = 6.66$$

For both conditions, the actual is less than the maximum permissible.

(3) Determine the total column area:

$$A = 3.67 + 2(0.5)6 = 9.67 \text{ in.}^2$$

(4) Calculate the moment of inertia for each principal axis:

$$I_x = 22.1 + 2\left[(0.05)6(3.25)^2\right]$$
$$= 85.5 \text{ in.}^4$$

$$I_y = 1.82 + 2\left[\frac{0.5(6)^3}{12}\right]$$

$$= 19.8 \text{ in.}^4$$

Note that $I_x = I_{(S\,6 \times 12.5)} + 2[I_{(plate)} + Az^2]$, but that $I_{(plate)}$ has been neglected (Art. 5.13); and that $I_y = I_{(S\,6 \times 12.5)} + 2[bd^3/12]_{(plate)}$.

(5) Calculate the radius of gyration for each axis:

$$r_x = \sqrt{\frac{85.5}{9.67}} = 2.97 \text{ in.}$$

$$r_y = \sqrt{\frac{19.8}{9.67}} = 1.43 \text{ in.}$$

(6) Determine the maximum slenderness ratio:

$$\frac{KL_x}{r_x} = \frac{1(16)12}{2.97} = 64.6$$

$$\frac{KL_y}{r_y} = \frac{1(8)12}{1.43} = 67.1$$

(7) Using the larger slenderness ratio, calculate the allowable average stress. (Either refer to an appropriate table or solve the applicable formula for this slenderness ratio.) Note that since $KL/r < C_c$, the formula for inelastic buckling governs, and

$$F.S. = \frac{5}{3} + \frac{3(67.1)}{8(126.1)} - \frac{(67.1)^3}{8(126.1)^3} = 1.85$$

$$F_a = \frac{\left[1 - \frac{(67.1)^2}{2(126.1)^2}\right]36,000}{1.85}$$

$$= 16,700 \text{ psi}$$

(8) Determine the maximum allowable axial load including the column weight, by multiplying the allowable stress by the total column area:

$$P = 16,700(9.67)$$

$$\approx 161,500 \text{ lb}$$

Example 2

A **W** 10×100 is used as a column with a height $L = 18$ ft. Using A242 high-strength, low-alloy steel, determine the maximum applied axial load permitted by the AISC Specification, $K_x = 2.0$ and $K_y = 1.0$.

Solution

(1) Referring to the AISC Manual, the area and radii of gyration are

$$A = 29.4 \text{ in.}^2, \qquad r_x = 4.60 \text{ in.}$$
$$r_y = 2.65 \text{ in.}$$

From Table 2, Part 1 of the AISC Manual,

the **W** 10×100 is in group 2, and from Table 1, $F_y = 50$ ksi.

(2) Determine the maximum slenderness ratio:

$$\frac{K_x L_x}{r_x} = \frac{2(18)12}{4.60} = 93.9$$

and

$$\frac{K_y L_y}{r_y} = \frac{1(18)12}{2.65} = 81.5$$

(3) Using the larger slenderness ratio, calculate the allowable average stress. This may be readily accomplished by reference to the tables of allowable compressive stress for steels of different yield points in Part 3 of the AISC Manual. Interpolating from the table, $F_a = 16,080$ psi.

(4) Determine the maximum allowable axial load:

$$P = 16,080(29.4)$$
$$= 473,000 \text{ lb}$$

(5) Subtract the weight of the column to determine the maximum allowable applied load:

$$P = 473,000 - 100(18) = 471,000 \text{ lb}$$

PROBLEMS

1. A **W** 8×31 of A36 Steel is used as a column with an unbraced height of 14 ft. $K = 1.0$. Determine the maximum axial load. (Answer given in Appendix G.)

2. A **W** 6×12 of A36 Steel is used as a column with an unbraced height of 12 ft. $K = 1.0$. Determine the maximum axial load.

3. A **W** 10×33 column of A242 Steel has an unbraced height of 16 ft for its major axis and 10 ft for its minor axis. Determine the maximum axial load if K for both axes is 1.0. (Answer given in Appendix G.)

4. A **W** 8×24 column of A36 Steel has an unbraced height of 14 ft. $K = 1.0$. It is desired that the maximum possible load be carried on this column and that intermediate bracing be used only for the minor axis. Determine the bracing interval for the minor axis and the maximum axial applied load.

5. A built-up channel and plate column of A36 Steel, similar to that shown in Fig. 6.3d, has an unbraced height of 13.5 ft. It is composed of two **C** 9×13.4 and two $\frac{1}{2} \times 9$-in. plates. The channels are so placed that their backs are 10 in. apart, and the overall size of the column is 10×10 in. $K = 1.0$. Determine the maximum axial load. (Answer given in Appendix G.)

6. A built-up plate and angle column of A36 Steel, similar to that shown in Fig. 6.3c, has an unbraced height of 20 ft for its major axis and 10 ft for its minor axis. It is composed of a $12 \times \frac{1}{2}$-in. web plate, four $6 \times 4 \times \frac{9}{16}$-in. angles, and two $13 \times \frac{5}{16}$-in.cover plates. Each pair of angles is set $\frac{1}{2}$ in. beyond the edge of the web plate with the 6-in. legs outstanding. K for both axis is 0.80. Determine the maximum axial load.

6.9
Column Design

The design of columns is an indirect process. Generally, the length of the column and the applied load it is to support are known, and the designer must select a rolled shape or built-up section in which the average (actual) stress does not exceed the allowable stress as given in the specifications. The two unknowns are A and r, and the value of one cannot be computed without knowing the other. The design, therefore, must be accomplished by trial and error. A trial section is selected and investigated by the method described in Arts. 6.6 and 6.7. If either the load-carrying capacity of the trial section is found to be less than that required, or it is so much greater than necessary that the section would be uneconomical, another trial section is selected using the first as a guide. This process is repeated until a satisfactory section is found. Steel handbooks contain tables of safe loads for different column sections and lengths. These tables are

a splended aid to design, but they should not be used until the underlying principles are understood. In Art. 6.14, safe-load tables are discussed and their use explained.

6.10
Summary: Column Design Procedure

The steps essential to the design of a column, when safe-load tables are not used, are as follows:

1. Assume a trial section. (The applied load, unbraced heights, and end conditions for each axis are known.)

2. Find the area and radii of gyration of the trial section from tables of properties for designing.

3. Compute the maximum slenderness ratio (KL/r).

4. Compute the allowable average stress by the appropriate formula as discussed in Art. 6.5.

5. Compute the allowable load on the column by multiplying the stress found in step 4 by the area of the section.

6. Compare the allowable load found in step 5 with the applied load plus the column weight. If the result is unsatisfactory, repeat the process until a satisfactory section is found.

6.11
Design and Investigation of Struts

Struts, including the lighter compression members of roof trusses, are designed in the same manner as columns. The more common strut cross sections are illustrated in Fig. 6.3(e), (f), and (g). In (e), two angles are shown in contact and connected by welds. In (f), the same angles are shown separated, the separation being accomplished by ring fillers and bolts or rivets. A single-angle strut is shown in (g).

Example 1

A 9-ft strut of A36 Steel is composed of two $5 \times 3\frac{1}{2} \times \frac{1}{2}$-in. angles with the 5-in. legs back to back. The angles are spaced $\frac{3}{8}$ in. apart as shown in Fig. 6.10; $K = 1.0$. Find the allowable load on the strut.

Solution

(1) From the AISC double-angle strut tables, for two unequal angles, long legs back to back,

$$A = 8.0 \text{ in.}^2$$

$$r_x = 1.58 \text{ in.}$$

$$r_y = 1.49 \text{ in.}$$

If tables are not available or do not include the particular combination of angles and spacing selected, r_x and r_y may be computed in a manner similar to that shown for built-up columns in Art. 6.7.

(2) The slenderness ratio is

$$\frac{KL}{r} = \frac{1(9)12}{1.49} = 72.5$$

(3) The allowable average stress is

$$F_a = \frac{\left[1 - \dfrac{(72.5)^2}{2(126.1)^2}\right]36,000}{\dfrac{5}{3} + \dfrac{3(72.5)}{8(126.1)} - \dfrac{(72.5)^3}{8(126.1)^3}}$$

$$= 16,100 \text{ psi}$$

(4) The allowable load on the strut is equal

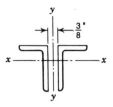

Figure 6.10/Example 1.

to the average stress times the area, or

$$P = F_a A = 16,100(8.0)$$

$$= 129,000 \text{ lb} = 129 \text{ kips}$$

It should be noted that when designing a single-angle strut, the least radius of gyration is about a diagonal axis, often referred to as the z-z axis (see properties tables for angles, Part 1, AISC Manual).

The design of struts composed of two angles is greatly facilitated by safe-load tables such as those discussed in Art. 6.14. Two-angle compression members will be discussed more fully in Chapter 9.

6.12
Loads on Columns

The design load for a given column in a building is generally made up of: (1) the live and dead load on the floor area immediately over the column being designed, or, in the case of a column supporting a roof, the dead, live, and environmental loads over the column; (2) the load transmitted by the column above, if one exists; and (3) the weight of the column itself. The portion of the column load contributed by the floor or roof that it supports is found by adding the reactions of the beams and girders or other members framing into it, or by multiplying the tributary area of the supported bay by the imposed per-square-foot live and dead loads. The design load for the column is taken as the total load at the base.

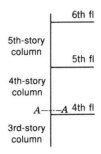

Figure 6.11/Fourth-story column.

Under certain conditions, and when the character of the occupancy is such that all floors in a building will not be subjected simultaneously to the full live load, most building codes permit a reduction.

The recommendations of the 1988 ASCE 7-88 standard "Minimum Design Loads for Building and Other Structures" (formerly ANSI A58.1) refer to any structural member—beam, girder, or column—having an influence area of 400 sq ft or more. Application of these code provisions to the design of floor beams and girders has been reserved for introduction in Chapter 13; application to the design of columns will be treated in the example below.

Columns are designated by the story through which they run. For example, a column between the fourth and fifth floors of a building, supporting the fifth floor, is known as a fourth-story column. The design load on a fourth-story column is the load at its base just above the fourth floor. This plane of load application is shown by the line *A-A* in Fig. 6.11.

Example 1

A steel-frame office building is laid out in bays 20 ft square. The floor live load is 70 psf, and the dead load, including an allowance for the weight of beams, girders, and columns, is 60 psf. The building com-

prises eight floors and a roof. The combined live and environmental load on the roof is 40 psf, and the dead load is 50 psf. Determine the design load on a typical interior fifth-story column (live-load reduction provisions of ASCE 7-88 will be used).

Solution

(1) Determine the total tributary area per floor supported by the typical column:

$$A = 20(20) = 400 \text{ sq ft}$$

The provision for live-load reduction applies.

(2) Calculate the total dead load per bay:

$$DL = 400(60) = 24,000 \text{ lb}$$

(3) Determine the reduced live load permitted for the floor area supported, and check to see if it exceeds the maximum allowable reduction. The influence area $A_I = 4(400) = 1600$ sq ft. Using Equation 1 from ASCE 7-88, the reduced live load is

$$L = L_0\left(0.25 + \frac{15}{\sqrt{A_I}}\right)$$

$$= 70\left(0.25 + \frac{15}{\sqrt{1600}}\right)$$

$$= 70(0.625)$$

$$= 43.75 \simeq 43.8 \text{ psf}$$

a 37.4 per cent reduction and thus less than the maximum 50 per cent permitted for members supporting more than one floor.

Since the fifth-story column supports three floors of 400 sq ft each, the total live load is

$$43.8(400)3 = 52,560 \simeq 52,600 \text{ lb.}$$

(4) Determine total load on the roof, with no live load reduction:

$$LL + EL = 400(40) = 16,000 \text{ lb}$$
$$DL = 400(50) = 20,000$$
$$W = \overline{36,000 \text{ lb}}$$

(5) The load P on a typical 5th-story column is as follows:

$$\text{Roof} = 36,000 \text{ lb}$$
$$\text{Live load} = 52,600$$
$$\text{Dead load} = 3(24,000) = 72,000$$
$$P = \overline{160,600 \text{ lb}}$$
$$\approx 161 \text{ kips}$$

6.13
Column Splices

Columns are usually spliced about 1 ft 6 in. above the floor level, the detail of the splice depending on the relative sizes of the members to be connected. In building construction, it is usual practice to mill the ends of columns, since most of the stress is transmitted by direct bearing. When the entire cross-sectional area of the upper column bears on the column below, the splice plates serve only to hold the columns in position or to transmit bending stresses (Fig. 6.12a, b, and d). Where full bearing cannot be assured, special splices, similar to those shown in Fig. 6.12(c) and (e), must be designed.

PROBLEMS

1. A column of A36 Steel has an unbraced height of 14 ft and must carry an axial load of 175 kips, including its own weight. Select the lightest-weight rolled column section that may

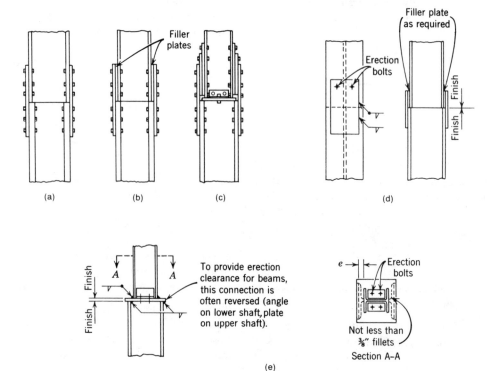

Figure 6.12/Typical column splices.

be used if K for both axes is 1.0. (Answer given in Appendix G.)

2. A column of A441 Steel has an unbraced height of 12 ft and must carry an axial load of 50 kips, including its own weight. Select the lightest-weight rolled column section that may be used if K for both axes is 1.0. (Answer given in Appendix G.)

3. A column of A36 Steel has an unbraced height of 12 ft. The applied axial load is 200 kips. Select the lightest weight 8-, 10-, and 12-in. rolled column sections that may be used. K for both axes is 1.0.

4. A double-angle strut (secondary member) of A36 Steel, 7 ft long, must sustain an axial load of 35 kips, including the strut weight. The angles are placed back to back with a $\frac{3}{8}$-in. separation between legs. Select the lightest-weight angles that may be safely used. Do not use angles of less than $\frac{1}{4}$-in. thickness. (Answer given in Appendix G.)

5. A column of A36 Steel has an overall height of 18 ft. There are bracing members framing perpendicularly into its web at 6-ft intervals. The applied axial load is 175 kips. Select the lightest-weight rolled column section that may be used. K for both axes is 1.0. (Answer given in Appendix G.)

6. A column of A588 Steel has a total height of 12 ft. A bracing member frames perpendicularly into its web, 3 ft 6 in. above the floor. The axial load, including the column weight, is 120 kips. Select the lightest-weight rolled section that may be used. K for the y-y axis is 1.0, and for the x-x axis, 1.2.

7. A column supports an axial load of 100 kips and has an unsupported height of 10 ft for both axes ($K_x = K_y = 1.0$).

(a) Select the lightest-weight structural steel pipe section (A501 Steel) that can safely be used.

(b) Redesign the column using structural tubing of A500, Grade B Steel. Use a square section having the same side dimension as the diameter of the pipe section selected in (a).

8. A steel frame building has ten floors and a roof. A typical interior bay is 24×24 ft. The live load on a typical floor is 80 psf, and the dead load, including an allowance for beams and girders, is 65 psf. Using the live-load reductions allowed by ASCE 7-88, determine the design load for a typical sixth-story column, neglecting the weight of the column. The combined live, environmental, and dead load on the roof is 110 psf. (Answers given in Appendix G.)

6.14 Safe-Load Tables— Columns and Struts

The AISC Manual provides tables which give the safe total axial load that columns and struts can carry for various effective unbraced heights. Such tables are very helpful to the designer. Once the total axial load to be carried has been determined and the effective unbraced length established, the column or strut is selected directly from the tables. Although these tables may be arranged differently in the various handbooks, a short period of study will generally reveal the method of selection.

Column safe-load tables of the type presented in the AISC Manual are most frequently encountered in practice, and include **W**, **M**, and **S** shapes and double angles, of 36- and 50-ksi yield-point grade steels, structural tubing of 46-ksi yield-point grade steel, and steel pipe of 36-ksi grade only. Table 6.1, extracted from the Ninth Edition of the AISC Manual, illustrates the make-up of a typical column safe-load table. It will be noted that this particular table is for 10-in. **W** shapes of A36 and F_y = 50-ksi steels, and that allowable axial (concentric) loads are given in kips.

The first vertical column of numbers refers to the effective unbraced length (height) with respect to the least radius of gyration. If the column is braced so that buckling will occur about the major axis, this table is not directly applicable.

Table 6.1
Typical Column Safe Load Table[†]

| F_y = 36 ksi |
| F_y = 50 ksi |

COLUMNS
W shapes
Allowable axial loads in kips

Designation	W10											
Wt./ft	60		54		49		45		39		33	
F_y	36	50	36	50	36	50	36	50	36	50	36	50
0	380	528	341	474	311	432	287	399	248	345	210	291
6	353	482	317	433	289	394	260	351	224	303	189	255
7	348	472	312	423	284	385	253	340	218	293	184	246
8	341	461	306	414	279	376	247	328	213	283	179	237
9	335	450	300	403	273	367	240	316	206	272	173	228
10	328	437	294	392	268	357	232	303	200	260	167	217
11	321	425	288	381	262	346	224	289	193	248	161	207
12	313	412	281	369	256	335	216	274	186	235	155	196
13	306	398	274	356	249	324	208	259	178	221	149	184
14	297	383	267	343	242	312	199	243	170	207	142	171
15	289	368	259	330	235	299	190	227	162	193	135	159
16	280	353	251	316	228	286	180	209	154	177	127	145
17	271	337	243	301	221	273	170	191	145	161	120	131
18	262	320	235	286	213	259	160	172	136	144	112	117
19	253	303	226	271	205	245	149	154	126	130	103	105
20	243	285	217	255	197	230	138	139	116	117	95	95
22	222	248	199	221	180	198	115	115	97	97	78	78
24	201	209	179	186	161	167	97	97	81	81	66	66
26	177	178	158	159	142	143	82	82	69	69	56	56
28	154	154	137	137	123	123	71	71	60	60	48	48
30	134	134	119	119	107	107	62	62	52	52	42	42
32	118	118	105	105	94	94	54	54	46	46	37	37
33	111	111	99	99	88	88	51	51	43	43		
34	104	104	93	93	83	83						
36	93	93	83	83	74	74						

Effective length in ft KL with respect to least radius of gyration r_y

Properties												
U	2.55	2.55	2.56	2.56	2.57	2.57	3.25	3.25	3.28	3.28	3.35	3.35
P_{wo} (kips)	99	138	83	116	73	101	79	109	64	89	55	77
P_{wl} (kips/in.)	15	21	13	19	12	17	13	18	11	16	10	15
P_{wb} (kips)	239	282	163	193	127	149	138	163	101	119	79	93
P_{fb} (kips)	104	145	85	118	71	98	86	120	63	88	43	59
L_c (ft)	10.6	9.0	10.6	9.0	10.6	9.0	8.5	7.2	8.4	7.2	8.4	7.1
L_u (ft)	31.1	22.4	28.2	20.3	26.0	18.7	22.8	16.4	19.8	14.2	16.5	11.9

Property	W10×60	W10×54	W10×49	W10×45	W10×39	W10×33
A (in.^{2})	17.6	15.8	14.4	13.3	11.5	9.71
I_x (in.^{4})	341	303	272	248	209	170
I_y (in.^{4})	116	103	93.4	53.4	45.0	36.6
r_y (in.)	2.57	2.56	2.54	2.01	1.98	1.94
Ratio r_x/r_y	1.71	1.71	1.71	2.15	2.16	2.16
B_x } Bending	0.264	0.263	0.264	0.271	0.273	0.277
B_y } factors	0.765	0.767	0.770	1.000	1.018	1.055
$a_x/10^6$	50.5	45.0	40.6	37.2	31.2	25.4
$a_y/10^6$	17.3	15.4	13.8	8.0	6.7	5.4
$F'_{ex}\,(K_xL_x)^2/10^2$ (kips)	200	198	196	194	189	182
$F'_{ey}\,(K_yL_y)^2/10^2$ (kips)	68.5	68.0	66.9	41.9	40.7	39.0

Note: Heavy line indicates Kl/r of 200.

[†]Courtesy of American Institute of Steel Construction: Manual of Steel Construction. Allowable Stress Design. Ninth Edition. p. 3–30.

Load listings in the table are terminated when the L/r for a given column exceeds 200. Properties are listed at the bottom of the table for ready reference. These properties are helpful in indirect ways. For example, notice the ratio r_x/r_y, which is used when the effective buckling length is different for each axis. When the buckling tendency is the same for both axes, the following expression is valid:

$$\frac{L_x}{r_x} = \frac{L_y}{r_y}$$

Solving for L_y,

$$L_y = \frac{L_x}{r_x/r_y}$$

L_y is the hypothetical length with respect to the minor axis, controlled by buckling about the major axis. Compare $L_y/(r_x/r_y)$ with the actual buckling length for the y-y axis and use the larger of the two when selecting loads from the table.

Other listed properties will be explained in subsequent articles dealing with eccentric loads on columns.

Example 1

Select a **W** column of A36 Steel to support an applied axial load of 130 kips if it has an effective unbraced height of 18 ft.

Solution

Knowing that the effective unbraced height (length) is 18 ft, enter Table 6.1 and select the lightest-weight section capable of supporting 130 kips. Select the **W** 10×39.

Example 2

Select a **W** column of A36 Steel to support an applied axial load of 195 kips if the effective unbraced height for the x-x axis is 24 ft, and for the y-y axis, 10 ft.

Solution

(1) Calculate the ratio of the effective unbraced heights:

$$\frac{KL_x}{KL_y} = \frac{24}{10} = 2.4$$

(2) Entering Table 6.1 with the unbraced height for the y-y axis (10 ft), a **W** 10×39 is tentatively selected because it will support 200 kips if the y-y axis is critical.

(3) Observe the ratio r_x/r_y for the trial shape selected. As the value shown for the **W** 10×39 is 2.16 (which is less than 2.4), the major axis is critical. Since the tabulated load is for a critical y-y axis, further investigation is necessary.

(4) Adjust the unbraced height for the x-x axis so that the corresponding tabulated loads will apply:

$$\frac{24}{2.16} = 11.1 \text{ ft}$$

In other words, an unbraced height of 11.1 ft for the y-y axis would result in the same KL/r ratio as that for the unbraced height of 24 ft for the x-x axis. Since the **W** 10×39 will carry only 193 kips for an unbraced length of 11 ft, it is not safe and a new trial section must be selected

(5) The next heavier section is a **W** 10×45. The ratio r_x/r_y is 2.15. Again, the major axis is critical; therefore,

$$\frac{24}{2.15} = 11.2 \text{ ft}$$

At 11 ft, the **W** 10×45 will support 224 kips, which is more than 195 kips plus the 1.08-kip weight of the column. The **W** 10×45, therefore, is safe.

The safe-load tables for struts included in the AISC Manual are for two angles, back

Table 6.2
Typical Strut Safe Load Table*

F_y = 36 ksi / F_y = 50 ksi	COLUMNS — Double angles — Allowable concentric loads in kips — Unequal legs — Long legs ⅜ in. back-to-back of angles

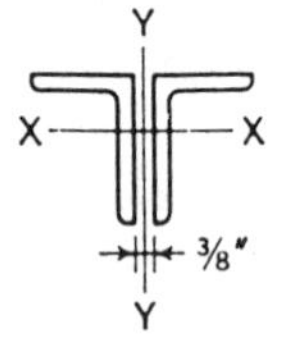

3 x 2½ — Effective length in ft KL with respect to indicated axis

X-X AXIS

Size	3 x 2½							3 x 2							
Thickness	⅜		¼		³⁄₁₆			⅜		⁵⁄₁₆		¼		³⁄₁₆	
Wt./ft	13.2		9.0		6.77			11.8		10.0		8.2		6.1	
F_y	36	50	36	50	36	50	KL	36	50	36	50	36	50	36	50
0	83	115	57	76	39	50	0	75	104	63	88	51	69	35	45
2	78	106	53	70	37	46	2	70	96	59	81	48	64	33	42
3	74	100	51	66	35	44	3	67	90	57	77	46	60	32	40
4	70	92	48	62	34	41	4	63	84	54	71	44	56	30	38
5	65	84	45	56	32	38	5	59	77	50	65	41	51	29	35
6	60	75	41	51	29	35	6	55	68	46	58	38	46	27	32
7	54	65	38	44	27	31	7	50	59	42	51	34	41	25	29
8	48	53	34	37	24	27	8	44	49	38	42	31	35	22	25
9	41	42	29	30	22	23	9	38	39	33	34	27	28	20	21
10	34	34	24	24	19	19	10	32	32	27	27	23	23	17	17
11	28	28	20	20	16	16	11	26	26	23	23	19	19	14	14
12	24	24	17	17	13	13	12	22	22	19	19	16	16	12	12
13	20	20	14	14	11	11	13	19	19	16	16	13	13	10	10
14	17	17	12	12	10	10	14	16	16	14	14	12	12	9	9
15	15	15	11	11	8	8	15	14	14	12	12	10	10	8	8
							16							7	7

Y-Y AXIS

KL	3 x 2½ ⅜ 36	50	¼ 36	50	³⁄₁₆ 36	50	KL	3 x 2 ⅜ 36	50	⁵⁄₁₆ 36	50	¼ 36	50	³⁄₁₆ 36	50
0	83	115	57	76	39	50	0	75	104	63	88	51	69	35	45
2	74	100	47	59	30	36	2	67	90	55	73	43	54	27	33
3	73	98	46	59	29	35	3	65	86	53	71	42	53	27	32
4	71	95	45	57	29	35	4	61	81	51	66	40	50	26	31
5	68	90	44	55	29	34	5	57	73	48	60	38	46	25	29
6	65	84	42	52	28	33	6	53	65	44	53	35	41	23	27
7	61	77	40	48	27	31	7	48	56	39	46	31	35	21	23
8	57	69	37	43	25	28	8	42	46	35	37	27	28	19	19
9	52	61	34	38	23	26	9	36	36	29	29	23	23	16	16
10	47	52	31	32	21	22	10	30	30	24	24	19	19	13	13
11	42	43	27	27	19	19	11	25	25	20	20	16	16	11	11
12	36	36	23	23	16	16	12	21	21	17	17	13	13	9	9
13	31	31	20	20	14	14	13	18	18	14	14	11	11	8	8
14	27	27	17	17	12	12	14	15	15	12	12	10	10	7	7
15	24	24	15	15	11	11	15	13	13	11	11				
16	21	21	13	13	10	10									
17	18	18	12	12	9	9									
18	16	16	11	11	8	8									
19	15	15													

Properties of 2 angles — ⅜ in. back-to-back

	⅜	¼	³⁄₁₆		⅜	⁵⁄₁₆	¼	³⁄₁₆
A (in.²)	3.84	2.63	1.99		3.47	2.93	2.38	1.80
r_x (in.)	0.928	0.945	0.954		0.940	0.948	0.957	0.966
r_y (in.)	1.16	1.13	1.12		0.917	0.903	0.891	0.879

Heavy line indicates Kl/r of 200.

†Courtesy of American Institute of Steel Construction: Manual of Steel Construction. Allowable Stress Design, Ninth Edition. p. 3–30.

to back, with a $\frac{3}{8}$-in. separation. Three series of tables are provided, one each for equal-leg angles, unequal-legs angles with short legs back to back, and unequal-leg angles with long legs back to back.

Table 6.2, taken from the AISC Manual, illustrates a typical strut safe-load table. It will be noted that this particular table is for two sizes of unequal-leg angles with long legs back to back, having a $\frac{3}{8}$-in. separation. Safe loads are tabulated for a different range of lengths in each case, and the unsupported length is shown for both axes. Properties listed at the bottom of the table are for the combination of the two angles.

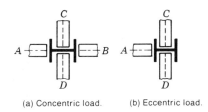

(a) Concentric load.   (b) Eccentric load.

Figure 6.13/Column loads.

Example 3

Design a double-angle strut of A36 Steel to carry a concentric load of 44 kips, including the strut weight. The unsupported length on the x-x axis is 4 ft, and on the y-y axis, 8 ft.

Solution

Referring to Table 6.2, two $3 \times 2\frac{1}{2} \times \frac{3}{8}$-in. angles with long legs back to back and separated by $\frac{3}{8}$ in. will support 70 kips as controlled by the x-x axis (4 ft), and 57 kips as controlled by the y-y axis (8 ft).

In order to determine if this is the lightest-weight angle combination that can be used, the more complete set of tables in the AISC Manual should be studied.

6.15
Columns with Eccentric Loads

All column loads thus far have been assumed to be concentric, i.e., applied along the axis of the column. This assumption is valid when the load is applied uniformly over the top of the column, or when beams

having equal reactions frame into the column opposite each other, as would be the case in Fig. 6.13a if the reactions of beams A and B were the same, and those of beams C and D were the same. If, however, beam B is omitted as shown in Fig. 6.13b, or if the reaction of B is considerably less than that of A in Fig. 6.13a, it is evident that the loads on the column no longer will be symmetrical and that the left column flange will be subjected to a greater unit stress than the right. This eccentric loading condition occurs frequently in wall columns of buildings, where a floor beam is supported on the interior face without a corresponding load on the exterior face. (Figure 6.14 simply illustrates one method of framing which may be used to lessen this eccentricity, or even to balance the loads if the total reaction of the two spandrel beams is nearly the same as that of the floor beam.)

In order to develop an expression that will account for the variation in stress over the column cross section due to the eccentric condition, consider once more the short compression block, this time with a load P eccentrically applied (Fig. 6.15). The distance e is the eccentricity, and c is the

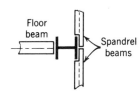

Floor beam — Spandrel beams

Figure 6.14/Exterior column load.

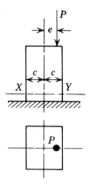

Figure 6.15/Short compression block.

distance from the axis of the block to the extreme fibers. The stress in any fiber, on any cross section of the block, such as X-Y, may be considered to be the sum of the average stress P/A and a stress caused by the moment Pe. To the right of the axis of the block, i.e., on the same side as P, this moment causes a compressive stress on the section, and to the left of the axis, a tensile stress. The unit stress at Y is equal to the average stress P/A, plus the extreme fiber stress Mc/I caused by the moment Pe. Substituting Pe for M, the intensity of stress at Y is expressed by the formula[5]

$$f_y = \frac{P}{A} + \frac{Pec}{I}$$

and the intensity of stress at X is

$$f_x = \frac{P}{A} - \frac{Pec}{I}$$

The above expressions are applicable to sections symmetrical about two axes such as rectangles, **W** sections, and **S** sections.

[5]The formulas developed in this article neglect stresses resulting from lateral deflection of the column, whether such deflection is caused by induced end-of-the-column bending moments or by the tendency to buckle under axial load. These latter effects are accounted for in Art. 6.17, where design procedures are developed for eccentrically loaded columns.

They may be stated in the general form

$$f = \frac{P}{A} \pm \frac{Pec}{I}$$

in which f is the unit stress at either edge of the section, depending on whether the plus or minus sign is used, and I is the moment of inertia in the direction of the eccentricity.

In the investigation of eccentrically loaded columns, maximum compression is usually the most critical, because seldom will the tensile stress on the far edge of the column due to the moment Pe be sufficient to counteract the direct compressive stress P/A. Where this does occur, it is of importance only when the column is to be spliced.

It is generally true in buildings that columns carry a direct axial load in addition to any eccentric loads that may exist. Where such is the case, a more convenient form of the expression is

$$f = \frac{P}{A} + \frac{P'ec}{I}$$

in which P is the total vertical load *including the eccentric load*, and P' is the eccentric load alone. The investigation of a column carrying an eccentric load is illustrated by the example which follows.

Example

A **W** 10×68 column, 13 ft long, supports the loads shown in Fig. 6.16. The point of application of the 50,000-lb load is assumed to be 8 in. from the axis of the column. Neglect the weight of the column. Assuming the effect of buckling and lateral deflection to be negligible, determine the maximum unit compressive stress.

Solution

(1) The total vertical load $P = 134,000 + 20,000 + 20,000 + 50,000 = 224,000$ lb. The eccentric load $P' = 50,000$ lb.

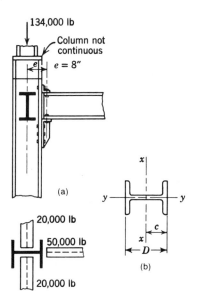

Figure 6.16/Example.

(2) The following properties of the **W** 10×68 section are found from the AISC Manual:

$$I_x = 394 \text{ in.}^4, \qquad r_x = 4.44 \text{ in.}$$
$$I_y = 134 \text{ in.}^4, \qquad r_y = 2.59 \text{ in.}$$
$$A = 20.0 \text{ in.}^2, \qquad d = 10.40 \text{ in.}$$
$$c = 5.20 \text{ in.}$$

(3) The maximum value of the unit compressive stress is

$$f = \frac{P}{A} + \frac{P'ec}{I}$$
$$= \frac{224,000}{20.0} + \frac{50,000(8)5.20}{394}$$
$$= 11,200 + 5280 = 16,480 \text{ psi}$$

This investigation will not, of course, reveal whether the column is safe unless there is a known allowable stress for this condition of loading, with which the 16,480 psi can be compared. This is discussed in Art. 6.17, where the AISC method of designing columns for combined axial and eccentric loadings is presented.

6.16
Induced Moments In Columns

Advances in arc-welding theory and methods have led to the practice of restraining beam ends against rotation where they are attached to columns. When that is done, the beam not only delivers an end reaction to the column, but induces a bending moment as well. In cases where the column is continuous above and below the floor in question, an end moment is induced in the column above as well as in the column below the beam concerned. The moment at the joint, caused by the restrained end of the beam, is resisted by all members rigidly framing into the joint, each receiving a portion of the bending moment in an amount proportional to its stiffness. This topic is covered in more detail in Chapter 10, "Continuous Beams and Frames," and is introduced here only to demonstrate how end-of-the-column bending moments are generated.

Figure 6.17 illustrates a design condition having a restrained-end beam and a discontinuous column above (i.e., none of the beam end moment is induced in the column above). The fiber stress in the column below the beam at any section *A-A* may

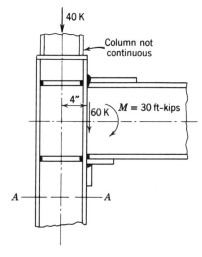

Figure 6.17/Induced moments.

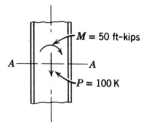

Figure 6.18/Combined condition.

be expressed as follows:[6]

$$f = \frac{P}{A} \pm \frac{P'ec}{I} \pm \frac{Mc}{I}$$

For the particular detail shown in Fig. 6.17, P is $40 + 60 = 100$ kips; P' is 60 kips; e is 4 in.; and M is 30 ft-kips.

This design condition may also be expressed as shown in Fig. 6.18, where the moment due to the eccentric load P' and the induced end-of-the-column moment have been combined. That is,

$$M = \frac{60(4)}{12} + 30 = 50 \text{ ft-kips}$$

and

$$f = \frac{P}{A} \pm \frac{Mc}{I}$$

where P again represents the *total* vertical load, as above, and M, the combined moments.

Since both expressions represent the same situation, the one chosen is simply a matter of preference on the part of the designer.

6.17
Design of Eccentrically Loaded Columns

As previously stated, the investigation of eccentrically loaded columns by the method shown in the example of Art. 6.15

[6]See footnote 5, Art. 6.15.

requires an allowable combined bending and axial stress for comparison with the actual stress computed. Some early specifications stipulate that this combined allowable stress be limited to that which would be permitted if only axial stress existed. This would appear to be too conservative, particularly in cases where bending stress is produced by gravity loading of beams framing to a column with a restraining connection similar to that shown in Fig. 6.17. The design procedure given here is based on separate values for allowable bending stress and allowable axial stress, and is sufficiently involved to require extended discussion of each related part.

The principal goal is to isolate actual bending stress and actual direct stress and to identify each with independently controlled allowable bending and direct stresses (F_a and F_b). The allowable stresses vary widely. The design approach is one of adjusting percentages, limiting the sum of axial-stress percentage (f_a / F_a) and bending-stress percentage (f_b / F_b) to 100 percent. The task is further complicated by the fact that actual bending stresses (f_b) are increased when an already bent column (having a bending moment) is subjected to an axial load. An amplification factor adjusts the bending-stress percentage for this condition. Finally, the bending stress (f_b) must be further adjusted to account for end conditions; a reduction factor (C_m) is used for this purpose.

The design procedure is begun by assuming a trial section and calculating the percentage of the allowable axial stress resulting from the loads, i.e. f_a / F_a. Next, the percentage of the allowable bending stress (f_b / F_b) is calculated. All values are established according to previously discussed procedures. The percentage of bending stress is further adjusted to allow for increased bending stress due to initial bend and for end conditions. This is accomplished by multiplying the bending-stress percentage by the amplification factor and the reduction factor. Then, the percentages are added together, and if the resultant value is less than 100 per cent, the design is considered safe.

A detailed explanation and derivation of applicable formulas needed for design is given in the paragraphs which follow. The nomenclature is as follows:

F_a = allowable axial unit stress.

F_b = allowable bending unit stress.[7]

f_a = actual axial unit stress developed.

f_b = actual bending unit stress developed.

The other symbols conform to those used in Art. 6.15 and 6.16. It should be noted from the definition of radius of gyration (Art. 6.4) that moment of inertia (I) may be expressed as Ar^2. Then, when investigating an eccentrically loaded column,

$$f_a = \frac{P}{A}$$

and

$$f_b = \frac{P'ec}{I} = \frac{P'ec}{Ar^2}$$

In the design, if a column is subjected to axial loading only, the required cross-sectional area (A_a) will be

$$A_a = \frac{P}{F_a}$$

If a column is subjected to bending action only, the required cross-sectional area (A_b) will be

$$A_b = \frac{P'ec}{F_b r^2}$$

If a column is subjected to both axial loading and bending action, the total required area becomes

$$A_a + A_b = A = \frac{P}{F_a} + \frac{P'ec}{F_b r^2}$$

---

[7]F_b is established as described in Arts. 5.6 and 5.7, and is entirely independent of the action and specifications related to the axial load.

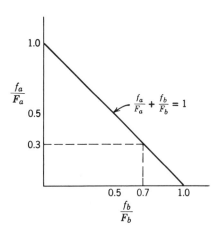

Figure 6.19/Fundamental interaction formula.

Dividing both sides of the equation by A,

$$1 = \frac{P}{F_a A} + \frac{P'ec}{F_b Ar^2}$$

This may be written as

$$\frac{P/A}{F_a} + \frac{P'ec/Ar^2}{F_b} = 1$$

or, by substitution,

$$\frac{f_a}{F_a} + \frac{f_b}{F_b} = 1$$

This is known as the fundamental interaction formula. It is a straight line when plotted graphically as shown in Fig. 6.19. One simple example of its use is as follows: when the ratio f_b/F_b is 0.7, the ratio f_a/F_a cannot exceed 0.3.

The fundamental interaction formula is made to account for the additional stresses caused by column buckling under an axial load, by reducing the value of F_a as the largest KL/r increases. However, the formula does *not* account for additional stresses from the lateral deflection caused by induced end moments.

End-Moment Stresses / This phenomenon can be described by referring to the sequential column loading shown in Fig. 6.20. The first, Fig. 6.20a, illustrates a

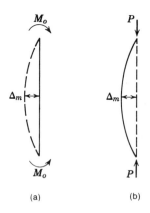

(a)   (b)

Figure 6.20/Effect of end moment.

vertical member having equal end moments M_o, deflecting to a shape shown exaggeratedly by the broken line. The maximum lateral deflection is Δ_m, and the moment at midheight is M_o. When the axial load P is applied to the already deflected shape (Fig. 6.20b), there will be an additional moment at midheight equal to $P\Delta_m$. This, in turn, causes more lateral deflection, causing more moment, and so on. Consequently, the final bending stress at midheight of the column will be the sum of the stresses caused by each action, or

$$f_b = \frac{M_o c}{I} + \frac{\Delta_m P c}{I}$$

The additional stresses caused by $P\Delta_m$ are very difficult to ascertain, often requiring the complex mathematical processes known as numerical integration. Such procedures and accompanying formulas are unrealistic for routine design application. However, some useful conclusions can be abstracted from the above-described structural action. It is seen that for a constant end condition such as that shown in Fig. 6.20 (equal end moments), the lateral deflection Δ_m will depend upon the slenderness ratio of the column with respect to the direction of bending. A large slenderness ratio permits a large lateral deflection. The corresponding bending stresses

from the deflection will increase with increasing values of the axial load P.

In order to simplify the design procedure, the AISC Specification recommends a method based upon the application of the interaction formula modified as necessary to agree with experimental test data.

The curves shown in Fig. 6.21 are typical of such test data. Columns having equal end moments (M_o) were tested to determine what additional axial load P could be applied before failure would occur. This was repeated for columns having different slenderness ratios, such ratios being determined with respect to the direction of the applied moment. These values, of varying combinations of P and M_o, were made dimensionless by dividing them by P_y and M_y, respectively. P_y is the axial load that would cause yielding if it alone occurred. M_y is the bending moment that would cause yielding if it occurred in the absence of an axial load. The similarity of the axial load ratios P/P_y and the axial stress ratios f_a/F_a should be apparent. The same similarity exists between the moment ratios M_o/M_y and the bending-stress ratios

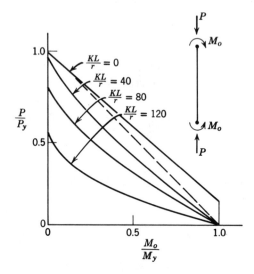

Figure 6.21/Test-data modification of interaction formula.

f_b/F_b. The straight-line interaction formula is shown by the broken line superimposed over these test-data curves in Fig. 6.21.

In all cases, except for columns having very small values of KL/r, the experimental data curves fall on the unsafe side of the straight-line interaction curve. The curves become increasingly more unsafe as KL/r increases.

The formula that produces curves closely approximating those shown in Fig. 6.21 is

$$\frac{P}{P_y} + \left[\frac{1}{1-\dfrac{P}{P_e}}\right]\frac{M_o}{M_y} = 1.0$$

where P_e is the elastic buckling load of the column (Art. 6.5), when buckling occurs about the same axis as the bending and is equal to

$$\frac{\pi^2 EI}{L^2} = F'_e A$$

The remaining terms in the formula may be interpreted as follows:[8]

$$P = f_a A, \qquad M_o = \frac{f_b I}{c}$$

$$P_y = F_a A, \qquad M_y = \frac{F_b I}{c}$$

Substituting these values into the basic curve formula

$$\frac{f_a A}{F_a A} + \left[\frac{1}{1-\dfrac{f_a A}{F'_e A}}\right]\frac{f_b\left(\dfrac{I}{c}\right)}{F_b\left(\dfrac{I}{c}\right)} = 1.0$$

which readily reduces to

$$\frac{f_a}{F_a} + \left[\frac{1}{1-\dfrac{f_a}{F'_e}}\right]\frac{f_b}{F_b} = 1.0$$

The term within the bracket is referred to as the *amplification factor*. F'_e is defined as the limiting Euler stress (divided by a factor of safety) and can be calculated as follows:[9]

$$F'_e = \frac{12\pi^2 E}{23(K_b L_b/r_b)^2}$$

where the slenderness ratio is measured with respect to the axis about which bending takes place, and F'_e is expressed in psi or ksi.

When F'_e is very large and/or the axial stress f_a is very small, the amplification factor is negligible. When significant, the amplification factor increases the ratio f_b/F_b and accounts for the additional bending stress illustrated in Fig. 6.20.

Reduction Factor / The basic column curves shown in Fig. 6.21 and the above-described amplification factor were based upon equal column end moments causing a single-curve deflection. These end conditions are the most severe. Any other combination of end conditions would permit a further refinement in the form of a reduction factor C_m, which should be multiplied by the amplified bending-stress ratio. Thus, the adjusted interaction formula specified by the 1989 AISC Specification is

$$\frac{f_a}{F_a} + \left[\frac{C_m}{1-\dfrac{f_a}{F'_e}}\right]\frac{f_b}{F_b} \leq 1.0$$

and must be satisfied for all points on a column between lateral supports when $f_a/F_a \geq 0.15$.

The reduction factor C_m needs further explanation. Its value is established by relative size and direction of the column end moments, and is never larger than 1.0. The 1989 AISC Specification establishes C_m for braced columns as follows:

$$C_m = 0.6 - 0.4 \frac{M_1}{M_2}$$

where M_1 is the smallest end moment and M_2 is the largest end moment.

The ratio M_1/M_2 is negative if the end moments act in the same direction, causing a single curve. The ratio M_1/M_2 is positive if the end moments are of opposite directions, causing a double-curve deflection.

Three end conditions frequently encountered in steel buildings are shown in Fig. 6.22.

For the case in which no bracing against sidsway buckling is provided, the effective length KL is larger than the story height. This must be accounted for in computing F_a by the selection of K (Art. 6.8). A safeguard is needed against the influence of additional moments caused by the axial

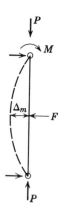

Figure 6.23/Transverse load on column.

force times the lateral story deflection. Therefore, the 1989 AISC Specification places the lower limit of C_m at 0.85 for this condition; i.e., C_m may be larger but not less than 0.85.

There is one further "combined loading"[10] condition, significantly different from those establishing the curves of Fig. 6.21. This is the case where the column receives transverse loads in addition to axial forces. Figure 6.23 illustrates such a condition. The magnitude of load and type of transverse loading will appreciably affect Δ_m. It also could influence the effect of end moment and frequently causes a larger moment between the ends. Since there is specific reference in the AISC Specification to the fact that f_b should be computed at the point under consideration, the larger M should be used. Under these conditions, C_m can be estimated conservatively as 1.0 or it can be computed by any rational analysis. The Commentary on the AISC Specification (Part 5 of the Manual) furnishes additional information for computing C_m, including a table listing six different situations and corresponding values of C_m.

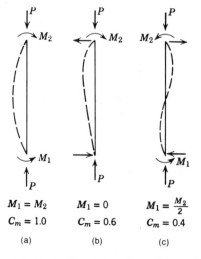

$M_1 = M_2$	$M_1 = 0$	$M_1 = \dfrac{M_2}{2}$
$C_m = 1.0$	$C_m = 0.6$	$C_m = 0.4$
(a)	(b)	(c)

Figure 6.22/Reduction factors for end-braced columns.

[10]"Combined loading" is a common structural term referring to a member subjected to the combined effect of moment and axial load.

When the axial load is small relative to the bending moment, the additional stresses caused by lateral deflection are negligible. This may be ascertained from a study of the column curves in Fig. 6.21. The AISC Specification permits the direct application of the straight-line interaction formula when $f_a/F_a \le 0.15$. Therefore, under such conditions, use

$$\frac{f_a}{F_a} + \frac{f_b}{F_b} \le 1.0$$

An additional criterion of safety needs to be established for all columns having combined loads. The previously discussed design formulas were applicable to positions near the center of a column, where buckling and lateral deflection are greatest. There is also the possibility (depending upon the slenderness ratio of a column unbraced in the plane of bending) that the combined stress computed at one end will exceed that at all points where lateral deflection is created by the end moments, even when the bending stress at these points has been modified. To provide for this case, the AISC Specification requires that the straight-line interaction formula be used, substituting $0.6F_y$ for F_a:

$$\frac{f_a}{0.6F_y} + \frac{f_b}{F_b} \le 1.0$$

For all columns, this condition should be satisfied at points where the column is braced in the plane of bending.

Summary of Procedure / Calculate the ratio of the direct axial stress to the allowable axial stress (f_a/F_a), where F_a is established by the maximum value of KL/r and is independent of the effect of bending moment. If the ratio is less than 0.15, satisfy the condition that

$$\frac{f_a}{F_a} + \frac{f_b}{F_b} \le 1.0$$

If the ratio f_a/F_a is greater than 0.15, satisfy the condition that

$$\frac{f_a}{F_a} + \left[\frac{C_m}{1 - \dfrac{f_a}{F'_e}}\right]\frac{f_b}{F_b} \le 1.0$$

In addition to the above, satisfy the following condition at points of lateral support in the plane of bending:

$$\frac{f_a}{0.6F_y} + \frac{f_b}{F_b} \le 1.0$$

Example 1

Using the above design procedure, the **W** 10×68 column selected in the example of Art. 6.15 will be investigated to determine whether the section is overstressed. The top and bottom of the column are assumed to be braced against translation but allowed to rotate. A36 Steel is specified. (Step 3 of the referenced example is repeated below, and subsequent operations are given step numbers in the same sequence.)

(3) The maximum value of the unit compressive stress is

$$f = \frac{P}{A} + \frac{P'ec}{I} = 11,200 + 5280 = 16,480 \text{ psi}$$

(4) The actual unit stresses developed are

$$f_a = 11,200 \text{ psi}, \qquad f_b = 5280 \text{ psi}$$

(5) The allowable axial stress is found in the same manner as discussed in Art. 6.5. If the unbraced height for both axes of the column is 13 ft and $K = 1.0$ for both axes, the maximum KL/r ratio will result from using the least r of the column, i.e.,

$$\frac{KL}{r_y} = \frac{1(13)12}{2.59} = 60.2$$

and the allowable axial stress may be calculated from the appropriate column formula, or be selected from the appropriate tables as is done here, i.e.,

$$F_a = 17,410 \text{ psi}$$

(6) Since the **W** 10×68 is compact for A36 Steel, the allowable bending stress depends upon the tendency of the column flange to buckle laterally (Art. 5.7). If lateral support is provided at the top and bottom only, the unsupported lateral distance is 13 ft. Reference to the 1989 AISC beam tables shows limits of $L_c = 10.7$ ft and $L_u = 34.8$ ft. Since $13.0 > 10.7 < 34.8$, $F_b = 0.6F_y = 22,000$ psi.

(7) Determine the ratio of axial stresses:

$$\frac{f_a}{F_a} = \frac{11,200}{17,410} = 0.643$$

Since $0.643 > 0.15$, the modified interaction formula must be satisfied.

(8) Determine the amplification factor; this is controlled by the slenderness ratio of the section with respect to the plane of bending:

$$\frac{K_b L_b}{r_b} = \frac{1(13)12}{4.44} = 35.1$$

$$F_e' = \frac{12\pi^2 \, 29,000,000}{23(35.1)^2}$$

$$= 121,200 \text{ psi}$$

$$1 - \frac{f_a}{F_e'} = 1 - \frac{11,200}{121,200} = 0.908$$

(9) Determine the reduction factor. Since column ends are braced against lateral displacement,

$$C_m = 0.6 - 0.4 \frac{M_1}{M_2} = 0.6$$

(10) Testing the stress ratios for stability (between ends of the column),

$$\frac{f_a}{F_a} + \left[\frac{C_m}{1 - \dfrac{f_a}{F_e'}} \right] \frac{f_b}{F_b} = \frac{11.2}{17.41} + \frac{0.6}{0.908} \left(\frac{5.3}{22} \right)$$

$$= 0.802$$

Since $0.802 < 1.0$, the section is adequate for stability.

(11) Testing the stress ratios at the top of the column where it is braced in the plane of bending,

$$\frac{f_a}{0.6F_y} + \frac{f_b}{F_b} = \frac{11.2}{22} + \frac{5.3}{22} = 0.750$$

Since this value is also less than 1.0, the column is considered safe and satisfies the AISC Specification.

From the above procedure, one can see that the design of eccentrically loaded columns is largely one of trial and error, i.e., first selecting a trial section and then testing the stress ratios. Tables will, of course, aid appreciably in expediting design. The "Numerical Values Tables" section of the AISC Specification (Part 5 of the AISC Manual) furnishes a table for values of F_e' for various KL/r ratios for all grades of steel. These tables are similar to those giving values of F_a. An alternative method of designing eccentrically loaded columns is shown in Art. 6.18.

For corner columns or unsymmetrical bays, another condition of eccentric loading may be encountered (Fig. 6.24). Beam A delivers an eccentric reaction to the column causing bending about the x-x axis, and beam B delivers an eccentric reaction to the column causing bending about the y-y axis. Should beam B frame directly to the web, the degree of eccentricity about the y-y axis can be very small and the effect negligible; should the connection be such

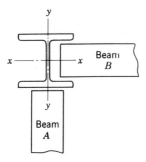

Figure 6.24/Corner column.

that the reaction acts at a distance from the web face, the effect can be quite significant. Following the design procedure just presented but adding one additional term to the stress-ratio formula, the safe criterion becomes

$$\frac{f_a}{F_a} + \frac{f_{bx}}{F_{bx}} + \frac{f_{by}}{F_{by}} \le 1.0.$$

also when

$$\frac{f_a}{F_a} > 0.15$$

the safe criterion becomes,

$$\frac{f_a}{F_a} + \left[\frac{C_{mx}}{1-\dfrac{f_a}{F'_{ex}}}\right]\frac{f_{bx}}{F_{bx}}$$

$$+ \left[\frac{C_{my}}{1-\dfrac{f_a}{F'_{ey}}}\right]\frac{f_{by}}{F_{by}} \le 1.0$$

Also when

$$\frac{f_a}{F_a} > 0.15$$

the safe criterion becomes

$$\frac{f_a}{0.6F_y} + \frac{f_{bx}}{F_{bx}} + \frac{f_{by}}{F_{by}} \le 1.0$$

at braced points, where the subscripts x

and y refer to the respective planes of bending.[11]

In determining the values for F_{bx} and F_{by}, the coefficient C_b can conservatively be taken as unity or, in the case of unbraced frames, it can be computed by the familiar formula

$$C_b = 1.75 + 1.05\left(\frac{M_1}{M_2}\right) + 0.3\left(\frac{M_1}{M_2}\right)^2 \le 2.3$$

This formula, its definition, and its application were introduced and described in Art. 5.6.

Example 2

A **W** 12×40 column has a total height of 12 ft. The y-y axis is braced at midheight. The column base is pinned, allowing free rotation, and is braced against translation. The top of the column is pinned and braced for the y-y axis, and is restrained against rotation but not braced against translation for the x-x axis. There is a 30-kip axial load, including the column weight; a moment about the x-x axis of 80 ft-kips at the top; and no moment at the bottom. Using A36 Steel, determine if the column is safe according to the 1989 AISC Specification.

Solution

(1) The following properties of the **W** 12×40 section are found from the AISC Manual:

$$A = 11.8 \text{ in.}^2, \qquad r_x = 5.13 \text{ in.}$$

$$S_x = 51.9 \text{ in.}^3, \qquad r_y = 1.93 \text{ in.}$$

---

[11]See also "Commentary on the Specification for the Design, Fabrication & Erection of Structural Steel for Buildings," Part 5 of the AISC Manual.

(2) Determine the actual unit axial stress and bending stress:

$$f_a = \frac{P}{A} = \frac{30}{11.8} = 2.54 \text{ ksi}$$

$$f_b = \frac{M}{S} = \frac{80(12)}{51.9} = 18.5 \text{ ksi}$$

(3) Determine the maximum slenderness ratio. The K value for the $x\text{-}x$ axis is 2.0, and for the $y\text{-}y$ axis, 1.0 (see Fig. 6.4); thus

$$\frac{K_x L_x}{r_x} = \frac{2(12)12}{5.13} = 56.1$$

$$\frac{K_y L_y}{r_y} = \frac{1(6)12}{1.93} = 37.3$$

(4) Using the larger slenderness ratio, determine the allowable unit axial stress. From the AISC tables,

$$F_a = 17.8 \text{ ksi}$$

(5) Determine the ratio of axial stresses:

$$\frac{f_a}{F_a} = \frac{2.54}{17.8} = 0.143$$

Since $0.143 < 0.15$, use the straight-line interaction formula to determine stability.

(6) Determine the maximum allowable unit bending stress. The **W** 12×40 is compact for A36 Steel. $L_c = 8.4$ ft and $L_u = 16.0$ ft. The brace at midheight establishes the length for lateral beam buckling at 6.0 ft. Since $6.0 < 8.4$, use $F_b = 0.66F_y = 24$ ksi.

(7) Test the stress ratios for the stability check:

$$\frac{f_a}{F_a} + \frac{f_b}{F_b} = \frac{2.54}{17.8} + \frac{18.5}{24} = 0.914$$

(8) Test the stress ratios at the top of the column where it is braced in the plane of

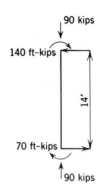

Figure 6.25/Example 3.

bending:

$$\frac{f_a}{0.6F_y} + \frac{f_b}{F_b} = \frac{2.54}{22} + \frac{18.5}{24} = 0.886$$

Since both stress-ratio checks give values less than 1.0, the column is considered safe.

Example 3

A **W** 14×48 column of A242 Steel, having a total length of 14 ft, is loaded as shown in Fig. 6.25. The 90-kip load includes an allowance for the column weight. The bending moments are applied relative to the major axis of the column. The top and bottom of the column are braced for both axes. The K values have been estimated as 1.0 for the minor axis and 0.65 for the major axis. Determine whether the column is safe according to the 1989 AISC Specification.

Solution

(1) The following properties of the **W** 14×48 section are found from the AISC Manual:

$$A = 14.1 \text{ in.}^2, \quad r_x = 5.85 \text{ in.}, \quad L_c = 7.2 \text{ ft}$$

$$S_x = 70.3 \text{ in.}^3, \quad r_y = 1.91 \text{ in.}, \quad L_u = 11.5 \text{ ft}$$

(2) Determine the actual unit axial stress and bending stress:

$$f_a = \frac{P}{A} = \frac{90}{14.1} = 6.38 \text{ ksi}$$

$$f_b = \frac{M}{S} = \frac{140(12)}{70.3} = 23.9 \text{ ksi}$$

(3) Determine the slenderness ratio for each axis:

$$\frac{K_x L_x}{r_x} = \frac{0.65(14)12}{5.85} = 18.7$$

$$\frac{K_y L_y}{r_y} = \frac{1(14)12}{1.91} = 88$$

(4) Using the largest slenderness ratio, determine the allowable unit axial stress (note that the section is in structural Group 1; therefore, $F_y = 50$ ksi). From the AISC tables,

$$F_a = 17.37 \text{ ksi}$$

(5) Determine the ratio of axial stresses:

$$\frac{f_a}{F_a} = \frac{6.38}{17.37} = 0.367$$

Since $0.367 > 0.15$, the modified interaction formula must be used to determine stability.

(6) Calculate the amplification factor:

$$F'_e = \frac{12\pi^2 \, 29{,}000}{23(18.7)^2}$$

$$= 427 \text{ ksi}$$

$$1 - \frac{f_a}{F'_e} = 1 - \frac{6.38}{427} = 0.985$$

(7) Determine the reduction factor:

$$C_m = 0.6 - 0.4\left(\frac{70}{140}\right) = 0.4$$

(8) Determine the allowable bending stress. The section could be checked for adequacy as a compact section, but, since the unsupported lateral length is 14 ft,

which is larger than L_c (7.2 ft), it does not satisfy the specification permitting $F_b = 0.66F_y$. Also, since 14 ft is greater than L_u (11.5 ft), the allowable bending stress could be less than $0.60F_y$ and must be established by the procedure described in Art. 5.7 and shown in Fig. 5.5. The moment gradient modifier is

$$C_b = 1.75 + 1.05\left(\frac{M_1}{M_2}\right) + 0.3\left(\frac{M_1}{M_2}\right)^2$$

$$= 1.75 + 1.05\left(\frac{70}{140}\right) + 0.3\left(\frac{70}{140}\right)^2$$

$$= 2.35$$

However, the upper limit of the modifier is set at 2.3. Therefore, use $C_b = 2.3$. Determine whether buckling or twisting controls L_u.

$$r_T \frac{d}{A_f} = 2.13(2.89) = 6.16 < \frac{62.62}{\sqrt{50}} = 8.86$$

Therefore, twisting controls and

$$L_{\text{eff.}} = C_b L_u = 2.3(11.5) = 26.45 \text{ ft}$$

Since 14 ft < 26.45 ft, $F_b = 0.6F_y = 30$ ksi.

(9) Testing the stress ratios for the stability check,

$$\frac{6.38}{17.37} + \frac{0.4}{0.985}\left(\frac{23.9}{30}\right) = 0.691$$

Inasmuch as $0.691 < 1.0$, the column is considered safe for stability.

(10) Test the stress ratios at the top of the column, where it is braced in the plane of bending:

$$\frac{f_a}{0.6F_y} + \frac{f_b}{F_b} = \frac{6.38}{30} + \frac{23.9}{30} = 1.01$$

Since $1.01 > 1.0$, the section is not safe according to the AISC Specification.

PROBLEMS

1. A **W** 8×31 column of A36 Steel has an unbraced height of 14 ft for both axes. It is braced at top and bottom so that $K_x = K_y = 1.0$. A 60-kip load is placed 2 in. from the face of the flange. Neglect the column weight. Calculate the sum of the stress ratios. Is the column safe? (Answer given in Appendix G.)

2. A **W** 14×43 of A36 Steel has an unbraced height of 10 ft for the y-y axis and 24 ft for the x-x axis. Bracing top and bottom is such that $K_x = K_y = 1.0$. There is a 90-kip load applied directly to the outside face of the column flange. Neglecting the column weight, calculate the sum of the stress ratios. Is the column safe?

3. A **W** 8×18 of A242 Steel has a total unbraced height of 14 ft. $K_x = K_y = 1.0$. Calculate the maximum applied load P that may be safely placed at the outside face of the column flange. Use $C_b = 1.75$. (Answer given in Appendix G.)

4. A square column of A36 Steel is fabricated by welding the toes of two $6 \times 6 \times \frac{5}{8}$-in. angles together. The unbraced height is 12 ft and $K_x = K_y = 1.0$.

(a) Calculate the maximum axial load the column can safely support, including its own weight.

(b) Calculate the maximum allowable applied load if it is to be applied to the outside of one of the faces.

5. A **W** 6×25 of A36 Steel serves as a column for a two-story building. It is continuous for the full two stories, and floor-to-floor heights are 11 ft 6 in. $K_x = K_y = 1.0$. The roof delivers an axial load of 18 kips to the column. The second floor delivers a total of 50 kips, of which 25 kips has an eccentricity of 8 in. (with respect to the x-x axis). Neglect the weight of the column. Calculate the sum of the stress ratios. Is the column safe? *Note:* Since the column is continuous, the bending moment is divided equally between the column above and the column below. (Answer given in Appendix G.)

6. A **W** 8×28 of A36 Steel serves as a column for a one-story building. It is pin-ended and braced for both axes. The loads applied at the top are as shown in Fig. 6.26. The unbraced height is 14 ft. Assume that the 30-kip load acts

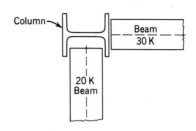

Figure 6.26

at the column flange, and the 20-kip load, 3 in. from the y-y axis. Neglecting the column weight, calculate the sum of the stress ratios. Is the column safe?

7. Select the lightest-weight **W** section of A36 Steel acting as a column that will safely support a vertical load of 100 kips, having an eccentricity of 6 in. with respect to the x-x axis. $K_x = K_y = 1.0$, and the unbraced height is 18 ft. (Answer given in Appendix G.)

8. Select the lightest-weight **W** section of A588 Steel, acting as a single-story column with an unbraced height for both axes of 20 ft, that will support loads as shown in Fig. 6.27. Top and bottom are pinned and braced so that $K_x = K_y = 1.0$. Neglect the column weight. (Answer given in Appendix G.)

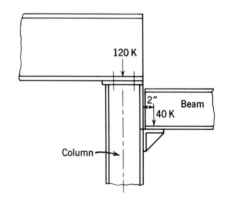

Figure 6.27

9. A single-story column has an unbraced height of 16 ft and must support an axial load of 200 kips, including its own weight. In addition, there is a steel beam welded to the column flange face that delivers to this flange a reaction of 30 kips and a moment of 80 ft-kips. The column is pinned and braced top and bottom for the *y-y* axis. The top is moment-resistant about the *x-x* axis but allowed to translate. The bottom is fixed against translation and rotation for the *x-x* axis. Select the lightest-weight 12 in. **W** section of A36 Steel that can be safely used. (*Note*: The column base has a moment equal to half the moment at the top, and opposite in direction.)

6.18
Equivalent Concentric Load

A common method for determining the unit bending stress in columns supporting eccentric loads is to consider the effect of eccentricity in terms of an equivalent concentric load; that is, the eccentric load is replaced by a concentric load of sufficient magnitude to produce a stress equal to the maximum stress produced by the eccentric load.

This method has many useful applications, one of which is to aid in approximating the first trial section of columns to be subsequently checked by the stress-ratio method. If P_{eq} is the equivalent concentric load, and A is the cross-sectional area of the column, the unit stress produced by P_{eq} may be equal to the unit bending stress produced by the eccentric load, i.e.,

$$\frac{P_{eq}}{A} = \frac{M}{S}$$

from which

$$P_{eq} = \frac{A}{S} M$$

In the foregoing equations, M is the bending moment caused by the induced beam end moment (Art. 6.16) or by the eccentric load ($P'e$). Also, S is the section modulus relative to the axis about which bending takes place.

The ratio A/S is called the bending factor and will be found tabulated in the AISC Manual. For convenience, bending factors are listed with the column safe-load tables rather than with the other properties of sections. The ratio A/S is symbolized by B, and there are two values for each column section, identified by subscript x or y, depending upon which axis of the column it pertains to. For the remainder of this discussion, reference will be made to the major *x-x* axis. Consequently, the expression for the equivalent concentric load becomes

$$P_{eq} = B_x M$$

The units for B_x are $1/\text{in.}$; therefore, M must be measured in in.-lb or in.-kips.

Applying this method to step (3) of the example in Art. 6.15, the listed value of B_x for the **W** 10×68 is 0.264, and

$$P_{eq} = 0.264(50,000)8 = 105,600 \text{ lb}$$

then

$$f = \frac{P + P_{eq}}{A} = \frac{224,000 + 105,600}{20.0}$$
$$= 16,480 \text{ psi}$$

This agrees with the result previously obtained. The equivalent-concentric-load method greatly facilitates design when safe-load tables are used. The object is to select a first trial section with as much accuracy as possible. However, each trial section must be checked by calculating the sum of the stress ratios as presented in Art. 6.17; therefore, accuracy and thoroughness in calculating this equivalent axial load are not warranted. The procedure that follows is considered adequate and allows for some judgment on the part of the designer. The procedure described in Section 3 of the AISC Manual is similar but more precise in that the checking of the stress ratios is automatically provided for.

It has been shown that the *stress* obtained by the previous method was also found using the equivalent concentric load method. However, the equivalent-concentric-load method produces results which approach an accurate design only when the allowable stress for the axial load approaches the allowable stress for the bending moment, or $F_a \rightarrow F_b$. From this it may also be concluded that the larger the axial load and the smaller the bending moment, the more nearly accurate will be the design resulting from use of the method.

The accuracy of the equivalent axial load can be improved by multiplying it by the ratio F_a / F_b. This, of course, requires knowing beforehand what these final values will be. The ratio may be small in cases where the influence of the axial buckling dominates; however, just the opposite is true in cases where the column is principally a bending-moment element. In all cases it can be conservatively taken as 1.0.

The usual procedure is to calculate the approximate total equivalent load based upon the requirements of the stability check (the modified interaction formula). The total equivalent load is the total axial load that is used in calculating f_a, plus the equivalent concentric load representing the effect of the eccentricity. The effect of the eccentricity in the stability check is measured by the bending-stress ratio f_b / F_b, which is further modified by the reduction factor C_m.

Similarly, the equivalent concentric load should be multiplied by C_m before adding its effect to that of the axial load. Therefore, the total equivalent axial load is

$$P_{\text{total}} = P_{\text{axial}} + P_{\text{eq}}$$

$$P_{\text{total}} = P_{\text{axial}} + B_x C_{mx} M_x \frac{F_a}{F_b}$$

The only factor not included in the above equation is the amplification factor. A further modification to allow for its influence is necessary.

By scanning the safe-load tables, a trial bending factor is selected which, when used and modified as described above, will produce a total load that can be compared with those given in the safe-load table.

For the loading condition shown in Fig. 6.16, $P = 134 + 20 + 20 + 50 = 224$ kips. The effective unbraced height for both axes is 13 ft, and A36 Steel is assumed. Scanning the AISC Manual safe-load tables for 10-in. columns (similar to Table 6.1), and using as a guide those loads greater than 224 kips and $KL = 13$ ft, an average B_x is approximately 0.266. The bracing is such that $C_m = 0.6$. The bending moment is $50(8) = 400$ in.-kips. It also may be estimated that $F_a / F_b = 0.8$. Consequently, the approximate equivalent total axial load is

$$P_{\text{total}} = 224 + 0.266(400)0.6(0.8) = 275 \text{ kips}$$

Entering the safe-load tables again with an effective unbraced height of 13 ft and an axial load of 275 kips, it is observed that the **W** 10×54 appears to be a borderline choice and that the **W** 10×60 is more reasonable. Both should be evaluated accurately by the stress-ratio method. In this instance, the **W** 10×54 proves not to be adequate.

The **W** 10×60 is selected for a trial section, and the modified interaction formula shows that the stress ratio is 0.91. The end-of-column check shows the stress ratio to be 0.85. Since both are less than 1.0, the selection is adequate.

It also may be observed that in this case the final F_a / F_b ratio turns out to be 0.79, while the amplification factor is $1/0.89$.

The equivalent-concentric-load method also could be applied to columns subjected to bending about both axes. To do so, however, requires calculating an equivalent concentric load for the x-x axis bending ($P_{\text{eq-}x}$) and another for the y-y axis ($P_{\text{eq-}y}$), and then adding both to the total axial load for comparison with loads listed in the safe-load tables.

6.19
Column Base Plates

Where a column is to be supported on a concrete footing, pier, or pedestal, a rolled steel slab is generally used to distribute the column load over the concrete support (Fig. 6.28). The bottom of the column and the surface of the steel plate under the column are planed so that the load is transmitted to the plate by direct bearing. The underside of the plate is not planed, but rests on cement grout approximately 1 in. thick on top of the footing. Angles may be used to fasten the column to the plate, and, by means of anchor bolts, the plate to the footing. The angles may be omitted when the plate is shop-welded to the column (Fig. 6.28c).

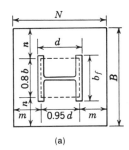

(a)

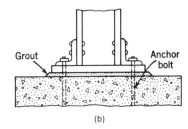

(b)

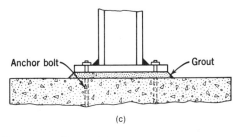

(c)

Figure 6.28/Steel slab column base plates.

The required area of plate is found by dividing the column load by the allowable unit bearing pressure on the concrete.

The allowable bearing pressure on concrete or masonry is set forth in model building codes. For example, the 1991 Uniform Building Code limits the bearing strength in concrete to $0.3f_c'$ (f_c' being the specified strength of the concrete). In addition, most building codes allow the bearing strength to be increased if the concrete surface is not entirely covered by the base plate. In many cases, even the smallest bearing area is more than adequate and the plate dimensions N and B are established by practical considerations.

It is assumed that the bearing pressure on the bottom of the plate is of uniform intensity and that the load from the column is uniformly distributed over a rectangular area on top of the plate. The dimensions of this "equivalent rectangle" are taken as 0.95 times the depth of the column section, and 0.8 times the column flange width. The portion of the plate projecting beyond the sides of the equivalent rectangle acts as an inverted cantilever with the maximum bending moment occurring at that side for which the overhang, either n or m, is greater (Fig. 6.28a). This is not an exact method of analysis, but the results obtained have proved satisfactory in practice, except in cases where the controlling value of m or n is very small, and then the AISC Manual should be consulted.

The following symbols are used in the derivation of the formulas:

P = total load on the column (lb).

N = length of plate (in.).

B = width of plate (in.).

d = depth of the column section (in.).

b_f = width of column flange (in.).

A = area of the plate (in.^{2}).

F_p = allowable unit bearing pressure on the concrete (psi).

f_p = actual unit bearing pressure on the underside of the plate ($f_p = P/NB$), in psi.

t = thickness of the plate (in.).

F_b = allowable extreme fiber stress in bending (psi).

The required area of the slab is $A = P/F_p$. The dimensions N and B are chosen so as to give this area, and the actual bearing pressure is calculated ($f_p = P/NB$).

Considering a strip of steel slab 1 in. wide, the bending moment on the strip at the edge of the equivalent rectangle is

$$M = f_p(m)\frac{m}{2} = \frac{f_p m^2}{2}$$

$$M = f_p(n)\frac{n}{2} = \frac{f_p n^2}{2}$$

The required section modulus is

$$S = \frac{M}{F_b} = \frac{f_p m^2}{2F_b}$$

$$S = \frac{M}{F_b} = \frac{f_p n^2}{2F_b}$$

But the section modulus is also equal to I/c, and the moment of inertia of a rectangle 1 in. wide and t_p inches thick is

$$I = \frac{1(t_p)^3}{12} = \frac{t_p^3}{12}$$

The value of c is $t/2$; therefore,

$$S = \frac{I}{c} = \frac{t_p^3}{12} \Big/ \frac{t_p}{2} = \frac{2t_p^3}{12t} = \frac{t_p^2}{6}$$

Equating I/c to M/f_b,

$$\frac{t_p^2}{6} = \frac{f_p m^2}{2F_b} \quad \text{and} \quad \frac{t_p^2}{6} = \frac{f_p n^2}{2F_b}$$

$$t_p^2 = \frac{6f_p m^2}{2F_b} = \frac{3f_p m^2}{F_b} \quad \text{or} \quad \frac{3f_p n^2}{F_b}$$

whichever is greater. The 1989 AISC Specification limits the extreme fiber stress in

bearing plates to $F_b = F_y$. Substituting this value in the above equations and simplifying gives

$$t_p = 2n\sqrt{\frac{f_p}{F_y}}, \qquad t_p = 2m\sqrt{\frac{f_p}{F_y}}$$

The operations involved in the design of steel slab column bases are illustrated by the example which follows.

Example

Design a rolled steel base plate for a **W** 14×90 column that transmits a load of 383,000 lb. A36 Steel is specified. The allowable unit bearing pressure on the concrete footing is 500 psi.

Solution

(1) The required area is

$$A = \frac{P}{F_p} = \frac{383,000}{500} = 766 \text{ in.}^2$$

(2) This area requires a slab 27.7 in. square. Working to the nearest inch, a 28×28-in. plate is selected, and the actual bearing pressure is

$$f_p = \frac{P}{NB} = \frac{383,000}{(28)^2} = 489 \text{ psi}$$

(3) The values of d and b for the column section are obtained from a handbook, and a sketch is made as shown in Fig. 6.29.

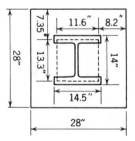

Figure 6.29/Base plate: example.

and,

$$n > m = 8.2 \text{ in.}$$

(4) Then

$$t_p = 2(8.2)\sqrt{\frac{489}{36,000}} = 1.91 \text{ in.}$$

(5) If the thickness is taken to the next greater $\frac{1}{4}$ in., the finished dimensions of the required plate are $28 \times 28 \times 2$ in.

6.20
Grillage Foundations

It is common practice to distribute column loads over the foundation bed, using spread footings. However, steel grillages are sometimes used for this purpose, and are of particular value in cases where extremely heavy loads must be supported on rock. Grillage foundations are made up of one or more layers of steel beams arranged as shown in Fig. 6.30 and encased in concrete. Pipe separators are placed near the ends of the beams and under points where concentrated loads occur, or welded diaphragms are used as spreaders. Sufficient space is left between the flanges

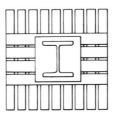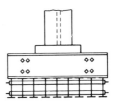

Figure 6.30/Grillage foundation.

of the beams in any one tier to permit proper placing of the concrete.

The unit bearing pressure on the bottom of the grillage is limited to the bearing capacity of concrete, inasmuch as a thin concrete mat is usually placed under the grillage, even where the supporting material is bedrock.

PROBLEMS

1–3. Rework Problems 7, 8, and 9 following Art. 6.17 using the equivalent-concentric-load method to arrive at the first trial section.

4. A **W** 10×39 column transmits a load of 122 kips. A base plate 16×16 in. is to be used. Determine the required thickness of the base plate if A36 Steel is specified. (Answer given in Appendix G.)

7
CONNECTIONS

7.1
Introduction

Chapters 2 through 6 dealt with the design and investigation of beams, girders, and columns as individual structural members. These members must be attached to one another to form a complete structure; the manner of doing so is the subject of this chapter. Considerable time and effort often are spent in establishing the most suitable framing arrangement and in selecting members with regard to size, shape, and economy, at the expense of adequate attention to connections and their design. This is evidenced by the fact that the more common structural failures occur most often in connections rather than in members.

Attention must be given to the total problem of bringing structural members together in a manner that is both satisfactory and consistent with assumptions made in the selection of individual members and the structure as a whole. To do this, a distinction must be made between the connection that is the total joint assembly and the connectors that are used to create that assembly. There are several types of connector that can be used either exclusively or in combination, and a number of types of connection. These will be discussed first, and then the detailed numerical design processes for both will be treated.

7.2
Types of Connector

The once predominant rivet is little used today, particularly in buildings. The two most common types of steel structure connector in use today are bolts and welds, and the former is but one in a family of

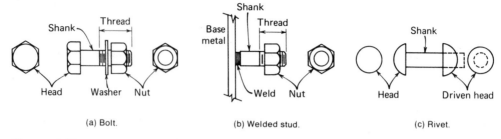

(a) Bolt.　　　(b) Welded stud.　　　(c) Rivet.

Figure 7.1/Fasteners.

mechanical fasteners that are referred to in the AISC Manual as "Bolts, Threaded Parts, and Rivets" (Fig. 7.1).

Bolts / Bolts (Fig. 7.1a) consist of a cylindrical body (shank) with an attached head. A portion of the body is threaded to receive a nut. Both head and nut may be square or hexagonal, and a washer may or may not be required. Bolts are installed with considerable ease, are relatively inexpensive, and can be checked visually for imperfections and damaged material, all of which are good reasons for their use. The 1989 AISC Specification lists four basic types of steel for use in the manufacture of bolts for structural connections; these are referred to by their ASTM designation. Type A307 bolts, which have no specified minimum yield point, were once the most frequently used in practice; however, currently, their use has been somewhat limited. This type of bolt has frequently been referred to as a common, plain, unfinished, rough, or machine bolt. With these bolts, there is some danger of the nut loosening; therefore, the AISC Specification prohibits their use in some major connections.

High-strength bolts are not so restricted; therefore, they have become the type most frequently specified for steel buildings where mechanical fasteners are to be used. The 1989 AISC Specification lists three types of material for these bolts—A325, A490, and A449. For general purposes,

the A325 bolt, with approximately a 90-ksi yield-point stress, is usually specified.[1] Its cost is about 15 per cent greater than the A307 bolt. A stronger high-strength bolt (A490, approximately 120 ksi yield) is available and may prove to be more suitable when a higher-strength material is to be connected by the fasteners. Its cost is about 35 per cent more than the A325. A449 bolts are used when a bolt diameter greater than $1\frac{1}{2}$ in. is required; consequently, they are seldom used in buildings.

Bolts may be installed by use of a pneumatic or similar impact wrench that can be adjusted to the degree of stress to be put on the bolt and threads. Other means of installation are described later in this chapter.

Welded Studs / Threaded fasteners, with one end welded to a steel member or plate (Fig. 7.1b), are being used more often because of their unique characteristics. The weld is accomplished automatically: one end of a threaded stud is inserted into an electrified hand tool (gun), and the other end is brought into contact with the base metal at the desired position. When contact is made and the gun triggered, the stud is immediately fused to the base metal. Welding guns, materials, types of stud, etc., vary with different manufacturers and generally are proprietary in nature; therefore,

---

[1]A325 bolts are available with weathering characteristics comparable to ASTM A242 and A588 Steel.

design data must be furnished by each manufacturer, and such data must be consulted when specifying a given type of stud. Typically, manufacturers use a stud material having a 50-ksi yield point and a corresponding design shear-stress capacity of 24 ksi.[2] The advantages of using welded stud fasteners are obvious—only half the normal number of holes are necessary, some positions are accessible for welding that are not accessible for inserting bolts, and it is not necessary to hold the head when tightening nuts. However, welded studs are more expensive than other types of mechanical fastener.

Rivets / Some fabricating shops continue to use rivets for certain purposes. A rivet consists of a cylindrical shank with a head at one end (Fig. 7.1c). The shank is made sufficiently long to extend through the parts to be connected and have enough metal remaining to form the second head when driven. Its diameter is slightly less than that of the hole prepared to receive it, so that when the rivet expands with heating, it can still be inserted into the hole. The formed end is backed up by a pneumatic jack or hammer with a die the same shape as the formed head. Once backed up, a second pneumatic hammer with a die in the shape of the desired second head is used to form that head and thus make the connection. The 1989 AISC Specification lists two basic types of steel for rivets, A502-1 and 502-2,3. For general purposes, Grade 1 (carbon steel) usually is specified, and when a higher-strength steel is desired, Grade 2 (carbon manganese steel).

Even though the cost of high-strength bolts is nearly twice that of rivets, the overall cost of riveted construction usually is more because of increased labor and equipment requirements.

Welds / The fourth type of connector is the weld. There are numerous welding procedures available, but the only procedure acceptable in structural work is fusion welding by the electric arc process. Even so, there are basically four different processes by which the fusion arc weld can be achieved. They are all covered in more detail in Art. 7.22 (Figs. 7.46 to 7.48 and Table 7.4), but for purposes of introduction, a typical manual arc-welding process will be described here.

With fusion welding, the base parts to be joined and additional metal in the form of a rod (filler metal) are heated to fusion temperature. Heat for this operation is provided by an electric arc. The arc is a sustained spark between a metal electrode or rod and the base metal (Fig. 7.2). At the instant the arc is formed, the temperature of the metal (electrode and base metal) at the tip of the electrode is brought to the melting point. At the tip of the electrode, small droplets of molten metal are formed, forced across the arc, and deposited in the crater of molten base metal. The force impelling the droplets of molten metal across the arc is actually strong enough to permit welding overhead. To create the arc, one terminal of a motor-generator is attached to the work—which may be an individual component being fabricated in the shop or an entire building frame being erected in the field—and the other terminal is fastened through a flexible cable to

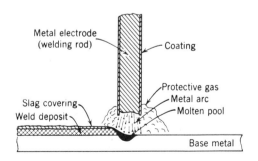

Figure 7.2/Manual, shielded-arc welding.

[2] Later in this chapter it is shown that most fasteners are used in such a manner that shear stress becomes the controlling design criterion.

the electrode. In ordinary manual welding, the electrode is gripped in a special insulated holder by the operator. With the generator running, the electrode is brought in contact with the work to strike the arc. It is then withdrawn slightly to provide the proper arc gap, i.e., one that is not so small as to cause melting through the parts or "spatter," or so large as to fail to melt the base metal to sufficient depth for adequate weld penetration. As the electrode metal melts and is deposited on the locally fused metal of the work, the operator must continually feed the electrode into the work to maintain a constant gap and to build the desired bead.

If the molten metal comes into direct contact with the air, there will be a chemical reaction, which results in a brittle metal, very susceptible to corrosion. Such action is prevented by using a coated electrode that produces large quantities of inert gas when heated that in turn shields the molten metal. This coating also produces a flux that enhances the fusion process and forms a slag coating over the weld, protecting it from the atmosphere while it is cooling. The slag, which is very brittle, is easily removed by tapping with a hammer and wire brushing; then the weld can be ground smooth and painted.

Coated rods are known as shielded arc electrodes and are available in a variety of structural steel grades. The ultimate tensile strength of a properly placed weld varies from 60 to 110 ksi, and the design shear strength, from 18 to 33 ksi. For example, the E70XX electrode frequently is specified for welding A36 Steel and has a design shear strength of 21 ksi. The E80XX electrode is sued for A572 Grade 65 Steel and has a design shear strength of 24 ksi.

There are other arc-welding processes, and their nature and use will be treated later in this chapter.

With large-size welds and/or with the higher-strength steels, a minimum preheat temperature of the parent steel is required, so that the mass of the parent metal will not dissipate the heat from the arc so rapidly that proper fusion will be prevented. Also, note should be made of the fact that shrinking, accompanying the cooling process, usually results in residual stresses in the weld and parent metal; therefore, care must be taken to assure sufficient strength to compensate for these stresses.

Combinations of the above-mentioned basic connectors (rivets, bolts, or welds) may be used in making structural connections. For example, one end of a connecting piece can be welded to a structural member in the fabricating shop, and the assembly can then be taken to the building site where it is bolted to another structural member. This is generally referred to as "shop-welded and field-bolted construction." Also, members often are temporarily bolted into position and then field-welded.

7.3
Types of Steel Construction

The AISC Specification recognizes three basic types of steel construction, each of which is related to the manner in which one member is attached to another. Consequently, there are three basic design assumptions upon which the size of members and types of connection are predicated.

Type 1, commonly designated as "rigid-frame" (continuous frame), assumes that beam-to-column connections have sufficient rigidity to hold the original angles between intersecting members virtually unchanged.

Type 2, commonly called "simple" framing (unrestrained, free-ended), assumes that, insofar as gravity loading is concerned, the ends of the beams and girders are connected for shear only and are free to rotate under gravity load.

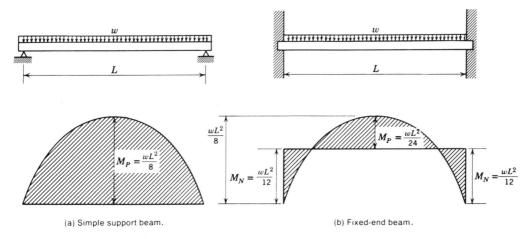

(a) Simple support beam.   (b) Fixed-end beam.

Figure 7.3/Consistency in design.

Type 3, commonly designated as "semi-rigid framing" (partially restrained), assumes that the connections of beams and girders possess a dependable and known moment capacity intermediate in degree between the complete rigidity of Type 1 and the complete flexibility of Type 2.

This simply means that the design of the members in a steel-frame building and the design of the connections joining the members must be carried out in accordance with the same basic assumptions. Members assumed to be simply supported or pin-connected to facilitate design cannot consistently be fastened together rigidly in the actual structure. Again, if all frame joints are considered rigid or continuous in the design analysis, they must be constructed to provide a reasonably complete degree of continuity.

If the fixed-end beam shown in Fig. 7.3b (ends completely restrained against any rotation) is designed to resist the given bending moments, and the connections are then designed to permit free-end rotation as in Fig. 7.3a, the beam will not be strong enough to resist the bending moment resulting from the actual, simply supported condition. In a like manner, if a connection is assumed to be pinned in its design (free rotation) as shown in Fig. 7.3a, and

then is detailed as a restrained end (Fig. 7.3b), the $wL^2/12$ bending moment induced at the end of the beam could cause the connection to fail.

Type 1 construction is unconditionally permitted by the AISC Specification. When Type 1 construction is used, the building frame is statically indeterminate. It is necessary that a careful stress analysis be made and that the normal unit working stress values allowed by the specification not be exceeded in the design of members and connections. Chapter 10 deals with elastic theory and the design of continuous beams and frames, and an example of Type 1 construction is presented in Chapter 13. The 1989 AISC Specification also permits an analysis based upon plastic design for Type 1 construction. Plastic design is treated in Chapter 11.

Type 2 construction is permitted if some structural system is used to develop the lateral forces on the frame, such as can result from wind and earthquake loading. One such system is the "wind bent," discussed in Chapter 9, in which a limited number of specific beam-to-column connections are selected and designed for moment resistance. In this kind of Type 2 construction, care must be taken to ensure that flexible connections have adequate in-

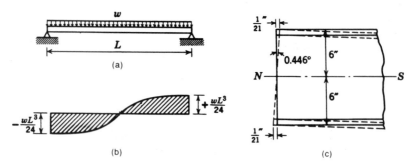

Figure 7.4/End rotation of simply supported beam.

elastic rotation capacity to avoid overstress of the connectors under combined gravity and wind loading.

Type 3 construction is permitted by the AISC Specification only upon evidence that the connections are capable of developing definite resisting moments without overstress. Designs of beam-to-column connections, together with experimentally determined bending resistances, are available from the American Institute of Steel Construction. Further discussion of this type of construction is not included in this text.

The design methods and procedures treated in this chapter are applicable to the analysis and design of both Type 1 and Type 2 construction.

In Type 2, the framing connections must be designed and detailed to permit free end rotation for gravity loads.

The angle of end rotation of a simply supported beam can be determined as readily as the maximum deflection. The basic theory for such a calculation was shown in Chapter 4, accompanying the discussion on deflection; and Example 1, following Art. 4.8, referred to any simply supported steel beam carrying a uniformly distributed load. Shown in the example was a method for developing the equation of the slope diagram. A sketch of the beam and the corresponding slope diagram is shown in Fig. 7.4 (a) and (b). In

the example, the largest angular rotation occurs at the ends and is

$$\theta = \pm \frac{wL^3}{24EI}$$

The angle is positive ($+$) for the right end and negative ($-$) for the left end. Both angles have the same magnitude. If the load w is 2 kips per ft and the span L is 18 ft, a **W** 12×40 would be an economical beam choice and would meet most design requirements. The corresponding rotation at the support would be[3]

$$\theta = \frac{wL^3}{24EI}$$

$$= \frac{2(18)^3 12^2}{24(29,000)310} = 0.00778 \text{ radian}$$

$$\theta = \frac{180}{\pi}(0.00778) = 0.446°$$

Fig 7.4c shows the end of the **W** 12×40 before loading and the resulting angle of rotation after the load is applied. In this example, the end plane of the beam is assumed to rotate about its neutral axis. Consequently, the horizontal movement at the top and bottom of the beam may be

[3] In the equation for θ, there are no resulting units. Consequently, θ must be measured in radians, and from geometry, it is known that one radian equals $\pi/180$ degrees.

found by trigonometry as follows:

$$\tan 0.446° = 0.00778$$
$$0.00778(6) = 0.047 \text{ in.} \simeq \tfrac{1}{21} \text{ in.}$$

In most design work, the actual angle of rotation is not significant and is seldom determined. However, Section J1.2 of the 1989 AISC Specification does stipulate that flexible beam connections shall accommodate end rotations of unrestrained (simple) beams. The reason for making such a calculation here is to show that a connection for Type 2 construction only need be flexible enough to permit this very small movement.

7.4
Common Connections

The different types of connection dealt with in practice are too numerous to list here, and the details of any given type will vary with the type of connector used as well. Each type of connector, i.e., rivet, bolt, or weld, has its own peculiar advantages and disadvantages; these will become apparent as they are studied in detail later.

Following are several examples of connections most frequently used in steel buildings, with the type of connector (bolt or weld) intermixed. The connections shown in Figs. 7.6 through 7.10 permit enough end rotation so that beams may be assumed to be simply supported and thus suitable for Type 2 construction as discussed in Art. 7.3.

Lap or Butt (Fig. 7.5) / These are generally the simplest type of connection to use when two members are in essentially the same plane. They can be arranged so that they will or will not require additional pieces of metal to complete the connection. The lap connection, or *splice*, may be in tension or compression.

Framing Angles (Fig. 7.6) / Structural members framing perpendicular to one another usually require the use of an angle to effect a connection. Bolts or welds may be used. Once the members are positioned it is customary to hold them temporarily in place with driftpins, bolts, or tack welds until the permanent connection can be made.

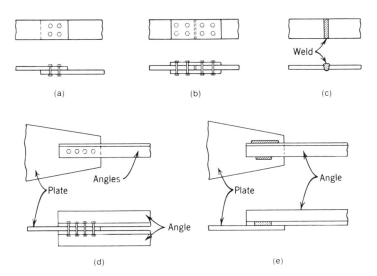

Figure 7.5/Lap and butt connections.

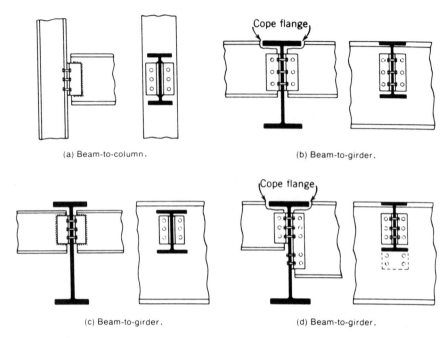

(a) Beam-to-column.

(b) Beam-to-girder.

(c) Beam-to-girder.

(d) Beam-to-girder.

Figure 7.6/Framing angles.

Shear Plates (Fig. 7.7) / A shear plate welded to the end of a beam may be used in place of the paired angles used as framing angles. Bolts are then used to connect the plate to the supporting member. Shear-plate connections require much closer fabricating tolerances for the beam, assuming that ends are cut parallel and to exact length.

Beam Seats (Figs. 7.8 and 7.9) / Beam seats provide a ledge or shelf for the beam to rest on while the permanent connection

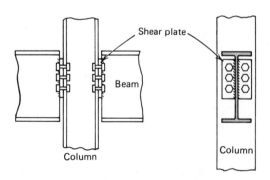

Figure 7.7/Beam-to-column end shear plate.

is made. The clip angle at the top of the beam provides lateral support at this point only and is assumed to carry no load. Flexible beam seats are the simplest and most desirable. Because the thickness of the seat angle provides the only resistance against bending in the outstanding leg, the outstanding leg must be stiffened when the load becomes too large.

Direct Web Connections (Fig. 7.10) / For beam-to-column connections, it is sometimes possible to secure the beam web directly to the column flange. The shelf angle is used to facilitate erection only, and is assumed to carry no load. Welds may be used as well as rivets or bolts.

Bracketed Connections (Fig. 7.11) / The bracketed type of connection is necessary whenever two members to be secured together do not intersect. It is not a desirable type of connection structurally, because a pronounced eccentricity is introduced both in the connection and in one of the members to be joined.

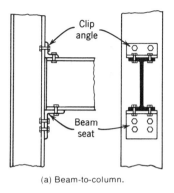

(a) Beam-to-column.

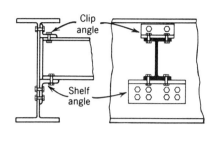

(b) Beam-to-girder.

Figure 7.8/Flexible beam seats.

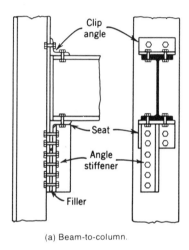

(a) Beam-to-column.

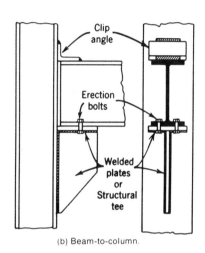

(b) Beam-to-column.

Figure 7.9/Stiffened beam seats.

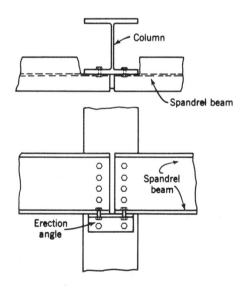

Figure 7.10/Spandrel connection.

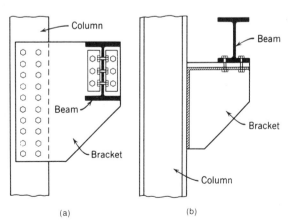

(a)　　　　　　　　(b)

Figure 7.11/Bracketed connections.

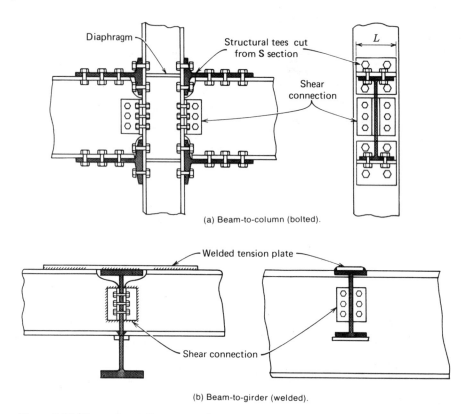

(a) Beam-to-column (bolted).

(b) Beam-to-girder (welded).

Figure 7.12/Moment-resisting connections.

Moment-Resisting Connections (Fig. 7.12) / Either bolts or welds can be used in moment-resisting connections. Any one of the previously discussed shear connections may be used to develop the vertical force. In addition, some means of developing the horizontal force at the beam flanges is required. These connections prevent rotation of one member with respect to the other; however, there may be some rotation of the entire joint, depending on the relative stiffnesses of the members (Chapter 10).

FASTENER CONNECTIONS

7.5
Kinds of Fastener Loads

Most connections are detailed and constructed so that the fastener itself is subjected to a load perpendicular to its shank (Fig. 7.13a). When this is the case, the fastener is designated as *shear type*; all the fasteners shown in Figs. 7.5 through 7.8 are of the shear type under normal loads. This designation will be used herein even though in some cases—e.g., friction fas-

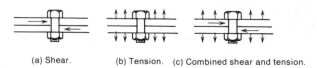

(a) Shear.   (b) Tension.   (c) Combined shear and tension.

Figure 7.13/Kinds of fastener loads.

teners—the shank itself is not in shear. Also, there are occasions when a fastener is actually in tension (Fig. 7.13b)—e.g., hanger-type connections. Only high-strength bolts in friction-type connections (Art. 7.8) or rivets are recommended for developing this kind of load. Figure 7.12a shows a bolted, moment-resisting connection. The bolts attaching the structural tee to the flange of the column are loaded in tension; the other bolts in the connection are in shear.

The third kind of load is a combination of shear and tension (Fig. 7.13c). Brackets and stiffened beam seats, such as are shown in Fig. 7.9a, load the fasteners in combined tension and shear.

7.6
Holes for Fasteners

Determining the size, type, and arrangement of holes for fasteners is an important part of the design procedure. Holes are usually made by punching, using standard

dies. However, if the material is too thick –i.e., larger than the diameter of the fastener to be used—the holes must be drilled or subpunched and reamed. Slotted holes require additional labor.

Figure 7.14a, shows the standard or most frequently used type of hole. Unless another type of hole is specified, this standard (STD) hole is assumed. It is round and only slightly larger ($\frac{1}{16}$ in.) than the fastener, allowing a reasonable tolerance for fabrication. Figure 7.14 shows four other types of hole that are allowed by the 1989 AISC Specification. Reference should be made to Table J3.1 in this specification for dimensions.

The principal reason for using oversized or slotted holes is to provide greater flexibility in assembly and "fitting up" the steel frame. Complicated geometry and/or long elements may make it advisable to specify one of these special types of hole. It should be noted that the NSL type (Fig. 7.14e) can be either a short slotted hole or a long slotted hole; however, the slot must be oriented so that the direction of load is

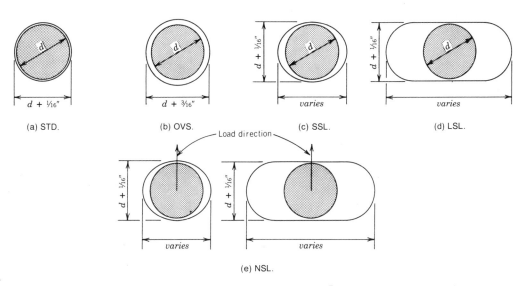

(a) STD. (b) OVS. (c) SSL. (d) LSL.

(e) NSL.

Figure 7.14/Fastener holes for diameters equal to or less than $\frac{7}{8}$ in. (a) Standard hole, (b) oversized hole, (c) short slotted hole, (d) long slotted hole, and (e) short or long slotted hole normal to load direction.

normal to the length of the slot. All slotted holes and oversized holes require use of washers.

Another important aspect of hole specification pertains to the strength design of the connection. Subsequent articles discuss *bearing-type* and *friction-type* connections. The friction type, referred to as "slip-critical" in the AISC specification, generally is more expensive and has lower strength capacity, yet must be used (with some exceptions) if other than STD or NSL holes are specified.

7.7
Bearing-Type Fasteners

A good example of a bearing-type fastener is the standard A307 bolt in a simple lap joint as shown in Fig. 7.15. The hole is slightly larger than the shank of the bolt (Art. 7.6), so that initially there is a loose fit. Even though the nut on the bolt is tightened to a snug fit, the small amount of friction is easily overcome with the application of the load (P), and the pieces "slip into bearing." Observe that the bearing surfaces are on opposite sides of the bolt shank for the full diameter of the bolt and thickness of the pieces that are connected. When using A307 bolts, they must

always be designed as bearing-type fasteners. The same is true for rivets. The degree of pre-tensioning accompanying the cooling of the rivet is unreliable. Consequently, all friction forces are neglected and the rivets are designed as bearing-type.

High-strength bolts (A325 or A490) may be used as bearing-type or as friction-type fasteners (Art. 7.8). If the connection is not considered to be slip-critical or does not subject the bolt to tension, a bearing-type connection may be used. Under these circumstances the bolts need only be tightened to a snug-tight condition, which can be attained with the full effort of a man using an ordinary spud wrench. Consequently, a slip is assumed to take place and the bolts are designed accordingly. The high-strength bolt in a bearing-type connection is the strongest and is more economical than a friction-type. Consequently, it is the most widely used of fastener connections.

7.8
Friction-Type Fasteners in Slip-Critical Connections

High-strength bolts (A325 or A490) have a high tensile-strength. This strength can be utilized by installing the bolt and nut in such a manner that the bolt shank is pretensioned to a very high degree. Such pretensioning provides a clamping effect between the parts being connected so that friction alone can transfer the shear load (Fig. 7.16). The applied force (P) will not exceed the summation of the frictional forces (F); therefore, in theory, the shank

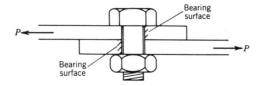

Figure 7.15/Bearing-type fastener.

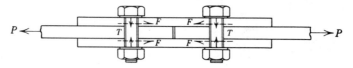

Figure 7.16/Friction-type connection using high-strength bolts.

of the fastener never comes in contact with the sides of the hole.

Slip-critical connections should always be used when no slip at service loads is desired. Also, if the loads are dynamic in nature (earthquake loading) or if the loads place the bolts in direct tension (Fig. 7.13b and c), use of a friction-type connection is required. Finally, it should be noted that oversized holes and most slotted holes require the use of friction-type connectors with washers. It is important to make these observations because the design strength of a friction-type connection is less than that of the bearing type[4] (of the same detail), and yet friction-type connections are more expensive.

The 1989 AISC Specification requires that the pre-tension be 70 per cent of the minimal tensile strength of the bolt. Since this stress capacity varies somewhat with the size of the bolt, the specification also lists the pre-tensioned load for each size bolt. For example, the $\frac{3}{4}$-in. A325 bolt must be tightened to 28 kips, while the same-size bolt of A490 material must be tightened to 35 kips. This pre-tensioning can be accomplished by several means, one of which is use of a calibrated impact wrench. Another is the so-called "turn of the nut" method, which is reliable, is least expensive, and consequently is most frequently used. In this method, the nut is initially made snug (as tight as possible) with ordinary spud wrenches, and then a long-handle wrench is used to achieve additional tightening, usually from one-third to three-quarters of a turn, but sometimes up to a full turn, depending upon the grade of steel and dimensions of the bolt. In addition, a variety of pre-tensioned indicators are used today to accomplish this pre-tensioning. Typical are patented washers

placed under the nut or bolt. The washers have protrusions that indicate when proper pre-tensioning is reached.

Finally there is the *alternate design bolt*. Fasteners of this type incorporate a design feature which indirectly indicates the bolt tension. Typically an extended tip end of the bolt shank will yield or twist off when the bolt reaches the proper tension.

The surface condition of the contact area undergoing friction (faying surface) is an additional consideration when designing friction-type connections. Usually the shop priming coat is held back from the connection areas, giving what is considered to be the standard condition. This is a clean mill scale condition, identified as a Class A surface. If special paint is used on the faying surfaces, the Class A surface may still be achieved, but the paint must first be tested in accordance with strict standards established by the Research Council on Structural Connections.

7.9 Failure of a Fastener Joint

The four principal types of failure with shear-loaded fastener joints are shown in Fig. 7.17; any clamping effect is neglected, and a bearing condition is assumed. The 1989 AISC Specification requires that friction-type connections be designed as bearing connections. The possibility of both shear and bearing failure must be considered in all cases, whereas failure in net section and end shear-out (Fig. 7.17c and d) are possible only when the joint is in tension. It is assumed that the fastener completely fills the hole, permitting a direct transfer of compression stress through the fastener. Except for unusually long fasteners, the possibility of failure by bending is ignored. For example, some bending will actually occur in the fastener shown in Fig. 7.17 and perhaps even more

---

[4]An exception to this is connections where special treatment of frictional surfaces is specified; however, even in these cases, the upper limit is the capacity set by the bearing type.

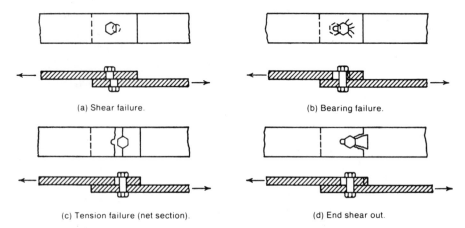

(a) Shear failure.   (b) Bearing failure.

(c) Tension failure (net section).   (d) End shear out.

Figure 7.17/Failure of a fastener joint.

in the fastener shown in Fig. 7.19, but with properly detailed connections within the size range normally encountered in building construction, it does not become critical and consequently is not investigated.

In these early examples (Figs. 7.17 through 7.25), fasteners are shown attaching one steel plate to another. It will be shown later that these fasteners also may be used to connect webs of beams, stems of tees, legs of angles, and the like (Figs. 7.5 through 7.12).

7.10
Strength of a Fastener in Shear

The joint shown in Fig. 7.18 is so constructed that the fastener tends to shear in one plane. In this case, the fastener is said to be in single shear and transfers the load (P) from one plate to the other. The joint

shown in Fig. 7.19 has two shearing planes; therefore, the fastener is in double shear and transfers the load (P) to the two outside plates.

The allowable strength of fastener connections is based upon the ultimate strength (Chapter 11) and ductility characteristics of the entire connection, verified by tests and adjusted by a factor of safety of 2. Consequently, the question of developed real stresses at design loads is not answered. However, the design procedure postulated by the AISC Specification comes from a concept of nominal stress which is developed from the actual service loads. These nominal stresses are always calculated from the nominal size of the shank of the fastener.

The resistance of a fastener to shear depends upon the cross-sectional area of the

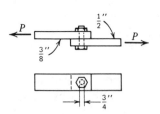

Figure 7.18/Fastener in single shear.

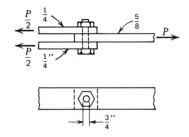

Figure 7.19/Fastener in double shear.

Table 7.1
Allowable Stress in Fasteners (ksi)

	Fastener Designation	Holes[a]	Allowable Shear F_v	Allowable Bearing[b,c] F_p	Allowable Tension F_t
Bearing-type connections	Bolts A307	STD NSL	10.0	1.2F_u or $\dfrac{L_e F_u}{2d}$ whichever is smaller	20.0
	Rivets 502-1	STD	17.5		23.0
	Rivets 502-2, 3	STD	22.0		29.0
	Bolts A325 N	STD NSL	21.0		44.0
	Bolts A325 X	STD NSL	30.0		44.0
	Bolts A490 N	STD NSL	28.0		54.0
	Bolts A490 X	STD NSL	40.0		54.0
Friction-type connections	Bolts A325 F	STD	17.0	1.2F_u or $\dfrac{L_e F_u}{2d}$ whichever is smaller	44.0
		OVS SSL	15.0		44.0
		LSL	12.0		44.0
	Bolts A490 F	STD	21.0		54.0
		OVS SSL	18.0		54.0
		LSL	15.0		54.0

[a]See Art. 7.6.
[b]F_u is the specified minimum tensile strength for the type of steel (see Numerical Values Tables in the 1989 AISC Specification; also see Fig. 7.20).
[c]Under special conditions an upper limit of $F_p = 1.5F_u$ can be used (see complete AISC Specification).

shank and the allowable unit shearing stress of the fastener. The 1989 AISC Specification establishes allowable shearing stresses for all fasteners except welded studs. These allowable stresses are listed in Table 7.1. Examination of Table 7.1 shows that high-strength bolts used in bearing-type connections have a larger allowable shear stress than in friction-type connections. Furthermore, larger stresses are permitted where the bolt threads are excluded from the shear plane (the X designation following the ASTM number). However, the usual case is the N designation, where the threads are not excluded from the shear plane. Also, the F designation represents a friction-type connection.

The shear stress shown for standard A307 bolts is valid for any position of the threads.

The symbol BV (bolt value) or RV (rivet value) is used to designate the strength capacity of a single bolt or rivet. For shear, this is determined by the product of the allowable shear stress and the area, i.e.,

$$BV_s = F_v A_v$$

in which

BV_s = bolt value in shear (kips).

F_v = allowable unit shearing stress (ksi) from Table 7.1.

A_v = nominal cross-sectional area ($\pi d^2/4$) of the shaft or shank (sq in.).

The strength of a fastener in double shear is, of course, twice the single-shear value.

Example

Find the shear bolt value for the following, using standard (STD) bolt holes:

(a) An A307 bolt, $\frac{3}{4}$-in. in diameter, and in single shear (Fig. 7.18)

(b) An A325 high-strength bolt in a slip-critical connection, $\frac{3}{4}$-in. in diameter, and in double shear (Fig. 7.19)

Solution (a)

(1) The area of the bolt is $\pi d^2 / 4$:

$$A_v = \frac{3.14(0.75)^2}{4} = 0.4418 \text{ in.}^2$$

(2) Select the allowable shear stress from Table 7.1. For A307, bearing type,

$$F_v = 10 \text{ ksi}$$

(3) BV_s (single shear) $= F_v A_v = 10(0.4418) = 4.42$ kips.

Solution (b)

(1) Area of $\frac{3}{4}$-in. bolt is 0.4418 in.^{2}.

(2) From Table 7.1 for A325 F,

$$F_v = 17.0 \text{ ksi}$$

(3) BV_s (double shear) $= 2F_v A_v = 2(17.0)0.4418 = 15.0$ kips.

The three standard fastener sizes most frequently used in steel buildings are $\frac{3}{4}$, $\frac{7}{8}$, and 1 in. Also, the Ninth Edition of the AISC Manual (Part 4) contains tables listing load values in kips for all rivets and threaded fasteners (excluding welded studs) in sizes from $\frac{5}{8}$ through $1\frac{1}{2}$ in. using the values shown in Table 7.1.

7.11
Strength of a Fastener in Bearing and End Shear-Out

The strength of a fastener in bearing must be investigated for both bearing and slip-critical connections. Once again, the design procedure calls for neglecting the clamping effect in the slip-critical connection.

The assumed bearing on a fastener is the force exerted on that fastener by the steel through which it passes. Although the bearing area is the cylindrical surface of contact between the fastener shank and the holed steel member, the area used in computations is the projected area of the surface, i.e., the area of a rectangle, the dimensions of which are the fastener diameter and the holed metal thickness. Here again, the nominal diameter of the fastener is used. Expressed in equation form, the bearing area is

$$A_b = dt$$

If two plates of different thickness are fastened together as shown in Fig. 7.18, it is evident that the bearing capacity will be determined by the thickness of the thinner plate. The area for a $\frac{3}{4}$-in. fastener in this case is

$$A_b = 0.75(0.375) = 0.281 \text{ in.}^2$$

The strength of a fastener in bearing is equal to the allowable bearing stress F_p times the bearing area, or

$$\text{BV}_b = F_p A_b$$

The quantity F_p requires further discussion.

Actual tests of fastener joints have revealed that neither the fastener nor the metal in contact with the fastener shank actually fails in bearing. The tests show that the bearing strength is a function of the tensile strength of the connected part, the spacing of the fasteners, and the distance from the fastener to the end of the material. The upper limit of the bearing strength has been established as $1.5F_u$, where F_u is the specified minimum tensile strength of the connected part. A table listing values of F_u for different grades of steel is furnished in Numerical Values Table 2 of the 1989 AISC Specification. For example, A36 Steel may have F_u varying from 58 to 80 ksi, while A588 Steel, Grade 50, has $F_u = 70$ ksi.

The maximum allowable bearing strength $(F_p = 1.5F_u)$ can be used only if the spacing of the fasteners and end distance (in the direction of the stress) meet limiting specifications described later in this article; otherwise F_p must be reduced. For example, the allowable bolt values based upon bearing (BV$_b$) listed in the Ninth Edition of the AISC Manual are based upon $F_p = 1.2F_u$. This is the allowable stress if, rather than meeting the detailed spacing requirements described later in this article, a standard spacing of $3d$ and a standard end distance of $1\frac{1}{2}d$ are used. Also, in all connections with a single bolt in the line of force, the specification sets the limit of F_p equal to $1.2F_u$ or a value based upon the formula

$$F_p = \frac{L_e F_u}{2d}$$

whichever is smaller. This formula is de-

rived later in this article. Most designers prefer to use these standards rather than meeting the special conditions that permit $F_p = 1.5F_u$.

Assuming the plates of the connection in Fig. 7.18 to be of A36 Steel with a minimum specified tensile strength of 58 ksi, the maximum bearing stress is

$$F_b = 1.2(58) = 69.2 \text{ ksi}$$

and the capacity of the fastener in bearing is

$$BV_b = 69.2(0.281) = 19.6 \text{ kips.}$$

In the joint shown in Fig. 7.19, the thickness of the thinnest plate is $\frac{1}{4}$ in., but the two $\frac{1}{4}$-in. plates act in the same direction; therefore, the load (P) must be divided between them. Assuming $\frac{3}{4}$-in. fasteners, the bearing area acting to the left of this joint is

$$A_b = (0.25 + 0.25)0.75 = 0.375 \text{ in.}^2$$

The bearing area of the $\frac{5}{8}$-in. plate on the right is

$$A_b = 0.625(0.75) = 0.469 \text{ in.}^2$$

It is apparent that the least total thickness in any one direction will establish the limiting capacity of the fastener in bearing.

Figure 7.20 shows three fasteners in a line parallel to the direction of the stress. The

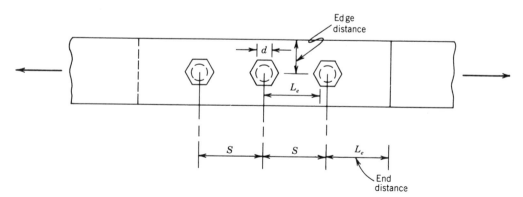

Figure 7.20/Spacing, end, and edge distances.

Table 7.2
Minimum End and Edge Distances

Fastener Diameter (in.)	At Sheared Edges (in.)	At Rolled or Gas-Cut Edges (in.)
$\frac{1}{2}$	$\frac{7}{8}$	$\frac{3}{4}$
$\frac{5}{8}$	$1\frac{1}{8}$	$\frac{7}{8}$
$\frac{3}{4}$	$1\frac{1}{4}$	1
$\frac{7}{8}$	$1\frac{1}{2}$	$1\frac{1}{8}$
1	$1\frac{3}{4}$	$1\frac{1}{4}$
$1\frac{1}{8}$	2	$1\frac{1}{2}$
$1\frac{1}{4}$	$2\frac{1}{4}$	$1\frac{5}{8}$
Over $1\frac{1}{4}$	$1\frac{3}{4}\times$ diameter	$1\frac{1}{4}\times$ diameter

fastener spacing is always taken as center to center of fasteners. The end distance is measured from the center of the fastener to the end of the member, parallel to the direction of the stress. This distance (L_e) also is seen to be the length from the center of one fastener to the nearest side of the next hole in the direction parallel to the stress. The edge distance is measured from the center of a fastener to the edge of the material, perpendicular to the direction of stress.

To prevent gaps and bulging of the material near its edge, a minimum edge distance is specified. This minimum edge distance also is the minimum end distance. Table 7.2 shows the 1989 AISC Specification values for these minimum distances, the variables being the diameter of the fastener and the edge condition of the material. This table is for standard holes only. Oversized and slotted holes require an incremental increase over these distances. (See the complete AISC Specification.)

As was stated earlier, the variables controlling the allowable bearing stress are the tensile strength of the part, the diameter of the fastener, and the spacing (or more specifically, L_e). Tests have show that there is a linear relationship between the ratios of these parameters, i.e.,

$$\frac{f_p}{F_u} = \frac{L_e}{d}$$

In this form, f_p is the developed bearing stress. By solving for f_p and using a factor of safety of 2, it then becomes the allowable bearing stress with the upper bound of $1.2F_u$:

$$F_p = \frac{L_e F_u}{2d} \le 1.2F_u$$

This equation can be further adjusted to represent the actual fastener load P and the required end distance L_e. Let $F_p = P/dt$; then

$$\frac{P}{dt} = \frac{L_e F_u}{2d}$$

and

$$L_e = \frac{2P}{tF_u}$$

Also, it is seen that the required center-to-center spacing is

$$S = \frac{2P}{tF_u} + \frac{d}{2}$$

These requirements for spacing, end distance, and bearing values are for standard holes. Oversized and slotted holes require an incremental increase over these distances. (See the complete AISC Specification.)

Another usual spacing of fasteners is 3 in.; it is adequate for a bearing stress of $1.5F_u$ for fastener sizes up to and including $\frac{3}{4}$ in. The 1989 AISC Specification recommends a minimum preferred spacing of $3d$ and an absolute minimum of $2\frac{2}{3}d$. The bearing values for $\frac{3}{4}$-, $\frac{7}{8}$-, and 1-in. fasteners for various material thicknesses are given in Part 4 of the Ninth Edition of the AISC Manual.

The limiting value for the capacity of any given fastener is the smallest value calculated for shear and bearing as described herein.

Example 1

Find the limiting bolt value for the $\frac{3}{4}$-in. bolt shown in Fig. 7.19. The parent metal is A36 Steel, and the bolts are of high-strength A325 Steel. The connection is of bearing type, with threads in the shear plane. Assume standard (STD) holes and end distances for maximum bearing stress.

Solution

(1) Determine the bolt value in shear. The bolt shank is in double shear, and the area of a $\frac{3}{4}$-in. bolt is 0.442 in.^{2}. From Table 7.1, $F_v = 21$ ksi; thus

$$\text{BV}_s \text{ (double shear)} = 2F_v A_v$$

$$= 2(21)0.442$$

$$= 18.6 \text{ kips}$$

(2) Determine the bolt value in bearing. The combined thickness of the two $\frac{1}{4}$-in. plates is $\frac{1}{2}$ in., which is less than the $\frac{5}{8}$-in. plate and therefore is controlling. The area in bearing is

$$0.5(0.75) = 0.375 \text{ in.}^2$$

Also

$$F_p = 1.2F_u = 1.2(58) = 69.6 \text{ ksi}$$

Therefore

$$\text{BV}_b \text{ (bearing)} = F_p A_b$$

$$= 69.6(0.375)$$

$$= 26.1 \text{ kips}$$

(3) Establish the limiting value. The limiting value will be the smaller of the capacities computed on the basis of shear and bearing, i.e.

$$\text{BV} = 18.6 \text{ kips}$$

These values (18.6 kips and 26.1 kips) can be obtained directly from the shear and bearing tables for fasteners in the Ninth Edition of the AISC Manual.

Example 2

Find the limiting bolt value for the 1-in. bolts shown in Fig. 7.21. The parent metal is A36 Steel and the bolts are A490. The connection is a bearing type with threads excluded from the shear plane, and the holes are standard. Determine the required end distance.

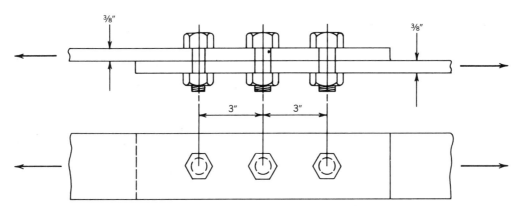

Figure 7.21/Example 2.

Solution

(1) Determine the bolt value in shear. The bolt shank is in single shear, and the area of a 1-in. bolt is 0.785 in.^{2}.

From Table 7.1, $F_v = 40$ ksi; therefore

$$BV_s = F_v A_v = 40(0.785) = 31.4 \text{ kips}$$

(2) Determine the bolt value in bearing. Assuming adequate spacing and end distance, the maximum bearing pressure is

$$F_p = 1.5F_u = 1.5(58) = 87 \text{ ksi}$$

The bearing pressure based upon fastener spacing is

$$F_p = \frac{L_e F_u}{2d}$$

where

$$L_e = 3 - \frac{1.125}{2} = 2.44$$

and

$$F_p = \frac{2.44(58)}{2(1)} = 70.8 \text{ ksi} \qquad \text{(governs)}$$

The bearing area is $A_v = 1(0.375) = 0.375$ in.^{2}. Consequently,

$$BV_b = F_p A_b = 70.8(0.375) = 26.6 \text{ kips}$$

(3) The limiting bolt value is the one in bearing—i.e., BV = 26.6 kips.

(4) Using $P = 26.6$ kips, calculate the required end distance:

$$L_e = \frac{2P}{tF_u} = \frac{2(26.6)}{0.375(58)} = 2.45 \text{ in.}$$

This end distance is larger than the minimum listed in Table 7.2 and hence must be specified.

7.12
Gross and Net Section

The unit stress in tension members is considered to be uniform throughout the entire cross section. Expressed in equation form,

$$f_t = \frac{P_t}{A}$$

where:

f_t = unit tensile stress (psi).

P_t = total force (lb).

A = total area resisting stress.

The presence of fastener holes reduces the effective area available to resist stress. Obviously, the strength of a tension member should be based upon the area available to resist stress after the holes are punched or reamed to full size. This is referred to as *net section*. Fastener holes are punched $\frac{1}{16}$ in. larger in diameter than the fastener. Punching holes in steel damages a small amount of the metal around the perimeter

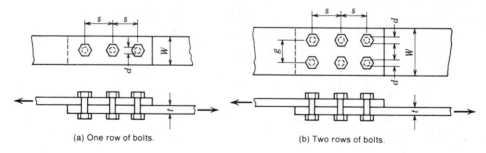

(a) One row of bolts.   (b) Two rows of bolts.

Figure 7.22/Net section.

of the hole. Consequently, the AISC Specification provides that in computing net section, the holes should be assumed $\frac{1}{16}$ in. larger in diameter than their punched dimension, or $\frac{1}{8}$ in. larger than the nominal fastener diameter.

Where one hole is involved, i.e., a single row of fasteners along the line of stress, the net section A is determined by multiplying the thickness of the member by its net width (width of member minus diameter of hole). For example, in Fig. 7.22a, the net section is

$$A_{net} = t(W - d)$$

When more than one row of holes are involved and holes are not staggered, but are as shown in Fig. 7.22b, the net section is

$$A_{net} = t(W - nd)$$

where n is the number of holes across the section (in Fig. 7.22b, $n = 2$).

The AISC Specification also requires that in no case shall the net section through a hole be considered to be more than 85 per cent of the corresponding gross section.

Fasteners may be placed in a variety of patterns, depending upon the available area and the shape of the members to be connected. A saving in material is achieved by keeping the fasteners in as compact a group as possible. However, as was shown in Art. 7.11, a lesser spacing can reduce the bearing capacity of a fastener. A row of fasteners generally refers to a line of fasteners placed parallel to the line of stress in a member. Figure 7.22a shows one row of fasteners, and (b), two rows. The *fastener pitch* is the center-to-center spacing of fasteners in a row. The symbol for this is s. The *fastener gage* (g) is another dimension, this time measured perpendicular to the line of stress in the joint. In a lap joint, the gage is the center-

to-center distance between rows of fasteners.

Figure 7.23 (a) and (b) show fasteners placed in staggered rows or in a zigzag pattern. Possible paths of tension failure are indicated by the dashed lines. It would seem reasonable to figure net section on the basis of the shortest path, because, with a constant plate thickness, that would involve the least area. However, experience has shown that failure does not generally occur in this manner. The area on the diagonal, between B and C in Fig. 7.23a, actually is capable of resisting higher stresses. The AISC Specification, therefore, recommends an empirical formula for determining net sections with staggered rows of fasteners which must be applied for all possible paths of tension failure:

$$A_{net} = t\left(W - nd + \sum \frac{s^2}{4g}\right)$$

where:

A = net area.

s = pitch.

g = gage.

d = diameter of hole.

The summation is over all diagonal paths of possible failure.

The least area resulting from applying the above formula is the net area on which the strength of such a tension member could be based.

This net section area (A_{net}), calculated as described here and applicable to splice and gusset plates only, cannot be greater than 85 per cent of the gross area. If 85 per cent of the gross area is less than the actual net area, then the former becomes the net area that must be used.

One further refinement must be considered. Frequently, a structural shape (tee, **W**, **S**, etc.) will be connected by only one

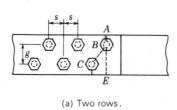

(a) Two rows.

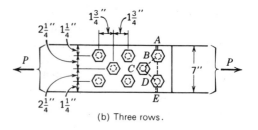

(b) Three rows.

Figure 7.23/Staggered bolts.

part of the cross-section—e.g., only the web of a **W** shape. Consequently, the entire area of the shape near the connection is not in uniform tension. To account for this, the 1989 AISC Specification requires that an additional coefficient (U) be used to reduce the net area to an *effective* net area, i.e.

$$A_e = U A_{net}$$

The coefficient U varies between 0.75 and 1.0, and the complete specification can be consulted. In particular, however, in all members (including flat bars) with bolted or riveted connections having only two fasteners per line in the direction of the stress, U should be 0.75.

Example

Determine the maximum permissible force P for the tension splice shown in Fig. 7.23b, as determined from the net section, if the steel is A36 and $\frac{3}{4}$-in. bolts are used. Neglect shear and bearing. The members are 7 in. wide and $\frac{1}{2}$ in. thick.

Solution

(1) Determine possible paths of tension failure. These are shown by the dashed lines *ABDE* and *ABCDE*.

(2) Applying the net section formula, determine the area for each path. Path

ABDE:

$$A_{net} = t(W - nd) = \tfrac{1}{2}[7 - 2(\tfrac{7}{8})]$$
$$= 2.63 \text{ in.}^2$$

Path *ABCDE*:

$$A_{net} = t\left(W - nd + \sum \frac{s^2}{4g}\right)$$

$$= 0.5\left[7 - 3(0.875)\right.$$

$$+ \frac{1.75^2}{4(2.25)} + \frac{1.75^2}{4(2.25)}\left.\right]$$

$$= 0.5[(7 - 2.63) + 0.34 + 0.34]$$

$$= 2.53 \text{ in.}^2$$

(3) Determine the maximum net section using 85 per cent of the gross section:

$$A_{max} = 0.85(0.5)7 = 2.98 \text{ in.}^2$$

(4) Determine the effective net area. In this example, using plates with more than two bolts in line, the coefficient U does not apply, and A_{max} is greater than A_{net}. Therefore, A_e is equal to the smallest A_{net}, i.e.,

$$A_e = 2.53 \text{ in.}^2$$

(5) Calculate the maximum tensile force based upon the effective net area:

$$P_t = A_e 0.5 F_u = 2.53(0.5)58 = 73.4 \text{ kips}$$

(6) Calculate the maximum tensile force based upon the gross area $A_g = 0.5(7) = 3.5$ in.^{2}:

$$P_t = A_g 0.6 F_y = 3.5(0.6)36 = 75.6 \text{ kips}$$

(7) The limiting tensile force is 73.4 kips.

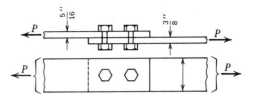

Figure 7.25/Single shear.

7.13
Design and Investigation of Lap-Type Fastener Connections

Lap splices, hangers, and similar types of connection usually have fasteners so positioned that the center of the fastener group coincides with the center of gravity of the material to be connected. Under these conditions, the connection is concentrically loaded. This is the case in Example 1 (Fig. 7.24) and nearly so in Example 2 (Fig. 7.26). The degree of eccentricity in Example 2 is so small that normally it is neglected in design.

The design of ordinary concentric connections is based on the assumption that the load is equally distributed among all the fasteners in the connection. Consequently, the number of fasteners required may be found by dividing the total load to be transmitted by the limiting value of one fastener in shear or bearing, whichever is smaller. Conversely, when investigating a connection to determine the safe load it will carry, compute the strength of one fastener as governed by shear or bearing and then multiply by the number of fasteners in the connection. Also, observe that

the load transfer shown in Fig. 7.24 is different from that shown in Fig. 7.25. In Fig. 7.24, two fasteners are used to transfer the load to the two outside plates, and two more are required to transfer the load from the outside plates back to the center plate. In Fig. 7.25, the fasteners transfer the load directly from one plate to the other. Again, it should be emphasized that if the members are in tension, net section also should be investigated to determine the safe load.

The 1989 AISC Specification requires that all connections be designed to support a minimum load of 6 kips. Furthermore, it is customary to use no less than two fasteners for any connection even though the computations indicate that one fastener is sufficient; this simply is a precaution against the obvious fact that failure for any reason would be total if only one fastener is used.

Summarizing:

(1) Investigate shear:

$$BV_s = F_v A_v$$

(2) Investigate bearing:

$$BV_b = F_p A_b$$

where

$$F_p = \text{the smaller of } 1.2\, F_u \text{ and } \frac{L_e F_u}{2d}$$

with d being the nominal diameter of the fastener, and L_e, the required length, given by $2P/tF_u$ (in which P is the force transmitted by the fastener).

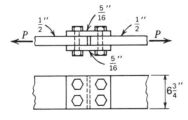

Figure 7.24/Example 1.

(3) Investigate gross section. The maximum tensile load on the part is

$$P_t \leq 0.6 F_y A_g$$

(4) Investigate effective net section. Determine the net area:

$$A_{net} = t(W - nd)$$

or

$$A_{net} = t\left(W - nd + \frac{s^2}{4g}\right)$$

or

$$A_{net} = 0.85 A_g$$

where

$$d = \text{diameter of the hole} + \tfrac{1}{16} \text{ in.}$$

Then, using the least A_{net} determined above, calculate the effective net section,

$$A_e = U A_{net}$$

and the maximum tensile load on the part,

$$P_t = 0.5 F_u A_e$$

Example 1

Determine the load P that the joint shown in Fig. 7.24 can transmit if $\tfrac{3}{4}$-in. A325 bolts are used and the parent steel is A36. The bolts are bearing type and the threads are not excluded from the shear planes. Standard holes are used. Calculate the required end distance for both inner and outer plates.

Solution 1

(1) Determine the value of one of the bolts in double shear. The shear area of one $\tfrac{3}{4}$-in. bolt is 0.4418 in.^{2}. From Table 7.1, the shear strength of the bolts is 21 ksi. Thus

$$\text{BV}_s(\text{double shear}) = 2 F_v A_v$$
$$= 2(21)0.4418$$
$$= 18.6 \text{ kips}$$

(2) Determine the value of one of the bolts in bearing. The two outside plates have a total thickness of $\tfrac{5}{8}$ in., compared to $\tfrac{1}{2}$ in. for the inner plate. Therefore, bearing on the inner plate controls, and the area in bearing is $(\tfrac{1}{2})\tfrac{3}{4} = 0.375$ in.^{2} so

$$\text{BV}_b(\text{bearing}) = 1.2 F_u A_b$$
$$= 1.2(58)0.375$$
$$= 26.1 \text{ kips}$$

(3) From steps (1) and (2), the limiting value for one fastener is governed by shear and is 18.6 kips. Two bolts transfer the load to the outside plates; therefore, the total load P that the bolts can transmit is

$$P = 2(18.6) = 37.2 \text{ kips}$$

(4) Investigating net section, again it is observed that the total thickness of the two $\tfrac{5}{16}$-in. plates is greater than that of the $\tfrac{1}{2}$-in. plate. The $\tfrac{1}{2}$-in. plate, therefore, will provide the lesser net section. The hole is taken as $\tfrac{1}{8}$ in. larger than the bolt size, and the net section is

$$A = t(W - nd) = 0.5[6.75 - 2(0.875)]$$
$$= 2.5 \text{ in.}^2$$

whereas

$$A_{max} = 0.85(6.75)0.5 = 2.87 \text{ in.}^2$$

Therefore, use the smaller value ($A_{net} = 2.5$ in.^{2}).

(5) The total tensile load (P_t), as governed by effective net section, is

$$P_t = 0.5(58)2.5 = 72.5 \text{ kips}$$

(6) The total tensile load (P_t), as governed by the gross section, is

$$P_t = 0.6 F_y A_g$$
$$= 0.6(36)0.5(6.75) = 72.9 \text{ kips}$$

(7) The limiting value for P is that established by bolt shear (step 3) and is 37.2 kips.

(8) Since each bolt develops 18.6 kips, the minimum end distance for the inner

plate is

$$L_e = \frac{2(18.6)}{0.5(58)} = 1.28 \text{ in.}$$

This is seen to be larger than that shown in Table 7.2 and therefore will be adopted, increasing the dimension to the nearest $\frac{1}{8}$ in., or $1\frac{3}{8}$ in.

(9) Each outside plate develops half the bolt value, or $18.6/2 = 9.3$ kips. Therefore, its minimum end distance is

$$L_e = \frac{2(9.3)}{0.31(58)} = 1.03 \text{ in.}$$

This is seen to be less than that shown for a sheared edge in Table 7.2; therefore, adopt a $1\frac{1}{4}$-in. end distance for the outer plates.

Solution 2

Note. Solution 1 demonstrates the general method for solving problems of this type. The following method, involving the use of fastener tables, is the one most frequently used in practice (and the one that will be used in subsequent examples where directly applicable).

(1) Determine the bolt value. Referring to the allowable shear load tables for "Bolts, Threaded Parts, and Rivets" (Ninth Edition of the AISC Manual), for $\frac{3}{4}$-in., A325 bolts, double-shear bearing type, with threads in the shear plane, $BV_s = 18.6$ kips. Referring to the AISC bearing load table for $F_u = 58$ ksi (A36 Steel) and $\frac{3}{4}$-in. bolts on $\frac{1}{2}$-in. plates, $BV_b = 26.1$ kips when spacing does not govern bearing strength. The smaller of the two, shear or bearing, limits the bolt-value capacity.

(2) Determine the load (P) that can be transmitted by the two bolts:

$$P = 2(18.6) = 37.2 \text{ kips}$$

(3) Investigate net section and end distance as shown in steps (1) through (9) of Solution 1.

Example 2

A typical strut–gusset-plate connection used in truss construction (Chapter 9) is shown in Fig. 7.26. The two angles that straddle the gusset plate carry a total tensile load of 62 kips. The parent steel (angles and gusset) is A572, Grade 65. Design the connection using $\frac{3}{4}$-in., A490, high-strength bolts in a friction-type connection using standard holes.

Solution

In Fig. 7.26a, the centroid of the angles, and consequently the line of action of the 62-kip load, is denoted by the symbol y. This distance is given in the AISC Manual as 1.09 in. for 2L $4 \times 4 \times \frac{1}{4}$. Therefore, the line of action for this size angle does not exactly coincide with the line of bolts located at a typical gage distance of $2\frac{1}{2}$ in.; however, the degree of eccentricity is sufficiently small to permit the assumption of a concentric load.

(1) Determine the bolt value based on shear:

$$BV_s = 18.6 \text{ kips}$$

(2) Determine the bolt value based on bearing. With the type of steel specified, $F_u = 80$ ksi. The bearing load tables in the Ninth Edition of the AISC Manual do not list values for this case; therefore, calculate the bearing area. The limiting area will be that for the $\frac{3}{8}$-in. plate:

$$A_b = 0.375(0.75) = 0.28 \text{ in.}^2$$

Assuming a 3-in. spacing, calculate L_e:

$$L_e = 3 - 0.44 = 2.56 \text{ in.}$$

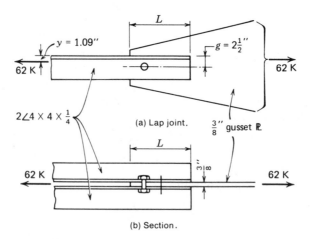

(a) Lap joint.

(b) Section.

Figure 7.26

The allowable bearing stress based upon spacing is

$$F_p = \frac{2.56(80)}{2(0.75)} \simeq 137 \text{ ksi}$$

However, the maximum allowable bearing stress is

$$F_p = 1.2(80) = 96 \text{ ksi}$$

Therefore, the bolt value in bearing is

$$BV_b = 0.28(96) = 26.9 \text{ kips}$$

which, because it is larger than the value based on shear, cannot be used.

(3) Calculate the number of bolts required:

$$n = \frac{62}{18.6} = 3.33; \quad \text{use four bolts}$$

(4) Check gross and net section. The maximum tensile force based on gross section is

$$P_t = 0.6 F_y A_g$$

$$P_t = 0.6(65)3.88 = 151 \text{ kips}$$

This is larger than the applied load of 62 kips; therefore, it is accepted. The net section is the gross area minus two $\frac{7}{8}$-in. holes in the $\frac{1}{4}$-in. thick material:

$$A_{net} = 3.88 - 2\left(\tfrac{7}{8}\right)\tfrac{1}{4} = 3.44 \text{ in.}^2$$

However, the maximum net area is limited to 85 per cent of the gross; therefore

$$A_{net} = 0.85(3.88) = 3.30 \text{ in.}^2$$

The maximum tensile load based upon effective net area is

$$P_t = 0.5 F_u A_e$$

$$P_t = 0.5(80)3.3 = 132 \text{ kips}$$

This is larger than the applied load of 62 kips and is accepted.

(5) Establish end distance and calculate the length of the connection. The bearing of one bolt against the $\frac{3}{8}$-in. plate is $62/4 = 15.5$ kips, and

$$L_e = \frac{2(15.5)}{0.375(80)} = 1.03 \text{ in.}$$

Use $1\frac{1}{4}$ in. (Table 7.2).

The bearing of one bolt against the $\frac{1}{4}$-in. angle is $15.5/2 = 7.75$ kips, and

$$L_e = \frac{2(7.75)}{0.25(80)} = 0.78 \text{ in.}$$

Use $1\frac{1}{4}$ in. (Table 7.2).

The total length of the connection is seen to be

$$L = 3(3) + 2\left(1\frac{1}{4}\right) = 11\frac{1}{2} \text{ in.}$$

PROBLEMS

1. Determine the maximum force (P) that the simple lap joint shown in Fig. 7.27 can safely resist if 1-in., A325 bolts are used in a friction-type connection with standard holes. The parent steel is A36. Calculate the minimum end distance permitted by the AISC Specification. (Answers given in Appendix G.)

resist when $\frac{7}{8}$-in., A307 bolts are used with standard holes. Also, determine the minimum end distance. The plates are of A36 Steel and gas-cut. (Answers given in Appendix G.)

4. Design a splice joint similar to that shown in Fig. 7.28 for a tensile force of 70 kips. All plates are to be $\frac{3}{8}$-in., A36 Steel. Use $\frac{5}{8}$-in., A325 high-strength bolts with threads in the shear plane. The maximum width of plate is limited to 10 in. Make a sketch of the joint showing all dimensions.

5. Two channels (**C** 8×11.5) are in tension and are attached to a structural tee (**WT** 9×30) which, in turn, is welded to a beam (Fig. 7.29). The total design load for the channels is 70 kips. Using $\frac{3}{4}$-in., bearing-type, A325 high-strength bolts, determine the number of bolts required, and show their arrangement and spacing for the connection between the channels and the structural tee. The bolts have threads in the shear plane, and all members are of A36 Steel.

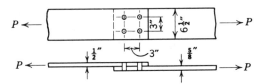

Figure 7.27

2. What changes would result in Problem 1 if 1-in., A502-1 rivets were used instead of high-strength bolts?

3. Determine the maximum force (P) that the butt-splice joint shown in Fig. 7.28 can safely

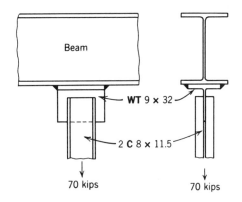

Figure 7.29

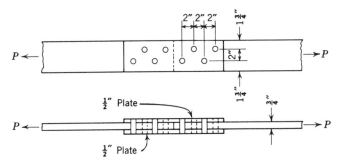

Figure 7.28

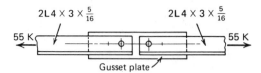

Figure 7.30

6. Design a tension splice between two sets of double-angle ties that straddle a gusset plate (Fig. 7.30). Use $\frac{3}{4}$-in., A325 bolts with threads in the shear plane. The angles are 2L4×3×$\frac{5}{16}$ and carry an axial load of 55 kips. All steel is A36. Select a gusset-plate thickness and spacing so that bearing will not control. Detail the connection. (Answer given in Appendix G.)

7.14
Framing Angles—
Fastener-Connected

The type of connection shown in Fig. 7.6 is the one most frequently used to connect floor beams to other beams or to girders and columns. The angles generally are used in pairs, and the legs that straddle the web of the member being supported are called connected legs because they usually are attached to the member in the fabricating shop before the assembly is transported to the building site. The other legs are called outstanding legs.

When the connected legs of framing angles are attached only to the beam web, enough end rotation (Art. 7.3) is permitted to enable the connection to be classified as a flexible type and thus appropriate for Type 2 construction (beams can be assumed to be simply supported). Furthermore, unless the framing angles are welded (Art. 7.30), any eccentricity is ignored in the design procedure, i.e., the beam reaction is treated as though it were concentric with the row(s) of fasteners that transfer the load from the web to the framing angles. The beam reaction also is treated as through it were concentric with the row(s) of fasteners transferring the load from the angles to the supporting member. The thickness of framing angles purposely is kept small to assure necessary flexibility.

The stresses that must be considered are: shear in the fastener; bearing on the beam webs, column flange, and framing angle; and gross shear on the vertical section through the beam web and framing angles. The limiting fastener value is determined as follows:

A.   Connected leg:

    1.   Double shear for fasteners through the beam web.

    2.   Bearing on the web.

B.   Outstanding leg:

    1.   Single shear for fasteners through the girder web or column flange.

    2.   Bearing on the girder web, column flange, or angle thickness.

Traditionally, there were three standard sizes of framing angles: $4 \times 3\frac{1}{2}$ in., 4×6 in., and 6×6 in. They were the most suitable for use of standard $\frac{3}{4}$-in. rivets in one or two vertical rows, with adequate driving clearances and standard gage dimensions. With the currently more common high-strength bolts, however, use of other sizes often is required to allow for nut, washer, bolt-projection, and impact-wrench clearances. Also, it is not unusual to find staggered lines of fasteners between the outstanding and connected legs to provide the needed clearances.

Confirming test data are incomplete for framing angles using two lines of bolts in a 6-in. leg. Consequently, until such data become available, it is recommended that bolts be limited to one line in the connected legs and one line in each outstanding leg.

The thickness of framing angles generally is so selected that bearing on the angle legs will not control the fastener load value. The thickness of the angles also can be determined by limiting the allowable shear to $0.40F_y$ on the gross area of the vertical section through the connecting angles. The angle sizes and fastener spacings shown in Fig. 7.31 are recommended unless further investigation indicates that increased clearances are required. It is customary to arrange fasteners symmetrically in beam connections, even though this may necessi-

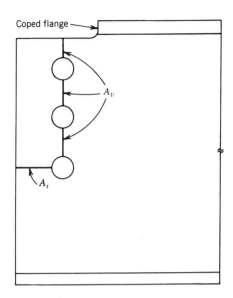

Figure 7.32/Block shear.

tate more fasteners than are actually required to resist the total stress.

When the top flange is coped as shown in Figs. 7.6 and 7.32, an investigation of web tear-out (referred to as "shear rupture") is required. The beam reaction cannot exceed the combined shear and tension strength along the net areas as shown in Fig. 7.32. In this case, the 1989 AISC Specification limits the shear strength (F_v) to 30 per cent of the minimum specified tensile strength (F_u). Expressed in equation form,

$$R = 0.30F_u A_v + 0.50F_u A_t$$

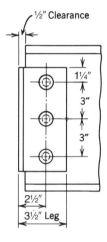

(a) Connected leg.

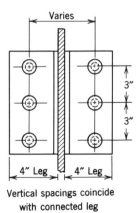

Vertical spacings coincide
with connected leg

(b) Outstanding leg.

Figure 7.31/Framing-angle details, $\frac{3}{4}$- and $\frac{7}{8}$-in. fasteners.

Example

Design the framing-angle connection for a **W** 16×40 girder framing to the flange of a **W** 10×33 column, both of A36 Steel. The end reaction is 42 kips. Use $\frac{3}{4}$-in. A325 bolts in a bearing-type connection with threads in the shear plane and standard holes.

Solution

(1) Determine the limiting bolt value for the connected leg:

$$BV_s = 2(21)0.442$$

$$= 18.6 \text{ kips}$$

$$BV_b = 1.2(58)0.75(0.305)$$

$$= 15.9 \text{ kips}$$

The limiting bolt value is 18.6 kips, and the minimum angle thickness is $0.5(0.305) = 0.153$ in.; use $\frac{1}{4}$ in.

(2) Calculate the number of bolts required in the connected leg:

$$n = \frac{42}{18.6} = 2.64$$

Three bolts are needed.

(3) Determine the limiting bolt value for the outstanding leg:

$$BV_s = 21(0.442)$$

$$= 9.28 \text{ kips}$$

$$BV_b = 1.2(58)0.75(0.25)$$

$$= 13.05 \text{ kips}$$

The limiting bolt value is 9.28 kips.

(4) Calculate the number of bolts required in the outstanding legs:

$$n = \frac{42}{9.28} = 4.53$$

Five bolts are needed, but use six for symmetry.

(5) Select a suitable arrangement for the bolts, and detail the joint according to the details shown in Fig. 7.31. The **W** 16×40 has a T depth of $13\frac{5}{8}$ in.; therefore, the length of angle is limited to $13\frac{5}{8}$ in. However, this is adequate for three fasteners in one row. Figure 7.33 shows one suitable arrangement, using three fasteners in each leg.

(6) Check shear on the net area of the angles:

$$F_v = 0.3(58) = 17.4 \text{ ksi}$$

There are two angles, each $8\frac{1}{2}$ in. long. The thickness required for shear is

$$t = \frac{42}{17.4\left[8.5 - 3\left(\frac{13}{16}\right)\right]2} = 0.20$$

The $\frac{1}{4}$-in. angle thickness is satisfactory.

The AISC Manual lists standard framed beam connections and their respective al-

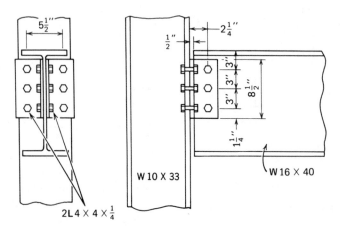

W 10 X 33

W 16 X 40

2L 4 X 4 X $\frac{1}{4}$

Figure 7.33/Framing angles.

lowable loads for various beam sizes, numbers and types of fasteners, sizes of fasteners, and thicknesses of framing angles. Part 4 of the AISC Manual lists standard framed beam connections using one row of fasteners in each leg.

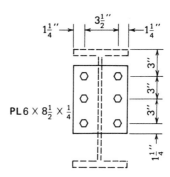

Figure 7.34/End shear plate.

7.15
End-Plate Shear Connections

Modern fabricating techniques have led to the development of end-plate shear connections (Fig. 7.7), which have proven economical if the beam reactions are in the light to medium range. These are flexible-type connections similar to framing angles (Art. 7.14) with a single plate replacing the paired angles. Precision in shop fabrication is required to assure that the plates, when welded to the beam web, are accurate in their length and exact in their orientation. End plates are treated in greater detail under welded connections (Art. 7.30); therefore, for purpose of the discussion here, it is assumed the plates are adequately welded to the beam.

The plate is attached to the supporting member by standard or high-strength bolts. To ensure adequate end rotation capacity, plate thickness is limited to $\frac{3}{8}$ in., plate depth is limited to $T = d - 2k$, and only two rows of bolts are permitted. The transverse spacing must be between $3\frac{1}{2}$ and $5\frac{1}{2}$ in., and the edge distance must be $1\frac{1}{4}$ in. (Fig. 7.34).

With the specifications described here, the design procedure assumes that there is no eccentricity and that the stresses are as follows:

1. Shear on bolts (single shear).

2. Bearing on end-plate thickness or thickness of supporting member, i.e., column flange or girder web.

Example

Design the end shear-plate connection for a **W** 14×34 beam framing to the flange of a **W** 8×28 column, both of A36 Steel. The end reaction is 43 kips. Use $\frac{3}{4}$-in., A325 bolts in a bearing-type connection with threads in the shear plane and standard holes.

Solution

(1) Determine the bolt value in single shear:

$$BV_s = 9.3 \text{ kips}$$

(2) Establish the plate thickness by determining a thickness and bolt spacing that will be sufficient to ensure that shear controls the bolt value. From bolt tables (bearing) for $\frac{3}{4}$-in. bolts, A36 Steel, and 3-in. spacing, a $\frac{3}{16}$-in. thickness gives a bolt value of 12.2 kips and could be adopted. However, to allow for proper welding to the beam web (Art. 7.30), a $\frac{1}{4}$-in. plate will be selected.

(3) Determine the number of bolts required:

$$n = \frac{43}{9.3} = 4.62; \qquad 6 \text{ bolts are needed}$$

(4) Check depth of plate with beam dimensions and other details. Depth of plate:

$$L = 2(3) + 2\left(1\tfrac{1}{4}\right) = 8\tfrac{1}{2} \text{ in.}$$

Allowable plate depth:

$$T = 12 \text{ in.}$$

Since $8\tfrac{1}{2} < 12$ in., accept. (See Fig. 7.34 for other details.)

The AISC Manual contains a safe-load table for standard end-plate shear connections for $\tfrac{3}{4}$- and $\tfrac{7}{8}$-in. bolts, using values of $F_y = 36$ and 50 ksi, and E70 electrodes.

7.16
Flexible Beam Seats

These connections (Fig. 7.8) also allow end rotation needed for Type 2 construction. One advantage to using beam seats is that a surface is provided on which the beam can rest during erection, thereby eliminating the need for driftpins or temporary erection bolts. The beam transfers its entire load through direct bearing to the seat angle; the top clip angle is used only to provide lateral support to the beam.

The seat angle has its connected leg attached to the supporting member (column or girder web) in the shop. A 4-in. outstanding leg (seat) with a minimum of two fasteners is usually sufficient for the attachment between seat and beam. Actually, there is no load on the fasteners in the outstanding leg, when considering the reaction of the beam. The beam must have a $\tfrac{1}{2}$-in. clearance from the supporting member, and frequently a $\tfrac{3}{4}$-in. clearance is used. A design length of bearing of $3\tfrac{1}{4}$ in., therefore, is all that should be relied upon. If this is not sufficient for web yielding (Art. 5.8) in the beam, a larger outstanding leg will have to be used.

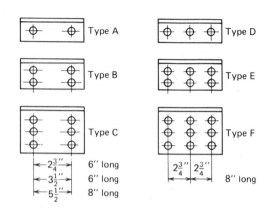

Figure 7.35/AISC standard beam seats.

The connected leg of the beam seat must be large enough to accommodate the number of fasteners required to transfer the beam reaction to the supporting member. Usually one, two, or three lines of fasteners, in two or three rows, are used. Six standard types are shown in Fig. 7.35, and are the types shown in the AISC Manual for which there are corresponding safe-load tables.

Types D, E, and F, of course, cannot be used on the flanges of columns, because the middle row of fasteners would interfere with the web of the column. Sizes other than those shown could, of course, be used as well. Fasteners are designed for the direct shearing force of the reaction, and because the seat is flexible, it is assumed that no additional stresses are developed in the fasteners due to the eccentricity of the reaction.

The thickness of the seat angle itself must be sufficient to ensure that the outstanding leg will not fail in bending. This is accomplished by limiting the maximum bending stress at the toe of the fillet to $0.75F_y$ (Fig. 7.36). The seat will act as a cantilever, having a length e_b equal to the distance from the outside of the seat angle fillet (the critical section) to R, which is located at the center of the length of bearing (N)

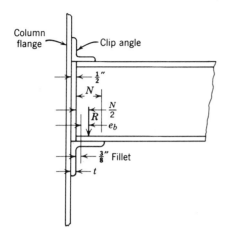

Column flange

Clip angle

$\frac{1}{2}''$

N

$\frac{N}{2}$

R

e_b

$\frac{3}{8}''$ Fillet

t

Figure 7.36/Flexible beam-seat design.

(Fig. 5.17a). This may be expressed as

$$e_b = \frac{1}{2} + \frac{N}{2} - t - \frac{3}{8} \text{ in.}$$

Then,

$$M = e_b(R)$$

The bending stress in the seat angle must not exceed $0.75F_y$; therefore

$$0.75F_y = \frac{M}{S} = \frac{e_b R}{bt^2/6}$$

where b = width of beam seat.

Beam seats usually have a width (b) equal either to the 6- or 8-in. length shown in Fig. 7.35, or to the column flange.

Example

Design a beam-seat connection similar to that shown in Fig. 7.36 for a **W 12×26** beam with a reaction of 21 kips and a **W 8×31** column, both of A36 Steel. Use $\frac{3}{4}$-in., A325 bolts in a bearing-type connection with threads in the shear plane and standard holes.

Solution

(1) Determine the end bearing length (N) required for web yielding:

$$N = \frac{R}{0.66F_y t_w} - 2.5k$$

$$= \frac{21}{0.66(36)0.23} - 2.5\left(\frac{7}{8}\right)$$

$$= 1.66 < 3\frac{1}{4} \text{ in.}; \qquad \text{use 4 in.}$$

(2) Calculate the eccentricity:

$$e_b = \frac{1}{2} + \frac{1.66}{2} - t - \frac{3}{8} = 0.955 - t$$

(3) There are two unknown values for the beam seat, namely the thickness (t) and the width (b). It is necessary to assume one unknown before evaluating the other. Some designers prefer to assume values for both and check by comparing f_{max} with $0.75F_y$. In the solution illustrated here, assume a width equal to that of the column flange, or 8 in.:

$$0.75F_y = \frac{e_b R}{bt^2/6}$$

$$0.75(36) = \frac{(0.955 - t)21}{8t^2/6}$$

Solving for t,

$$0.75(36)8(t^2) = (0.955 - t)21(6)$$
$$216t^2 = 120 - 126t$$
$$t^2 + 0.58t = 0.56$$

Completing the square,

$$t^2 + 0.58t + \left(\frac{0.58}{2}\right)^2 = 0.56 + \left(\frac{0.58}{2}\right)^2$$

$$(t + 0.29)^2 = 0.64$$
$$t + 0.29 = \pm 0.80$$
$$t = 0.51 \text{ in.};$$

$$\text{use } \tfrac{5}{8} \text{ in.}$$

(4) Determine the limiting bolt value:

$$BV_s = 9.3 \text{ kips}$$

$$BV_b\left(\tfrac{7}{16} \text{ in.}\right) = 22.8 \text{ kips}$$

$$BV = 9.3 \text{ kips}$$

(5) Calculate the number of bolts required in the connected leg:

$$n = \frac{21}{9.3} = 2.26; \qquad 4 \text{ bolts are needed}$$

(6) Detail the connection. Use an $8 \times 4 \times \tfrac{5}{8}$-in. seat angle, with four $\tfrac{3}{4}$-in., A325-N bolts in a pattern similar to Type B (Fig. 7.32).

PROBLEMS

1. Design a framing-angle connection between a beam (**W** 12×53) and the flange of a column (**W** 10×49), both of A36 Steel. The beam reaction is 42 kips. Use $\tfrac{7}{8}$-in., A325 high-strength bolts, in a bearing-type connection with threads in the shear plane. (Answer given in Appendix G.)

2. Two **W** 14×22 beams, each having a reaction of 40 kips, frame into a **W** 16×40 girder in a manner similar to that shown in Fig. 7.6b. All steel is A588, Grade 50. Determine the size of A490 bolts required in a bearing-type connection with threads in the shear plane and standard holes. Use three bolts in the connected leg and six bolts in the outstanding legs (Answer given in Appendix G.)

3. Design an end shear-plate connection to field bolt a **W** 12×45 beam to the flange of a **W** 10×45 column, both of A36 Steel. The beam reaction is 24 kips. Use A307 standard bolts.

4. Design a flexible beam seat for a **W** 10×15 beam of A36 Steel and having an end reaction of 20 kips, framing to an A36 Steel **W** 8×24 column. Use $\tfrac{7}{8}$-in., A307 standard bolts.

5. Two **W** 10×22 beams, having reactions of 26 kips each, frame opposite each other to the web of a **W** 18×35 girder. The top flanges of the beams and girders are flush. All steel is A36. Design beam seats using $\tfrac{3}{4}$-in., A325 bolts, bearing type, with threads in the shear plane. (Answer given in Appendix G.)

6. A C 10×15.3 spandrel beam is connected to the flange of a **W** 8×35 column in a manner similar to that shown in Fig. 7.10. Both are of A36 Steel. In this direct connection, only three fasteners are to be used. The spandrel reaction is 24 kips. Determine the bolt diameter required for each of the seven different bolt specifications (A307; A325-X, -N, and -F; and A490-X, -N and -F). Use 3-in. spacing and standard holes.

7.17
Eccentric Fastener Connections

In all the connections thus far considered, the fasteners were arranged symmetrically about the line of action of the force. Under such conditions, the stress in a joint is uniformly distributed among the fasteners. It frequently happens, however, that beams are offset from column centerlines to such an extent that they cannot be directly connected to the column flange. Where this occurs, it is necessary to use some form of eccentric connection such as that shown in Fig. 7.37 (a) or (b), where the offset load (indicated here diagrammatically) frames into the plate or bracket (Fig. 7.11). There is a distinct difference in the design of the connection depending on whether the load acts in or outside a plane passed through the shear area of the fastener shanks.

When the eccentric load acts in the plane of fastener shears, as shown in Fig. 7.37a, the plate rotates, producing additional shear on the fasteners. When the eccentric load acts outside the plane of fastener shears, as in Fig. 7.37b, the contact surfaces between members separate, placing the fastener in tension as well as shear. The structural phenomenon described here is assumed to take place in all eccentric

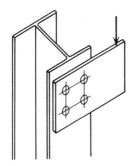 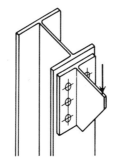

(a) Load in plane of fastener shears.   (b) Load outside plane of fastener shears.

Figure 7.37/Eccentric fastener connections.

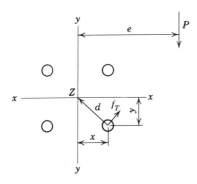

Figure 7.38/Torque stress.

fastener connections and is valid for all types of fastener.

7.18
Load in the Plane of Fastener Shears

The rotating moment caused by the eccentric load acting in the plane of fastener shears is referred to as a torque. This torque is equal to the product of the load P and the eccentric distance e, measured from the centroid of the fastener group (Z), perpendicular to the line of action of the load (Fig. 7.38).

The total stress in each fastener is made up of two parts: the ordinary uniform stress (load divided by number of fasteners), and the torque stress, which varies with the

distance of the fastener from the centroid of the group. The joint should be so proportioned that the resultant of these two components on any fastener does not exceed the maximum permissible fastener value as determined by shear or bearing.

The stresses in the fasteners produced by the torque may be determined as demonstrated below. Letting the four-bolt group, shown diagrammatically in Fig. 7.38, represent any fastener group where all fasteners are the same size,

P = load to be carried.

Z = centroid of the bolt group.

e = the eccentric distance of the load P.

x, y = coordinates of *any* bolt referred to Z as the origin.

d = polar distance of any bolt from the origin ($d = \sqrt{x^2 + y^2}$).

f_0 = torque stress on a bolt at a *unit* distance from Z.

f_T = torque stress on any bolt.

The torque stress on any bolt is proportional to the distance of the bolt from the centroid of the group, and is equal to its distance from Z multiplied by the torque stress on a bolt a unit distance from Z, or

$$f_T = f_0 d = f_0 \sqrt{x^2 + y^2}$$

The moment of resistance of this stress for

any bolt is

$$f_T d = f_0 d^2 = f_0 \left(\sqrt{x^2 + y^2} \right)^2$$
$$= f_0 (x^2 + y^2)$$

The total resisting moment of all the bolts in the group equals the applied torque, or

$$Pe = f_0 \left(\Sigma x^2 + \Sigma y^2 \right)$$

$$f_0 = \frac{Pe}{\Sigma x^2 + \Sigma y^2}$$

When f_0 has been determined from the above equation, the torque stress f_T on any bolt is found by multiplying this value by the distance of the bolt from the origin. This stress acts in a line *perpendicular* to a line drawn from the bolt to the center of gravity of the group, as indicated in Fig. 7.38.

To determine the resultant total stress on any fastener, it is first necessary to resolve the torque stress into its vertical and horizontal components. The uniform stress on the fastener (vertical load divided by the number of fasteners) is then added to the vertical component of the torque stress to obtain the total vertical component. The resultant of this vertical component and the horizontal component of the torque stress is found by the use of the familiar equation

$$R = \sqrt{V^2 + H^2}$$

or it may be determined graphically. It is this resultant total stress R that must be equal to or less than the maximum permissible fastener value, as determined by shear or bearing. Since all fasteners of a group will not be stressed equally, care must be exercised to investigate that fastener having the maximum resultant stress. After fully examining a fastener group, one may establish the fact that this critical fastener (having maximum resultant stress) is the one farthest from the centroid of the group and nearest the applied eccentric load.

Example

Determine the resultant stress on bolt B in the eccentric connection shown in Fig. 7.39. Compare this stress with the allowable value if $\frac{3}{4}$-in., A307 bolts are used. The A36 Steel plate and column flange are each $\frac{3}{8}$ in. thick.

Solution

(1) The center of gravity of the bolt group is at Z. Compute the value of f_0:

$$\Sigma x^2 = 8 \left(2\frac{3}{4} \right)^2 = 60.5 \text{ in.}^2$$
$$\Sigma y^2 = \left[4(6)^2 \right] + \left[4(2)^2 \right] = 160 \text{ in.}^2$$

$$f_0 = \frac{Pe}{\Sigma x^2 + \Sigma y^2}$$

$$= \frac{10,000(9)}{220.5} = 408 \text{ lb per in.}$$

(2) Compute the torque stress (f_T) on bolt B. The distance from B to Z is

$$d = \sqrt{x^2 + y^2} = \sqrt{(2.75)^2 + 6^2} = 6.6 \text{ in.}$$

and the torque stress is

$$f_T = f_0 d = 408(6.6) = 2690 \text{ lb} \nwarrow$$

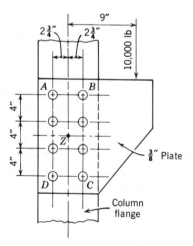

Figure 7.39

(3) Resolve the torque stress into its vertical and horizontal components. (It should be borne in mind that the direction of the torque stress is at right angles to the line joining *B* and *Z*.) The horizontal component is

$$2690(6.0/6.6) = 2450 \text{ lb},$$

and the vertical component is

$$2690(2.75/6.60) = 1120 \text{ lb}.$$

(4) The uniform vertical stress on the bolt is

$$\frac{10,000}{8} = 1250 \text{ lb} \uparrow$$

The total vertical component is

$$1250 + 1120 = 2370 \text{ lb} \uparrow$$

The resultant stress is

$$R = \sqrt{V^2 + H^2} = \sqrt{(2370)^2 + (2450)^2}$$
$$= 3410 \text{ lb} \ \nwarrow$$

(5) Compare this resultant stress with the bolt value. From the AISC Manual,

$$BV_s = 4.4 \text{ kips}$$
$$BV_b = 19.6 \text{ kips}$$

The resultant stress obviously is well within the bolt value as determined by single shear. An excellent exercise would be to compute the total resultant stress for other bolts in the group such as *A*, *D*, and *C*.

This procedure can be used safely to determine the adequacy of eccentric connections. The Ninth Edition of the AISC Manual refers to this procedure as the "elastic method."

Laboratory tests have shown the results to be conservative. Furthermore, use of this design procedure does not result in unduly large or expensive connections. However, test results also have shown that the factor of safety does not remain constant. Consequently, the AISC describes two alternative design methods.

One such method is called the *ultimate-strength method* and is based upon principles similar to those described in Chapter 11. However, its use is limited to vertical lines of fasteners in connections subjected to eccentric vertical loads. For other directional loads and/or fastener patterns an "Alternate Method 2" procedure is described, which is a variation of the ultimate-strength method. Refer to the AISC Manual for further details and examples.

7.19
Load Outside the Plane of Fastener Shears

Connections supporting loads falling outside the plane of fastener shears generally are of the type shown in Fig. 7.37b, employing a structural tee; the type shown in Fig. 7.40, where a plate is sandwiched between two angles; or the stiffened beam seat shown in Fig. 7.9. The fasteners connected to the column flange are subjected to a load outside the shear plane.

The eccentric load tends to separate the bracket from the column flange at the top and press the bracket against the flange at the bottom. Therefore, with no initial prestress, beginning at the top and moving downward, the fasteners are subjected to

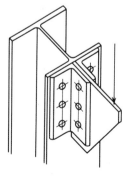

Figure 7.40/Plate and angle bracket.

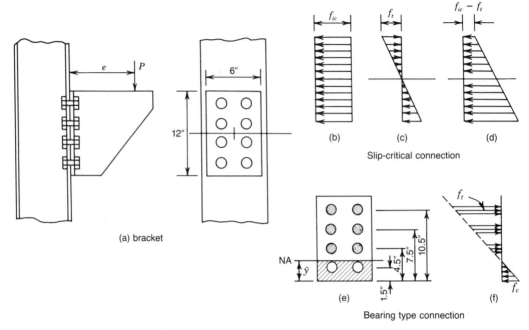

Figure 7.41/Bolts in tension and shear.

a decreasing tensile force. The load also places the fasteners in direct shear.

It was stated earlier that rivets are seldom used in today's buildings and that the use of A307 bolts in tension is not recommended. Therefore, this discussion will be limited to high-strength bolts subjected to tension and shear. Articles 7.7 and 7.8 discussed in detail the advantage that high-strength bolts may have when used either in bearing-type connections, where they are installed in a snug fit, or in slip-critical connections, where they are installed pre-stressed to 70% of their tensile capacity. The distribution of stresses, and consequently the design procedure, depends upon which type is used.

In slip-critical connections, the initial pre-stress of the bolts clamps the plates tightly together. This initial compression stress between the plates (f_{ic}) can be calculated as the sum of all the pre-stressed bolt forces (ΣT_b) divided by the contact area of the plates (Fig. 7.41a and b). When the

eccentric force is applied, it tends to separate the top of the plates, thereby relieving the clamping effect and reducing the initial pressure exerted by the bolts. It also increases the pressure at the bottom. The tension and compression stress pattern caused by the eccentric load varies linearly from the center of the connection to a maximum at the top and bottom (Fig. 7.41c). This is similar to the bending-stress distribution in a beam, but this distribution occurs only if the initial pre-stress at the top (f_{ic}) is not exceeded. Finally, the resulting compressive stress will be the sum of the two stresses as shown in Fig. 7.41d. The moment of inertia (I) of the contact surface can be calculated and the bolt stress can be determined by the familiar beam formula

$$f_t = \frac{Pey}{I}$$

As long as the initial pre-stress f_{ic} is not

overcome by the eccentric stress f_t, the connection is adequate, and the 1989 AISC Specification states that the shear component can be neglected. This is not the case when there is a direct shear and tensile stress due to an axially applied tensile force. Under these circumstances the allowable stresses shown in Table 7.1 must be multiplied by a reduction factor $1 - f_t A_b / T_b$, where f_t is the average tensile stress due to the direct axial loads

The stresses in a bearing-type connection are quite different from the slip-critical type. Refer to Figure 7.41 (e) and (f). The top of the bracket tends to open up, and the bottom is compressed. The cross-hatched area in Fig. 7.41e represents the compressed area. Note that the bolts in the compression area are ineffective. The bolts that are shaded are in tension. In addition, all bolts are in shear.

The neutral axis passes through the centroid of the shaded and cross-hatched areas. Once the location of the neutral axis is established, the moment of inertia of the shaded and cross-hatched areas can be calculated. And finally, the tensile stresses in the bolts can be determined by the beam formula

$$f_t = \frac{My}{I}$$

In addition to these tensile stresses, the fasteners are subjected to a shear stress. All fasteners are effective in shear, and it is assumed to be evenly divided between them. Consequently

$$f_v = \frac{P}{nA_s}$$

where n is the number of bolts and A_s is the nominal area of the bolt shank.

The connection is considered to be adequate if the shear stresses do not exceed the allowable ones shown in Table 7.1 and the tensile stresses do not exceed the allowable ones calculated from the formulas shown in Table 7.3.

Example

For the connection shown in Fig. 7.41, assume that $P = 40$ kips and $e = 7$ in. Determine if $\frac{3}{4}$-in. bolts will be adequate if the bolts are A325 in a bearing-type connection, with threads in the shear plane, and spaced 3 in. vertically. Would the same bolts be adequate in a friction-type connection?

Solution

(1) Calculate the location of the neutral axis. Assuming it lies just above the bottom two bolts, at an unknown distance ($\bar{y}$) up from the bottom of the bracket, and equating the first moment of the shaded bolt area (about the neutral axis) to the first moment of the cross-hatched area, the following equation can be developed (see Fig. 7.41e; each bolt has an area

Table 7.3
Allowable Tensile Stress (F_t) for Bolts in Bearing-Type Connections, (ksi)

Bolt Type	Threads Not Excluded from Shear Plane	Threads Excluded from Shear Plane
A325	$\sqrt{(44)^2 - 4.39 f_v^2}$	$\sqrt{(44)^2 - 2.15 f_v^2}$
A490	$\sqrt{(54)^2 - 3.75 f_v^2}$	$\sqrt{(54)^2 - 1.82 f_v^2}$

equal to 0.44 in.^{2}):

$$2(0.44)[(10.5 - \bar{y}) + (7.5 - \bar{y})$$
$$+ (4.5 - \bar{y})(\bar{y} - 1.5)] = \frac{6\bar{y}^2}{2}$$

This can be simplified to

$$\bar{y}^2 + 0.59\bar{y} = 6.16$$

Completing the square,

$$\bar{y}^2 + 0.59\bar{y} + \left(\frac{0.59}{2}\right)^2 = 6.16 + \left(\frac{0.59}{2}\right)^2$$

$$(\bar{y} + 0.30)^2 = 6.25$$

$$\bar{y} = 2.5 - 0.3$$

$$\bar{y} = 2.2 \text{ in.}$$

(2) Compute the moment of inertia of the shaded and cross-hatched areas with reference to the neutral axis. Neglect the moment of inertia of the bolt area about its own centroid:

$$I = 2(0.44)[(8.3)^2 + (5.3)^2 + (2.3)^2 - (0.7)^2]$$
$$+ \frac{6(2.2)^3}{3}$$

$$I = 111 \text{ in.}^4$$

(3) Calculate the tensile stress in the top bolts:

$$f_t = \frac{(40)(7)(8.3)}{111} = 21 \text{ ksi}$$

(4) Calculate the average shear stress:

$$f_v = \frac{40}{8(0.44)} = 11.4 \text{ ksi}$$

(5) Select the correct formula from Table 7.3, and calculate the allowable tensile stress:

$$F_t = \sqrt{(44)^2 - 4.39(11.4)^2} = 37 \text{ ksi}$$

(6) Compare actual to allowable stress. Since 21 < 37 and 11.4 < 17, the design is acceptable for a bearing connection.

(7) Determine the pre-stress force in the $\frac{3}{4}$-in. A325 bolt:

$$T_b = (0.7)(90)(0.44) = 28 \text{ kips}$$

This value can also be obtained from Table J3.7 in the Ninth Edition of the AISC manual.

(8) Calculate the moment of inertia of the contact surface:

$$I = \frac{6(12)^3}{12} = 684 \text{ in.}^4$$

(9) Determine the initial compressive stress on the contact surface due to the pre-stress in all the bolts:

$$f_{ic} = \frac{8(28)}{(6)(12)} = 3.11 \text{ ksi}$$

(10) Calculate the tensile stress at the top of the plate:

$$f_t = \frac{7(40)(6)}{864} = 1.91 \text{ ksi}$$

Since 1.91 < 3.11, it is adequate for a slip-critical connection.

Some designers prefer to use stiffened beam seats rather than flexible beam seats, particularly when designing for beams having large reactions. The stiffener can be another angle, placed as shown in Fig. 7.42, and it is good practice to use an angle thickness equal to or greater than that of the web of the beam being supported.

Since the seat is not flexible, the reaction is assumed to have a greater eccentricity than for flexible seats. Referring to Fig. 7.42, the beam reaction is assumed to act at $N/2$, measured from the toe of the outstanding leg of the seat angle. The resulting eccentricity is

$$e_s = W - \frac{N}{2}$$

where e_s is the distance from R to the

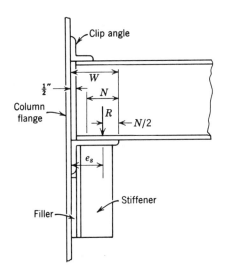

Figure 7.42/Stiffened beam seat.

plane of the fastener shears. The design of the fasteners follows the same procedure as that described for the connection in Fig. 7.41, and the AISC Manual provides safe-load tables for a variety of stiffened beam seats.

7.20
Moment-Resisting Connections— Fastener-Connected

The main reason for using moment-resisting connections in buildings is to accommodate lateral forces (Chapter 9). However, even when they are used solely for this purpose, there also will be negative moments at the beam ends due to the gravity (vertical) loads (Chapter 10). To resist these end moments, some way of developing the horizontal force in the top and bottom flanges must be provided. The value of the total horizontal force can be determined by dividing the end moment by the depth of the beam, i.e.,

$$F_t = \frac{M}{d}$$

Figure 7.12a shows one way in which this

horizontal force can be developed. The force in the beam flange is transferred to the web of the tee by single shear in the fasteners. Then this force is transmitted to the flange of the tee by bending in the flange, and finally to the column flange by tension in the other fasteners. The stiffening diaphragms on the column may or may not be required. Other details also are possible.

In the design of this kind of connection, it is assumed that the vertical beam reaction is resisted entirely by the shear connection and that the horizontal forces are resisted entirely by the tees. The design nearly always is accomplished through a trial-and-error procedure, adjusting and refining the details as the various requirements become known. One disadvantage of using fasteners (as opposed to welds) in this kind of connection is the difficulty of providing adequate space for fastener heads and assembly clearances. For this reason, detailed scale drawings should be prepared during design to be sure that the assembly actually can be fabricated.

Structural tees cut from **S** sections frequently are used because of their thicker webs and flanges. Beveled washers are used under nuts on sloping flange surfaces, and the tension flange and fasteners must be capable of resisting the additional prying forces caused by distortion from bending (Fig. 7.43). The prying force (Q) is assumed to be a line force acting uniformly along the edge of the flange over the distance L shown in Fig. 7.12a. Consequently, the tension fastener must be designed for the combined load ($F + Q$). The bending in the flange varies as shown in Fig. 7.43a.

Either M_1 or M_2 can be critical. Once the value of Q has been determined, the bending moments can be calculated.

However, the 1989 AISC Specification recommends a procedure based upon ultimate strength for calculating Q. Refer to

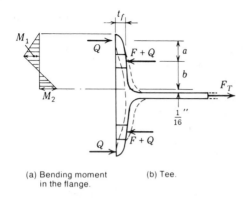

(a) Bending moment
in the flange.

(b) Tee.

Figure 7.43/Prying forces.

the Ninth Edition of the AISC Manual for details on determining the value of Q and complete design of this type of connection.

PROBLEMS

1. A $\frac{5}{16}$-in. angle is bolted to the flange of a **W** 14×90 column, as shown in Fig. 7.44. Both members are of A36 Steel. Determine which bolt is critical and what the maximum total stress will be when the 8500-lb load is applied. Use $\frac{7}{8}$-in. A325-N bolts, in standard holes. Is the connection safe? (Answer given in Appendix G.)

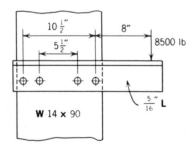

Figure 7.44/Problem 1

2. Determine the maximum permissible load (P) that can be supported by the $\frac{3}{8}$-in. plate connected to the **W** 10×22 column flange

shown in Fig. 7.45. Both members are of A36 Steel. Use $\frac{3}{4}$-in. A490-F bolts in oversize round holes.

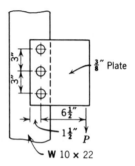

Figure 7.45/Problem 2

3. A bracket similar to that shown in Fig. 7.11a has two rows of eight bolts each. The distance between rows is $5\frac{1}{2}$ in., and the distance between the column centerline and the beam web is 12 in. Members are of A36 Steel. The beam reaction is 50 kips. Select a bracket thickness based on critical shear, and determine the size of A325-X bolts needed at a 3-in. spacing, in standard holes. (Answer given in Appendix G.)

4. A bracket similar to that shown in Fig. 7.40 has a total of nine A325-F bolts in standard holes. Three of these bolts connect the $\frac{3}{8}$-in. plate and $\frac{3}{8}$-in. angles, and six connect the angles to the column flange. The angles are 3×3 in. and 12 in. long. The load is 10 kips and is applied 14 in. from the face of the column flange. The gage distance (spacing) on the column flange is 5 in., and the gage distance between the flange and the line of bolts attaching the plate to the angles is 3 in. The vertical spacing is 4 in. Determine the size of bolts required if all bolts are to be of the same size.

5. A stiffened beam seat, similar to that shown in Fig. 7.42, must support a **W** 16×67 beam with a reaction of 50 kips. The seat angle is an L $5 \times 3 \times \frac{1}{4}$ in. (6 in. long), with the 5-in. leg outstanding. All steel is A36 and the bolts are A325-N, in standard holes. There are two rows of three bolts each, and the gage distance for bolt spacing is 3 in. (vertical). The filler plate between the vertical stiffener angle and the flange is $\frac{1}{4}$ in. thick and 6 in. wide, extending

2 in. below the centerline of the bottom two bolts. Determine the bolt size required.

WELDED CONNECTIONS

7.21
General

An introduction to welding was given in the latter part of Art. 7.2 dealing with types of connector. It would be advisable to review that discussion before proceeding further.

Design for welding, as discussed here, is applicable to members and joints of ordinary building frames in which the loads to be carried are essentially static in nature. Special members, such as crane girders in industrial buildings and framing for elevator machinery or vehicular ramps, are subjected to dynamic loading. Adequate provision for impact loading in building members usually can be made by increasing the assumed total load to be carried in accordance with the AISC Specification. However, a thorough understanding of the behavior of materials under fatigue conditions is essential to the intelligent design of structures subjected to dynamic loadings which cause many repetitions of maximum design stress.

In addition to fatigue, another danger to be avoided is lamellar tearing, which can occur when very thick welds are used in attaching large members. Lamellar tearing has been described as the failure of steel in its through-thickness dimension—the z direction, if the x and y axes are oriented in the plane of rolling. Failure occurs in the parent metal and not the weld deposit. Presently, there are no detailed specifications to guard against this type of failure. Until more is known about lamellar tearing and detailed specifications are adopted, it is best to avoid welds over $1\frac{1}{2}$ in. thick.

As a general rule, this can be easily accomplished in building design as will be shown in subsequent articles.

With welded connections, it is sometimes possible to attach one member directly to another without the use of any additional connecting parts, such as are necessary when using fasteners. Welded connections, therefore, are customary in rigid construction (Chapter 10), where the intent is to develop full member strength. Where this is not the intent, as in Type 2 construction (Art. 7.3), it is difficult to achieve a connection allowing free rotation without use of connecting parts such as angles and plates to transmit loads. Consequently, welded connections between beam and girder and between beam and column in such construction are in many respects similar to fastener connections except that welds replace the fasteners. Design procedures for such connections will be described in Art. 7.30.

7.22
Welding Processes

A basic manual arc-welding process was described in Art. 7.2. In the past, there were only two different welding processes available: the shielded-metal arc and the submerged arc. The manual shielded-arc process (Art. 7.2 and Fig. 7.2) is used both in the field and in the fabricating shop. The submerged-arc process (described below) is limited to shop fabrication. With the emergence of newer steels and advancing technology, however, new and improved welding techniques have been developed; and the AISC now also recognizes the gas-metal-arc and flux-cored-arc processes and the electroslag-and-electrogas process.

The submerged-arc welding process (Fig. 7.46) is an important shop procedure and

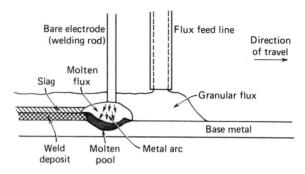

Figure 7.46/Submerged-arc welding.

is usually automatic or semiautomatic, using a machine. A bare electrode is pushed through a mound of flux and the metal arc is created. Since the flux is simply dropped on the metal, the process is effectively limited to flat work. However, because this process results in deeper weld penetration, the AISC allows a greater value for the actual weld size.

The gas-metal-arc welding process (Fig. 7.47) also can be automatic or semiautomatic and can be used either in the shop or in the field. It differs from the submerged-arc process in that the flux is replaced by a stream of gas. This permits welding in all positions (horizontal, vertical, and overhead), but field use is limited by the possibility that wind will interrupt the gas stream.

Flux-cored-arc welding (Fig. 7.48) is similar to gas-metal-arc welding except that the electrodes are tubular and contain flux that is deposited on the weld. This process too may be automatic or semiautomatic for shop or field welding.

The electrogas-and-electroslag welding process is a special method using either the gas-metal arc or the flux-cored arc. It is a fully automatic method employing a copper shoe to contain the welding components. The weld must be deposited in a vertical path, and coolants are used. This process is limited to shop use and usually to larger welds.

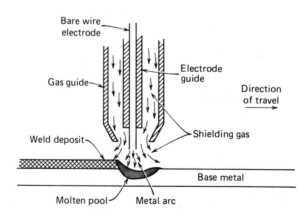

Figure 7.47/Gas-metal-arc welding.

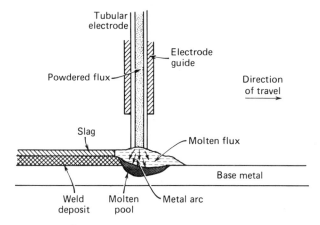

Figure 7.48/Flux-cored-arc welding.

Each of these four welding processes requires special types of electrodes, bearing special designations. There also are special strength grades for each process. The AISC Specification recommends a matching base metal and strength grade of electrode. However, the lower-strength-grade electrodes may be used with higher-strength matching metal, provided the lower stress value of the weld is used in design. Conversely, a higher-strength-grade electrode can be used with a lower-strength-grade steel if the electrode is assigned the matching lower stress-grade value.

Table 7.4
Electrodes: Allowable Stresses and Matching Steels

Welding Process and Electrode Grade				*Allowable F_v for Deposited Weld Metal (ksi)*	*ASTM Designation for AISC and AWS "Matching Base" Metals*
Manual Shielded-Metal Arc	*Submerged Arc*	*Gas-Metal Arc*	*Flux-Cored Arc*		
E60XX or E70XX	F6X-EXXX or F7X-EXXX	E70S-X or E70U-1	E60T-X	18.0 or 21.0	A36, A53 Grade B, A500, A501, A529 A570 Grades D and E, A709 Grade 36
E70XX	F7X-EXXX	E70S-X or E70U-1	E70T-X	21.0	A242, A441, A572 Grades 42–55; A588; and A709 Grades 50 and 50W
E80XX	F8X-EXXX	Grade E80S	Grade E80T	24.0	A572 Grades 60 and 65
E100XX	F10X-EXXX	Grade E100S	Grade E100T	30.0	A514, over $2\frac{1}{2}$ in. thick
E110XX	F11X-EXXX	Grade E110S	Grade E110T	33.0	A514, $2\frac{1}{2}$ in. thick and under

Key: E, electrode; F, flux; S, bare solid electrode; U, coated solid electrode; T, flux-cored electrode; 60, 70, 80, 100, 110, min. tensile strength; X, design specification numbers.

The basic strength of most welds corresponds to its shear capacity, as will be illustrated in the next article. The six strength grades of electrodes for each welding process, along with matching base metals and specific electrode designations, are listed in Table 7.4.

The values shown in Table 7.4 agree with the 1989 AISC Specification provision limiting the allowable stress to 0.30 times the nominal tensile strength of the weld metal (kips per square inch), when it is observed that the number following the E- or F-designation of the electrode represents the tensile strength of the weld metal (e.g., an E80XX has a tensile strength of 80 ksi, and $0.30(80) = 24$ ksi).

The listed values (F_v) for the electrodes are to be used only when the corresponding matching steels are used. When this is not the case, the lesser value of F_v must be used. For example, E60 electrodes can be used with $F_v = 18$ ksi for the deposited weld metal for both A500, Grade A base metals and A36 Steel base metals, whereas the E70 electrode can be used with $F_v = 21$ ksi for A36 Steel but must be used with an $F_v = 18$ ksi when the base metal is A500, Grade A.

For design purposes, usually only the selected electrode for the shielded-metal-arc process is specified. If the contractor or fabricator decides to select another welding process, the appropriate change in electrode designation is assumed but should be verified by the designer. For example, suppose an E70XX electrode is specified, and the fabricator decides to use the flux-cored-arc process: the substitute electrode should be the E70T-X. This practice of specifying only the manual shielded-metal-arc electrode will be used in all examples and problems presented herein.

7.23
Types and Strengths of Welds

Most welds used to join structural elements in buildings are of two general types: *groove* and *fillet*.

Groove Welds / Groove welds are used in joints between two abutting parts lying in approximately the same plane (Fig. 7.49). They are classified according to the method of grooving or preparing the base metal before weld metal is deposited. The square butt weld (Fig. 7.49a) requires no preparation; however, for complete penetration, it is limited to plates not more than $\frac{1}{4}$ in. thick. For greater thicknesses, it is necessary to weld from both sides (or use a back-up bar on one side) and/or to groove the parts that are to be welded. Nine forms of groove weld are recognized in practice, three of which are shown in Fig. 7.49. (See Fig. 7.52 for others.)

Complete-penetration butt welds have an effective thickness or throat dimension equal to the thickness of the thinner member joined. Consequently, when the full

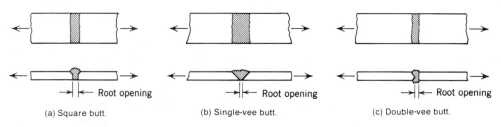

(a) Square butt. (b) Single-vee butt. (c) Double-vee butt.

Root opening Root opening Root opening

Figure 7.49/Groove welds.

width of the member is welded, the full strength of the member is developed in the weld. It is important that the matching steels and electrodes meet the AISC Specification.

When the weld is made from both sides, the root of the initial layer of weld metal (or bead) must be thoroughly chipped out on the reverse side before welding is started on that side. This removal of slag also is required when making more than one pass to build up the size of weld.

Fillet Welds / These are shown in Fig. 7.50. A fillet weld has a cross section which is approximately triangular. The weld size is designated D in Fig. 7.50 (b), (c), and (d). This type of weld generally is used to join two surfaces at right angles to each other, and is the type most frequently used for structural connections. Unless clearly noted otherwise, such welds are assumed to have equal legs, i.e., 45° faces, and the throat dimension is the size multiplied by

$\sin 45°$. When an unequal-leg fillet weld is necessary for a special design condition, the design throat dimension is the least dimension from the root to the face.

The strength of a fillet weld is determined by the throat dimension; therefore, small fillet welds are most economical. This is true because the throat dimension is proportional to the leg size, while the amount of weld metal varies approximately as the square of the leg size. Fillet welds of $\frac{1}{8}$-, $\frac{3}{16}$-, $\frac{1}{4}$-, and $\frac{5}{16}$-in. size can be made in one pass (one progression of the electrode along the axis of the weld). Larger welds are made in several passes. As with groove welds, the surface of each weld must be thoroughly chipped and wire-brushed to remove all slag before the next pass.

Fillet welds may be loaded in two different ways. Figure 7.50c shows two fillets, each having its axis parallel to the direction of stress in the members. Each weld transfers stress from one member to the other by means of shear parallel to its axis. The

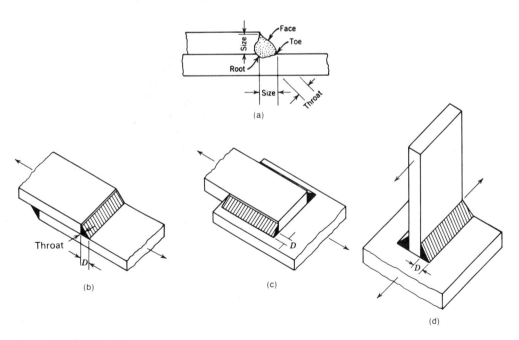

Figure 7.50/Fillet welds.

shearing stress is assumed to be uniformly distributed over the length of the weld.[5] Figure 7.50b shows two fillets, each having its axis perpendicular to the direction of stress in the members. Welds of this type fail through the throat as a result of the combined effect of shear and tension (or compression). Since tests show the strength of such welds in tension or compression to be greater than that for shear, it is conservative but reasonable to limit the strength to that controlled by shear alone.

Consequently, for design purposes, it is generally assumed that the strength per lineal inch of fillet weld is the shearing strength, regardless of the direction of load on the weld. The allowable shearing stress for various fillet welds is shown in Table 7.4. Since the critical section is through the throat of the fillet, the strength per lineal inch (F) is equal to the leg size times $\sin 45°$, multiplied by the allowable shear stress.

Therefore, for the 18,000-psi-strength weld,

$$F = D(\sin 45°)18{,}000$$

$$= 12{,}700D \text{ lb per lineal in.}$$

and for the 21,000-psi-strength weld,

$$F = 14{,}800D \text{ lb per lineal in.}$$

Some designers prefer to calculate shear strength of welded connections "per $\frac{1}{16}$ in. of fillet weld size." Using this method, the strengths are:

For the 18,000 psi-strength weld,

$$F = 800 \text{ lb per lineal in. per } \tfrac{1}{16} \text{ in. of } D.$$

For the 21,000 psi-strength weld

$$F = 930 \text{ lb per lineal in. per } \tfrac{1}{16} \text{ in. of } D.$$

The above processes also could be applied to the higher stress levels listed in Table 7.4. However, for the applications shown herein, only E60 and E70 electrodes will be referenced.

In order to compensate for irregularities in weld deposit and the tapered shape at the ends of fillet weld passes, caused by starting and stopping the weld, the welder should always make the actual length $2D$ greater than the computed length. However, the length shown on detailed drawings is always the net or computed length.

Tests have also shown that the ends of fillet welds sustain greater than average shears. Consequently, where possible, fillets should be continued around corners a short distance. This is referred to as a return, and is usually equal to $2D$ (Fig. 7.51). Returns are not generally shown on detail drawings but are executed by the welder wherever feasible.

Fillet welds made by the submerged-arc process are permitted a larger capacity because of their deep penetration. The 1989 AISC Specification allows the throat of the weld to be the size of the weld when the size is equal to or less than $\frac{3}{8}$ in. Thus, its capacity is $18D$ kips per in. for the E60 electrode. For submerged-arc welds larger than $\frac{3}{8}$ in., the specification allows 0.11 in.

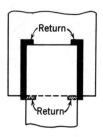

Figure 7.51/Weld returns.

---

[5]Although this assumption is not valid in the range of elastic stresses, it greatly simplifies design and produces a safe solution.

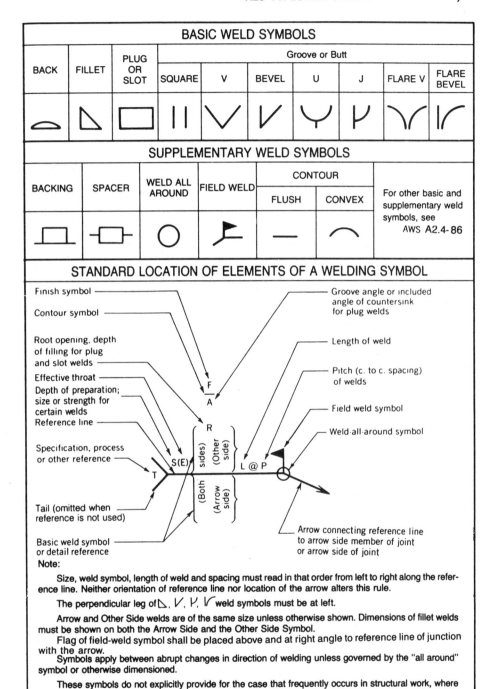

Figure 7.52/Welded joints—standard symbols. Courtesy of American Institute of Steel Construction; *Manual of Steel Construction*: *Allowable Stress Design*, p. 4-155.

to be added to the computed throat dimension.

7.24
Designation of Welds

In addition to type, there are numerous other features, such as size, exact location, and finishes, that must be described to completely identify a weld. To convey this information from the designer to the fabricator and welder, standard welding symbols are used. These symbols provide the means for putting complete welding information on drawings. In practice, designers may need only a few of the symbols available. They may, therefore, elect to use only such parts of the scheme as suits their need. More detailed information may be found in "Welding Symbols and Information for Their Use," published by the American Welding Society.

The complete welding symbol contains the following elements:

Reference line

Arrow

Basic weld symbol (type of weld)

Dimensions and other data

Supplementary symbols

Finish symbols

Tail

Specification, process, or other references

Figure 7.52 illustrates the basic weld symbols, i.e., types of weld. It also shows symbols for supplementary information, and the location of all elements in a complete welding symbol.

With this system, the *joint* is the basic reference. Any joint, the welding of which is indicated by a symbol, will always have an *arrow side* and an *other side*. Accord-

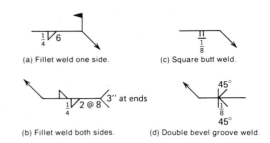

(a) Fillet weld one side. (c) Square butt weld.

(b) Fillet weld both sides. (d) Double bevel groove weld.

Figure 7.53/Examples of weld designation.

ingly, the words *arrow side*, *other side*, and *both sides* are used to locate the weld with respect to the joint. The tail of the symbol is used for designating the welding specification, procedures, or other supplementary information necessary to make the desired weld. The notation placed in the tail to indicate this information will usually be established by the individual designer. If no notation is used, the tail of the symbol may be omitted.

Figure 7.53 shows three specific examples. The horizontal line in each case is the reference line. The triangular symbol below the line (Fig. 7.53a) indicates a fillet weld positioned at the arrow side of the joint. Observe that the left side of the triangle always is vertical and that next to the vertical side, the size of weld is given. In this case, the size is $\frac{1}{4}$ in., and it is 6 in. long. If the length dimension is omitted, the weld is for the full width of the member to be joined. The flag indicates a field weld.

Figure 7.53b shows staggered, intermittent fillet welds on both sides of the joint. Weld lengths are 2 in. each (except at the ends), and they are spaced 8 in. center to center.

Figure 7.53c shows a flush-ground groove weld with square edges and a $\frac{1}{8}$-in. root opening.

Figure 7.53d shows a double bevel cut at 45° and a weld placed with a root opening of $\frac{1}{8}$ in.

7.25 Maximum and Minimum Size of Welds

The AISC Specification limits the size of a fillet weld so that its total strength may be developed without overstressing the adjacent base metal. If there are welds opposite each other on both sides of a plate, each fillet size is limited by one-half the plate shear strength.

The AISC Specification allows $0.40F_y$ in shear on the gross area of beam webs. It was previously shown (Art. 7.23) that the strength of the 18,000-psi-strength fillet is $F = 12.7D$ kips per inch. Letting t represent the thickness of the member being welded, its strength per lineal inch is $0.40F_y t$. If the member is welded on both sides,

$$F = 12.7D = \tfrac{1}{2}(0.40F_y)t$$

from which

$$D_{max} = 0.0157F_y t$$

A similar formula may be derived for the 21,000-psi-strength fillet, i.e.,

$$D_{max} = 0.0135F_y t$$

The fillet size is further restricted when the weld is to be placed along edges of connected parts. The maximum size may be equal to the part thickness when the part is less than $\tfrac{1}{4}$ in. thick, and to $\tfrac{1}{16}$ in. less than the part thickness when the part is $\tfrac{1}{4}$ in. or more thick.

A small weld placed on a thick member is undesirable. The heat generated in depositing the small weld is not enough to appreciably expand the base metal. Consequently, as the weld cools and tries to contract, it is prevented from doing so by the stable base metal. Initial stresses (residual stresses) in the weld metal are thereby induced. To avoid this as well as other undesirable metallurgical effects, the

Table 7.5
Minimum Sizes for Fillet Welds

Material Thickness of Thicker Part Joined (in.)	Minimum Size of Fillet Welds (in.)
To $\tfrac{1}{4}$ inclusive	$\tfrac{1}{8}$
Over $\tfrac{1}{4}$ to $\tfrac{1}{2}$	$\tfrac{3}{16}$
Over $\tfrac{1}{2}$ to $\tfrac{3}{4}$	$\tfrac{1}{4}$
Over $\tfrac{3}{4}$	$\tfrac{5}{16}$

AISC recommends fillet-weld minimum sizes in relation to plate or member thickness (Table 7.5).

The AISC Specification also sets minimum effective lengths for fillet welds. When necessary for strength, the minimum effective length is four times the weld size. Also, if longitudinal fillet welds are used alone in end connections of flat bar tension members, the length of each fillet weld must be not less than the perpendicular distance between them.

7.26 Design of Simple Welded Lap-Type Connections

Once a given type weld is selected (this is usually governed by the connection detail), the size and length must be established for the desired strength and type of electrode. At first there may seem to be an infinite number of possible solutions to a given design problem, but further investigation generally reveals that code and other limitations reduce the number to relatively few.

Example 1

The connection of a double-channel hanger to the lower flange of a $W\,36 \times 194$ beam, both of A36 Steel, is shown in Fig. 7.54. A plate 8 in. wide is selected to

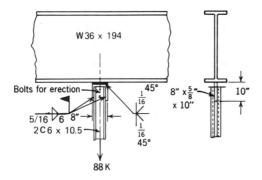

Figure 7.54/Example 1.

provide a 1-in. extension on each side for field welding the channels. This plate is joined to the beam with a double-bevel butt weld. The full 8-in. length of the weld cannot be counted as effective, unless the ends are cut down to solid metal and side welds applied to furnish weld reinforcement similar to that provided at the faces, or unless short extension bars are used to eliminate reduction of weld due to crater effects by continuing the full weld section beyond the ends. Normally, it is more economical to assume an effective weld length somewhat shorter than the total length in order to avoid special welding. The effective throat section required is

$$\frac{88,000}{22,000} = 4 \text{ in.}^2$$

Using an effective length of 7 in., the throat thickness required is $\frac{4}{7} = 0.57$ in.; use a $\frac{5}{8}$-in. plate and a complete-penetration weld ($\frac{5}{8}$-in. throat dimension). The applicable welding symbol (Fig. 7.54) shows the type of groove, root opening, and bevel angle. Where only one member at a joint is grooved, the arrow is broken and points to that member; therefore, the arrow is drawn to the plate rather than to the beam.

Vertical fillet welds along each side of the double channel member (Fig. 7.54) will be used to field-connect the member to the $\frac{5}{8}$-in. plate. No portion of the load can be considered to be transferred by the bolts, which are used during erection to hold members in position. Each weld must carry

$$\frac{88,000}{4} = 22,000 \text{ lb}$$

The minimum size of weld for the $\frac{5}{8}$-in. plate is $\frac{1}{4}$ in. (Table 7.5). Welds are to be placed on each side of the plate. Using E60 electrodes, the maximum weld size is

$$D = 0.0157(36)\tfrac{5}{8} = 0.353 \text{ in.}$$

Adopting a $\frac{5}{16}$-in. weld,

$$F = 12.7D = 12.7\left(\tfrac{5}{16}\right) = 3.97 \text{ kips per in.}$$

Length required = $22/3.97 = 5.54$; use 6 in.

The applicable welding symbol shown in Fig. 7.54 indicates a fillet weld, the size and length of the welds, that welds are required on both sides of the joint at each of two points, and that these welds are to be made in the field.

Example 2

When the connected member is not symmetrical, welds should be proportioned to avoid eccentricity if possible. In Fig. 7.55,

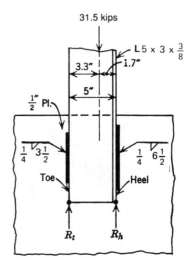

Figure 7.55/Example 2.

an angle strut with a total stress of 31.5 kips is welded to a $\frac{1}{2}$-in. gusset plate. The action line of the load coincides with the gravity axis of the angle, which is 1.7 in. from the back of the 5-in. connected leg. By taking moments about a point on the weld action line of R_h, the amount of weld resistance (R_t) along the toe can be determined:

$$R_t(5) - 31,500(1.7) = 0$$

$$R_t = \frac{31,500(1.7)}{5} = 10,700 \text{ lb}$$

As the total resistance of the two welds must equal the applied load. R_h can be calculated as

$$R_h = 31,500 - 10,700 = 20,800 \text{ lb}$$

The angle must be connected for at least 10,700 lb along the toe and 20,800 lb along the heel. The maximum size of weld along the toe of the angle is $\frac{3}{8} - \frac{1}{16} = \frac{5}{16}$ in.; use a $\frac{1}{4}$-in. weld. For a $\frac{1}{4}$-in. weld, produced from E60 electrodes, the length required is

$$\frac{10,700}{4(800)} = 3.34; \quad \text{use } 3\frac{1}{2} \text{ in.}$$

The length of $\frac{1}{4}$-in. weld required along the heel of the angle is

$$\frac{20,800}{4(800)} = 6.5; \quad \text{use } 6\frac{1}{2} \text{ in.}$$

It should be pointed out that generally codes no longer requires this proportioning of welds and that the total length ($3\frac{1}{2} + 6\frac{1}{2} = 10$ in.) may be divided equally between heel and toe (5 in. along the heel and 5 in. along the toe).

PROBLEMS

1. A **W** 6×15 supports a total tensile load of 56 kips and is connected to the bottom flange of a **W** 16×40 girder by two plates as shown in Fig. 7.56. Determine the required plate thickness if a full-penetration butt weld is used for

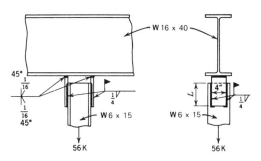

Figure 7.56/Problem 1.

the connection between the plates and the bottom flange of the girder. Assume the effective length of the butt weld to be 3 in. Using a $\frac{1}{4}$-in. fillet between the plates and the **W** 6×15, calculate the length of overlap L required. Members are of A36 Steel, and electrodes are E60. (Answer given in Appendix G.)

2. Two angles, $5 \times 3 \times \frac{1}{2}$ in., straddle a $\frac{7}{16}$-in. plate. They carry a total axial load of 85 kips and are fillet-welded to the plate along the toe and heel of each angle. Determine the maximum permissible size of weld and, using this size, calculate length of welds required. Use A36 Steel and E70 electrodes. The 5-in. legs of both angles are in contact with the plate.

3. A $6 \times \frac{1}{2}$-in. plate is welded to another member, both of A36 Steel, as shown in Fig. 7.57. Determine dimension L necessary to provide a total length of weld sufficient to resist the load shown. Use E60 electrodes. (Answer given in Appendix G.)

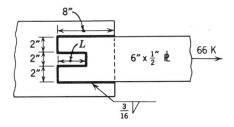

Figure 7.57/Problem 3.

4. Design a welded connection for two $6 \times 6 \times \frac{3}{8}$-in. angles straddling a $\frac{5}{8}$-in. plate, if the angles carry a total axial load of 90 kips. Steel is A572, Grade 65, and electrodes are E70.

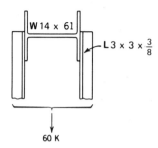

Figure 7.58/Problem 5.

5. Two $3 \times 3 \times \frac{3}{8}$-in. angle hangers are welded to the flanges of a **W** 14×61 as shown in Fig. 7.58. Design the welds to support a total load of 60 kips on the hangers. Steel is A242 and electrodes are E60.

7.27
Eccentric Welded Connections

All the welded connections discussed thus far have been loaded concentrically, i.e., the line of action of the load passes through the centroid of the weld shape, thus justifying the assumption that all portions of the weld are equally stressed. In some cases it was found necessary to balance the welds in order to achieve a concentrically loaded connection (Art. 7.26). This balancing cannot be done if all welds have to be placed on one side of the line of action of a load. When this occurs, the weld shape or connection is said to be eccentrically loaded.

If the plate receiving load in Fig. 7.37a were welded to the column flange instead of being bolted, the connection still would be eccentric, with the load acting in the plane of welds. The welds would be subjected to a torque, causing additional shearing stresses in the welds.

If the tee-bracket bolts shown in Fig. 7.37b were replaced by welding the vertical edges of the bracket to the column flange, this connection would still be eccentric, with the load acting outside the plane of welds. The top of the bracket would tend to pull away from the flange. In this case the weld shape may be thought of as being subjected to a load causing shear, and to a moment causing a stress perpendicular to the shear.

7.28
Load in the Plane of Welds

The following discussion is valid for any pattern of welds in one plane.

Figure 7.59a shows a weld in the form of two vertical lines. The eccentric load P is in the plane of welds. In (b) of the same figure, the eccentric load is replaced by its equivalent, a direct load and a torque ($T = Pe$). The direct load P is assumed to be distributed uniformly throughout the entire weld length. Therefore, designating as F_1 the force per lineal inch of weld resulting from the direct load,

$$F_1 = \frac{P}{2L}$$

where L is the length of one weld line.

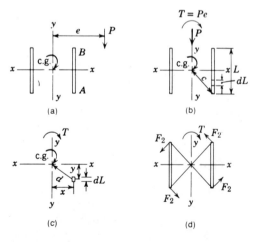

Figure 7.59/Eccentricity in the plane of welds.

The denominator in the above equation is always the total length of weld, and the direction of the force F_1 is always opposite to that of the load P.

In addition to the force F_1, there is another force F_2 resulting from the torque. An elemental portion of the weld, dL, is shown isolated in Fig. 7.59c. This may be any portion of the weld length, located a distance d from the center of gravity of the welds and having coordinates x and y.

Let:

F_0 = force per inch of weld at a unit distance from the c.g. of the welds.

F_2 = force per inch of weld at the element.

dT = resisting moment of the element.

From the stress-strain relationship applicable to an elastic material, a unit force is proportional to its distance from the centroid. Consequently,

$$F_2 = F_0 d = F_0 \sqrt{x^2 + y^2}$$

The total force on this elemental length is $F_2\, dL$, and it has a resisting moment of

$$dT = F_2(dL)d$$

Replacing F_2 by its equivalent shown above,

$$dT = F_0(x^2 + y^2)\, dL$$

The resisting moment of the complete weld is the sum of the effects of all the elemental portions, and this, in turn, must be equal to the applied torque in order to establish equilibrium:

$$T = \sum \left[F_0(x^2 + y^2)\, dL \right]$$

and since F_0 is constant for any weld pattern,

$$F_0 = \frac{T}{\sum x^2\, dL + \sum y^2\, dL}$$

Each term of the denominator in the above formula should be recognized as the moment of inertia of the total line of welds with respect to the x or y axis. Consequently,

$$F_0 = \frac{T}{I_x + I_y}$$

The force F_0 represents a unit force per inch, located a unit distance from the center of gravity of the welds. The maximum force in the welds occurs at the point furthest from the center of gravity, and its direction is perpendicular to a line connecting this point to the c.g. Letting c (Fig. 7.59b) represent this distance, and F_2 represent the maximum force,

$$F_2 = \frac{Tc}{I_x + I_y}$$

This is the general formula for the maximum force resulting from a torque. The welds shown in Fig. 7.59 are such that the maximum force F_2 occurs at each end, and their respective directions are indicated in (d) of this figure.

The above demonstration shows that the effect of an eccentric load on a weld connection, where the load acts in the plane of welds, is a force resulting from the direct load F_1 and a force resulting from the torque F_2, each of which has a determined direction. The total effect, therefore, is the vectorial sum of F_1 and F_2. For a safe design, the maximum vectorial sum of F_1 and F_2 should not exceed the strength of the weld F as determined by its size.

The location of the critical total force (vectorial sum of F_1 and F_2) may be determined by applying a general rule of thumb similar to that for bolts. The critical point occurs at the point furthest from the c.g. of the welds and nearest to the line of action of the eccentric load. Applying this rule to the detail shown in Fig. 7.59a locates the critical point at A and B.

Example

A 9-in. channel is welded to the flange of a column as shown in Fig. 7.60. Both members are of A36 Steel. Determine the size of E60-electrode weld required to resist the eccentric 12-kip load.

Solution

(1) Determine the location of the center of gravity of the 17 lineal inches of weld:

$$\bar{x} = \frac{2(4)2}{17} = 0.94 \text{ in.}$$

(2) Calculate I_x and I_y of the weld line:

$$I_x = \frac{9^3}{12} + 2\left[4(4.5)^2\right] = 223 \text{ in.}^3$$

$$I_y = 9(0.94)^2 + 2\left[\frac{4^3}{12} + 4(1.06)^2\right]$$

$$= 7.96 + 19.66 = 27.6 \text{ in.}^3$$

(3) Calculate the distance from the c.g. to the critical point of the weld. The critical

point is at A or B, both having a distance

$$c = \sqrt{x^2 + y^2} = \sqrt{(3.06)^2 + (4.5)^2}$$

$$= 5.45 \text{ in.}$$

(4) Determine the maximum force resulting from the torque:

$$F_2 = \frac{Tc}{I_x + I_y} = \frac{12(14.06)5.45}{223 + 27.6}$$

$$= 3.66 \text{ kips per in.}$$

(5) Determine the average force resulting from the direct load:

$$F_1 = \frac{P}{\text{total weld length}}$$

$$= \frac{12}{17} = 0.71 \text{ kips per in.}$$

(6) Calculate the vectorial sum of the forces F_1 and F_2. This may be accomplished by a graphical solution or mathematically by resolving F_2 into vertical and horizontal components:

(a) The *graphical solution* is shown in Fig. 7.61a for both points A and B.

(b) The *mathematical solution* for point A is shown in Fig. 7.61b. Resolve the force F_2 into horizontal and vertical components. From simple proportions established through similar triangles (oriented at right angles in the figure),

$$F_{2V} = \frac{3.06}{5.45}(3.66) = 2.06$$

(vertical component)

$$F_{2H} = \frac{4.50}{5.45}(3.66) = 3.02$$

(horizontal component)

The total vertical force at A is

$$F_{2V} + F_1 = 2.06 + 0.71 = 2.77 \text{ kips per in.}$$

The total horizontal force at A is

$$F_{2H} = 3.02 \text{ kips per in.}$$

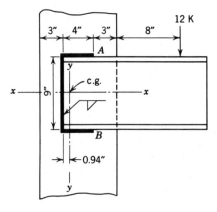

Figure 7.60/Example 1.

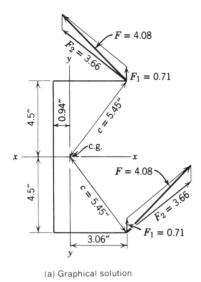

(a) Graphical solution.

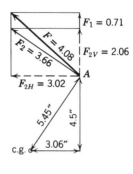

(b) Mathematical solution.

Figure 7.61/Solution to Example 1.

Therefore, the total vectorial sum of the forces at A is

$$F = \sqrt{(2.77)^2 + (3.02)^2} = 4.08 \text{ kips per in.}$$

(7) Knowing the maximum force per inch of weld produced by the loading, the required size of weld necessary to resist this force, using E60 electrodes, is

$$12.7D = F = 4.08$$
$$D = \frac{4.08}{12.7}$$
$$= 0.321; \quad \text{use } \tfrac{3}{8}\text{-in. weld}$$

As with fasteners, the procedure described here can safely be used to design eccentric welded connections. Tests have shown the results to be conservative. Furthermore, use of this design procedure does not result in unduly large or expensive connections. However, test results have shown that the factor of safety does not remain constant. Consequently, the AISC permits an alternate design procedure based on ultimate-strength design (Chapter 11). The Ninth Edition of the AISC Manual describes this alternate procedure and fur-

nishes tables of coefficients for various weld patterns that simplify that procedure.

7.29
Load Outside the Plane of Welds

The following discussion is valid for any pattern made by welding in one plane. Figure 7.62 shows a weld consisting of two vertical weld lines and one horizontal weld line. An eccentric load P is shown outside the plane of welds.

The eccentric load P in Fig. 7.62a is replaced by its equivalent, a concentric load and a moment as show in (b). The new concentric load P is assumed to be uniformly distributed over the weld length. Designating F_1 as the force per lineal inch of weld length resulting from the direct load,

$$F_1 = \frac{P}{\text{total weld length}}$$

F_1 is opposite in direction to the load P, in the plane of welds.

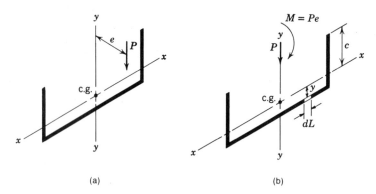

Figure 7.62/Eccentricity outside the plane of welds.

In addition to the force F_1, there is another force F_2 resulting from the moment. F_2 is not a uniform force, but varies with its location from the x-x axis. The moment forces are zero at the x-x axis (passing through the centroid of the welds), they are tensile for all weld portions above the axis, and they are compressive for all portions below the axis. It should be noted also that these forces always act perpendicular to the plane of the welds.

Let:

F_0 = moment force at a unit distance from the x-x axis.

dL = an elemental length of weld a distance y from the x-x axis.

dM = moment about the x-x axis of the force on the element dL.

Then, the force per inch of weld at a distance y from the x-x axis is $F_0 y$, and the total force on the element is $F_0(y)dL$. Also,

$$dM = F_0(y)\,dL(y)$$

The resisting moment of the entire weld group is the sum of the effects of all the elemental portions and, to establish equilibrium, must be equal to the applied moment M. Therefore,

$$M = \sum \left[F_0(y^2)\,dL \right]$$

or, since F_0 remains constant,

$$F_0 = \frac{M}{\sum y^2\,dL}$$

The term $\sum y^2\,dL$ should be recognized as the moment of inertia of a line (representing the weld pattern) with respect to the x-x axis. The maximum force occurs at the furthest point from the x-x axis. Consequently, letting c represent this distance,

$$F_2 = \frac{Mc}{I_x}$$

The combined effect of the concentric force and the moment will be the vectorial sum of the forces F_1 and F_2. As explained earlier, these forces act at right angles to one another. Consequently,

$$F = \sqrt{F_1^2 + F_2^2}$$

For a safe design, this maximum force should not exceed the strength of the weld as established by the size of the weld.

Example

Figure 7.63 shows a tee bracket, fillet-welded along its entire depth to a column flange. Calculate the required weld size to safely resist the 80-kip eccentric load. Steel is A36 and electrodes are E60.

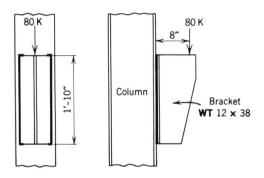

Figure 7.63/Example.

Solution

(1) Calculate the average vertical force per unit length of weld:

$$F_1 = \frac{80}{2(22)} = 1.82 \text{ kips per in.}$$

(2) The x-x axis, passing through the centroid of the welds, is located at mid-depth.

(3) Determine the moment of inertia of the welds with reference to the x-x axis:

$$I_x = 2\left(\frac{22^3}{12}\right) = 1770 \text{ in.}^3$$

(4) Calculate the maximum moment force on the welds:

$$F_2 = \frac{Mc}{I_x} = \frac{8(80)11}{1770} = 3.98 \text{ kips per in.}$$

(5) Determine the vectorial sum of F_1 and F_2:

$$F = \sqrt{F_1^2 + F_2^2} = \sqrt{(1.82)^2 + (3.98)^2}$$
$$F = 4.38 \text{ kips per in.}$$

(6) Calculate size of weld necessary to develop this maximum force.

$$12.7D = 4.38$$
$$D = 0.345; \quad \text{use } \tfrac{3}{8}\text{-in. weld}$$

PROBLEMS

1. The back of a **C** 8×11.5 is welded to the flange of a **W** 10×33 column as shown in Fig. 7.64. The weld is a $\frac{3}{16}$-in. fillet placed around the full perimeter as shown. Determine the maximum beam reaction R, located 4 in. from the edge of the column flange. Use A36 Steel and E60 electrodes. (Answer given in Appendix G.)

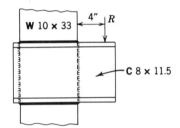

Figure 7.64/Problem 1.

2. A $\frac{3}{4}$-in. plate is cut and welded to the flange of a column as shown in Fig. 7.65. Determine the size of fillet weld required to support the 30-kip eccentric load shown. Use A242 Steel and E70 electrodes. (Answer given in Appendix G.)

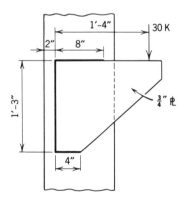

Figure 7.65/Problem 2.

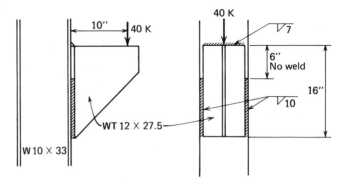

Figure 7.66/Problem 3.

3. Determine the size of fillet weld required for the connection shown in Fig. 7.66. All steel is A36. Use E70 electrodes.

7.30
Welded Flexible Beam Connections

Standard flexible beam connections are sometimes referred to as shear connections. Three common types were discussed in Art. 7.4, i.e., beam seats, end plates, and framing angles (Figs. 7.6 through 7.9). Design procedures for these connections, using fasteners, were described in Arts. 7.14, 7.15, and 7.16. Here, the design procedure using welds will be treated.

Beam Seats / Beam seats may be either flexible or stiffened as discussed in Art. 7.16. Flexible seats usually are 6 or 8 in. wide, and their design is essentially the same as that described for fastener connections. Of particular importance are the angle thickness and the calculated eccentricity of the load (Fig. 7.36). The attachment of the seat to the supporting member (Fig. 7.67) is achieved by welds along the vertical edges of the angle only. This type of weld is eccentrically loaded, with the load falling outside the plane of welds. The design of the welds follows the procedure described in Art. 7.29, and Part 4 of

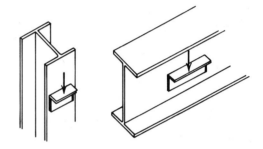

Figure 7.67/Welded beam seats.

the AISC Manual contains safe-load tables for several standard conditions.

End Plates / This type of beam connection is shown in Fig. 7.7. A rectangular plate is welded to the end of the beam. Current practice calls for bolting the plate to the supporting member, the design of which was covered in Art. 7.15.

The length of the plate (depth) is made less than the depth of the beam so that all the welding will be on the web. This limitation, plus the further stipulation that the plates be in a $\frac{1}{4}$- to $\frac{3}{8}$-in. thickness range, assures adequate flexibility and end rotation necessary for Type 2 construction. No eccentricity is considered in either the welds or the bolts.

The AISC Code requires the weld to be composed of two fillet welds on each side of the beam web. The code further specifies that the welds not be returned across the web at the top or bottom of the end plate. Therefore, the effective weld length is the depth of the plate minus twice the weld size. Since the weld is placed on both sides of the beam web, the theoretical size limitation should be imposed. Article 7.25 described this limited weld size to be

$$D = \begin{cases} 0.0154 F_y t_w & \text{for E60 electrodes} \\ 0.0135 F_y t_w & \text{for E70 electrodes} \end{cases}$$

However, actual design procedures permit larger weld sizes *provided* their capacity is reduced by the ratio of the thickness supplied to the thickness required. The AISC Manual furnishes a safe-load table for several standard end plates, shear connections for $\frac{3}{4}$- and $\frac{7}{8}$-in. bolts using $F_y = 36$ and 50 ksi, and E70 electrodes.

Example 1

Design an end-shear-plate connection for a **W** 12×40 beam having a reaction of 32 kips. The supporting column is a **W** 10×33. All steel is A36, and E60 electrodes are specified. Bolts are A325N in standard holes and at 3-in. spacing.

Solution

(1) Make a first estimate of the number of bolts, thickness of end plate, and size of bolts required. Try the $\frac{1}{4}$-in. minimum shear-plate thickness. Using A325N bolts, bearing is a consideration, and in this example the bolts will be in single shear. Try $\frac{5}{8}$-in. bolts, having a capacity of 6.4 kips per bolt (bearing is 13.1 kips per bolt and is not controlling):

$$\text{number of bolts} = \frac{32}{6.4} = 5.0$$

Use 6 bolts, because an even number is required.

(2) Calculate the maximum depth of plate:

$$L = T = 9\tfrac{1}{2} \text{ in.}$$

Use an $8\frac{1}{2}$-in.-deep plate (standard).

(3) Try a $\frac{3}{16}$-in. weld on each side of the beam web and determine the nonadjusted weld capacity. The effective length of weld is $8\frac{1}{2} - 2(\frac{3}{16}) = 8.13$ in.:

$$F = 2(12.7)\tfrac{3}{16} = 4.76 \text{ kips per in.}$$

Capacity = 4.76(8.13) = 38.7 kips.

(4) Check the required beam web thickness for the $\frac{3}{16}$-in. weld, compare with the actual web thickness, and use the reduction ratio if necessary:

$$t_w = \frac{\frac{3}{16}}{0.0154(36)} = 0.338 \text{ in. required}$$

The actual t_w is 0.295 in., and the reduction ratio is

$$\frac{0.295}{0.338}$$

(5) Calculate the reduced capacity of the weld and compare with the beam reaction:

$$\frac{0.295}{0.338}(38.7) = 33.8 > 32 \text{ kips} \qquad \text{OK}$$

Framing Angles / Framing-angle connections were shown in Fig. 7.6 and are presented in more detail in Fig. 7.68 The framing angles (used in pairs) straddle the web of the beam, and the fillet weld is

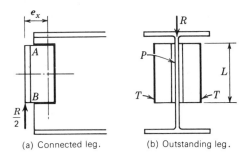

(a) Connected leg.   (b) Outstanding leg.

Figure 7.68/Welded framing angles.

placed along the top, bottom, and vertical sides, of the angle. These welds occur on each side of the web of the beam and consequently are limited in size.

The outstanding legs of the angles are welded to the supporting member only along the vertical edge of the angle (toe). This is done to assure the necessary flexibility needed for Type 2 construction. The angles usually have 3-in. legs and the necessary thickness for the size weld being used. The length of the angles is governed by strength requirements but is never more than the depth of the beam minus $2k$ and never less than one-half the depth.

Frequently, the outstanding leg is the field connection, and mechanical fasteners (bolts) are used instead of welds. The connection thus becomes "shop-welded and field-bolted," which is the type most often used in steel construction. In order to allow for bolting, the outstanding leg must have a minimum width of 4 in.

From Fig. 7.68a, it is seen that the weld on the connected leg is eccentrically loaded, with the load placed in the plane of welds. Each angle carries one-half the beam reaction, and the weld must be designed to carry the axial load ($R/2$) as well as the torque ($R/2$ times e_x). Points A and B are critical. The design process was described in Art. 7.28. Since this weld is placed on both sides of the beam web, the theoretical size limitation should be imposed (Art. 7.25). However, actual design procedures permit larger weld sizes *provided* their capacity is reduced by the ratio of the thickness supplied to the thickness required.

The outstanding legs welded to the supporting member are shown in Fig. 7.68b. It may appear that the welds are axially loaded, since the line of action of the reaction seems to pass through the centroid of the two vertical welds. However, experience has shown that because of the

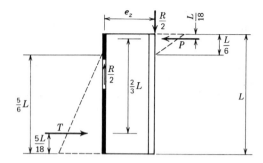

Figure 7.69/Force equilibrium in outstanding leg.

long weld length involved and the flexible action desired, there is some eccentricity. The tops of the angles bear inward, creating a pressure on the web of the beam in the vicinity of P (Fig. 7.68b). At the same time, the bottom of the welds (T) shears outward.

For design purposes, it is satisfactory to assume that the pressure near the top of the angles occurs over a distance (Fig. 7.69), $\frac{1}{6}$ the length of the weld (L). This pressure develops the total force (P), and the unit stress varies linearly from a maximum at the top to zero at $L/6$. These assumptions place the force P at $\frac{1}{3}(L/6)$, or $L/18$, down from the top.

The weld for the remaining distance ($\frac{5}{6}L$) develops the counterforce T and has a linear stress distribution from zero to a maximum at the bottom of the weld. This places the force T at $\frac{5}{18}L$ up from the bottom. Consequently, the internal couple between T and P has a moment arm of $2L/3$. This internal couple opposes the applied moment ($R/2)e_z$. From this,

$$T\tfrac{2}{3}L = \frac{R}{2}e_z$$

$$T = \frac{3Re_z}{4L}$$

The maximum unit horizontal force in the weld at the very bottom is twice the average, or

$$F_2 = \frac{2T}{\frac{5}{6}L}$$

To this value must be added, vectorially, the unit vertical force

$$F_1 = \frac{R/2}{L}$$

Thus the maximum unit force becomes

$$F = \sqrt{F_1^2 + F_2^2}$$

It is this maximum unit force that determines the size of the weld.

Example 2

Design a framing-angle connection between a **W 18×60** beam having an end reaction of 40 kips, and a **W 10×49** column. All steel is A36. Use E70 electrodes. The end of the beam is to be set $\frac{1}{2}$ in. from the face of the column (a standard procedure).

Solution

(1) Assume a 3×3-in. angle, 10 in. long. The T dimension of the beam is $15\frac{1}{2}$ in., which is greater than the 10-in. angle length and therefore satisfactory.

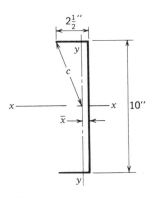

Figure 7.70/Solution to Example 2.

(2) Make a sketch of the weld shape for the connected leg, and calculate pertinent properties for determining unit axial stress and unit torque stress (Fig. 7.70):

Total length = 15 in.

$$\bar{x} = \frac{2(2.5)1.25}{15} = 0.42 \text{ in.}$$

$$c = \sqrt{(2.08)^2 + (5)^2} = 5.42 \text{ in.}$$

$$I_x = [2(2.5)5^2] + \frac{(10)^3}{12}$$

$$= 125 + 83.3 = 208 \text{ in.}^3$$

$$I_y = \frac{2(2.5)^3}{12}$$

$$+ [2(2.5)0.83^2] + 10(0.42)^2$$

$$= 2.60 + 3.44 + 1.76 = 7.8 \text{ in.}^3$$

$$I_x + I_y = 208 + 7.8 \approx 216 \text{ in.}^3$$

(3) Determine the unit axial force in the weld:

$$F_1 = \frac{20}{15} = 1.33 \text{ kips per in.}$$

(4) Determine the maximum unit torque force in the weld and convert to horizontal and vertical components:

$$F_2 = \frac{20(2.58)5.42}{216} = 1.29 \text{ kips per in.}$$

$$F_{2V} = \frac{2.08}{5.42}(1.29) = 0.50$$

$$F_{2H} = \frac{5}{5.42}(1.29) = 1.19$$

(5) Calculate the vectorial sum of the unit axial force and the unit torque force:

$$F = \sqrt{(1.33 + 0.5)^2 + (1.19)^2}$$

$$= 2.18 \text{ kips per in.}$$

(6) Determine the required size of weld, and check with the web thickness and required angle thickness:

$$D = \frac{2.18}{14.8}$$
$$= 0.15 \text{ in.}$$

($\frac{3}{16}$-in. weld required);

$$D_{max} = 0.0135(36)0.415$$
$$= 0.20 \text{ in.} \quad \text{OK}$$

Required angle thickness $= \frac{3}{16} + \frac{1}{16} = \frac{1}{4}$ in.

(7) Calculate the unit axial force in the outstanding leg:

$$F_1 = \frac{20}{10} = 2 \text{ kips per in.}$$

(8) Determine the total spreading shear force on the bottom of the angles and the resulting maximum unit force:

$$T = \frac{3(20)3}{4(10)} = 4.5 \text{ kips}$$

$$F_2 = \frac{2(4.5)}{\frac{5}{6}(10)} = 1.08 \text{ kips per in.}$$

(9) Calculate the vectorial sum of the unit shears:

$$F = \sqrt{(2)^2 + (1.08)^2} = 2.27 \text{ kips per in.}$$

(10) Determine the required weld size, and check angle thickness:

$$D = \frac{2.27}{14.8} = 0.15 \text{ in.}$$

($\frac{3}{16}$-in. weld required). Required angle thickness $= \frac{3}{16} + \frac{1}{16} = \frac{1}{4}$ in.

(11) Summarizing,

Angle size $= 3 \times 3 \times \frac{1}{4}$-in., 10 in. long

Weld size of connected leg $= \frac{3}{16}$ in.

Weld size of outstanding leg $= \frac{3}{16}$ in.

The AISC Manual furnishes safe-load tables for a few standard framing-angle connections, some of which are all welded and others of which have a combination of welds and fasteners. These tables are prepared using the ultimate-strength method (Chapter 11) for the connected leg and the more traditional vector-analysis method for the outstanding legs.

7.31
Welded Moment-Resisting Connections

As noted earlier in this chapter, the principal reason for using moment-resisting connections in buildings is to resist the effect of lateral forces such as wind and earthquake. Consequently, they are used most frequently between main beams and columns, creating a "rigid frame" (Chapter 10). However, even though they are used principally to resist lateral loads, the vertical gravity load will develop negative bending moments at the ends of the beams.

Figure 7.12a shows a bolted, beam-to-column, moment-resisting connection. The structural tees could be replaced by rectangular plates, field-welded to the flanges of both beams and column (Fig. 7.71a). It also is possible to omit tees and plates altogether and field-weld flange to flange directly, as shown in Fig. 7.71b. However, this procedure requires exactness in fabrication (allowing only a $\frac{1}{4}$-in. clearance) that may become too costly. The plate and flange welds must be designed to develop a horizontal force equal to the end moment divided by the depth of the beam, or $F = M/d$.

Usually the top plate is made narrower than the beam flange, so that the weld can be placed on top of the flange. The bottom plate, however, is made wider than the flange to permit placing the weld on top of the plate, thereby avoiding overhead welding.

The vertical reaction of the beam is transferred to the column through either a

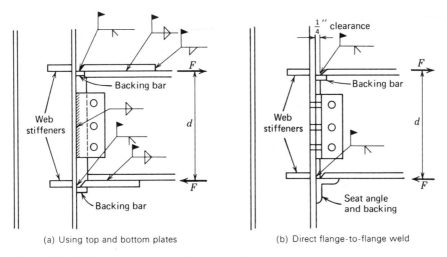

(a) Using top and bottom plates   (b) Direct flange-to-flange weld

Figure 7.71/Welded moment-resisting connections.

standard framing-angle connection (Fig. 7.71b) or a shear plate, shop-welded to the column flange (Fig. 7.71a). The field connection usually is bolted as shown, but also may be welded.

Moment-resisting end connections reduce the positive moment near the center of the beam, which may permit a reduction in beam size (Fig. 7.3). As a consequence, moment-resisting connections often are used for this purpose even when beams frame into girders as shown in Fig. 7.12b.

With moment-resisting beam-to-column connections, column web stiffeners may be required. In Fig. 7.71, only one beam is framed to the column. When beams are connected to both column flanges and stiffeners are required, the stiffeners should extend continuously between the flanges of the column. The 1989 AISC Specification (Section K1.8) contains requirements for column web stiffeners. These requirements are based on test data and ultimate-strength design principles (Chapter 11); therefore, the force from the beam flange (or flange plate) must be factored upward to provide a factor of safety. When the computed force derives from gravity live, environmental, and dead loads,

this factor is specified to be $\frac{5}{3}$. The flange force is designated P_{bf}, and $P_{bf} = \frac{5}{3}F$, where F is the flange force calculated on the basis of elastic behavior.

The force P_{bf} is transmitted through the column flange into the column web. The AISC Specification limits the length of the web over which this force can be developed at the yield stress of the web. This length has been set at five times the column dimension k, plus the thickness of the beam flange (or flange plate). Therefore, if web stiffeners are not to be used,

$$P_{bf} \leq t_w(t_f + 5k)F_{yc}$$

where F_{yc} is the yield stress of the column web.

When this condition is not met, web stiffeners, having an area A_{st} and stressed to the yield point F_{yst}, must make up the difference. Expressed in equation form,

$$A_{st}F_{yst} \geq P_{bf} - t_w(t_f + 5k)F_{yc}$$

This criterion for design is based on the compression force from the beam flange (which is usually controlling), but is applied to the tension flange as well. Occasionally, the tension flange is controlling and must be checked.

Section K1.8 of the 1989 AISC Specification stipulates that when the thickness of the column flange (t_f) is less than $0.4\sqrt{P_{bf}/F_{yc}}$, a pair of stiffeners is required opposite the tension force. This implies that when t_f is equal to this ratio, the tensile capacity of the column flange is equal to the applied force and no web stiffeners are required. Using P_{fb} as the column-flange strength, substituting it for P_{bf} in the above expression, and solving for its value,

$$t_f = 0.4\sqrt{\frac{P_{bf}}{F_{yc}}}$$

$$= 0.4\sqrt{\frac{P_{fb}}{F_{yc}}}$$

$$t_f^2 = 0.16\frac{P_{fb}}{F_{yc}}$$

$$P_{fb} = \frac{t_f^2 F_{yc}}{0.16}$$

When the beam-flange tensile force (P_{bf}) exceeds this column-flange strength, the web stiffener must be designed to make up the difference, i.e.,

$$P_{bf} - \frac{t_f^2 F_{yc}}{0.16}$$

One additional check is needed to determine if stiffeners are required. This check is based on buckling of the column web behind the compression flange force. When the column web depth clear of fillets (d_c) is more than

$$\frac{4100t_w^3\left(\sqrt{F_{yc}}\right)}{P_{bf}}$$

stiffeners are required.

Example

Design a moment-resisting connection similar to that shown in Fig. 7.71a. The beam is a **W** 18×35, and the column, a **W** 10×49. The beam has an end reaction of 22 kips and an end moment of 90 ft-kips. All steel is A36. Use E70 electrodes and A325N bolts in standard holes and at 3-in. spacing.

Solution

(1) Calculate the horizontal force that must be transmitted by the top and bottom plates:

$$F = \frac{90(12)}{17.7} \simeq 61 \text{ kips}$$

(2) Determine the area required for the top and bottom plates. The allowable stress in tension is

$$F_t = 0.6F_y \simeq 22 \text{ ksi}$$

$$A_p = \frac{61}{22} \simeq 2.77 \text{ in.}^2$$

(3) Design the top plate and its weld requirements. Since the beam flange is 6 in. wide, select a plate width of $4\frac{1}{2}$ in. Then, the required thickness of this plate is

$$t_A = \frac{2.77}{4.5} \simeq 0.62 \text{ in.}; \qquad \text{use } \tfrac{5}{8} \text{ in.}$$

Use a complete-penetration, single-vee groove weld between the plate and column flange, which requires a backing bar as shown in Fig. 7.71. The minimum fillet-weld size for the plate is $\frac{1}{4}$ in. The $\frac{1}{4}$-in. weld has a strength of $14.8(\frac{1}{4}) = 3.7$ kips per in.; therefore, the total length of weld required is

$$L = \frac{61}{3.7} \simeq 16.5 \text{ in.}$$

Distribute the 16.5 in. as follows: 4.5 in. on the end and 6 in. along each side of the plate.

(4) Design the bottom plate and its weld requirements. Select a $7\frac{1}{2}$-in. wide plate to permit welding between the edge of the flange and the plate. The required thickness of the plate is

$$t_B = \frac{A_p}{7.5} = \frac{2.77}{7.5} \simeq 0.37 \text{ in.}; \quad \text{use } \tfrac{3}{8} \text{ in.}$$

Using a $\frac{1}{4}$-in. weld, the total length required is the same as for the top plate, i.e., $16\frac{1}{2}$ inches. Place the weld $8\frac{1}{2}$ in. on each side.

(5) Design the shear plate for the vertical reaction. Select $\frac{5}{8}$-in. bolts. BV = 6.4 kips, and the number required is

$$\frac{22}{6.4} \simeq 3.44; \quad \text{use 4 bolts}$$

The total length of the plate is

$$3(3) + 2(1\tfrac{1}{4}) = 11\tfrac{1}{2} \text{ in.}$$

The allowable unit shear stress is $0.4F_y \simeq 14.5$ ksi; therefore, the required plate thickness is

$$\frac{22}{11.5(14.5)} \simeq 0.13 \text{ in.}$$

Therefore, use a $\frac{1}{4}$-in. min. plate thickness. A fillet weld will be used on both sides of the plate to connect it to the column flange. The thickness of the flange is 0.56 in., which requires a minimum fillet-weld size of $\frac{1}{4}$-in.; however, this exceeds the $0.0135(36)\frac{1}{4} = 0.122$-in. maximum allowable size. Therefore, a reduced weld strength is needed. This reduced strength is

$$\frac{0.122}{0.25}(14.8)0.25 = 1.81 \text{ kips per in.}$$

The required length is

$$\frac{22}{2(1.81)} = 6.08 \text{ in.}; \quad \text{use 7 in.}$$

(6) Investigate the need for column stiffeners (AISC Specification, Section K1). Assuming the end moment of 90 ft-kips results from gravity live and dead loads, the load factor is $\frac{5}{3}$. Thus

$$P_{bf} = \tfrac{5}{3}F = \tfrac{5}{3}(61) \simeq 102 \text{ kips}$$

The capacity of the column web opposite the compression plate is

$$t_w(t_f + 5k)F_{ye} = 0.34(\tfrac{3}{8} + 5\tfrac{3}{16})36 = 68 \text{ kips}$$

Since $102 > 68$, web stiffeners are required and will be placed opposite both the tension and compression flange plates.

(7) Calculate the required area of the web stiffeners:

$$A_{st} = \frac{102 - 68}{36} = 0.94 \text{ in.}^2$$

The width of the compression flange plate is $7\frac{1}{2}$ in. (step 4). The AISC Specification requires the width of the stiffener plate to be not less than

$$\frac{7.5}{3} + \frac{0.34}{2} = 2.67 \text{ in.}$$

Use a 3-in. plate width. The required thickness is

$$t_{ts} = \frac{0.94}{3} = 0.31 \text{ in.}$$

The AISC Specification requires the minimum stiffener thickness to be $t_b/2$. Using the thicker flange plate (top),

$$\frac{0.625}{2} = 0.31 \text{ in.}$$

Use a $\frac{3}{8}$-in.-thick stiffener plate.

(8) The required length of the stiffener (along the web of the column) will depend,

in part, on the weld, even though the AISC Specification states that, in this case, the length does not have to be greater than one-half the depth of the column. Try a length of $4\frac{1}{2}$ in.

(9) Check the width-thickness ratio (AISC Specification, Section B5):

$$\frac{4.5}{0.375} = 12 < \frac{95}{\sqrt{f_y}} = 15.8 \qquad \text{OK}$$

(10) Design the welds for the stiffener. The net compression force on the stiffener is

$$102 - 68 = 34 \text{ kips}$$

The net tensile force in the stiffener is

$$P_{bf} = \frac{t_f^2 F_{yc}}{0.16}$$

and

$$102 - \frac{0.56^2(36)}{0.16} = 31.4 \text{ kips}$$

Using 34 kips for the design of the welds, the force per plate is $\frac{34}{2} = 17$ kips.

(a) Weld to column web. The minimum weld size is $\frac{3}{16}$ in. The maximum (full strength) of the weld, based on thickness of column web and welding both sides, is

$$D_{\max} = 0.0135(0.34)36 \approx 0.17 \text{ in.}$$

Adopt a $\frac{3}{16}$-in. weld on both sides of the plate, but reduce its strength. The regular strength of the $\frac{3}{16}$-in. weld is $0.1875(14.8) = 2.78$ kips per in. The reduced strength is

$$\frac{0.17}{0.19}(2.78) = 2.49 \text{ kips per in.}$$

The length of weld required is

$$l = \frac{17}{2(2.49)} = 3.41 \text{ in.}$$

Use a $3\frac{1}{2}$-in.-length weld along the $4\frac{1}{2}$-in. length of the plate.

(b) Weld to column flange. The minimum weld size is $\frac{1}{4}$ in. (based on the column flange thickness). The maximum fillet weld for welding both sides of the $\frac{3}{8}$-in. plate is

$$D_{\max} = 0.0135(0.375)36 = 0.18 \text{ in.}$$

Use the $\frac{1}{4}$-in. weld size, but reduce the capacity of the weld:

$$\frac{0.18}{0.25}(14.8)0.25 = 2.66 \text{ kips per in.}$$

The required length is

$$l = \frac{17}{2(2.66)} = 3.2 \text{ in.}$$

Increase the width of the stiffener plate to $3\frac{1}{2}$ in., and use two $\frac{1}{4}$-in. fillet welds along the entire length.

7.32
Ductile Moment-Resisting Connections

It was stated earlier that one major reason for using moment-resisting connections is to resist lateral forces such as those produced by earthquakes. Chapter 8 will treat earthquake-resistant design, and it will be pointed out that under certain conditions the steel frame must consist of *ductile moment-resisting connections*. Not only must such connections meet more restrictive specifications in order to be classified as ductile, but the ultimate-strength procedure also must be taken into consideration in their design.

PROBLEMS

1. Design a welded flexible beam seat for a **W** 14×22 beam having an end reaction of 26 kips. The seat is attached to the flange of a **W** 10×49 column. Allow for $\frac{3}{4}$-in. clearance between the end of the beam and the column flange. Use A36 Steel and E60 electrodes. (Refer to Art 7.16 for information related to design of seat angles.)

2. (a) Determine the maximum allowable beam reaction for the end-plate shear connection shown in Fig. 7.72. All steel is A36 and E60 electrodes are to be used.

(b) Same as (a), except use E70 electrodes. Check answers with the AISC Manual. (Answers given in Appendix G.)

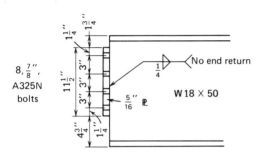

Figure 7.72/Problem 2.

3. Determine the maximum allowable beam reaction for the all-welded framing-angle connection shown in Fig. 7.73. Use A36 Steel and E70 electrodes. (Answer given in Appendix G.)

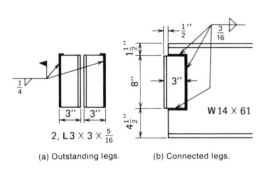

(a) Outstanding legs. (b) Connected legs.

Figure 7.73/Problem 3.

4. The framing-angle connection shown in Fig. 7.74 is shop-welded and field-bolted. Determine the maximum reaction for each beam using A36 Steel and E70 electrodes. (Answer given in Appendix G.)

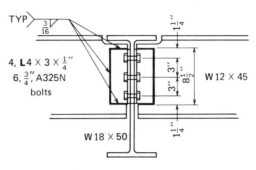

Figure 7.74/Problem 4.

5. Determine the moment and shear capacity of the connection shown in Fig. 7.75. All steel is A36; E60 electrodes and $\frac{3}{4}$-in., A325N bolts are to be used.

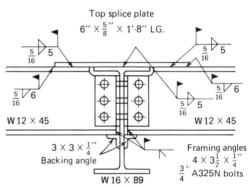

Figure 7.75/Problem 5.

8

LATERAL LOADS

8.1
Introduction

The term *lateral load* is applied to a horizontal load, as distinct from a gravity load that acts vertically downward. Lateral loads on a building are not uncommon. They may result from soil or water pressure, snow drifting on one side, or such special circumstances as a heavy vehicle braking to a stop on the floor of a parking garage. However, of concern in this chapter are lateral loads caused by wind and earthquake.

All buildings must possess some ability to develop resistance to lateral loads. Buildings with steel frames or steel elements may utilize those frames or elements to provide this resistance; if not, then it must be provided by other structural elements such as walls and floors. Although the discussions and procedures that follow are general enough to apply to any building type, emphasis will be placed on buildings with steel frames or steel elements.

For general design purposes, it must be assumed that wind and/or earthquake loads can come from any horizontal direction, and that lateral wind load can create an uplift force on the roof. It is customary to consider only two building axes, one transverse and one longitudinal (Fig. 8.1). Then, if the structure is deemed safe and adequate in these two directions treated separately, it is considered to be safe for loads from any direction.[1] It should be added that lateral loads are applied concurrently with all dead loads and such other gravity loads as will produce the

[1]Some codes make an exception to this procedure when designing for earthquake loads. This exception pertains to corner columns and to irregular framing systems.

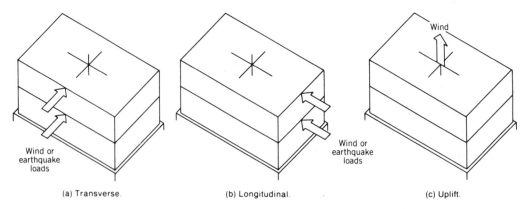

(a) Transverse.   (b) Longitudinal.   (c) Uplift.

Figure 8.1/Load direction and building axes.

most unfavorable load combination. How-ever, earthquake loads and wind loads need not be applied at the same time. (See Art. 9.4, "Combining Loads.")

8.2
Development of Lateral Loads

The development of lateral loads on a building must be discussed in the context of the architectural solution and/or the type of construction used. The only thing that wind loads and earthquake loads have in common is that they are considered to be basically horizontal in direction. Conse-quently, a structural scheme must be de-vised to transfer these horizontal loads downward to the foundation. In the case of wind, the exterior envelope of the build-ing usually delivers the horizontal load to each floor level and at the roof. In the case of earthquakes, building weights are as-sumed to be collected at each floor level and at the roof and are then converted to equivalent horizontal loads. Therefore, in the case of both wind and earthquake, it is the horizontal load at each level that must be transferred downward to the founda-tion.

There are two distinct types of structural element used to transfer these loads, namely, horizontal and vertical. The hori-zontal element (Fig. 8.2) does not neces-sarily have to be positioned in a horizontal plane (e.g., a sloping roof could be used). Its principal purpose is to transfer the lateral load in a horizontal direction to the vertical element. Trussed floors (cross-bracing) and diaphragms usually are used for this purpose, and at times both are required in the same building.[2]

Figure 8.2a illustrates a trussed floor with lateral loads applied in one direction only. The triangulated center portion acts like a truss (Chapter 9) lying on its side and spanning between the vertical elements lo-cated at each end.

Figure 8.2b illustrates a floor diaphragm, with lateral loads applied in one direction only. The entire floor acts like a deep beam lying on its side, spanning between the two vertical elements located at each end. A diaphragm is an element stressed principally in shear, but tension and com-pression forces must be accounted for as well (see Art. 8.3).

The vertical element transfers the horizon-tal load vertically downward to the foun-dation (Fig. 8.3). Vertical elements may be placed anywhere within the building, but

[2]Occasionally, a series of closely spaced beams can function as the horizontal element in the transverse direction, if the vertical elements are distributed throughout the building as shown in Fig. 8.23.

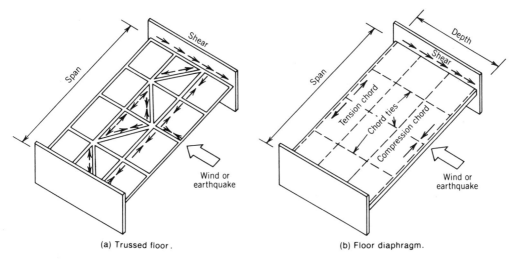

(a) Trussed floor.

(b) Floor diaphragm.

Figure 8.2/Horizontal structural elements.

their location is a prime factor in the design of the horizontal element and has a significant effect on the load it must develop. Since load must be considered from two directions, a minimum of three vertical elements are required for the stability of any given building.

Figure 8.3a illustrates two rigid frames (Chapter 10) located in the exterior walls of a building. Lateral loads are shown in one direction only, and the frames transfer the horizontal load downward to the foundation. Figure 8.3b shows two vertical trusses (cross-bracing) placed in the exterior walls, and Fig. 8.3c shows the vertical elements as shear walls. The rigid frame and cross-bracing frequently are constructed from steel, while the shear wall generally is either of reinforced concrete or reinforced masonry. Chapter 9 contains a detailed discussion and examples of vertical elements consisting of braced frames.

The building in Example 2, Art. 8.7 (Fig. 8.23) appears to be an exception to the manner in which horizontal loads are developed. In this example, only wind in the transverse direction is considered, and the horizontal load development is achieved by the girts and purlins. The load from

wind on the surface of the building is transferred directly to the gable frames by the purlins and girts (sometimes bending is about the minor axis). No cross-bracing or diaphragm is used. However, when considering wind in the longitudinal direction, one or the other would be required to transfer the loads to the cross-bracing in the side walls.

The location of the vertical elements in a building is a critical consideration and is usually dictated by architectural requirements. Figure 8.4 shows the minimum three vertical elements located in three of the four exterior walls. The lateral load is shown in one direction only, and in this case it is parallel to wall A. Under such conditions, it should be apparent that there is a tendency for rotation in a horizontal plane, and the structure could collapse without the existence of both wall B and wall C. Horizontal torsion from lateral loads always should be investigated.

Example Cross-Braced Roof

A rectangular, one-story building having overall dimensions in plan of 60×100 ft is shown in Fig. 8.5. The main framing grid is

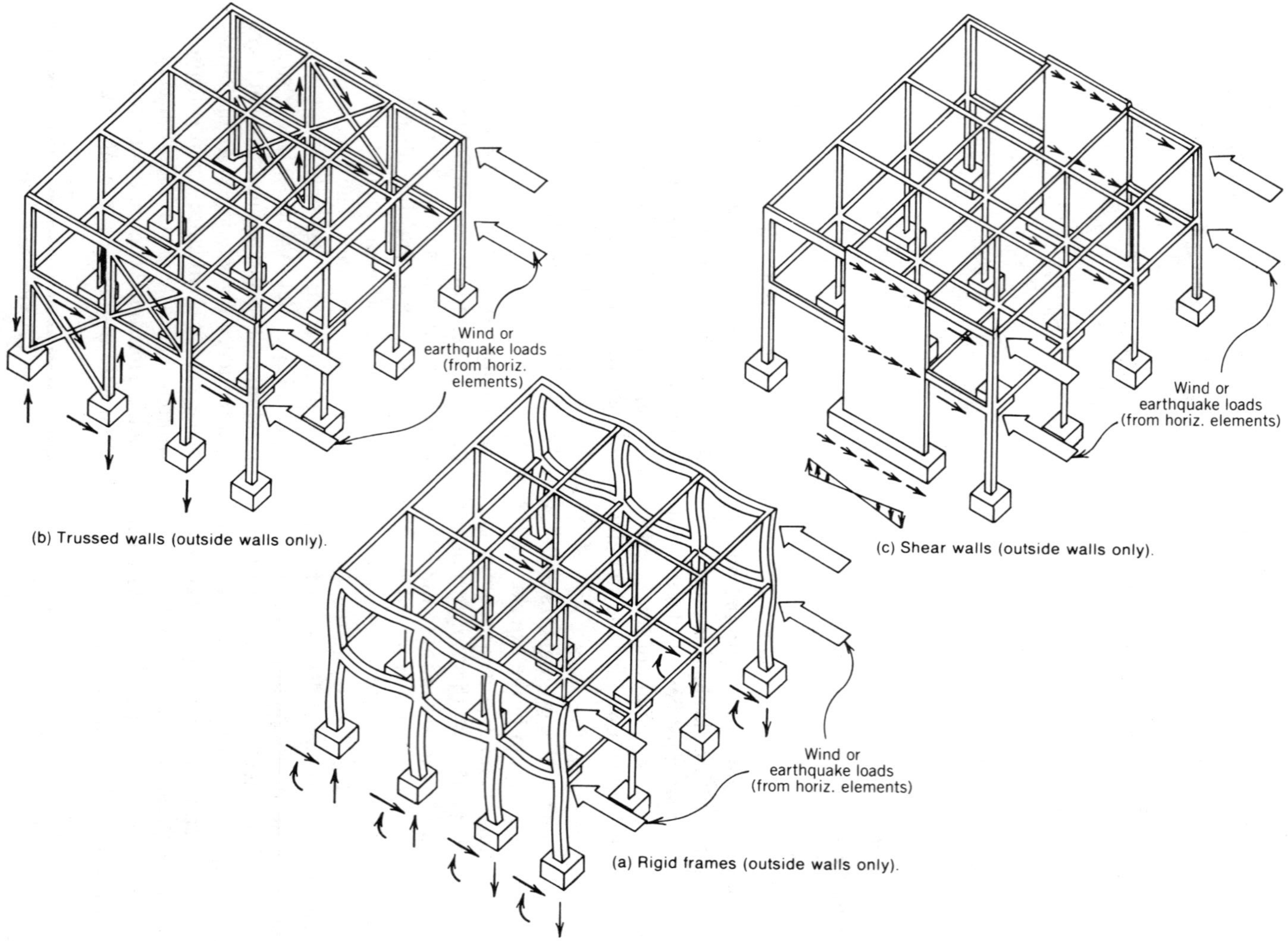

(c) Shear walls (outside walls only).

Wind or earthquake loads (from horiz. elements)

(a) Rigid frames (outside walls only).

Wind or earthquake loads (from horiz. elements)

(b) Trussed walls (outside walls only).

Wind or earthquake loads (from horiz. elements)

Figure 8.3/Vertical structural elements.

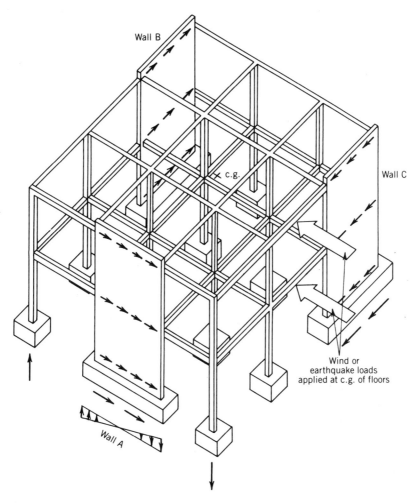

Wall B

c.g.

Wall C

Wind or
earthquake loads
applied at c.g. of floors

Wall A

Figure 8.4/Building torsion.

20×25 ft. The complete structure required for gravity loads is not shown. Several systems could match this grid system. In this example, it is assumed that roof purlins (small beams) span the 25-ft direction at a spacing of 5 ft o.c. so that one purlin occurs along the eaves line and along each grid line in the longitudinal direction. It is also assumed that main beams, and girders or trusses, occur along the grid lines in the transverse direction. For example, *DE* and *JK* are purlins, and *JD* and *KE* are main beams or trusses.

A lateral load of 200 plf will be applied in a transverse direction as shown along the back eaves line in Fig. 8.5. This load could come from either wind or earthquake. The overall scheme for developing this lateral load will be the vertical elements shown at the ends of the building and the horizontal element consisting of cross-bracing along the center bay. The design force calculations for the horizontal element loading shown will be illustrated; however, it must be borne in mind that a similar solution is required for the lateral load applied in the

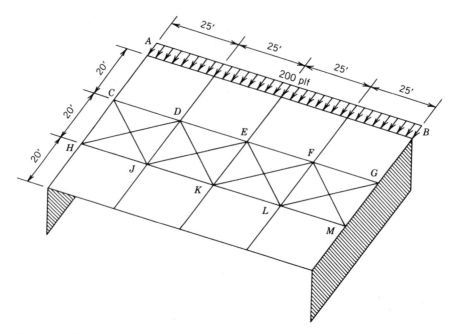

Figure 8.5/Example of cross-braced roof.

opposite direction. Also, although not shown, the resisting structural scheme for the longitudinally applied load would need to be determined.

The cross-bracing members commonly are flexible angles, or cables or rods with turnbuckles, that extend continuously from grid corner to grid corner. They cross one another but do not actually intersect. All other intersections are assumed to be pinned connections. Under these conditions, one of each pair of cross-ties becomes an effective tension component; the other (the compression component) is not considered to be effective. This actually will be the case if the ties are cables and nearly so for other types of cross-ties.

Solution

(1) Calculate the horizontal reactions of the eaves purlin. The pinned-end construction allows for a simply supported beam solution in the horizontal direction; therefore

$$R = \frac{wL}{2} = \frac{0.2(25)}{2} = 2.5 \text{ kips}$$

(2) Determine the axial load on each strut that transfers the horizontal load to the cross-bracing system. Each internal strut transmits two purlin reactions, i.e., $2(2.5) = 5$ kips. The exterior strut transmits one purlin reaction, or 2.5 kips.

(3) Redraw the horizontal element showing the strut loads (calculated in step 2), the effective cross-ties (delete the cross-tie in compression), and the remaining components (along grid lines) that form a triangulated truss spanning between the two vertical elements. This step is shown in Fig. 8.6a. In this example, members *CJ*, *DK*, *FK*, and *GL* are the tension cross-ties. Purlins that function as compression struts are *CD*, *DE*, *EF*, and *FG*. Purlins that

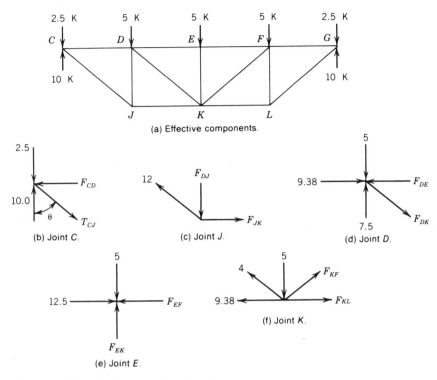

(a) Effective components.

(b) Joint C.

(c) Joint J.

(d) Joint D.

(e) Joint E.

(f) Joint K.

Figure 8.6/Solution of cross-braced roof.

perform as tension members are *JK* and *KL*. Components *DJ*, *EK*, and *FL* are in compression and are formed by the main beams or portions of the vertical truss that supports the purlins.

(4) Determine the horizontal reactions from the horizontal element to the vertical elements. In this example, each vertical element develops one-half the total lateral load, or 10 kips.

(5) Calculate the axial load in each component of the truss system. Every component is pin-ended, and loads occur only at the intersection of the components. Consequently, there is no bending in the components, and each is a two-force member. This means each member is either in tension or in compression, which is consistent with the assumptions made earlier. Therefore, the line representing each member

(Fig. 8.6a) also is the line of action of the forces developed by the members.

Two ways of determining the force in a member are the *graphical method* and the *algebraic method of joints*. Both methods are developed in Chapter 9. The algebraic method will be used here.

Proceeding in the horizontal plane from joint to joint and applying the principles of static equilibrium—i.e., $\Sigma V = 0$ and $\Sigma H = 0$—the algebraic sum of all forces in both the transverse and longitudinal directions must equal zero. Beginning at a joint that has only two unknown forces and then proceeding to other joints where there are only two unknown forces, use an arrow directed away from a joint to indicate a member in tension and an arrow toward the joint to indicate a member in compression.

Joint C is shown in Fig. 8.6b. The length of the cross-tie is

$$L = \sqrt{(25)^2 + (20)^2} \simeq 32 \text{ ft}$$

From geometry

$$\sin \theta = \tfrac{25}{32}$$
$$\cos \theta = \tfrac{20}{32}$$

Applying $\Sigma V = 0$ in the transverse direction, it is observed that the transverse component of T_{CJ} is $10 - 2.5 = 7.5$ kips. From geometry it is observed that this component is the product of $\cos \theta$ and T_{CJ}, i.e.,

$$(\cos \theta) T_{CJ} = \tfrac{20}{32} T_{CJ} = 7.5$$
$$T_{CJ} = 12 \text{ kips} \quad (\text{tension})$$

The longitudinal component of T_{CJ} is $(\sin \theta) T_{CJ}$. Applying $\Sigma H = 0$ in the longitudinal direction,

$$F_{CD} = (\sin \theta) T_{CJ}$$
$$= \tfrac{25}{32}(12) = 9.38 \text{ kips} \quad (\text{compression})$$

Proceed to joint J and apply the tie force, calculated as 12 kips (Fig. 8.6c). There remain only two unknowns, and the components of $T_{JC} = 12$ kips have already been calculated. Therefore,

$$F_{JK} = 9.38 \text{ kips} \quad (\text{tension})$$
$$F_{DJ} = 7.50 \text{ kips} \quad (\text{compression})$$

Proceed to joint D (Fig. 8.6d). The vertical (transverse) component of F_{DK} is $(\cos \theta) F_{DK}$ and must be equal to $7.5 - 5 = 2.5$ kips, i.e.,

$$\tfrac{20}{32} F_{DK} = 2.5$$
$$F_{DK} = 4 \text{ kips} \quad (\text{tension})$$

The horizontal component of F_{DK} is $(\sin \theta) F_{DK}$, which is

$$\tfrac{20}{32}(4) = 3.13 \text{ kips}$$

Consequently,

$$F_{DE} = 9.38 + 3.13 = 12.5 \text{ kips}$$
$$(\text{compression})$$

Proceed to joint E (Fig. 8.6e). It should be apparent that

$$F_{EF} = 12.5 \text{ kips} \quad (\text{compression})$$

and

$$F_{EK} = 5 \text{ kips} \quad (\text{compression})$$

Proceed to joint K (Fig. 8.6f). This step is used to confirm that the forces in the scheme are symmetrical:

$$F_{KF} = 4 \text{ kips} \quad (\text{tension})$$
$$F_{KL} = 9.38 \text{ kips} \quad (\text{tension})$$

(6) Summarize the design forces. For the lateral loads shown, the cross-ties in the outer bays (adjacent to the vertical elements) develop a force of 12 kips, and the inner cross-ties, a force of 4 kips. These become the design forces if the 200-plf load results from wind. On the other hand, if the lateral load is caused by an earthquake, most codes require that the developed load be multiplied by 2 or 3. This means a design force of $2.0(12) = 24$ kips and $2.0(4) = 8$ kips. It should be noted also that codes generally permit a 33 per cent increase in the allowable stress when the developed forces include the short-time effect of wind or earthquake.

The remaining members of this horizontal element system (purlins and beams) must be designed in such a manner as to include the effect of the computed axial loads in combination with other gravity loads. The various load combinations are described in Chapter 9.

8.3 Diaphragms

A diaphragm is a frequently used and very effective and efficient horizontal element. It spans horizontally, carrying the lateral load to the vertical elements (Fig. 8.2b). The number and location of these vertical elements bear an important relationship to the design of the diaphragm. The diaphragm has been described as a deep beam lying on its side and spanning between the vertical elements.

For design purposes, all diaphragms must be classified as either rigid or flexible. Four very common types of deck construction that can be classified as rigid diaphragms are (1) reinforced concrete, (2) reinforced gypsum concrete, (3) steel decks with concrete fill, and (4) precast concrete components that are properly attached to one another. Steel decks without concrete fill may be either flexible or rigid, depending upon their thickness, their attachment to one another, and the manner of connection to supporting elements. Two very common types of deck construction that can be classified as flexible diaphragms are (1) some steel decks without concrete infill, similar to those shown in Figs. 5.42 and 8.10, and (2) plywood.

A rigid diaphragm acts more like a flat plate than a deep beam, and it transmits lateral load to the vertical elements in proportion to their relative stiffnesses. Torsion may be included; in fact, if the vertical elements are positioned in a manner that requires torsion action to maintain lateral stability, then a rigid diaphragm is required. A rigid diaphragm must be designed for both shear and bending due to forces acting in its own plane. Flexible diaphragms, on the other hand, cannot transmit loads by torsion. They are typically very flexible compared to vertical elements and consequently transmit loads by acting as a simple or continuous beam spanning between the vertical elements.

Figure 8.7 (a) and (b) show plans of a building having overall dimensions L_1 and L_2. Lateral loads are shown along the floor line (L_1) in one direction only. Three vertical elements are shown as heavy lines along sides AB, AC, and CD. The uniform load is summed up as $F = wL_1$ and, for purposes of stability, can be considered as concentrated, acting as shown. This entire lateral load must be transmitted to, and be resisted by, the vertical element in

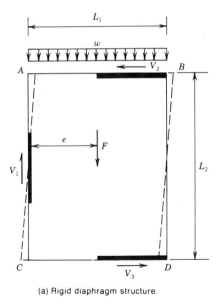

(a) Rigid diaphragm structure.

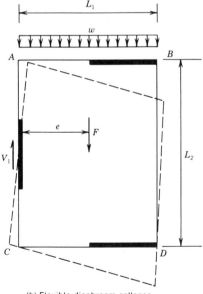

(b) Flexible diaphragm collapse.

Figure 8.7/Rigid and flexible diaphragms.

wall AC, so that $V_1 = F$. It should be apparent that this condition creates a torsion (eF), and that the diaphragm will tend to rotate as indicated by the dashed lines in Fig. 8.7a. The presence of the other two vertical elements in walls AB and CD will stabilize this rotational tendency by developing the lateral shears in each, as shown, i.e.,

$$V_2 = V_3 = \frac{eF}{L_2}$$

Figure 8.7b is intended to illustrate the same condition as that shown in fig. 8.7a, except that a flexible diaphragm is present. The same torsional tendency exists, but with a flexible diaphragm no parallel forces are developed in the vertical elements located in walls AB and CD. The diaphragm will deform as shown by the dashed lines, and collapse is possible. Stability without torsion requires a minimum of two vertical elements in the direction under consideration. Under these conditions, the flexible diaphragm transmits loads by an action similar to that of continuous or simple beams.

The maximum span-depth proportions (4:1) for the most rigid of flexible diaphragms is shown in the right span of Fig. 8.8a. Continuous beam action best describes the function of the diaphragm. Deformations take place similar to that shown by the dashed lines—observe the reverse curvature. In addition, the maximum force in the vertical element occurs at B and is larger than $w(L_1 + L_2)/2$ because of continuity. Consequently, the maximum shear in the diaphragm occurs along element B.

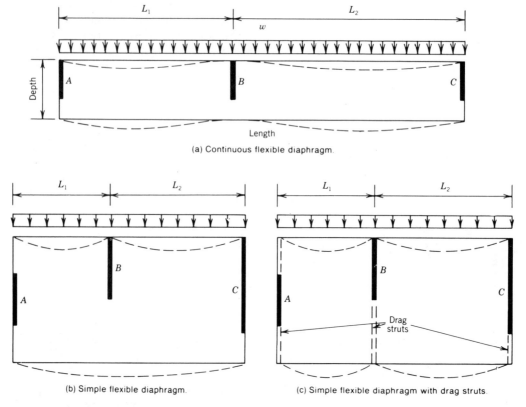

(a) Continuous flexible diaphragm.

(b) Simple flexible diaphragm.

(c) Simple flexible diaphragm with drag struts.

Figure 8.8/Flexible diaphragms.

A more frequently encountered small span-depth proportion is that shown in Fig. 8.8 (b) and (c). In this case, the assumption of continuous beam action is not justified for the flexible diaphragm. Under the conditions shown, a simple beam action is assumed, and the lateral loads are distributed to the vertical elements in proportion to the tributary area of the diaphragm. The middle element (B) will develop $w(L_1 + L_2)/2$.

Note that in Fig. 8.8b, the dashed lines showing deflections appear incompatible with one another. This condition should be avoided and can be by using a *collector*, or *drag strut*, as shown in Fig. 8.8c. A drag strut is a linear tension or compression member placed in the plane of the diaphragm, to which the diaphragm is attached. The drag strut is parallel to, and anchored to, the vertical element. Proper placement of collectors greatly reduces shear stress in the diaphragm.

Comparing a diaphragm to a beam lying on its side could be misleading. There are two main reasons for this: first, the span-depth ratio is very small (Fig. 8.2b), so at times the depth will be larger than the span—particularly when investigating lateral loads in the longitudinal direction of the building (Fig. 8.1b); second, flexible diaphragms usually consist of a number of materials and independent pieces that must be tied together to make the entire diaphragm function as a unit, whereas a conventional beam usually is homogeneous.

The small span-depth ratio means that the chord forces are very small in comparison to the deck shear forces, which are very large. The principal means of distributing forces in diaphragms is through shear action. This means that the deflection due to shear cannot be neglected as it is in regular beam design. Shear deflection in diaphragms is usually larger than flexural deflection.

Since most roof and floor constructions that classify as flexible diaphragms consist of many independent parts, special care

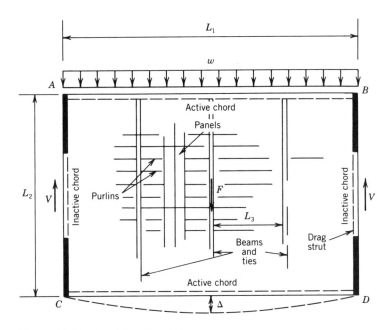

Figure 8.9/Design data—example.

must be taken to tie these parts together. The many deck pieces—that is, panels (Fig. 8.9) making up the web—must be attached to one another and stiffened by the supporting purlins. In addition, the panels must be properly attached to chords around the full perimeter of the deck. Examination of failures has shown that the chords (usually at the top of walls or frames) tend to pull away from the deck. Consequently, chord ties are required that function independently of the deck. Finally, the diaphragm must be properly attached to the vertical element and to collectors, so that they in turn can transmit the horizontal load down to the foundation. These design and detailing considerations, as they pertain to the diaphragm

shown in Fig. 8.9, will be treated next, and formulas will be developed that specifically apply to the configuration and arrangement of elements shown. It should be noted, however, that care must be taken in applying these procedures when conditions are other than those shown.

The building of Fig. 8.9 has a rectangular plan of dimensions L_1 and L_2. Lateral loads in only one direction will be investigated—those applied uniformly along the edge of the deck from A to B. Vertical elements are symmetrically placed in the outer walls along lines AC and BD. A wide-rib metal deck, similar to that shown in Figs. 5.43 and 8.10, will be used to carry the vertical load to the purlins. This type of roof or floor construction can function

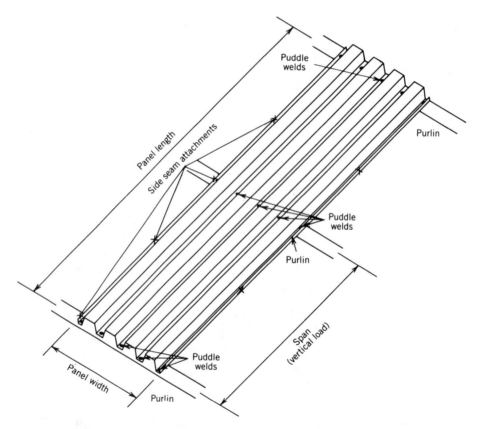

Figure 8.10/Light-gage steel deck panel.

as either a flexible or a rigid diaphragm, transferring the lateral load (w) to the vertical elements.

Metal decking is produced by a number of manufacturers; consequently, a wide variation in physical properties and design values is available. The thickness of the metal is measured in gages—16-, 18-, 20-, and 22-gage being the ones most widely used —and some are made of more than one layer of metal. Panels usually are 24, 30, or 36 in. wide with lengths in 2-in. increments up to a maximum of 30 ft. They are either galvanized or painted. Each type has its own vertical load-carrying capacity, and span-load tables are published and made readily available by each manufacturer. In a like manner, manufacturers publish tables containing design data for use of their decking as a diaphragm (Table 8.1). Of particular concern in this case are shear values and the degree of flexibility (the amount of deformation under load) for each variation in specification. These design data generally are accepted by building regulatory authorities when accompanied by supporting test results. Usually a factor of safety of 2.5 is required, and a and safe metal-deck diaphragm design will require reference to the manufacturer's tables and specifications.

A metal-deck diaphragm must have three or more puddle welds per panel width, placed where the surface of the metal-deck rib contacts the supporting joist or purlin. In addition, deck panels should be attached to one another by *button punch* side-seam welds, or screws placed at intervals not exceeding 36 in. (Fig. 8.10). The shear capacity and degree of flexibility depend upon the number of end welds per panel and the type and spacing of the side-seam attachments. Some manufacturers provide design data for self-drilling, cadmium-plated stainless-steel screws for this side-seam attachment. In addition, the

flexibility factor depends upon the ratio of the span for vertical load to the total length of the panel. This ratio is given the symbol R. The flexibility factor is the average deflection in micro-inches (in the plane of the diaphragm) of the deck per foot of distance from the support stressed with a shear of one pound per foot. A complete set of shear and flexibility tables is much too large to include here; however, a representative sample of such tables, taken from the military tri-service manual "Seismic Design for Buildings," is shown as Table 8.1.[3] Panels 1 and 2 are wide-rib panels, 24 in. wide, made from a single sheet of steel. Panel 3 is made from two sheets of metal, and panel 4 from one narrow-ribbed sheet, 30 in. wide. Only the button punch at 24-in. intervals is shown for the side seam attachment for the wide-rib panels. In addition, the design constants are for $1\frac{1}{2}$-in.-deep galvanized panels only. It should be apparent that this table represents only a portion of the data needed for design.

Referring to Fig. 8.9, the total lateral load developed by the diaphragm is $F = wL_1$, and its line of action passes through the center of the diaphragm, which in turn is midway between the vertical elements located at walls AC and BD. Consequently, the vertical element in each side wall develops one-half of F, or

$$V = \tfrac{1}{2}F$$

The drag struts along the edges AC and BD may be inactive chords. They will not function as active chords unless the lateral load is applied in the longitudinal direction, perpendicular to lines AC and BD. The deck is attached to this inactive chord along its total length, and the average run-

---

[3]Available from Superintendent of Documents, U.S. Government Printing Office, Washington, D.C. 20402.

Table 8.1
Steel Deck Diaphragms

Section	End Welds	Seam Fastening	Gage		Allowable Shear (q_D) and Flexibility Factor (F)[a] — Span (L_v)						
					4'-0"	5'-0"	6'-0"	7'-0"	8'-0"	9'-0"	10'-0"
1.	3	Button Punch @24" o.c.	16	q_D	1260	1030	870	760	680	620	560
				F	5.7+34.7R	7.0+27.8R	8.3+23.1R	9.6+19.8R	11+17.4R	12+15.4R	14+13.9R
			18	q_D	900	740	630	550	500	450	410
				F	8.1+67.8R	9.9+54.2R	12+45.2R	13+38.7R	15+33.9R	17+30.1R	19+27.1R
			20	q_D	520	430	370	320	290	260	240
				F	13+161R	15+129R	18+107R	21+91.9R	23+80.4R	26+71.4R	28+64.3R
			22	q_D	340	280	240	210	190	180	160
				F	17+278R	20+222R	23+185R	27+159R	30+139R	32+123R	35+111R
2.	5	Button Punch @24" o.c.	16	q_D	1650	1340	1130	980	870	790	720
				F	5.0+8.68R	6.1+6.94R	7.3+5.79R	8.5+4.96R	9.8+4.34R	11+3.86R	13+3.47R
			18	q_D	1220	990	840	730	660	580	520
				F	7.1+17.0R	8.6+13.6R	10+11.3R	12+9.69R	14+8.48R	15+7.54R	17+6.78R
			20	q_D	700	560	470	410	360	320	290
				F	11+40.2R	13+32.1R	16+26.8R	18+23.0R	21+20.1R	23+17.9R	26+16.1R
			22	q_D	450	370	310	270	240	220	200
				F	15+69.4R	18+55.5R	21+46.3R	24+39.7R	27+34.7R	30+30.9R	32+27.8R

Section 1 profile dimensions: 2 3/8", 3 3/8", 1 1/2", 2'-0"

Section 2 profile dimensions: 2 3/8", 3 3/8", 1 1/2", 2'-0"

Table 8.1
(Continued)

Allowable Shear (q_D) and Flexibility Factor (F)[a]

Section	End Welds	Seam Fastening	Gage		Span (L_v)						
					4'-0"	5'-0"	6'-0"	7'-0"	8'-0"	9'-0"	10'-0"
3.	3	Button Punch @24" o.c.	18-18	q_D	1580	1280	1080	930	830	750	680
				F	3.5+3.98R	4.4+3.18R	5.3+2.65R	6.2+2.28R	7.2+1.99R	8.3+1.77R	9.4+1.59R
			16-16	q_D	1990	1610	1360	1170	1040	930	850
				F	2.5+2.81R	3.1+2.25R	3.8+1.87R	4.5+1.61R	5.2+1.41R	6.0+1.25R	6.9+1.13R
			16-18	q_D	1920	1550	1310	1130	1000	900	820
				F	2.9+3.34R	3.6+2.67R	4.4+2.22R	5.2+1.91R	6.1+1.67R	7.0+1.49R	8.0+1.34R
			20-20	q_D	1180	960	810	690	600	530	480
				F	5.6+5.92R	6.9+4.73R	8.2+3.94R	9.7+3.38R	11+2.96R	13+2.63R	14+2.37R
4.	6	1½" Seam Weld @18" o.c.	18	q_D	990	890	820	760	710	680	650
				F	5.7+17.0R	5.5+13.6R	5.4+11.3R	5.3+9.68R	5.2+8.47R	5.1+7.53R	5.0+6.78R
			20	q_D	710	640	590	550	520	490	460
				F	8.5+40.2R	8.1+32.1R	7.8+26.8R	7.6+23.0R	7.3+20.1R	7.1+17.9R	6.9+16.1R
			22	q_D	480	420	380	350	330	310	300
				F	11+69.4R	10+55.5R	9.7+46.3R	9.3+39.7R	8.9+34.7R	8.6+30.9R	8.3+27.8R

[a] From *Seismic Design for Buildings*, NAVFAC P-355, 1982, p. 5-25.

Note. The gages for multiple-sheet decks are designated with the gage of the flat sheet first and fluted sheet second. All deck sections are made from galvanized sheets.

ning shear force is

$$q_D = \frac{V}{L_2} \quad \text{(plf)}$$

A deck with given specifications is selected to provide this shear strength without the usual one-third increase generally allowed for short-time loads. The side-seam attachment to this inactive chord must develop this same shear strength as well.

The active chord forces (tension and compression) are computed as

$$T = C = \frac{M_{max}}{L_2} = \frac{w(L_1)^2}{8L_2} \quad \text{(kips or lb)}$$

Chord ties prevent the active chords from pulling away from the deck. Judgment is necessary, depending upon the source of the load w, to determine the force developed by the ties. A conservative procedure would be to assume that the tie force is wL_3.

The transverse girder supporting the purlins can function as the tie if it is continuous and connected to the chords.

The end of the deck panels must be connected to the active chords to develop the perpendicular running shear, which is calculated as

$$v = \frac{VQ}{I} \quad \text{(plf)}$$

This shear is similar to the horizontal shear between the flange and web of a plate girder when loaded in the normal manner. Q is the first moment of the flange area about the center of the diaphragm, and I is the moment of inertia of the entire diaphragm relative to the longitudinal axis through the center of the diaphragm. It is difficult to establish their values, since the deck is an irregular web. However, since Q is small compared to I, the shear is small

and seldom controlling. The controlling shear is that for the lateral load in the longitudinal direction (not shown), and is

$$q_D = \frac{w_L L_2}{2L_1}$$

where w_L represents the longitudinal load (not shown).

Flexible diaphragms must be checked for excessive deflection, shown as Δ in Fig. 8.9. The maximum deflection occurs at the center of span L_1 and consists of both flexural deflection and shear deflection, i.e.,

$$\Delta = \Delta_f + \Delta_s$$

The flexural deflection is calculated by the usual procedure (i.e., as was described in Chapter 4), and will be

$$\Delta_f = \frac{5}{384} \left(\frac{wL_1^4}{EI} \right)$$

where I generally is established by considering chords only, i.e.,

$$I = 2A_c \left(\frac{L_2}{2} \right)^2$$

The shear deflection can be approximated by the formula

$$\Delta_s = \frac{M}{A_{web}G}$$

where G is the shear modulus of elasticity (11,200 psi for steel), and A_{web} is the thickness of metal times the length L_2. In Fig. 8.9

$$M = \frac{wL_1 L_1}{8} = \frac{VL_1}{4}$$

and $V = q_D L_2$. Therefore,

$$\Delta_s = \frac{q_D L_1 L_2}{4A_{web}G}$$

This shear deflection formula does not take into consideration the slip between adjacent panels or local buckling distortion. Consequently, each panel manufacturer furnishes a value representing the flexibility factor for each panel specification. These factors are listed in the tables along with shear strengths (Table 8.1). The flexibility factor has been defined as the deflection in micro-inches for a unit length developing a unit shear force:

$$\Delta_s = \frac{F q_D L}{10^6}$$

where F is the flexibility factor, q_D is the average shear force, and L is the distance measured in inches from the vertical element to the place where the deflection is to be determined. In Table 8.1, R is the ratio of the vertical load span to the overall length of the panel.

One additional problem is that of establishing an allowable limit for maximum total deflection. Some codes are very specific in this regard, while others permit judgment on the part of the designer. A major concern is the potential for damage to walls and frames that are attached to the diaphragm. It should be apparent that wall assemblies consisting of wood or steel can sustain more deflection (perpendicular to the plane of the wall) without serious damage than can walls made from brittle materials such as masonry. One frequently used formula for the maximum deflection of masonry walls is

$$\Delta_{max} = \frac{H^2 f_c}{0.01 E t}$$

where

H = height of wall in feet.

f_c = compression stress of wall material (psi).

E = modulus of elasticity of wall material (psi).

t = thickness of wall in inches.

The International Conference of Building Officials, which publishes the Uniform Building Code (see Chapter 1), specifies span-depth limits and overall maximum spans of diaphragms for various flexibility-factor (F) ranges in UBC Research Report No. 2078 (1980). For example:

If $F > 150$, the diaphragm cannot be used for masonry or concrete walls; its use is limited to flexible walls, and the diaphragm span-depth ratio is limited to $2:1$.

If $150 > F > 70$, the maximum diaphragm span is limited to 200 ft for masonry or concrete walls, and the span-depth ratio is limited to $2\frac{1}{2}:1$ if deflections are not calculated; there is no limit on diaphragm span if flexible walls are used, but the span-depth ratio cannot exceed $3:1$.

If $10 < F < 70$, the diaphragm can be either flexible or rigid; the maximum span is limited to 400 ft if masonry or concrete walls are used, and the span-depth ratio is limited to $2\frac{1}{2}:1$ if deflections are not calculated; and there is no limit on diaphragm span if flexible walls are used, but the span-depth ratio cannot exceed $4:1$. (These are the proportions used in Fig. 8.8a.)

Example Diaphragm Roof

A one-story rectangular building having overall plan dimensions of 80×100 ft is shown in Fig. 8.11. The height is 11 ft. A 500-plf lateral load is applied in one direction as shown. The complete gravity-load-carrying system is not shown; however, a metal deck is used to span the 5-ft spacing between purlins. The purlins span $33\frac{1}{3}$ ft

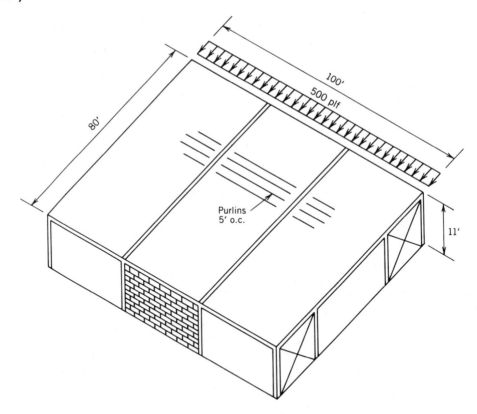

Figure 8.11/Example.

to the main girders. All vertical elements (for lateral-load design) are in the outside walls. Design the diaphragm for the lateral load.

Solution

(1) Calculate and locate the total lateral load:

$$F = 0.5(100) = 50 \text{ kips}$$

The line of action is midway between the vertical elements.

(2) Determine the running shear to be developed in the diaphragm. Total shear in each vertical element is $V = F/2 = 25$ kips. The inactive chord is assumed con-

tinuous for the 80 ft. Therefore,

$$q_D = \frac{25,000}{80} = 313 \text{ plf}$$

(3) Make a trial deck selection for further investigation. Enter Table 8.1 with a vertical span of 5 ft and a shear force of 313 plf. It is observed that a 20-gauge, single-sheet deck having three puddle welds and button punched at 24-in. intervals has a shear capacity of 430 plf. (This is a $1\frac{1}{2}$-in deep galvanized deck and, although undetermined in this example, must be adequate to carry the vertical loads imposed as well.)

(4) Determine the spacing of puddle welds along the marginal edge (inactive chords). Each puddle weld must have an effective diameter of $\frac{1}{2}$ in. and, as is usually the

case, will be made using E70 electrodes. The strength of each weld is 32,000 times the total thickness of the metal sheets. The selected metal gauge (20) has a thickness of 0.0359 in. Therefore, the required spacing is

$$s = \frac{32,000(t)}{q_D} = \frac{32,000(0.0359)}{313} = 3.67 \text{ ft}$$

Most codes limit maximum spacing to one-third the vertical span; therefore,

$$s_{max} = \frac{5(12)}{3} = 20 \text{ in.}$$

(5) Design the active chords. The chord forces are

$$T = C = \frac{0.5(100)^2}{8(80)} = 7.81 \text{ kips}$$

This is a very small force, so the elements used for its development would depend principally upon the construction used at the eaves. In this example, assume an angle with its vertical leg anchored to the vertical supporting element and the deck welded to its horizontal leg. With such a detail, angle buckling will not be a consideration. The tension area required for A-36 Steel is

$$A = \frac{7.81}{22} = 0.36 \text{ in.}^2$$

Use a $4 \times 3 \times \frac{1}{4}$-in. angle with lengths welded end to end for full continuity.

(6) Investigate deflection of the diaphragm. Calculate the moment of inertia of the two angle chords:

$$I = 2(1.69)[40(12)]^2 = 779,000 \text{ in.}^4$$

Determine the maximum flexural deflection at the center of the span:

$$\Delta_f = \frac{5(0.5)100^4(12)^3}{384(29,000)779,000} = 0.05 \text{ in.}$$

Calculate the shear deflection using the flexibility factors from Table 8.1.

Assume three panel lengths to cover the 80-ft distance. They must fit the 5-ft purlin spacing; therefore, use two at 25 ft and one at 30 ft. The maximum R is $\frac{5}{25} = 0.2$. From Table 8.1, the selected deck has a flexibility factor

$$F = 15 + 129R$$

Therefore, $F = 15 + 129(0.2) = 40.8$, and the maximum shear deflection is

$$\Delta_s = \frac{40.8(313/2)50(12)}{10,000,000} = 0.39 \text{ in.}$$

and

$$\Delta_f + \Delta_s = 0.05 + 0.39 = 0.44 \text{ in.}$$

Figure 8.11 shows a masonry wall at the point of this maximum deflection. Using the ultimate strength of masonry units as 1350 psi,

$$f_c = 0.33 f_m' = 0.33(1350) = 446 \text{ psi}$$
$$E = 1000 f_m' = 1,000(1350) = 1,350,000 \text{ psi}$$

The maximum allowable deflection at the top of the masonry wall is

$$\Delta = \frac{H^2 f_c}{0.01 Et} = \frac{11^2(446)}{0.01(1,350,000)8} = 0.50 \text{ in.}$$

Since $0.44 < 0.50$ in., the selected diaphragm is satisfactory.

(7) Design the chord ties. The maximum force in each tie is $0.5(100)/3 = 16.7$ kips. The girder supporting the purlins could be used to develop this tie force; however, frequently the position of the girder is sufficiently outside the plane of the diaphragm that it cannot be attached to the chords. Under these circumstances, an angle resting on top of the purlins can be used, i.e.,

$$A = \frac{16.7}{22} = 0.76 \text{ in.}^2$$

Use a $2\frac{1}{2} \times 2 \times \frac{1}{4}$ in. angle, made continuous from chord to chord.

8.4
Wind—Introduction

Wind creates a dynamic loading on a building. These changing loads act in every direction and last from a fraction of a second to several minutes; they can range from small to destructive in magnitude. For design purposes, it is necessary to convert this dynamic chaos into some orderly pattern of equivalent static loading. However, under certain conditions, equivalent static loads are unrealistic, and actual wind-tunnel testing is advisable.

Although wind can come from any direction and follow a variety of paths from horizontal to inclined and even vertically upward or downward on the face of a building, it generally is accepted that wind should be treated as a horizontal movement of the air mass. This movement, however, can cause pressures and forces on structures that act in any direction.

The process of calculating an equivalent static wind load begins with establishing a basic wind velocity in miles per hour (mph). This velocity is then converted to a wind pressure in pounds per square foot (psf). Next, a variety of modifying coefficients are applied to arrive at a combined design pressure (psf) that is imposed on specific parts or the whole of a building.

From model analyses and from studies of wind damage to buildings, it has long been known that, in addition to exerting a positive pressure against a windward wall (Fig.

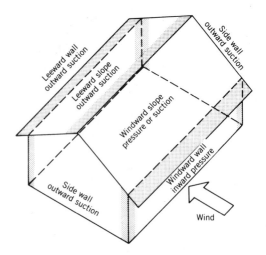

Figure 8.12/Wind on exterior surfaces.

8.12), wind can simultaneously create a suction on the leeward wall and on walls parallel to the wind. In addition, there is external suction on flat roofs, on the leeward slopes of gable roofs, and on slopes parallel to the wind; and there is suction, turning to pressure, on windward slopes, with the pressure increasing as the roof slope becomes steeper. Furthermore, because buildings are not completely airtight, and because they may be partially open or subject to being left open or broken open by flying debris, the interior may also be subjected to pressure or suction depending on whether openings occur primarily in the windward or leeward walls (Fig. 8.13). It is necessary to investigate the combina-

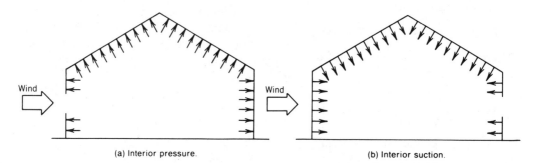

(a) Interior pressure.

(b) Interior suction.

Figure 8.13/Wind on interior surfaces.

tion of exterior forces (Fig. 8.12) and either internal pressure or internal suction (Fig. 8.13) in order to determine the critical design load.

Primary attention will be given here to determining wind-load values at the eaves, ridges, and rakes of one-story buildings, or at the floor and roof lines of multistory buildings. These are the horizontal-element loads that must be transmitted to vertical elements (Art. 8.2). Building codes refer to these elements as "main wind-force resisting systems" or "primary frames and systems." However, it should be emphasized that individual building components such as purlins, girts, mullions, and surface cladding also must be properly designed for the imposed wind loads, and in this regard, localized surface areas on buildings can be subjected to large "buildup pressures" that are in excess of the average pressures used for mainframe design. These local areas generally are roof and wall corners, overhangs, leeward ridges, and eaves (shown shaded in Fig. 8.12). It often is local failure in these high-pressure areas that starts a progressive failure of a building. Building codes require special treatment of such individual components.

The three articles that follow, all relating to wind, treat the procedures for determining basic wind speed and for converting this wind speed to design pressures. It should be noted, however, that such detailed procedures are seldom justified in preliminary design. An often-used approximate formula for determining a wind speed for purposes of preliminary design is

$$p = 0.0033V^2$$

where V = wind speed in mph and p = design pressure in psf.

Using this formula, a wind speed of 77 mph gives a design pressure of 20 psf. Some designers consider this value a minimum and adjust it upwards for areas with recorded wind speeds in excess of 77 mph. It should be emphasized that this approximate procedure does not always produce a safe design.

8.5
Wind Speed

Winds can reach velocities of 200 to 250 mph; however, it is unrealistic to design all structures to resist winds of this magnitude. Extreme winds usually are associated with cyclonic disturbances, principally tornadoes and hurricanes, with tornadoes accounting for the higher velocities. Designing for tornadoes is a special topic and, as such, is not covered in this text—nor is design for tornadoes covered in most building codes. Because wind speeds associated with hurricanes generally are less than those with severe tornadoes and usually affect a much wider geographic area, it is more reasonable to establish an adequate design procedure based on hurricane-force winds. Hurricanes (severe tropical cyclones affecting the U.S. land mass) develop over warm areas in the Atlantic Ocean, Caribbean Sea, Gulf of Mexico, and the northeastern Pacific Ocean. Because coastal, estuary, and large exposed inland water surges, as well as direct wind and wind- and water-borne debris resulting from hurricanes, can have devastating consequences for the survivability of communities and their inhabitants, it is not infrequent that such communities will have individually fashioned building-code requirements or will need to be treated in an individual manner.

As was pointed out in Chapter 1, most building codes in the medium- to smaller-sized communities in the United States are to a greater or lesser degree adaptations of, or are based upon, one or the other of the nationally recognized model building codes, usually the one that is most commonly adopted in that region. The large cities, as a general rule, have their own building code, but even the provisions of these codes are increasingly coming into harmony with the model codes, and vice versa. As noted above, however, geophysical conditions do differ among code-issuing jurisdictions, and development patterns can have a powerful influence on what will be required to achieve even comparable levels of public health and safety.

Where hurricanes are concerned, the South and Southeast have had the most experience. The *Southern Standard Building Code* is the predominant model code in this area. Where tornadoes are concerned, the central and southcentral areas of the country have had the most experience. In these areas, all three of the extant model codes have their adherents (the *National Building Code* is no longer being promulgated). The far West has had the most experience with damaging earthquakes, and it is the *Uniform Building Code* (UBC) that is most used in this area.

The American Society of Civil Engineers ASCE 7-88, *Minimum Design Loads for Buildings and Other Structures*, is a voluntary standard that is finding increasing acceptance by the model-code bodies and code-issuing jurisdictions. It is not a building code; it contains recommended requirements for adoption by the model codes and/or building-code-issuing jurisdictions.

Although it might be more appropriate to reference the Southern Standard Building Code when dealing with hurricane-force winds, the Uniform Building Code will be referenced because it does treat extreme winds and currently is one of the more appropriate model codes to reference for earthquake design—a topic that is also treated in this chapter. ASCE 7-88 also will be referenced because of its potential for more universal adoption as a voluntary

standard by both the model-code bodies and code-issuing jurisdictions. Actual code provisions should, of course, always take precedence.

Both the UBC and ASCE 7-88 require local investigation to establish basic wind speeds exceeding their recommendation. In addition, ASCE 7-88 contains a special coefficient (I) that accounts for the relative importance of a structure in relation to its occupancy and exposure to hurricane-force winds. The UBC also contains an I factor, but it is not related to hurricanes. Table 8.2 shows the I values for both references.

The velocity of wind is a function of time. Peak wind speeds last for only one or two seconds and are not used in design. However, an allowance for longer gusts is made. Gusts are not fully effective unless they last long enough to engulf the structure. Of significance in this discussion and in the codes is the *fastest-mile wind speed*. This is defined as the average speed of 1 mile of air passing an anemometer. For example, a wind speed of 120 mph means that a mile of air mass passed the anemometer in 30 seconds.

Wind velocity is a function of the height above the ground and ground conditions. Rough, irregular ground and man-made obstructions on the ground retard movement of the air near the ground and reduce wind velocity. At some height above the ground, however, the movement of air is not affected by ground conditions. This height is called the *gradient height*, and the velocity is referred to as the *gradient wind speed*.

It is generally agreed that for building design purposes, the variation in wind speed, for varying ground conditions, can be expressed by the power-law formula

$$V_z = V_g \left[\frac{z}{z_g} \right]^{1/\alpha}$$

Table 8.2
Values of the Importance Factor _I_ for Wind

ASCE 7-88[a]			
		I	
Occupancy Description	*Category*	*100 mi or more from Hurricane at Oceanline*	*At Hurricane Oceanline*
All buildings except those listed below	I	1.00	1.05
Buildings where assembly is for more than 300 persons	II	1.07	1.11
Essential facilities such as hospitals, fire & communication buildings	III	1.07	1.11
Buildings that represent low hazard to life	IV	0.95	1.00
1991 UBC[b]			
Occupancy Description	*Category*	*I*	
Essential facilities such as hospitals, fire & communication buildings	I	1.15	
Hazardous facilities, buildings containing toxic or explosive substances	II	1.15	
Special facilities where assembly is for more than 300 persons (schools, 250 persons; colleges, 500 persons, etc.)	III	1.00	
Standard occupancy	IV	1.00	

[a]Courtesy of the American Society of Civil Engineers, Minimum Design Load For Buildings and other Structures, ASCE 7-88, 1988.

Reproduced from the 1991 edition of the Uniform Building Code, Copyright © 1991, with permission of the publisher, The International Conference of Building Officials.

where:

V_z = wind speed at any height z.

V_g = gradient wind speed.

z = height above ground.

z_g = gradient height.

$1/\alpha$ = coefficient based upon ground conditions.

Research by Professor A. G. Davenport of the University of Western Ontario has established values and profiles as shown in Fig. 8.14. Included are four degrees of ground roughness. Exposure A represents the most severe roughness and exposure D the least—i.e., over calm water. Many building codes have adopted these expo-

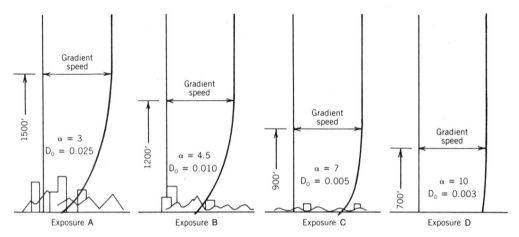

Figure 8.14/Gradient height and gradient speed.

sure categories; they are more fully defined in Art. 8.7.

Records of wind-speed data are sufficient to produce a U.S. map with wind-speed contours. However, it is necessary to adjust the contours to account for some probability of recurrence. The generally accepted map is that shown in Fig. 8.15. This map is for fifty-year mean recurrence intervals. The contour values on the map are the equivalent to reasonably smooth terrain (Exposure C) at a height of 33 ft above ground. With this map as a base, equations and coefficients are applied to account for changing heights, other exposures, gusts, and building configurations. Observe that the minimum basic wind speed is 70 mph.

8.6
Wind Pressure

A wind speed in mph can be converted to wind pressure in psf. In so doing, some assumptions are necessary. Wind possesses kinetic energy by reason of the mass and velocity of the moving air. It is assumed that the temperature of the air is 15°C and that the pressure is 760 mm of mercury. Under these conditions, the density of the air can be taken as 0.0765 lb per cubic

foot. The kinetic energy in ft-lb for 1 cu ft of air mass in motion is

$$E = \tfrac{1}{2}mV^2$$

where m is the air mass. To use gravitational units (lb) the mass can be expressed as

$$m = \frac{W}{g}$$

where W is the weight and g is the universal acceleration due to gravity, or 32.2 ft per sec.[2]. In addition, it is desirable to keep the units of V in mph. Therefore, V must be multiplied by 5280 ft per mile and divided by 3600 sec per hour. Making these substitutions in the kinetic energy equation and using the symbol q to represent wind pressure,

$$q = \frac{1}{2}\left(\frac{0.0765}{32.2}\right)\left(\frac{5280}{3600}\right)^2 V^2$$

$$q = 0.00256V^2$$

Thus, it is seen that a wind speed of 100 mph converts to a wind pressure

$$q = 0.00256(100)^2$$

$$= 25.6 \text{ psf}$$

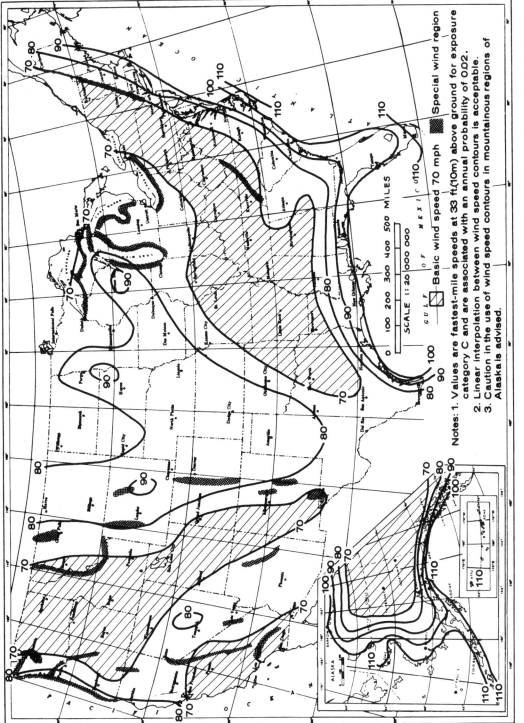

Figure 8.15/Basic wind speed (miles per hour). (This figure is reprinted with permission from ASCE 7-88, Minimum Design Loads for Buildings and Other Structures, copyrighted 1988 by the American Society of Civil Engineers, 345 East 47th Street, New York, NY 10017.

Notes: 1. Values are fastest-mile speeds at 33 ft(10m) above ground for exposure category C and are associated with an annual probability of 0.02.
2. Linear interpolation between wind speed contours is acceptable.
3. Caution in the use of wind speed contours in mountainous regions of Alaska is advised.

☐ = Basic wind speed 70 mph   ▨ Special wind region

8.7
Wind Codes and Coefficients

Most codes accept and use the wind-pressure formula q as derived in the previous article; however, basic wind speeds may differ from area to area and thus for different codes. If this were not the case, the speed could be taken from the map of Fig. 8.15 for actual design purposes. It was stated, however, that these map speeds were for a height 33 ft above a terrain of Exposure C. Consequently, they must be adjusted for other heights and other terrain exposures. This adjustment usually is accomplished by multiplying the wind pressure by a *velocity-pressure exposure coefficient*, K_Z.

A brief description of the exposure categories is as follows:

Exposure A. Centers of large cities and very rough terrain.

Exposure B. Suburban areas, towns, city outskirts, wooded areas, and rolling terrain.

Exposure C. Flat, open country and grassland.

Exposure D. Flat, unobstructed coastal areas directly exposed to wind blowing over bodies of water.

Since the basic wind-speed map was prepared for Exposure C at a height of 10 meters (33 ft), it is necessary to adjust its values for other heights and exposures. This is done by applying the *velocity pressure exposure coefficient*, K_Z.

To arrive at the correct formula for the velocity-pressure exposure coefficient K_Z, it is necessary to place in the power-law formula (Art. 8.5) the values of $\alpha = 7$ and $z_g = 900$ (Exposure C) as follows:

$$V_z = \left[V_{33} \left(\frac{900}{33} \right)^{1/7} \right] \left(\frac{z}{z_g} \right)^{1/\alpha}$$

The factor within the brackets converts the basic wind velocity to gradient wind velocity, whereas the factor $(z/z_g)^{1/\alpha}$ determines the wind velocity for any height and exposure. It should be borne in mind that the desired expression is for a pressure coefficient rather than a velocity coefficient. Also, as was shown in Art 8.6, pressures vary as the square of the velocity. Consequently, in deriving the formula for K_Z, it becomes necessary to square both factors in the velocity formula, i.e.,

$$K_Z = \left(\frac{V_z}{V_{33}} \right)^2 = \left(\frac{900}{33} \right)^{2/7} \left(\frac{z}{z_g} \right)^{2/\alpha}$$

This simplifies to

$$K_Z = 2.58 \left(\frac{z}{z_g} \right)^{2/\alpha}$$

Table 8.3 was prepared using this formula. Another modification must be made for gusts. This is accomplished through use of a *gust response factor*, G_z. This factor takes into account the additional loading effect due to wind turbulence over the fastest-mile wind. Variables are the exposure category (α value), height (z), and surface drag coefficient (D_0) for the various exposures. Empirical formulas for this coefficient were developed through the research efforts of J. W. Vellozzi, E. Cohen, and A. G. Davenport. For buildings that are not sensitive to dynamic amplifications, the formula for calculating the gust response factor is

$$G_Z = 0.65 + 3.65 T_Z$$

T_Z is defined as the turbulence intensity factor and can be determined by the formula

$$T_Z = \frac{2.35(D_0)^{1/2}}{(z/30)^{1/\alpha}}$$

where D_0 and α are obtained from their respective exposure categories shown in

Table 8.3
Velocity Pressure Exposure Coefficient, K_Z

Height in Feet z	Exposure A $\alpha = 3.0$ $z_g = 1500$ ft.	Exposure B $\alpha = 4.5$ $z_g = 1200$ ft.	Exposure C $\alpha = 7.0$ $z_g = 900$ ft.	Exposure D $\alpha = 10.0$ $z_g = 700$ ft.
0–15	0.12	0.37	0.80	1.20
20	0.15	0.42	0.87	1.27
25	0.17	0.46	0.93	1.32
30	0.19	0.50	0.98	1.37
35	0.21	0.54	1.02	1.42
40	0.23	0.57	1.06	1.46
45	0.25	0.60	1.10	1.49
50	0.27	0.63	1.13	1.52
55	0.28	0.66	1.16	1.55
60	0.30	0.68	1.19	1.58
70	0.33	0.73	1.24	1.63
80	0.37	0.77	1.29	1.67
90	0.40	0.92	1.34	1.71
100	0.42	0.86	1.38	1.75
110	0.45	0.89	1.42	1.78
120	0.48	0.93	1.45	1.81
130	0.51	0.96	1.48	1.84
140	0.53	0.99	1.52	1.87
150	0.56	1.02	1.55	1.90
160	0.58	1.05	1.58	1.92
180	0.63	1.11	1.63	1.97
200	0.67	1.16	1.68	2.01
250	0.78	1.28	1.79	2.10
300	0.88	1.39	1.88	2.18
350	0.98	1.49	1.97	2.25
400	1.07	1.58	2.05	2.31
450	1.16	1.67	2.12	2.36
500	1.24	1.75	2.18	2.41

Note. Shading indicates heights and coefficients that correlate with the 1991 UBC.

Fig. 8.14. Table 8.4 was prepared using these formulas.

There still are other coefficients that are used for building areas (inside and out) and for special components of buildings.

The adjusted wind speed, multiplied by the applicable coefficients, gives the design pressure (p). The National Institute of Standards and Technology has been instrumental in establishing *building pressure*

Table 8.4
Gust Response Factors, G_z

Height in Feet z	Exposure A $\alpha = 3.0$ $D_0 = 0.025$	Exposure B $\alpha = 4.5$ $D_0 = 0.010$	Exposure C $\alpha = 7.0$ $D_0 = 0.005$	Exposure D $\alpha = 10.0$ $D_0 = 0.003$
0–15	2.36	1.65	1.32	1.15
20	2.20	1.59	1.29	1.14
25	2.09	1.54	1.27	1.13
30	2.01	1.51	1.26	1.12
35	1.94	1.48	1.24	1.11
40	1.88	1.46	1.23	1.11
45	1.83	1.43	1.22	1.10
50	1.79	1.42	1.21	1.10
55	1.76	1.40	1.21	1.09
60	1.73	1.39	1.20	1.09
70	1.67	1.36	1.19	1.08
80	1.63	1.34	1.18	1.08
90	1.59	1.32	1.17	1.07
100	1.56	1.31	1.16	1.07
110	1.53	1.29	1.15	1.06
120	1.50	1.28	1.15	1.06
130	1.48	1.27	1.14	1.06
140	1.46	1.26	1.14	1.05
150	1.44	1.25	1.13	1.05
160	1.43	1.24	1.13	1.05
180	1.40	1.23	1.12	1.04
200	1.37	1.21	1.11	1.04
250	1.32	1.19	1.10	1.03
300	1.28	1.16	1.09	1.02
350	1.25	1.15	1.08	1.02
400	1.22	1.13	1.07	1.01
450	1.20	1.12	1.06	1.00
500	1.18	1.11	1.06	1.00

Note. Shading indicates heights and coefficients that correlate with the 1991 UBC.

coefficients (C_p). Figure 8.16 is an excerpt from the Department of the Navy's publication Military Handbook 1002/2, Loads, 1988. These are essentially the same coefficients used by ASCE 7-88. It should be pointed out that these building coefficients are to be used only for the main wind-force-resisting system affecting the exterior surface of the building; the minus sign means the direction of the pressure is away

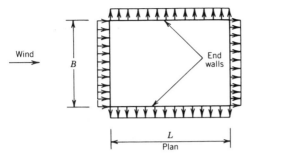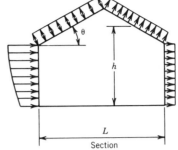

External Wall Pressure Coefficients (C_p)

Surface	L/B	C_p
Windward wall	All values	0.8
Leeward wall	0–1	−0.5
	2	−0.3
	≥ 4	−0.2
End wall	All values	−0.7

External Roof Pressure Coefficients (C_p)

Wind Direction	h/L	Windward							Leeward
		Angle θ Degrees							
		0	10–15	20	30	40	50	≥ 60	
Normal to Ridge	≤ 0.3	−0.7	0.2 and −0.9	0.2	0.3	0.4	0.5	0.01θ	−0.7 for all values of h/L
	0.5	−0.7	−0.9	−0.75	−0.2	0.3	0.5	0.01θ	
	1.0	−0.7	−0.9	−0.75	−0.2	0.3	0.5	0.01θ	
	≥ 1.5	−0.7	−0.9	−0.9	−0.9	−0.35	0.21	0.01θ	
Parallel to Ridge	h/B or $h/L \le 2.5$	−0.7							−0.7
	h/B or $h/L > 2.5$	−0.8							−0.8

Figure 8.16/Building pressure coefficients.

from the surface, whereas the implied plus sign means pressure is toward the surface.

Some regulations require the investigation of internal as well as external pressures. The ASCE 7-88 specifies this requirement only for single-story buildings. The internal coefficient is referred to as the combined gust and pressure coefficient and is designated by the symbol GC$_{pi}$. It usually has the value of ± 0.25. If unusually large wall openings are encountered, then the values change to a $+0.75$ and a -0.25.

The basic wind speed (V) is taken from Fig. 8.15, and the velocity pressure (q) is calculated. This velocity pressure is then multiplied by the velocity-pressure exposure coefficient (K_z) to obtain the appropriate pressure (q_z) for the given height and exposure. One further refinement accounts for the criticality of building occupancy and hurricane exposure, and is termed the I factor. Expressed in equation form when using ASCE 7-88,

$$q_z = 0.00256 K_z (IV)^2$$

I factors may be obtained from Table 8.2 and are seen to vary from 0.95 to 1.11. The velocity pressure exposure coefficient may be selected from Table 8.3 or calculated from the formula

$$K_z = 2.58 \left(\frac{z}{z_g} \right)^{2/\alpha}$$

where:

z = height above ground.

z_g = gradient height (Fig. 8.14).

α = power coefficient (Fig. 8.14).

The four exposure categories A, B, C, and D are described in greater detail in ASCE 7-88; category A is seldom used.

The height increment (z value) to be used in determining K_z depends upon the nature of the problem under investigation. Usually 5 ft is a minimum, but as z be-

comes large, the change in K_z is very small. In Example 3 which follows, a story height of 10 ft is used for the increment change in z. This procedure is best illustrated by use of the example.

Application of additional procedures described in ASCE 7-88 depends on whether the structure is (1) a regular building, (2) a structure other than a building, or (3) a building or structure that is considered to be flexible or wind-sensitive. A flexible structure is further defined as one in which the height-to-width ratio is greater than 5, or one in which the fundamental frequency of vibration is less than 1 Hz (cycle per second). These can be best understood by studying ASCE 7-88.

A further division in the procedure depends on whether the load is applied to the main wind-force-resisting system or to the components and cladding. In the case of main frames of regular buildings, the q_z-value is multiplied by a *gust response factor* (G_h), and finally by a pressure coefficient for the building (C_p). Thus, the design pressure becomes

$$p = q_z G_h C_p$$

The gust response factor depends upon the height and the exposure category. Values of G_h can be obtained from Table 8.4 or be derived from formulas. The height to be used in selecting G_h is the mean height of the roof.

The building pressure coefficient C_p depends upon the surface location in the building under investigation (see Fig. 8.16). The coefficients in the table pertain to exterior wall and roof surfaces for main frames. ASCE 7-88 also contains tables of coefficients for components and parts.

Summary of ASCE 7-88 Procedures for Main Frames in Regular Buildings

1. Establish the basic wind speed (Fig. 8.15).

2. Determine the importance coefficient I (Table 8.2).

3. Calculate q_z as a function of K_z from

$$q_z = 0.00256 K_z (IV)^2$$

4. Select the various heights (z) to be used, depending on the problem being solved, and determine the corresponding K_z value from Table 8.3. One z value always to be used is equal to h, which is the mean height of the roof. This step provides various K_z values and a K_h value.

5. Calculate q_z and q_h by multiplying corresponding values from steps 3 and 4. This will provide wind velocity pressures at all z levels.

6. Select the gust response factor (G_h) from Table 8.4, using $z = h$.

7. Determine the appropriate external pressure coefficients from Fig. 8.16, and multiply them by the velocity pressure (step 5) and the gust response factor (step 6). This step determines all external force effects on the building.

8. Determine the appropriate combined gust and internal pressure coefficient and multiply it by the velocity pressure (g_h) found in step 5.

9. Combine the effects of the external forces and the internal effect. These will determine the critical loading.

The procedure thus described is used in the example problems that follow, with some small variations required for each example. Similar procedures are used for determining loads on components or cladding and are so stipulated in ASCE 7-88.

Summary of 1991 UBC Procedure for Primary Frames / The 1991 edition of the UBC specifies a simplified version of ASCE 7-88. The same basic wind-speed map is used and must be modified when local conditions so dictate. This wind speed is then converted to velocity pressure in a manner similar to that shown in Table 8.5a. The UBC refers to this as the stagnation pressure (q_s), and the values agree with those derived using the wind velocity-pressure formula developed in Art. 8.6. This velocity pressure is then converted to design pressures by the application of three coefficients, i.e.,

$$P = C_e C_q q_s I$$

The coefficient C_e is a combined height, exposure, and gust factor obtained from Table 8.5b. Only three exposures are used (B, C, and D), and they correspond reasonably well with the exposure descriptions given in ASCE 7-88. The factor C_q is the pressure coefficient of the building obtained from Table 8.5c. Observe that there is separate treatment of primary frames, elements and components, and other structures.

The I factor in the design pressure formula is designated by UBC as an importance factor. A value of either 1.0 or 1.15 must be used (see Table 8.2).

The application of the UBC requirements will be illustrated in Examples 1 and 3 which follow.

Example 1/Single Story

A gymnasium intended for public assembly of more than 300 people is constructed of steel columns, continuous from floor to eaves, and steel trusses (Fig. 8.17). These truss-column bents are 30 ft apart and span 90 ft. The clear height is 23 ft, and the truss depth, 7 ft. The roof deck and purlins are not shown, but the horizontal element is in the plane of the roof deck. All exterior sides have vertical girts extending from the floor to eaves. Cross-bracing occurs between two bents on each long side of the building. The building is located in Omaha, Nebraska, and has a site exposure of Category *B* and minimal exterior openings.

(a) Calculate the design pressures and eaves loading due to wind according to the 1991 UBC. Assume the eaves load to be uniform even though the wall girts are spaced 10 ft apart.

(b) Calculate the design pressures and eaves loading due to wind according to ASCE 7-88. Assume the eaves load to be uniform, even though the wall girts are spaced 10 ft apart.

Solution (a) (UBC)

(1) Determine the basic wind speed. Referring to Fig. 8.15, interpolation between the contours gives a basic wind speed of 85 mph.

(2) Determine the wind stagnation pressure from Table 8.5a, interpolation again being required:

$$q_s = 18.6 \text{ psf}$$

(3) Select the combined height, exposure, and gust factor from Table 8.5b. Use increments of 5 ft for the windward side. The C_e for roof and leeward surfaces is

Table 8.5
Wind Coefficients from 1991 UBC[a]

(a) Wind stagnation pressure q_s at standard height of 30 ft

Basic Wind Speed (mph)	Pressure q_s (psf)
70	12.6
80	16.4
90	20.8
100	25.6
110	31.0
120	36.9
130	44.3

(b) Combined height, exposure, and gust-factor coefficient C_e

Height above Average Level of adjoining ground (feet)[b]	C_e		
	Exposure D	Exposure C	Exposure B
0–15	1.39	1.06	0.62
20	1.45	1.13	0.67
25	1.50	1.19	0.72
30	1.54	1.23	0.76
40	1.62	1.31	0.84
60	1.73	1.43	0.95
80	1.81	1.53	1.04
100	1.88	1.61	1.13
120	1.93	1.67	1.20
160	2.02	1.79	1.31
200	2.10	1.87	1.42
300	2.23	2.05	1.63
400	2.34	2.19	1.80

Table 8.5
(Continued)

(*c*) Pressure coefficients C_q

Structure or Part Thereof	Description	C_q
1. Primary frames and systems	Method 1 (normal-force method) Walls:	
	Windward wall	0.8 inward
	Leeward wall	0.5 outward
	Roofs:[c]	
	Wind perpendicular to ridge	
	Leeward roof or flat roof	0.7 outward
	Windward roof:	
	Slope < 2:12	0.7 outward
	Slope 2:12 to < 9:12	0.9 outward or 0.3 inward
	Slope 9:12 to 12:12	0.4 inward
	Slope > 12:12	0.7 inward
	Wind parallel to ridge and flat roofs	0.7 outward
	Method 2 (projected-area method) On vertical projected area:	
	Structures 40 ft or less in height	1.3 horizontal any direction
	Structures over 40 ft in height	1.4 horizontal any direction
	On horizontal projected area[c]	0.7 upward
2. Elements and components not in areas of discontinuity[d]	Wall elements:	
	All structures	1.2 inward
	Enclosed and unenclosed structures	1.2 outward
	Open structures	1.6 outward
	Parapet walls	1.3 inward or outward
	Roof elements:[e] Enclosed and unenclosed structures:	
	Slope < 7:12	1.3 outward
	Slope 7:12 to 12:12	1.3 outward or inward
	Open structures:	
	Slope < 2:12	1.7 outward
	Slope 2:12 to 7:12	1.6 outward or 0.8 inward
	Slope > 7:12 to 12:12	1.7 outward or inward

Table 8.5
(Continued)

(c) Pressure coefficients C_q

Structure or Part Thereof	Description	C_q
3. Elements and components in areas of discontinuities[d,f,h]	Wall corners[i]	1.5 outward or 1.2 inward
	Roof eaves, rakes or ridges without overhangs:[i]	
	Slope < 2:12	2.3 upward
	Slope 2:12 to 7:12	2.6 outward
	Slope > 7:12 to 12:12	1.6 outward
	For slopes less than 2:12— Overhangs at roof eaves, rakes or ridges, and canopies	0.5 added to values above
4. Chimneys, tanks and solid towers	Square or rectangular	1.4 any direction
	Hexagonal or octagonal	1.1 any direction
	Round or elliptical	0.8 any direction
5. Open-frame towers[e,f]	Square and rectangular:	
	Diagonal	4.0
	Normal	3.6
	Triangular	3.2
6. Tower accessories (such as ladders, conduit, lights and elevators)	Cylindrical members:	
	2 in. or less in diameter	1.0
	Over 2 in. in diameter	0.8
	Flat or angular members	1.3
7. Signs, flagpoles, lightpoles, minor structures[f]		1.4 any direction

[a]Reproduced from the *Uniform Building Code*, 1991 Edition, Copyright 1991, with permission of the publisher, the International Conference of Building Officials.

[b]Values for intermediate heights above 15 ft may be interpolated.

[c]For one story or the top story of multistory open structures, an additional value of 0.5 shall be added to the outward C_q. The most critical combination shall be used for design. For definition of open structures, see Section 2312.

[d]C_q values listed are for 10-sq-ft tributary areas. For tributary areas of 100 sq ft, the value of 0.3 may be subtracted from C_q, except for areas at discontinuities with slopes less than 7:12, where the value of 0.8 may be subtracted for C_q. Interpolating may be used for tributary areas between 10 and 100 sq ft. For tributary areas greater than 1000 sq ft, use primary-frame values.

[e]For slopes greater than 12:12, use wall-element values.

[f]Local pressures shall apply over a distance from the discontinuity of 10 ft or 0.1 times the least width of the structure, whichever is smaller.

[g]Wind pressures shall be applied to the total normal projected area of all elements on one face. The forces shall be assumed to act parallel to the wind direction.

[h]Discontinuities at wall corners or roof ridges are defined as discontinuous breaks in the surface where the included interior angle measures 170° or less.

[i]Load is to be applied on either side of discontinuity but not simultaneously on both sides.

based upon the mean roof height (in this example, 30 ft):

0–15 ft,	$C_e = 0.62$
15–20 ft,	$C_e = 0.67$
20–25 ft,	$C_e = 0.72$
25–30 ft,	$C_e = 0.76$

(4) Determine the importance factor. Since the gymnasium is expected to accommodate more than 300 persons, $I = 1.00$ (Table 8.2).

(5) Select the pressure coefficients for the main wind-resistant structure. These coefficients are the same for the transverse and longitudinal directions. From Table 8.5c,

Windward wall,	$C_q = 0.8$ inward
Leeward wall,	$C_q = 0.5$ outward
Roof (flat),	$C_q = 0.7$ outward

(6) Calculate the wall and roof pressures as the product of q_s and the coefficients.

Windward wall:

0–15 ft,	$P = 0.62(0.8)(18.6)(1.0)$
	$= 9.23$ psf
15–20 ft,	$P = 0.67(0.8)(18.6)(1.0)$
	$= 9.97$ psf
20–25 ft,	$P = 0.72(0.8)(18.6)(1.0)$
	$= 10.71$ psf
25–30 ft,	$P = 0.76(0.8)(18.6)(1.0)$
	$= 11.31$ psf

Leeward wall:

$$P = 0.76(0.5)(18.6)(1.0) = 7.07 \text{ psf}$$

Roof:

$$P = 0.76(0.7)(18.6)(1.0) = 9.90 \text{ psf}$$

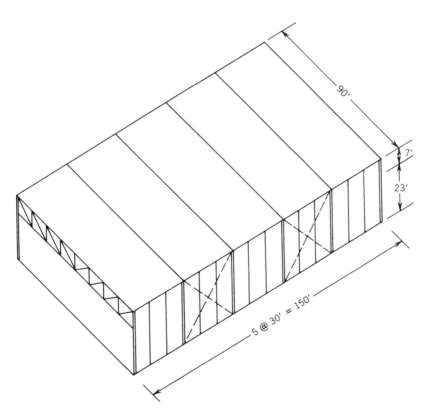

Figure 8.17/Example 1: Single story.

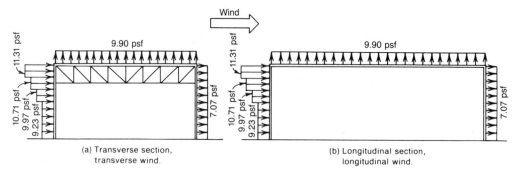

(a) Transverse section,
transverse wind.

(b) Longitudinal section,
longitudinal wind.

Figure 8.18/Design pressures, UBC.

(7) Sketch the design pressures caused by transverse wind: longitudinal wind will generate the same pressures. These pressures are shown in Fig. 8.18 (a) and (b).

(8) Calculate the eaves loading for the horizontal element in the plane of the roof deck. Vertical girts span from floor to eaves. For the leeward wall, one-half the load is developed at the eaves and at the foundation. The windward wall, with a staggered loading, requires proportioning the loads:

Leeward wall:

$$w = 7.07(30/2) = 106 \text{ plf}$$

Windward wall:

$$9.23(15)(7.5/30) = 34.61$$
$$9.97(5)(17.5/30) = 29.08$$
$$10.71(5)(22.5/30) = 40.16$$
$$11.31(5)(27.5/30) = 51.84$$
$$\overline{w = 155.69}; \quad \text{use 160 plf}$$

The diaphragm loads are shown in Fig. 8.19.

Solution (b) (ASCE 7-88)

(1) Determine the basic wind speed. Interpolating in Fig. 8.15, $V = 85$ mph.

(2) Determine the importance coefficient. Referring to Table 8.2, with a Type II

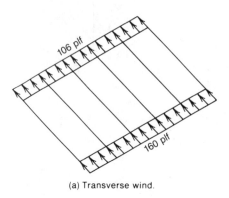

(a) Transverse wind.

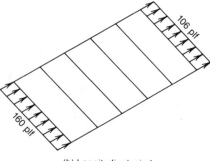

(b) Longitudinal wind.

Figure 8.19/Diaphragm loads, UBC.

building classification,

$$I = 1.07$$

(3) Select the velocity-pressure exposure coefficients (K_z) from Table 8.3. Use z at 5-ft increments:

0–15 ft,	$K_z = 0.37$
15–20 ft,	$K_z = 0.42$
20–25 ft,	$K_z = 0.46$
30 ft,	$K_h = 0.50$

(4) Calculate the wind velocity pressure for each height range:

$$q_z = 0.00256 K_z [1.07(85)]^2 = 21.18\, K_z$$

0–15 ft,	$q_z = 21.18(0.37) = 7.84$ psf
15–20 ft,	$q_z = 21.18(0.42) = 8.90$ psf
20–25 ft,	$q_z = 21.18(0.46) = 9.74$ psf
30 ft,	$q_h = 21.18(0.50) = 10.59$ psf

(5) Select the gust response factor (G_h) from Table 8.4. Use $z = 30$ ft:

$$G_h = 1.51$$

(6) Determine the internal pressure coefficient, and calculate the internal pressure to be used for all heights:

$$GC_{pi} = \pm 0.25$$

$$p = GC_{pi} q_h = \pm 0.25(10.59)$$

$$= \pm 2.65; \quad \text{use 2.7 psf}$$

It should be emphasized that this 2.7 psf is either internal suction or internal pressure.

(7) Select the appropriate external pressure coefficient from Fig. 8.16, and calculate and sketch the design pressures for a transverse wind:

$$L/B = 90/150 < 1.0$$

Windward wall,	$C_p = 0.8$
Side walls,	$C_p = -0.7$
Leeward wall,	$C_p = -0.5$
Roof (flat),	$C_p = -0.7$

Design pressures $p = q_z G_h C_p \pm 2.7$; therefore,

Windward wall:

0–15 ft,	$p = 7.84(1.51)0.8 \pm 2.7$
	$= 9.5 \pm 2.7$ psf
15–20 ft,	$p = 8.90(1.51)0.8 \pm 2.7$
	$= 10.8 \pm 2.7$ psf
20–25 ft,	$p = 9.74(1.51)0.8 \pm 2.7$
	$= 11.8 \pm 2.7$ psf
30 ft,	$p = 10.59(1.51)0.8 \pm 2.7$
	$= 12.8 \pm 2.7$ psf

Leeward wall (all heights):

$$p = 10.59(1.51)(-0.5) \pm 2.7$$

$$= -8.0 \pm 2.7 \text{ psf}$$

Side walls:

$$p = 10.59(1.51)(-0.7) \pm 2.7$$

$$= -11.2 \pm 2.7 \text{ psf}$$

Roof (flat):

$$p = 10.59(1.51)(-0.7) \pm 2.7$$

$$= -11.2 \pm 2.7 \text{ psf}$$

The calculated design surface pressures are shown in Fig. 8.20.

(8) Repeat step (7) for longitudinal wind:

$$L/B = 150/90 = 1.67$$

It is seen from Fig. 8.16 that all pressure coefficients remain the same except those for the leeward walls, where $C_p = -0.37$. Thus

$$p = 10.59(1.51)(-0.37) \pm 2.7$$

$$= -5.9 \pm 2.7 \text{ psf}$$

The calculated design surface pressures are shown in Fig. 8.21.

(9) Calculate and sketch the eaves loading on the diaphragm in each direction.

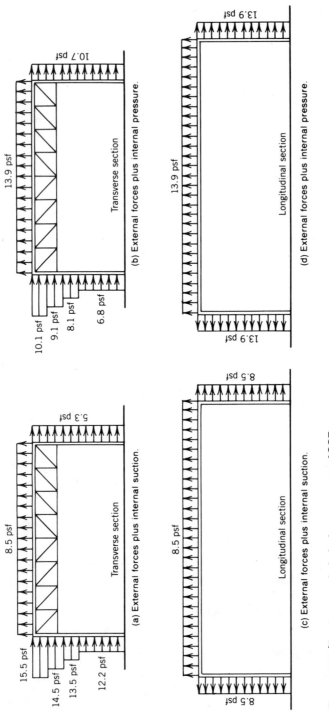

8.5 psf

15.5 psf
14.5 psf
13.5 psf
12.2 psf

5.3 psf

Transverse section

(a) External forces plus internal suction.

13.9 psf

10.1 psf
9.1 psf
8.1 psf
6.8 psf

10.7 psf

Transverse section

(b) External forces plus internal pressure.

8.5 psf

8.5 psf

8.5 psf

Longitudinal section

(c) External forces plus internal suction.

13.9 psf

13.9 psf

13.9 psf

Longitudinal section

(d) External forces plus internal pressure.

Figure 8.20/Transverse wind, design pressures, ASCE.

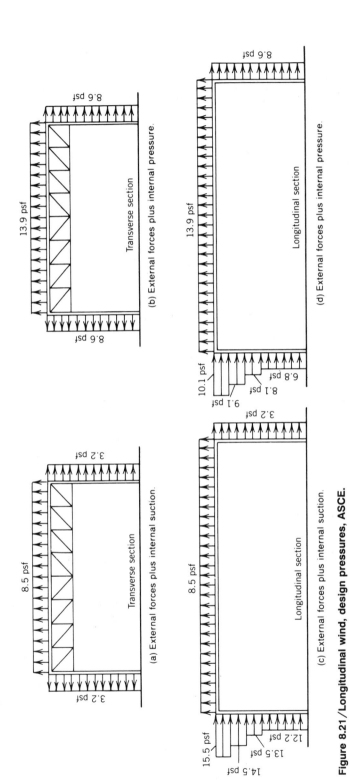

(a) External forces plus internal suction.

Transverse section

8.5 psf

3.2 psf

3.2 psf

(b) External forces plus internal pressure.

Transverse section

13.9 psf

8.6 psf

8.6 psf

(c) External forces plus internal suction.

Longitudinal section

8.5 psf

3.2 psf

15.5 psf

14.5 psf

13.5 psf

12.2 psf

(d) External forces plus internal pressure.

Longitudinal section

13.9 psf

8.6 psf

10.1 psf

9.1 psf

8.1 psf

6.8 psf

Figure 8.21 / Longitudinal wind, design pressures, ASCE.

289

Vertical wall girts span from floor to eaves as simple beams. Under uniform wall pressures, one-half the total wall load is developed at the eaves. The variable wall loading on the windward walls require calculating eaves reactions from the stepped loading, as in the procedures described for simple beams in Chapter 2.

Transverse Wind / The longitudinal section in Fig. 8.20 (c) and (d) shows the outward side-wall pressures canceling each other. Therefore, there is no net lateral load.

The transverse section in Fig. 8.20 (a) and (b) shows a larger windward load at (a) than at (b), but the reverse is true on the leeward walls. Although the net result will be the same, it is necessary to calculate wall reactions at the eaves on each side in order to determine attachment requirements.

Transverse frame of Fig. 8.20a: The eaves reaction for the leeward wall is

$$w = 5.3(30/2) = 79.5; \quad \text{use 80 plf}$$

The eaves reaction for the windward wall is

$$12.2(15)\frac{7.5}{30} = 45.75$$

$$13.5(5)\frac{17.5}{30} = 39.38$$

$$14.5(5)\frac{22.5}{30} = 54.38$$

$$15.5(5)\frac{27.5}{30} = \frac{71.04}{210.55}; \quad \text{use 211 plf}$$

Transverse frame of Fig. 8.20b: The eaves reaction for the leeward wall is

$$w = 10.7(30/2) = 160.5; \quad \text{use 161 plf}$$

The eaves reaction for the windward wall is

$$6.8(15)\frac{7.5}{30} = 25.50$$

$$8.1(5)\frac{17.5}{30} = 23.63$$

$$9.1(5)\frac{22.5}{30} = 34.13$$

$$10.1(5)\frac{27.5}{30} = \frac{46.29}{129.55}; \quad \text{use 130 plf}$$

The diaphragm loads from transverse wind are shown in Fig. 8.22 (a) and (b). Observe that the total load for each case is 291 plf.

Longitudinal Wind / The transverse section shown in Fig. 8.21 (a) and (b) shows the outward side-wall pressures canceling each other; therefore, there is no net lateral load. Each eaves load must be calculated for the longitudinal section in Fig. 8.21 (c) and (d). Observe that the windward wall forces are the same as those from a transverse wind. Therefore, the eaves load for (c) is $w = 211$ plf, and the eaves load for (d) is $w = 130$ plf. The leeward-wall eaves load is $w = 3.2(30/2) = 48$ plf for (c) and $w = 8.6(30/2) = 129$ plf for (d).

The diaphragm loads from the longitudinal wind are shown in Fig. 8.22 (c) and (d). Observe that the total load for each case is 259 plf.

Example 2/Gable Frames

Five single-story gable frames, spanning 64 ft and spaced 28 ft apart, frame a building 64×112 ft (Fig. 8.23). The eaves height is 20 ft and the ridge height 32 ft. This is a nonessential industrial-type building located in an Exposure C area at Corpus Christi, Texas. Horizontal girts on the walls and purlins along the roof are spaced fairly close together, and exterior openings are minimal. Determine the loading due to transverse wind on a typical interior frame using ASCE 7-88.

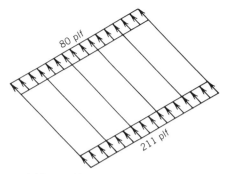

(a) External forces plus internal suction. Transverse wind.

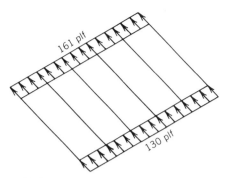

(b) External forces plus internal pressure. Transverse wind.

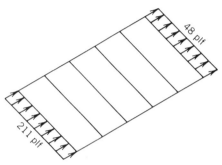

(c) External forces plus internal suction. Longitudinal wind.

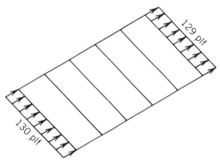

(d) External forces plus internal pressure. Longitudinal wind.

Figure 8.22/Diaphragm loads, ASCE.

Solution

(1) Determine the basic wind speed. Interpolating from the map of Fig. 8.15,

$$V = 95 \text{ mph}$$

(2) Determine the importance factor. From Table 8.2 for a building occupancy of Type I, $I = 1.05$ for the hurricane oceanline.

(3) Select velocity-pressure coefficients (K_z) from Table 8.3:

0–15 ft, $K_z = 0.80$
15–20 ft, $K_z = 0.87$
$h = 26$ ft, $K_h = 0.94$ (mean roof height)

(4) Calculate the wind velocity pressure for each height range using the formula $q_z = 0.00256 K_z(IV)^2$. For each height range, the value $0.00256[1.05(95)]^2 = 25.47$

remains constant; therefore,

0–15 ft, $q_z = 25.47(0.80) = 20.38$ psf
15–20 ft, $q_z = 25.47(0.87) = 22.16$ psf
$h = 26$ ft, $q_h = 25.47(0.94) = 23.94$ psf

(5) Select the gust response factor from Table 8.4, using a height of 26 ft:

$$G_h = 1.27$$

(6) Determine the internal pressure coefficient and calculate the internal pressure to be used for all heights (there are only minor openings in the building):

$$GC_{pi} = \pm 0.25$$

$$p = GC_{pi}q_h = \pm 0.25(23.94)$$

$$= \pm 5.99; \quad \text{use } \pm 6.0 \text{ psf}$$

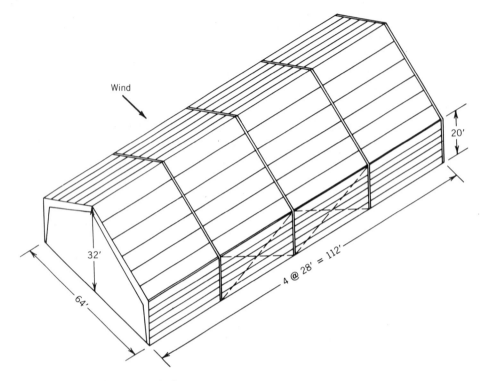

Figure 8.23/Example 2: Gable frames.

(7) Select the external pressure coefficients from Fig. 8.16:

Windward wall,

$$C_p = 0.8$$

Leeward wall,

$$L/B = 64/112 < 1.0$$

$$C_p = -0.5$$

Windward roof,

$$h/L = 26/64 = 0.41$$

$$\theta = \arctan(12/32) \approx 20°$$

Interpolating between 0.2 and −0.75,

$$C_p = 0.2 + \left[\frac{0.41-0.30}{0.2}(-0.95) \right] = -0.32$$

Leeward roof, $C_p = -0.7$.

(8) Calculate the design pressures for the external forces for each segment:

Windward wall,

$$0\text{–}15 \text{ ft}, \quad p = 20.38(1.27)0.8 = 20.71 \text{ psf}$$
$$15\text{–}20 \text{ ft}, \quad p = 22.16(1.27)0.8 = 22.51 \text{ psf}$$

Leeward wall,

$$p = 22.16(1.27)(-0.5) = -14.07 \text{ psf}$$

Windward roof,

$$p = 23.94(1.27)(-0.32) = -9.73 \text{ psf}$$

Leeward roof,

$$p = 23.94(1.27)(-0.7) = -21.28 \text{ psf}$$

(9) combine the external pressure for each segment determined in step (8) with the internal pressures (±6.0) found in step (6). These pressures are transferred to the

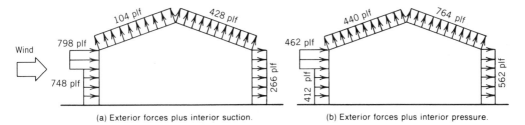

(a) Exterior forces plus interior suction.

(b) Exterior forces plus interior pressure.

Figure 8.24/Wind loads on gable frames.

frames by the girts and purlins. Multiply the combined pressures by 28 ft, the spacing between the frames. Sketch the loading on the frame (Fig. 8.24). The frames must be designed to safely develop each of these loading patterns.

Example 3/Multistory Building

A four-story, nonessential office building is constructed of steel frames placed on grid lines as shown in Fig. 8.25. Wall panels containing fixed windows form the ex-

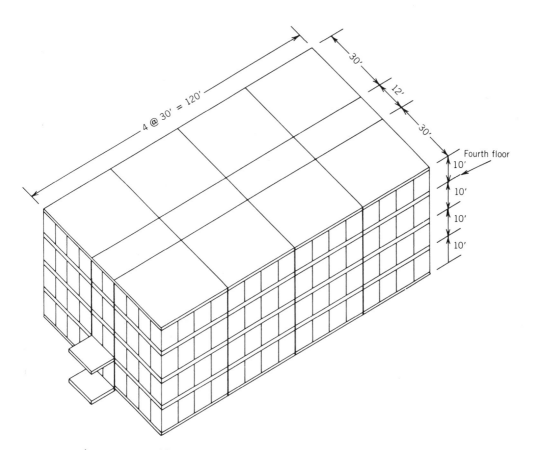

Figure 8.25/Example 3: multistory.

terior envelope of the building and span vertically from floor to floor. Diaphragms are developed at each floor and at the roof level. This is not considered to be a wind-sensitive building. It is located in an Exposure C area at Salt Lake City, Utah. Investigate for transverse wind only.

(a) Calculate the diaphragm load at the fourth floor using the 1991 UBC.

(b) Calculate the diaphragm load at the fourth floor using ASCE 7-88.

Solution (a) (UBC)

(1) Determine the basic wind speed from Fig. 8.15:

$$V = 70 \text{ mph}$$

(2) Select the stagnation pressure from Table 8.5a:

$$q_s = 12.6 \text{ psf}$$

(3) Select the combined height, exposure, and gust coefficient from Table 8.5b. The height under investigation is from the mid third floor (25 ft) to the mid fourth floor (35 ft). Use 30 ft for the windward wall and 40 ft for the leeward wall:

Windward wall,	$C_e = 1.23$
Leeward wall,	$C_e = 1.31$

(4) Determine the importance factor, Table 8.2:

$$I = 1.0$$

(5) Determine the pressure coefficients for the diaphragms from Table 8.5c:

Windward wall,	$C_q = 0.8$ inward
Leeward wall,	$C_q = 0.5$ outward

(6) Calculate the pressures on the walls:

Windward wall,

$$P = 1.23(0.8)(12.6)(1.0) = 12.4 \text{ psf}$$

Leeward wall,

$$P = 1.31(0.5)(12.6)(1.0) = 8.25 \text{ psf}$$

(7) Calculate the diaphragm load by multiplying the pressures by the 10-ft wall height:

Windward wall,	$w = 12.4(10) = 124$ plf
Leeward wall,	$w = 8.25(10) \approx 83$ plf

This diagphragm load is shown in Fig. 8.26.

Solution (b) (ASCE)

(1) Determine the basic wind speed from Fig. 8.15:

$$V = 70 \text{ mph}$$

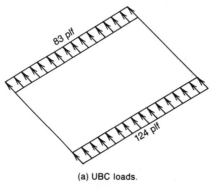

(a) UBC loads.

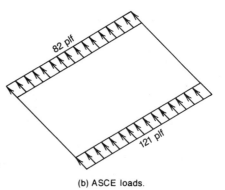

(b) ASCE loads.

Figure 8.26/Fourth-floor diaphragm loads.

(2) Determine I from Table 8.2. The office building is of occupancy category I; therefore,

$$I = 1.0$$

(3) Calculate the velocity pressure exposure coefficient from Table 8.3. The height under investigation is from the mid third floor (25 ft) to the mid fourth floor (35 ft). Use 30 ft for the windward wall:

$$z = 30 \text{ ft}, \qquad K_z = 0.98$$

$$h = 40 \text{ ft}, \qquad K_z = 1.06$$

(4) Calculate the velocity pressures:

$$q_z = 0.00256(0.98)\left[1.0(70)\right]^2 = 12.29 \text{ psf}$$

$$q_h = 0.00256(1.06)\left[1.0(70)\right]^2 = 13.30 \text{ psf}$$

(5) Select the gust response factor from Table 8.4:

$$G_h = 1.23$$

(6) Determine the pressure coefficients for both walls from Fig. 8.16:

Windward wall, $\qquad C_p = 0.8$
Leeward wall, $\qquad C_p = -0.5$

(7) Calculate the pressure on the walls:

Windward wall,

$$p = 12.29(1.23)0.8 = 12.09 \text{ psf}$$

Leeward wall,

$$p = 13.30(1.23)(-0.5)$$
$$= -8.18 \text{ psf}$$

(8) Calculate the diaphragm load by multiplying the pressures by the 10-ft wall height:

Windward wall, $\quad w = 12.09(10) \approx 121 \text{ plf}$
Leeward wall, $\quad w = -8.18(10) \approx -82 \text{ plf}$

This diaphragm load is shown in Fig. 8.26b.

PROBLEMS

1. Redo Example 1, Fig. 8.17, with the following changes: the truss spans 120 ft, the eaves height is 35 ft, and the building location is in Tampa, Florida. Calculate the loading along the eaves for the transverse direction only.

(a) Use ASCE 7-88. (Answers given in Appendix G.)

(b) Use the 1991 UBC. (Answers given in Appendix G.)

2. Redo Example 2, Fig. 8.23, according to the provisions of the 1991 UBC.

3. The building is the same as that used in Example 3, Fig. 8.25.

(a) Calculate the diaphragm loading at the roof according to the provisions of ASCE 7-88.

(b) Calculate the diaphragm loading at the roof according to the provisions of the 1991 UBC.

8.8
Earthquakes—Introduction

Earthquakes occur somewhere several times each day; fortunately, however, most of them are small and are sensed only by instruments. Earthquake occurrences are not uniformly distributed over the globe. They occur principally in bands or strips that follow identifiable paths over the surface of the earth. Earth science has produced convincing evidence that the earth's crust is composed of irregular pieces or plates that are "floating" and thus constantly moving with respect to one another. Although the rate of movement of these plates is small, it is measurable and significant over geological time. It is the movement of these plates that is believed to be the principal cause of earthquakes.

Two adjacent plates moving with respect to one another can cause a buildup of strains in the crustal structure. When these strains become large enough, a sudden

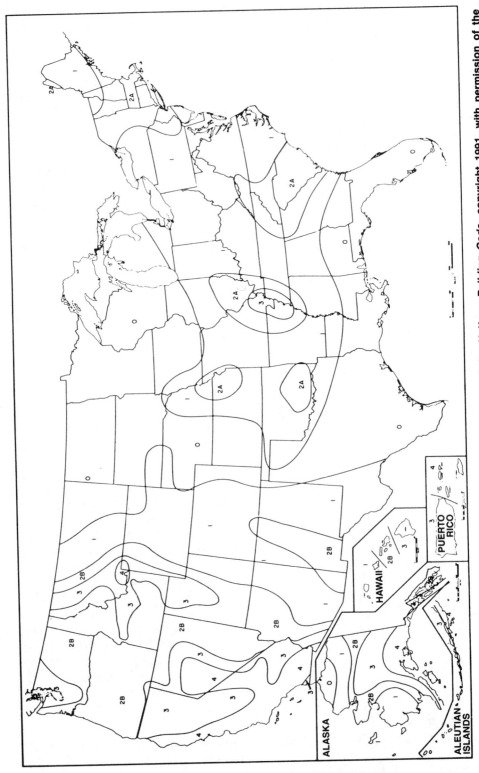

Figure 8.27/UBC seismic-risk map. Reproduced from the 1991 edition of the *Uniform Building Code*, copyright 1991, with permission of the publisher, the International Conference of Building Officials.

ALASKA

ALEUTIAN ISLANDS

HAWAII

PUERTO RICO

296

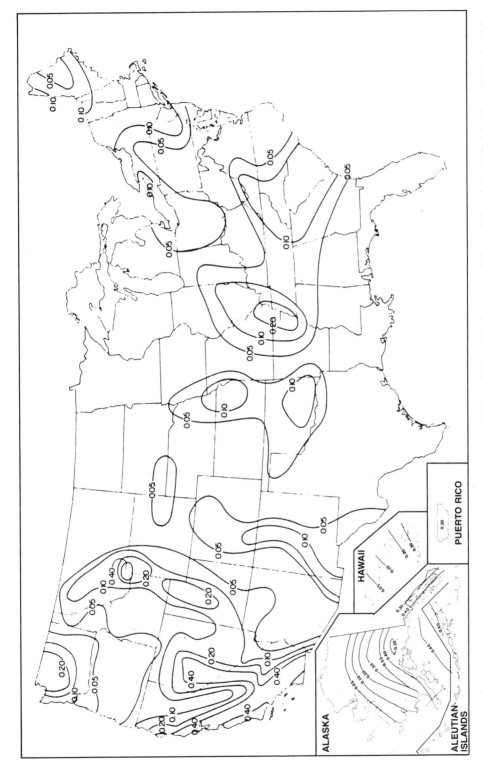

Figure 8.28 / NEHRP contour map for effective peak acceleration (EPA) coefficient A_a, Courtesy of the Building Seismic Safety Council, 1991 edition of the NEHRP Recommended Provisions.

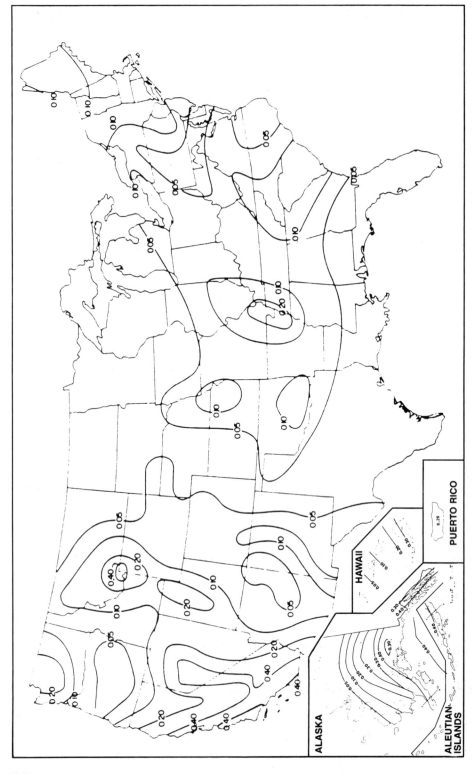

Figure 8.29/NEHRP contour map for effective peak velocity-related acceleration (EPV) coefficient A_v, Courtesy of the Building Seismic Safety Council, 1991 edition of the NEHRP Recommended Provisions.

fracture takes place. This fracture is recognized as the earthquake. Earth scientists believe that 90 per cent of all earthquakes take place in the vicinity of plate boundaries.

Some earthquakes are of sufficient magnitude or intensity to cause severe damage to, and even total collapse of, some buildings. The earthquake hazard is a function of geographic location. Figure 8.27 is the map adopted by the 1991 Uniform Building Code (UBC). It contains contours dividing the U.S. into zones of approximately equal danger; the larger the number, the greater the danger. Figures 8.28 and 8.29 are the maps adopted by the 1991 National Earthquake Hazards Reduction Program (NEHRP) "Provisions for the Development of Seismic Regulations for New Buildings." They contain contours representing the effective peak acceleration and the effective peak velocity of potential ground motion. The larger the number, the greater the danger. The use of these maps in design procedures is discussed in Art. 8.11.

Large earthquakes can produce many damaging effects. For example, there can be ground surface fractures, landslides, soil liquefaction, avalanches, and even large-water-body tsunamis. Statistically, however, the most damaging effect is ground motion or vibration, because it covers such a large geographic area. Strong vibrations can occur in any direction, even vertically. The current practice, however, is to consider only horizontal (lateral) movement, but to consider that such movement can occur in any direction. (See Art. 8.1 and Fig. 8.1.)

Similar to wind, ground vibrations have a dynamic affect on a building, but, unlike wind, there are no externally imposed forces. The weight of the building provides static inertia which must be overcome before the motion of the building follows the motion of the ground. The distortions from these "racking" forces can be severe enough to overstress portions or the whole of a structure, resulting in permanent damage or total collapse. Under certain conditions, it is not only advisable but a code requirement to undertake a detailed dynamic analysis of the structure to accurately evaluate these potential distortions and their stress resultants. In order to avoid the resonance effect, the natural period of vibration of the structure should be quite different from the predominant period of vibration of the ground.

Up until recently, a dynamic analysis of a structure for earthquake vibrations was

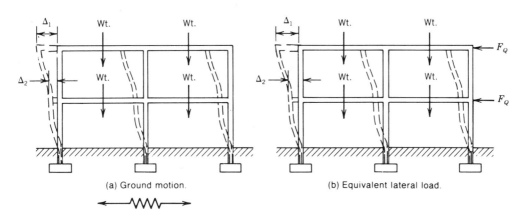

(a) Ground motion.

(b) Equivalent lateral load.

Figure 8.30/Equal structure deflection.

seldom attempted. However, recent advances in structural engineering and in computer programs have made this task more practicable. Consequently, regulatory codes can now require a design procedure based upon a dynamic anlaysis. But this requirement is usually made only under special circumstances that will be discussed later. This text does not cover dynamic analysis.

The alternative to the dynamic analysis is called the *equivalent-lateral-force procedure*, and the remainder of this chapter covers this procedure as it relates to steel buildings.

The current practice is to assume static lateral forces that would cause the same distortion as that caused by the anticipated ground vibrations. A helpful way to visualize earthquake effects is to assume that the ground grips the base of the building, with the rest of the building cantilevering out of the ground (Fig. 8.30a). The ground vibration from the earthquake shakes the base, and the cantilevered portion of the building above responds by distorting accordingly. Earthquake loads, then, are assumed to be those static lateral forces that will cause the same distortion as the anticipated ground vibrations (Fig. 8.30b).

8.9
Design Procedure and Codes

The 1989 AISC Specification does not address the issue of earthquake design, except to allow a one-third increase in allowable stress for elastic design or a reduced load factor of 1.3 for plastic design (Chapter 11). Therefore, it is necessary to refer to other standards and/or regulations.

Most regulatory jurisdictions adopt (with modifications) one of the three model building codes available in the U.S. and

cited in Chapter 1. These three model codes are: (1) the Uniform Building Code (UBC) in the western part of the U.S., (2) the Standard Building Code in the southern and southeastern part, and (3) the Basic Building Code in the east and midwest. Historically the UBC has been more active in promoting seismic safety, and its current version (1991) has resulted from refinement and improvement over the last fifty years. Consequently, the UBC will be used here, in both discussion and examples.

In addition, the National Earthquake Hazards Reduction Program (NEHRP) "Recommended Provisions for the Development of Seismic Regulations for New Buildings" will also be referenced and used in the examples. The NEHRP is not a building code as such, but recommended provisions for a wide range of regulations to improve seismic safety, including new and existing buildings, urban and community planning, equipment and non-structural components, etc. These provisions are based in large measure on recommendations of the Structural Engineers Association of California (SEAOC) and the Applied Technology Council (ATC), which is also a California-based group. Both the SEAOC and the ATC have had a major influence on the development of the earthquake standards contained in the UBC.

The other two model building codes, Basic and SSBC, have adopted (or are in the process of adopting) many of the new building provisions contained in NEHRP. Also, the ASCE 7-88 (referenced earlier for wind design, Art. 8.5) is currently undergoing a revision scheduled for publication in 1995 and will have its earthquake portion comply with NEHRP. The 1991 Edition of the NEHRP is referenced here and is available from the U.S. Government Printing Office.

The general procedure in earthquake design is first to calculate the overall base shear, which is the basic total lateral load on the building. Then, this base shear is distributed "back" to the vertical and horizontal elements according to specific rules. Once the loading in the elements is determined, their design may be accomplished in several ways. NEHRP states that the strength of members and connections subjected to seismic forces in combination with other prescribed loads shall be determined by the LRFD procedure (Chapter 11) except for some modification in load factors. It also states that an alternative procedure for strength design is the allowable-stress-design (ASD) method, also with modifications.

The UBC relates the strength design procedures to specific detailing requirements and to seismic zones. Zones 3 and 4 have more strict criteria, but in general the UBC permits strength design to be accomplished by either the LRFD method or the ASD method (which also includes plastic design, Chapter 11). The codes must be consulted in detail, as certain structural elements are more critical and require larger safety factors. For example, some bracing elements are required to be designed for 3 times the determined seismic forces.

8.10
Equivalent Lateral Load
or Base Shear

The basis for establishing the equivalent static lateral load is Newton's second law of motion, i.e.,

$$F = ma$$

where F is the unbalanced force causing the mass m to have an acceleration a. The mass can be obtained by dividing the weight by the acceleration due to gravity, or approximately 32 ft per sec^2. Expressed

in equation form,

$$F = \left(\frac{W}{g}\right)a$$

It is convenient to alter this equation to

$$F = \left(\frac{a}{g}\right)W$$

In earthquake design, F is the equivalent static lateral load of a building or building element having a weight W. The quantity a/g is referred to as the *seismic factor*. Conceivably, an infinitely rigid structure subjected to an earthquake movement having an acceleration of 16 ft/sec^2 would have a seismic factor of $16/32 = 0.5$. Notice that the seismic factor has no units. It is defined as a dimensionless number which, when multiplied by the weight, produces an equivalent static lateral load.

In arriving at the seismic factor, both the UBC and NEHRP require that the degree of probability of a large earthquake, the type of building occupancy, the flexibility of the structure, the type of structural system, and the physical site characteristics be taken into consideration. Symbols are used to represent these considerations, and the seismic factor is then expressed as the product of a number of coefficients without units. This seismic factor, multiplied by the appropriate weight (W), becomes the lateral force F. If W is the entire building weight, the lateral force is considered to be the base shear V. The two referenced documents have a slightly different form and definition of these considerations and symbols as follows:

UBC:

$$V = \left[\frac{ZIC}{R_w}\right]W$$

where

$$C = \frac{1.25S}{T^{2/3}} \leq 2.75$$

NEHRP:

$$V = [C_s]W$$

where

$$C_s = \frac{1.2 A_v S}{RT^{2/3}} \quad \text{or} \quad C_s = \frac{2.5 A_a}{R}$$

whichever is smaller.

The new quantities entering the above expressions—i.e., Z, I, C, R_w, S, T, A_v, A_a, and R—are collectively referred to as *seismic force factors*; along with W, they are discussed individually in the following article.

8.11
Seismic Force Factors

Although the formulas for base shear look quite different, they frequently produce reasonably close values. Each code, however, maintains its own way of treating these considerations because of cross-referencing with its many other aspects, including some that are non-structural. The discussion here relates only to the structural aspects of each consideration and further limits the discussion to those features most appropriate for steel buildings.

Geography—Z, A_v, A_a Factors / The UBC seismic-risk map, Fig. 8.27, initially partitioned the U.S. into five zones. Zone 0 represents no risk, whereas Zone 4 represents the greatest risk. Zone 2 covered such a large area that it was subsequently subdivided into zones 2A and 2B. Both the probability of a large earthquake and the frequency of occurrence were considered in establishing the boundaries. The UBC assigns a numerical value, a *Z factor*, for each zone, as shown in Table 8.6, and its value should be used in the base-shear formula.

The NEHRP map for the effective peak acceleration V_a (Fig. 8.28) shows values

Table 8.6
UBC Seismic Zone Factor Z [a]

Zone	Z
1	0.075
2A	0.15
2B	0.20
3	0.30
4	0.40

[a] Excerpted from the 1991 edition of the Uniform Building Code, copyright 1991, with permission of the publisher, The International Conference of Building Officials.

from 0.05 to 0.4. These numbers are related to the maximum expected peak acceleration in percentages of g, the universal acceleration of gravity (32 ft per sec²). Consequently they are numbers without units. The contours on the velocity-related acceleration map (Fig. 8.29) are also without units. Maximum expected peak velocities of 1.5 in. per sec would correlate with an acceleration of $0.05g$, and 12 in. per sec would correlate with accelerations of $0.40g$. Intermediate contours are also shown on the map. These percentages should be used in each of the two formulas for establishing C_s for base-shear calculations. It should also be noted that the maps in the NEHRP publication are very large colored maps showing the boundaries of every county in the United States and identifying each with an A_v and A_a. In the absence of these detailed maps, it is suggested that linear interpolation between the contours on smaller maps be used.

Occupancy—I Factor / The I factor occurs only in the UBC base-shear formula. NEHRP does consider occupancy, but does not use it in establishing force levels. This difference in the treatment of occupancy is the largest discrepancy between the two documents and warrants separate discussions.

The UBC lists four occupancy categories as follows:

Group I, essential facilities, $I = 1.25$.

Group II, hazardous facilities, $I = 1.25$.

Group III, special-occupancy facilities, $I = 1.00$.

Group IV, standard-occupancy facilities, $I = 1.00$.

Essential facilities are those that are expected to function and operate during and immediately after an earthquake. Examples are hospitals, police and fire stations, communication facilities, etc. Hazardous facilities are those that contain explosive or toxic substances. Special-occupancy facilities are educational or medical buildings or any public facilities with a capacity of over 300 persons. Standard-occupancy facilities are those that do not fit into the other categories.

The NEHRP provision has only three occupancy categories, referred to as seismic-hazard exposure categories, and has a grouping number just the opposite of the UBC's. Group I consists of standard-occupancy buildings that do not fit into the other groups. Group II contains buildings having special occupancy, such as schools, jails, buildings for public assembly of more than 300 persons, etc. Group III contains the essential facilities that must continue operation during and after an earthquake.

As stated earlier, NEHRP does not use the occupancy category in establishing the base shear. Instead, it considers the combined effect of the value of A_v from Fig. 8.30 and the exposure-group number. With these two parameters a *seismic performance category* of A, B, C, D, or E is established according to Table 8.7. This performance category is used for establishing critical features for detailing, structural analysis, and design. The use of this performance category in the design procedure will be shown later.

Period of Vibration—T Factor / Every structure has a fundamental period of vibration, which is defined as the length of time (in seconds) for the structure to complete one full cycle of vibratory motion. This can range from less than a tenth of a second for a short, stiff building to several seconds for a tall, flexible one, and it can be different for each orthogonal axis.

There is no simple or direct way of accurately calculating this period of vibration. Perhaps the procedure offering the most credibility is one based upon dynamic analysis, where the eigen-values are extracted. However, even this method requires assumptions that are seldom met in the actual building.

The value of T also can be calculated by procedures based upon static analysis.

Table 8.7
NEHRP Performance Categories[a]

	Seismic-Hazard Exposure Group		
Range of A_v	*I*	*II*	*III*
$A_v < 0.05$	A	A	A
$0.05 \le A_v < 0.10$	B	B	C
$0.10 \le A_v < 0.15$	C	C	C
$0.15 \le A_v < 0.20$	C	D	D
$0.20 \le A_v$	D	D	E

[a]Reprinted from the 1991 NEHRP Recommended Provisions, courtesy of the Building Seismic Safety Council.

However, once again, story drifts (lateral deflection) must be calculated, and this requires basic assumptions that have significant influence on the accuracy of the results. One way of calculating T from static mechanics is to use the Rayleigh formula

$$T = 2\pi \sqrt{\left(\sum_{i=1}^{n} w_i \delta_i^2 \right) \div \left(g \sum_{i=1}^{n} f_i \delta_i \right)}$$

Explanations of the symbols in this formula are given in Fig. 8.31.

A building consists of weights assumed to be collected at the center of each floor and the roof ($w_1, w_2, w_3, \ldots$). A base shear (V) is assumed or estimated. This base shear is redistributed to each floor ($f_1, f_2, f_3, \ldots$) according to the procedure described in Art. 8.13. These lateral forces causes the structure to deflect at each floor level ($\delta_1, \delta_2, \delta_3, \ldots$). It should be noted that in the Rayleigh formula g is the acceleration due to gravity; if 384 in. per sec^2 is used as the value for g, then δ_i must be measured in inches. The units for weights and forces in the formula can be either pounds or kips, but must be consistent.

The difficulty with using this formula should be apparent—the value of V must be known beforehand. Consequently, other formulas are used to provide an approximate and conservative value of T. The Rayleigh formula then is used as a refinement or a check on the adequacy of the first estimated value of T. This method is used in the example in Art. 8.13 which follows.

Both documents provide an approximate conservative approach in establishing this basic period. It can be estimated from the formula

$$T = C_t (h_n)^{3/4}$$

where h_n is the height in feet above the

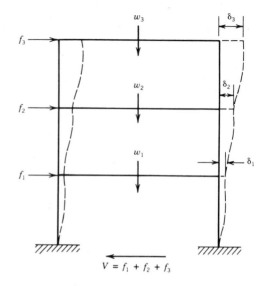

Figure 8.31/Forces and deflections for calculating T.

base to the highest level, and

$C_t = 0.035$ for steel moment-resisting frames.

$C_t = 0.030$ for reinforced concrete frames and eccentrically braced frames.

$C_t = 0.02$ for all other buildings.

Site soil—S Factor / The type of soil at the building site has a strong influence on vibrations in the structure caused by an earthquake. With thorough site investigation and research it is usually possible to determine a characteristic site period similar to that for a building structure. If the two periods match one another, there is danger from resonance with the structure. NEHRP provision contains a section in its appendix devoted exclusively to methods for determining site-structure response.

However, both UBC and NEHRP contain an approximate conservative method of dealing with the site consideration. Table 8.8 shows four types of soil classification. The description shown is the precise description given in the NEHRP. The wording in the UBC is slightly different, but the

Table 8.8
Site Soil Factor[a]

Soil Profile Type	Description	Site Coefficient S
S_1	A soil profile with either (1) rock of any characteristic, either shale-like or crystalline in nature, that has a shear-wave velocity greater than 2500 ft per sec, or (2) stiff soil conditions where the soil depth is less than 200 ft and the soil types overlying the rock are stable deposits of sands, gravels, or stiff clays.	1.0
S_2	A soil profile with deep cohesionless or stiff clay conditions where the soil depth exceeds 200 ft and the soil types overlying rock are stable deposits of sands, gravels, or stiff clays.	1.2
S_3	A soil profile containing 20 to 40 ft in thickness of soft to medium-stiff clays with or without intervening layers of cohesionless soils.	1.5
S_4	A soil profile characterized by a shear-wave velocity of less than 500 ft per sec containing more than 40 feet of soft clays or silts.	2.0

[a]Reproduced from the 1991 edition of the NEHRP Recommended Provisions, courtesy of the Building Seismic Safety Council.

general types remain the same. Initially there were only the first three types, but the 1985 Mexico City earthquake emphasized the need for the fourth type. The UBC further states that soil type S_4 need only be used if thorough geotechnical data prove its presence or if the building official requires it to be used. Also, the UBC states that if soil information is not known, type S_3 should be used.

Structural Systems— R Factor / This factor represents the effectiveness of the structural system and includes consideration of both the lateral load-carrying system and the gravity load-carrying system. UBC and NEHRP essentially agree on the descriptions and classification of the systems, but assign different numerical coefficients, as shown in Table 8.9. Observe that the numerical factor used by NEHRP is called R and that used by the UBC is called R_w. The main reason why NEHRP's value of R is less than the UBC's is that NEHRP's procedure produces strength-

level forces, while the UBC is intended to produce service-level forces.

Each code contains a larger and more complete list of structural systems, their R values, height limitations, and, in some cases, lateral-deflection limitations. Table 8.9 is a selected listing of only those systems most appropriate for steel buildings.

Figure 8.32 illustrates four basic structural systems, showing both gravity loads and lateral loads. The bearing-wall system refers to gravity loads. However, when lateral loads occur, the bearing walls also function as shear walls. If the walls are damaged as a consequence of lateral loads, they may lose their ability to support the gravity loads. Steel buildings may have bearing walls that are used in place of some columns.

Figure 8.32b shows a building frame in which there is an essentially complete space frame which carries the gravity loads. This frame is considered to be a simple frame, which means the connections are flexible and are incapable of developing

Table 8.9
Basic Structural Systems and Lateral Load-Resisting Factors

| Description | | *NEHRP*[a] | | | | | | *UBC*[b] | |
| | | | *Height Limitation*[c] (ft) | | | | | | *Height Limitation*[c] (ft) |
	R	*Category* A, B	C	D	E		R_w		*in Zones* 3 & 4
Bearing wall:									
Reinforced concrete shear walls	$4\frac{1}{2}$	NL		NL	160	100	6		160
Reinforced masonry shear walls	$3\frac{1}{2}$	NL		NL	160	100	6		160
Steel concentrically braced frames	4	NL		NL	160	100	6		160
Steel building frame:									
Reinforced concrete shear walls	$5\frac{1}{2}$	NL		NL	160	100	8		240
Reinforced masonry shear walls	$4\frac{1}{2}$	NL		NL	160	100	8		160
Steel concentrically braced frames	5	NL		NL	160	Dual	8		160
Steel eccentrically braced frames	7	NL		NL	160	100	10		240
Steel moment-resisting frames:									
Ordinary moment-resisting frames	$4\frac{1}{2}$	NL		NL	160	100	6		160
Special moment-resisting frames	8	NL		NL	NL	NL	12		NL
Dual system with ordinary moment-									
resisting frames for 25% of strength:									
Steel with concentrically braced frames	5	NL		NL	160	100	6		160
Reinforced concrete shear walls	6	NL		NL	160	100	6		160
Reinforced masonry shear walls	5	NL		NL	160	100	6		160
Dual system with special moment-									
resisting frames for 25% of strength:									
Steel concentrically braced frames	6	NL		NL	NL	NL	10		NL
Reinforced concrete shear walls	8	NL		NL	NL	NL	12		NL
Reinforced masonry shear walls	$6\frac{1}{2}$	NL		NL	NL	NL	8		160

[a]Reprinted from the 1991 edition of the NEHRP Recommended Provisions, courtesy of the Building Seismic Safety Council.

[b]Reproduced from the 1991 edition of the Uniform Building Code, Copyright 1991, with permission of the publisher, The International Conference of Building Officials.

[c]NL = no limit

resistance to bending moments. Consequently, the frame, by itself, is incapable of resisting any lateral load. Therefore, some type of bracing or shear wall must be used with this frame system.

The moment-resisting frame system shown in Figure 8.32c resists both the gravity loads and the lateral loads. The moment-resisting connections, in effect, substitute for the bracing or shear walls of the simple frame system. The connections may be one of two types, ordinary or special. *Special moment-resisting frames* (SMRF) are distinct from *ordinary moment-resisting frames* (OMRF) in that they conform to more strict design criteria, such that they can deform and still maintain their strength. This requirement makes them energy-absorbing, which is particularly desirable in earthquake design.

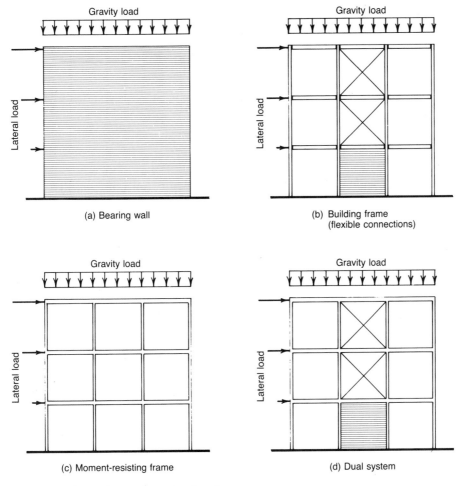

Figure 8.32/Basic types of structural systems.

Figure 8.32d, the dual system consists of a moment-resisting frame, either OMRF or SMRF, along with either shear walls or bracing. The whole system must meet the following criteria:

1. There is an essentially complete space frame which carries the gravity loads.

2. The combined action of moment-resisting frame and bracing or shear walls resist the lateral roads. The frame is considered to be a back-up to the bracing and walls. The moment-resisting frame must be designed to carry at least 25% of the total base shear.

3. The two parts of the dual system must be designed to resist the total lateral force in proportion to their relative stiffnesses.

One additional distinction should be made with respect to lateral load-carrying systems. Refer to Figure 8.33, where two different bracing systems are shown. The normal bracing in Figure 8.33a is referred to as concentric. The bracing meets at the joint between girder and column. Eccentric bracing (Fig. 8.33b) meets the girder at a location several feet from the girder-column connection. In so doing, it forms a plastic link or fuse to absorb energy and

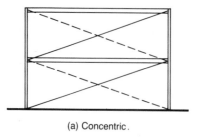

(a) Concentric.

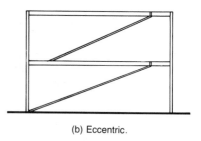

(b) Eccentric.

Figure 8.33/Bracing.

therefore becomes an improved earthquake bracing system.

The complete code should be consulted with regard to use and restrictions for all of these systems as well as other systems that are not listed. For example, Table 8.9 shows many systems that cannot be used for buildings over 160 ft high.

Weight—W Factor / The symbol W in the seismic force equation for the total base shear represents the total weight of all permanent construction above the base (level at which the shear is calculated) plus, under some circumstances, additional weight due to use or environmental factors.

All of the gravity dead loads, including all building elements plus permanent equipment, should be included when determining the total weight W for base-shear calculations. Also, the codes prescribe that 25 per cent of the live load for warehouses and storage facilities is to be included in W. Floor live load is seldom, if ever, included. One exception is a live-load allowance for interior partitions (such an allowance often is made for design purposes when the precise makeup and location of partitions cannot be established). A minimum of 10 psf is used under these circumstances.

The question of whether or not to include snow load requires judgement. It seems reasonable, for certain locations which consistently have heavy snow cover over extended periods of time, that all or part of the snow load should be included in determining the weight for seismic-force calculations. All of the codes conclude that, when the design snow load is equal to or less than 30 psf, none of it need be included. For snow loads greater than 30 psf, some portion of the weight should be included, and the local building official should decide the amount.

Summary / Structural design for earthquake resistance, following firm regulatory codes, has proven to be very good in most cases. Where significant structural damage has occurred, it usually can be associated with buildings with obvious structural irregularities. Consequently, all codes contain specific regulations involving irregularities. These irregularities pertain to the overall configuration of the building and can be described as being either vertical or in plan. The intent is not to prohibit irregular buildings, but to specify certain restrictions and consequences in design when they do occur.

Both UBC and NEHRP list five types of irregularities in each category as follows:

Vertical structural irregularities:

A, 1. Stiffness irregularity—soft story
B, 2. Weight (mass) irregularity
C, 3. Vertical-geometry irregularity
D, 4. In-plane discontinuity in vertical lateral force-resisting element
E, 5. Discontinuity in capacity—weak story

Plan structural irregularities:

A,1.   Torsion irregularity
B,2.   Reentrant corners
C,3.   Diaphragm discontinuity
D,4.   Out-of-plane offsets
E,5.   Nonparallel systems

The detail descriptions and limiting requirements associated with irregularities specified in the codes must be consulted.

It was stated earlier that although the equivalent static lateral load could be used in most design circumstances, there are conditions which require a detailed dynamic analysis. The flowcharts in Figures 8.34 and 8.35, further explain how this is decided.

Figure 8.34 was prepared following the provision of the UBC. The process begins at the top, concerning the seismic zone. Zone 1 proceeds directly to the static analysis. Zone 2 proceeds to the occupancy category, and if the occupancy is IV, then it proceeds to the static analysis. All other occupancy categories or locations in zones 3 and 4 must proceed to the irregular-building consideration.

Figure 8.35 was prepared following the provision of the NEHRP. The process begins at the top by considering jointly the coefficient A_v from map 2 (Figure 9.29) and the NEHRP occupancy group (grouping is different from the UBC). Based upon these two considerations, the building is placed in one of the five seismic performance categories according to Table 8.7. Buildings in categories D and E must proceed next to the building-irregularity consideration.

It is apparent in the static analysis that the procedure for calculating the total base shear is an empirical process that is subject to changing rules established by various building regulating and advisory authorities.

Once the base shear has been established, it is necessary to determine what part of it

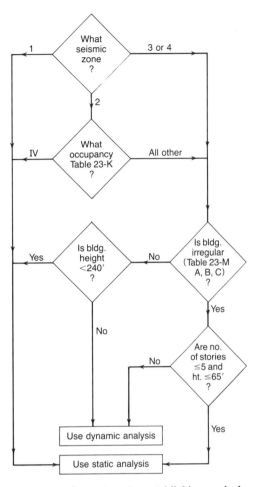

Figure 8.34/Flowchart for establishing analysis procedure (based upon UBC 1991 Seismic Provisions).

is developed by each vertical and horizontal element. This procedure can be compared to that of determining the loads on a beam after the reactions have been established. The sum of loads developed by each vertical element equals the base shear.

Procedures for distributing the base shear to all vertical and horizontal elements of multistory buildings require additional consideration of the dynamic response of the building. Articles 8.12 and 8.13 cover these topics. Calculating element loads for one-story buildings is much easier and is

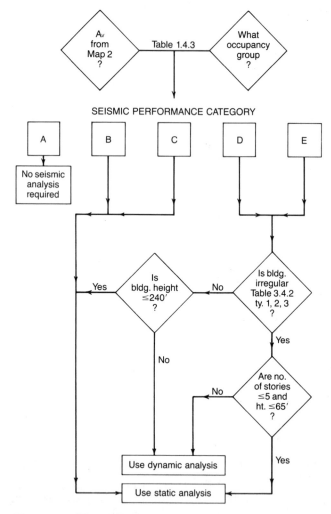

Figure 8.35/Flowchart for establishing analysis procedure (based upon NEHRP provisions).

accomplished by use of ordinary structural mechanics, as will be shown in Example 2 which follows.

Example 1/Base Shear

A three-story, steel-framed office building is shown in Fig. 8.36. Columns are placed so as to provide a bay size of 20×30 ft for the entire building. Beams and girders occur in both directions and are connected to the columns with standard moment-resisting connections. Purlins are used but are not shown.

The building is located in Salt Lake City, Utah. The soils investigation report describes the soil at the site of the proposed building as a mixture of gravel and sand for a depth of over 200 ft.

All exterior walls are glass and prefabricated metal panels that are estimated to weigh an average 12 psf of wall area.

Floor and roof construction and the respective weights of the elements of each are as follows:

Roof	
Built-up roofing	6 psf
Steel deck	2
Steel structure	4
Ceiling	5
Miscellaneous	3
Total	20 psf

Floors	
Flooring	2 psf
Steel deck and lightweight fill	30
Steel structure	6
Ceiling	5
Partitions	8
Total	51 psf

The deck construction is such that it functions as a diaphragm. The vertical elements are steel frames at each grid line.

(a) Calculate the base shear in each direction, using the 1991 NEHRP requirements.

(b) Calculate the base shear in each direction, using the 1991 UBC requirements.

Solution (a) (Base Shear, NEHRP)

(1) Calculate the building weight (no snow load):

Roof weight = 60(120)0.02 ≈ 144 kips
3rd-floor weight = 60(120)0.051 ≈ 367
2nd-floor weight = 60(120)0.051 ≈ 367
Exterior wall weight
= 30[2(120) + 2(60)]0.012 ≈ 130
Total W ≈ 1008 kips

(2) Determine the effective peak acceleration from the map (Fig. 8.28): $A_a = 0.20$.

(3) Determine the effective peak velocity-related acceleration from the map (Fig. 8.29): $A_v = 0.20$.

(4) Calculate the fundamental period of vibration of the building:

$$T = C_t(h_n)^{3/4}$$
$$= 0.035(30)^{3/4} = 0.45 \text{ sec}$$

(both directions)

(5) Determine the site coefficient from Table 8.8. The soil profile is classified as S_2; thus $S = 1.2$.

(6) The structure has been described as a steel ordinary moment-resisting frame (OMRF). Consequently, the R-value from Table 8.9 is 4.5.

(7) Calculate the regular seismic factor C_s:

$$C_s = \frac{1.2 A_v S}{R T^{2/3}}$$
$$C_s = \frac{1.2(0.20)(1.2)}{4.5(0.45)^{2/3}} = 0.11$$

(8) Check the maximum value for C_s:

$$C_s = \frac{2.5 A_a}{R}$$
$$C_s = \frac{2.5(0.20)}{4.5} = 0.111$$

(9) Calculate the base shear:

$$V = C_s W$$
$$V = 0.11(1008) = 111 \text{ kips}$$

Observe that the base shear of 111 kips is valid for both the transverse and longitudinal axes of the building. This will not always be the case. The values for T and R could be different for each axis and thus could result in a different base shear for each axis.

It also should be pointed out that the computed base shear is considered to be conservative. A more refined analysis *might* result in a smaller base shear. To make this more refined investigation, it would be necessary to distribute the base shear to each floor level (Art. 8.13), calcu-

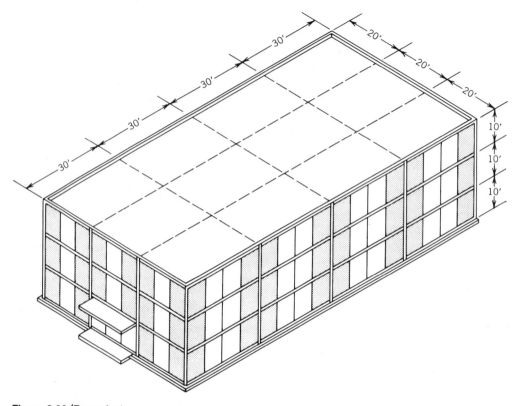

Figure 8.36/Example 1.

late each floor lateral displacement (Fig. 8.31), and then recalculate the building period by the Rayleigh formula. If this new value of T was larger than that calculated in step (4) (0.45 sec), a smaller base shear coefficient (C_s) would be calculated in Step (7). However, NEHRP places an upper limit on this refinement.

Solution (b) (Base Shear, UBC)

(1) Calculate the building weight W (same as Solution (a), step 1):

$$W = 1008 \text{ kips}$$

(2) Determine the zone value. Referring to the UBC earthquake zone map (Fig. 8.27), the building is located in Zone 3.

(3) Determine the Z value from Table 8.6:

$$Z = 0.30$$

(4) Determine the importance factor. The building is classified as standard occupancy: $I = 1.0$.

(5) Determine the site coefficient from Table 8.8. The soil profile is S_2. Therefore $S = 1.2$.

(6) Calculate the fundamental period of vibration for the building:

$$T = C_t (h_n)^{3/4}$$

$$= 0.035(30)^{3/4} = 0.45 \text{ sec}$$

(both directions)

(7) Calculate the base-shear coefficient C, and check it against the maximum value of 2.75:

$$C = \frac{1.25S}{T^{2/3}}$$

$$C = \frac{1.25(1.2)}{(0.45)^{2/3}} = 2.55 < 2.75$$

(8) The building structure has been described as a steel ordinary moment-resisting frame (OMRF). Therefore, from Table 8.9, $R_w = 6$.

(9) Calculate the base shear:

$$V = \left[\frac{ZIC}{R_w} \right] W$$

$$V = \left[\frac{0.3(1.0)2.55}{6} \right] 1008 = 129 \text{ kips}$$

The 129-kip base shear is valid for both the transverse and longitudinal directions of the building. This will not always be the case, since T, R_w, and C could be different for each axis.

Example 2/One-Story Building

A single-story, light industrial building is shown in Fig. 8.37. The interior structure is composed of steel purlins and steel frames with flexible connections (non-moment-resisting). The exterior walls are symmetrical for each axis and are constructed of reinforced masonry that is 10 in. thick and weights 62 psf of wall surface. The masonry walls are bearing type for the gravity loads and shear type for the lateral loads. The building location, site information, and roof construction are the same as given in Example 1.

Calculate the base shear, roof diaphragm loads, and design loads for the shear walls. Follow the specifications of either NEHRP or the 1991 UBC (they will give the same results).

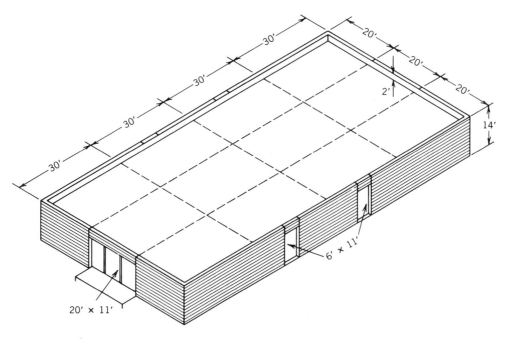

Figure 8.37/Example 2.

Solution

(1) Calculate the roof weight:

$$W_R = 144 \text{ kips}$$

See Example 1, Solution (a) step (1)).

(2) Calculate the weight of the masonry walls. Make separate calculations for long side walls and end walls:

Side walls, $2[14(120) - 2(6)11]0.062 = 192$ kips.

End walls, $2[14(60) - 20(11)]0.062 = 77$ kips.

(3) Determine the effect of the site location in Salt Lake City. From Figs. 8.27, 8.28, and 8.29, $A_a = 0.20$, $A_v = 0.02$, zone 3, $Z = 0.3$.

(4) Calculate the fundamental period of vibration:

$$T = C_t(h_n)^{3/4}$$

$$T = 0.20(14)^{3/4} = 0.14 \text{ sec}$$

$$\text{(both directions)}$$

(5) The importance factor for the UBC is 1.0.

(6) Determine the site coefficient from Table 8.8 with a soil profile of S_2: $S = 1.2$.

(7) Determine the coefficient for the structure from Table 8.9. This building is a bearing-wall type with masonry shear walls. Thus for NEHRP, $R = 3.5$; for UBC, $R_w = 6$.

(8) Calculate the seismic factor for the UBC:

$$C = \frac{1.25S}{T^{2/3}}$$

$$C = \frac{1.25(1.2)}{(0.14)^{2/3}} = 5.56 > 2.75$$

$$\frac{ZIC}{R_w} = \frac{0.3(1.0)2.75}{6} = 0.14$$

(9) Calculate the seismic factor for

NEHRP:

$$C_s = \frac{1.25A_v S}{RT^{2/3}}$$

$$C_s = \frac{1.25(0.20)(1.2)}{3.5(0.14)^{2/3}} = 0.32$$

$$\max C_s = \frac{2.5A_a}{R} = \frac{2.5(0.20)}{3.5} = 0.14$$

(10) The seismic factor of 0.14 is the same for both codes and for both axes.

(11) Calculate the base shear. The total weights of the roof and masonry walls are

$$W = 144 + 192 + 77 = 413 \text{ kips}$$
$$V = 0.14(413) = 58 \text{ kips}$$

This step is not necessary and consequently seldom executed for one-story buildings having heavy shear walls.

(12) Calculate the design load on the roof diaphragm. This step requires an investigation of the action and function of the masonry shear walls. Walls function as shear walls only when the loads are in a direction parallel to the walls. Loads perpendicular to the walls tend to overturn them; consequently, these walls require a supporting lateral structure. Figure 8.38 illustrates two methods of providing laterally supporting structures for walls. In each case, the earthquake vibration under consideration is perpendicular to the plane of the wall. This creates (in effect) the lateral wall load shown. Figure 8.38a shows the wall spanning between the top of the foundation and the eaves of the diaphragm. The diaphragm spans between the two end shear walls; therefore, the weight of the top half of the wall is assumed to be in the plane of the diaphragm, and the bottom half, applied directly to the foundation.

Figure 8.38b shows the wall spanning between two vertical elements, which in this case are columns. These columns could be pilasters, cross walls, or buttresses. In the case of columns or pilasters, however, they must be attached to both the foundation

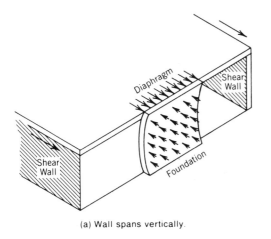

(a) Wall spans vertically.

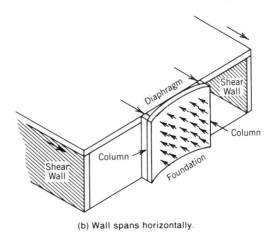

(b) Wall spans horizontally.

Figure 8.38/Lateral structure for walls.

and the diaphragm to transfer the wall load. Cross walls or buttresses could develop the lateral resistance without the aid of the diaphragm.

In this example, the lengths of the side walls are not the same as those of the end walls, making it necessary to perform a separate investigation for each building axis. For the transverse direction, the parapet depth plus one-half the weight of the two side walls is assigned to the diaphragm level:

$$W_T = 144 + \left(\frac{2 + 12/2}{14}\right)192 = 254 \text{ kips}$$

For the longitudinal direction, the parapet depth plus one-half the depth of the two end walls is assigned to the diaphragm:

$$W_L = 144 + \left(\frac{2 + 12/2}{14}\right)77 = 188 \text{ kips}$$

The lateral load on the roof diaphragm is determined by multiplying the seismic factors by the associated weights in each direction. The symbol F_Q is given to this total roof lateral load, and

$$F_{QT} = 0.14(254) = 36 \text{ kips}$$

and

$$F_{QL} = 0.14(188) = 26 \text{ kips}$$

For diaphragm design purposes, the total lateral load is assumed to be uniformly distributed over the length of the diaphragm; therefore,

$$w_T = \frac{36,000}{120} = 300 \text{ plf}$$

and

$$w_L = \frac{26,000}{60} = 433 \text{ plf}$$

The roof diaphragm is designed for strength and flexibility between the shear walls using these loads according to the procedures described in Art. 8.3.

(14) Calculate the design loads for the shear walls. Two loads for each direction are calculated, one at the top and one at the bottom of the walls.

Transverse direction: For $F_{QT} = 36$ kips, each shear end wall develops one-half the load at the top, or

$$R = 36/2 = 18 \text{ kips}$$

The shear at the bottom must include an allowance for the weight of the end walls;

therefore, use the seismic factor that was calculated for the entire building. Then, the shear load at the bottom is

$$V = 18 + 0.14\left(\frac{77}{2}\right) = 23.4 \text{ kips}$$

Each end wall has two segments of the same dimensions and stiffness; therefore, each segment is designed to develop one-half the shear load, or 11.7 kips.

Longitudinal direction: For $F_{QL} = 26$ kips, each shear side wall develops one-half the load at the top, or

$$R = 26/2 = 13 \text{ kips}$$

The shear at the bottom must include an allowance for the weight of the side walls, using the seismic factor for the building, i.e.,

$$V = 13 + 0.14\left(\frac{192}{2}\right) = 26.4 \text{ kips}$$

Each side wall consists of three segments that are of different lengths, and therefore of different stiffnesses. The total shear load (26.4 kips) is developed by each wall segment in proportion to its relative stiffness. The formula for calculating relative stiffnesses for piers of constant thickness and material, and fixed at the top and bottom, is

$$K_{rel} = \frac{1}{0.1(h/d)^3 + 0.3(h/d)}$$

where:

h = height of the pier.
d = width of the pier.

The middle segment of one side wall has dimensions 11 ft (high)×24 ft (wide), so

$$\frac{h}{d} = \frac{11}{24} = 0.46.$$

Therefore,

$$K_{rel} = \frac{1}{0.1(0.46)^3 + 0.3(0.46)} = 6.77$$

Each end segment of one side wall has dimensions 11 ft (high)×42 ft (wide), so

$$\frac{h}{d} = \frac{11}{42} = 0.26$$

Therefore,

$$K_{rel} = \frac{1}{0.1(0.26)^3 + 0.3(0.26)} = 12.54$$

The sum of the stiffnesses is

$$\sum K_{rel} = 6.77 + 2(12.54) = 31.85$$

Proportion the total shear wall load of 26.4 kips as follows:

Middle segment,

$$V = \frac{6.77}{31.85}(26.4) = 5.6 \text{ kips}$$

Each end segment,

$$V = \frac{12.54}{31.85}(26.4) = 10.4 \text{ kips}$$

Checking,

$$2(10.4) + 5.6 = 26.4 \text{ kips}$$

PROBLEMS

1. Redo Example 1 (Fig. 8.34) with both of the following changes:

(a) Increase the height to five equal stories.

(b) Use special moment-resisting connections.

(Answers given in Appendix G.)

2. Redo Example 2 (Fig. 8.35) for UBC with all of the following changes:

(a) Increase the height from 14 ft to 16 ft.

(b) Change the exterior walls to regular stone concrete having a weight of 150 pcf.

(c) Change the side-wall openings from a 6-ft to a 10-ft width.

(Answers given in Appendix G.)

8.12
Force Distribution—Single Story

The base shear, established by the procedures described in the preceding article, is the total lateral force that the structure must be designed to resist. This base shear (for each direction) is proportioned to each horizontal and vertical resisting moment that is so positioned in the structure that it is able to develop some of the load. For one-story buildings, the load distribution is determined by using the ordinary principles of structural mechanics associated with static loads. The variables to consider are (1) type of horizontal element, (2) direction of horizontal load, (3) position of vertical elements, and (4) relative rigidities of the vertical elements.

Example 2 in the preceding article will serve as a good introduction to this topic. It was shown how the diaphragm developed nearly all of the base shear (only the lower part of the walls being omitted). The diaphragm delivered its load equally to both vertical elements positioned in the exterior walls. Also, these shear walls picked up additional lateral load due to their own weight and transmitted it all to the foundation.

Vertical elements need not always occur in the exterior walls of a building. For example, refer to the plan of the one-story building shown in Fig. 8.39a where the

lateral load (F_Q) is applied in the transverse direction only; vertical elements occur along lines AB and CD. There are two segments of the vertical element along line AB. In order to illustrate the function of the diaphragm, let it be assumed that the lateral load (F_Q) is 100 kips, occurring at the center of the roof, which implies that F_Q results from a uniform load along the length of the building.

A *rigid* diaphragm roof deck (Art. 8.3) would deflect as shown in Fig. 8.39b. This is similar to a continuous beam. The force in the vertical element along line CD can be determined by taking moments of the forces about any point on line AB and applying the condition of equilibrium, i.e., $\Sigma M = 0$, or

$$15(100) - 50V_{CD} = 0$$
$$V_{CD} = 30 \text{ kips}$$

Similarly, taking moments about a point along line CD,

$$35(100) - 50V_{AD} = 0$$
$$V_{AD} = 70 \text{ kips}$$

The vertical element along line CD occurs for only 15 ft of the 35-ft width of the building. It must be designed to develop the full 30-kip diaphragm load plus any additional lateral loads. The vertical element along line AB consists of two segments, each having different dimensions.

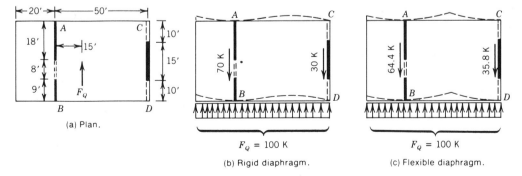

(a) Plan.

(b) Rigid diaphragm. $F_Q = 100$ K

(c) Flexible diaphragm. $F_Q = 100$ K

Figure 8.39/Plan distribution.

The combined load the two segments must be designed to develop is 70 kips.

The roof deck could be designed and constructed as a *flexible* diaphragm (Art. 8.3). Under these conditions it would deflect as shown in Fig. 8.39c. The loads developed by vertical elements will be according to tributary areas for each element. Converting the 100 kips to a uniform load,

$$w = \frac{100}{70} = 1.43 \text{ kips per ft}$$

and the load for each vertical element is

$$V_{AB} = 1.43(20 + 25) = 64.4 \text{ kips}$$

$$V_{CD} = 1.43(25) = 35.8 \text{ kips}$$

These values are different from those developed using a rigid diaphragm. The vertical element along line *CD* can be a shear wall, a rigid frame, or a cross-braced frame, and must be designed to resist either the 35.8-kip or the 30-kip load. Figure 8.40a shows this element as a cross-braced frame, and the load from a rigid diaphragm. Only the tension arm of the cross-bracing is considered effective. The noneffective compression brace is shown as a dashed line. Joints are assumed pinned. The horizontal strut along the roof lines is designed for a compression force of 30 kips. The angle the tension brace makes with the horizontal is given by

$$\tan \alpha = \tfrac{9}{15}$$

$$\alpha = 31°$$

Using the principle that the sum of the forces at a joint must be equal to zero, it is determined that the tension force in the brace is

$$T = \frac{30}{\cos 31°} = 35 \text{ kips}$$

and the compression force in the column is

$$C = 30 \tan 31° \simeq 18 \text{ kips}$$

The vertical element along line *AB* that consists of two segments may be braced frames (similar to line *CD*), rigid frames, or shear walls. Each segment contributes to the development of the total force. Cross-braced frames are shown in Fig. 8.40b for the rigid-diaphragm load of 70 kips. The compression diagonals are assumed to be noneffective and are shown as broken lines. Observe that the lateral deflection of each segment will be the same. Consequently, the proportion of the total load developed by each segment will depend upon the relative rigidity of each. This observation remains valid (within the elastic range of the material) whether the element consists of braced frames, rigid frames, or shear walls.

As used here, the term *ridigity* refers to the resistance against deflection for a given load. The three types of vertical elements are shown in Fig. 8.41. The variables for each type are the height (*h*) and the length (*L*). Formulas are given for each that relate the deflection (Δ) caused by the load (*P*) that is resisted by the rigidity of the system (*R*). The rigidity of each system is

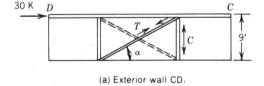

(a) Exterior wall CD.

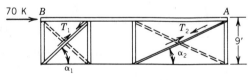

(b) Interior wall AB.

Figure 8.40/Vertical elements.

shown as the quantity within the brackets. The rigidity is sometimes referred to as the stiffness of the system. Inasmuch as these types are seldom mixed for any one line of vertical elements, the absolute formulas for R (shown in brackets) can be simplified to produce only relative values of rigidity for any one type. This is the procedure usually used in design.

For example, for a series of cross-braced frames similar to that shown in Fig. 8.41a, having constant cross-sectional areas and moduli of elasticity for each corresponding member, the relative rigidity formula reduces to

$$R_B = \frac{1}{L\left[1 + r^3 + \left(1 + r^2\right)^{3/2}\right]}$$

where r is h/L. Applying this formula to the two segments of braced frames shown in Fig. 8.40b will give their relative rigidities as follows:

Segment at B, $h/L = 9/9 = 1$:

$$R_B = \frac{1}{9\left[1 + 1^3 + \left(1 + 1^2\right)^{3/2}\right]} = 0.023$$

Segment at A, $h/L = 9/18 = 0.5$:

$$R_B = \frac{1}{18\left[1 + 0.5^3 + \left(1 + 0.5^2\right)^{3/2}\right]} = 0.022$$

Observe that the rigidities are very nearly the same in spite of the fact that one segment is twice the length of the other. This is typical.

Using these values of relative rigidity, the total 70-kip load can be proportioned to each segment as follows:

Segment at B,

$$\frac{R_B}{\Sigma R_B}(70) = \frac{0.023}{0.045}(70) = 35.8 \text{ kips}$$

Segment at A,

$$\frac{R_B}{\Sigma R_B}(70) = \frac{0.022}{0.045}(70) = 34.2 \text{ kips}$$

The procedure described here can be used for rigid frames and shear walls as well as for braced frames. However, in cases where the braced frame or rigid frame is made of steel, that is seldom done. This is because steel is a ductile material and earthquake loads occur over a very short period of time. Steel does not fail suddenly; instead, it yields (still developing its yield load) and permits other structural elements to share in developing the load. This is basically an ultimate-strength-design approach (Chapter 11).

Accepting this yield (some permanent distortion will occur), all that is necessary is that the combined strength capacity of the segments be equal to the total load. Applying this principle to the braced frames shown in Fig. 8.36b

$$T_1 \cos \alpha_1 + T_2 \cos \alpha_2 = 70 \text{ kips}$$

This type of reasoning for ductile steel cannot be applied to structures made from brittle materials. Once the strength capacity of masonry or concrete is reached, it fails suddenly and has essentially no remaining strength. If more than one segment is relied on to develop loads and one segment fails, the remaining segments likely will also fail. This phenomenon is known as progressive failure. Consequently, the procedure described for proportioning loads to segments according to their relative rigidities must always be followed for wood, masonry, and concrete shear walls.

Refer to the formula for the absolute rigidity of shear walls, shown in Fig. 8.41c. If the wall thickness remains constant and there is a consistent modulus of elasticity, the formula can be adjusted to represent

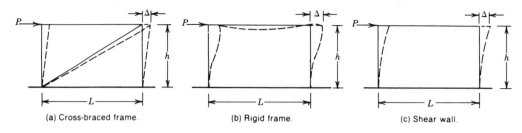

(a) Cross-braced frame.

$$P = R_B \Delta$$

$$P = \left[\frac{E}{\sum \dfrac{SUL_1}{A}} \right] \Delta$$

where:

S = stress in each member caused by the load P.

U = stress in each member caused by unit load at P.

L_1 = Length of each member.

A = Area of each member.

(b) Rigid frame.

$$P = R_F \Delta$$

$$P = \left[\frac{17EI_c}{h^3} \right] \Delta$$

where:

I_c = Moment of inertia of one column

Assumptions:

I is same for both columns

I/L of the beam remains equal to I/h of columns.

(c) Shear wall.

$$P = R_S \Delta$$

$$P = \left[\frac{E}{h \left(\dfrac{h^2}{3I} + \dfrac{3}{A} \right)} \right] \Delta$$

where:

t = thickness of wall

$$I = \frac{tL^3}{12}$$

$$A = tL$$

Assumption:

The shear modulus is 40% of elastic modulus.

Figure 8.41/Rigidities of vertical elements.

relative values as follows:[4]

$$R_S = \frac{1}{(h/L)^3 + 0.75(h/L)}$$

In order to illustrate the significant dif-f rence between the frames and shear walls, let it be assumed that the two segments of the frame shown in Fig. 8.37 are replaced by two segments of shear wall. The segment of B has an $h/L = 1$ and a relative rigidity

$$R_S = \frac{1}{1^3 + 0.75(1.0)} = 0.571$$

The segment at A has an $h/L = 0.5$ and a relative rigidity

$$R_S = \frac{1}{0.5^3 + 0.75(0.5)} = 2.0$$

[4]This formula is slightly different from that used for the shear walls in Example 2 in Art 8.10. The reason for this difference is that in this example the top of the wall is assumed free (unrestrained), while in Example 2 it is assumed that the top is fixed against rotation.

Observe that the 9×18-ft segment at A is $2.0/0.571 = 3.5$ times more rigid than the 9×9-ft segment at B. This is an abrupt turnabout from the result for cross-braced frames.

Finally, the proportioning of the 70 kips to each wall segment is determined as follows:

Segment at A,

$$\frac{0.571}{2.571}(70) = 15.5 \text{ kips}$$

Segment at B,

$$\frac{2.0}{2.571}(70) = 54.5 \text{ kips}$$

8.13 Force Distribution—Multistory Buildings

The load distribution for multistory buildings is quite different from that for single-story buildings. The reason for this difference is that the load distribution is based upon structural dynamics.

The load distribution of the base shear for multistory buildings is based upon the structural response of the building due to ground vibrations. For a dynamic analysis to be performed, the structure must be modeled into a form appropriate for application of dynamic theory. A most frequently used model consists of *lumped masses* at each floor level and at the roof, with elastic connections between each pair of adjacent masses. The elastic connection represents the structure of the vertical element between the floors.

A partial section through a multistory building is shown in Fig. 8.42a. The lumped mass consists of the weight of the floor and one-story height of walls, partitions, columns, and, in some cases, part of a sustained live load. The R for the structure shown in Fig. 8.42b represents the vertical elements, such as rigid or braced frames or shear walls. All weights, properties, and dimensions must be known (or assumed) before the modeling can be completed.

The ground vibration sets these lumped masses in motion, and the main purpose of the dynamic analysis is to describe their relative motion. This must be done with respect to time and, for practical reasons, will almost always involve the use of computers. The output of the computer pro-

gram is the time-history response of the distorted structure.

Structures vibrate according to their characteristic modes of vibration. A mode is described as a basic displacement shape of the distorted structure. A model of a three-story structure is shown in Fig. 8.43a. Three basic modes of vibration are shown in (b), (c), and (d). The first mode is the dominant one and is referred to as the fundamental mode. This mode also possesses the longest period of vibration of the three. The actual deflected shape at any one time is the summation of a certain percentage of each mode. However, the most severe displacement at any one time most closely resembles the shape of the fundamental mode. Consequently, codes usually require that the equivalent static lateral loads should be applied at each floor level in such a manner as to cause a deflected shape similar to the shape of the fundamental mode. A load distribution according to this rule generates a shear envelop that is similar in shape to that caused by the superposition of the forces associated with each mode.

This load distribution is commonly referred to as the inverse triangle loading. The forces vary linearly from zero at the bottom to a maximum at the top. However, it must be emphasized that this exact triangular loading occurs only where there is a uniform mass distribution. To explain

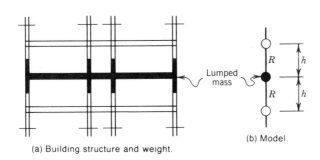

(a) Building structure and weight.

(b) Model.

Figure 8.42/Model for dynamic analysis.

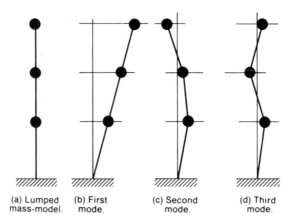

(a) Lumped mass-model. (b) First mode.   (c) Second mode.   (d) Third mode.

Figure 8.43/Mode shapes.

further, refer to Fig. 8.44a where the building section represents equal weights (W) and equal floor heights (h) for each floor. The theoretical model is shown in Fig. 8.44b, where all lumped masses are the same and the rigidity between the masses (R) is the same. The fundamental mode is shown in Fig. 8.44c, with a base shear of V. This base shear is distributed in a uniformly varying pattern, beginning with $\frac{1}{10}$ of V at the second floor and ending with $\frac{4}{10}$ of V at the roof level. Figure 8.44c shows the shear-force envelope, be-

ginning with the $\frac{4}{10}$-V shear force at the top floor and successively adding the story shear for each floor below.

Generally, building codes adopt these principles but modify the load distribution formulas to allow for tall buildings and for variations in story weights and story heights. Pure cantilever deflections, and buildings having large height-to-width ratios (aspect ratios), have a fundamental mode that departs slightly from a straight line. The two codes, UBC and NEHRP, have different procedures to account for

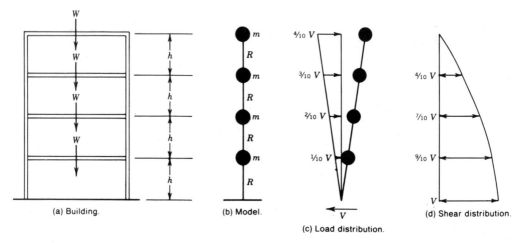

(a) Building.   (b) Model.   (c) Load distribution.   (d) Shear distribution.

Figure 8.44/Equal mass distribution.

this small variation. Each procedure will be explained here, and each will be used in the example problem which follows.

The UBC accounts for this variation by requiring a portion of the base shear to be applied at the top under certain conditions. This need only be applied in cases where the period of vibration is larger than 0.7 sec. The value of this additional top force (F_t) is $0.07TV$, but does not have to exceed $0.25V$. The remaining portion of the base shear $(V - F_t)$ is distributed to each floor and roof level according to the inverse triangle loading if the story heights and story weights remain constant. Usually this is not the case. The following formula is valid for either uniform masses or a variation in weights and heights:

$$F_x = \frac{(V - F_t)w_x h_x}{\sum_{i=1}^{n} w_i h_i}$$

This formula requires further explanation. The subscript x refers to the level for which the lateral force is calculated, i.e., F_x. The value of w_x is the gravity load at level x, and h_x is the height above the base of level x. The denominator in the formula is the sum of all the floor weights times their respective heights above the base. It is seen that the force at any level is the modified base shear $(V - F_t)$ multiplied by a dimensionless fraction calculated as $w_x h_x / \sum w_i x_i$. A summary of the load distribution to the vertical elements for a four-story building is shown in Fig. 8.45.

The NEHRP procedure is to modify the force at each level rather than applying an additional force at the top. This need be done only if the fundamental period of vibration of the building is larger than 0.5 sec. An exponent k for the height level h_k is inserted in the force level F_x formula as follows:

$$F_x = \left[\frac{w_x h_x^k}{\sum_{i=1}^{n} w_i h_i^k} \right] V$$

The exponent k is 1.0 for buildings with periods equal to or less than 0.5 sec, and 2.0 for buildings with periods equal to or larger than 2.5 sec. For buildings with periods between 0.5 and 2.5 sec, k is determined by linear interpolation between 1 and 2.

When making this adjustment for NEHRP procedures, the force F_x can be used for

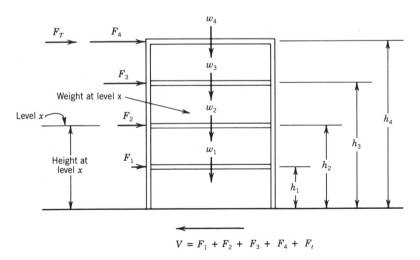

$$V = F_1 + F_2 + F_3 + F_4 + F_t$$

Figure 8.45/Load distribution—vertical elements.

the design of the diaphragm. When using the UBC procedure, the F_x for the vertical element may have to undergo further adjustment to establish a realistic load for the diaphragm. In the UBC method the load distribution to the vertical elements cannot be used for the design of the horizontal elements, because the distribution was established in a pattern that required duplication of the deflection of the fundamental mode of vibration. In doing so, larger forces were assigned to the top floors and much smaller forces to the bottom floors. These forces at the lower floors are unrealistically low in view of the dynamic movement taking place. The loading on any diaphragm (horizontal element) at any level should represent the acceleration acting at that level on the mass at that level. To accomplish this, the 1991 UBC provides an empirical formula for calculating

minimum design loads in diaphragms. The results do not correlate well with the base shear and so constitute a discrepancy that requires explanation.

It must be kept in mind that these earthquake loads are unreal. No lateral loads are actually imposed on the structure during an earthquake; there is just the distortion due to the ground vibrations, and the loads are selected for each element that would duplicate the maximum distortion in that element.

The diaphragm loading formula is

$$F_{p_x} = \frac{F_t + \sum_{i=x}^{n} F_i}{\sum_{i=x}^{n} w_i} w_{px}$$

where:

F_{px} = the diaphragm load, but does not have to be greater than $0.3ZIw_{px}$, nor can it be less than $0.14ZIw_{px}$.

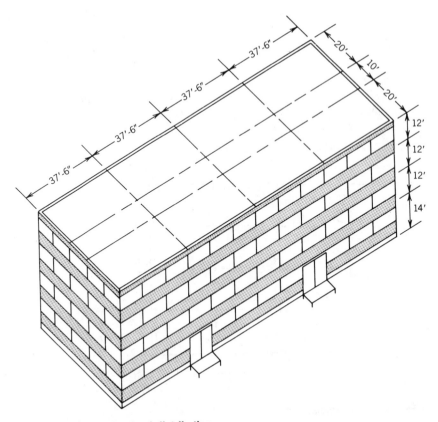

Figure 8.46/Example—load distribution.

w_{px} = the gravity load imposed on the diaphragm at level x.

$\sum_{i=x}^{n} F_i$ = the sum of all the lateral loads at level x and above. (These are the lateral loads—inverse triangle—at each level that are assumed to act on the vertical elements.)

$\sum_{i=x}^{n} w_i$ = the sum of all the gravity loads at level x and above.

Example/Load Distribution

A four-story, steel-framed building is shown in Fig. 8.46. The steel frames conform to the grid lines and have ductile moment-resisting connections, making them special frames. Each floor deck as well as the roof deck is a rigid diaphragm. Building weights are as follows:

Roof construction, including ceilings	48 psf
Floor construction, including ceilings	68 psf
Interior partition weight (per floor area)	12 psf
Exterior wall weight (average per wall area)	20 psf

In order to make this example applicable under the provisions of both NEHRP and the UBC, the seismic factor will be assumed to be 0.084.

Calculate the lateral loads per frame in the transverse direction only (both directions actually would require investigation). Also, calculate the diaphragm design loads for the transverse direction according to the UBC.

Solution

(1) Determine all the gravity loads contributing to the base shear and itemize them according to the level to which they

are assigned:

Roof construction	50(150)0.048 =	360 kips
Floor construction	50(150)0.068 =	510
Floor partitions	50(150)0.012 =	90
Typical walls	12(400)0.020 =	96
Roof walls	6(400)0.020 =	48
Second-floor walls	13(400)0.020 =	104

Level 4 (roof), $w = 360 + 48$		= 408 kips
Level 3, $w_3 = 510 + 90 + 96$		= 696
Level 2, $w_2 = 510 + 90 + 96$		= 696
Level 1, $w_1 = 510 + 90 + 104$		= 704
		$W = \overline{2504}$ kips

(2) Calculate the total base shear:

$$V = 0.084(2504) = 210 \text{ kips}$$

(3) Calculate the period of vibration:

$$T = 0.035(50) = 0.66 \text{ sec}$$

Since $0.66 < 0.7$ sec, the UBC procedure has $F_t = 0$. Since $0.66 > 0.5$, the NEHRP requires that the exponent k be calculated by interpolation:

$$k = \frac{0.66 - 0.5}{2.0} = 1.08$$

(4) The UBC procedure is described below, followed by changes required when using NEHRP procedures. Calculate the portion of $V - F_t = 210$ kips that is applied at each level. This step is accomplished by completing Table 8.10 as follows:

(a) Place the gravity weights at each level determined in step (1) in column (2).

(b) Place the height of each level above the base in column (3).

(c) Determine the product of columns (2) and (3) at each level, and add the total at the bottom of column (4).

(d) Divide each value in column (4) by the total of column (4), observing that these fractions should total 1.0.

(e) Determine the value of F_x at each level by multiplying the value in column (5) by $V - F_t$, and enter the

Table 8.10
Distribution of Base Shear

(1)	(2)	(3)	(4)	(5)	(6)	(4a)	(5a)
Level	w_x (kips)	h_x (ft)	$w_x h_x$ (kip-ft)	$\dfrac{w_x h_x}{\Sigma w_i h_i}$	$F_x = \dfrac{(V - F_t)w_x h_x}{\Sigma w_i h_i}$	$w_x h_x^k$ (kip-ft)	$\dfrac{w_x h_x^k}{\Sigma w_i h_i^k}$
4 (roof)	408	50	20,400	0.28	58.8	27,896	0.28
3	696	38	26,448	0.35	73.5	35,381	0.36
2	696	26	18,096	0.24	50.4	23,485	0.24
1	704	14	9,856	0.13	27.3	12,173	0.12
Σ			74,800	1.00	210.0	98,935	1.00

results in column (6). When using the NEHRP procedure, $F_t = 0$, and columns (4) and (5) are modified as shown in columns (4a) and (5a) (Table 8.10) by including the exponent k. Although there are large changes in column (4a), the percentages in column (5a) remain essentially the same. This is because the period of 0.66 sec is very close to 0.5 sec. The values in column (5) will be used to complete this example.

(5) Convert the lateral load F_x and the gravity load w_x at each level to their respective values per frame. Since the diaphragms are rigid and there are five

frames of equal rigidity, each value is divided by five. Show these frame values on a sketch, and determine the shear envelope (Fig. 8.47).

Many designers would feel that the investigation at this stage constitutes only a preliminary analysis. Whether this is so depends in part on the accuracy of the values used in the analysis, in particular, the value of the seismic factor. Perhaps a more refined and reduced seismic factor could be justified. In order to determine if such a reduction is possible, it is necessary to establish the size of the beams and columns in the frame and calculate the lateral displacement of each floor due to the lateral

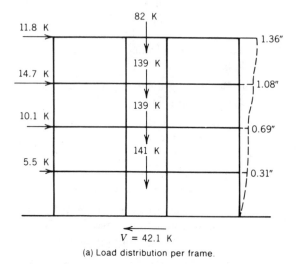

(a) Load distribution per frame.

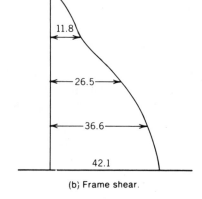

(b) Frame shear.

Figure 8.47/Example—load distribution.

load as currently calculated. This step could be accomplished manually by elastic structural theory or through the use of computer program such as is furnished and described in Appendix E. To demonstrate this refinement, a computer run was made, and the resulting story displacements (relative to the base) are noted in Fig. 8.47a.

From the data shown in Fig. 8.47a, the fundamental period of vibration (T) is calculated by using the Rayleigh formula (Art. 8.10), i.e.,

$$T = 2\pi \sqrt{\sum_{i=1}^{n} \left[w_i \delta_i^2 \right] \div g \sum_{i=1}^{n} \left[f_i \delta_i \right]}$$

$$\sum w_i \delta_i = 1.41(0.31)^2 + 139(0.69)^2$$

$$+ 139(1.08)^2 + 82(1.36)^2 \simeq 380$$

$$\sum f_i \delta_i = 5.5(0.31) + 10.1(0.69)$$

$$+ 14.7(1.08) + 11.8(1.36) \simeq 41$$

$$T = 2\pi \sqrt{380/384(41)} = 0.98 \text{ sec}$$

Since $0.98 > 0.66$ sec, a refinement could be made in the base shear. It is necessary to check the code upper limits for T and recalculate the seismic factor.

(6) Using the values calculated in step (4), determine the design load for the diaphragm at each level. This step is accomplished by completing Table 8.11 as fol-

lows:

(a) Place the frame lateral load at each level determined in step (4) in column (2).

(b) Place the progressive summation of the values in column (2) in column (3).

(c) The values in column (4) are the gravity weights at each level, as determined in step (1).

(d) Place the progressive summation of the values in column (4) in column (5).

(e) The value at each level in column (3) is divided by the corresponding value in column (5) and placed in column (6).

(f) A check is required on maximum and minimum diaphragm factors. Assuming Zone 3 ($Z = \frac{3}{4}$) and a nonessential building ($I = 1.0$), the diaphragm factors are

Max $= 0.30ZI = 0.3(0.75)1 = 0.225$

Min $= 0.14ZI = 0.14(0.75)1 = 0.105$

Since the maximum is larger than the values in column (6), it is not used. However, the minimum exceeds those at levels 1 and 2; therefore, it is placed in column (7).

(g) Multiply the appropriate value from column (6) or (7) by the corresponding level diaphragm weight in column (4), and place in column (8).

The diaphragm design load F_{px} as shown in Table 8.11 is applied to the entire length

Table 8.11
Diaphragm Load

(1)	(2)	(3)	(4)	(5)	(6) $\dfrac{F_t + \Sigma F_i}{\Sigma w_i}$	(7) Min. Max.	(8)
Level	F_i	ΣF_i	w_i	Σw_i			F_{px}
4 (roof)	58.8	58.8	408	408	0.144	—	58.8
3	73.5	132.3	696	1104	0.120	—	83.5
2	50.4	182.7	696	1800	0.102	0.105	73.1
1	27.3	210.0	704	2504	0.084	0.105	73.9

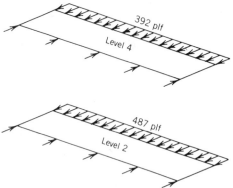

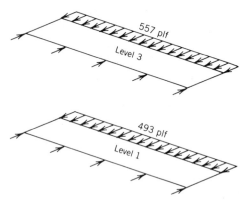

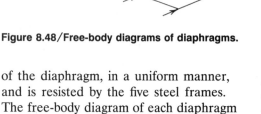

Figure 8.48/Free-body diagrams of diaphragms.

of the diaphragm, in a uniform manner, and is resisted by the five steel frames. The free-body diagram of each diaphragm is shown in Fig. 8.48.

Complete design for earthquake effects and/or wind loads requires more investigation than is presented in this chapter. Components or individual parts of a building frequently have a local loading requirement which is different from that utilized when considering the building as a whole. The emphasis in this chapter has been on whole buildings and main structural systems rather than on their individual parts.

PROBLEMS

The building shown in Fig. 8.46 is assumed to be a four-story office building having *all* floor-to-floor heights of 11 ft. Other dimensions, weights, and properties remain the same as described in the example. Rigid diaphragms are used throughout. However, the main steel frames consist of standard moment-resisting connections. Calculate loadings in the transverse direction only for:

1. A building site that has very loose and deep soil located in Charleston, South Carolina. Use the provisions of NEHRP.

2. A building site to be located in Los Angeles, California, for which there are no geotechnical data. Use the provisions of the 1991 UBC.

9

LOW-RISE AND INDUSTRIAL-TYPE BUILDINGS

9.1
Introduction

Principal attention in this chapter will be given to analysis and design of roof trusses and braced frames using statically determinate methods, and to the use of these elements in low-rise and industrial-type buildings.

Statically indeterminate frame analysis will be treated separately in Chapter 10 (conventional method), Chapter 11 (plastic design), and Chapter 12 (computer method).

9.2
Loads on External Building Elements—General

The general subject of loads on buildings and building elements was introduced in Chapter 1. It was pointed out that, although there is general agreement on the types of load to be considered in building design, there is much less agreement on how to classify, ascertain, specify, and apply these loads. In general, however, the loads to be considered are as follows:

Dead loads. The weight of permanent (fixed) materials of construction, including the structural elements themselves, and equipment.

Live loads. Occupancy loads (i.e., people, vehicular and other moving equipment, furniture, stored fluids and materials, movable equipment and partitions); loads incidental to construction, maintenance, and repair; and impact loads.

Environment-related loads. Wind; earthquake; snow; rain, including forces and effects due to ponding of water; weight and pressure of soil and soil water; and self-straining forces and effects (expansion and contraction) resulting from temperature and moisture changes, creep, and movement due to differential settlement.

Not all of these loads will necessarily apply to any given building, nor will they all be additive, and certainly not all will be applied externally to building element, i.e., to roofs and exterior walls.

In the case of roofs, the dead load carried by the structural members supporting the roof will consist principally of the weight of the roof deck and its covering, and the weight of the structural members themselves.[1] The dead load on exterior load-bearing walls or on load-bearing piers integral with walls will be the load brought to them by the roof-supporting structural members plus their own weight. The dead load on columns to which a weather envelope system is attached will be the load brought to them by the roof-supporting structural members, the weight of the columns, and the attached weather-envelope elements. Where the roof-supporting structural element is a truss or the trussed element of a braced frame, the roof dead load generally is carried to the trusses by rafters and/or purlins, which in turn add their own weight to the dead load.

Live loads on external building elements will depend to a great extent on the intended function and design of the given building. A flat roof, for example, might be called upon to support a number of loads associated with occupancy—if, for instance, it also is intended to serve as a terrace or for parking. All roofs, whether flat, pitched, or gabled, need to support loads incidental to construction, maintenance, and repair. And in the case of truss-supported or braced-frame-supported roofs where the truss or the trussed element of a braced frame is exposed on the exterior, it will be necessary to provide for the possibility that the trusses will be used to support unplanned concentrated loads, such as would result from hoisting with a block and tackle hung from the lower chord or a panel point (a minimum of a 2-kip load-carrying capability usually is specified). Live loads on exterior walls, other than unusual impact loads, are less likely; however, the possibility should be investigated.

Environment-related loads are those that derive from the geophysical setting of the building. Most act on external building elements. Wind and earthquake loads were dealt with in Chapter 8 as lateral loads, i.e., in the context of loads that either act horizontally on roofs and walls or can be converted to a horizontal equivalent and then transferred downward through vertical structural elements to the building foundation. Snow and rain also are important environmentally related factors in load determinations and will be treated in a succeeding article. If there is reason to believe that self-straining forces and effects will be significant, they, too, should be dealt with individually; however, these forces and effects do not necessarily apply only to external building elements.

9.3
Snow and Rain Loads

The proper design snow load for roofs will depend on a number of factors, geographic location being the most obvious. However, roof exposure and configuration also are important—snow tends to slide

---

[1]Trusses frequently support suspended ceilings, and at times suspended floors as well. Ceilings may be hung from the purlins or from a light framework supported by the lower chord of the trusses. In the first instance, the ceiling weight reaches the truss through the purlins resting on the top chord; in the latter, the loads are generally considered to be applied at the lower-chord panel points.

off steep and smooth roof surfaces and to accumulate on rough and relatively flat roofs.

Snow is more likely to be blown off roofs when exposed to wind, and to build to depths greater than normal ground surface accumulations where drifting can occur and where accumulated snow can slide from higher to lower roof levels or fall from adjacent natural or man-made structures or foliage. Snow is more likely to accumulate on roofs of unheated buildings and superinsulated buildings. Finally, it is important to assess the likelihood and possible extent of added load that could be created by rain on snow and by ponding of snowmelt and/or rainwater.

Freshly fallen dry snow may weigh as little as 5 lb per cu ft, whereas wet snow can weigh 12 lb per cu ft. It may be necessary to estimate the maximum cumulative depth as well as make assumptions as to the type of snow that is likely in order to establish a sound design load. Also, a sequence of fresh snow, partial melting, freezing, and further snow accumulation is not uncommon in some climates. In such cases, a conservative estimate of probable load would be prudent. In many areas, of course, the local building-code authority establishes the design load.

Maps of the ground snow load for the contiguous 48 states and a table for various locations in Alaska are given in ASCE 7-88, "Minimum Design Loads for Buildings," where it is pointed out that there are many areas where local variations make mapping impossible, while in others the values given may not be appropriate because they do not take into account all high elevations. The ASCE 7-88 procedure for determining the appropriate design snow load is to:

1. Determine the psf ground snow load for the specific geographic location and investigate local weather records and ex-

perience where there is reason to believe that the actual load could exceed the map or table value given.

2. Calculate the flat-roof snow load, using the following formulas:

$$p_f = \begin{cases} 0.7 C_e C_t I p_g & \text{for the contiguous states} \\ 0.6 C_e C_t I p_g & \text{for Alaska}^2 \end{cases}$$

where:

p_f = psf flat roof snow load.

C_e = exposure factor (to account for wind effects).

C_t = thermal factor (to account for effect of structure heating).

I = importance factor for building occupancy (to account for the relative consequences of structure failure).

p_g = psf ground snow load (as discussed above).

3. Consider the slope of the roof, if any, using the follwing formula (provided the many exceptions do not apply):

$$p_s = C_s p_f$$

where:

p_s = psf sloped roof design snow load acting on the horizontal projection of the roof surface.

C_s = slope factor (to account for warm and cold roofs, and whether or not the surface is slippery).[3]

p_f = psf flat roof snow load.

4. Consider unbalanced loads for hip and gable roofs—i.e., uniform snow load on the leeward side only of roofs with slopes between 15 and 70° —using

---

[2]Note that minimum values for the flat-roof snow load are given in ASCE 7-88.

[3]Note that in ASCE 7-88, portions of curved roofs where the slope exceeds 70° are to be considered free of snow, and that no snow-load reduction due to slope is to be applied to multiple folded-plate, sawtooth, and barrel-vault roofs.

the following formula:[3]

$$p_l = 1.5 p_s / C_e$$

5. Consider snow drifts on lower roofs of the structure and from adjacent structures and terrain features, where the ground snow load (p_g) exceeds 10 psf.

6. Consider snow drift from roof projections longer than 15 ft.

7. Consider snow sliding from higher to lower roof surfaces.

8. Consider unloaded portions, i.e., the effect of removing half the snow load from any portion of a loaded area.

9. Consider extra loads from rain on snow in areas where intense rains are likely to occur in conjunction with snow.

10. Consider effects of ponding (i.e., deflections and subsequent increase in load) from rain on snow and/or snowmelt.

It should be noted that, as stated earlier, the possibility of snow drifting against walls, and against roofs that extend to near the ground, also must be investigated.

The following example will illustrate the use of this procedure.

Example

Determine the snow load on a building having a simple pitched roof supported by the truss of Fig. 9.3 (Art. 9.9). The building is located in Springfield, Illinois, where the ground snow load (p_g) is 20 psf. There is no shelter from the wind, and the building is heated in the interior. It will have a slippery roof surface, more than 300 people in its undivided interior, and eaves that are well above the ground surface.

Solution

(1) Determine the flat-roof snow load:

$$p_f = 0.7 C_e C_t I p_g$$

where the exposure factor $C_e = 0.8$ for windy areas with the roof exposed on all sides with no shelter afforded by the terrain, higher structures, or trees; the thermal factor $C_t = 1.0$ for heated structures; and the importance factor $I = 1.1$ for a structure that can have more than 300 people in its undivided interior. Therefore

$$p_f = 0.7(0.8)1.0(1.1)20 = 12.32 \text{ psf}$$

(2) Consider the effect of roof slope,

$$p_s = C_s p_f$$

where the roof-slope factor[4] $C_s \approx 0.47$ for a slope $37°$ and $p_f = 12.32$ psf:

$$p_s = 0.47(12.32) = 5.79 \text{ psf}$$

(3) Consider an unbalanced snow load:

$$p_l = 1.5 p_s / C_e \qquad \text{for the leeward slope}$$
$$p_l = 1.5(5.79)/0.8 = 10.86 \text{ psf}$$

(4) Drifting, sliding snow and ponding need not be considered.

The design snow load, therefore, would be either 5.79 psf applied over the entire roof surface, or the unbalanced 10.86 psf applied to either roof slope (considering that wind can come from either direction), whichever produces the greatest stress.

9.4
Combining Loads

As noted earlier, there is no universal agreement on the way in which loads should be combined. In Chapter 8, it was pointed out that there is reasonably good agreement that wind and earthquake loads need not be applied simultaneously. The

---

[4]ASCE 7-88 contains graphs for determining the roof-slope factor C_s for warm and cold roofs. For unobstructed, slippery surfaces of warm roofs, C_s is 1.0 for zero slope, decreases uniformly from 1.0 to 0 at a slope of $70°$, and remains 0 for slopes greater than $70°$. Expressed in equation form, $C_s = 1.0 -$ (slope/70), which for a slope of $37°$ is 0.47.

AISC 1989 Specification includes the snow load in its definition of live loads. It also states that, in the absence of specific requirements by other codes, the ASCE 7-88 should be used for load combinations.

ASCE 7-88 stipulates that the design load is to be whichever of the following combinations produces the most unfavorable effect, recognizing that the effect may occur when one or more of the contributing loads is not acting:

D	= dead load
$D + G$	= dead load plus variable gravity load, i.e., live, snow, and rain loads
$D + (W$ or $E)$	= dead load plus wind or earthquake
$D + G + (W$ or $E)$	

It also is stated in ASCE 7-88 that the total combined loads may be multiplied by the following factor (for those loads of principal interest here):

$$D + 0.75[G + (W \text{ or } E)]$$

9.5 Roof Trusses

A complete trussed roof structure consists of a roof deck, a system of rafters and purlins that support the deck, and roof trusses that carry the purlins and span the distance between supporting walls, piers, or columns. The area of the roof between two trusses is called a bay.

Purlins are the horizontal beams which span the distance between trusses. In some instances the deck is supported directly on purlins; in others, rafters or subpurlins, perpendicular to and supported by purlins, carry the deck.

The components of a roof truss are generally a top chord, a lower chord, and a web system. Although it is perhaps most common to have an inclined top chord and a horizontal bottom chord as shown in Fig. 9.1, there are numerous variations. For

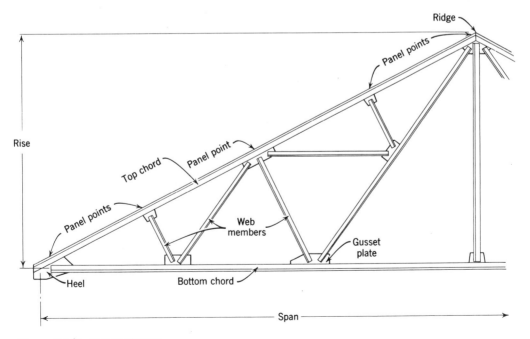

Figure 9.1/Typical roof truss.

example, the top chord can be horizontal to permit a flat roof, or the lower chord can be inclined, or *cambered*, to increase headroom under the truss. And trusses of a type similar to that shown in Fig. 9.1 can be inverted.

With the typical truss of Fig. 9.1, the top joint or peak supports the roof ridge; the end joints are called heels. The joints at which web members intersect chords are called panel points, and the distance between these joints is a panel length. The plates used to connect members are known as gusset plates. Members are generally connected to the gusset plates or to one another by either bolts or welds.

The rise or height of a truss is measured vertically from peak to horizontal lower chord. The rise of a truss with cambered lower chord is measured from peak to a horizontal line passing through supports.

The pitch is the ratio of rise to span length; e.g., a truss having a rise of 10 ft and a span of 40 ft has a pitch of $\frac{10}{40}$, or $\frac{1}{4}$. This should not be confused with the slope, which is the ratio of the rise to one-half

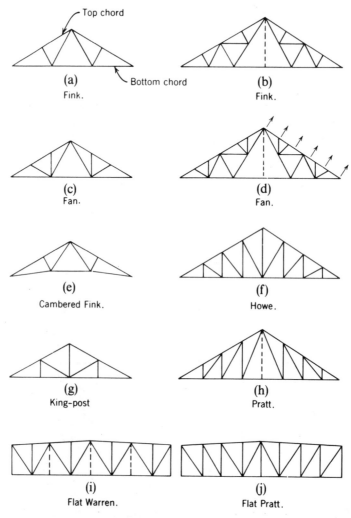

Figure 9.2/Types of roof truss.

the span. In the case mentioned above, the slope is $\frac{10}{20}$, or $\frac{1}{2}$. These two terms are often used interchangeably and thus erroneously, especially by workmen in the field. A common way of expressing the slope of a roof is to give the amount of rise per foot of run; e.g., a slope of $\frac{1}{2}$ is the same as a rise of 6 in. in 12 in. or a slope of "6 in 12." Slope and pitch are frequently determined by architectural considerations or by the capabilities of roofing materials; in the absence of any governing factor, a rise equal to $\frac{1}{4}$ the span will generally prove economical.

9.6
Types of Roof Truss

There are a great many steel truss forms used in building construction. For average spans, the Fink truss, as shown in Fig. 9.2 (a) and (b), is among the more common. To construct a Fink truss as shown in Fig. 9.2a, perpendiculars to and from the center of the top left- and right-hand chords are constructed and extended until they meet the lower chord. The remaining two web members are drawn from these intersections to the peak as shown. The number of panels in half the truss may be increased to four by subdividing each panel into two, as shown in Fig. 9.2b. This process may be repeated again, resulting in eight panels. The dashed line at the center indicates the position of a sag rod which often is added for the sake of stiffness but theoretically is under no stress.

A disadvantage of this type of truss is that the number of panels can be increased only by doubling the previous number. A modification of the Fink truss that permits greater flexibility in number of panels is the fan truss shown in Fig. 9.2 (c) and (d); here the web members do not intersect the top chord at right angles.

Figure 9.2 illustrates several other frequently used types of roof truss as well.

Those shown in Fig. 9.2 (i) and (j) are considered to be flat trusses even though their top chords are slightly pitched.

Where trusses support purlins at panel points only, a desirable top-chord panel length is about 8 ft. This would result in a maximum span of about 30 ft for the truss shown in Fig. 9.2 a, 40 ft for c, 55 ft for (b), and 75 ft for (d). These maximum spans are, of course, only approximate and may be varied to meet actual conditions.

9.7
Spacing of Trusses

Roofing systems involve so many variables that no simple rule can be given for the selection of the most economical truss spacing. It is usually desirable, however, to use bays of approximately equal size over any one portion or wing of a building so that as many trusses as possible will be identical. For spans up to about 70 ft, a spacing of 15 to 20 ft is generally satisfactory for loads approximating 30 psf. In a great many cases, the location of windows, piers, etc., will also be significant factors in the selection of a truss spacing.

9.8
Application of Loads—General

Except for the dead weight of the truss and any loads suspended from the lower chord, all loads are brought to the truss by the purlins. There is usually a purlin connection at every top-chord panel point, and not infrequently, one or more purlins are supported between panel points. In the first instance, the top chord acts as an ordinary compression member, while in the second, bending is combined with direct stress. For the present, discussion will be confined to trusses receiving loads at panel points only.

The area of roof surface supported by one purlin is equal to the panel length (distance between purlins) multiplied by the length of a bay (distance between trusses). Tables giving the weights of various building materials used for deck and roofing will be found in most handbooks (Art. 1.3).

All components of the dead load except the truss weight can be determined directly from the roof specifications and layout, and the truss weight itself can be approximated for ordinary spans and bay lengths by using a value of 2 to 5 psf of supported area. For spans of 75 ft and less, this weight will generally represent a small part of the total load to be carried; it should, nevertheless, be considered. The weight of the truss itself is actually a distributed load, but for the sake of simplicity it is assumed to be concentrated at top-chord panel points and distributed equally among them. After the truss has been designed, the actual truss weight should be computed and compared with the assumed weight. If the two are not in reasonable agreement, a new weight should be assumed, using the weight of the first truss as a guide, and the design repeated.

The computation of wind and earthquake loads was treated in Chapter 8, and snow and rain loads in Art. 9.4. Combining loads was treated in Art. 9.5, and in Art. 9.2 it was noted that ponding and self-straining forces also need to be considered where they can be significant; they will not be considered to be significant in the example that follows. Nevertheless, each component member of the truss must be designed to safely withstand the maximum force to which it will be subjected. Depending on the type of truss and the nature of the loads, this force may produce either tension or compression in any given member. It is possible for a member to be in tension under one loading condition and in compression under another. Such a

member is said to have a reversal of stress and obviously must have adequate strength in both tension and compression. It remains to be determined which of the above recommended combinations of loading will produce the maximum force (tensile or compressive) in any given member.

In Part 1, Section A5.2 of the 1989 AISC Specification, it is stated that allowable stresses may be increased $\frac{1}{3}$ above the values otherwise provided, when produced by wind or seismic loading acting alone or in combination with the design dead and live loads, provided the required section computed on this basis is not less than that required for the design dead and live load and impact (if any), computed without the $\frac{1}{3}$ stress increase, nor less than that required by Section K4 (repeated stress variation) where applicable. This means that a member generally can carry wind or earthquake loads $\frac{4}{3}$ greater in magnitude than dead and snow and rain loads.

Except in those cases where the dead load is small and the wind or earthquake load large, it has been found that stress reversal in roof trusses supported on masonry walls is quite rare. Because of this, some designers prefer to devise an "equivalent" vertical loading to simulate the combined effect of all possible loadings. When this is done, only one truss analysis is necessary. A discussion of equivalent vertical loads will not be attempted here, because this approach is too arbitrary, and ultimately a detailed analysis must in any case be made for each combination of loading to assure a safe design.

9.9
Panel Loads

With the more frequently used roof trusses, where top chord panels are of equal length and there is only one slope between heel and peak, vertical panel loads due to dead and snow loads (plus rain where applicable) are equal except at the heel, where the value is one-half that of a full panel load. To determine the panel load resulting from snow, the *horizontal* distance between panel points is multiplied by the truss spacing, and this product is multiplied by the maximum snow load expressed in psf of horizontal projection. To determine the panel load resulting from dead load, the panel length is multiplied by the spacing of trusses, and this product is multiplied by the unit dead load of the deck construction. To this must be added the weight of purlins and the truss itself. The weight of the truss is computed as explained in the previous article and distributed among the panel points. Attention is called to the fact that both the snow and the dead panel-point loads act vertically downwards.

Equivalent static wind loads are assumed to act perpendicular to the roof surfaces, either inward or outward, and are calculated as shown in Chapter 8 (see Art. 8.7, Example 2, "Gable Frames"). Panel loads resulting from wind frequently are different for the windward and leeward slopes of a truss.

Example

The typical interior Fink truss shown in Fig. 9.3 rests on piers incorporated within masonry walls, and the eave height is 20 ft. Trusses are spaced 18 ft on centers. The roof deck is of galvanized corrugated metal weighing 6 psf. All purlins are **W** 10×15 and span between trusses at each panel point. The structure is an industrial type, located in Springfield, Illinois, as in Art. 9.3, and has only minor openings. Calculate all possible panel point loads for: (a) snow load, (b) dead load, (c) wind load.

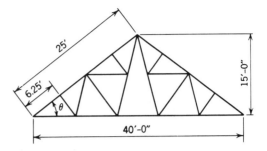

Figure 9.3/Fink truss.

Solution

(a) *Snow Load*

The snow loads are as determined in the example in Art. 9.3, i.e.,

5.79 psf applied over the entire roof surface.

10.86 psf in the unbalanced condition applied to the leeward roof surface (wind from the left).

(1) Determine the horizontal projected area supported by one purlin:

$$A = 5(18) = 90 \text{ ft}^2$$

(2) Calculate the snow panel-point loads for both conditions by multiplying the area by the unit snow load:

$$P = 90(5.79) = 521 \text{ lb; use } 0.52 \text{ kips}$$
$$P = 90(10.86) = 977 \text{ lb;} \qquad \text{use } 0.98 \text{ kips}$$

(unbalanced condition)

These loadings are shown in Fig. 9.4 (a) and (b).

(b) *Dead Load*

(1) Determine the panel area supported by one purlin:

$$A = 6.25(18) = 112.5 \text{ ft}^2$$

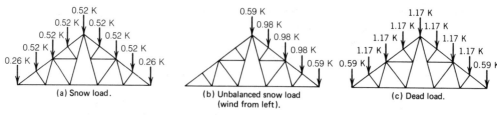

(a) Snow load.

(b) Unbalanced snow load (wind from left).

(c) Dead load.

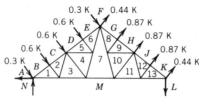

(d) Wind load. External forces plus internal suction (wind from left).

(e) Wind load. External forces plus internal pressure (wind from left).

Figure 9.4/Panel-point loads (example).

(2) Calculate the roof deck load supported by one purlin:

$$112.5(6) = 675 \text{ lb}$$

(3) Add to this the weight of each purlin: $15(18) = 270$, or

$$675 + 270 = 945 \text{ lb}$$

(4) Since this is a short span, a light truss will suffice; therefore, estimate the weight to be 2.4 psf of supported area. Trusses are spaced 18 ft apart and span 40 ft; therefore, $A = (18)40 = 720 \text{ ft}^2$. The estimated weight of each truss in $(720)2.4 \simeq 1730$ lb. This weight, divided equally between the panel points, is $1730/8 \simeq 216$ lb; use 220 lb. Then,

$$P = 945 + 220 = 1165; \quad \text{use } 1.17 \text{ kips}$$

This loading is shown in Fig. 9.4c.

(c) *Wind Load*

(1) Basic wind speed $V = 70$ mph, from Fig. 8.15.

(2) Determine the importance factor. From Table 8.2 for a building classification I,

$$I = 1.00$$

(3) Select the velocity-pressure coefficient from Table 8.3 for a mean roof height

$h = 20 + 15/2 = 27.5$ ft, and an unsheltered Exposure C. Interpolating between 0.93 and 0.98,

$$K_z = 0.955 \simeq 0.96$$

(4) Calculate the velocity pressure:

$$q_h = 0.00256 K_z (IV)^2$$
$$= 0.00256(0.96)[1.00(70)]^2$$
$$= 12.04 \text{ psf}$$

(5) Select the gust response factor for 27.5 ft from Table 8.4:

$$G_h \simeq 1.27$$

(6) For normal wall openings, the internal-pressure coefficient is ± 0.25. Calculate the internal pressure:

$$p = GC_{pi}q_h = \pm 0.25(12.04) = \pm 3.01 \text{ psf}$$

(7) Select the external-pressure coefficient from Fig. 8.16 (normal to ridge):

Windward:

$$h/L = 27.5/40 = 0.688$$
$$\theta = \arctan(15/20) = 37°$$

Interpolating between 30° and 40° for $h/L = 0.688$,

$$C_p = 0.15$$

Leeward:

$$C_p = -0.70$$

(8) Calculate the external design pressures,

$$p = q_h G_h C_p$$

Windward:

$$p = 12.04(1.27)0.15 = 2.29 \text{ psf}$$

Leeward:

$$p = 12.04(1.27)(-0.70) = -10.7 \text{ psf}$$

(9) Combine the external pressure from step (8) with the internal pressure from step (6).

Windward: External force plus internal suction,

$$2.29 + 3.01 = 5.30 \text{ psf}$$

External force plus internal pressure,

$$2.29 - 3.01 = -0.72 \text{ psf}$$

Leeward: External force plus internal suction,

$$-10.7 + 3.01 = -7.69 \text{ psf}$$

External force plus internal pressure,

$$-10.7 - 3.01 = -13.71 \text{ psf}$$

(10) Calculate panel loads.

$$\text{Panel area} = 112.5 \text{ ft}^2$$

External force plus internal suction:

Windward:

$$5.30(112.5) = 596 \text{ lb}$$
$$\simeq 0.60 \text{ kips}$$

Leeward:

$$-7.69(112.5) = -865 \text{ lb}$$
$$\simeq -0.87 \text{ kips}$$

External force plus internal pressure:

Windward:

$$-0.72(112.5) = -81 \text{ lb}$$
$$\simeq -0.08 \text{ kips}$$

Leeward:

$$-13.71(112.5) = -1542 \text{ lb}$$
$$\simeq -1.54 \text{ kips}$$

These loads are shown in Fig. 9.4 (d) and (e).

9.10
Truss-Analysis Assumptions

An exact determination of stresses in a loaded truss would be very difficult and would involve complex procedures; in fact, so many variables are involved that an exact analysis is seldom attempted except by the computer method discussed in Chapter 12 and Appendix D. Experience has shown, however, that *exact* stress values for truss members are not necessary in order to arrive at a reasonable and safe design. Consequently, assumptions are made that permit an easy solution.

It is assumed that all component members of a truss lie in one plane, one which is also the plane of loading. Most truss constructions satisfy this assumption reasonably well. In making a truss sketch, a single line represents a member, and subsequent analysis is based on the assumption that this line coincides with the centroid axis of that member. Consequently, the centroid axes of all members meeting at a joint intersect at a common point. The final layout of the truss is dimensioned accordingly. It is further assumed that all joints are hinges, that is, no bending moment can occur or be transferred through a joint. This assumption is made even though the top chord may actually be continuous from heel to ridge, and connections made by attaching each member to a gusset plate with welds or bolts, which creates a joint that offers some degree of restraint.

In conclusion, if component members and loads are assumed to lie in a common plane, joints are assumed to be pinned, and loads are assumed to be applied only at panel points, bending throughout the truss is eliminated. Each member, then, from joint to joint, is in effect a "two-force" member, i.e., it is in either tension or compression, and the stress resultant developed is axial and acts along the centroid axis of the member.

Ordinarily, trusses are designed only for primary stresses which result from the static loading condition. This assumes no joint movement. Actually, as the loads are applied, the truss deflects and the joints are displaced. This displacement of joints also produces stress in the members, i.e., secondary stresses. Such deflections and secondary stresses can be computed; however, inasmuch as the resultant stress in any member is normally less than 8 percent of the total, and deflections are frequently of no great consequence, they will be neglected in the discussion that follows.

Truss analysis for other than vertical loads, such as those due to wind, requires an assumption as to the proportioning of the horizontal force component to the reactions. One way to deal with such lateral loads is to assume that all horizontal force is resisted by one reaction, the other reaction resting on frictionless rollers. This may require two stress solutions, one for wind from the left and one for wind from the right. Another way to deal with the same problem is to assume that the horizontal forces are distributed equally between reactions.

9.11
Bow's Notation

As explained earlier, the loads on a truss are concentrated loads, and each member develops a stress resultant acting along its centroid axis. *Bow's notation* provides a systematic method for lettering spaces between external forces and numbering spaces between members, which in turn enables each force and each member to be easily identified.

A scale diagram of the truss having been made, the loads and reactions are calculated and shown on the truss diagram. Starting to the left of the left reaction, spaces between forces are lettered alphabetically in a clockwise direction. In cases where it is advantageous to indicate reactions by horizontal and vertical components, the space between components is also lettered. Each triangular space within the truss is given a number. Usually the triangular spaces are placed in numerical order starting at the left and progressing to the right.

With this system of letters and numbers each external force can be identified by two successive letters and each member (or stress resultant within the member) can be identified with either a letter and a number or two numbers.

The application of Bow's notation is shown in Fig. 9.4d. The panel load at the left heel is identified as *A-B*, and the right reaction as *L-M*. The top chord, just to the left of the ridge, is *E*-6. In this example, the left reaction is assumed to take all the lateral thrust; therefore, it is shown as having vertical and horizontal components, *M-N* and *N-A*, respectively.

9.12
Graphical Vector Analysis

Member stress resultants are sometimes referred to as *bar stresses*. Most trusses lend themselves to an easy graphical solution for the bar stresses. The graphical solution is based on definitions and principles from mechanics, a review of which follows.

An external force or a bar stress can be shown graphically as a scaled line, called a vector (Fig. 9.5). The magnitude of the external force or bar stress is indicated by

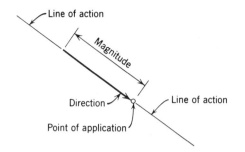

Figure 9.5/Vector.

the length of the vector, and the direction by an arrow. By extending the vector in both directions the line of action may be shown. A force or bar stress can act anywhere along its line of action; usually the point of application is known.

When two forces (or two bar stresses, or a force and a bar stress) meet at a point, their algebraic sum (resultant) can be determined by constructing a parallelogram, as shown in Fig. 9.6a. Each vector is laid off in its proper direction and at a scaled length equal to its magnitude. The resultant is the diagonal of the parallelogram, and one point on its line of action is the point where the two original forces meet. The magnitude of the resultant is determined by scaling its length. If more than two forces meet at a point, the resultant of all the forces can be determined by constructing successive parallelograms, each

new parallelogram containing the previous resultant as one force and another of the original force vectors as the other force.

The resultant of two concurrent forces (or two bar stresses, or a force and a bar stress) may also be determined by constructing a triangle as shown in Fig. 9.6b. The tail of one vector is placed at the head of the other vector, and the resultant is a line drawn from the tail of the first vector to the head of the second. Notice, however, that a resultant determined in this manner is not placed in its proper position, i.e., it is not shown passing through its true line of action.

If more than two vectors meet at a point, the resultant can be determined by constructing a polygon as shown in Fig. 9.6c. Each vector is drawn with its tail at the head of another vector. This being done, the resultant is the line drawn from the tail of the first vector to the head of the last.

If two or more forces (and/or bar stresses) meet at a point and are in a state of equilibrium, their vectorial sum is equal to zero, i.e., there is no resultant, and consequently the triangle (or polygon, as the case may be) must close (Fig. 9.6d).

Figure 9.7a shows three forces meeting at a common point. Only force *A-B* is known. However, the lines of action of forces *B-C*

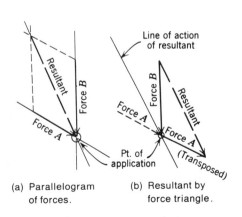

(a) Parallelogram of forces.

(b) Resultant by force triangle.

(c) Resultant by force polygon.

(d) Vector equilibrium, closed force triangle.

Figure 9.6/Graphic vector analysis.

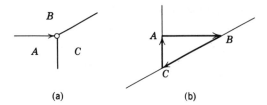

Figure 9.7/Solving for two unknowns.

and *C-A* are as shown. If the three forces are in a state of equilibrium, the magnitudes and directions of the two unknown forces can be determined by constructing a closed force triangle as follows (Fig. 9.7b):

1. Lay off the known vector *A-B* to a selected scale; label it as shown.

2. At one end of the known vector (point *B* in this case), construct a line parallel to the line of action of one of the unknown forces, e.g., force *B-C*.

3. At the other end of the known vector construct a line parallel to the line of action of the other unknown force *A-C*.

4. The intersection of the two construction lines establishes the closed triangle, and the magnitudes of the unknown forces can be determined by scaling their respective lengths. The directions of the forces *B-C* and *C-A* are determined by always joining a tail to a head, so that the triangle closes.

9.13
Stress Analysis of a Truss

Four common methods of determining the stresses in the members of a roof truss are the algebraic method of sections; the algebraic method of joints introduced in Art. 8.2 and used later in this chapter; computer analysis—treated in Chapter 12; and the graphical method of successive joints. The graphical method will be used here. A description of the method of sections and

discussions of the theory on which it is based will be found in textbooks on statics and strength of materials.

From the assumptions discussed in Art. 9.9, the lines shown on a truss diagram, which ordinarily represent the center of gravity location of members, are also used to represent stress resultants (or bar stresses). The lines of action of all members of a truss are known once the dimensions and type of truss are established; therefore, the line of action of each bar stress also is known. If external forces are present at a joint, their magnitudes and directions are determined as discussed in Arts. 9.8 through 9.11. Reactions are also stress resultants; however, they can be thought of as external forces and be determined in the usual manner. The bar stresses, then, represent the only unknowns, and if only two unknowns occur at a joint, they can be determined in accordance with the principles discussed in Art. 9.12.

Figure 9.8a represents part of a loaded truss. Examination shows that the heel has only two unknowns, designated as *B*-1 and 1-*X*. This heel joint is in static equilibrium, as are all other joints in the truss. Conse-

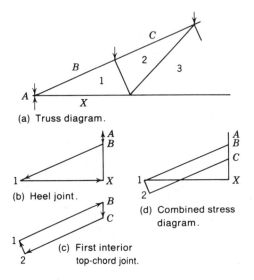

(a) Truss diagram.

(b) Heel joint.

(c) First interior top-chord joint.

(d) Combined stress diagram.

Figure 9.8/Graphical method of joints.

quently, all of the forces and bar stresses at this joint must make a closed force polygon. Figure 9.8b shows the closed force polygon of all forces and bar stresses at the heel joint. This polygon is constructed as follows:

The external forces *X-A* and *A-B* are laid off to scale; their resultant is *X-B*. At point *B*, a line is constructed parallel to the line of action of *B*-1. At point *X*, a line is constructed parallel to the line of action of 1-*X*. The intersection of these two construction lines establishes the location of point 1, and the polygon is closed. The magnitude of *B*-1 and of 1-*X* is obtained by scaling the diagram, and their directions are as shown.

Moving to the right, the next interior joint on the bottom chord cannot be solved, because three unknown bar stresses exist (1-2, 2-3, and 3-*X*). However, the next interior joint on the top chord can be solved, as there are only the two unknowns *C*-2 and 2-1 (bar stress 1-*B* previously having been established). The stress diagram for this joint (Fig. 9.8c) is constructed in the same manner as for the heel joint. Bar stresses *C*-2 and 2-1 are scaled from the force polygon.

This process could be repeated for each successive joint in the truss. However, by observation it is seen that line *B*-1 in Fig. 9.8b is identical with line *B*-1 in Fig. 9.8c, except for the directions indicated. Therefore, the two can be combined in one diagram as shown in Fig. 9.8d. Further investigation will show that each successive joint has one line (bar stress) in common; therefore, a combined stress diagram can be constructed for the entire truss, so long as direction is not indicated.

The stress analysis procedure for an entire truss, employing the combined diagram, is illustrated in Fig. 9.9, and is as follows:

1. Draw a diagram of the truss to a convenient scale, and show the loads and positions of the reactions.

2. Compute the reactions.

3. Using Bow's notation, letter or number each space. (Steps 1, 2, and 3 are shown in Fig. 9.9a.)

4. Draw the force polygon for the external forces to some convenient scale, plotting them in order around the truss diagram in a clockwise direction. For a system of vertical loads, the polygon is a straight line as shown in Fig. 9.9b. In this case, *A-B* = 2000 lb, *B-C* = 4000 lb, *C-D* = 4000 lb, *D-E* = 4000 lb, and *E-F* = 2000 lb. The reactions *F-G* and *G-A*, each of which equals 8000 lb, close the polygon.

5. Select a joint, such as the left heel of the truss, where there are only two unknowns. In the truss shown in Fig. 9.9a, the two unknowns are the stresses in members *B*-1 and *G*-1; their line of action, however, is known. The force polygon for this joint is then drawn as follows:

- Through *B* on the stress diagram, draw a line parallel to *B*-1 on the truss diagram.
- Through *G* (stress diagram), draw a line parallel to *G*-1 (truss diagram).
- The intersection of these two lines locates the point 1, and completes the force polygon for this joint.
- The magnitude of the stress in the member *B*-1 is obtained by scaling the length of the line *B*-1 on the stress diagram.

6. Select the next joint to the right where there are only two unknown forces and proceed as above. For the truss in Fig. 9.9a, the next joint should be that formed by the intersection of member 2-1 and the lower chord. Then, proceed as follows:

- Through 1 (stress diagram), draw a line parallel to 1-2 (truss diagram).
- Through *G* (stress diagram), draw a line parallel to 2-*G* (truss diagram).

Point 2 will coincide with 1, since *G*-1 and 2-*G* have the same line of action, i.e., the stress in 2-1 is zero. The stresses in the

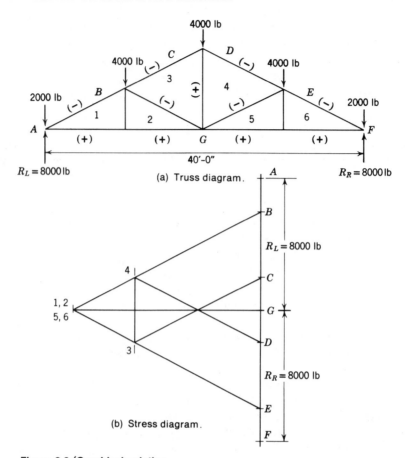

(a) Truss diagram.

(b) Stress diagram.

Figure 9.9/Graphical solution.

remaining members are found in a similar manner and are recorded on the truss diagram.

7. After the magnitude of the stress in each member is found, it is necessary to determine which members are in tension and which are in compression. A negative sign usually denotes compression, and a positive sign tension. To determine the sign of the stress in a member, read the letters and/or numbers (Bow's notation) clockwise about one of its joints in the truss diagram. For the member B-1, use the left heel joint (Fig. 9.9a) and read clockwise from B to 1. On the stress diagram, the direction of B to 1 is downward and to the left, or *toward* the joint in the truss diagram. This indicates that the

member is in compression. Using the heel joint again to find the nature of the stress in member 1-G, read clockwise 1 to G on the truss diagram. On the stress diagram, 1 to G is toward the right, or *away* from the joint in the truss diagram. This indicates that the member is in tension. The sign of the stress in each of the remaining members is found similarly and recorded on the truss diagram. If this method is consistently applied, the sign of the stress can always be found.

Further study will show that a given percentage increase or decrease in all vertical panel loads will produce a proportionate increase or decrease in internal bar stresses. Consequently, only one stress diagram is necessary for vertical loads. The bar stresses for other vertical loads can be determined by simple proportions.

Wind loads that act perpendicular to the top chord require a separate stress diagram. The resulting stresses in each member are then combined as explained in Art. 9.8. The maximum value obtained from the three load combinations is the stress used for design.

9.14
Procedure for the Design of a Roof System

The procedure for the design of roof purlins and trusses is as follows:

1. Select the type of roof truss best suited for the span, interior function and design, and roof deck and overing to be used. Make a line drawing of the truss to scale.

2. Calculate the psf weight of the roof covering and deck. Multiply this unit dead-load increment by the panel length to obtain the per-foot uniform dead load supported by each purlin (half this amount for heel purlins, and for ridge purlins if two are used). Make an allowance for the dead load of the purlin itself.

3. Determine the psf live load per foot of horizontal projection, and convert it to a uniform load per foot acting on the purlin as in step 2 above.

4. Determine the psf balanced snow load (and rain load, if any) per foot of horizontal projection, and convert it to a uniform load per foot acting on the purlin as in step 2 above.

5. Determine the psf wind loads (both transverse and longitudinal) acting normal to the roof surface, and calculate the per-foot uniform load supported by each purlin as in step 2 above.

6. Ascertain the most unfavorable combination of loads (Art. 9.4).

7. Design the purlins for the most unfavorable load combination.

8. Estimate the weight of the truss, and, adding this to the dead loads determined in step 2 above, calculate the panel-point dead loads.

9. Determine the panel-point loads due to both balanced and unbalanced snow loads.

10. Determine the panel-point loads due to wind for each loading condition using the data from step 5 above.

11. Draw the dead-load stress diagram.

12. Draw the unbalanced-snow-load stress diagram.

13. Draw the wind-load stress diagrams.

14. Prepare a stress analysis table (stresses due to live load and balanced snow load are found by simple proportions, using the dead-load stresses as the reference).

15. Ascertain the critical stress for each member based upon the most unfavorable load combinations (Art. 9.4).

16. Prepare a design table.

17. Design compression members.

18. Design tension members.

19. Design joints.

20. Design end bearing and anchorage.

21. Design bracing.

22. Prepare a design drawing.

The design of a truss-supported roof will be illustrated in the example that follows.

Example

Note. Because of the unusual length of the design procedure involved, and the fact that load computations already have been amply demonstrated, certain of the loading conditions will be given. Also, because of the need for a considerable number of amplifying statements, the various steps in the procedure will be treated as separate articles (9.15 through 9.30). The number of each step will be shown in parentheses after the article title; however, these step numbers will not correspond to those above, because the given loading conditions eliminate the initial steps in the design procedure. Thereafter, the format thus far followed in this text for examples will be resumed.

Given that a rectangular building having brick masonry side and end walls is to be constructed, design a truss-supported roof system.

Piers to receive trusses are to be spaced 16 ft on centers along the side walls, and the trusses must span 70 ft between side-wall pier centers. A Fink truss, having an 8-panel top chord on a rise of 12 ft 6 in., is to be used. The roofing is to be tile having a weight, including bedding material, of 15 psf (this heavy-weight material being selected here principally to develop sufficient stress in the members to ensure that other than all minimum-size members will be required, thus improving the example for illustrative purposes).

The decking is to consist of light-weight, precast concrete slabs supported on structural tees welded to the purlins in a manner similar to that shown in Fig. 9.27; weight, 20 psf. The design should conform to the 1989 AISC Specification for A36 Steel.

Although $\frac{1}{4}$-in.-thick sections for truss members would be adequate, a minimum section of two angles, $2\frac{1}{2} \times 2 \times \frac{5}{16}$ in., will be selected for this example. Gusset plates are to be $\frac{3}{8}$ in. in thickness. All connections and joints will be welded.

It should be noted that tee sections are frequently used for top and bottom chords in welded trusses. This enables double-angle web members to straddle the webs of the tees. Also, when the chord members (tees) are deep, it is frequently possible to

use tee web members as well—the web portion of the tee web members being cut back, and the outstanding flange being slotted so as to enable the web members to be fitted over the tee chord web. Then, welds from web flange to chord web and, if need be, from web flange end to chord flange can be made.

It will be assumed that critical purlin stresses will result from a combination of dead and snow loads and that critical truss member stresses will be determinable from a combination of dead, snow, and transverse wind loads.

9.15
Line Drawing of Truss (Step 1)

Figure 9.10a shows the truss drawn to scale, with loads and Bow's notation, which are actually added later.

9.16
Design Dead Load (Step 2)

Determine the total dead load per square foot of roof surface:

$$\begin{array}{r} \text{Tile roofing} = 15 \text{ psf} \\ \text{Roof deck} = \underline{20} \\ \text{Total} = 35 \text{ psf} \end{array}$$

The panel length obtained from the scale drawing is approximately 9.3 ft. The uniform load each purlin supports is 9.3(35) = 326 lb per ft. Assume that each purlin contributes 14 lb per ft to the dead load. The total purlin dead load, then, is 326 + 14 = 340 lb per ft.

9.17
Design Snow Load (Step 3)

The snow load will be taken as 30 psf of horizontal projection. The horizontal projection supported by each purlin is $\frac{35}{4} =$

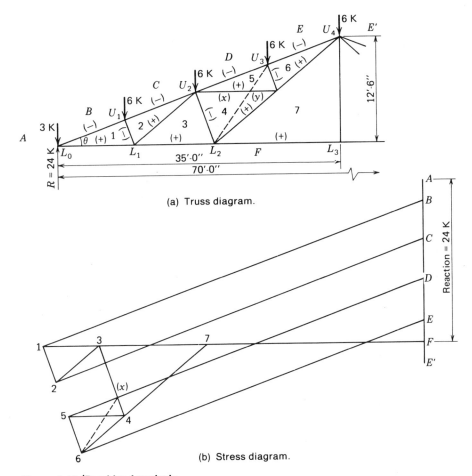

(a) Truss diagram.

(b) Stress diagram.

Figure 9.10/Dead-load analysis.

8.75 ft. The snow load supported by each purlin is then 8.75(30) = 263 lb per ft.

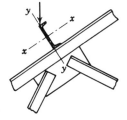

Figure 9.11/Purlin orientation.

9.18
Purlin Design (Step 4)

Purlins may be classified in one of two ways for purposes of design: as members free to bend in any direction and as members fixed laterally. It is evident that the channel purlin loaded as shown in Fig. 9.11 is subjected to unsymmetrical bending and is therefore of the first type. Design of members subjected to unsymmetrical bending is treated in detail in Art. 5.15.

When the roof deck is of such a nature that it provides lateral support in a direction parallel to the roof surface, the vertical load on the purlin may be resolved into normal and parallel components. It then may be assumed that the parallel component is carried by the deck, and the purlin need be designed for the normal component only. When the roof deck is not sufficiently rigid, lateral support can be provided by placing tie rods at the center or third-points of the purlin span. These rods (usually $\frac{5}{8}$ or $\frac{3}{4}$ in. in diameter) run from heel purlins to the ridge. Since the stress in the rod is greatest at the ridge, it is frequently necessary to increase the size of the ridge purlin in order to provide for the concentrated load due to the vertical component of the tie-rod stresses.

Tie rods may also be used with rigid roof decks to provide lateral support during erection. They are seldom necessary on roofs having a slope of less than 3 in 12.

In the problem under consideration, the welded tees and roof deck will be assumed to furnish adequate lateral support, i.e., both sides of the roof-deck system are secured at the ridge. The purlin, therefore, carries a total vertical load of $340 + 263 = 603$ lb per ft (dead load plus snow load). This load is converted into components parallel to and perpendicular to the roof surface using either a graphical or a mathematical method. Finding the angle of the roof truss,

$$\tan \theta = \frac{12.5}{35} = 0.357, \quad \text{or} \quad \theta = 19.7°$$

the parallel component P is $603 \sin 19.7° = 203$ lb per ft; the force N, normal to the roof surface, is $603 \cos 19.7° = 568$ lb per ft (Fig. 9.12).

Figure 9.12/Purlin load.

Because the component P will be resisted by the tees and roof deck, the purlin may be designed as a simple beam carrying a uniform load of 568 lb per ft on a 16-ft span. The maximum moment at midspan for a uniformly loaded beam is

$$M = \frac{wl^2}{8} = \frac{0.568(16)^2}{8} = 18.2 \text{ kip-ft}$$

and the required section modulus for an A36 Steel beam having adequate lateral support is

$$S = \frac{M}{F_b} = \frac{18.2(12)}{24} = 8.72 \text{ in.}^3$$

From the AISC Manual "Allowable Stress Design Selection Table" the **M** 12×10 is seen to be the lightest-weight section. However, it is decided in this case (for a 16-ft span) to use a maximum depth of 8 in. and to use a channel shape. Consequently, a **C** 8×13.75 is selected.

9.19
Panel-Point Loads for Dead and Snow Loads (Step 5)

Estimating the weight of the truss to be 4 psf of supported area, the total truss weight is

$$4(16)70 = 4480 \text{ lb}$$

Assuming this weight is evenly distributed among the eight panels, the weight per panel is

$$\frac{4480}{8} = 560 \text{ lb}$$

The full dead load at interior panel points then is

$$16(340) + 560 = 6000 \text{ lb}; \quad \text{use 6.0 kips}$$

The full dead load at heel panel points is

$$\tfrac{1}{2}(6000) = 3000 \text{ lb}; \quad \text{use 3.0 kips}$$

The total snow load at interior panel points

is

$$16(263) = 4210 \text{ lb}; \quad \text{use 4.2 kips.}$$

The total snow load at heel panel points is

$$\tfrac{1}{2}(4.2) = 2.1 \text{ kips}$$

9.20
Panel-Point Loads for Wind (Step 6)

The angle of roof slope has been determined to be $19.7°$, and external force plus internal pressure is found to be the controlling wind condition, as follows:

External forces:

Windward side = 12 psf   (suction)

External forces:

Leeward side = 9 psf   (suction)

Internal forces:

Both sides = 7 psf   (pressure)

Internal forces:

Both sides = 6 psf   (suction)

Panel area = $16(9.3) = 149 \text{ ft}^2$; therefore,

Windward side:

$$\frac{(12+7)(149)}{1000} = 2.8 \text{ kips} \quad \text{(outward)}$$

Leeward side:

$$\frac{(9+7)(149)}{1000} = 2.4 \text{ kips} \quad \text{(outward)}$$

This loading condition is shown in Fig. 9.13a. It will be assumed that each reac-

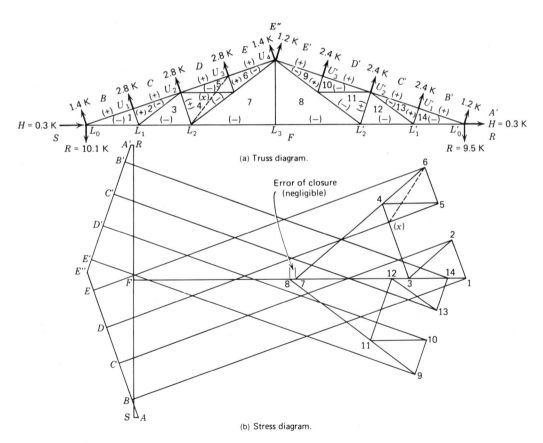

(a) Truss diagram.

(b) Stress diagram.

Figure 9.13/Wind-load analysis.

tion will develop one-half the total lateral horizontal force due to wind.

9.21
Dead-Load Stress Analysis (Step 7)

With the loads and reactions recorded on the truss diagram (Fig. 9.10a), the stress diagram may be drawn. Inasmuch as the truss and loads are symmetrical, it is necessary to draw the diagrams for only half the truss. The notation consisting of the letters A to E may be considered as repeated on the right half of the truss by the addition of primes ('). The stress diagram is constructed as explained in Art. 9.12 until the joint U_2 is reached. Here there are three unknowns, the stresses in members D-5, 4-5, and 3-4.

The unknowns at this joint may be reduced temporarily to two by replacing members 4-5 and 5-6 by the member x-y, shown dashed in Fig. 9.10a. On the stress diagram (Fig. 9.10b), a line is drawn through D parallel to D-5 on the truss diagram, and another through 3 on the stress diagram parallel to 3-4 on the truss diagram, thus locating x. Through x, on the stress diagram, a dashed line is drawn parallel to x-y on the truss diagram, and another through E on the stress diagram parallel to E-6 on the truss diagram (same as E-y). The intersection locates point 6, which is also the point y. The line E-6 represents the true stress in member E-6.[5]

Member x-y is now removed, and 4-5 and 5-6 are replaced. Knowing the position of point 6 in the stress diagram, the stresses in the remaining members are found in the usual manner. This method is often

referred to as the *substitute-member method*.

When the panel loads on a Fink truss are equal and symmetrically placed, as is the case in this example, a simple method not requiring the substitute member may be used for constructing the stress diagram. In Fig. 9.10b it is observed that points 1, 2, 5, and 6 lie on a straight line perpendicular to B-1. When the point 1 has been determined, points 2, 5, and 6 are located by the intersection of this perpendicular with lines C-2, D-5, and E-6. When the panel loads are not equal this method cannot be used, as points 1, 2, 4, and 5 will no longer lie on the same straight line.

The fact that the lines 5-4, 3-4, and 6-7 intersect at a common point will serve as a check on the accuracy of the construction of the stress diagram. Additional checks will become evident from a study of Fig. 9.10 and the bar stresses tabulated in the stress table. The signs of the stresses in the members are determined as explained in Art. 9.12, and the values are scaled from the stress diagram.

9.22
Wind-Load Stress Analysis (Step 8)

The truss diagram is drawn to scale (Fig. 9.13a). Panel-point loads due to wind from the left are shown on the diagram. Since the loadings are not symmetrical, the entire truss is shown, and the stress diagram will have to be constructed for both left and right halves of the truss.

Reactions are determined by applying the laws of statics, together with the given assumption that the horizontal component of the loads is divided equally into each reaction.

The stress diagram (Fig. 9.13b) is constructed in a manner similar to that for dead load (Art. 9.21), except that the full

[5]It is evident that the stress in member E-6 is not affected by any change in the form of the truss to the left of the joint U_3 so long as the positions of the loads remain the same.

diagram must be drawn. Also, to minimize drawing errors, it is advisable to start at one heel joint and construct the diagram up to and including the middle member 7-8, and then start at the other heel joint, closing at the center. The error of closure should be minimal and should be indicated on the diagram as shown in Fig. 9.13b.

It should be noted that due to the assumption that the horizontal component of the loads is divided equally into each reaction, wind coming from the right would cause identical stresses to wind from the left, if the truss were turned end for end. This would not be true if the horizontal reaction components were not equal, for example, if a roller were used at one reaction.

9.23
Stress-Analysis Table (Step 9)

A stress-analysis table (Table 9.1) is prepared, and the stress in each member due to each loading condition is recorded therein. For the reasons stated in Arts. 9.21 and 9.22, only half the members need be identified. The members are usually grouped: top chord, bottom chord, and web members.

The stresses due to dead load are recorded as they are scaled from the stress diagram. The stresses due to snow load are directly proportional, i.e., the snow-load stress in each member is obtained by multiplying the dead-load stress by the ratio

$$\frac{\text{Snow panel-point load}}{\text{Dead panel-point load}} = \frac{4.2}{6.0} = 0.70$$

Stresses due to wind load left are recorded as scaled directly from the stress diagram for the member listed. Stresses due to wind load right are recorded as scaled from the stress diagram for the counterpart member in the other half of the truss.

Table 9.1
Stress Analysis Table (kips)

Member	Dead Load	Snow Load	Wind Load Left	Wind Load Right	Dead Load + Snow Load	$\frac{3}{4}$ of [Dead Load + Wind Load (L)]	$\frac{3}{4}$ of [Dead Load + Wind Load (R)]	$\frac{3}{4}$ of {Dead Load + $\frac{3}{4}$ of [Snow Load + Wind Load (L)]}	$\frac{3}{4}$ of {Dead Load + $\frac{3}{4}$ of [Snow Load + Wind Load (R)]}
B-1	−62.4	−43.7	+26.1	+25.1	−106.0	−27.2	−28.0	−56.7	−57.3
C-2	−60.3	−42.6	+26.1	+25.1	−102.5	−25.7	−26.4	−54.5	−55.1
D-5	−58.1	−40.7	+26.1	+25.1	−98.8	−24.0	−24.8	−51.8	−52.4
E-6	−56.0	−39.2	+26.1	+25.1	−95.2	−22.4	−23.2	−49.4	−49.9
F-1	+57.8	+40.5	−24.4	−23.1	+98.3	+25.1	+26.1	+52.6	+53.1
F-3	+50.2	+35.1	−20.3	−19.1	+85.3	+22.4	+23.3	+46.0	+46.7
F-7	+33.2	+23.2	−11.7	−11.7	+56.4	+16.1	+16.1	+31.4	+31.4
F-8	+33.2	+23.2	−11.7	−11.7	+56.4	+16.1	+16.1	+31.4	+31.4
1-2	−5.7	−4.0	+2.8	+2.4	−9.7	−2.2	−2.5	−5.0	−5.2
2-3	+8.5	+6.0	−4.1	−4.0	+14.5	+3.3	+3.4	+7.4	+7.5
3-4	−11.3	−7.9	+5.6	+4.8	−19.2	−4.3	−4.9	−9.8	−10.2
4-5	+8.5	+6.0	−4.1	−4.0	+14.5	+3.3	+3.4	+7.4	+7.5
5-6	−5.7	−4.0	+2.8	+2.4	−9.7	−2.2	−2.5	−5.0	−5.2
6-7	+25.4	+17.8	−12.5	−11.5	+43.2	+9.7	+10.4	+22.0	+22.6
4-7	+17.0	+11.9	−8.2	−7.5	+28.9	+6.6	+7.1	+14.8	+15.2
7-8	0.0	0.0	0.0	0.0	0.0	0.0	0.0	0.0	0.0

In other words, the stress in member B-1 resulting from wind left is the same as the stress in member B'-14 resulting from wind right.

The remaining columns in the table are determined from the probable loading conditions, as noted in the column headings (see Art. 9.4). Since the last four columns in the table pertain to loadings which include some wind, the total load (for design purposes) is multiplied by $\frac{3}{4}$, the result being the same as multiplying the allowable stress by $\frac{4}{3}$. (See Art. 9.8.)

A member must be designed to safely resist the maximum tension and/or compression it is likely to be subjected to. It so happens that, in this example, the maximum stress (design stress) for *all* members results from dead plus snow load. This is not uncommon for this type of truss with the proportions as shown. Of course, other trusses, or even this same truss with different proportions and/or loadings, may well require different load combinations, which, in turn, result in different stress distributions. Some truss members may even be subjected to a reversal of stress (Art. 9.8).

9.24
Design Table (Step 10)

At this point, it is desirable to prepare a table in which design information may be recorded: The form of Table 9.2 is only one of many types that might have been chosen. Up to this point in the solution of the present problem, enough information has been obtained to fill out the first three columns. As the members are designed, the required section is entered in the fifth column. When all members have been designed, slight variations in sections are eliminated to simplify the truss. In trusses

of this type and span, it is usual practice to use one section for the entire top chord, the waste in material being offset by savings in fabrication.

The lengths recorded in the third column usually refer to the length between joints of a designated member. For compression members, this is the design length along which buckling takes place and consequently could dictate the allowable stress. Joint movement perpendicular to the plane of the truss also must be considered. Such movement depends upon the bracing between trusses. For example, the purlins prevent joints in the top chord from moving perpendicular to the plane of the truss. The joint where member 4-7 meets member 6-7 is not so restricted, and one continuous member is usually used between L_2 and U_4 (Fig. 9.13a). It is necessary, therefore, to investigate the slenderness ratios for both lengths when fulfilling the requirement set forth in Section B5.7 of the 1989 AISC Specification, or whenever these members are in compression.

Section B5.7 of the AISC Specification sets the maximum slenderness ratio of compression members to $KL/r = 200$. K for all truss elements is 1.0. This, then, requires a minimum r of $L/200$ for all compression elements. This section of the specification also sets the maximum slenderness ratio of tension members to 300. In earlier specifications it was limited to 240. Further examination shows that limiting slenderness ratios for tension members is not necessary for structural integrity, but is recommended only for practical reasons. Therefore it is somewhat arbitrary. For this example a maximum value of 240 is used.

The L to be used for bottom-chord members depends upon the bottom-chord bracing. In this problem it will be assumed that bracing perpendicular to the plane of the truss will be attached at joints L_2 and L'_2.

Table 9.2
Design Table

Member	Design Stress (kips)	Length (ft & in.)	Min r (in.)	Required Section	Adopted Section	Weight (lb)
B-1	−106.0	9-3½	0.56	2L6×4×$\frac{1}{2}$ ⊤⊤	2L6×4×$\frac{3}{8}$ ⊤⊤	301
C-2	−102.5	9-3½	0.56	do.	do.	301
D-5	−98.8	9-3½	0.56	2L5×5×$\frac{3}{8}$ ⊤⊤	do.	301
E-6	−95.2	9-3½	0.56	do.	do.	301
F-1	+98.3	9-10 / 19-8	0.49 / 0.98	2L4×3½×$\frac{5}{16}$ ⌐⌐	2L4×3½×$\frac{5}{16}$ ⌐⌐	151
F-3	+85.3	9-10	0.49	2L4×3×$\frac{5}{16}$ ⌐⌐	do.	151
F-7	+56.4	15-4 / 30-8	0.77 / 1.53	2L3½×3×$\frac{5}{16}$ ⌐⌐	2L3½×3×$\frac{5}{16}$ ⌐⌐	202
1-2	−9.7	3-4	0.20	2L2½×2×$\frac{5}{16}$ ⊤⊤	2L2½×2×$\frac{5}{16}$ ⊤⊤	30
2-3	+14.5	9-10	0.49	do.	do.	88
3-4	−19.2	6-7½	0.40	do.	do.	60
4-5	+14.5	9-10	0.49	do.	do.	88
5-6	−9.7	3-4	0.20	do.	do.	30
6-7	+43.2	9-10 / 19-8	0.49 / 0.98	2L2½×2×$\frac{5}{16}$ ⌐⌐	2L2½×2×$\frac{5}{16}$ ⌐⌐	88
4-7	+28.9	9-10	0.49	do.	do.	88
7-8	0.0	12-6	0.50	L3×3×$\frac{5}{16}$	L3×3×$\frac{5}{16}$	77

Total weight = 2(2180) + 77 = 4437 lb

9.25
Selection of Compression Members (Step 11)

As stated in Art. 9.14, the top chord (from heel to ridge) of this truss will be composed of one double-angle member. (The design of double-angle struts was discussed in Art. 6.11.) For purposes of illustration, member B-1 of the truss will be treated first. The largest load which must be carried by any top-chord member is 106 kips. Therefore, a strut must be selected that will carry 106 kips for a length of 9.3 ft. Two unequal-leg angles with the long legs back to back (⊤⌐) will be selected so as to equalize the radii of gyration for both axes as nearly as possible (Fig. 9.14).

The AISC Manual provides double-angle-strut safe-load tables for A36 Steel; therefore, scanning the table for unequal-leg angles with long legs back to back, it is seen that the lightest-weight combination

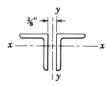

Figure 9.14/Top chord.

is composed of $6 \times 4 \times \frac{1}{2}$-in. angles at 32.4 lb per ft. The safe load for the y-y axis is 145 kips for 10 ft. Since member B-1 must carry only a 106-kip load at 9.3 ft, the section is adequate.

The remaining members in the top chord could be designed in a similar manner; however, they would all carry a lesser stress. It is noted, for example, that a section slightly lighter in weight could be used for members D-5 and E-6, but it is more practical to make the top chord one continuous member. Therefore, the $6 \times 4 \times \frac{1}{2}$-in. double-angle section is adopted for the entire top chord. The remaining compression members in the truss are similarly designed, remembering that the minimum section has been set at two $2\frac{1}{2} \times 2 \times \frac{5}{16}$-in. angles.

9.26
Design of Tension Members (Step 12)

The strength of a member in tension depends on the area of its least cross section. With bolted double-angle tension members—which are not of direct concern in this example—the least area is obtained when a cross section is taken through the bolt holes (Fig. 9.15). This is called the net area and was discussed in detail for fasteners in Art. 7.12.

The procedure for design is as follows.

1. Find the required area by dividing the total stress to be carried by the allowable unit tensile stress. This may be expressed by the formula $A = P/F_t$, where A is in square inches, P is in kips, and F_t is the allowable axial tensile stress in kips per square inch.

2. Select two angles that have a combined gross cross-sectional area somewhat larger than that found in step 1.

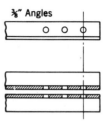

Figure 9.15/Net section.

3. Determine the net area by the method described in Art. 7.12. The net area will need to be equal to or slightly larger than the required area found in step 1. If the two are not in reasonable agreement, another trial section will have to be selected.

The above net-area procedure is, of course, not necessary for welded members; however, whether or not the truss is bolted, or welded as is the case in this example, the selected section should preferably have a slenderness ratio not exceeding 240 (not a code requirement).

The lower chord member F-1 will be used to illustrate the design (or selection) of tension members to be welded. The total stress in this member was found to be 98.3 kips.

1. The required area for A36 Steel is

$$A = \frac{P}{F_t} = \frac{98.3}{22} = 4.47 \text{ in.}^2$$

2. Referring again to the AISC safe-load tables for unequal-leg angles, it is seen that two $4 \times 3\frac{1}{2} \times \frac{5}{16}$-in. angles have an area of 4.49 in.^{2}.

Member F-1 has a length of 9 ft 10 in.; consequently, it needs to have a minimum radius of gyration of

$$r = \frac{L}{240} = \frac{118}{240} = 0.49 \text{ in.}$$

Member F-3 has the same length and therefore would need the same minimum radius of gyration. For the sake of simplicity, however, one continuous member is usually used for both F-1 and F-3. But, if the joint L_1 is not braced laterally and is free to move in a direction perpendicular to the plane of the truss (or should the bracing be disrupted), the full member must then have a minimum radius of gyration (perpendicular to the plane of the truss), i.e.,

$$r = \frac{L}{240} = \frac{2(118)}{240} = 0.98 \text{ in.}$$

To provide this required r, the trial section of two $4 \times 3\frac{1}{2} \times \frac{5}{16}$ in. angles will be placed with the short legs back to back ($\lrcorner\llcorner$), and $r_y = 1.86$ in. for a $\frac{3}{8}$-in. gusset plate.

It is common practice to brace the other joints laterally; as previously stated, bracing will be provided at joints L_2 and L'_2. Consequently, the remaining bottom-chord members (F-7 and F-8) must also provide a greater r perpendicular, rather than parallel, to the truss. And it is for this reason that the bottom-chord members are frequently composed of unequal-leg angles with short legs back to back.

The remaining tension members of the truss are designed (or selected) in a similar manner. One angle, $3 \times 3 \times \frac{5}{16}$ in., is used for the 12-ft 6-in. member 7-8. This section is selected as being the lightest-weight angle, $\frac{5}{16}$ in. thick, which furnishes the desired radius of gyration of 0.50 in. for a secondary member, i.e.,

$$r = \frac{L}{300} = \frac{150}{300} = 0.50 \text{ in.}$$

Notice, from the AISC "Dimensions and Properties" table, that the smallest radius of gyration of this angle is 0.589 in. (Z-Z axis). Then

$$P = 22(1.78) = 39.2 > 6 \text{ kips}$$

(AISC minimum allowable capacity is 6 kips.)

The safe-load tables which are available for various angle sizes and arrangements greatly facilitate the design process. Such tables occasionally include the statement that angles must be connected to the gusset plates by both legs in order to develop full strength. However, this is usually not necessary, and where the ability of the angle to transmit the load is in question, selecting a slightly larger section will generally suffice.

When all the members have been designed, the fifth column of Table 9.2 is filled in, and the weight of the truss is computed. The weight of the truss, exclusive of welds, gusset plates, bracing, etc., is found to be approximately 4437 lb. The assumed weight was about 4480 lb. Therefore, assuming 350 lb for the weight of welds, gussets, etc., or 8 per cent, there will be a discrepancy between the actual and assumed weight of only 300 lb, which is close enough.

9.27
Design of Joints (Step 13)

The lines of action of all truss members entering a joint should meet at a common point (Art. 9.10). If this cannot be arranged, eccentricity will result. With angle members, such eccentricity cannot help but occur, because the centroid axis of such members does not lie in the plane of either leg.

With bolted trusses, this eccentricity is further accentuated by the fact that very often one cannot install bolts in a plane passing through the centroid axis of the member—doing so would bring them too close to the outstanding leg. Therefore, bolts are usually placed along the standard gage lines (AISC Manual, Part 1) of the leg in contact with the gusset plate, and additional connectors added to compensate for the resulting eccentricity.

The same centroid-axis problem, of course, exists with welded members. However, when welding, the same problem of working space does not generally occur, and members generally can be brought into closer alignment. Plug or slot welds then can be made, so long as the hole diameter or width can be achieved. With fillet welds not in holes or slots, but on the heel or toe of the connected angle leg, not even this concern exists. However, as pointed out in Chapter 7, balancing of welds due to the lack of section symmetry should be considered. The AISC Specification does not actually call for such balancing, and certainly with small, lightly loaded members, it is difficult to justify. Nevertheless, for this example, welds will be balanced to further illustrate the technique.

Regardless of the type of connectors used —i.e., welds as used in this example, or rivets or bolts—two general possibilities must be considered in truss joint design: do the members simply extend from one joint to the other, or are they continuous through the joint?

The first case is illustrated by the heel joint L_0. The stress in the top-chord member B-1 and the stress in the lower-chord member F-1 are shown in Fig. 9.16. B-1 is composed of two $6 \times 4 \times \frac{1}{2}$-in. angles, and F-1 is composed of two $4 \times 3\frac{1}{2} \times \frac{5}{16}$-in. angles. As stated earlier, all gusset plates are $\frac{3}{8}$ in. thick. In this case, the welds on each member must transfer the full load of that member to the gusset plate.

The second case is illustrated by the design of joint U_1. The stresses in members

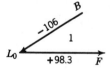

Figure 9.16/Joint L_0.

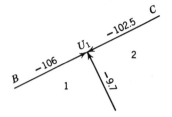

Figure 9.17/Joint U_1.

meeting at this joint are shown in Fig. 9.17. Member 1-2 is composed of two $2\frac{1}{2} \times 2 \times \frac{5}{16}$ in. angles. In this case, the welds on member 1-2 must transfer the full load to the gusset; but the welds on the continuous top-chord member need only transfer to the gusset the difference between the stress in B-1 and in C-2, or $106 - 102.5 = 3.5$ kips.

Returning to the general discussion of joints, shop-applied fillet welds will be placed along the heel and toe of each angle member in sufficient length to develop the design stress—those recommended by the 1989 AISC Specification should be used.

The minimum-size fillet weld (AISC Specification, Section J2) for parts up to $\frac{1}{2}$-in. thickness is $\frac{3}{16}$ in. The maximum-size fillet weld along the toe of an angle leg which is $\frac{1}{4}$ in. or more thick (AISC Specification) cannot exceed (except under prescribed circumstances) $\frac{1}{16}$ in. less than the thickness of the angle leg. For $\frac{5}{16}$-in. angle legs the maximum fillet weld size then is

$$\tfrac{5}{16} - \tfrac{1}{16} = \tfrac{1}{4} \text{ in.}$$

Members vary from $\frac{1}{2}$ to $\frac{5}{16}$ in. thick; however, the bracing is only $\frac{1}{4}$ in. thick. Therefore, a $\frac{1}{16}$-in. reduction will be made for consistency, and the $\frac{3}{16}$-in. fillet weld used throughout. The minimum effective length of welds shall not be less than the perpendicular distance between them, which in this case is the length of the leg in contact with the gusset plate. Further, the mini-

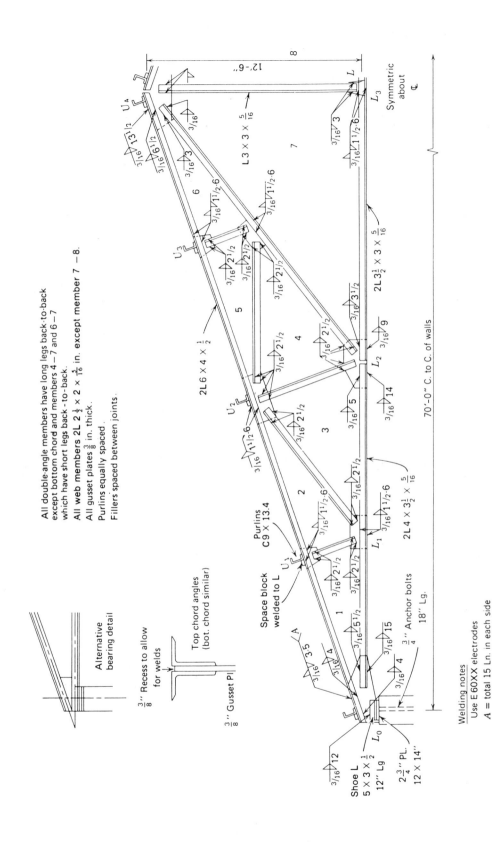

All double-angle members have long legs back-to-back except bottom chord and members 4–7 and 6–7 which have short legs back-to-back.

All web members 2L $2\frac{1}{2} \times 2 \times \frac{5}{16}$ in. except member 7–8.

All gusset plates $\frac{3}{8}$ in. thick.

Purlins equally spaced.

Fillers spaced between joints.

Alternative bearing detail

$\frac{3}{8}''$ Recess to allow for welds

Top chord angles (bot. chord similar)

$\frac{3}{8}''$ Gusset Pl

Purlins C9 × 13.4

Space block welded to L

A

$L_3 \times 3 \times \frac{5}{16}$

$2L 3\frac{1}{2} \times 3 \times \frac{5}{16}$

$2L 6 \times 4 \times \frac{1}{2}$

$2L 4 \times 3\frac{1}{2} \times \frac{5}{16}$

$\frac{3}{4}''$ Anchor bolts 18'' Lg.

Shoe L $5 \times 3 \times \frac{1}{2}$ 12'' Lg

$2\frac{3}{4}''$ PL. 12 × 14''

L_0

12'-6"

Symmetric about ℄

70'-0" C. to C. of walls

Welding notes

Use E 60XX electrodes

A = total 15 Ln. in each side

Figure 9.18/Design drawing.

357

mum weld length should be at least 4 times the nominal size of the weld.

Gusset plates are recessed $\frac{3}{8}$ in. from the outstanding legs of the top and bottom chords (see detail, Fig. 9.18). This is done to provide welding surfaces between the backs of the angles and gussets, while maintaining a smooth, uninterrupted surface at the top and bottom chord for the placement of purlins. The minimum length of intermittent welds along this surface is $1\frac{1}{2}$ in.

The calculation of weld lengths for the joint L_0 follows.

The strength of a $\frac{3}{16}$-in. E60-electrode fillet weld is $0.80(3) = 2.4$ kips per in. (Art. 7.23). Top-chord weld to gusset:

$$\text{total length needed} = \frac{106}{2(2.4)} = 22 \text{ in.}$$

As stated near the outset of this article, welds need not be balanced, but they will be here for illustrative purposes. From properties-of-section tables, determine the distance from the centroid of the angle to the toe and to the heel of the leg that will receive the welds. The distances are 4.01 and 1.99 in., respectively. To balance, the welds should be inversely proportional to these distances:

$$\frac{1.99}{4.01} = \frac{\text{weld length along angle toe}}{\text{weld length along angle heel}}$$

$$= \frac{\text{use } 7\frac{1}{2} \text{ in.}}{\text{use } 15 \text{ in.}}$$

Bottom-chord weld to gusset:

$$\text{total length needed} = \frac{98.3}{2(2.4)} = 20.5 \text{ in.}$$

Balancing welds,

$$\frac{0.93}{2.57} = \frac{\text{toe weld length}}{\text{heel weld length}} = \frac{\text{use } 5\frac{1}{2} \text{ in.}}{\text{use } 15 \text{ in.}}$$

Calculation of weld lengths for the chord member at joint U_1 follows.

As previously noted, the difference between bar stresses (members *B*-1 and *C*-2) is 3.5 kips. Therefore, for top-chord weld to gusset,

$$\text{total length needed} = \frac{3.5}{2(2.4)} = 0.73 \text{ in.}$$

The controlling minimum-weld-length criterion here, then, is the perpendicular distance between welds, or a minimum of 6 in. The minimum $1\frac{1}{2}$-in. intermittent welds also will be used at other joints.

The design of other joints is accomplished in a similar manner.

9.28
End Bearing and Anchorage (Step 14)

The lengths of weld required in the upper and lower chords at the joint L_0 were determined in the preceding article. The length of weld on the shoe angle (Fig. 9.18) is made long enough to transmit the end reaction. In this case, the end reaction due to dead and snow load is $24 + 16.8 = 40.8$ kips, and the length required is

$$\frac{40.8}{2(2.4)} = 8.50; \quad \text{use } 8\frac{1}{2} \text{ in.}$$

The full 12-in. length of the shoe angle will be used.

When roof trusses rest on a masonry wall or pier, a sole plate is usually secured to the underside of the shoe angles and a bearing plate rests on the masonry. These plates are made somewhat wider than the shoe angles in order to provide holes for the anchor bolts, and the minimum thickness of each plate is commonly $\frac{1}{2}$ in. They are designed in the same manner as beam bearing plates (Art. 5.16), each plate resisting one-half the bending moment developed at a vertical section through the edge of the fillet on a shoe angle. The area of the bearing plate must be sufficient to distribute the end reaction over the masonry.

If the allowable bearing pressure (F_p) on the masonry is 250 psi, the required area of the plate in this example is

$$A = \frac{R}{F_p} = \frac{40,800}{250} = 163 \text{ in.}^2$$

A 12×14 in. plate is selected. The actual bearing pressure on the underside of the plate is

$$\frac{40,800}{12(14)} = 243 \text{ psi}$$

The critical moment section through the angle fillet edge occurs 1 in. from the centerline of the gusset (allowing $\frac{1}{4}$ in. for the fillet radius). The bending moment for a strip 1 in. wide is

$$243(6)\frac{6}{2} = 4374 \text{ in.-lb}$$

Since each plate is considered to resist half of this moment, the required section modulus for one plate is

$$S = \frac{M}{F_b} = \frac{2187}{27,000} = 0.081 \text{ in.}^3$$

The required thickness of each plate (Art. 5.16) is

$$t = \sqrt{6S} = \sqrt{6(0.081)} = 0.70 \text{ in.}$$

Two 12×14-in. plates, each $\frac{3}{4}$ in. thick, are satisfactory.

Two $\frac{3}{4}$-in. round anchor bolts approximately 18 in. long are generally provided to fasten the truss to the masonry. The slotted holes in the sole plate provide for movement due to expansion and contraction.

Figure 9.18 also shows an alternative end-bearing detail in which the lower-chord angles act as shoe angles. It will be noted from the figure that the lower chord has been dropped and that the lines of action of the upper and lower chords and the end reaction do not meet at a common point. This, of course, produces a moment in the connection, which may be compensated for in light trusses by making the gusset plate larger and providing additional lengths of weld. There must also be sufficient weld in the lower chord to transmit the end reaction. It is better to avoid this type of end detail in heavy trusses, but where it must be used, the joint should be specially designed to resist the stresses due to eccentricity.

9.29
Design of Bracing (Step 15)

Roof trusses resting on masonry walls usually have diagonal bracing, consisting of $\frac{3}{4}$-in. round rods placed in the plane of the top chord in alternate bays (Fig. 9.19a). In addition, the lower chord frequently is braced with one or two lines of continuous struts, depending on the span (Fig. 9.19b). These struts may consist of angles, channels, or light beams, and can be selected on the basis of a maximum L/r of 200 (Art. 9.24). In this problem, the trusses are 16 ft apart and require struts with a least radius of gyration

$$r = \frac{L}{200} = \frac{16(12)}{200} = 0.96 \text{ in.}$$

From a table of properties of double-angle struts, two $3\frac{1}{2} \times 2\frac{1}{2} \times \frac{1}{4}$-in. angles, with long legs vertical and $\frac{3}{8}$ in. back to back, would be satisfactory. Because of the indefinite nature of the forces which the bracing is called upon to resist, an accurate stress analysis is difficult to model. The designer must rely upon his judgment and experience (see also Chapter 8).

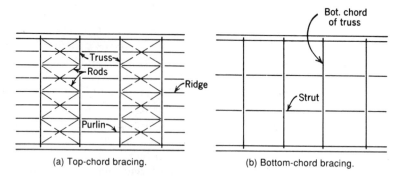

(a) Top-chord bracing. (b) Bottom-chord bracing.

Figure 9.19/Truss bracing.

9.30
The Design Drawing (Step 16)

The size of members, type and lengths of weld at each joint, thickness of gusset plates, and all other details determined by the calculations are recorded on the design drawing shown in Fig. 9.18. It should be noted that because of the gusset-plate size required to adequately receive members, it is often desirable to space out welds, even though more weld is used than is required by the calculation is doing so, i.e., use intermittent welds. Also, one should consider the use of all single-angle members and the placement of welds at angle ends to reduce gusset sizes.

The shop drawings of the fabricator are made from the design drawings and give complete details such as exact weld lengths and spacing, size of gusset plates, length of members, etc.

9.31
Trusses with Loads between Panel Points

It frequently happens that the limiting span of the roof deck is such that purlins must be placed between panel points of the top chord. The lower chord also is often called upon to support loads between panel

points. The chord must then act as a beam in addition to carrying either a direct compression or tension load, i.e., it is subjected to both bending and direct stress.

The stress diagram for the axial stress in the members is drawn in the same manner as explained in Art. 9.12, except that the panel loads must include the load brought to the truss by the intermediate purlins. For example, if the actual top-chord loading on a truss is as shown in Fig. 9.20, the panel load at U_1 for the stress diagram is the sum of 1000 lb due to the purlins framing at the joint, 500 lb from the intermediate purlin to the left, and 500 lb from the intermediate purlin to the right, or a total of 2000 lb. When the panel loads have been determined in this manner, the stress diagram is drawn as previously explained.

In the design of members subjected to bending and direct stress, however, a trial section must be assumed and then checked for adequacy by one of the following methods, as may be appropriate (see also Art. 6.17).

1. Bending and direct tension: Members must satisfy the formula

$$\frac{f_a}{0.6F_y} + \frac{f_b}{F_b} \leq 1.0$$

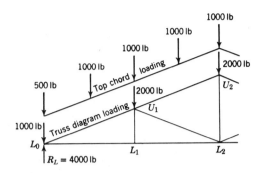

Figure 9.20/Bending in top chord.

where:

$$f_a = \frac{P}{A} \text{ (actual tensile stress).}$$

$$f_b = \frac{Mc}{I} \text{ (actual tensile bending stress).}$$

F_b = allowable maximum tensile bending stress.

In addition, members must satisfy limiting conditions for bending only, between points of lateral support.

2. Bending and direct compression members must satisfy the following requirements:

(a) When $f_a / F_a \leq 0.15$,

$$\frac{f_a}{F_a} + \frac{f_b}{F_b} \leq 1.0$$

(b) When $f_a / F_a > 0.15$,

$$\frac{f_a}{F_a} + \frac{C_m f_b}{\left(1 - \dfrac{f_a}{F'_e}\right) F_b} \leq 1.0$$

where:

$$f_a = \frac{P}{A} \text{ (actual axial compressive stress).}$$

$$f_b = \frac{Mc}{I} \text{ (actual compressive bending stress).}$$

F_a = allowable axial stress based on KL/r.

F_b = allowable compressive bending stress.

$$F'_e = \frac{12\pi^2 E}{23(KL_b / r_b)^2}$$

(L_b = unbraced length in plane of bending; r_b = corresponding radius of gyration; $K = 1.0$).

$C_m = 1.0$, or a lesser value determined by Table C-H1.1 in the 1989 AISC Manual.

In addition, at points braced in the plane of bending, the member must satisfy the basic interaction formula.

The design of such members will be treated in Arts. 9.33 and 9.34.

9.32
Bending Moments for Continuous Members

This topic is discussed in detail in Chpater 10. However, some understanding of what happens to members that are continuous over supports is necessary before one can cope with the design of continuous-chord members which support loads between panel points. The concepts underlying continuity and bending under restraint are illustrated in Fig. 9.21. Figure 9.21a represents a single beam resting on three supports and carrying equal loads at the centers of the two spans. If one imagines the beam to be cut over the middle support as shown in Fig. 9.21b, the result will be two simple beams. Each of these simple beams will deflect as shown. However, when the beam is made continuous over the sup-

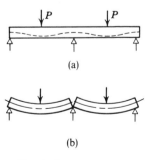

Figure 9.21/Continuity.

port, the deflection curve has a shape similar to that indicated by the dashed line in (a).

It is evident that there is no bending moment developed over the center support in Fig. 9.21b, while there must be a moment over the support in Fig. 9.21a. A study of (a) reveals that there is tension in the bottom of the member at both midspans, and in addition that there is tension in the top of the member over the center support. In other words, the member shown in (a) has a positive bending moment at the middle of each span and a negative moment over the support. It will be shown in Chapter 10 that the positive bending moment at midspan for a continuous member such as shown in (a) is less than that for the simple spans in (b).

The values of the bending moments in continuous beams cannot be found by the usual equations of equilibrium. Additional equations which involve the elasticity of the material are required; therefore, the continuous beam is described as "statically indeterminate." Bending-moment formulas can be developed for various typical conditions of continuous beam loading and restraint, and these are reproduced in many handbooks. For example refer to "Beam Diagrams and Formulas" in Part 2 of the AISC Manual, and in Appendix B of this text. The designer should understand these derivations and exercise judgement in their use, bearing in mind that the actual conditions under which the structure is built may not duplicate the theoretical conditions on which the formulas are based.

9.33
Combined Bending and Direct Tension

The design of members to resist combined bending and tensile stresses is illustrated by the example which follows.

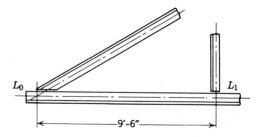

Figure 9.22/Combined tension and bending.

Example

Design the lower chord of a roof truss similar to that shown in Fig. 9.22, carrying a direct tensile load of 46,000 lb and a uniform load over both spans, including the weight of the member, of 600 lb/ft. The distance between panel points is 9 ft 6 in. The member is to be composed of two angles placed $\frac{3}{8}$ in. back to back and made continuous over the first interior panel point. (From tabulated beam diagrams, the positive moment between supports is $0.07wL^2$, and over the interior support the negative moment is $0.125wL^2$.) Use A36 Steel. The arrangement is indicated diagrammatically in Fig. 9.22.

Solution

(1) Assume a trial section of two $6 \times 3\frac{1}{2} \times \frac{3}{8}$-in. angles placed with the long legs vertical (Fig. 9.23). The following properties are found from the AISC Manual:

$$A = 6.84 \text{ in.}^2$$
$$I_x = 25.7 \text{ in.}^4$$
$$c_1 = 2.04 \text{ in.}$$
$$c_2 = 3.96 \text{ in.}$$

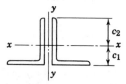

Figure 9.23/Tension and bending.

(2) The direct stress (P) is 46,000 lb, and the actual unit stress due to axial tension is

$$f_a = \frac{P}{A} = \frac{46,000}{6.84} = 6730 \text{ psi}$$

(3) The value of the positive bending moment is

$$0.07wL^2 = 0.07(600)9.5^2(12)$$

$$= 45,500 \text{ in.-lb}$$

(4) The value of the negative bending moment is

$$0.125wL^2 = 0.125(600)9.5^2(12)$$

$$= 81,200 \text{ in.-lb}$$

(5) The maximum tensile bending stress between supports is

$$f_b = \frac{Mc_1}{I} = \frac{45,500(2.04)}{25.7} = 3610 \text{ psi}$$

The maximum tensile bending stress over the interior support L_1 is

$$f_b = \frac{Mc_2}{I} = \frac{81,200(3.96)}{25.7} = 12,500 \text{ psi}$$

(6) The maximum allowable tensile bending stress (F_b) is taken as $0.6F_y = 22,000$ psi, because longitudinal supports bringing the load to the lower chord are assumed to provide lateral support.

(7) The member must satisfy the formula

$$\frac{f_a}{0.6F_y} + \frac{f_b}{F_b} \le 1.0$$

$$\frac{6730}{22,000} + \frac{12,500}{22,000} = 0.306 + 0.568$$

$$= 0.874 < 1.0$$

If bolts are to be used to connect this member at panel points, it will need to be further checked for adequacy in net section.

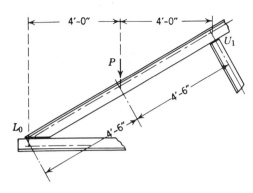

Figure 9.24/Combined compression and bending.

9.34
Combined Compression and Bending

The design of top-chord members to resist combined bending and direct compression is illustrated by the example which follows.

Example

Design a top-chord member composed of two angles with long legs $\frac{3}{8}$ in. back to back, carrying a direct compression load of 65,000 lb, and a 3500-lb concentrated load (P) in each panel. The member is continuous over U_1. From tabulated beam diagrams, the positive moment between supports is $0.156PL$, and over the interior support the negative moment is $0.187PL$. Purlins will be considered as furnishing lateral support to the top-chord member at the points where they are attached to the chord. The arrangement and dimensions are shown in Fig. 9.24. Use A36 Steel. Neglect weight of angles.

Solution

(1) Assume a trial section of two angles, $7 \times 4 \times \frac{1}{2}$ in. (Fig. 9.25). The following

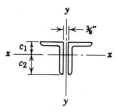

Figure 9.25/Compression and bending.

properties are found from the AISC Manual:

$$A = 10.5 \text{ in.}^2$$
$$I_x = 53.3 \text{ in.}^4$$
$$c_1 = 2.42 \text{ in.}$$
$$c_2 = 4.58 \text{ in.}$$
$$r_x = 2.25 \text{ in.}$$
$$r_y = 1.57 \text{ in.}$$

(2) The direct stress (P') is 65,000 lb, and the actual unit stress is

$$f_a = \frac{P'}{A} = \frac{65}{10.5} = 6.19 \text{ ksi}$$

(3) The value of the positive bending moment is

$$0.156PL = 0.156(3500)8(12)$$
$$= 52,400 \text{ in.-lb.}$$

and

$$f_b = \frac{Mc_1}{I} = \frac{52.4(2.42)}{53.3} = 2.38 \text{ ksi}$$

(compression)

(4) The value of the negative bending moment is

$$0.187PL = 0.187(3500)8(12)$$
$$= 62,800 \text{ in.-lb}$$

and

$$f_b = \frac{Mc_2}{I} = \frac{62.8(4.58)}{53.3} = 5.40 \text{ ksi}$$

(compression)

(5) Determine the value of F_a:

$$\frac{KL}{r_y} = \frac{1.0(4.5)12}{1.57} = 34.4$$

$$\frac{KL}{r_x} = \frac{1.0(9)12}{2.25} = 48.0$$

From the AISC Manual, Part 3, Table C-36, $F_a = 18.53$ ksi.

(6) Determine whether 1989 AISC Formula H1.1 or H1.3 applies:

$$\frac{f_a}{F_a} = \frac{6.19}{18.53} = 0.33 > 0.15$$

Use Formula H1.1.

(7) The allowable compressive bending stress (Art. 5.13) is governed by lateral support. The points where stress and bending moment are maximum (midspan and interior support) have a purlin giving lateral support at these points. Therefore, $F_b = 22$ ksi.

(8) To obtain F_e'

$$\frac{KL_b}{r_b} = \frac{1.0(9)12}{2.25} = 48.0$$

From AISC Table 8, $F_e' = 64.81$ ksi. Then the check at midpoint is

$$\frac{f_a}{F_a} + \frac{C_m f_b}{\left(1 - \dfrac{f_a}{F_e'}\right)F_b} \leq 1.0$$

$$\frac{6.19}{18.53} + \frac{1.0(2.38)}{\left(1 - \dfrac{6.19}{64.81}\right)22} = 0.334 + 0.120$$

$$= 0.454 < 1.0$$

(9) Checking at the ends where the bending stress is larger but no buckling occurs

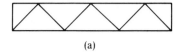 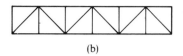

(a)   (b)

Figure 9.26/Warren truss with parallel chords.

about either axis,

$$\frac{f_a}{0.6F_y} + \frac{f_b}{F_b} \leq 1.0$$

$$\frac{6.19}{22.00} + \frac{5.40}{22.00} = 0.28 + 0.245$$

$$= 0.525 < 1.0$$

A lighter-weight section obviously should be sought; first try two $6 \times 4 \times \frac{1}{2}$-in. angles.

9.35
Parallel-Chord Trusses

Trusses with parallel chords are frequently used for flat-roof construction (and for the support of floors over long spans). One of the chief advantages of parallel-chord trusses over plate girders for such purposes is the comparatively clear space provided for pipes, conduits, ventilating ducts, and the like. The depth of such trusses is frequently established by architectural considerations; however, in the absence of other governing factors, a depth of $\frac{1}{8}$ to $\frac{1}{12}$ the span will be found to be a reasonable starting pont for achieving an economical design.

Figure 9.26a illustrates a Warren truss with parallel chords. The distance between joints on the upper or lower chords may be reduced by the addition of vertical web members as shown in (b) of the same figure. Roof trusses of this same general type may also be built with the top chord slightly inclined, as shown in Fig. 9.2 (i) and (j). They are particularly applicable where a roof slope is required to ensure proper draingage. Purlins can be connected directly to the truss, without the necessity for blocking to establish the slope. Here again, a depth at the center of $\frac{1}{8}$ to $\frac{1}{12}$ the span length will usually prove economical. The depth at the wall may be varied to suit the end conditions.

9.36
Roof-Deck Construction

Precast slabs of gypsum or lightweight concrete are only two of the great variety of roof-deck systems available. Some of these can span up to 7 ft and are therefore supported directly on the purlins. Others, which span lesser distances, must be supported on subpurlins, generally T bars spaced 30 in. on centers and welded or clipped to the main purlins (Fig. 9.27). For those roofing materials requiring it, a 2-in. coating of "nailing concrete" may be applied over the slabs to provide a nailing surface. Slabs should be securely clipped to their supporting members. Complete details for design, such as available slab thicknesses, permissible spans, weight of

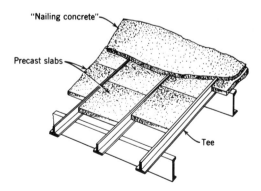

Figure 9.27/Precast slabs supported on tee subpurlins.

construction, etc., may be found in the literature of the manufacturers.

Ribbed steel deck panels (Fig. 8.10), covered with an insulating material and a membrane roofing, are also frequently used, the deck being either clipped or welded to the purlins. Also, of course, there are numerous other products and concepts which should be explored before deciding upon the most suitable type of construction.

PROBLEMS

1. Compute the snow panel-point loads for the fan truss shown in Fig. 9.2c. The span of the truss is 38 ft, and it has a vertical rise of 10 ft. Trusses are spaced 14 ft on centers. The snow load is 30 psf of horizontal projection. (Answers given in Appendix G.)

2. Find the stress in the members of the king-post truss shown in Fig. 9.2g, if the span is 38 ft, the rise is 12 ft, and the top-chord panel loads are 8000 lb each, applied vertically.

3. Find the stress in the members of the fan truss shown in Fig. 9.2c, if the span is 40 ft, the rise is 8 ft, and the top-chord panel loads are 4000 lb each, applied vertically. (Answers given in Appendix G.)

4. Find the stress in the members of the Pratt truss shown in Fig. 9.2j, if the span is 64 ft, the height is 8 ft, and the top-chord panel loads are 6000 lb each, applied vertically. Top and bottom chords are parallel.

5. Design an 8-panel Warren roof truss of the type shown in Fig. 9.2i, using verticals at each panel pont. The truss is to rest on masonry piers with the span from center to center of bearing being 70 ft. The depth of the truss at midspan is to be 10 ft, and the depth at each end, 8 ft. The trusses are spaced 22 ft apart. Interior top-chord panel loads due to construction (including the truss weight) are 4000 lb each, and the interior lower-chord panel loads due to a suspended ceiling are 2500 lb each. The snow load is 30 psf of horizontal projection. Bottom chord bracing is at every other

panel point. Make a design drawing of the truss. Use A36 Steel. Design for a combination of live and dead load (no wind load).

6. Using the example problem given in Art. 9.31 (Fig. 9.20), design the top-chord member subjected to direct compression and bending, using two channels. Base the design on the 1989 AISC Specification, using A242 Steel. Assume that all purlins provide point lateral support. Span of truss is 64 ft, rise is 10 ft., and $\frac{3}{8}$-in. gusset plates are used. (Answer given in Appendix G.)

9.37
One-Story Braced Frame Construction—General

The discussion of roof trusses in the preceding articles was confined to those resting on load-bearing walls or walls with integral stable piers. As noted at the outset of this chapter, these walls and/or piers are inherently stable against racking forces. Stability of the total structure is achieved by longitudinal bracing between trusses as well as the roof deck; by end panels; and by securing trusses to the walls or piers.

For many structures, such as mill buildings, the trusses are supported on exterior columns. The exterior wall surface then consists of thin curtain walls affixed to the columns, or walls supported directly on foundations, but not necessarily imparting a dependable degree of racking resistance to the total structure.

Under such conditions, each frame, consisting of a truss and supporting columns, lacks lateral stability. Needed frame stability is usually achieved by means of a knee brace, as shown in Fig. 9.28a, or by making supporting end columns act as end posts when trusses of the type shown in Fig. 9.28b are used. The resulting frame, in either case, is known as a bent and is statically indeterminate for lateral loads.

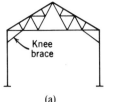
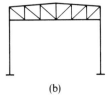

(a)   (b)

Figure 9.28/Transverse bent.

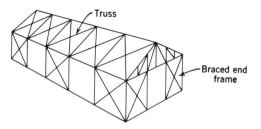

Figure 9.31/Braced frame, type 2.

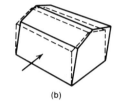

(a)   (b)

Figure 9.29/Racking.

may be indeterminate; however, approximate methods of analysis based on statically determinate principles and tempered with experience and judgment usually substitute for precise mathematical analyses. (See also Art. 8.2.)

9.38
Braced Frame—Types

Two typical braced frames are shown in Figs. 9.30 and 9.31, and principal members and components are identified.

Although it is an oversimplification, one may view the transverse bents of Fig. 9.30a as providing lateral bracing; the side-wall portals (or the side-wall cross-bracing in Fig. 9.30b), struts, and girts as providing longitudinal bracing; and the cross-bracing in the upper and/or lower chords as providing diagonal bracing. Actually, because the various members and components are

When several bents are spaced longitudinally to form a building, the entire resulting building frame must then be made stable against racking. Racking failure may occur along any axis; however, it is generally sufficient to consider only two (Fig. 9.29), i.e., longitudinal and lateral. The building frame is unstable both during and after erection; therefore, bracing squares the building during erection and provides needed stability after the building is completed.

The analysis of braced-frame elements will be treated in the articles which follow, as well as bracing systems. Such structures

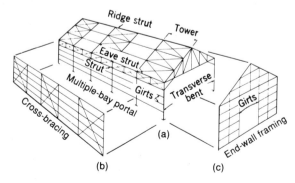

Figure 9.30/Braced frame, type 1.

brought together as a single structure, each assists the others in resisting imposed loads. Member stresses will vary with the nature and direction of loading as well as with the juxtaposition of members and components.

The transverse bents of Fig. 9.28 are stable against sidesway, but not stable when subjected to forces normal or diagonal to the plane of their span. A stable structural substructure can be produced, however, by introducing bracing between two bents in the plane of side walls, upper and/or lower chords, and by using longitudinal struts between bents at points of cross-bracing connection. Such a stable element is known as a tower or braced bay (Fig. 9.30a). Each building frame should have at least two such elements, and it is customary to so brace each fourth or fifth bay in long structures and alternate bays in shorter structures. At times, bracing in every bay may be justified.

Longitudinal struts run from end to end of the building to prevent vibration, and are of constant cross section. Ridge and eave struts are always provided and, as previously mentioned, should be located wherever cross-bracing connects to columns or chords. Purlins often substitute for struts in the top chord, and girts, which are used to receive the exterior cladding, may also be designed to serve as struts.

The end wall of a structure using towers may be framed simply to receive exterior cladding (Fig. 9.30c).

The multiple-bay portal of Fig. 9.30a is useful when more than a normal amount of free area is desired for placement of fenestration and entrances. However, full cross-bracing is more efficient and is therefore preferred: the framing is simpler and the column moments are less. The use of a side-wall girt along the line of the knee-brace connection or the lower-chord connection to column is preferred by some designers (Fig. 9.32b). Placement of cross-bracing in interior rather than exterior

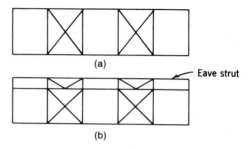

Figure 9.32/Side-wall cross-bracing.

panels, as shown in Fig. 9.32a, is often preferred in long structures because accumulated temperature movements tend to distort outer columns when the bracing is placed only in end bays.

It is also customary to provide vertical cross-bracing between truss chords to assist in truss erection as well as to impart added stability. Such bracing may be placed in alternate bays, so that trusses may be erected in pairs, or in all bays. Generally, one set of such longitudinal bracing is placed on the building centerline, and if the structure is wide, in two or even three vertical, longitudinal planes.

All cross-bracing should be put in initial tension so that there will be no frame movement under load before the bracing members accept stress. To accomplish this, bracing members are generally cut 0.02 in. short for every 10 ft of length, and then drifted into position for connection. Those diagonals included in the same direction are assumed to act in unison and only in tension—when forces are applied from the opposite direction, the opposite diagonals will act.

A second type of braced frame is shown in Fig. 9.31. Instead of transverse bents, trusses are landed on columns, and end walls (in this instance braced end frames) are used to impart lateral stability to the entire structure. At least two panels of cross-bracing must be used in the end frame, and are generally required to permit the placement of doors.

Because there are no towers, and the interior trusses, together with their supporting columns, are not laterally stable in themselves, bracing in a horizontal or nearly horizontal position is needed to carry lateral loads to the end walls. This is generally accomplished by providing cross-bracing between lower chords in each bay, and/or between upper chords, to form one or more longitudinal trusses from end to end of the building. In other words, these longitudinal horizontal and/or inclined trusses will have their horizontal reactions in the end walls or end braced frames.

To resist longitudinal forces, full cross-bracing is shown in alternate bays in the plane of side walls. The use of struts, girts, and vertical and longitudinal cross-bracing is as previously described. The entire braced structure is thus capable of resisting diagonal forces as well.

The degree to which any structure is braced and the specific pattern used, including consideration of possible diaphragm action of sheathing, are usually matters of individual choice. They depend upon the designer's appraisal of the forces to be resisted and the manner in which he feels they may most efficiently be carried to foundations (see also Art. 8.2). And, of course, the designer will need to consider building function and desired architectural characteristics. Generally speaking, bracing is relatively inexpensive in relation to the structure as a whole; therefore, one should not attempt to effect economies in this area of design without considerable experience.

9.39
Braced Frame—Component Analysis

Two of the more basic components of braced-frame buildings are the transverse bent and the portal—the latter being helpful in the analysis of multistory buildings as well (see Chapter 13). An approximate method of analysis will be presented first for a single portal, then for a multiple portal and a transverse bent.

Applicable to both bents and portals are assumptions regarding column restraints. A simple portal consisting of two columns, each pin-ended at its base, with a girder between them (Fig. 9.33a), has four reaction components and is thus indeterminate to the first degree. An assumption is needed to eliminate one unknown. If it is assumed that the girder will not shorten significantly under direct stress, the two columns will then deflect horizontally an equal amount Δ (Fig. 9.33b), and the applied load P positioned as shown must be divided equally to produce equal deflections, i.e., $H_1 = H_2 = P/2$. It remains to determine the values of V_1 and V_2 from the laws of statics.

A simple portal of two columns with fixed bases and a girder between columns (Fig. 9.34a) has six reaction components and is thus indetermineate to the third degree. If restraints are assumed to be equal at the top and bottom of columns, there is an inflection point at the midheight of each column, and that portion above the points of inflection may be treated as a free body having the same characteristics as the portal of Fig. 9.33. To permit analysis by the laws of statics, it is necessary both to assume a location for the points of inflection and that the horizontal shear is divided equally between the two columns ($H_1 = H_2$). It is of extreme importance, then,

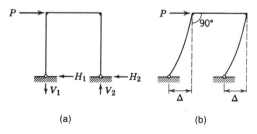

Figure 9.33/Pin-ended base: assumptions.

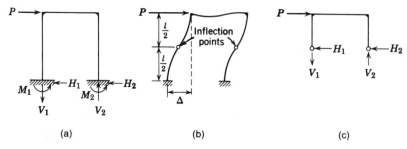

Figure 9.34/Fixed base: assumptions.

that the degree of fixity of the columns at both ends be correctly appraised. If both ends are fully fixed, the maximum column moment will be but one-half that of a column pin end connected at the base only. If there is a difference between degrees of fixity, the point of inflection will be nearer the less restrained end.

9.40
Portals

Two illustrative examples will now be presented to show the essential stress calculations for the two conditions described above, i.e., pin-end and fixed bases.

Example 1

A portal is to be constructed as shown in Fig. 9.35a. Column bases are to be pin-connected. Determine the stress in all members caused by the horizontal force of 2000 lb.

Solution

Note. The following conventions will be adopted for indicating the sign of stresses in the members:

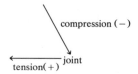

(1) Assuming the total shear is divided equally between columns,

$$H_1 = H_2 = 1 \text{ kip}$$

(2) Taking moments about the base of the lefthand column to determine V_2,

$$\sum M = 0, \qquad V_2 = \frac{2(24)}{16} = 3 \text{ kips} \uparrow$$

Then, by summation of the vertical stress resultants,

$$\sum V = 0, \qquad 3 - V_1 = 0$$
$$V_1 = 3 \text{ kips} \downarrow$$

(3) Isolating the left-hand column and the members framing into it (Fig. 9.35b), calculate the stress in the members. There are three unknowns ($\overline{B\text{-}2}$, $\overline{1\text{-}2}$, and $\overline{C\text{-}1}$) and two points of zero moment (the top and base of the column); therefore, by taking moments about the top, two unknowns will be eliminated for the first calculation:

$$\sum M = 0, \qquad -1(24) + \overline{C\text{-}1}_H(8) = 0$$
$$8\overline{C\text{-}1}_H = 24$$
$$\overline{C\text{-}1}_H = 3 \text{ kips} \rightarrow$$

All angles are 45°; therefore,

$$\overline{C\text{-}1} = \frac{3}{0.707} = 4.24 \text{ kips} \nearrow \text{ (tension)}$$

Taking the sum of the vertical components

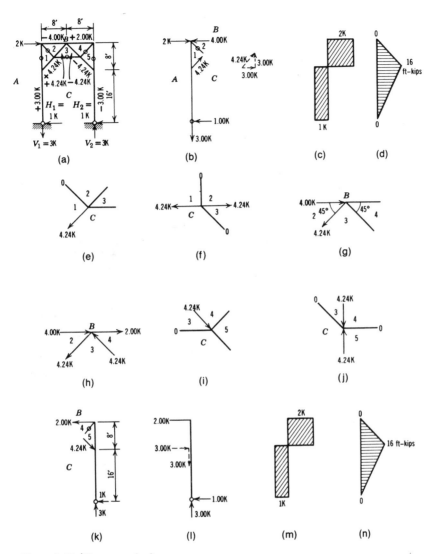

Figure 9.35/Stress analysis.

(from the base upwards),

$$\sum V = 0, \qquad -3+3\pm\overline{1\text{-}2}_V=0$$

$$\overline{1\text{-}2}=0$$

Taking the sum of the horizontal components (from the base upwards),

$$\sum H = 0, \qquad -1+3+2-\overline{B\text{-}2}=0$$

$$\overline{B\text{-}2}=4 \text{ kips} \leftarrow (\text{compression})$$

To check the accuracy of the calculations

thus far, take moments about the column base:

$$\sum M = 0$$

$$+3(16)+2(24)-4(24)=0$$

$$+96-96=0$$

(4) Using the horizontal force and stress components, the shear diagram (Fig. 9.35c) may be constructed for the left-hand column, and from shear areas, the moment

diagram (Fig. 9.35d). Then

$$V_{max} = -2 \text{ kips}$$
$$M_{max} = 16 \text{ ft-kips}$$

(5) Stresses in the remaining members will be found using the *method of joints*. Joint $\overline{1\text{-}2\text{-}3\text{-}C}$ (Fig. 9.35e) is rotated to the position in Fig. 9.35f, and the horizontal components are added algebraically:

$$\Sigma H = 0, \qquad -4.2 + \overline{2\text{-}3} = 0$$

$$\overline{2\text{-}3} = 4.24 \text{ kips} \rightarrow (\text{tension})$$

The stress in $\overline{C\text{-}3}$ is 0. Next, joint $\overline{2\text{-}B\text{-}4\text{-}3}$ (Fig. 9.35g) is analyzed. $\Sigma V = 0$; therefore, it is apparent that the stress in $\overline{3\text{-}4}$ is 4.24 kips compression (Fig. 9.35h). Then,

$$\Sigma H = 0, \qquad +4 - 3 - 3 + \overline{B\text{-}4} = 0$$

$$\overline{B\text{-}4} = 2 \text{ kips} \rightarrow (\text{tension})$$

For joint $\overline{3\text{-}4\text{-}5\text{-}C}$ (Fig. 9.35i) rotated as shown in Fig. 9.35j, it is apparent from

$\Sigma V = 0$ that the stress in $\overline{C\text{-}5}$ is 4.24 kips compression.

(6) It remains to analyze the right-hand column (Fig. 9.35 k and l):

$$\Sigma V = 0, \qquad +3 - 3 = 0$$
$$\Sigma H = 0, \qquad -2 + 3 - 1 = 0$$

The shear and moment diagrams are shown in Fig. 9.35 (m) and (n).

Example 2

The portal of Example 1 is to be constructed as shown in Fig. 9.36a. Column bases are fixed. Determine the stress in all members caused by the horizontal force of 2000 lb.

Solution

Note. The same sign convention adopted for Example 1 will be used. It should be

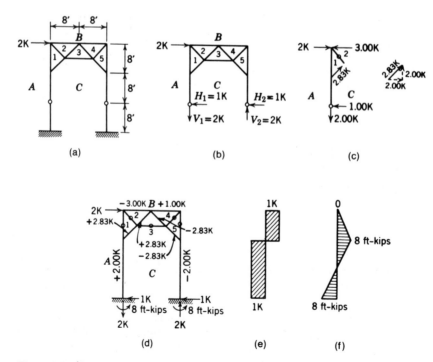

(a)   (b)   (c)

(d)   (e)   (f)

Figure 9.36/Stress analysis.

noted that a column which carries a large direct stress, as would be the case if the portal columns were also the transverse bent columns and thus carried a portion of the roof load, may well be fixed against rotation without special anchorage. So long as the moment does not reverse the pressure beneath the bearing plate and the plate is restrained against lateral movement, the column may be assumed to be fixed. Analysis for possible stress reversal should be made, however, and should a stress reversal be found, special anchor bolts, adequate for the load and under initial tension to avoid any movement, would be necessary to resist the uplift force (Fig. 6.28).

(1) It will be assumed that the point of inflection occurs at the midheight of columns, and that the total shear is divided equally between columns (Fig. 9.36b), or

$$H_1 = H_2 = 1 \text{ kip}$$

(2) Isolating that portion of the portal above the point of inflection, and taking moments about the inflection point of the left-hand column to determine V_2,

$$\sum M = 0, \qquad V_2 = \frac{2(16)}{16} = 2 \text{ kips} \uparrow$$

Then, by vertical summation,

$$\sum V = 0, \qquad 2 - V_1 = 0$$
$$V_1 = 2 \text{ kips} \downarrow$$

(3) Isolating the left-hand column and the members framing into it, calculate the stress in the members (Fig. 9.36c). Take moments about the top of the column:

$$\sum M = 0, \qquad -1(16) + \overline{C\text{-}1}_H(8) = 0$$
$$8\overline{C\text{-}1}_H = 16$$
$$\overline{C\text{-}1}_H = 2 \text{ kips} \rightarrow$$

All angles are $45°$; therefore,

$$\overline{C\text{-}1} = \frac{2}{0.707} = 2.83 \text{ kips} \nearrow \text{ (tension)}$$

Taking the sum of the vertical forces (from the base upwards),

$$\sum V = 0, \qquad -2 + 2 \pm \overline{1\text{-}2}_V = 0$$
$$\overline{1\text{-}2} = 0$$

Taking the sum of the horizontal components (from the base upwards),

$$\sum H = 0, \qquad -1 + 2 + 2 - \overline{B\text{-}2} = 0$$
$$\overline{B\text{-}2} = 3 \text{ kips} \leftarrow \text{ (compression)}$$

To check the accuracy of the calculation thus far, take moments about the column base:

$$\sum M = 0, \qquad +2(8) + 2(16) - 3(16) = 0$$
$$+48 - 48 = 0$$

(4) Using the horizontal components, the shear diagram (Fig. 9.36e) may be constructed for the left-hand column, and from shear areas, the moment diagram (Fig. 9.36f).

(5) The stresses in the remaining members and in the right-hand column are found in the same manner as in Example 1.

(6) The moment at the base may be checked by taking moments at the base. For the left-hand column,

$$\sum M = 2(16) + 2(24) - 3(24) = 8 \text{ ft-kips}$$

which agrees with the value obtained by plotting the moment diagram from shear areas.

9.41
Multiple Portals

A multiple portal of the type shown in Fig. 9.37 has horizontal and vertical reaction components and a moment at each column base, or a total of 12 unknowns. The struc-

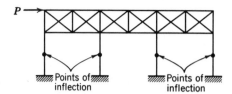

Figure 9.37/Multiple portal.

ture is thus highly indeterminate, and several assumptions must be made if the structure is to be subjected to a determinate method of analysis.

The first assumption is that for fixed bases, the points of inflection will occur at the midheight of columns, the same as for the portal of Example 2. This implies that the location of points of inflection is set by considered judgment of the relative fixities of the columns.

Second, it is assumed that the shear is distributed to the columns in proportion to the moment of inertia of the cross sections of the respective columns. Thus, if all columns are alike, the shear is divided equally among the columns. If, however, the moment of inertia of interior columns is twice that of exterior columns, load P will be divided by 6, and then $\frac{1}{3}P$ will be resisted by each interior column and $\frac{1}{6}P$ by each exterior column. This assumption permits the use of $\Sigma H = 0$ to resolve 4 of the 12 unknowns.

Column moments may be found by applying the horizontal shear at the point of inflection and then taking moments either about the base or about the bottom of the knee brace. The moments resisted by the interior columns are thus related to the horizontal shears, and from $\Sigma M = 0$, another 4 unknowns may be resolved.

It remains to consider the vertical reaction components in each column. The division of direct stress varies in proportion to the distance of the particular column from the centroid of the portal system. In this sense, that portion of the portal system above a horizontal line drawn through the points of inflection acts much as a beam, the neutral axis of which passes vertically through the centroid of the portal system. This is shown diagrammatically in Fig. 9.38. The trussed portion of the portal system above the knee brace must be sufficiently stiff to permit the assumption of equality of end conditions and a straight-line variation in direct column stress. This having been assured, the flexure formula may be applied:

$$f_a = \frac{My}{I}$$

where:

f_a = column direct unit stress.

M = overturning moment or Px.

I = moment of inertia = ΣAy^2

$\quad$ (A = column area).

Assuming, for the multiple portal of Figs. 9.37 and 9.38, that interior columns have

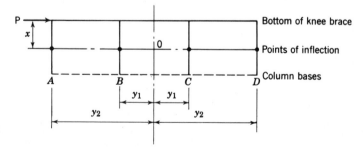

Figure 9.38/Multiple portal: assumptions.

twice the cross-sectional area ($A = 2$, or twice an assumed unity) and moment of inertia of exterior columns,

$$I = \sum Ay^2 = 2\left(y_1^2 + y_1^2\right) + 1\left(y_2^2 + y_2^2\right)$$
$$= 4y_1^2 + 2y_2^2$$

But $y_2 = 3y_1$ for the limited case shown of equal span portals; therefore,

$$I = 4y_1^2 + 2y_2^2$$
$$= 4y_1^2 + 2(3y_1)^2$$
$$= 22y_1^2$$

The stress f_a in columns B and C is

$$f_a = \frac{My}{I} = \frac{Px(y_1)}{22y_1^2}$$

The stress f_a in columns A and D is

$$f_a = \frac{My}{I} = \frac{Px(y_2)}{22y_1^2}$$

Thus

$$RA_V, RD_V = f_a(A) = 1\left[\frac{Px(y_2)}{22y_1^2}\right]$$

$$RB_V, RC_V = f_a(A) = 2\left[\frac{Px(y_1)}{22y_1^2}\right]$$

The procedure may now be applied to an example problem.

Example

Determine the approximate stress in all members of the multiple portal shown in Fig. 9.39. The column bases are fixed, and points of inflection may be assumed to occur at midheight of columns. Interior columns have twice the area and moment of inertia of exterior columns.

Solution

Note. Exterior column areas are taken as unity.

(1) The horizontal force of 20 kips will be divided among the columns in proportion to their ability to resist shear:

$$\frac{20}{6} = 3.33 \text{ kips}$$
$$RA_H = RD_H = 3.33 \text{ kips}$$
$$RB_H = RC_H = 2(3.33) = 6.67 \text{ kips}$$

Checking,

$$\sum H = 0$$
$$3.33 + 6.67 + 6.67 + 3.33 - 20.00 = 0$$
$$20 - 20 = 0$$

(2) The column stresses may be determined using the formulas developed in the preceding discussion (the portal of Fig. 9.39 satisfies the limiting conditions set

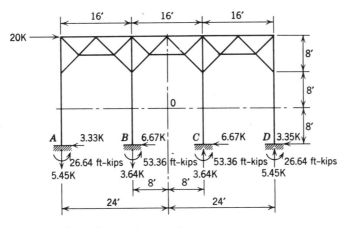

Figure 9.39/Reactions and moments.

forth). The stress f_a in columns B and C is

$$\frac{Px(y_1)}{22y_1^2} = \frac{20(16)8}{22(8)^2} = 1.82 \text{ kips/ft}^2$$

The stress f_a in columns A and D is

$$\frac{Px(y_2)}{22y_1^2} = \frac{20(16)24}{22(8)^2} = 5.45 \text{ kips/ft}^2$$

(3) The vertical reaction components are

$$RA_V = RD_V = 1\left[\frac{Px(y_2)}{22y_1^2}\right] = 1\left[\frac{20(16)24}{22(8)^2}\right]$$

$$= 5.45 \text{ kips}$$

$$RB_V = RC_V = 2\left[\frac{Px(y_1)}{22y_1^2}\right]$$

$$= 2\left[\frac{20(16)8}{22(8)^2}\right]$$

$$= 3.64 \text{ kips}$$

Checking,

$$\sum M_0 = 0$$

$$20(16) - 5.45(24) - 3.64(8)$$

$$- 3.64(8) - 5.45(24) = 0$$

$$320 - 320 = 0$$

(4) Moments at the column base may be found by taking moments about the respective points of inflection of the horizontal reaction component at the base (the same as for a cantilever beam):

$$MA = MD = 3.33(8) = 26.64 \text{ ft-kips}$$

$$MB = MC = 6.67(8) = 53.36 \text{ ft-kips}$$

Moment diagrams may be constructed the same as for the single portal. The moments at the knee brace and column base are equal for each column.

(5) Determine the stress in the trussed members, beginning with the left-hand col-

umn. (Using Bow's notation, re-letter and number the portal system, Fig. 9.40a.) Taking moments about the top of the column (Fig. 9.40b),

$$\sum M = 0, \qquad 3.33(16) - \overline{E\text{-}1}_H(8) = 0$$

$$8\overline{E\text{-}1}_H = 53.28$$

$$\overline{E\text{-}1}_H = 6.66 \text{ kips} \rightarrow$$

All angles are $45°$; therefore,

$$\overline{E\text{-}1} = \frac{6.66}{0.707} = 9.42 \text{ kips} \nearrow \text{(tension)}$$

Taking the sum of the verticals (from the base upwards),

$$\sum V = 0, \qquad -5.45 + 6.66 - \overline{1\text{-}2}_V = 0$$

$$\overline{1\text{-}2}_V = 1.21 \text{ kips} \downarrow$$

Then,

$$\overline{1\text{-}2} = \frac{1.21}{0.707} = 1.71 \text{ kips} \searrow \text{(tension)}$$

and

$$\sum H = 0$$

$$-3.33 + 6.66 + 1.21 + 20.00 - \overline{B\text{-}2} = 0$$

$$\overline{B\text{-}2} = 24.50 \text{ kips} \leftarrow \text{(compression)}$$

Moving to joint $\overline{E\text{-}1\text{-}2\text{-}3}$ (Fig. 9.40c),

$$\sum V = 0$$

$$+\overline{1\text{-}2}_V - \overline{E\text{-}1}_V \pm \overline{2\text{-}3}_V = 0$$

$$+1.21 - 6.66 + \overline{2\text{-}3}_V = 0$$

$$\overline{2\text{-}3}_V = 5.45 \text{ kips} \uparrow$$

$$\overline{2\text{-}3} = \frac{5.45}{0.707} = 7.71 \text{ kips} \nearrow \text{(tension)}$$

$$\sum H = 0$$

$$-\overline{E\text{-}1}_H - \overline{1\text{-}2}_H + \overline{2\text{-}3}_H \pm \overline{3\text{-}E} = 0$$

$$-6.66 - 1.21 + 5.45 + \overline{3\text{-}E} = 0$$

$$\overline{3\text{-}E} = 2.42 \text{ kips} \rightarrow \text{(tension)}$$

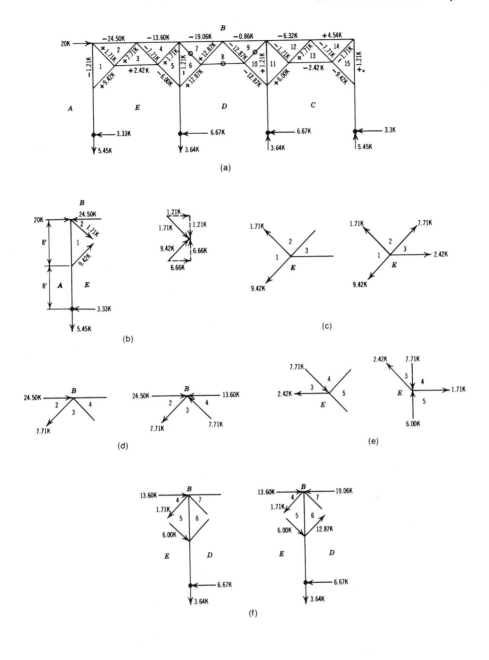

Figure 9.40/Member stresses.

Joint $\overline{3\text{-}2\text{-}B\text{-}4}$ (Fig. 9.40d):

$$\sum V = 0$$

$$-\overline{3\text{-}2}_V + \overline{4\text{-}3}_V = 0$$

$$-5.45 + \overline{4\text{-}3}_V = 0$$

$$\overline{4\text{-}3}_V = 5.45 \text{ kips} \uparrow$$

$$\overline{4\text{-}3} = \frac{5.45}{0.707}$$

$$= 7.71 \text{ kips} \nwarrow \text{ (compression)}$$

$$\sum H = 0$$

$$+\overline{2\text{-}B} - \overline{4\text{-}3}_H - \overline{3\text{-}2}_H \pm \overline{B\text{-}4} = 0$$

$$+24.50 - 5.45 - 5.45 - \overline{B\text{-}4} = 0$$

$$\overline{B\text{-}4} = 13.60 \text{ kips} \leftarrow \text{ (compression)}$$

Joint $\overline{E\text{-}3\text{-}4\text{-}5}$ (Fig. 9.40e):

$$\sum V = 0$$

$$+\overline{E\text{-}3}_V - \overline{3\text{-}4} \pm \overline{5\text{-}E} = 0$$

$$+1.71 - 7.71 + \overline{5\text{-}E} = 0$$

$$\overline{5\text{-}E} = 6.00 \text{ kips} \uparrow \text{ (compression)}$$

$$\sum H = 0$$

$$-\overline{E\text{-}3}_H \pm \overline{4\text{-}5} = 0$$

$$-1.71 + \overline{4\text{-}5} = 0$$

$$\overline{4\text{-}5} = 1.71 \text{ kips} \rightarrow \text{ (tension)}$$

Column $\overline{D\text{-}E}$ (Fig. 9.40f): Taking moments about the top of the column,

$$\sum M = 0$$

$$+6.67(16) - 4.24(8) \pm 8\,\overline{6\text{-}D}_H = 0$$

$$+106.72 - 33.92 + 8\,\overline{6\text{-}D}_H = 0$$

$$8\,\overline{6\text{-}D}_H = 72.80$$

$$\overline{6\text{-}D}_H = 9.10 \text{ kips} \rightarrow$$

$$\overline{6\text{-}D} = \frac{9.10}{0.707}$$

$$\overline{6\text{-}D} = 12.87 \text{ kips} \nearrow \text{ (tension)}$$

$$\sum V = 0$$

$$-3.64 - \overline{E\text{-}5}_V + \overline{6\text{-}D}_V - \overline{5\text{-}4}_V \pm \overline{7\text{-}6}_V = 0$$

$$-3.64 - 4.24 + 9.10 - 1.21 \pm \overline{7\text{-}6}_V = 0$$

$$\overline{7\text{-}6}_V = 0$$

$$\overline{7\text{-}6} = 0$$

$$\sum H = 0$$

$$-6.67 + \overline{E\text{-}5}_H + \overline{6\text{-}D}_H - \overline{5\text{-}4}_H$$

$$+\overline{4\text{-}B} \pm \overline{B\text{-}7} = 0$$

$$-6.67 + 4.24 + 9.10 - 1.21$$

$$+13.60 - \overline{B\text{-}7} = 0$$

$$\overline{B\text{-}7} = 19.06 \text{ kips} \leftarrow \text{ (compression)}$$

The remainder of the Fig. 9.40a member stresses are determined in a like manner.

The various members of the trussed portion of the portal system are designed in a manner similar to that for truss members treated in earlier articles.

PROBLEM

Determine the stress in all members of the portals shown in Fig. 9.41. Construct shear and moment diagrams for both windward and leeward columns for each of the single portals. (Partial answers given in Appendix G.)

9.42
Transverse Bent

A transverse bent is shown in Fig. 9.42a. As stated in Art. 9.37, it is essentially a truss, supported on columns and braced for lateral stability by knee braces between the lower chord and the column.

The general discussion in Art. 9.38 regarding the action of bents and portals, including establishment of points of inflection in

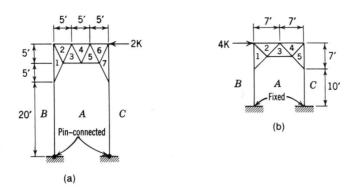

(a)

(b)

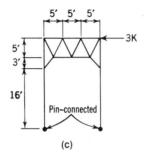

(c)

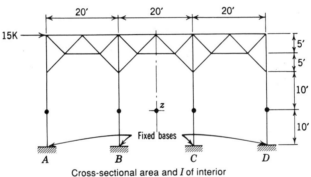

Cross-sectional area and I of interior
columns are twice those of exterior columns

(d)

Figure 9.41/Problems.

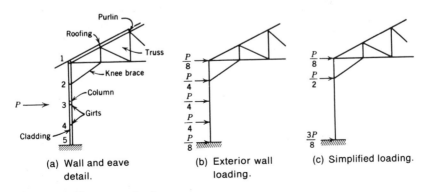

(a) Wall and eave
detail.

(b) Exterior wall
loading.

(c) Simplified loading.

Figure 9.42/Transverse bent.

379

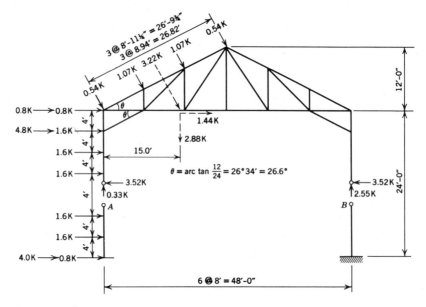

$$\theta = \text{arc tan } \frac{12}{24} = 26°34' = 26.6°$$

Figure 9.43/Transverse bent.

supporting columns, as well as the general method of analysis developed in Art. 9.39 for single portals, applies equally to the transverse bent.

The lateral loading depends largely on the manner of supporting the exterior wall sheathing material. In Fig. 9.42a, the exterior wall sheathing is supported on longitudinal girts, which in turn transmit the dead load of that material, as well as wind loads, to the column. Let it be assumed that this column is fixed at the base and continuous through the bottom of the knee brace to the eave. The point of inflection would then be located at the midheight of the column. For purposes of simplifying analysis, it is well to assume that the loads brought to the column by the girts, which are located between the base and knee brace, are acting at points of connection of the bent members (Fig. 9.42c).[6] The girt load at 3 is therefore moved to 2, and the load at 4 is moved to 5 (it is also not uncommon practice to apply one-half the

[6]From Art. 8.6, it should be noted that using the ASCE 7-88 wind-loading procedure, the loading of Fig. 9.42b could depend on the height to the eave.

total lateral load at the top of the column at 1, and one-half at the base).

It must also be remembered that it is the transverse bents which support the roof loads, not the longitudinal portals between bents (Fig. 9.30). For the present, only the lateral or wind load will be considered. Before the column is designed, however, the very significant gravity loads brought by the truss to the column would also need to be considered.

Returning to the wind load, it will be recalled from Chapter 8 and earlier articles in this chapter that, in the absence of positive shielding, it is customary to assume that the wind can come from any direction. Therefore, it is customary to design both columns for the most critical combination of external force and internal pressure or suction.

Example

Given the wind loading shown in Fig. 9.43, determine the stresses due to wind in the transverse bent columns and in the members framing into the columns. Columns are fixed at their bases and fully restrained by the knee brace.

It should be noted here that with the proportions and dimensions as shown in Fig. 9.43 (θ, h, and L) the actual roof wind loading (by wind only) would be outward if only external effects were considered and the building entirely enclosed. However, other possible loading combinations could result in force loads similar to those shown.

Solution

(1) Isolate that portion of the bent above the points of inflection. Those wall loads occurring between the column base and the knee brace will be temporarily divided equally between the knee brace and base as shown. For purposes of analysis, the resultant of the roof load is carried to the bottom chord of the truss and there separated into its horizontal and vertical components. The horizontal shear is divided equally between the columns:

$$\Sigma H = 0.80 + 4.80 + 1.44 = 7.04$$

$$\frac{\Sigma H}{2} = \frac{7.04}{2} = 3.52 \text{ kips}$$

$$RA_H = RB_H = 3.52 \text{ kips}$$

Take moments about A to find RB_V:

$$\Sigma M = 4.8(10) + 0.8(14)$$

$$+ 1.44(14) + 2.88(15)$$

$$= 122.56$$

$$RB_V = \frac{122.56}{48} = 2.55 \text{ kips} \uparrow$$

Then,

$$\Sigma V = 0, \qquad -2.88 + 2.55 + RA_V = 0$$

$$RA_V = 0.33 \text{ kips} \uparrow$$

(2) Isolate the left-hand column (Fig. 9.44a) and determine stress in the members. (Bow's notation will be used.) Taking moments about the top of the column,

$$\Sigma M = 0$$

$$-4.8(4) + 3.52(14) \pm \overline{0\text{-}D}_H(4) = 0$$

$$-19.20 + 49.28 - \overline{0\text{-}D}_H(4) = 0$$

$$\overline{0\text{-}D}_H = 7.52 \text{ kips} \rightarrow$$

$$\overline{0\text{-}D} = \frac{7.52}{0.894} = 8.41 \text{ kips} \nearrow \text{(tension)}$$

$$\overline{0\text{-}D}_V = 0.447(8.41) = 3.76 \text{ kips} \uparrow$$

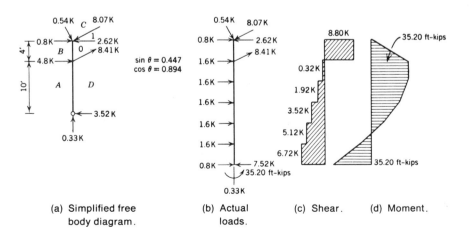

(a) Simplified free body diagram.

(b) Actual loads.

(c) Shear.

(d) Moment.

Figure 9.44/Left-hand column.

Taking the sum of the vertical components,

$$\sum V = 0$$

$$+0.33 + 3.76 - [0.894(0.54)] \pm \overline{C\text{-}1}_V = 0$$

$$3.61 - \overline{C\text{-}1}_V = 0$$

$$\overline{C\text{-}1}_V = 3.61 \text{ kips} \downarrow$$

$$\overline{C\text{-}1} = \frac{3.61}{0.447} = 8.07 \text{ kips} \swarrow \text{ (compression)}$$

$$\overline{C\text{-}1}_H = 0.894(8.07) = 7.22 \text{ kips} \leftarrow$$

Determine the stress in $\overline{1\text{-}0}$ by a summation of the horizontal components:

$$\sum H = 0$$

$$-3.52 + 4.8 + 7.52 + 0.8 - 7.22$$

$$+ [0.447(0.54)] \pm \overline{1\text{-}0} = 0$$

$$2.62 - \overline{1\text{-}0} = 0$$

$$\overline{1\text{-}0} = 2.62 \text{ kips} \leftarrow \text{ (compression)}$$

(3) Return the girt loads to their original positions on the full left-hand column (Fig. 9.44b), and construct the shear and moment diagrams (Fig. 9.44c and d). At the top of the column,

$$\sum H = +0.8 + 0.24 - 7.22 - 2.62$$

$$= -8.80 \text{ kips}$$

At the knee brace,

$$\sum H = +1.6 + 7.52 = 9.12 \text{ kips}$$

$$9.12 - 8.8 = 0.32 \text{ kips}$$

The moment diagram may be constructed from the shear areas. This moment may be checked by taking the sum of the moments

about the column base (Fig. 9.44b):

$$\sum M = +1.6(4) + 1.6(8) + 1.6(12)$$

$$+1.6(16) + 1.6(20) + 7.52(20)$$

$$+0.8(24) - 2.62(24) - 7.22(24)$$

$$+0.24(24)$$

$$= 271.36 - 236.16$$

$$= 35.20 \text{ ft-kips}$$

Having determined the stress in members framing into the left-hand column, the remaining member stresses, as well as the stresses in the right-hand column, can be found graphically and/or algebraically. The known member stresses, in addition to the remaining roof loads, are treated as external truss loads. Of course, dead and live gravity loadings would also need to be investigated, as was the case with trusses discussed in previous articles.

Note. Had the column bases not been fully fixed against rotation, the point of inflection would be nearer the base (Art. 9.39). It is not infrequent that column-base connections are such that only partial fixity may be assumed. In such instances a point of inflection $\frac{1}{3}$ the distance between base and knee brace, measured from the base, would be a reasonable assumption.

Also, the roof-deck system and bracing between bents would be treated as discussed in previous articles on trusses.

PROBLEMS

1. Complete the calculation of member stresses for the bent of the above example (Fig. 9.43).

2. Determine the stress in all members of the transverse bent of the above example (Fig. 9.43), if the point of inflection is assumed to occur 16 ft up from the base. All other stipulations remain the same.

3. Determine the stress due to wind in all members of the transverse bent in Fig. 9.45a. Column bases are fixed, and bents are spaced

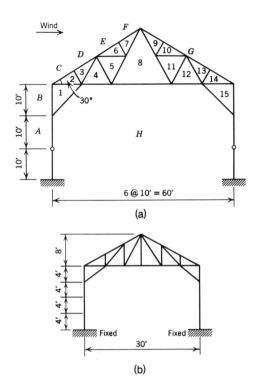

(a)

(b)

Figure 9.45/Problems.

20 ft apart. Draw the shear and moment diagrams for both columns, and indicate maximum shears and moments. Use ASCE 7-88, assuming a basic wind speed of 70 mph, Exposure C, importance factor $I = 1.0$, and openings in excess of 20 per cent.

4. Determine the stress due to wind in all members of the transverse bent in Fig. 9.45b; draw the shear and moment diagrams for both columns, and indicate maximum shear and moment for design in each case. Bents are spaced 16 ft apart. The structure is located in an area having a basic wind speed of 90 mph, Exposure C, and importance factor $I = 1.0$; openings are minimal. Use ASCE 7-88.

9.43
Portal Method: Multistory, Multibay Construction

The principles developed in Arts. 9.40 and 9.41 for an approximate method of analy-sis of simple and multiple portals and bents for lateral wind loads can be extended to multistory, multibay portal wind bents.

The assumptions on which the portal method of analysis for such wind bents are based are as follows.

1. A bent of a frame acts as a series of independent (simple) portals.

2. The point of inflection of each column is at story midheight.

3. The point of inflection of each beam is at its midlength.

4. The horizontal shear on any plane is divided equally over the aisles—an exterior column taking one-half the shear of an interior column.

5. Floors remain plane and level.

6. Beam and column length changes may be neglected.

7. The wind load is resisted entirely by the steel frame.

Referring to Fig. 9.46, assume that the bent shown is but one in a series of east-west bents spaced 20 ft apart in a building which has its longest dimension in the north-south direction. This bent, and all others in the east-west direction, must resist a wind load of 30 psf applied over the entire vertical surface. Assume a story height of 12 ft and the existence of a parapet wall extending 3 ft above the roof level; then the roof-level panel load is

$$30(6+3)(10+10) = 5400 \text{ lb}$$

and the 2nd-floor panel load is

$$30(12)(10+10) = 7200 \text{ lb}$$

The wind load at the base will be transmitted directly into foundations.

Using the assumptions noted above:

Column shear at any story is divided equally between the aisles, and exterior

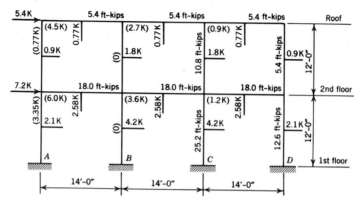

Figure 9.46/Portal method of analysis.

columns take half as much shear as interior columns (assumption 4). Then, for the shear in 2nd-story columns:

Total shear in 2nd story
$$= 5400 \text{ lb} = 5.4 \text{ kips}$$

Shear in each aisle
$$= 5400/3 = 1800 \text{ lb} = 1.8 \text{ kips}$$

Shear in exterior columns
$$= 1800/2 = 900 \text{ lb} = 0.9 \text{ kips}$$

Shear in interior columns
$$= 1800 \text{ lb} = 1.8 \text{ kips}$$

For the shear in 1st-story columns:

Total shear $= 5400 + 7200 = 12{,}600 \text{ lb}$
$$= 12.6 \text{ kips}$$

Shear in each aisle
$$= 12{,}600/3 = 4200 \text{ lb} = 4.2 \text{ kips}$$

Shear in exterior columns
$$= 4200/2 = 2100 \text{ lb} = 2.1 \text{ kips}$$

Shear in interior columns
$$= 4200 \text{ lb} = 4.2 \text{ kips}$$

These values are recorded in Fig. 9.46 at the midheights of the columns.

Beam direct compressive stress in floor beams that distribute the panel load as shear to the column on the right of each beam (Fig. 9.46) is found by deducting from the panel load that portion of the panel load taken as shear by the column on the left of each beam. For the direct

stress in the roof beam of aisle *AB*,

stress = panel load − shear in column *A*
$$= 5400 - 900 = 4500 \text{ lb}$$
$$= 4.5 \text{ kips}$$

For the direct stress in the roof beam of aisle *BC*,

stress = panel load
    − shear in columns *A* and *B*
$$= 5400 - (900 + 1800) = 2700 \text{ lb}$$
$$= 2.7 \text{ kips}$$

For the direct stress in the roof beam of aisle *CD*,

stress = panel load
    − shear in columns *A*, *B*, and *C*
$$= 5400 - (900 + 1800 + 1800)$$
$$= 900 \text{ lb}$$
$$= 0.9 \text{ kips}$$

These and similar values for the 2nd floor are shown in Fig. 9.46 in parentheses immediately under the respective beams.

Column bending moments are equal at the top and bottom of each column and are equal to the column shear multiplied by one-half the story height. The bending moment in each interior column is twice that in an exterior column, because the shears are twice as large; therefore, for the bending moment in 2nd-story columns *A* and D,

$$M = 900(6) = 5400 \text{ ft-lb} = 5.4 \text{ ft-kips}$$

For the bending moment in 2nd-story columns B and C,

$$M = 1800(6) = 10,800 \text{ ft-lb} = 10.8 \text{ ft-kips}$$

These values are recorded parallel to the columns—only those for columns C and D being shown in Fig. 9.46. The bending moments for the 1st-story columns are determined in the same manner.

Beam bending moments at the ends of floor beams are equal to the sum of the moments in exterior columns immediately above and below the floor under consideration. Moments in beams of the same floor are equal and are independent of the width of the aisle. Bending moment in roof beams:

$$M = 5400 + 0 = 5400 \text{ ft-lb} = 5.4 \text{ ft-kips}$$

Bending moment in 2nd-floor beams:

$$M = 5400 + 12,600 = 18,000 \text{ ft-lb}$$
$$= 18.0 \text{ ft-kips}$$

These values are shown directly above the respective beams (Fig. 9.46).

Beam shears are computed from beam end moments, i.e., by dividing the end moment by one-half the beam span. Since moments in the beams of any one floor are equal, the beam shears are also equal when aisles are of the same width:

Shear in roof beams
$$= 5400/7 = 770 \text{ lb} = 0.77 \text{ kips}$$
Shear in 2nd floor beams
$$= 18,000/7 = 2580 \text{ lb} = 2.58 \text{ kips}$$

These values are shown perpendicular to and beneath the respective beams at the assumed point of inflection, i.e., at midspan (Fig. 9.46).

Column direct stresses are computed directly from beam shears. Since beam shears are equal for any one floor when aisles are of equal width, there is no direct stress in the interior columns of this bent. With

wind from the left (Fig. 9.46), the direct stress in column A is tension, and in column D, compression. The values are as follows: For columns A and D in the 2nd story,

$$P = 770 \text{ lb} = 0.77 \text{ kips}$$

For columns A and D in the 1st story,

$$P = 770 + 2580 = 3350 \text{ lb} = 3.35 \text{ kips}$$

These values are shown in parentheses, parallel to and to the left of columns A and B, and although not shown, are the same for columns D and C, respectively.

The approximate portal method of analysis will be used in Chapter 13 as an aid to design of a simplified, though larger, building frame.

9.44
Long-Span Roof Construction

When a roof must span a distance of 150 ft or more, which is well beyond the capability of the braced-frame construction presented herein, either a multiple-bay system with interior columns must be used, or some form of arch or rigid-frame construction is usually necessary.

In this country, structural steel arches are probably the most commonly used. As in Europe, however, reinforced-concrete arches of long span are achieving increased acceptance, and thin shells and space frames are likely to be used to an ever greater extent as well. Nevertheless, the three-hinged steel arch truss (Fig. 9.47) is still used for the roofs of buildings such as exhibition halls, armories, and gymnasiums, which may or may not be of one-story construction. The general shape of the truss is usually established by the architectural features of the structure. The chord and web members may be built up of channels or plates and angles, similar to those in ordinary roof trusses; however, S

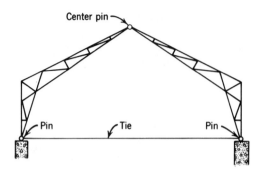

Figure 9.47/Three-hinged arch.

beams and **W** sections are frequently more satisfactory.

Inasmuch as a great deal of shop work is required in the fabrication of trusses of this type and as the truss itself will usually be quite heavy, the distance between trusses should be large. For spans 100 ft and over, a spacing of 22 to 40 ft is customary, the greater spacing being more economical for the longer spans. Such a spacing necessitates the use of framed trusses for purlins; however, the purlin trusses may also be made a part of the bracing system between arches.

In recent years, steel rib arches and rigid frames fabricated from wide flange sections have been widely adopted for long-span construction. Bents of this type are essentially two-hinged arches with the outward thrust at the base of the columns taken by tie rods under the floor. As stated at the outset of this chapter, detailed discussion of the design of rigid frames has been set aside for Chapter 10.

10

CONTINUOUS BEAMS AND FRAMES

10.1 Introduction

All structural members in a building may be classified as follows:

1. Beams (transverse load carriers).

2. Columns or struts (direct compressive load carriers).

3. Ties (direct tensile load carriers).

4. Those which carry combined bending and axial loads.

Their primary purpose is to frame space, and to achieve this purpose, they may be assembled and connected in a variety of ways. The simplest form for space enclosure is a beam resting on a wall; if the span is great, a truss composed of struts and ties can replace the beam (Fig. 10.1). And should walls supporting beams or trusses be undesirable, they can be replaced by columns. However, should connections between columns and beams, or columns and trusses, be flexible (Type 2 construction, Art. 7.3), an unstable assembly will result and, under any unbalanced loading condition, would collapse (Fig. 10.2). Collapse of this nature can be prevented and the structure rendered stable by bracing, i.e., using steel members or panel walls (Chapter 9).

Another way to provide this needed stability is to use rigid column-to-beam or column-to-truss connections. Such connections must be sufficiently rigid to prevent rotation of one member with respect to the other, and when they are, a structural element results which is essentially rigid also. This element is generally referred to as a *bent* or *rigid frame* (Fig. 10.3). Rigid frames may be designed in a variety of shapes, depending upon architectural and

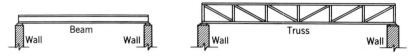

Figure 10.1/Stable beam and truss.

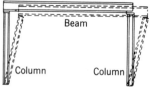

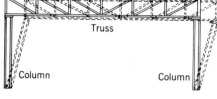

Figure 10.2/Unstable frames.

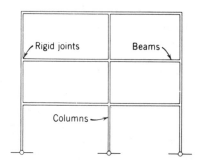

Figure 10.3/Rigid frame or bent.

structural needs, e.g., the three-story, three-column bent shown in Fig. 10.4. Such elements, however, cannot be analyzed by the laws of statics alone; they are therefore called statically indeterminate.

The beam was selected in previous chapters as the basis for gaining an understanding of structures in static equilibrium and became a tool available for their analysis. So too will the beam be used to introduce the study of statically indeterminate structures.

For coplanar, parallel force systems, such as a simply supported beam carrying vertical loads (Fig. 10.5a), it has been shown that two of the equations of statics, $\Sigma V = 0$ and $\Sigma M = 0$, are sufficient to solve for two unknown forces—in beams, these are usually the reactions. In the absence of additional equations, any more unknown forces are *redundant* and render a beam statically indeterminate, the *degree of indeterminacy* being measured by the number of redundant reaction components, i.e., V, H, and M.

Any support that is restrained has a moment as well as a force. Thus, a cantilever beam has two unknowns: the moment in the wall and the vertical reaction at the wall. The simple cantilever is, nevertheless, statically determinate because these two unknowns can be found using the two equations $\Sigma V = 0$ and $\Sigma M = 0$.

Members may be continuous *over* as well as *through* supports. Figure 10.5b illustrates a beam resting on three supports. The beam is *continuous over* the middle support. Had, for example, the middle support been restrained (as shown in Fig. 10.4, where beams are column-supported), the beam would be continuous *through* this support.

Figure 10.4/Stable bent.

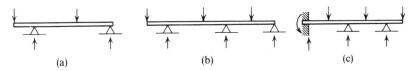

Figure 10.5/Continuity.

Figure 10.5b shows a beam having three unknown reactions; this beam is statically indeterminate to the first degree. Figure 10.5c shows a beam having four unknowns —three vertical reactions and the moment at the fixed end. This beam is statically indeterminate to the second degree.

As before, beam analysis requires that reactions be determined and that shear and bending moment diagrams be constructed. However, shear and bending moment diagrams can be drawn only when the external reactions are known. Consequently, the analysis of indeterminate beams begins with the calculation of redundant reactions, followed by a static solution of the remaining reactions.

The magnitude of redundant reactions may be found by various methods; however, deflections resulting from the loads are the tools of analysis common to all methods. This chapter will deal principally with one method of analysis known as *moment distribution*. Although other methods may lead to an easier solution in some cases, it is beyond the scope of this text to treat more than this one method in detail. The concepts basic to the other methods will be presented, however.

10.2
Method of Consistent Deformation

The sole purpose of this article will be to show why deflections are effective analytical tools. The method of consistent deformation is basic to indeterminate analysis. Essentially, it involves the removal of redundants to create a determinate struc-

ture, calculation of the determinate-structure deflections at the points of redundant forces, and then calculation of the magnitudes of the redundant forces necessary to prevent these deflections. In other words, the determinate structure produced by the removal of redundants is allowed to deflect. And because there can be no deflection at a reaction, the magnitude of the force necessary to bring this deflection back to zero must also be the magnitude of the redundant reaction originally removed. This procedure is demonstrated in the illustrative example which follows.

Example

A **W** 10×22 beam is 40 ft long and rests on three supports, 20 ft apart. This continuous beam supports three concentrated loads of 2 kips each, as shown in Fig. 10.6a. Neglect the weight of the beam. Find the magnitude of the reactions, shears, and moments.

Solution

(1) Make the beam statically determinate by removing redundants. It is seen that the beam is statically indeterminate to the first degree, i.e., there is only one redundant. Any one of the three supports may be considered to be the redundant. In this case, the middle support will be so designated, and temporarily removed (Fig. 10.6b).

(2) Calculate the deflection at the point where the support was removed. The beam is now simply supported with concentrated

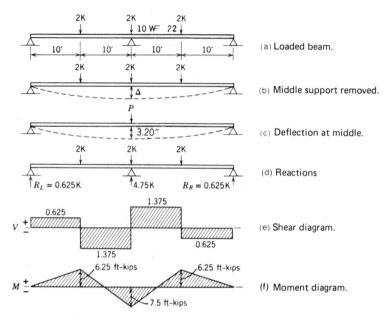

Figure 10.6/Method of consistent deformation.

loads at the quarter-points. From Fig. 4.15, the deflection at midspan is

$$\Delta = \frac{19}{384}\left(\frac{PL^3}{EI}\right)$$

$$= \frac{19(2)1000(40)^3 12^3}{384(29,000,000)118} = 3.20 \text{ in.}$$

(3) Calculate the magnitude of a single force at this point, which will produce this same deflection (Fig. 10.6c). This would result in a simply supported beam with a single concentrated load at midspan. From Fig. 4.15, the deflection at midspan is

$$\Delta = \frac{PL^3}{48EI} = 3.20 \text{ in.}$$

Solving for P,

$$\frac{48EI(3.20)}{L^3} = \frac{48(29,000,000)118(3.20)}{40^3(12)^3}$$

$$P = 4753 \text{ lb} \simeq 4.75 \text{ kips}$$

This force of 4.75 kips acting upward at the middle support will prevent the deflection at this point initially calculated.

Consequently, 4.75 kips is the magnitude of the redundant, or reaction, at the middle support.

(4) Determine the magnitude of the remaining unknown reactions. This can be accomplished by applying the equations of static equilibrium: $\Sigma M = 0$ and $\Sigma V = 0$. Taking moments about the left support (Fig. 10.6d),

$$2(10) + 2(20) + 2(30)$$

$$- 4.75(20) - 40R_R = 0$$

$$R_R = 0.625 \text{ kips}$$

From symmetry, or $\Sigma V = 0$,

$$R_L = 0.625 \text{ kips}$$

(5) Draw the shear and moment diagrams (Fig. 10.6 e and f).

The beam just analyzed was selected because of its simplicity. The same method of analysis can be applied to any indeterminate structure but becomes increasingly more complicated for higher degrees of indeterminacy and/or unsymmetrical loading conditions.

It should be noted that in order to determine beam deflections, the moment of inertia and thus the beam section must be known. This means that the size must be either given or assumed. Inherent in the analysis and design of indeterminate structures is the need to make successive approximations, i.e., the size of components must first be assumed, the structure analyzed, the size modified as required, the structure reanalyzed, etc. Fortunately, the size first assumed is usually sufficiently close so that for design purposes further refinement is not necessary. This subject will be treated in greater detail later in this chapter.

10.3
Fixed End Moments

The beam shown in Fig. 10.7a simply rests on its supports. There is *no* restraint at the supports. When the load P is applied, the beam will deflect and the ends of the beam will rotate (θ_A and θ_B).

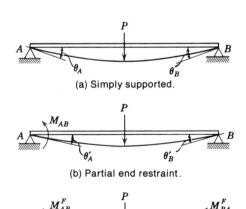

(a) Simply supported.

(b) Partial end restraint.

(c) Fixed ends.

Figure 10.7/End restraint.

Assume that the end at A is partially restrained so that angle θ_A' is less than θ_A (Fig. 10.7b). This partial restraint (or *fixity*) is represented by the moment M_{AB} as shown. Various degrees of restraint, for various reasons, may occur either at one end or at both ends. However, the condition which results in both ends being fully restrained so that there is no rotation at the supports (Fig. 10.7c) is most important.

Such a beam is said to have fixed ends, and the restraining moments M_{AB}^F and M_{BA}^F are called fixed-end moments. It should be noted that fixed-end moments are independent of any adjacent beams or members.

The magnitude of the fixed-end moments for six different conditions of loading is shown in Fig. 10.8 (see also Appendix B). Moments for combinations of the loadings shown may be obtained by simply adding them together. For example, assume a beam has a uniformly distributed load of intensity w, plus two equal concentrated loads at the third-points. From Fig. 10.8,

$$M_{AB}^F = M_{BA}^F = \frac{wL^2}{12} + \tfrac{2}{9}PL$$

The fixed-end moments shown in Fig. 10.8, or those for any loading condition, may be calculated by applying the area-moment principles (Art. 4.2).

For example, the beam shown in Fig. 10.9 has a concentrated load P acting at midspan. The moment and deflection diagrams also are shown. Since the load is symmetrically placed, the fixed-end moments are equal, and

$$M_{AB}^F = M_{BA}^F$$

Note also that the fixed-end moments are negative whereas the moment at midspan is positive.

The bending moment diagram for this beam also may be drawn in parts as shown in Fig. 10.9b. If restraint did not exist at

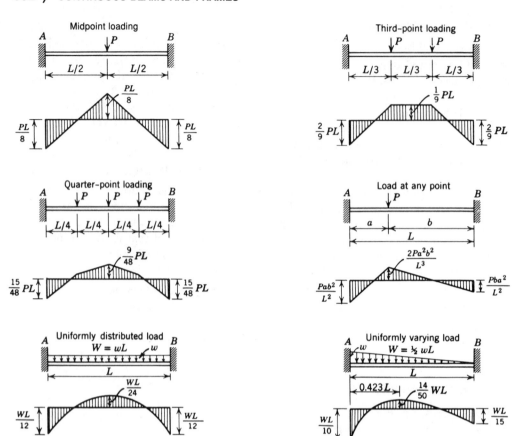

Figure 10.8/Fixed-end moments.

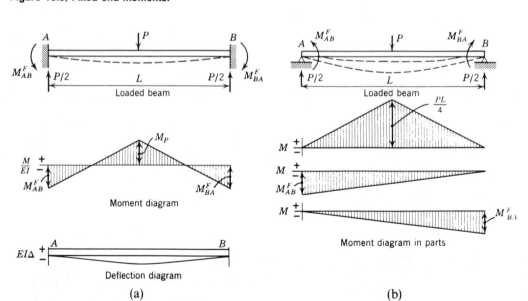

(a)

(b)

Figure 10.9/Fixed-end moment.

the ends, the moment diagram would be that for a simply supported beam. The restraining moments at the ends may be thought of as acting independently and thus diminishing to zero at opposite ends. The tangents drawn to the elastic curve at points A and B coincide; therefore, the angle subtended between them is zero. From the first of the area-moment principles, this angle is equal to the area of the M/EI diagram between the two points of tangency.

From the moment diagram as drawn in parts (Fig. 10.9b),

$$+\frac{PL}{4}\left(\frac{L}{2}\right) - M_{AB}^F\left(\frac{L}{2}\right) - M_{BA}^F\left(\frac{L}{2}\right) = 0$$

Since $M_{AB}^F = M_{BA}^F$,

$$+\frac{PL^2}{8} - M_{AB}^F(L) = 0$$

and

$$M_{AB}^F = M_{BA}^F = \frac{PL}{8}$$

The positive moment at midspan may now be calculated from the conditions of static equilibrium. It is also equal to $PL/8$ in this example.

Another example is the unsymmetrically loaded beam shown in Fig. 10.10a. The beam is statically indeterminate to the second degree. By choosing M_{AB}^F and M_{BA}^F as the redundants, the fixed-end beam may be broken into three separate simple beams as shown in Fig. 10.10 (b), (d), and (f): the first is a simple beam supporting the load P and deflecting according to the broken line as indicated; the second is a simple beam, with the moment M_{AB}^F applied, deflecting upward as the broken line indicates; and the third is a simple beam with the moment M_{BA}^F applied. In each simple-beam diagram, lines are shown drawn tangent to the elastic curve at points A and B, identifying the end rotation angles θ. The bending moment diagram for

each separate simple beam is shown directly below the beam diagram, i.e., Fig. 10.10 (c), (e), and (g).

The initial problem (Fig. 10.10a) has fully restrained ends; therefore, by definition, the slope at the fixed ends must be zero. Applying this observation to the three separate simple-beam conditions,

$$\theta_A = \theta_{A_1} + \theta_{A_2}$$
$$\theta_B = \theta_{B_1} + \theta_{B_2}$$

Applying trigonometry and the second area-moment principle (Fig. 10.10 b and c),

$$\tan\theta_A = \frac{BB'}{L}$$

$$= \frac{\left[\frac{Pab}{L}\left(\frac{b}{2}\right)\frac{2b}{3}\right] + \left[\frac{Pab}{L}\left(\frac{a}{2}\right)\left(b+\frac{a}{3}\right)\right]}{L}$$

$$= \frac{2Pab^3 + 3Pa^2b^2 + Pa^3b}{6L^2}$$

$$= \frac{Pab}{6L}(a+2b)$$

$$\tan\theta_B = \frac{AA'}{L}$$

$$= \frac{\left[\frac{Pab}{L}\left(\frac{a}{2}\right)\frac{2a}{3}\right] + \left[\frac{Pab}{L}\left(\frac{b}{2}\right)\left(a+\frac{b}{3}\right)\right]}{L}$$

$$= \frac{2Pa^3b + 3Pa^2b^2 + Pab^3}{6L^2}$$

$$= \frac{Pab}{6L}(2a+b)$$

Similarly, from Fig. 10.10 (d) and (e),

$$\tan\theta_{A_1} = \frac{B_1B_1'}{L} = \frac{M_{AB}^F\left(\frac{L}{2}\right)\frac{2L}{3}}{L} = \frac{M_{AB}^F L}{3}$$

$$\tan\theta_{B_1} = \frac{A_1A_1'}{L} = \frac{M_{AB}^F\left(\frac{L}{2}\right)\frac{L}{3}}{L} = \frac{M_{BA}^F L}{6}$$

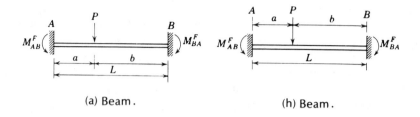

(a) Beam.   (h) Beam.

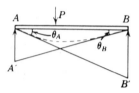

(b) Simple beam part.

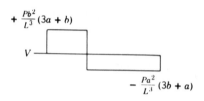

(i) final shear.

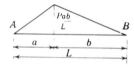

(c) Moment (simple).

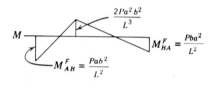

(j) Final moment.

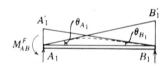

(d) Left reaction part.

(e) Moment (left reaction).

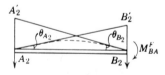

(f) Right reaction part.

(g) Moment (right reaction).

Figure 10.10/Fixed-end moment by parts.

From Fig. 10.10 (f) and (g),

$$\tan \theta_{A_2} = \frac{B_2 B_2'}{L} = \frac{M_{BA}^F \left(\dfrac{L}{2}\right)\dfrac{L}{3}}{L} = \frac{M_{AB}^F L}{6}$$

$$\tan \theta_{B_2} = \frac{A_2 A_2'}{L} = \frac{M_{BA}^F \left(\dfrac{L}{2}\right)\dfrac{2L}{3}}{L} = \frac{M_{BA}^F L}{3}$$

Substituting in the original equations, and using the rule that for very small angles $\theta = \tan \theta$ and $\tan \theta_{A_1} + \tan \theta_{A_2} = \tan \theta_A$, then

$$\frac{M_{AB}^F L}{3} + \frac{M_{BA}^F L}{6} = \frac{Pab}{6L}(a + 2b)$$

and since $\tan \theta_{B_1} + \tan \theta_{B_2} = \tan \theta_B$,

$$\frac{M_{AB}^F L}{6} + \frac{M_{BA}^F L}{3} = \frac{Pab}{6L}(2a + b)$$

from which

$$M_{AB}^F = \frac{Pab^2}{L^2} \quad \text{and} \quad M_{BA}^F = \frac{Pba^2}{L^2}$$

The beam diagram of Fig. 10.10h may now be drawn, and the vertical reactions of A and B computed. The shear and moment diagrams are shown in Fig. 10.10 (i) and (j).

An excellent exercise would be to calculate fixed-end moments for other conditions of loading and verify them by comparison with those given in Fig. 10.8 and Appendix B.

PROBLEMS

1. A W 12×26 is continuous for two 12-ft spans and supports a uniformly distributed load of 3 kips per ft. Determine the magnitude of reactions, using the method of consistent deformation. (Answers given in Appendix G.)

2. A W 10×26 spans 18 ft. It is fixed at one end and simply supported at the other. There is one concentrated load of 10 kips at midspan. Determine the magnitude of reactions, using

the method of consistent deformation. (It is suggested that the reaction at the simply supported end be taken as the redundant.)

3. Calculate the fixed-end moments for a 26-ft beam which carries a uniformly distributed load of 2 kips per ft and a concentrated load of 12 kips acting at midspan. (Answers given in Appendix G.)

4. Calculate the fixed-end moments for a 24-ft beam which carries a uniformly distributed load of 3 kips per ft, a concentrated load of 10 kips acting 8 ft from one end, and a concentrated load of 20 kips acting 8 ft from the other end.

10.4
Sign Convention

In preceding chapters, a positive moment was assumed to cause a beam to bend concave upward, and a negative moment, concave downward. This convention was particularly suited to the analysis and design of statically determinate beams. However, it will lead to confusion when analyzing continuous beams and frames; therefore, a new sign convention will be introduced here.

To specify a moment, one must know whether it tends to produce clockwise or counterclockwise rotation and whether it acts on a beam or on a joint. Figure 10.11 shows a loaded cantilever beam. Moment M_A is acting on the support; for equilibrium, moment M_B represents the effect of the support acting on the end of the beam. The sign convention may be expressed as follows: *A moment acting on a joint is positive if it tends to rotate the joint clockwise.* In Fig. 10.11, then, M_A is positive and M_B is negative. For downward loads

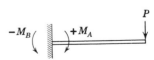

Figure 10.11/Sign convention.

on a horizontal member, the fixed-end moment is always positive to the right of a joint and negative to the left.

10.5
Moment Distribution—
Mathematical Procedure

The moment-distribution method of analyzing indeterminate structures centers on the direct determination of bending moments at a joint or support. Once moments are known, the magnitude of reactions may be determined from statics.

The moment distribution method was developed by the late Hardy Cross and is considered by many architects and engineers to be one of the more significant developments in structural theory thus far in the 20th century. Although the conceptual basis for the moment distribution method is actually quite simple, the mechanics of application may at first seem complicated. Therefore, the step-by-step mathematical procedure will be outlined before the method is applied to actual examples. The underlying theory will be explained in subsequent articles. This is the order of study recommended by Hardy Cross when he said: "...learn the elementary process of moment distribution... accepting its theorems temporarily as proved. Practice simple moment distribution restricted to prismatic members[1] and apply it...."

Moment distribution, as described in this article, is applicable only if the following conditions are met:

1. All members are initially straight.

2. All members are of uniform cross section throughout their entire length (support to support).

3. There is no displacement of the supports (joints).

[1]Beams of constant cross section.

The procedure may be thought of as a series of successive approximations. First, all spans are assumed to have fixed ends; therefore, no rotation takes place at joints or supports. Fixed-end moments are calculated. Then, one by one, each joint is unlocked and allowed to rotate to balance out the moments to the left and right. As a joint is allowed to rotate, any unbalanced moment will be distributed to adjacent members according to their relative stiffnesses—this stiffness being measured in terms of I/L for each member. Any induced moment at one end of a member produces an effect at the opposite fixed end of that member. This effect is one-half the value of the distributed moment. Thus, when this one-half value is *carried over* to the opposite end, it tends to create lack of balance at that end, requiring the process to be repeated again and again until the *carry-over* moments are negligible.

To eliminate the chance of errors as much as possible, some orderly way of presenting the work must be adopted. The tabulation shown in Fig. 10.12 is recommended.

Example 1

A beam of constant cross section is to be continuous over three supports and loaded as shown in Fig. 10.12. Using the moment-distribution method, determine the magnitude of the bending moments at A, B, and C.

Solution

(1) Prepare the base lines for the tabulation. There are two spans and three joints (supports in this problem). These lines are drawn heavy in Fig. 10.12.

(2) Calculate the relative stiffness factor for each member. This has been defined as I/L, and the symbol K' is used here to

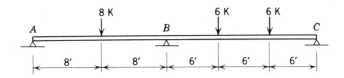

Stiffness factor			0.063		0.056		
Distribution factor	0.0	1.0	0.53	0.47	1.0	0.0
Fixed-end moments	0.0	+ 16.0	− 16.0	+ 24.0	− 24.0	0.0
Balance	0.0	− 16.0	− 4.2	− 3.8	+ 24.0	0.0
Total	0.0	0.0	− 20.2	+ 20.2	0.0	0.0
Carry-over		− 2.1	− 8.0	+ 12.0	− 1.9	
Balance		+ 2.1	− 2.1	− 1.9	+ 1.9	
Total		0.0	− 30.3	+ 30.3	0.0	
Carry-over		− 1.1	+ 1.1	+ 1.0	− 1.0	
Balance		+ 1.1	− 1.1	− 1.0	+ 1.0	
Total	0.0	0.0	− 30.3	+ 30.3	0.0	0.0

Figure 10.12/Moment distribution.

denote its value.[2] Since a beam of constant cross section is to be used, the value of I will remain constant:

$$K'_{AB} = \frac{I}{L} = \frac{I}{16} \approx 0.063I$$

$$K'_{BC} = \frac{I}{L} = \frac{I}{18} \approx 0.056I$$

Enter these values above the base line as shown in Fig. 10.12.

(3) Determine the distribution factors (DF) for each support. The distribution factor for any given member is equal to its K' value divided by $\Sigma K'$ of all members framing into the joint. Considering point B,

$$\Sigma K' = 0.063I + 0.056I = 0.119I$$

$$DF_{BA} = \frac{K'_{AB}}{\Sigma K'} = \frac{0.063I}{0.119I} \approx 0.53$$

$$DF_{BC} = \frac{K'_{BC}}{\Sigma K'} = \frac{0.056I}{0.119I} \approx 0.47$$

Notice that the sum of the distribution

[2] The symbol K' used here should not be confused with the symbol K used to denote the effective length factor for columns (Art. 6.4).

factors at a support must equal unity $(0.53 + 0.47 = 1.00)$. Considering point A,

$$\Sigma K' = 0 + 0.063I$$

$$DF_{AB} = \frac{K'_{AB}}{\Sigma K'} = \frac{0.063I}{0.063I} = 1.0$$

For point C,

$$DF_{CB} = \frac{K'_{CB}}{\Sigma K'} = \frac{0.056I}{0.056I} = 1.0$$

Enter these distribution factors directly above the base line as shown in Fig. 10.12.

(4) Assuming *all* joints fully fixed, calculate the fixed-end moments. From Fig. 10.8,

$$M^F_{AB} = + \frac{PL}{8} = \frac{8(16)}{8} = + 16 \text{ ft-kips}$$

$$M^F_{BA} = - \frac{PL}{8} = - 16 \text{ ft-kips}$$

$$M^F_{BC} = + \tfrac{2}{9}PL = \tfrac{2}{9}(6)18 = + 24 \text{ ft-kips}$$

$$M^F_{CB} = - 24 \text{ ft-kips}$$

Enter these values in the tabulation on the line labeled "Fixed-end moments."

(5) Balance the moments at each joint. Joint B is subjected to a positive moment

of 24.0 and a negative moment of 16.0. Consequently, it is out of balance by $24.0 - 16.0 = +8.0$ ft-kips. When the assumed fixity is released, the joint will rotate clockwise and the unbalanced moment of $+8.0$ ft-kips will be apportioned to members AB and BC according to their distribution factors:

Member BA receives

$$0.53(+8.0) = +4.2 \text{ ft-kips}$$

Member BC receives

$$0.47(+8.0) = +3.8 \text{ ft-kips}$$

However, since this is an applied moment, the induced moment which establishes equilibrium will be opposite in sign. Enter -4.2 and -3.8 on the "Balance" line in the tabulation. Add the moments on each side of the support:

$$M_{BA} = -16.0 - 4.2 = -20.2 \text{ ft-kips}$$
$$M_{BC} = +24 - 3.8 = +20.2 \text{ ft-kips}$$

This joint is now in equilibrium. Proceed to another joint and repeat the above process. At A, the joint is out of balance an amount $+16.0$, so the moment necessary to place it in balance is -16.0. All of this balancing moment goes to the right of the joint, which has a distribution factor of 1.0. The sum of the moments about A is 0.0. Repeat this step for joint C.

(6) Carry over one-half the distributed moments to the opposite end. In balancing joint B, a moment of -3.8 was applied to the right of the joint. This affects the moment on the far end, so carry over $\frac{1}{2}(-3.8) = -1.9$ to the left of joint C (shown as an arrow). This is done for every joint when an induced moment results from the balancing step.

(7) Balance each joint again as in step (5). The carry-over moments upset the balance of each joint once more. For example,

$$M_{BA} = -20.2 - 8.0 = -28.2$$
$$M_{BC} = +20.2 + 12.0 = +32.2$$
$$\text{out-of-balance moment} = +4.0$$

A moment of -4.0 now must be multiplied by the distribution factors 0.53 and 0.47, and added to the joint. Again, the total is computed and observed to be in balance.

These steps, carry-over, balance, and total, are repeated over and over until the carry-over moment is negligible. Each total represents one full cycle. The total following the final balancing step represents the moment in each member at a joint. As seen from this example, the moments at A and C are zero, and at B, 30.3 ft-kips.

Example 2

Make a complete analysis of the continuous beam shown in Fig. 10.13. The beam has a constant moment of inertia.

Solution

For a detailed explanation of steps (1) through (6), see Example 1.

(1) Calculate stiffness factors. These are the ratios I/L for each member having two supports. The cantilevered portion of the beam (extreme right end) has a stiffness of zero, since the adjacent spans do not affect its moment. The stiffness factor at a fully fixed end may be considered infinite, since no rotation is allowed to take place.

(2) Calculate the distribution factors at each joint.

(3) Determine fixed-end moments from Fig. 10.8.

(4) Balance moments at each joint.

(5) Total moments after balancing.

(6) Carry over one-half the induced moments to opposite ends, and repeat steps (4), (5), and (6) until carry-over moments are negligible. The final moments are shown as the last totals in the tabulation (Fig. 10.13).

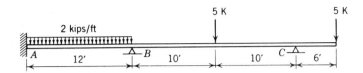

Stiffness factor	$K'=\infty$	$K'=0.084$		$K'=0.050$		$K'=0$
Distribution factor	1.0	0.0	0.63	0.37	1.0	0.0
Fixed-end moments	0.0	+24.0	−24.0	+12.5	−12.5	+30.0
Balance	−24.0	0.0	+7.3	+4.2	−17.5	0.0
Total	−24.0	+24.0	−16.7	+16.7	−30.0	+30.0
Carry-over		+3.7	0.0	−8.8	+2.1	
Balance	−3.7	0.0	+5.5	+3.3	−2.1	0.0
Total	−27.7	+27.7	−11.2	+11.2	−30.0	+30.0
Carry-over		+2.8	0.0	−1.1	+1.7	
Balance	−2.8	0.0	+0.7	+0.4	−1.7	0.0
Total	−30.5	+30.5	−10.5	+10.5	−30.0	+30.0
Carry-over		+0.3	0.0	−0.8	+0.2	
Balance	−0.3	0.0	+0.5	+0.3	−0.2	
Total	−30.8	+30.8	−10.0	+10.0	−30.0	+30.0

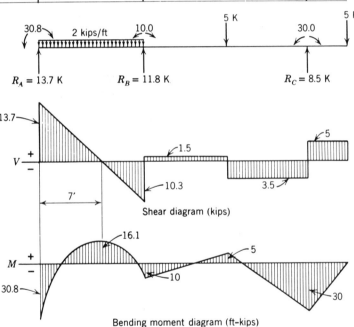

Figure 10.13/Moment distribution.

(7) Determine the reactions. This can be accomplished by applying the laws of statics. One possible procedure is described here. Cut the beam at point B. Considering only that portion of the beam which lies to the right, the moment acting on the cut section is −10.0 ft-kips. Taking moments of all forces to the right of the cut section (using point B as the center of moments) and equating to zero,

$$\sum M_B = 0$$
$$5(10) + 5(26) - 20(R_C) - 10.0 = 0$$
$$R_C = 8.5 \text{ kips}$$

Still dealing with the cut section at point B, but treating the portion to the left as a free body, the moment acting on the cut section is $+10.0$ ft-kips. Hence

$$\sum M_B = 0$$

$$-2(12)6 + 10.0 + 12(R_A) - 30.8 = 0$$

$$R_A = 13.7 \text{ kips}$$

The moment at A is -30.8 ft-kips. Using A as the center of moments for the entire beam,

$$\sum M_A = 0$$

$$2(12)6 + 5(22) + 5(38) - 8.5(32)$$
$$- 12(R_B) - 30.8 = 0$$

$$R_B = 11.8 \text{ kips}$$

Checking the calculated reactions,

$$\sum V = 0$$

$$2(12) + 5 + 5 = 8.5 + 13.7 + 11.8$$

$$34 = 34$$

(8) Construct the shear and moment diagrams (Fig. 10.13).

PROBLEMS

1. A beam of constant cross section is loaded as shown in Fig. 10.14. Find the bending moments at A and B, using the moment-distribution method. (Answers given in Appendix G.)

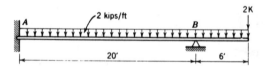

Figure 10.14

2. A beam of constant cross section is continuous for two spans and is loaded as shown in Fig. 10.15. Find the bending moments at A, B, and C, using the moment-distribution method.

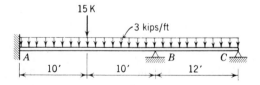

Figure 10.15

3. A beam of constant cross section is continuous for two spans and carries 4 kips per ft uniformly distributed over its full length. One span is 18 ft and the other is 12 ft. Only simple supports are used. Calculate the reactions.

4. Make a complete analysis of the beam shown in Fig. 10.16. The moment of inertia from B to D is twice that from A to B. (Answers given in Appendix G.)

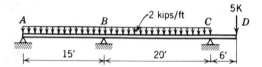

Figure 10.16

10.6
Stiffness and Carry-Over Factors

If a moment M_{BA} is applied at B as shown in Fig. 10.17a, the end will rotate

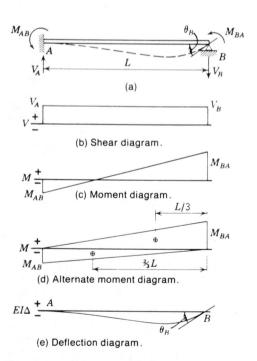

Figure 10.17/Stiffness and carry-over.

through an angle θ_B, and a moment M_{AB} will be induced at the fixed end A. Notice that the induced moment has the same sign as the applied moment. The ratio between applied and induced moments depends on the physical properties of the beam.

The angle θ_B through which the beam end rotates varies with the magnitude of the applied moment and the beam stiffness, and the ratio between applied moment and angle is the stiffness factor for the beam.

Figure 10.17 (b) and (c) show the corresponding shear and bending moment diagrams for a beam of constant cross section (prismatic). The method employed here for determining stiffness and carry-over factors will make use of the area-moment principles as detailed in Chapter 4, Art. 4.2. As shown in Fig. 10.17d, the moment diagram may be simplified. The deflection diagram is shown in Fig. 10.17e, and illustrates in an exaggerated manner the shape of the elastic curve.

The tangent to the elastic curve at A passes through B. Thus, the first moment of the bending-moment-diagram area about B is equal to zero (second rule, Art. 4.2). Expressing this in equation form,

$$EI\Delta = \left(+ M_{BA}\frac{L}{2} \right)\frac{L}{3}$$

$$+ \left(- M_{AB}\frac{L}{2} \right)\frac{2L}{3} = 0$$

$$\frac{M_{BA}L^2}{6} - \frac{M_{AB}L^2}{3} = 0$$

$$\frac{M_{AB}}{M_{BA}} = \frac{1}{2}, \quad \text{the carry-over factor.}$$

The extension of the tangent at A and the tangent to the elastic curve at B form the angle θ_B. Applying the first rule (Art. 4.2), this angle is equal to the area of the bending moment diagram between A and B, or

$$EI\theta_B = + M_{BA}\frac{L}{2} - M_{AB}\frac{L}{2}$$

However, since $M_{AB} = \frac{1}{2}M_{BA}$,

$$EI\theta_B = + M_{BA}\frac{L}{2} - \frac{M_{BA}}{2}\left(\frac{L}{2}\right)$$

$$\theta_B = \frac{M_{BA}L}{4EI}$$

The stiffness factor was defined as the ratio between applied moment and corresponding angle, i.e.,

$$\frac{M_{BA}}{\theta_B} = \frac{4EI}{L}$$

The quantity $4EI/L$ may be thought of as the *absolute* stiffness factor. For most problems, only the *relative* stiffness factors, which vary as the ratio I/L, are needed, i.e.,

$$K'_{abs} = \frac{4EI}{L}$$

$$K'_{rel} = \frac{I}{L}$$

10.7
Moment Distribution—Theory

Once the mechanics of moment distribution have been practiced and understood, the theoretical basis for locking and releasing joints may be explained quite simply.

Figure 10.18 shows three members meeting at a common point B, where they are rigidly connected to one another. The opposite end of each member is fully restrained. Apply an unbalanced moment M_u to joint B. This moment tends to distort the members as shown by the dashed lines. In order to establish equilibrium, each member does its share in resisting this unbalanced moment. The joint B rotates, and, since all the members meeting at this joint are continuous, each member will rotate through the same angle θ. In Art. 10.6 it was shown that the relationship between the moment and angle of rotation

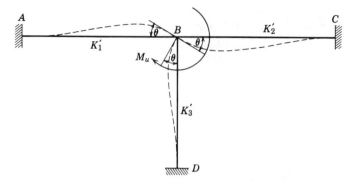

Figure 10.18/Theory of moment distribution.

for this condition[3] is

$$M = \frac{4EI}{L}\theta$$

Since 4, E, and θ are the same for all members, the portion of the unbalanced moment resisted by each member is in proportion to its ratio I/L, and is opposite in sign to the unbalanced moment. As previously stated, the ratio I/L is called the stiffness factor and is designated by the symbol K'.

Consequently,

induced moment in BA

$$= -M_u\left(\frac{K_1'}{K_1' + K_2' + K_3'}\right)$$

induced moment in BC

$$= -M_u\left(\frac{K_2'}{K_1' + K_2' + K_3'}\right)$$

induced moment in BD

$$= -M_u\left(\frac{K_3'}{K_1' + K_2' + K_3'}\right)$$

and

the sum of induced moments

$$= -M_u = -M_u\left(\frac{K_1' + K_2' + K_3'}{K_1' + K_2' + K_3'}\right)$$

The condition of equilibrium, $\Sigma M = 0$, is

---

[3]"This condition" refers to an angle of rotation at one end of a beam, the far end being fully fixed.

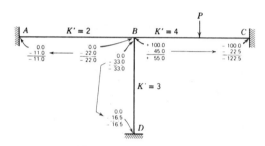

Figure 10.19/Moment distribution.

thus established. The quantity $K'/\Sigma K'$ is referred to as the distribution factor. Observe that the sum of the distribution factors of all members continuous at a joint must equal unity.

Reconsider this same frame as shown in Fig. 10.19. The unbalanced moment is removed and a load is placed on member BC. If the joint at B is assumed fixed before the load is applied, all joints are fixed and any loads will produce fixed-end moments. This is not the final moment, because joint B is artificially locked. A correction must be added to each fixed-end moment, a correction made by removing the artificial restraint.

Assume the load and span of member BC to be such as to produce fixed-end moments of $M_{BC}^F = +100$ and $M_{CB}^F = -100$. All other fixed-end moments are zero. Joint B is now subjected to an unbalanced moment of $+100$, and, when the artificial restraint is removed, it will rotate in a clockwise direction. This rotation induces balancing moments in each of the members as follows:

$$M_{BA} = -100\left(\frac{K'}{\Sigma K'}\right)$$

$$= -100\left(\frac{2}{2+4+3}\right) = -22$$

$$M_{BD} = -100\left(\frac{3}{2+4+3}\right) = -33$$

$$M_{BC} = -100\left(\frac{4}{2+4+3}\right)$$

$$= -44.4; \quad \text{use} -45$$

The total induced moments at $B = -100$.

These induced moments are the correction values to be added to the original fixed-end moments. Consequently, the final moments at B are

$$M_{BC} = +100-45 = +55$$

$$M_{BA} = \quad 0-22 = -22$$

$$M_{BD} = \quad 0-33 = -33$$

The sum of all the moments at B equals zero, and the joint is balanced.

It was shown in Art. 10.6 that if a moment is induced at one end of a beam which is fixed at its other end, one-half the induced moment (same sign) is carried over to the fixed end. This is the case with each in-

duced moment at B resulting from balancing that joint. Therefore,

$$M_{AB} = \tfrac{1}{2}(-22) = -11.0$$

$$M_{DB} = \tfrac{1}{2}(-33) = -16.5$$

$$M_{CB} = -100 + \tfrac{1}{2}(-45) = -122.5$$

are the final moments at the fully fixed ends (Fig. 10.19). If the above beam ends were not fully fixed, a further balancing step would be necessary.

Assume that joint A is not fixed, but continuously attached to another member AE, as shown in Fig. 10.20. When artificially fixed, it receives a moment of -11 from the balancing of joint B. Joint A must then be released and balanced as follows:

$$M_{AE} = +11\left(\frac{1}{1+2}\right) = +3.7$$

$$M_{AB} = -11 + 11\left(\frac{2}{1+2}\right)$$

$$= -11 + 7.3 = -3.7$$

Also, one-half the induced moment of $+3.7$ at AE must be carried over to the fixed end E, and one-half the induced moment of $+7.3$ at AB must be carried back to B. Joint B again has been assumed fixed while balancing the moments at A. Although not shown, the cycle must be repeated again at B, with $+3.7$ as the unbalanced moment.

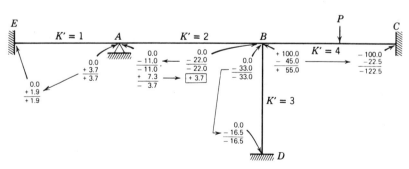

Figure 10.20/Moment distribution.

The general procedure for moment distribution then is as follows.

1. Assume all ends fixed, and compute the fixed-end moments.

2. Select a joint, release the restraint, and balance all moments.

3. Temporarily fix that joint again.

4. Select another joint, and repeat the same procedure (this must be done for all joints and is referred to as one cycle).

5. Carry over one-half the induced moments to the beam's opposite end and repeat the cycle.

When there is no longer a carry-over moment, or its value is insignificant, the process is completed and the final moments have been determined.

The treatment of beams having ends initially pinned *or* fully fixed is often a source of confusion. The discussion which follows should help in this regard.

Consider the beam shown in Fig. 10.21a. The beam is simply supported (pinned) at *A* and fully fixed at *B*. In step (1), it is assumed that both ends are fixed, and the fixed-end moments are calculated (Fig. 10.21b). In step (2), joint *A* is released,

and, since it is pinned, it can resist no moment. Therefore, the entire unbalanced moment of $+100$ must be offset by an induced moment of -100, so that the final moment at *A* will be zero (Fig. 10.21c). One-half of this induced moment is carried over to *B*. Joint *B* is never unbalanced, because it was fixed originally. It will take all moments carried over to it and never send any back (Fig. 10.21d).

10.8
Simplified Treatment of Pinned End

The moment distribution method may be simplified when one or more ends of a continuous beam or frame are pinned. This simplified treatment concerns only that portion of the structure between the pinned end and the first interior support. In Art. 10.6, the stiffness factor was defined as the ratio between an applied moment at a joint and the corresponding angle of rotation of the member at that joint. It was further shown that this ratio was

$$\frac{M_{BA}}{\theta_B} = \frac{4EI}{L}$$

when the far end of the member under consideration was fully *fixed*.

Now, consider the beam shown in Fig. 10.22. A moment M_{BA} is applied at one end, the far end being pinned. The broken line represents the deflected beam, and the moment M_{BA} produces the angle θ_B.

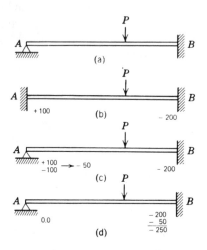

Figure 10.21/Moment distribution (fixed and pinned ends).

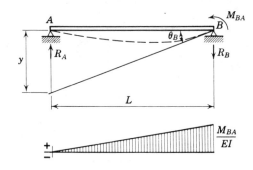

Figure 10.22/Stiffness factor (pinned ends).

A tangent to the elastic curve is constructed at B. Applying the second area-moment rule, the length y is equal to the first moment of the M/EI diagram about A, or

$$y = \frac{1}{2}\left(\frac{M_{BA}}{EI}\right)L\left(\tfrac{2}{3}L\right) = \frac{M_{BA}L^2}{3EI}$$

The tangent of θ_B is y/L and, for small angles θ_B, may be considered equal to y/L. Therefore,

$$\theta_B = \frac{y}{L} = \frac{M_{BA}L^2}{3EIL}$$

or

$$\frac{M_{BA}}{\theta_B} = \frac{3EI}{L}$$

Comparing the stiffness factors of a beam having a pinned end to one having fixed ends, it is seen that the former is $\tfrac{3}{4}$ the latter, i.e.,

$$\frac{3EI}{L} \div \frac{4EI}{L} = \frac{3}{4}$$

In order to use this factor $\tfrac{3}{4}$ properly in the moment-distribution procedure, the beam must be considered to be initially pinned at one end, and the fixed-end moment at the opposite end computed on that basis. This type of beam is often referred to as a *propped cantilever*. Figure 10.21a shows such a propped cantilever, and, from the moment distribution shown, it is evident that the actual fixed-end moment of 250 equals the moment produced

when both ends are fixed plus one-half the fixed-end moment from the opposite end, i.e.,

$$200 + \frac{100}{2} = 250$$

The moment-distribution procedure incorporating the simplified treatment of pinned ends is as follows.

1. Calculate the relative stiffness factors, i.e., I/L for all members not pinned. Members with one end pinned have a relative stiffness factor of $\tfrac{3}{4}(I/L)$.

2. Determine the distribution factor for each joint, based on the relative stiffness factors calculated in step (1).

3. Calculate all fixed-end moments, including the adjustment necessary for those members acting as propped cantilevers. The moments at the pinned ends are zero.

4. Release each joint where applicable, and balance the moments.

5. Carry-over $\tfrac{1}{2}$ the induced moments for all beams except those having pinned ends.

6. Repeat steps (4) and (5); usually three cycles are sufficient when following the steps as described herein.

Example

Find the moments at A, B, and C for the continuous beam shown in Fig. 10.23. The moment of inertia remains constant.

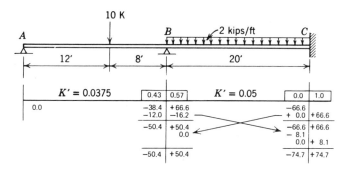

Figure 10.23/Simplified treatment of pinned ends.

(a) Simple beam. (b) Continuous beam.

Figure 10.24/Movement of supports.

Solution

(1) Calculate the relative stiffness factors:

$$K'_{AB} = \frac{3}{4}\left(\frac{I}{20}\right) \approx 0.038I$$

$$K'_{BC} = \frac{I}{20} = 0.050I$$

(2) Determine the distribution factors:

$$DF_{BA} = \frac{0.038I}{0.088I} \approx 0.43$$

$$DF_{BC} = \frac{0.050I}{0.088I} \approx 0.57$$

(3) Calculate the fixed-end moments:

$$M^F_{BC} = - M^F_{CB} = \frac{wL^2}{12} = \frac{2(20)^2}{12}$$

$$= 66.67 \text{ ft-kips}$$

$$M^F_{BA} = -\frac{Pab^2}{L^2} + \left[-\frac{1}{2}\left(\frac{Pba^2}{L^2}\right)\right]$$

$$= -\frac{10(8)12^2}{20^2} + \left[-\frac{1}{2}\left(\frac{10(12)8^2}{20^2}\right)\right]$$

$$= -28.8 - 9.6 = -38.4 \text{ ft-kips}$$

(4) Record the values found in steps (1) through (3) in their appropriate places on the tabulation shown in Fig. 10.23. Release the artificial restraint at the supports, and balance the moments.

(5) Carry over the induced moments. There is no carry-over to A, because A is pinned and no artificial restraint has been applied.

(6) Repeating step (4) completes the moment-distribution procedure, because

there are no resulting induced moments to carry over.

Final moments are as shown in the figure.

10.9
Movement of Supports

Joints or supports may move with respect to one another. This translation can result from a variety of conditions, e.g., differential ground settlement, changes in length of members, deflections of frames. Stresses in simply supported, determinate beams are not affected by this translation. However, bending moments and shears are introduced when settlement occurs in continuous beams (Fig 10.24).

The ends of the beam shown in Fig. 10.25a are fully fixed, allowing no rotation. Consider, however, that the support B moves a distance Δ as shown by the broken line. Fixed-end moments and reactions will be induced by this translation, as is shown in Fig. 10.25b. It is evident that $M^F_{AB} = M^F_{BA}$, $R_L = R_R$, and $M^F_{AB} + M^F_{BA} = R_L L = R_R L$.

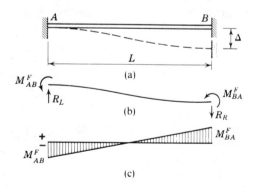

Figure 10.25/Movement (both ends fixed).

The bending moment diagram is shown in Fig. 10.25c.[4] Applying the second area-moment rule, it is seen that the distance Δ is equal to the first moment of the area of the M/EI diagram about B, or

$$
\begin{aligned}
EI\Delta &= \frac{+ M_{BA}^F}{2}\left(\frac{L}{2}\right)\left[\frac{1}{3}\left(\frac{L}{2}\right)\right] \\
&\quad - \frac{M_{AB}^F}{2}\left(\frac{L}{2}\right)\left[\frac{L}{2} + \frac{2}{3}\left(\frac{L}{2}\right)\right] \\
&= \tfrac{1}{24}\left(M_{BA}^F L^2\right) - \tfrac{5}{24}\left(M_{AB}^F L^2\right) \\
&= -\tfrac{1}{6}\left(M_{AB}^F L^2\right)
\end{aligned}
$$

or

$$
M_{AB}^F = -\frac{6EI}{L^2}\Delta
$$

Now consider the propped cantilever beam shown in Fig. 10.26a. Support B is initially pinned; therefore, the moment in the beam at this point is always zero. Support A is fully fixed, allowing no rotation. Support B settles an amount Δ. This translation will cause the reactions and moments shown in Fig. 10.26b. From the figure it is evident that $R_L = R_R$, and that $M_{AB}^F = R_R(L)$. The corresponding bending moment diagram is shown in Fig. 10.26c. Applying the second area-moment rule, it is seen that the distance Δ is equal to the first moment of the area of the M/EI diagram about B, or

$$
EI\Delta = -M_{AB}^F\left(\frac{L}{2}\right)\tfrac{2}{3}L = \frac{-M_{AB}^F L^2}{3}
$$

and

$$
M_{AB}^F = -\frac{3EI\Delta}{L^2}
$$

From the two formulas just derived, the magnitude of the fixed-end moments may

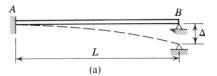

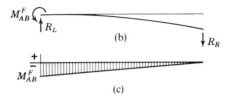

Figure 10.26/Movement of support (one end pinned).

be determined for any known relative translation of joints. To determine final fixed-end moments, these may be added algebraically to those caused by loads. Members not having fixed ends but which are continuous with other members require the additional procedure of moment distribution to produce final end moments.

Example 1

Calculate the fixed-end moments for the beam shown in Fig. 10.27. Support B moves vertically downward $\frac{1}{2}$ in. The moment of inertia of the beam is 400 in.[4]. Neglect the dead load of the beam.

Solution

Select the appropriate formula from Fig. 10.8 and from those derived in Art. 10.9, and solve for the fixed-end moments. (Care

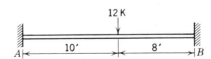

Figure 10.27

must be taken with units and signs, Art. 10.4.)

$$M_{AB}^F = +\frac{Pab^2}{L^2} + \frac{6EI\Delta}{L^2}$$

$$= +\frac{12(10)8^2}{18^2} + \frac{6(29,000)400(0.5)}{18^2(12)^3}$$

$$= +23.7 + 62.2 = 85.9 \text{ ft-kips}$$

$$M_{BA}^F = -\frac{Pba^2}{L^2} + \frac{6EI\Delta}{L^2}$$

$$= -\frac{12(8)10^2}{18^2} + \frac{6(29,000)400(0.5)}{18^2(12)^3}$$

$$= -29.6 + 62.2 = +32.6 \text{ ft-kips}$$

Example 2

The continuous beam shown in Fig. 10.23 is a **W** 14×43. Find the final moments at A, B, and C, if there is also a 1-in. vertical settlement at B.

(Two solutions to this problem will be shown.)

Solution (a)

(1) Determine moments at A, B, and C resulting from the loads only (no settlement). This was done in the example problem following Art. 10.8 and is shown in Fig. 10.23.

(2) Select the proper formulas, and calculate the fixed-end moments caused by the settlement:

$$M_{AB}^F = 0$$

$$M_{BA}^F = +\frac{3EI\Delta}{L^2} = \frac{3(29,000)428(1.0)}{20^2(12)^3}$$

$$= +53.9 \approx +54 \text{ ft-kips}$$

$$M_{BC}^F = M_{CB}^F = -\frac{6EI\Delta}{L^2}$$

$$= -\frac{6(29,000)428(1.0)}{20^2(12)^3}$$

$$= -107.7 \approx -108.0 \text{ ft-kips}$$

(3) Using the moment-distribution method, determine the moments at A, B, and C caused by the settlement alone. This step is shown in Fig. 10.28.

(4) Add (algebraically) the moments resulting from the loads (step 1), and the moments resulting from the settlement (step 3):

$$M_{AB} = 0$$

$$M_{BA} = -50.4 + 77.2 = +26.8 \text{ ft-kips}$$

$$M_{CB} = -74.7 - 92.6 = -167.3 \text{ ft-kips}$$

Solution (b)

(1) Select the appropriate formulas, and calculate the fixed-end moments caused by

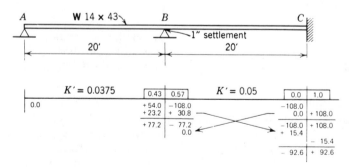

Figure 10.28

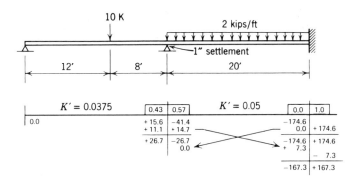

Figure 10.29

the loads and the settlement:

$$M_{AB}^F = 0$$

$$M_{BA}^F = -\frac{P}{L^2}\left(a^2 b + \frac{b^2 a}{2}\right) + \frac{3EI\Delta}{L^2}$$

$$= -38.4 + 54.0 = +15.6 \text{ ft-kips}$$

$$M_{BC}^F = +\frac{wL^2}{12} - \frac{6EI\Delta}{L^2}$$

$$= +66.6 - 108.0 = -41.4 \text{ ft-kips}$$

$$M_{CB}^F = -\frac{wL^2}{12} - \frac{6EI\Delta}{L^2}$$

$$= -66.6 - 108.0 = -174.6 \text{ ft-kips}$$

(2) Using the moment-distribution method, determine final moments resulting from the combined effect of loads and settlement. This step is shown in Fig. 10.29.

PROBLEMS

1. Repeat the process of moment distribution for Problems 2, 3, and 4 following Art. 10.5, using the simplified treatment of pinned ends. (Answers given in Appendix G.)

2. The continuous beam shown in Fig. 10.30 is a **W** 12×45. Loads are as shown. Calculate the

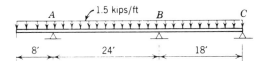

Figure 10.30

final moments at A, B, and C, if the support at A settles 1 in. vertically.

3. The continuous beam shown in Fig. 10.31 is a **W** 16×36. Loads are as shown. Calculate the final moments at the supports if both exterior supports settle $\frac{1}{2}$ in. and both interior supports settle $1\frac{1}{2}$ in. (Answers given in Appendix G.)

Figure 10.31

4. The propped cantilever beam shown in Fig. 10.32 is a **W** 12×35. Loads are as shown. The support at B moves vertically *upward* $\frac{3}{4}$ in. Sketch the shear and bending moment diagrams.

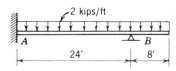

Figure 10.32

10.10
Building Frames

The nature and function of frames were discussed in Art. 10.1. How they act struc-

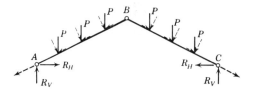

Figure 10.33/Thrust.

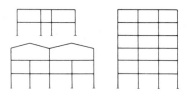

Figure 10.34/Single story frames.

turally and how stability is achieved to prevent collapse were also described. Frequently, frames are but an assemblage of horizontal and vertical members—the vertical members acting principally as columns and the horizontal members as beams or girders. However, these members need not always be vertical or horizontal.

For example, the roofs of buildings are frequently pitched, and the supporting members of the building frame are inclined.

Special consideration, however, must be given to *thrust* when dealing with inclined members in building frames. The existence of thrust is apparent in Fig. 10.33.[5] The inclined members *AB* and *BC* are parts of a frame. The end connections are of a flexible type; therefore, the ends are shown pinned. The vertical *P* forces are held in equilibrium by the vertical reactions R_V. However, there is a tendency for points *A* and *C* to move outward, caused by the cumulative tangential components of the *P* loads. This thrust is resisted by the horizontal reactions R_H, at *A* and *C*. It is imperative that the joints at *A* and *C* be able to develop needed resistance to this thrust; otherwise, the frame must be modified or some joints fixed against relative rotation.

The shapes and proportions of frames are nearly limitless. However, if frames are to be used to create space and enclosure,

their ultimate shape should be very much in mind during the early architectural planning stages. Figures 10.34 and 10.35 illustrate several basic shapes. Single-story frames, deriving stability from one or more rigid connections, are usually analyzed by one of the so-called exact mathematical procedures to be described in this chapter. Large multistory frames, or "high-rise" building frames, which derive stability in whole or in part from rigid connections, are usually analyzed first by approximate mathematical procedures (Art. 9.43) and then (following a preliminary design) a more detailed analysis by computer.

Spacing of frames, sometimes referred to as "bay spacing," depends on a great number of factors, such as size of building; occupancy; floor, wall, and roof construction; etc. In the absence of any particularly important factor to dictate the spacing of frames, the following suggestions are offered, based on greatest potential for economy of material. Generally speaking, the greater the span, the greater may be the spacing:

Span of Frames	Spacing of Frames
30 to 40 ft	16 ft
40 to 60 ft	18 ft
60 to 100 ft	20 ft
Over 100 ft	$\frac{1}{5}$ to $\frac{1}{6}$ the span

Building frames generally may be classified

[5]In order to simplify the figures, the members of all frames are shown as a heavy single solid line rather than the more realistic double line as was shown with beams.

Figure 10.35/Multistory frames.

Figure 10.36/Determinate frames.

either as three-hinged arches or rigid frames. Various forms of the three-hinged arch are shown in Fig. 10.36; these are statically determinate. That is to say, the reactions may be calculated from equations based on the laws of statics. It is the presence of the "extra" hinge (other than those at supports) that provides the additional condition equation necessary for solution by statics alone. Two examples are presented at the end of this article to illustrate the analysis of this type of frame.

Rigid frames are indeterminate, and the remainder of this chapter will be devoted to their analysis. The degree of external indeterminacy is established by the number of reaction components in excess of those needed for stability. Figure 10.37 shows six variations of the rigid frame, all of which are indeterminate. In both Fig. 10.37 (a) and (b), one horizontal component could be removed from the reactions

and the structure would remain stable; hence, both frames are indeterminate to the first degree. The frames shown in Fig. 10.37 (c) and (f) have three redundant reaction components. The frame in Fig. 10.37e is indeterminate to the fifth degree, and that shown in (d) is indeterminate to the first degree.

The design of building frames requires three distinct steps, the first being to determine reactions. This is easily accomplished in the case of the three-hinged arch, as will be illustrated in the two examples which follow. If the frame is indeterminate, however, determination of the reactions is considerably more involved—this will be covered in the remaining articles of this chapter.

The second step is to calculate maximum shears and bending moments. The best way to accomplish this is by sketching both shear and bending moment diagrams. From the sketches, the position of maximum values will be evident, and the work can be easily verified and checked. Included in this second step of the analysis is the determination of the direct (axial) load value each member must resist. This value is equal to the reaction components or difference in shears between members.

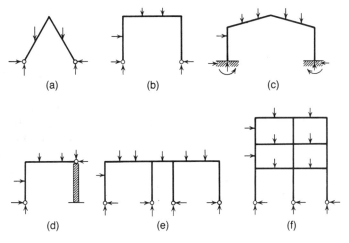

Figure 10.37/Indeterminate frames.

The third step is the actual design or proportioning of a member so as to be of sufficient size and of a shape to permit it to safely resist the stresses resulting from shear, bending moment, and direct load. End connections are then designed and detailed in conformance with the assumptions used in the analysis.

Example 1

Determine the maximum value of shear bending moment and axial load in the two members forming the frame shown in Fig. 10.38a.

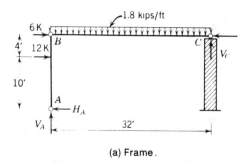

(a) Frame.

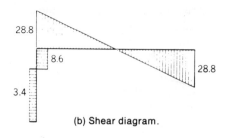

(b) Shear diagram.

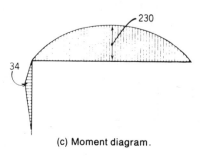

(c) Moment diagram.

Figure 10.38/Three-hinged arch.

Solution

By observation, this frame is a three-hinged arch and is therefore statically determinate.

(1) Determine reactions. Reactions may be calculated by taking moments about A, B, or C of a part or the whole of the frame. At A, B, and C the moment is zero, because ends are pinned.

$\Sigma M_B = 0$ (*BA* only):

$$H_A(14) - 4(12) = 0$$
$$H_A = 3.4 \text{ kips}$$

$\Sigma M_B = 0$ (*BC* only):

$$1.8(32)16 - V_C(32) = 0$$
$$V_C = 28.8 \text{ kips}$$

$\Sigma M_A = 0$ (entire frame):

$$12(10) + 6(14) + 1.8(32)16$$
$$- 28.8(32) - H_C(14) = 0$$
$$H_C = 14.6 \text{ kips}$$

$\Sigma M_C = 0$ (entire frame):

$$-1.8(32)16 - 12(4) + 3.4(14)$$
$$+ V_A(32) = 0$$
$$V_A = 28.8 \text{ kips}$$

Check:

$$\Sigma V = 0; \qquad 1.8(32) = 28.8 + 28.8$$
$$\Sigma H = 0; \qquad 12 + 6 = 3.4 + 14.6$$

(2) Sketch the shear diagram. This is shown in Fig. 10.38b. It is necessary to adopt a system for referencing the top and bottom of members, in order to be consistent with the sign convention for shear and bending moment previously adopted. To accomplish this, it is suggested that the inside of the frame be compared to the bottom of a beam. This convention is used for both shear and bending moment diagrams.

(3) Sketch the bending moment diagrams and calculate maximum values (Fig. 10.38c).

Summarizing: Member AB must be designed for both a bending moment of 34 ft-kips and a direct load of 28.8 kips. A maximum shear of 8.6 kips is also indicated. Member BC has a maximum bending moment of 230 ft-kips, a maximum axial load of 14.6 kips, and a maximum shear of 28.8 kips.

Example 2

Determine the maximum value of shear, bending moment, and axial load in the two members forming the frame shown in Fig. 10.39a. The uniform load is 1.5 kips per ft of horizontal projection. This is consistent with usual practice when referring to roof loads. It should be noted that this is the same as a vertical load of 0.93 kips per ft acting along the 30.6-ft length of the member, i.e., $0.93(30.6) = 1.5(19)$.

Solution

(1) Determine reactions:

$\Sigma M_A = 0$ (entire frame):

$$1.5(38)19 + 10(24) - V_C(38) = 0$$
$$V_C = 34.8 \text{ kips}$$

$\Sigma M_C = 0$ (entire frame):

$$-1.5(38)19 + 10(24) + V_A(38) = 0$$
$$V_A = 22.2 \text{ kips}$$

$\Sigma M_B = 0$ (AB only):

$$-1.5(19)\tfrac{19}{2} + 22.2(19) - H_A(24) = 0$$
$$H_A = 6.3 \text{ kips}$$

$\Sigma M_B = 0$ (BC only):

$$1.5(19)\tfrac{19}{2} - 34.8(19) + H_C(24) = 0$$
$$H_C = 16.3 \text{ kips}$$

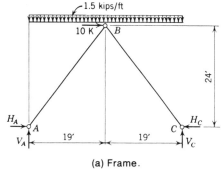

(a) Frame.

(b) Reactions.

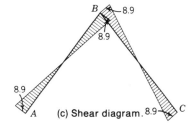

(c) Shear diagram.

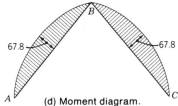

(d) Moment diagram.

Figure 10.39/Three-hinged arch.

Checking:

$$\Sigma H = 0; \qquad 10 = 16.3 - 6.3$$
$$\Sigma V = 0; \qquad 1.5(38) = 34.8 + 22.2$$

(2) Resolve the reaction components into components normal and parallel to the members. This is accomplished by similar triangles and is shown in Fig. 10.39b. The computations follow.

Support at A: parallel to AB,

$$\frac{24}{30.6}(22.2) + \frac{19}{30.6}(6.3) = 21.3 \text{ kips}$$

Normal to *AB*,

$$\frac{19}{30.6}(22.2) - \frac{24}{30.6}(6.3) \approx 8.9 \text{ kips}$$

Support at *C*: parallel to *CB*,

$$\frac{24}{30.6}(34.8) + \frac{19}{30.6}(16.3) = 37.4 \text{ kips}$$

Normal to *CB*,

$$\frac{19}{30.6}(34.8) - \frac{24}{30.6}(16.3) \approx 8.9 \text{ kips}$$

(3) Sketch shear diagram (Fig. 10.39c).

(4) Sketch bending moment diagram (Fig. 10.39d).

Summarizing:

Member *AB*: shear = 8.9 kips, bending moment = 67.8 ft-kips, axial load = 21.3 kips.

Member *BC*: shear = 8.9 kips, bending moment = 67.8 ft-kips, axial load = 37.4 kips.

It should be noted that the 10-kip horizontal load at *B* does not cause bending in the members; it tends only to put *AB* in tension and *BC* in compression. Also, the maximum bending moment in each member can be computed on the basis of the horizontal projected load. Knowing that the maximum moment will occur at the center of *AB*,

$$M = \frac{wL^2}{8} = \frac{1.5(19)^2}{8} = 67.7 \text{ ft-kips}$$

PROBLEMS

1. Determine the reactions, sketch the shear and bending moment diagrams, and calculate maximum shear and moment values for the frame shown in Fig. 10.40. Establish shear, bending-moment, and axial-load design values for members *AB* and *BC*. (Answers given in Appendix G.)

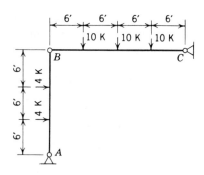

Figure 10.40/Problem 1.

2. Determine shear, bending-moment, and axial-load design values for the frame shown in Fig. 10.41. The support at *A* is a pin. The support at *D* is a pin on rollers; therefore, the reaction has only a vertical component.

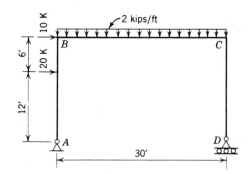

Figure 10.41/Problem 2.

3. Sketch the shear and bending moment diagrams for the frame shown in Fig. 10.42, and establish shear, bending-moment, and axial-

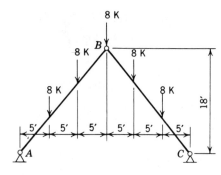

Figure 10.42/Problem 3.

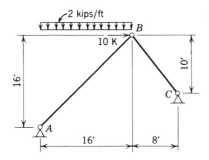

Figure 10.43/Problem 4.

load design values. (Answers given in Appendix G.)

4. Determine design values for all members of the frame shown in Fig. 10.43. The uniform load on the horizontal projected length of member *AB* is 2 kips per ft.

10.11
Analysis of Rigid Frames

Indeterminate frames can be analyzed using the moment-distribution method in a manner similar to that applied to continuous beams. After bending moments at supports and joints have been determined, reaction components can be calculated and shear and bending moment diagrams drawn.

Frequently, more than two members meet at a joint. In such cases, the unbalanced moment is distributed to the ends of *all* members in proportion to their relative stiffness factors. The work may be tabulated in many ways. The particular procedure shown in the examples which follow is both convenient and easy to read.

For purposes of analysis, rigid frames may be divided into two classes: Those that have joint translation, and those that do not. The frames shown in Fig. 10.44 have no joint translation. Each joint is held in position either by a support or by another member. Any joint movement resulting from a change in length of a member is considered negligible.

The frames shown in Fig. 10.45, however, do have joint translation, or what is commonly referred to as sidesway. A study of these examples will reveal that classification is dependent on the configuration of the frame and/or the type of loading.

The moment distribution process is straightforward for those cases where no sidesway is involved, and the true value of bending moments can be determined directly. Frames which do have sidesway require an additional step, which is a correction of the bending moments calculated on the basis of no sidesway. Frames having

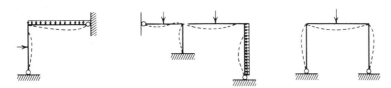

Figure 10.44/Frames having no sidesway.

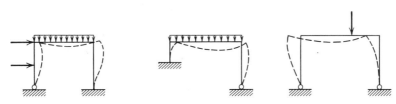

Figure 10.45/Frames having sidesway.

sidesway are treated in detail in Arts. 10.13 and 10.14.

Example 1

Analyze the rigid frame loaded as shown in Fig. 10.46a. The moment of inertia of the horizontal member is three times that of each column. Sketch the bending moment diagram.

Solution

Since the frame is symmetrical and the loading is symmetrical, no sidesway occurs.

(1) Determine the relative stiffness of each member. Use the simplified treatment for

pinned ends:

$$K'_{AB} = K'_{CD} = \frac{3}{4}\left(\frac{I}{12}\right) \simeq 0.063I$$

$$K'_{BC} = \frac{3I}{26} \simeq 0.115I$$

(2) Calculate the distribution factors:

$$DF_{BA} = DF_{CD} = \frac{0.063I}{0.178I} = 0.35$$

$$DF_{BC} = DF_{CB} = \frac{0.115I}{0.178I} = 0.65$$

(3) Determine the fixed-end moments:

$$M^F_{BC} = -M^F_{CB} = \tfrac{15}{48}(20)26 \simeq 162 \text{ ft-kips}$$

(4) Using the moment-distribution method, determine final moments at B and C (Fig. 10.46b).

(5) Calculate the end reactions. Determine the value of bending moments at other points necessary to sketch the bending moment diagram (Fig. 10.46c).[6]

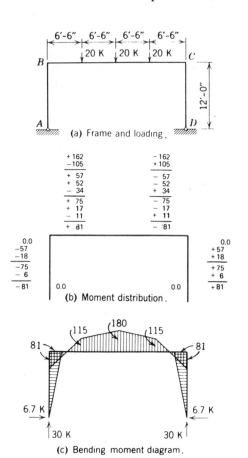

(a) Frame and loading.

(b) Moment distribution.

(c) Bending moment diagram.

Figure 10.46/No sidesway.

Example 2

Analyze the rigid frame loaded as shown in Fig. 10.47a. The moment of inertia of member BC is twice that of the other members.

Solution

Since the frame is such that no sidesway can occur, the procedure is the same as that used in Example 1.

---

[6] When studying frames, it sometimes becomes convenient to draw the moment diagram on the tension side of the frame. However, in order to maintain conformity with other portions of this text, that method will *not* be illustrated.

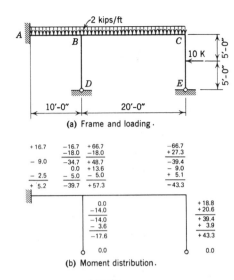

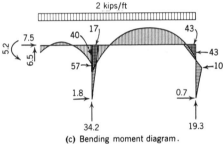

(c) Bending moment diagram.

Figure 10.47/No sidesway.

(1) Relative stiffness factors:

$$K'_{AB} = \frac{I}{10}$$

$$K'_{BD} = K'_{CE} = \frac{3}{4}\left(\frac{I}{10}\right)$$

$$K'_{BC} = \frac{2I}{20}$$

(2) Distribution factors:

$$DF_{BA} = 0.36$$

$$DF_{CB} = 0.57$$

$$DF_{BC} = 0.36$$

$$DF_{CE} = 0.43$$

$$DF_{BD} = 0.27$$

(3) Fixed-end moments:

$$M^F_{AB} = -M^F_{BA} = \frac{2(10)^2}{12} = 16.7 \text{ ft-kips}$$

$$M^F_{BC} = -M^F_{CB} = \frac{2(20)^2}{12} = 66.7 \text{ ft-kips}$$

$$M^F_{CE} = \frac{10(10)}{8}(1.5) = 18.8 \text{ ft-kips}$$

(4) Moment distribution: See Fig. 10.47b, where two cycles are illustrated.

(5) End reactions are determined and the bending moment diagram is drawn (Fig. 10.47c).

10.12
Analysis and Design
of Single-Story Frames

The design of a rigid frame involves the selection of shapes and proportioning of sections so that the actual stresses resulting from shear, bending moment, and direct load do not exceed the allowable stresses. The joints which are assumed rigid in the analysis must be made so by detailing them in such a manner that there will be no relative rotation when they are finally assembled in the field.

Although analysis precedes actual design, a preliminary estimate first must be made of the relative size of members. The analysis requires, first, that all joints be assumed held in position so that no relative translation takes place. Bending moments, resulting from joint rotation (moment distribution), which are in turn caused by loads placed between joints, are then calculated. Next, a sidesway correction is made to account for the forces used to establish the assumed condition of no relative joint translation. These corrections are applied to the previously calculated moments and the analysis completed by the use of statics. The example which follows illustrates the first stages in this procedure. The sideway correction and sizing of members will be illustrated in the article following.

Example

Analyze and design a typical interior rigid frame for the building shown in Fig. 10.48. Column bases are to be pinned, and roofing, roof deck, and purlins weigh 28 psf. No earthquake load is to be assumed. Use A36 Steel.

Solution

(1) Determine all loads acting on the frame. For purposes of this example, it will be assumed that the combined dead, live, and environmental (i.e., wind, snow, rain, etc.) loads impose a net downward load of 68 psf on the roof, and that there is a lateral load of 20 psf (wind only) from the left, acting on the wall. (See Chapters 8 and 9 for a more extensive treatment of such loads and their computation.)

Each purlin supports 6×18 ft of roof surface and delivers this as a concentrated load to the frame. However, the concentrated loads are sufficiently close together to permit an assumption that the load is uniformly distributed on the frame (this will greatly simplify the analysis). Therefore, the roof load per lineal foot of frame is

$$w = 68(18) \simeq 1220 \text{ plf}$$

Use 1.3 kips per ft, allowing 80 lb per ft for the weight of the horizontal spanning member (beam).

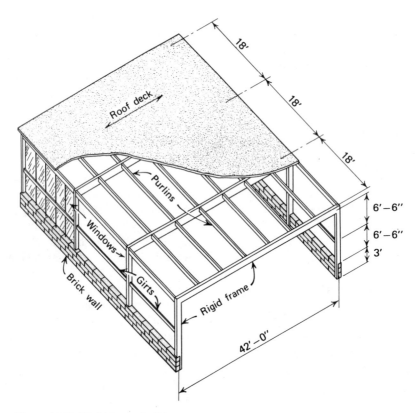

Figure 10.48/Rigid frame building.

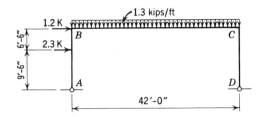

Figure 10.49/Typical frame—design loading.

The lateral (wind) load on the brick wall will be transmitted directly to the foundation and thus will not affect the frames. Wind load on the windows, however, will be transmitted through the girts and eave purlins to the frames as concentrated loads. For one frame, the maximum exterior wind load is:

Girt:

$$20(18)6.5 = 2340 \text{ lb}; \qquad \text{use } 2.3 \text{ kips}$$

Eave purlin:

$$\frac{2300}{2} = 1150 \text{ lb}; \qquad \text{use } 1.2 \text{ kips}$$

These loads are shown acting on the frame in Fig. 10.49.

The detailed analysis shown here and in Art. 10.13 is based on full live, dead, and environmental load on the roof, plus full wind load on one side. Before making a final decision on the size of the beam and column, an analysis would need to be made for the full roof load acting alone, i.e., no wind on the frame. This is discussed in the latter portions of steps (13) and (14) of Art. 10.13.

(2) Estimate the relative size of the horizontal and vertical members (beam and columns) constituting the frame. This must be done on the basis of experience. If the beam were simply supported, it would have a maximum moment of $wL^2/8 = 287$ ft-kips. If it is assumed that this moment is reduced $\frac{1}{4}$ by the effect of the rigid frame,

the approximate design moment would be 215 ft-kips. From the AISC Beam Charts, it is seen that beams in the **W** 21 group will be adequate.

The column size is somewhat more difficult to estimate. The axial load will be approximately $1.3(21) = 27$ kips, but there also will be an appreciable bending moment induced by the beam and wind loads. Estimate this bending moment to be $\frac{1}{3}$ the design moment for the beam, or 72 ft-kips. Then, scanning the safe-load tables and applying the B_x factor, columns in the lighter-weight **W** 14 group appear adequate.

Scanning the tables once more for moments of inertia in the beam and column groups selected, an approximate I of 1600 in.^{4} for the beam and 400 in.^{4} for the column seems reasonable. Consequently, assume a ratio of 4 to 1.

(3) Calculate stiffness factors and distribution factors:

$$K'_{AB} = \frac{3}{4}\left(\frac{I}{16}\right) \simeq 0.047I$$

$$K'_{BC} = \frac{4I}{42} \simeq 0.095I$$

$$DF_{BA} = DF_{CD} = \frac{0.047I}{0.142I} \simeq 0.33$$

$$DF_{BC} = DF_{CB} = \frac{0.095I}{0.142I} \simeq 0.67$$

(4) Calculate fixed-end moments on the basis of no relative joint translation:

$$M^F_{AB} = M^F_{DC} = 0$$

$$M^F_{BA} = -\frac{(2.3)6.5(9.5)^2}{(16)^2}$$

$$-\frac{1}{2}\left[\frac{(2.3)9.5(6.5)^2}{(16)^2}\right]$$

$$= -7.0 \text{ ft-kips}$$

(propped cantilever), and

$$M_{BC}^F = - M_{CB}^F = \frac{1.3(42)^2}{12} = 191 \text{ ft-kips}$$

(5) Make a sketch of the frame, tabulate the fixed-end moments in their appropriate positions, and balance the joints (moment distribution). This step is shown in Fig. 10.50a.

(6) Using the bending moments determined in step (5) calculate the horizontal force at each reaction (Fig. 10.50b). Notice that when checking $\Sigma H = 0$, there is a resultant force of 2.9 kips acting to the right; this is the magnitude of the force needed at B to prevent joint translation. It is frequently referred to as the artificial joint restraint (AJR) and in this example acts to the left. The vertical components of the reactions at this point could also be computed, but are not helpful in the still incomplete analysis; the moments at B

and C need to be corrected for the effect of sidesway.

10.13
Design of Rigid Frame—Sidesway

The rigid-frame analysis in the preceding article was based on the assumption that no sidesway took place. In step (6) it was shown that to make this assumption valid, a force (AJR) of 2.9 kips acting to the left at point B was necessary. This is a fictitious force and does not actually exist. Consequently, the bending moments in the frame at B and C must be found when this force is removed.

These moments are calculated by assuming a joint translation (any amount), determining bending moments induced by this translation, calculating the force necessary to produce the translation, and—by pro-

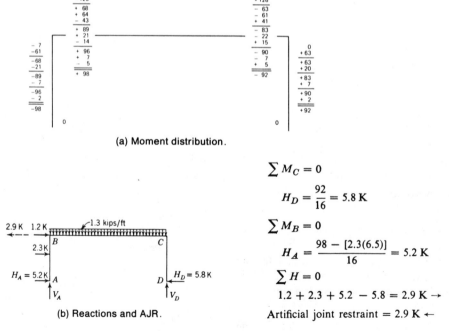

(a) Moment distribution.

(b) Reactions and AJR.

$$\sum M_C = 0$$

$$H_D = \frac{92}{16} = 5.8 \text{ K}$$

$$\sum M_B = 0$$

$$H_A = \frac{98 - [2.3(6.5)]}{16} = 5.2 \text{ K}$$

$$\sum H = 0$$

$$1.2 + 2.3 + 5.2 - 5.8 = 2.9 \text{ K} \rightarrow$$

Artificial joint restraint = 2.9 K ←

Figure 10.50/Frame solution (no joint translation).

portions of the assumed and actual forces —determining the actual value of the moments. This procedure is treated in detail in the following steps and is a continuation of the analysis begun in the preceding article.

Solution

(7) Assume that point B moves to the right a distance Δ. This joint movement occurs when the AJR is removed. A sketch of the deflected frame shows that point C also moves to the right the same distance (Fig. 10.51a). Any relative displacement caused by a change in length of any given member is negligible. It was shown in Art. 10.9 that the fixed-end moment caused by

joint translation is

$$M^F = \frac{6EI\Delta}{L^2}, \qquad \text{both ends fixed}$$

$$M^F = \frac{3EI\Delta}{L^2}, \qquad \text{one end pinned}$$

In the solution of this problem, it is not necessary to calculate the actual displacement Δ, but only to determine the relative value of the fixed-end moments when a displacement takes place. In this problem, since E, I, Δ, and L are the same for both BA and CD, and since both have one pinned end, $M_{BA}^F = M_{CD}^F$.

(8) Take any value fixed-end moment in the proportions determined in step (7), and allow the joints to rotate (moment

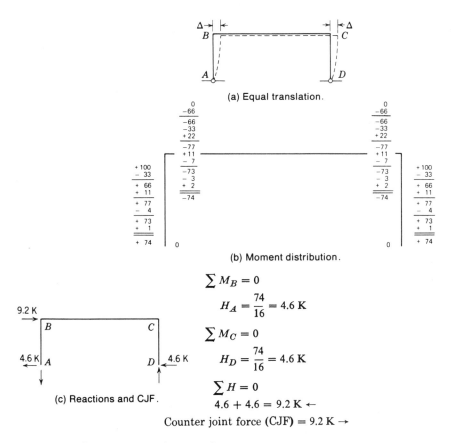

(a) Equal translation.

(b) Moment distribution.

(c) Reactions and CJF.

$$\sum M_B = 0$$

$$H_A = \frac{74}{16} = 4.6 \text{ K}$$

$$\sum M_C = 0$$

$$H_D = \frac{74}{16} = 4.6 \text{ K}$$

$$\sum H = 0$$

$$4.6 + 4.6 = 9.2 \text{ K} \leftarrow$$

Counter joint force (CJF) = 9.2 K →

Figure 10.51/Frame solution (sidesway).

distribution):

$$M_{AB}^F = M_{DC}^F = M_{BC}^F = M_{CB}^F = 0$$

$$M_{BA}^F = M_{CD}^F = 100 \text{ ft-kips} \quad \text{(assumed)}$$

The moment distribution is shown in Fig. 10.51b.

(9) Calculate the horizontal components of the reactions resulting from the moments determined from step (8). From these reaction components, the value of the force necessary to induce the assumed moments can be calculated. This is frequently referred to as a counter joint force (CJF) (Fig. 10.51c).

(10) From the proportions, determine the bending moments when the artificial joint restraint is released. The artificial joint restraint was found to be 2.9 kips←. When this is released, it will have the same effect as placing an equal force of 2.9 kips→ on the frame. The final bending moments shown in Fig. 10.51b were caused by a force of 9.2 kips→. Consequently, the bending moments caused by the 2.9 kips→ will be

$$M_{BA} = M_{CD} = \frac{2.9}{9.2}(+74) = +23 \text{ ft-kips}$$

$$M_{BC} = M_{CB} = \frac{2.9}{9.2}(-74) = -23 \text{ ft-kips}$$

(11) Correct the bending moments calculated on the basis of no joint translation (step 5, Art. 10.11), by adding to them the bending moments which result from removal of the artificial joint restraint (step 10):

$$M_{AB} = M_{DC} = 0$$

$$M_{BA} = -98 + 23 = -75 \text{ ft-kips}$$

$$M_{BC} = +98 - 23 = +75 \text{ ft-kips}$$

$$M_{CB} = -92 - 23 = -115 \text{ ft-kips}$$

$$M_{CD} = +92 + 23 = +115 \text{ ft-kips}$$

These are the final bending moments at the joints in the frame.

(12) Determine the reactions, and sketch the shear and bending moment diagrams (Fig. 10.52).

(13) Design the beam. The maximum bending moment for the beam is 189 ft-kips. Purlins provide lateral support; therefore, $L_u = 6$ ft. Assuming an allowable bending stress of 24 ksi, the required section modulus is

$$S = \frac{M}{F_b} = \frac{189}{24}(12) = 94.5 \text{ in.}^3;$$

$$\text{use 95 in.}^3$$

The lightest-weight section furnishing this section modulus is a **W** 18×55 ($S = 98.3$ in.^{3}). This section is compact for A36 Steel and has an allowable L_c length of 7.9 ft. Consequently, $F_b = 24$ ksi. The actual bending stress is

$$f_b = \frac{M}{S} = \frac{189(12)}{98.3} = 23.1 \text{ ksi} \quad \text{(safe)}$$

Checking shear,

$$f_v = \frac{V}{A_w} = \frac{28}{18.11(0.39)} = 3.96 \text{ ksi}$$

$$\text{(safe)}$$

This beam also carries a direct load of 7.2 kips; however, it is proportionately small and will be neglected.

The **W** 18×55 meets the AISC Specification with respect to strength requirements in both bending and shear and would appear adequate. However, further examination with respect to deflection is needed. This aspect of design often involves judgment on the part of the designer, since limiting deflections are seldom specifically established.

Using the $F_y/800$ suggested depth-span guide (1989 AISC Specification), the required depth is

$$\frac{36}{800}(42)12 = 22.7 \text{ in.}$$

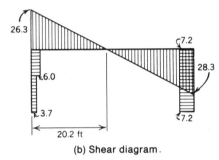

(a) Final reactions.

$$\sum M_C = 0$$

$$H_D = \frac{115}{16} = 7.2 \text{ K}$$

$$\sum M_B = 0$$

$$H_A = \frac{75 - [2.3(6.5)]}{16} = 3.7 \text{ K}$$

$$\sum M_A = 0$$

$$V_D = \frac{2.3(9.5) + 1.2(16) + 1.3(42)21}{42} = 28.3 \text{ K}$$

$$\sum M_D = 0$$

$$V_A = \frac{1.3(42)(21) - 2.3(9.5) - 1.2(16)}{42} = 26.3 \text{ K}$$

(b) Shear diagram.

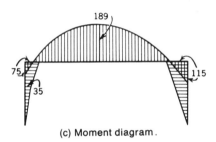

(c) Moment diagram.

Figure 10.52/Final analysis.

if the section is fully stressed (Art. 5.4). Therefore, the **W** 18×55 should not be used. Either a larger 18-in. beam should be selected (thereby reducing the bending stress) or a deeper beam.

Assuming the end restraint developed by rigidly attaching the beam to the column reduces the 42-ft span to an effective 38-ft simple span, and limiting the maximum total-load, simple-span deflection to $L/240$,

$$\Delta_{\max} = \frac{38(12)}{240} = 1.9 \text{ in.}$$

The required moment of inertia is

$$I = \frac{5wL^4}{384E\Delta} = \frac{5(1.3)38^4(12)^3}{384(29,000)1.9} = 1110 \text{ in.}^4$$

A **W** 21×57 meets the strength requirement and also has a moment of inertia of 1170 in.[4]. Use the **W** 21×57 for the beam.

The AISC Specification also permits an increase in the allowable stress when loading includes the effect of wind. Since the **W** 21×57 was selected primarily on the basis of deflection requirements, this special allowance should not alter the design. It also should be noted that the **W** 21×57

is adequate for bending and shear when the full roof load is acting alone in the absence of any wind load.

(14) Design the columns. These columns must be designed for the most severe condition which could occur. For column CD, the most severe condition occurs when the wind is from the left (Fig. 10.52). Column AB would be subjected to the same forces when the wind was from the right; therefore, design conditions are the same for both columns. These design conditions are: shear = 7.2 kips; bending moment = 115 ft-kips; direct load = 28.3 kips; unsupported height = 16 ft for the major axis and 6.5 ft for the minor axis. The effective-length factor[7] (K) for the minor axis is assumed to be 1.2, and for the major axis, 2.0.

After making a few trial selections, a **W** 12×40 is chosen for detailed investigation.

The maximum slenderness ratios are

$$\frac{K_y L_y}{r_y} = \frac{1.2(6.5)12}{1.93} = 48.5$$

$$\frac{K_x L_x}{r_x} = \frac{2.0(16)12}{5.13} = 74.9$$

From tables, the maximum allowable axial stress is

$$F_a = 15.91 \text{ ksi}$$

Since design conditions include wind load, this stress (using the AISC Specification) can be increased $\frac{1}{3}$:

$$F_a = 1.33(15.91) = 21.2 \text{ ksi}$$

[7]See Art. 6.8; K should not be confused with K', the symbol used to denote *stiffness factor* in the discussion on moment distribution.

The actual axial stress is

$$f_a = \frac{P}{A} = \frac{28.3}{11.8} = 2.4 \text{ ksi}$$

and

$$\frac{f_a}{F_a} = \frac{2.4}{21.2} = 0.113$$

Since $f_a / F_a < 0.15$, the straight-line interaction formula should be used for the stability check.

The actual bending stress is

$$f_b = \frac{M}{S} = \frac{115(12)}{51.9} = 26.6 \text{ ksi}$$

The **W** 12×40 is a compact section and has an allowable $L_c = 8.4$ ft > 6.5 ft. Therefore, the allowable bending stress is 24 ksi. Increasing this stress $\frac{1}{3}$ for wind,

$$F_b = 1.33(24) = 32 \text{ ksi}$$

Checking the sum of the stress ratios,

$$\frac{f_a}{F_a} + \frac{f_b}{F_b} = \frac{2.4}{21.2} + \frac{26.6}{32} = 0.94 < 1.0$$

(safe)

Also,

$$\frac{f_a}{0.6F_y} + \frac{f_b}{F_b} = \frac{2.4}{28.8} + \frac{26.6}{32} = 0.91 < 1.0$$

(safe)

Shear is negligible.

Before finally adopting this section, it should be checked for stresses resulting from gravity loads alone, i.e., without wind and the $\frac{1}{3}$ increase in allowable stress for wind. This will be found to be controlling, and the final adopted section is **W** 12×45. This conclusion should be verified.

Summary / Use **W** 21×57 for the beam. Use **W** 12×45 for the columns. Then

$$\frac{I_{beam}}{I_{col}} = \frac{1170}{350} = 3.34$$

which is reasonably close to the originally assumed ratio of 4:1 therefore, no further refinement is needed.

10.14
Gable Frames

The procedure for the analysis and design of gable frames is similar to that illustrated for the rigid frame in Art. 10.13. A gable frame is a special form of the rigid frame. The analysis, however, is considerably more complex than for rectangular rigid frames. This is particularly so when moment distribution is employed. Frequently, the designer will resort to the use of special handbooks, computer programs, or moment coefficients rather than undertake a time-consuming analysis. For the sake of completeness, however, an example will be presented showing the analysis of a gable frame. Each step is explained in detail. It is recommended that the entire example be read through, then each step in the process be studied.

Example

The rigid gable frame shown in Fig. 10.53 is loaded as indicated. Column bases are

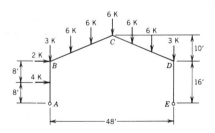

Figure 10.53/Rigid gable frame (example).

assumed pinned. All members have the same moment of inertia. Determine the reactions and sketch the bending moment diagram for the frame. The following analysis is both lengthy and cumbersome. Consequently, current procedures frequently include computer programs. The example in Appendix F solves this frame by a computer program.

Solution

(1) Determine relative stiffness factors and distribution factors for each member of the frame. Use the factor $\frac{3}{4}$ for pinned ends. Then,

$$K'_{AB} = K'_{ED} = \frac{3}{4}\left(\frac{I}{16}\right) \simeq 0.047I$$

$$K'_{BC} = K'_{CD} = \frac{I}{26} \simeq 0.038I$$

$$DF_{BA} = DF_{DE} = \frac{0.047I}{0.085I} = 0.55$$

$$DF_{BC} = DF_{DE} = \frac{0.038I}{0.085I} = 0.45$$

$$DF_{CB} = DF_{CD} = \frac{0.038I}{0.076I} = 0.50$$

(2) Compute fixed-end moments. Because joints B, C, and D are assumed to be fully fixed, loads acting at these points do not enter the calculation. For the first analysis, therefore, these loads are removed (Fig. 10.54a); their effect will be considered later.

$$M^F_{AB} = M^F_{ED} = M^F_{DE} = 0$$

$$M^F_{BA} = 1.5\left[-\frac{4(16)}{8}\right] = -12 \text{ ft-kips}$$

$$M^F_{BC} = -M^F_{CB} = \tfrac{2}{9}(6)24 = 32 \text{ ft-kips}$$

$$M^F_{CD} = M^F_{DC} = 32 \text{ ft-kips}$$

(3) Assuming that there is no joint translation, allow each joint to rotate, and balance the moments (moment distribution). This step is shown in Fig. 10.54b.

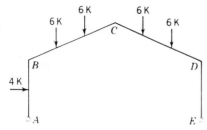

(a) Loads producing fixed-end moments.

(b) Moment distribution.

$$\sum M_D = 0 \qquad\qquad \sum M_B = 0$$

$$H_E = \frac{18}{16} = 1.1 \text{ K} \qquad H_A = \frac{4(8) - 23}{16} = 0.6 \text{ K}$$

$$\sum M_D = 0$$

$$V_A = \frac{6[8 + 16 + 32 + 40] + 4(8) - [1.1 + 0.6]16}{48} = 12.1 \text{ K}$$

$$\sum M_B = 0$$

$$V_E = \frac{6[8 + 16 + 32 + 40] + [1.1 + 0.6]16 - 4(8)}{48} = 11.9 \text{ K}$$

$$\sum M_C = 0 \text{ (right half)}$$

$$AJR_D = \frac{6[8 + 16] + 1.1(26) - 11.9(24) - 38}{10} = 15.1 \text{ K}$$

$$\sum M_C = 0 \text{ (left half)}$$

$$AJR_B = \frac{-6[8 + 16] - 4(18) + 0.6(26) + 12.1(24) + 38}{10} = 12.8 \text{ K}$$

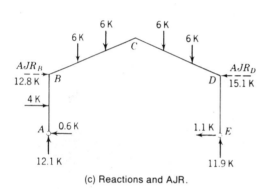

(c) Reactions and AJR.

Figure 10.54/Analysis—no joint translation.

(4) Calculate the reactions and the artificial joint restraint at B and D (Fig. 10.54c). These are the forces required to sustain the original assumption that there is no joint translation.

(5) Sidesway. Assume a side lurch which produces a displacement at B, equal to the displacement at C and in the same direction (Fig. 10.55a). From geometry, it is seen that the displacement of joint C is the same as for B and D, but that the relative displacement between B and C, and between C and D, is zero. Consequently, there will be no fixed-end moments in members BC and CD due to this

assumed displacement, and

$$M_{BA}^F = M_{DA}^F = \frac{3EI\Delta}{L^2}$$

Assume an arbitrary value of 40 ft-kips for M_{BA}^F and M_{DA}^F, the moment at the ends of the other members being zero. Allow joints to rotate to get balanced end moments (Fig. 10.55b).

(6) With the end moments found in step (5), and applying the conditions of statics, calculate the reactions and the counter joint forces F_{B-1} and F_{D-1} applied at B and D, respectively. These are the forces necessary to produce the side lurch associ-

ated with the end moments. The calculations are shown in Fig. 10.55c.

(7) Spread. Assume joints B and D move an equal distance in opposite directions (Fig. 10.56a). The displacement of point C relative to points B and D is as shown and may be calculated from geometry (Fig. 10.56b). The slope of BC is $1/2.4$, making the hypotenuse 2.6. From similar triangles,

$$\sin \alpha = \frac{1}{2.6} = \frac{\Delta}{\Delta_{BC}}$$

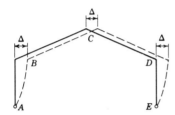

(a) Assumed joint translation.

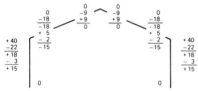

(b) Moment distribution.

(c) Reactions and CJF.

$$\sum M_B = 0$$

$$H_A = \frac{15}{16} = 0.94 \text{ K}$$

$$\sum M_D = 0$$

$$H_E = \frac{15}{16} = 0.94 \text{ K}$$

$$\sum H = 0$$

$$F_{B-1} = 0.94 \text{ K}$$

$$F_{D-1} = 0.94 \text{ K}$$

Figure 10.55/Analysis—sidesway.

from which

$$\Delta_{BC} = 2.6\Delta$$

(8) Assume any fixed-end moment values for the members which are commensurate with the relative end displacements and end conditions. It was shown earlier in this chapter that for fixed ends,

$$M^F = \frac{6EI\Delta}{L^2}$$

and that for pinned ends,

$$M^F = \frac{3EI\Delta}{L^2}$$

The ends of the members at B, C, and D are assumed fixed. Consequently, if 20 ft-kips is assumed for M^F_{BA} and M^F_{DE}, the other fixed-end moments will be 2(2.6)20, or 104 ft-kips. Care should be exercised in assigning the right sign to these fixed-end moments. The joints now may be allowed to rotate to get balanced end moments (moment distributions). (See Fig. 10.56c.)

(9) From the balanced end moments found in step (8), and applying the conditions of statics, calculate the reactions at A and E, and the counter joint forces F_{B-2} and F_{D-2} necessary at points B and D, respectively, to produce the initially assumed spread (Fig. 10.56d).

(10) With the same spread, under the same assumed conditions, calculate the vertical force F_{C-2} necessary at point C (Fig. 10.56e). It should be noted that either F_{C-2} applied at point C, or the forces F_{B-2} and F_{D-2}, will produce spread resulting in the end moments calculated in step (8).

(11) The artificial joint restraints, AJR_B and AJR_D, prevent both sidesway and spread. Calculate the proportions of each that prevent sidesway and spread. This step is shown in Fig. 10.57a. Assign the symbol x to components preventing

(a) Assumed joint translation.

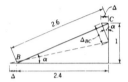

(b) Relative joint translation (B and C).

(c) Moment distribution.

(d) Reactions and CJF at B and D.

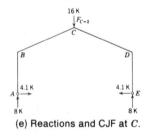

(e) Reactions and CJF at C.

Figure 10.56/Analysis—spread.

$$\sum M_D = 0; \qquad H_E = \frac{66}{16} = 4.1 \text{ K}$$

$$\sum M_B = 0; \qquad H_A = \frac{66}{16} = 4.1 \text{ K}$$

Horiz. forces at B and D to cause spread:

$$\sum M_C = 0 \text{ (right half)}$$

$$F_{D-2} = \frac{4.1(26) + 85}{10} = 19.2 \text{ K} \rightarrow$$

$$\sum M_C = 0 \text{ (left half)}$$

$$F_{B-2} = \frac{4.1(26) + 85}{10} = 19.2 \text{ K} \leftarrow$$

Vert. force at C to cause spread:

$$\sum M_C = 0 \text{ (right half)}$$

$$V_E = \frac{4.1(26) + 85}{24} = 8 \text{ K}$$

$$\sum M_C = 0 \text{ (left half)}$$

$$V_A = \frac{4.1(26) + 85}{24} = 8 \text{ K}$$

$$\sum V = 0$$

$$F_{C-2} = 8 + 8 = 16 \text{ K}$$

sidesway and the symbol y to components preventing spread. The components x for sidesway are equal at B and D and are applied in the same direction. The components y for spread are equal at B and D and are applied in opposite directions.

(12) Determine the proportions of the actual force at B causing sidesway and spread (Fig. 10.57b). This procedure is the same as that described for the AJRs in step (11).

(13) Correction factors. Absolute numbers which, when multiplied by the assumed balanced bending-moment values, will result in true values based on an actual loading condition are termed *correction factors*. The effect of removing the AJR will be the same as an equal and opposite force applied at the same point. This is also true of the components x and y. Consequently, the part of the AJR for sidesway, 1.15 kips, when divided by the

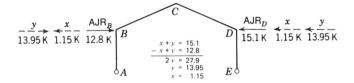

(a) Proportions of the AJR preventing sidesway (x) and spread (y)

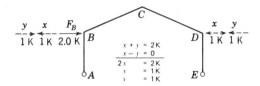

(b) Proportions of the force at B causing sidesway (x) and spread (y)

Figure 10.57/Components for sidesway and spread.

Table 10.1
Summary of End Moments (ft-kips)

Location of moment		A-B	B-A	B-C	C-B	C-D	D-C	D-E	E-D
Arbitrary sidesway		0	+15	−15	0	0	−15	+15	0
Arbitrary spread		0	−66	+66	+85	−85	−66	+66	0
Loads between ends of members no joint translation		0	−23	+23	−38	+38	−18	+18	0
Correction for AJR	$(+1.22)M$ from arbitrary sidesway	0	+18	−18	0	0	−18	+18	0
	$(+0.73)M$ from arbitrary spread	0	−48	+48	+62	−62	−48	+48	0
Effect of horiz. force at B (2^k)	$(+1.06)M$ from arbitrary sidesway	0	+16	−16	0	0	−16	+16	0
	$(−0.05)M$ from arbitrary spread	0	+3	−3	−4	+4	+3	−3	0
Effect of vert. force at C (6^k)	$(+0.38)M$ from arbitrary spread	0	−25	+25	+32	−32	−25	+25	0
Final moments		0	−59	+59	+52	−52	−122	+122	0

counter joint force for sidesway, F_{B-1}, becomes a correction factor which when multiplied by the end moments found in step (5) will give the end moments generated by the 1.15-kip force. The correction factors are as follows:

Effect of removing the AJR:

$$\text{sidesway} = \frac{1.15 \rightarrowtail}{0.94 \rightarrowtail} = +1.22$$

$$\text{spread} = \frac{13.95 \longleftrightarrow}{19.20 \longleftrightarrow} = +0.73$$

Effect of the actual horizontal force at B:

$$\text{sidesway} = \frac{1.00 \rightarrowtail}{0.94 \rightarrowtail} = +1.06$$

$$\text{spread} = \frac{1.00 \rightarrowtail\!\leftarrow}{19.20 \longleftrightarrow} = -0.052$$

Effect of the actual force at C:

$$\text{spread} = \frac{6\downarrow}{16\downarrow} = +0.375$$

(14) Final end moments. The final end moment at any point will be the algebraic

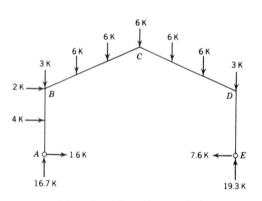

(a) Final reactions (from statics)

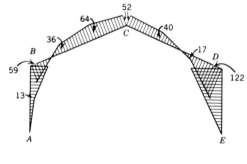

(b) Bending moment diagram

$$\sum M_D = 0; \qquad H_E = \frac{122}{16} = 7.6 \text{ K}$$

$$\sum M_B = 0; \qquad H_A = \frac{59 - 4(8)}{16} = 1.6 \text{ K}$$

$$\sum M_A = 0;$$

$$V_E = \frac{4(8) + 2(16) + 6[8 + 16 + 24 + 32 + 40] + 3(48)}{48} = 19.3 \text{ K}$$

$$\sum M_E = 0$$

$$V_A = \frac{6[8 + 16 + 24 + 32 + 40] + 3(48) - 2(16) - 4(8)}{48} = 16.7 \text{ K}$$

Check:

$$\sum H = 0; \quad 4 + 2 + 1.6 = 7.6 \text{ K}$$

$$\sum V = 0; \quad 6(5) + 2(3) = 16.7 + 19.3 \text{ K}$$

Figure 10.58/Final analysis.

sum of:

(a) Balanced end moment caused by loads between ends of members and no joint translation.

(b) Effect of removing the artificial joint restraint:

(1) Correction factor times sidesway moment.

(2) Correction factor times spread moment.

(c) Effect of the horizontal force at *B*:

(1) Correction factor times sidesway moment.

(2) Correction factor times spread moment.

(d) Effect of the vertical force at *C*; correction factor times spread moment.

The above step is shown executed for each end moment in Table 10.1.

(15) From the final end moments found in step (14), compute the reactions at *A* and *E*. Notice that the frame stands in equilibrium without the use of artificial joint restraints. Construct the bending moment diagrams (Fig. 10.58).

10.15
Rigid Frames—Conclusion

The design of rigid frames is not complete until all joints are designed and detailed to achieve a condition compatible with the design conditions. One member can be rigidly attached to another using welds, and additional plates are usually employed to prevent local buckling.

Special note should be made of the fact that the method of analysis described in this chapter is applicable only to members of constant cross section. It is frequently desirable to "build up" members where maximum bending moments occur. This

"built-up" portion is frequently referred to as a haunch. Rigid frames using haunched members are nearly always analyzed by the use of coefficients, handbooks, or computer. For more information on this subject, the reader is referred to "Single Span Rigid Frames in Steel," a pamphlet prepared by the American Institute of Steel Construction. (See also Chapter 11 for an introduction to plastic design.)

PROBLEMS

1. The rigid frame shown in Fig. 10.59 has a constant moment of inertia. Calculate the moments at *A*, *B*, and *C*, and all reactions. Sketch the shear and bending moment diagrams. (Answers given in Appendix G.)

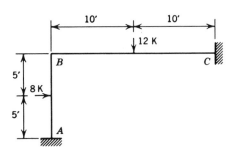

Figure 10.59

2. Determine the reactions and sketch the shear and bending moment diagrams for the rigid frame shown in Fig. 10.60. Column bases are pinned.

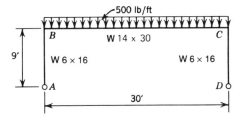

Figure 10.60

3. Determine the reactions and sketch the shear and bending moment diagrams for the rigid frame shown in Fig. 10.61. Column bases are assumed pinned, and the moment of inertia is constant for all members. (Answers given in Appendix G.)

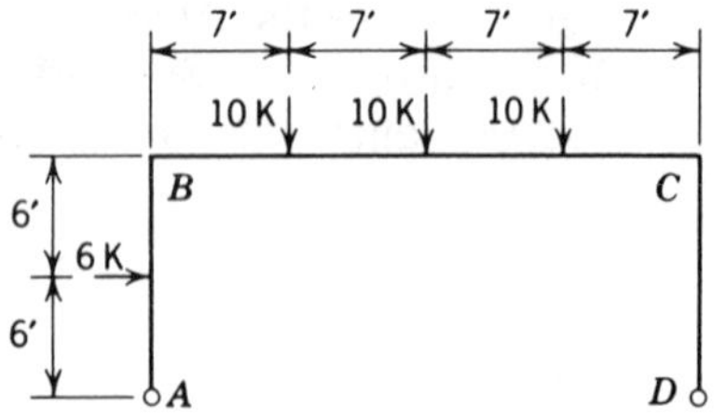

4. Determine the reactions and sketch the shear and bending moment diagrams for the rigid frame shown in Fig. 10.62. Moments of inertia for all members are equal. The base of the column at A is pinned, while that at D is fully fixed.

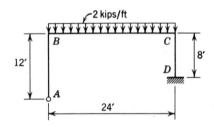

Figure 10.61

Figure 10.62

11

LRFD AND PLASTIC DESIGN

11.1
Introduction

The principal intent of Chapters 1 through 10 has been to present general principles regarding the structural analysis and design of buildings, and to utilize the most prevalent analytical and design procedures in doing so, i.e., elastic analysis and elastic (allowable-stress) design (ASD). This is quite reasonable, because using elastic analysis and ASD does provide protection against permanent structural damage, with an adequate, if somewhat obscure, factor of safety.

The underlying assumption in the ASD methodology is that the maximum stress will nowhere exceed the yield stress of the material. However, with the greater availability of higher-strength steels and concomitant greater reliance on continuity in structures, coupled with continuing efforts to improve the performance and efficiency of steel structures, additional analytic and design tools have emerged—specifically, plastic analysis, and LRF (load- and resistance-factor) and plastic (ultimate-strength) design. LRFD, which was introduced in Chapter 1 and referred to in Chapter 8, is based on plastic-design concepts, and can be used in conjunction with either elastic or plastic analysis procedures. Thus, in a sense LRFD could be considered a transition between elastic and full plastic (ultimate-strength) design. There is every reason to believe that both LRFD and plastic design procedures will see ever greater acceptance and use in the future.

Therefore, although the theory of elasticity has not lost its legitimacy as a basis for design, there is need to present these alternate procedures as well.

With these approaches, the ultimate strength of the structure as a whole becomes of increasing importance. This recognition of ultimate strength is not new. It will be remembered that in Euler's original paper (1757), the ultimate strength of columns was treated. Utilizing ultimate strength as an approach to design does not presuppose that the yield stress will be exceeded at any point within the structure under expected lifetime loads; on the contrary, every precaution is taken to prevent such an occurrence because the result would be permanent injury to the structure. The distinction is that with elastic design, the factor of safety is based on the difference between the maximum actual stress at predicted loads and the initial yield stress that would occur if loading were allowed to increase. With plastic design a portion of the reserve strength of the structure after initial yield stress has been reached is utilized as part of the factor of safety. *However, this reserve strength is significant only if the structure being designed is statically indeterminate.*

It is important that the terms and phrases used in this alternative design approach be clearly understood.

Ultimate Strength / This is the condition which occurs when a beam, an element of a structure, or the structure as a whole, reaches its load-carrying capacity. In this situation, any additional increment in load, no matter how small, would cause fracture or collapse.

Plastic Theory / Steel has a predictable structural behavior in the stress range in excess of the yield point, i.e., the plastic range. Within this range, stress is no longer proportional to strain and permanent deformation takes place. Plastic design implies a design procedure based upon behavior of the structure or element stressed within the plastic range of the material.

There are two principal aspects of plastic theory. The first involves the stress pattern at a single cross section and the developed stress resultant. (Article 11.2 elaborates on this aspect for a simple case of bending.) The second aspect involves the element as a whole—e.g., the entire length of a series of continuous beams, the entire frame, the structure as a whole—where a possible redistribution of bending moments, shears, and direct loads after the structure has begun to yield can be taken into account. This second aspect is referred to as *plastic analysis* (Art. 11.4).

Limit States / This term describes performance measures that can be used to control a design. For example, the maximum load on a beam can be established by limiting bending stress, shear stress, deflection, buckling tendency, and the like. In the case of LRFD, there are two types of limit states: strength and serviceability. Both are important, but obviously strength considerations, because they affect safety and load-carrying capacity, are paramount and are those dealt with in codes.

In the strength category are the onset of yielding, formation of a plastic hinge (Art. 11.3), formation of a plastic mechanism (Art. 11.5), overall frame or member stability, lateral-torsional buckling, tensile fracture, fatigue cracks, deflection instability, alternating plasticity, and excessive deformation. In the serviceability category are unacceptable elastic deflections and drift, unacceptable vibrations, and permanent deformations.

Any of these criteria could establish the load; therefore, in a sense, even the conventional (ASD) design procedure treated in the body of this text is a form of limit design. However, in an analytical sense, limit states are more generally associated with the general design process which includes ultimate strength as one of the limiting criteria. In these processes (LRFD and plastic design) it is necessary to consider all ways in which the structure could fail or collapse in order to isolate that type of failure requiring least load.

LRFD and Plastic Design / Structural design based on ultimate strength using plastic theory can provide a saving in material if the structure is adequately braced against local and lateral buckling and the serviceability aspect is not controlling. *Since the design is based upon higher stresses, more attention needs to be focused on buckling. Also, since smaller sections may result, special care must be taken to prevent excessive deflections.* This is why the ultimate-strength design process can never completely replace design based upon the elastic theory.

This introduction to the subject of LRFD and plastic design will be limited to certain aspects of bending and direct stress. For a more thorough treatment, including shear as well as direct stress and bending, the reader is referred to "Commentary on Plastic Design in Steel" by the American Society of Civil Engineers, or "Plastic Design in Steel" by AISC.

11.2
Plastic Theory

Because steel is a ductile material, it is able to withstand deformation under load without fracture. Figure 11.1a shows the

normal stress-strain curve for structural steel, whereas at (b), that portion of (a) contained within the broken lines is replotted on an expanded horizontal scale. In (b), the curve is further altered for illustrative purposes by changing the slightly curved lines to straight lines.

When the yield point, designated F_y, is first reached, it has a corresponding strain ϵ_y. Any slight increase in stress beyond this point will cause the steel to behave plastically. However, its ductility allows it to be strained, or stretched, 15 times its yield-point strain without a noticeable increase in stress. This is frequently referred to as a large flow of the material at a constant stress.

If this phenomenon is considered in a member in bending (Fig. 11.2a), the stress patterns illustrated at (b), (c), (d), and (e) are those resulting from successively higher bending moments. From elastic theory, stress is proportional to distance from the neutral axis, and triangular stress patterns are formed such as those shown in Fig. 11.2 (b) and (c). At (c), the outermost fiber has reached the yield point, and the corresponding stress resultant or yield bending moment is $M_y = F_y(S)$. Any additional moment will cause the outermost fiber to flow but at the same time maintain the yield stress. Consequently, the adjacent fibers undergo an increase in strain with corresponding increase in stress up to a maximum of F_y. This condition is shown in Fig. 11.2d. For still larger moments, the

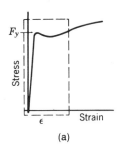

(a)

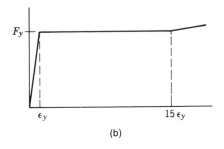

(b)

Figure 11.1/Stress-strain relationship.

process continues until the entire section is stressed up to F_y as shown in Fig. 11.2e. This is the maximum stress on the entire section. In this stressed condition, a so-called plastic hinge is formed, and the corresponding moment is referred to as the plastic moment (M_p). It is important to observe that in this state, the stressed beam will appear to rotate easily with any tendency to increase the moment; however, the beam will be able to sustain the plastic moment as it rotates.

Since the stress is constant (though of opposite sign) in both the tension and compression regions, the areas and internal forces must be equal. Expressed in equation form,

$$F_{\text{tension}} = F_{\text{compression}} = F_y \frac{A}{2}$$

where A = total area of the beam.

This internal force is located at the centroid of each of the respective tensile and compressive areas. Therefore, the plastic moment is the product of this internal force and the perpendicular distance between their lines of action, or

$$M_p = F_y \frac{A}{2} 2\bar{y}$$

This expression also may be shown as

$$M_p = F_y \left[2\left(\frac{A\bar{y}}{2} \right) \right]$$

The quantity within the parentheses will be recognized as the static moment of one-half the cross section with respect to the neutral axis. This quantity was introduced in this text under the discussion of horizontal shear. It was referred to as Q; therefore,

$$M_p = 2QF_y$$

The quantity $2Q$ is called the plastic section modulus, and the symbol Z denotes its value. Values of Z for various steel shapes are listed in Part 2 of the AISC Manual, immediately following the listing of the elastic section modulus (S).

$$M_p = ZF_y$$

Z will always be larger than S by an amount depending upon the actual dimensions of the beam. The ratio of Z to S is referred to as the shape factor (u):

$$u = \frac{Z}{S}$$

For **S** and **W** shapes, the value of u will generally be found to be from 1.02 to 1.2, and generally very close to 1.12.

Not only must the full plastic moment (M_p) be developed, but additional strain, permitting rotation, must be allowed to take place as well, and both must be achieved without local buckling or twisting. These requirements will be discussed below in greater detail. As a consequence,

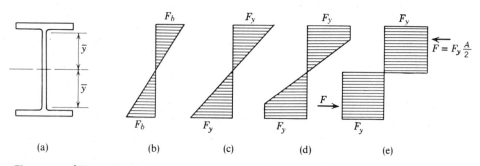

Figure 11.2/Elastic-plastic stress sequence in bending.

both rigid width-thickness limits and special criteria to prevent flange buckling are established. These restrictions are more severe than those governing design by the elastic method.

Section N7 of the 1989 AISC ASD Specification lists maximum ratios $b_f/2t_f$ (width to thickness of flanges for single web shapes) for different yield-stress grades of steel. This ratio for A36 Steel is 8.5. The same section of the specification provides a formula for limiting the depth-thickness ratio of webs—i.e., $d/t \le 412/\sqrt{F_y}$ — further modified when the beam must resist an axial load as well as a bending moment.

Section N9 of the ASD Specification describes the lateral bracing criteria. The appropriate formula for bracing dimensions depends on the ratio of the applied end moment (smallest) to the plastic moment of the beam (M/M_p). If a single curvature results in a ratio of negative sign and a double curvature results in a ratio of positive sign, the value of the M/M_p ratio ranges from -1.0 to $+1.0$. When $M_p \ge -1.0$ but < -0.5, the maximum unbraced length is

$$l_{cr} = r_y \left(\frac{1375}{F_y} \right)$$

and when $M/M_p \ge -0.5$ but $< +1.0$, the unbraced length is limited to

$$l_{cr} = r_y \left(\frac{1375}{F_y} + 25 \right)$$

where r_y is the radius of gyration of the member about its weak axis.

___

Example 1

A **W** 12×26 has two $\frac{3}{8} \times 12$-in. flange plates welded to it to make a built-up section as shown in Fig. 11.3. All steel is A36.

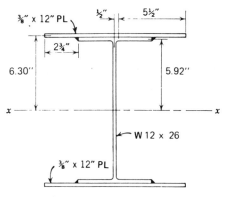

Figure 11.3/Built-up W beam.

(a) Check to be sure that width-thickness ratios conform to Section N9 of the 1989 AISC Specification.

(b) Calculate that bending moment about the x-axis which would cause first yield (M_y). Calculate the plastic bending moment M_p.

(c) Assuming a linear moment gradient of 0 to M_p (i.e., $M/M_p = 0$), determine at what intervals lateral support would be required.

Solution (a)

(1) Checking the combined flange projection,

$$\frac{W}{t} = \frac{5.5}{0.38 + 0.375} = 7.28 < 8.5 \qquad \text{OK}$$

(2) Checking the plate overhang,

$$\frac{W}{t} = \frac{2.75}{0.375} = 7.33 < 8.5 \qquad \text{OK}$$

(3) Checking the web depth-thickness ratio,

$$\frac{412}{\sqrt{F_y}} = \frac{412}{\sqrt{36}} = 68.7$$

$$\frac{d}{t} = \frac{12.97}{0.23} = 56.4 < 68.7 \qquad \text{OK}$$

Solution (b)

(1) Neglecting the moment of inertia of the plate about its own center of gravity, calculate the moment of inertia of the built-up section:

$$I_x = I_w + 2(A_p d^2)$$

$$= 204 + 2[0.375(12)6.30^2] = 561 \text{ in.}^4$$

(2) $\quad S = \dfrac{I}{c} = \dfrac{561}{6.49} = 86.4 \text{ in.}^3$

(3) $\quad M_y = SF_y = 86.4(36) \approx 3110 \text{ in.-kips}$

(4) $\quad Q = A_{pl} y_1 + A_f y_2 + A_w y_3$

$$= 4.5(6.30) + 0.38(6.49)5.92$$

$$+ 0.23(5.73)\frac{5.73}{2}$$

$$= 46.7 \text{ in.}^3$$

(5) $\quad Z = 2Q = 2(46.7) = 93.4 \text{ in.}^3$

(6) $\quad M_p = ZF_y = 93.4(36) \approx 3362 \text{ in.-kips}$

Solution (c)

(1) Calculate the radius of gyration for the axis perpendicular to bending:

$$I_y = I_{\text{beam}} + 2I_{\text{plate}}$$

$$= 17.3 + 2(0.375)\frac{12^3}{12} = 125.3 \text{ in.}^4$$

$$r = \sqrt{\frac{I}{A}} = \sqrt{\frac{125.3}{16.65}} = 2.74 \text{ in.}$$

(2) Referring to Section N9 of the 1989 AISC ASD Specification,

$$\frac{M}{M_p} = \frac{0}{M_p} = 0$$

Therefore,

$$l_{cr} = 2.74\left(\frac{1375}{36} + 25\right) = 173 \text{ in.} = 14.4 \text{ ft}$$

Example 2

Find the first yield moment and the plastic moment for a **W** 14×34 made from A36 Steel. Also, calculate the shape factor and the maximum unbraced length assuming a moment slope gradient from 0 to M_p.

Solution

(1) Referring to the AISC ASD Manual, select the following properties of the **W** 14×34:

$$S_x = 48.6 \text{ in.}^3, \qquad Z_x = 54.6 \text{ in.}^3,$$
$$r_y = 1.53 \text{ in.}$$

(2) $M_y = SF_y = 48.6(36) = 1750 \text{ in.-kips.}$

(3) $M_p = ZF_y = 54.6(36) = 1966 \text{ in.-kips.}$

(4) $u = Z/S = 54.6/48.6 = 1.12.$

(5) The maximum unbraced length is

$$\frac{M}{M_p} = 0$$

Therefore,

$$l_{cr} = 1.53\left(\frac{1375}{36} + 25\right) = 96.7 \text{ in.} = 8.1 \text{ ft}$$

PROBLEMS

1. A built-up section of channels and plates is constructed as shown in Fig. 11.4. A36 Steel is specified, and bending is about the x axis.

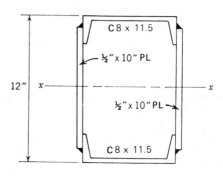

C 8 x 11.5

½" x 10" PL

12" x ———|———————|——— x

½" x 10" PL

C 8 x 11.5

Figure 11.4/Built-up box beam.

(a) Does the beam meet the width-thickness requirements necessary for a plastic moment?

(b) Calculate the maximum elastic bending moment it could develop, if laterally supported.

(c) Find the bending moment that would cause first yield.

(d) Calculate the plastic moment.

(e) Determine the maximum unbraced length allowed, assuming a moment gradient from 0 to M_p.

(Answers given in Appendix G.)

2. Find the plastic moment and shape factor in reference to the major axis for a **W** 16×36. Use A36 Steel. Calculate the plastic moment for the minor axis. (Answers given in Appendix G.)

3. A **W** 14×22 of A36 Steel is used as a simple beam for a 24-ft span. There is complete lateral support. Calculate the total uniform load (W) that would cause a plastic moment to be developed.

11.3
LRFD

The concept of balancing the effects of loads on a structure with the ability of the structure to withstand those effects is quite simple, even if the manner of dealing with them is not. Simply stated, the load effects can be viewed as representing the "required strength," whereas the structure resistance to those effects can be viewed as the "design strength." In the two design approaches cited above (ASD and LRFD), these relationships are expressed as follows: for ASD,

$$\sum Q_i \leq \frac{R_n}{\text{F.S.}}$$

and for LRFD

$$\sum \gamma_i Q_i \leq \phi R_n$$

The common symbols in these equations,

as defined by AISC, are:

Q_i = nominal load effect (i.e., effect of forces or moments), where i = type of load (e.g., dead, live, wind, earthquake, as defined earlier, and as ascertained by applying the load determination methods set forth in the applicable code or ASCE 7-88).

R_n = nominal resistance (i.e., nominal strength).

The other symbols reflect the differences in the two approaches to dealing with the major uncertainties in the assumptions.

In the case of ASD, unfactored loads (service loads) are equated with the resistance, to which a factor of safety (F.S.) has been applied. The left side of the equation thus represents the "required strength," and the right, the "design strength." From Chapter 3 it can be seen that on dividing the appropriate load effects (e.g., moment) by the appropriate section property (in this case, the section modulus), the result is f_b, the computed or developed bending stress. The right side of the equation is the yield stress of the steel being used with the F.S. applied, resulting in F_b, the permitted bending stress.

In the case of LRFD, both sides of the equation are factored—a load factor (γ_i) being applied to the left side of the equation, and a resistance factor (ϕ) to the right side. The load factors in the AISC ASD and LRFD Specifications—one for each possible load combination—are those recommended in ASCE 7-88, and are probability-based. The resistance factors contained in the AISC LRFD Specification, which are different for each limit state, are based on test results, with a lower factor representing a lesser uniformity in results, and chosen with the overall goal of uniformity of reliability. Each limit state has a associated value of R_n.

The two approaches—ASD and LRFD—will not give dramatically different results, because experience gained in the application of ASD over the years were a consideration in establishing LRFD parameters.

The yield stresses permitted, range from 36 ksi for A36 Steel to 100 ksi for A514 Steel; however, not all shapes and plate thicknesses are available for the various yield stresses. The LRFD Manual provides tables similar to that contained in the ASD Manual (Art. 1.5). With both of these approaches it is acceptable to utilize either elastic or plastic analysis; however, in the case of the latter, only steels in yield strengths up to and including 65 ksi (ASTM A514 Steel being excluded) are permitted.

Referring again to the inequality $\Sigma\gamma_i Q_i \leq \phi R_n$, the factored load combinations (ASCE 7-88 and the AISC Specification) are:

$1.4D$,

$1.2D + 1.6L + 0.5(L_r \text{ or } S \text{ or } R)$,

$1.2D + 1.6(L_r \text{ or } S \text{ or } R) + (0.5L \text{ or } 0.8W)$,

$1.2D + 1.3W + 0.5L + 0.5(L_r \text{ or } S \text{ or } R)$,

$1.2D + 1.5E + (0.5L \text{ or } 0.2S)$,

$0.9D - (1.3W \text{ or } 1.5E)$,

where the numbers are the load factors, and the letters represent the type of load: D = dead load; L = live load; L_r = roof live load; S = snow load; R = rain load (except ponding); W = wind load; and E = earthquake load.[1] The expression $\Sigma\gamma_i Q_i$ then is the maximum value attained by these combinations.

Typical values of ϕ are:

$\phi_t = 0.90$ for tensile yielding in the gross section.

$\phi_t = 0.75$ for tensile fracture in the net section.

$\phi_c = 0.85$ for compression.

$\phi_b = 0.90$ for flexure (bending).

$\phi_v = 0.90$ for shear yielding.

An example of a beam design in LRFD for flexure and shear will help illustrate the approach.

Example 1

A simple beam supports a flat roof area measuring 7 ft 0 in. between beams × a span of 21 ft 6 in. The dead load is 70 psf, the roof live load 20 psf, the snow load 20 psf, and the rain load 5 psf. Select the most economical **W** section of A36 Steel.

Solution

Assume a section weight of 30 plf, and compute the plf loads:

$$D = (70)(7.0) + 30 = 520 \text{ plf}$$
$$L_r = (20)(7.0) = 140 \text{ plf}$$
$$S = (20)(7.0) = 140 \text{ plf}$$
$$R = (5)(7.0) = 35 \text{ plf}$$

Calculating the load combinations,

a. $\qquad\qquad 1.4(520) = 728 \text{ plf}$

b. $\quad 1.2(520) + 0 + 0.5(140) = 694 \text{ plf}$
$\quad 1.2(520) + 0 + 0.5(140) = 694 \text{ plf}$
$\quad 1.2(520) + 0 + 0.5(35) = 642 \text{ plf}$

c. $\quad 1.2(520) + 1.6(140) + 0 = 848 \text{ plf}$
$\quad 1.2(520) + 1.6(140) + 0 = 848 \text{ plf}$
$\quad 1.2(520) + 1.6(35) + 0 = 680 \text{ plf}$

d. $1.2(520) + 0 + 0 + 0.5(140) = 694 \text{ plf}$
$\quad 1.2(520) + 0 + 0 + 0.5(140) = 694 \text{ plf}$
$\quad 1.2(520) + 0 + 0 + 0.5(35) = 642 \text{ plf}$

e. $\quad 1.2(520) + 0 + 0.2(140) = 652 \text{ plf}$

f. $\qquad\qquad 0.9(520) - 0 = 468 \text{ plf}$

The critical load for design is 848 plf.[2]

---

[1] See ASCE 7-88 for limitations and other types of load, e.g., limitations for garages and places of public assembly, and loads due to fluids and self-restraining forces.

[2] Obviously, as one gains experience, it becomes easier to save time by eliminating the unlikely combinations.

This roof beam supports and is embedded in a base concrete deck which gives it continuous lateral (braced) support; therefore

$$M_{req} = \frac{wL^2}{8} = \frac{0.848(21.5)^2}{8} = 49 \text{ ft-kips}$$

The design strength $(\phi_b R_n)$ is

$$\phi_b M_n \geq 49.0 \text{ kip-ft}$$

where $\phi = 0.90$, and $M_n = 49/0.90 = 54.4$ ft-kips.

The AISC load- and resistance-factor design specification gives the formula $M_n = M_p = Z_x F_y$ for bending about the major axis when $L_b \leq L_p$, where M_p will be recognized as the plastic moment, L_b is the laterally unbraced length (in.) for full plastic bending capacity under uniform moment, and L_p is the limiting laterally unbraced length (in.) for full plastic bending capacity under uniform moment. Then

$$Z_x = \frac{54.4(12)}{36} = 18.1 \text{ in.}^3$$

Referring to the AISC "Plastic Design Selection" table (ASD Manual), select the least-weight section (in boldface type) that will provide this section modulus. This is the **W** 12×16 with $Z_x = 20.1$ in.^{3}.

Checking for shear (no stiffeners), the design shear strength is $\phi_v V_n$, where $\phi = 0.90$, and $V_n = 21.6$ ksi$\times (dt_w)$:

$$\text{required shear} = (0.848 \text{ plf})\frac{21.5}{2}$$
$$= 9.1 \text{ kips}$$

For the **W** 12×16, $d = 11.99$ in. and $t_w = 0.220$ in. Then the design shear strength is

$$\phi_v V_n = 0.90(21.6)11.99(0.22)$$
$$= 51.3 > 9.1 \text{ kips} \quad \text{OK}$$

Example 2

Design a **W** column of A36 Steel with an unbraced height of 18 ft. The column sup-

ports an axial load of 175 kips, composed of 100 kips dead load (including the column weight), and 75 kips reduced live load. $K = 1.0$ (Fig. 6.4).

Solution

Compute the factored load:

$$1.2D + 1.6L = 1.2(100) + 1.6(75)$$
$$= 120 + 120 = 240 \text{ kips}$$

The required strength then is 240 kips. From the AISC ASD Manual "Column Axial Load" tables, select a **W** 8×48 as a trial section; then $A = 14.1$ in.^{2}, $r_x = 3.61$ in., $r_y = 2.08$ in. The AISC LRFD Specification provides a value $\phi = \phi_c = 0.85$ if $Kl/r \leq 133.7$. Then

$$\frac{Kl}{r_x} = \frac{1.0(18)12}{3.61} = 59.8$$
$$\frac{Kl}{r_y} = \frac{1.0(18)12}{2.08} = 103.8 < 133.7$$

Therefore

$$\phi_c = 0.85$$

From the LRFD specification, the design compressive strength is $\phi_c P_n$, where the nominal axial strength $P_n = A_g F_{cr}$, A_g being the gross area of the section in in.^{2}, and F_{cr} being the critical stress in ksi. The specification defines λ_c as the column slenderness parameter, with the boundary between elastic and inelastic instability being $\lambda_c = 1.5$. Then

$$\lambda_c = \frac{Kl}{r\pi}\sqrt{\frac{F_y}{E}}$$

For $\lambda_c \leq 1.5$,

$$F_{cr} = (0.658^{\lambda_c^2}) F_y$$

and for $\lambda_c > 1.5$,

$$F_{cr} = \frac{0.877}{\lambda_c^2} F_y$$

Then

$$\lambda_c = \frac{1.0(18)12}{2.08(3.14)} \sqrt{\frac{36}{29,000}} = 1.16 < 1.5$$

and

$$F_{cr} = (0.658^{\lambda_c^2}) F_y$$
$$= (0.658^{1.16^2})36 = 20.5 \text{ ksi}$$

The design compressive strength then is

$$\phi_c P_n = (\phi_c F_{cr}) A_g$$
$$0.85(20.5)14.1 = 246 > 175 \text{ kips}$$

The **W** 8×48 is too large. Try the **W** 10×39, $A_g = 11.5$ in.^{2}, and $r_y = 1.98$ in.:

$$\frac{Kl}{r} = \frac{1.0(18)12}{1.98}$$
$$= 109 < 133.7$$
$$\phi_c = 0.85$$
$$\lambda = 0.0112 \frac{Kl}{r}$$
$$= 0.0112(109)$$
$$= 1.22 < 1.5$$
$$F_{cr}(0.658^{1.22^2})36 = 19.3 \text{ ksi}$$
$$\phi_c P_n = (\phi_c F_{cr}) A_g$$
$$= 0.85(19.3)11.5$$
$$= 189 > 175 \text{ kips}$$

Assuming the column is checked for any moment due to wind and/or earthquake, the **W** 10×39 is adequate. It also is apparent that the savings in member size using LRFD over ASD are significant.

11.4
Application of Plastic Analysis

It was stated earlier that a design based upon ultimate strength could provide a saving if the reserve strength (after the initial yield stress was reached) contributed to the overall factor of safety. With plastic theory, there is a nonlinear relationship between stress and ultimate load. Therefore, the factor of safety must

be accounted for by increasing the actual load to an ultimate design load. This is done by multiplying the actual load by a load factor (LF).

The AISC ASD Specification, Chapter N, "Plastic Design," is based on the premise that continuous beams designed for ultimate load should have the same margin of safety as simply supported beams designed according to elastic theory. This implies that no reduction in strength will be permitted for simply supported beams and that such beams serve as the standard for determining the load factor.

For simply supported beams subjected to a uniformly distributed load (Fig. 11.5a), there are three intensities of load to be considered, i.e.,

$$W_1 = w_1 L, \qquad W_2 = w_2 L, \qquad W_3 = w_3 L$$

The corresponding bending moment diagrams are shown in Fig. 11.5b. The maximum moment for all three occurs at the center. Let M_e equal the maximum elastic moment allowed by the code, and F_b, the allowable bending stress. Then

$$M_e = F_b S = \frac{W_1 L}{8}$$

If the load is allowed to increase, the beam will behave elastically until reaching a load W_2 that will cause first yield stress. This will occur at the center of the beam; therefore,

$$M_y = F_y S = \frac{W_2 L}{8}$$

Any further increase in load will create an elastic-plastic stress pattern similar to that shown in Fig. 11.2d. The maximum load (W_3) will form a plastic hinge at the center of the beam. No further load can be applied; the beam has reached ultimate strength. Therefore,

$$M_p = F_y Z = \frac{W_3 L}{8}$$

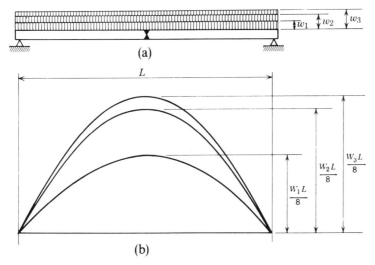

Figure 11.5/Simple beam—ultimate load.

The difference between M_e and M_p represents the margin of safety in this beam; therefore, $M_p/M_e = F_y Z/F_b S$ establishes the load factor.

It has been shown that $F_b = 0.66F_y$ for adequately braced beams, and that the ratio Z/S equals u and has a mean value of 1.12. Therefore,

$$\text{LF} = \frac{M_p}{M_e} = \frac{F_y}{0.66F_y}(1.12) = 1.70$$

This value of the load factor is used for the plastic design of all continuous beams and frames when only live and dead loads are considered. However, the load factor is reduced to 1.3 when the effects of wind or earthquake are added to the live and dead load.

It has been emphasized that ultimate strength design could provide a saving in material only if the structure to which it is applied is statically indeterminate. Simply supported beams are statically determinate, and no savings in material could result.

The 1989 AISC ASD Specification requires that the factored load be 1.7 times the given live load and dead load or 1.3 times these loads acting in conjunction with 1.3 times any specified wind or earthquake loads.

Example

A simply supported beam spans 30 ft and carries a uniform load of 1 kip per ft, including an allowance for its own weight. It has complete lateral support, and A36 Steel is to be used. Find the lightest-weight beam permissible, using both conventional design and ultimate-strength design. Consider both bending and shear.

Solution

(1) Designing according to the conventional ASD method,

$$M_{\text{max}} = \frac{wL^2}{8} = \frac{1(30)^2}{8} = 113 \text{ ft-kips}$$

$$S = \frac{M}{F_b} = \frac{113(12)}{24} = 56.5 \text{ in.}^3$$

From the AISC "Allowable Stress Design Selection" table select **W** 18×35 as the lightest-weight beam section.

(2) Designing according to the ultimate-strength method, the load factor must be taken into account, i.e.,

$$w = 1.7(1) = 1.7 \text{ kips per ft}$$

$$M_{max} = \frac{wL^2}{8} = \frac{1.7(30)^2}{8} = 191 \text{ ft-kips}$$

$$Z = \frac{M}{F_y} = \frac{191(12)}{36} = 63.7 \text{ in.}^3$$

From the AISC "Plastic Design Selection Table," select a **W** 18×35 as the lightest-weight beam section.

(3) Check for shear. Beam and girder web shear seldom controls the design, which is consistent with the working-stress design procedure (elastic design). Nevertheless, it should be checked. In regions of high shear, such as connections to columns, it could become critical and web doubler plates or web diagonals could be required. Section N5 of the 1989 AISC Specification limits the factored shear as follows:

$$V_u = 0.55 F_y t_w d$$

This implies an equivalent shear yield stress of $0.55 F_y$.

In this example, the **W** 18×35 can develop a maximum shear load

$$V_u = 0.55(36)0.30(17.70)$$
$$= 105 \text{ kips}$$

which is seen to be considerably larger than the factored shear load

$$V_u = R_u = \frac{1.7(30)}{2} = 25.5 \text{ kips}$$

(4) Check the plastic width-thickness ratios.

Flange:

$$\frac{b}{2t_f} = \frac{6.0}{2(0.425)} = 7.06 < 8.5 \qquad \text{OK}$$

Web:

$$\frac{d}{t_w} = \frac{17.7}{0.30} = 59.2 < \frac{412}{\sqrt{36}} = 68.7 \qquad \text{OK}$$

11.5
Fixed-End Beams

Because fixed-end beams are statically indeterminate, the ultimate-strength design method will result in economies. This is true because more than one plastic hinge must be formed before ultimate load is achieved. The reserve strength is furnished by the redundancy.

The fixed-end beam shown in Fig. 11.6 would require three plastic hinges before "mechanism" and thus ultimate load. A distinction must be made here between a

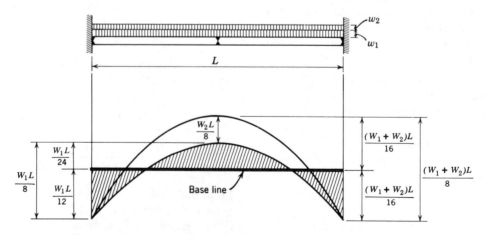

Figure 11.6/Fixed-end beam—ultimate load.

structure and a *mechanism*. A structure must be stable and non-moving (under static loads), whereas mechanism is defined as that state where a structure has reached its structural limit and any additional load would result in movement. The hinges needed for mechanism would occur at each fixed end and at the center of the beam. In order to understand the manner in which these hinges are formed, consider the uniform load w to be slowly increasing in magnitude. When w is small, the beam will behave elastically until w_1 is reached, giving a total load $W_1 = w_1 L$. The corresponding moment diagram is shown as the shaded area in Fig. 11.6b. Maximum moment occurs at the ends; therefore, first yield is reached at these points of maximum bending, and a plastic hinge is formed. Thus,

$$M_p = \frac{W_1 L}{12}$$

from which

$$W_1 = \frac{12 M_p}{L}$$

The beam still has an additional load-carrying capacity. The hinges maintain their developed plastic moment but allow the ends to rotate when additional load is applied. Only the ends are stressed in the plastic range; the remainder of the beam behaves elastically. The additional load ($W_2 = w_2 L$) the beam can carry is that load causing an additional bending moment at the center in a manner similar to that of a simply supported beam. This is so because the plastic end hinges offer no additional rotational resistance. If the beam is of constant size through its entire length, the moment developed at the center by the first load (W_1) is $W_1 L / 24$, which is observed to be one-half the ultimate capacity (M_p) that has already been developed at the support. Consequently, the load W_2 is limited to that amount required

to develop the remaining half of M_p, i.e.,

$$\frac{M_p}{2} = \frac{W_2 L}{8}$$

from which

$$W_2 = \frac{4 M_p}{L}$$

Of interest here is the total load at ultimate strength, which is equal to $W_1 + W_2$, i.e.,

$$W_1 + W_2 = \frac{12 M_p}{L} + \frac{4 M_p}{L} = \frac{16 M_p}{L}$$

or

$$M_p = \frac{(W_1 + W_2) L}{16}$$

This moment would be developed at both ends *and* at the center (Fig. 11.6b).

Observe this final ultimate-strength bending moment diagram carefully. It is seen that the simple-beam moment, having a maximum value of $WL/8$, is divided evenly between positive moment at the center and negative moment at the ends; namely, $WL/16$ each. This characteristic is typical and is valid for any type of symmetrical loading for fixed-end beams. (It will need to be used in solving the problems given at the end of this article.)

When designing a beam similar to that shown in this example, for uniform load, the formulas used are as follows:

$$M_p = F_y Z = \frac{LF(W) L}{16}$$

or

$$Z = \frac{LF(W) L}{16 F_y}$$

in which W is the actual load the beam must support.

Example 1

A **W** 12×30 is used as a fixed-end beam carrying a uniform load over its full 25-ft span. The beam has complete lateral support. Use A36 Steel. Find the total load (W_1) that would cause first yield. Find the total load (W_2) that represents the ultimate strength of the beam.

Solution

(1) Referring to the AISC Manual,

$$S_x = 38.6 \text{ in.}^3, \qquad Z_x = 43.1 \text{ in.}^3$$

(2) When first yield is reached, the beam has performed within its elastic range. Therefore,

$$M_y = SF_y = 38.6(36) = 1390 \text{ in.-kips},$$

$$= 116 \text{ ft-kips}$$

and

$$M_{\max} = \frac{W_1 L}{12}$$

from which

$$W_1 = \frac{12 M_y}{L} = \frac{12(116)}{25} \simeq 56 \text{ kips}$$

(3) The ultimate strength of the beam is limited by its plastic moment:

$$M_p = ZF_y = 43.1(36) = 1550 \text{ in.-kips},$$

$$= 129 \text{ ft-kips}$$

and

$$M_{\max} = \frac{W_2 L}{16}$$

from which

$$W_2 = \frac{16 M_p}{L} = \frac{16(129)}{25} = 82.6 \text{ kips}$$

Example 2

Find the lightest-weight beam that can carry a uniform load of 2 kips per ft for a 28-ft span. Each end of the beam is fully restrained, and there is complete lateral support. Use A36 Steel. Neglect shear and

deflection. Compare the beam section selected with the one selected by the conventional method.

Solution

(1) Using the ultimate-strength method, the required plastic section modulus is found, i.e.,

$$Z = \frac{LF(W)L}{16 F_y} = \frac{1.7(56)28(12)}{16(36)} = 55.5 \text{ in.}^3$$

(2) Referring to the AISC "Plastic Design Selection Table," a **W** 18×35 is selected.

(3) Using the elastic method, the required section modulus is

$$S = \frac{WL}{12 F_b} = \frac{56(28)12}{12(24)} = 65.3 \text{ in.}^3$$

(4) Referring to the AISC "Allowable Stress Design Selection Table," a **W** 18×40 is selected.

PROBLEMS

1. Find the lightest-weight A36 Steel section that can support a concentrated load of 20 kips at midspan, if the total span is 30 ft and both ends are fully restrained. There is complete lateral support. Neglect shear, deflection, and any effect of the beam dead load. (Answer given in Appendix G.)

2. A fixed-end beam has a span length of 24 ft and supports two equal concentrated loads of 12 kips each, located at the third-points. A36 Steel is specified, and there is complete lateral support. Design the beam. Neglect shear, deflection, and beam dead load. (Answer given in Appendix G.)

3. An **S** 12×35, A36 Steel section is used as a fixed-end beam with a span length of 26 ft. There is complete lateral support, and the beam supports a uniform load of 1 kip per ft, including the beam weight. Find the value of an additional concentrated load that can safely be applied at midspan. Neglect shear and deflection.

11.6
The Propped Cantilever

Another type of indeterminate beam that can prove more economical if designed by the ultimate-strength method is the propped cantilever. Such a beam requires the development of two plastic hinges before mechanism can occur. Also, unlike the fixed-end beam, there is no symmetry in the bending moment diagram. The location of the first plastic hinge to develop depends on the elastic behavior of the beam (usually it will be located at the fixed end). Additional load then can be applied until the second plastic hinge, having the same value of M_p as the first, develops. The loading that produces the two plastic hinges is the ultimate load for the beam. The process of design can best be ex-plained by the use of an illustrative problem.

Example 1

The propped cantilever shown in Fig. 11.7a carries a single concentrated load (P) at midspan. The beam is an A36, **W** 12×22, and there is complete lateral support. Neglect shear, beam dead load, and deflection.

(a) Find the maximum value of P allowed, using the elastic design method.

(b) Find the value of P causing first yield.

(c) Find the value of P causing ultimate failure.

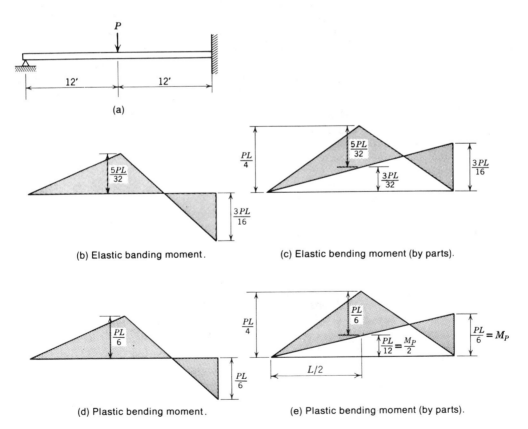

(a)

(b) Elastic banding moment.

(c) Elastic bending moment (by parts).

(d) Plastic bending moment.

(e) Plastic bending moment (by parts).

Figure 11.7/Propped cantilever, concentrated load.

Solution (a)

(1) Referring to the AISC Manual, the elastic section modulus is 25.4 in.^{3}. Therefore, the maximum permissible bending moment is

$$M_b = SF_b = 25.4(24) \approx 610 \text{ in.-kips}$$
$$= 50.8 \text{ ft-kips}$$

(2) The elastic bending moment diagram is sketched (Fig. 11.7b); the maximum moment occurs at the support and is

$$M_{max} = \frac{3PL}{16}$$

The procedure for this elastic solution was presented in Chapter 10. An alternative method of sketching the bending moment diagram (by parts) is shown in Fig. 11.7c. The simple-beam diagram is sketched first, having a maximum value of $PL/4$ at midspan. The negative fixed-end moment ($3PL/16$), diminishing linearly to zero at the simple end, is superimposed on the first diagram. The area bounded by the two diagrams represents the actual bending moment diagram and is shaded. This method will be extremely useful in the ultimate-strength design process described later.

(3) Setting the maximum elastic bending moment equal to the elastic design capacity of the beam, one can then solve for the load as follows:

$$M_{max} = \frac{3PL}{16} = 50.8 \text{ ft-kips}$$
$$P = \frac{16(50.8)}{3(24)} = 11.3 \text{ kips}$$

Solution (b)

(1) The moment developed by the **W** 12× 22 at yield stress is

$$M_y = SF_y = 25.4(36) = 914 \text{ in.-kips}$$
$$\approx 76 \text{ ft-kips}$$

(2) Setting this yield-stress moment equal to the maximum value from the moment diagram and solving for the load P,

$$M_{max} = \frac{3PL}{16} = 76 \text{ ft-kips}$$
$$P = \frac{16(76)}{3(24)} = 16.9 \text{ kips}$$

Solution (c)

(1) The plastic moment developed by the **W** 12×22 at yield stress is

$$M_p = ZF_y = 29.3(36) = 1055 \text{ in.-kips,}$$
$$= 88 \text{ ft-kips}$$

(2) The plastic bending moment diagram is drawn as shown in Fig. 11.7d. The maximum moment value occurs both at the support and at midspan and is equal to $PL/6$. To arrive at this conclusion, draw the diagram by parts as shown in Fig. 11.7e. This simple-beam diagram is sketched first, with maximum value at midspan ($PL/4$). The straight line representing the negative moments, from zero at the simple support to M_p at the fixed end, is drawn on top of the simple-beam moment diagram, and its value at midspan is equal to $M_p/2$. Therefore, the net moment at midspan is $PL/4 - M_p/2$. Since both plastic hinges (positive at midspan, negative at the fixed end) must have the same value, the net value at midspan can be equated to M_p at the fixed end:

$$\frac{PL}{4} - \frac{M_p}{2} = M_p$$

from which $M_p = PL/6$.

(3) Setting this plastic-moment value caused by the loads equal to the capacity of the beam, the load can be determined as follows:

$$M_p = \frac{PL}{6} = 88 \text{ ft-kips}$$
$$P = \frac{6(88)}{24} = 22 \text{ kips}$$

However, the design ultimate-strength load must include a load factor. Therefore,

$$P = \frac{22}{LF} = \frac{22}{1.7} = 12.9 \text{ kips}$$

A design formula for this particular propped cantilever problem is

$$Z = \frac{(LF)\,PL}{6F_y}$$

The design procedure for a propped cantilever carrying a uniform load is much the same. However, the computations are considerably more involved, as will be illustrated by the second solution in the example which follows. Consequently, a semigraphical trial-and-error procedure is recommended. This method will be described first.

Example 2

Design a propped cantilever beam of A36 Steel, as shown in Fig. 11.8a. Use the

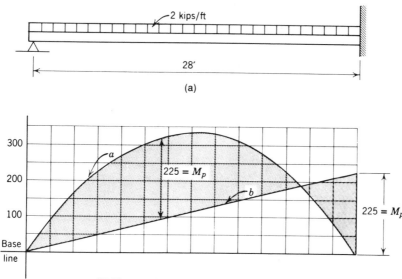

(a)

(b) Moment diagram: semigraphical solution.

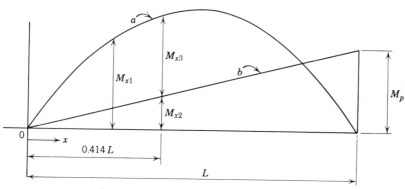

(c) Moment diagram: mathematical solution.

Figure 11.8/Propped cantilever beam.

ultimate strength method of design. Assume complete lateral support, and neglect shear, deflection, and beam weight.

Solution (Using the Semigraphical Procedure)

(1) The increased load necessary for ultimate-strength design is

$$w = LF(w) = 1.7(2) = 3.4 \text{ kips per ft}$$

(2) Construct the simple-beam moment
diagram to scale. A quadruled graph paper will improve the accuracy, requiring
calculation of the numerical value of the
bending moment at only a few points.

$$M(\text{center point}) = \frac{3.4(28)^2}{8} = 333 \text{ ft-kips}$$

$$M(\text{quarter-point}) = 47.6(7) - \frac{3.4(7)^2}{2}$$
$$= 250 \text{ ft-kips}$$

This diagram is shown as curve a in Fig.
11.8b.

(3) Superimpose the straight line b on this
simple-beam diagram, to represent the
negative-moment effect of the fixed end.
The final position of this straight line is
determined from a trial-and-error process
as follows: The left end is placed at the
origin, and the right end is adjusted so
that its vertical ordinate will be equal to
the maximum vertical ordinate between
curve a and line b. The maximum difference in ordinate will occur to the left of
the centerline of the beam. In this example, the maximum plastic bending moment
is approximately 225 ft-kips. The shaded
portion of this diagram represents the final
bending moment diagram.

(4) Determine the required plastic section
modulus needed to develop M_p:

$$M_p = 225(12) = ZF_y$$

$$Z = \frac{225(12)}{36} = 75 \text{ in.}^3$$

(5) Referring to the AISC "Plastic Design
Selection Table," select the **W** 18×40 as
the lightest-weight beam section.

Solution (Using the Mathematical Procedure)

(1) The increased load necessary for ultimate-strength design is

$$1.7(2) = 3.4 \text{ kips per ft}$$

(2) Determine the reactions. It will be recalled that for an elastic analysis, the reaction at the fixed end will be larger than at
the simple end. However, after the plastic
hinge at the support is developed and rotation takes place, the load will be redistributed until, at failure, both reactions
will be equal:

$$R = 3.4(14) = 47.6 \text{ kips}$$

(3) Sketch the simple-beam bending moment diagram, and develop the formula
for its curve. The origin for the variable
distance x is at the simple end. Let M_{x1}
represent the simple-beam moment. This
is curve a in Fig. 11.8c. Then,

$$M_{x1} = -\frac{3.4x^2}{2} + 47.6x$$

where M_{x1} is the vertical distance from
the base line to the curve a, for any value
of x.

(4) Superimpose the straight line b, varying from zero at the origin to M_p at the
fixed end. The equation of this line, letting
M_{x2} represent the vertical distance from
the base line to line b, is

$$M_{x2} = \frac{M_p x}{L} = \frac{M_p x}{28}$$

(5) The formula for M_{x3}, representing the
vertical distance between curve a and line
b, is

$$M_{x3} = M_{x1} - M_{x2}$$
$$= -\frac{3.4x^2}{2} + 47.6x - \frac{M_p x}{28}$$

(6) In the ultimate-strength design solution, it is seen that M_{x3} must be equal to M_p; therefore,

$$M_{x3} = M_p = -\frac{3.4x^2}{2} + 47.6x - \frac{M_p x}{28}$$

Solving the above formula for M_p,

$$M_p = \frac{-1.7x^2 + 47.6x}{1 + 0.036x}$$

(7) It is now necessary to establish the value of x, i.e., the location where M_p is maximum. This is accomplished by taking the first derivative of M_p with respect to x and setting it equal to zero:

$$\frac{dM}{dx} = \frac{(1+0.036x)(-3.4x+47.6)}{(1+0.036x)^2} = 0$$

Simplifying,

$$x^2 + 53.4x = 754$$

And solving for x by completing the square,

$$x^2 + 53.4x + 26.7^2 = 754 + 26.7^2$$
$$(x + 26.7)^2 = 1468$$
$$x = \pm 38.3 - 26.7$$
$$x = 11.6 \text{ ft.}$$

(8) The value of M_p can be determined by substituting the value 11.6 in either equation developed in step (6), e.g.,

$$M_p = \frac{-1.7(11.6)^2 + 47.6(11.6)}{1 + 0.036(11.6)}$$
$$= 226 \text{ ft-kips}$$

(9) Determine the required plastic modulus needed to develop M_p:

$$M_p = 226(12) = ZF_y$$

$$Z = \frac{226(12)}{36} = 75.3 \text{ in.}^3$$

(10) Referring to the AISC "Plastic Design Selection Table," select the **W** 18×40 as the lightest-weight beam section. It can be seen from the illustrative problems that

the procedure for designing propped cantilevers involves first establishing the simple-beam moment diagram, and then superimposing the straight-line variation of the effect of the fixed-end moment such that the maximum value of M_p occurs both at the fixed end (negative) and at some other position (positive moment) near the midspan. The process used can be either semigraphical or direct mathematical. Usually the semigraphical process is employed when the uniform load is significant.

It is also helpful to observe that when designing propped cantilever beams carrying uniform loads only, the maximum *positive* plastic moment will always occur at $0.414L$ from the simple end (Fig. 11.8c). The formula for M_p will also prove useful. It can be developed from the equation established in step (5) by substituting the following identities:

$$M_{x3} = M_p, \qquad x = 0.414L$$
$$3.4 = w, \qquad 28 = L$$
$$47.6 = \frac{wL}{2}$$

That is,

$$M_p = -\frac{w(0.414)^2 L^2}{2}$$
$$+ \frac{wL}{2}(0.414)L - \frac{M_p(0.414)L}{L}$$

and

$$1.414 M_p = -0.0858 wL^2 + 0.207 wL^2$$
$$M_p = \frac{0.1212}{1.414} wL^2$$
$$= 0.0858 wL^2$$

PROBLEMS

1. Design a propped cantilever beam of A36 Steel, carrying a concentrated load of 20 kips at midspan. The beam span is 28 ft. Assume complete lateral support. Neglect shear, deflection, and beam weight. (Answer given in Appendix G.)

2. Design a propped cantilever beam of A36 Steel, carrying two equal concentrated loads of 12 kips at the third points. The span length is 27 ft. Assume complete lateral support. Neglect shear, deflection, and beam weight. (Answers given in Appendix G.)

3. A **W** 18×46 of A36 Steel is used as a propped cantilever beam having a span of 26 ft. It carries a uniformly distributed load of 1.5 kips per ft, including the beam weight. Determine the additional concentrated load at midspan the beam could carry if it had complete lateral support. Neglect shear and deflection.

11.7
Continuous Beams

The design of continuous beams is accomplished in a manner similar to that discussed in the previous articles. The real loads are multiplied by the load factor (1.7), and the corresponding bending moment diagram is constructed such that maximum values (M_p) are located at the plastic hinges necessary for mechanism. As was illustrated for fixed-end and propped

cantilever beams, the total bending is proportioned to positive and negative components in the most favorable manner. It is the maximum bending moment from this process which establishes the required plastic section modulus (Z) and thus the beam size.

For cases where a single-size beam is used for all spans, and all spans are equal and equally loaded with symmetrically placed loads, it can be observed that the end span is controlling. In fact, under these conditions, interior spans have plastic hinges formed only at the supports and thus do not reach ultimate strength.

Example 1

Design the three-span continuous beam shown in Fig. 11.9, using A36 Steel. Neglect shear, deflection, and beam weight. Assume complete lateral support.

Solution

(1) Calculate the increased load to allow for a factor of safety:

$$P = 10(1.7) = 17 \text{ kips}$$

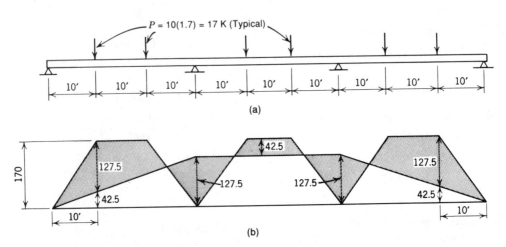

Figure 11.9/Continuous beam—concentrated load.

(2) Determine the simple-beam moment:

$$M = P\left(\frac{L}{3}\right) = 17(10) = 170 \text{ ft-kips}$$

(3) Determine the value of the maximum moment at the two plastic hinges, one at the interior support and the other at 10 ft from the simple end:

$$M_p = 170 - \frac{M_p}{3}$$

$$= \tfrac{3}{4}(170) = 127.5 \text{ ft-kips}$$

(4) Construct the moment diagram for the continuous beam (Fig. 11.9b). It will be observed that the moment at midspan of the interior span is

$$M = 170 - 127.5 = 42.5 \text{ ft-kips}$$

(5) Determine the required plastic section modulus:

$$Z = \frac{M_p}{F_y} = \frac{127.5(12)}{36} = 42.5 \text{ in.}^3$$

(6) Select the lightest-weight beam section from the AISC "Plastic Design Selection Table". This is a **W 16×26**.

Example 2

Design the three-span continuous beam shown in Fig. 11.10, using A36 Steel. The 1.5 kip-per-ft load includes an allowance for beam weight. Neglect shear and deflection, and assume complete lateral support.

Solution

(1) The real load of 1.5 kips per ft is first multiplied by the load factor to allow for a factor of safety:

$$w = 1.5(1.7) = 2.55 \text{ kips per ft}$$

(2) Determine the simple-beam moment:

$$M = \frac{wL^2}{8} = \frac{2.55(26)^2}{8} \simeq 216 \text{ ft-kips}$$

(3) The plastic moment for the end span occurs at the first interior support and at $0.414L$, or 10.76 ft from the simple end. Determine the value of M_p for the end span (see Art. 11.6):

$$M_p = 0.0858WL = 0.0858(2.55)26^2$$

$$= 148 \text{ ft-kips}$$

(4) Construct the moment diagram for the continuous beam (Fig. 11.10b). It will be observed that the moment at midspan of the interior span is

$$M = 216 - 148 = 68 \text{ ft-kips}$$

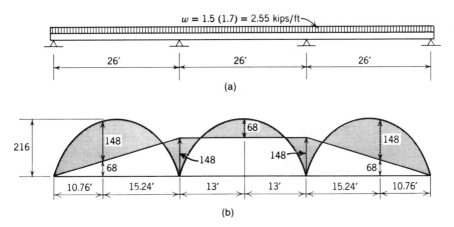

Figure 11.10/Continuous beam—uniform load.

(5) Determine the required plastic section modulus:

$$Z = \frac{M_p}{F_y} = \frac{148(12)}{36} = 49.3 \text{ in.}^3$$

(6) Referring to the AISC "Plastic Design Selection Table," select the **W 16×31** as the lightest-weight beam section.

Unequal span lengths and variations in load complicate the design procedure, but the principles remain the same. The location of maximum moments and plastic hinges is not always readily apparent. The recommended procedure is the semi-graphic process introduced in Art. 11.6.

The procedure is as follows: Draw the beam lengths and positions of reactions and loads to some convenient scale. Calculate the bending moments based upon a series of independent beams (all simply supported). Draw the bending moment diagram to scale. Then, superimpose over the simple-beam bending moment diagrams a new base line dividing the diagrams into positive and negative areas, and locate this base line so as to create the smallest possible maximum ordinate. The trial-and-error method accompanied by

some mathematical calculations will usually prove to be the most rapid means of solution.

Example 3

Investigate the three-span continuous beam shown in Fig. 11.11. Construct the bending moment diagram according to the principles established for the ultimate-strength process.

Solution

(1) Draw spans and locations of loads to some convenient scale (Fig. 11.11a).

(2) Calculate the maximum simple-span moment for each span:

Left span:

$$M_{max} = 10.5\left(\frac{20}{3}\right)$$

$$= 70 \text{ ft-kips}$$

Center span:

$$M_{max} = \frac{1.1(24)^2}{8}$$

$$= 79.2 \approx 80 \text{ ft-kips}$$

Right span:

$$M_{max} = \frac{15(16)}{4}$$

$$= 60 \text{ ft-kips}$$

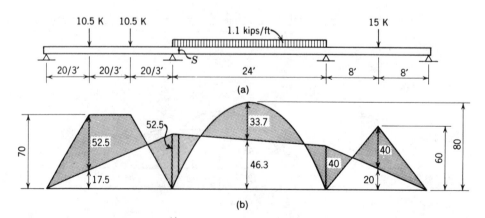

(a)

(b)

Figure 11.11/Continuous beam—various spans and loads.

(3) To some convenient scale, construct the diagram variations in the bending moment based upon a series of independent simply supported beams (Fig. 11.11b).

(4) Superimpose the straight line representing the negative moment effect, thereby dividing the diagram into positive and negative areas. Starting with the left span, it is seen that if 52.5 ft-kips is used for the continuous end, the straight line will result in a negative value under the left concentrated load as follows:

$$M = 52.5\left(\tfrac{1}{3}\right) = 17.5 \text{ ft-kips}$$

and the net positive moment at this same point is

$$M_p = 70 - 17.5 = 52.5 \text{ ft-kips}$$

Any value other than 52.5, established at the continuous support, would result in a negative or positive bending-moment value larger or less than the 52.5. Consequently, the position shown is most favorable.

In like manner, the right end-span moment will be divided into 40 ft-kips for both the negative value at one end and a net positive value at midspan. The bending diagram for the interior span will be as shown in Fig. 11.11b.

Observe that the left-end span is controlling. If a beam section of one size were used throughout, it would need to have a Z-value adequate to develop the 52.5 ft-kip moment. Under such a condition, the right end span would not have reached a mechanism, and the center span would have developed only one plastic hinge, and thus would not be at ultimate strength.

This points up the final consideration in designing continuous beams, i.e., variation of beam sizes and location of beam splices. Consider further the bending moment diagram of Fig. 11.11. It has been shown that the left end span must have a sufficient plastic section modulus to develop a moment of 52.5 ft-kips, and, in doing so, it

will be at ultimate load. The right end span must have a Z-value adequate to develop only 40 ft-kips and thus ultimate load. The interior span is considerably more involved. It must have a Z-value adequate to develop the 52.5 ft-kips at the left end, but, since only one plastic hinge would have been developed, the beam would not be at ultimate strength. By providing a Z-value less than that needed to develop 52.5 ft-kips, the superimposed base line for the end span would be lowered (thereby increasing the net positive moment), and the end span would be underdesigned. If, however, the left end-span beam, capable of developing a moment of 52.5 ft-kips, were extended over the support and spliced to the interior beam at point S, the remainder of the interior beam could be reduced in size, since it would have to develop a moment of only 40 ft-kips.

The design of continuous beams of varying size can best be illustrated with examples.

Example 4

A three-span continuous beam of A36 Steel supports a uniform load of 2 kips per ft, including an allowance for the dead weight of the beam (Fig. 11.12). The total length of the beam, $28(3) = 84$ ft, is composed of three separate beams spliced as described below. Assume complete lateral support, and neglect shear and deflection.

(a) Execute the design so that end-span beams cantilever over the support and are spliced to the interior span at point a.

(b) Execute the design so that the interior beam cantilevers over the supports and is spliced to the end-span beams at point b.

Solution (a)

(1) Sketch the beam spans to some convenient scale (Fig. 11.12a).

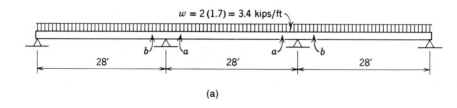

(a)

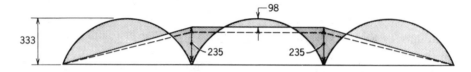

(b) Solution when splice occurs at point a.

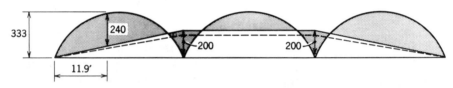

(c) Solution when splice occurs at point b.

Figure 11.12/Continuous beam—spliced.

(2) Calculate the increased load to allow for the load factor:

$$w = 2(1.7) = 3.4 \text{ kips per ft}$$

(3) Calculate the maximum positive moment based upon a series of simple spans:

$$M = \frac{wL^2}{8} = \frac{3.4(28)^2}{8} = 333 \text{ ft-kips}$$

(4) Determine the moments causing plastic hinges in the end span, and superimpose the straight negative-moment variation lines for end spans and interior span. (This is the broken line in Fig. 11.12b.) Then

$$M_p = 0.0858WL = 0.0858(3.4)28^2$$

$$= 229 \text{ ft-kips}$$

(5) Determine the required plastic section modulus for the end span:

$$Z = \frac{M_p}{F_y} = \frac{229(12)}{36} = 76.3 \text{ in.}^3$$

(6) Select the section for the exterior span from the AISC "Plastic Design Selection Table". Select a **W** 18×40 ($Z = 78.4$ in.^{3}).

(7) The exterior beam is slightly overdesigned because of the limited availability of shapes ($Z = 78.4$ versus $Z = 76.3$ required). Therefore, calculate the plastic moment for the selected section:

$$M_p = ZF_y = \frac{78.4(36)}{12} = 235 \text{ ft-kips}$$

(8) Adjust the superimposed negative-moment variation line to account for the actual plastic moment resistance for the selected section. (This is shown as the solid line in Fig. 11.12b.)

(9) Calculate the net positive moment remaining at midspan of the interior span:

$$M = 333 - 235 = 98 \text{ ft-kips}$$

(10) Determine the plastic section modulus necessary for the interior span:

$$Z = \frac{M_p}{F_y} = \frac{98(12)}{36} = 32.7 \text{ in.}^3$$

(11) Select the section for the interior span from the AISC "Plastic Design Selection Table". Select a **W** 18×35. Although the **W** 14×22 would be adequate, the 18-in.-deep section is selected so as to maintain a constant 18-in. depth (approximately) and to simplify the beam splice.

Solution (b)

Steps (1) through (3) are identical with Solution (a).

(4) Consider the interior span first. Superimpose the most favorable base line, and determine M_p. The most favorable base line divides the simple-span moment evenly between negative and positive moments:

$$M_p = \frac{333}{2} = 166.7 \text{ ft-kips}$$

(This line is shown as the broken line in Fig. 11.12c.)

(5) Find the required plastic section modulus:

$$Z = \frac{M_p}{F_y} = \frac{166.7(12)}{36} = 55.5 \text{ in.}^3$$

(6) Select the lightest-weight beam section: a **W** 18×35 ($Z = 66.5$ in.^{3}).

(7) Since the beam is slightly overdesigned, determine its plastic resisting moment and adjust the superimposed base line (the solid line in Fig. 11.12c):

$$M_p = ZF_y = \frac{66.5(36)}{12} \simeq 200 \text{ ft-kips}$$

(8) Find the maximum positive moment for the end span which corresponds to the adjusted straight-line variation effect of the 200-ft-kip negative moment. To accomplish this, develop an equation representing this positive moment at any distance x from the left support. The left reaction is approximately 47.5 kips. Therefore,

$$M = -\frac{3.4x^2}{2} + 47.5x - \frac{x}{28}(200)$$
$$= -1.7x^2 + 40.36x$$

The position of the maximum moment (M) from the above equation is determined by taking the first derivative with respect to x and setting it equal to zero:

$$\frac{dM}{dx} = -3.4x + 40.36 = 0$$
$$x = \frac{40.36}{3.4} = 11.9 \text{ ft}$$

Determine the value of M_p located 11.9 ft from the left reaction:

$$M_p = -1.7(11.9)^2 + 40.36(11.9)$$
$$= 240 \text{ ft-kips}$$

(9) Find the required plastic section modulus for the end span:

$$Z = \frac{M_p}{F_y} = \frac{240(12)}{36} = 80 \text{ in.}^3$$

(10) Select the lightest-weight 18-in. beam section furnishing this section modulus: a **W** 18×46.

PROBLEMS

1. A two-span continuous beam supports two 20-kip concentrated loads, one each at third-points, as shown in Fig. 11.13. Assume that one beam section is to be used over the total length (60 ft), and that both top and bottom flanges

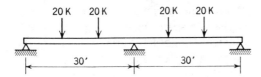

Figure 11.13

have complete lateral support. A36 Steel is specified. Neglect the beam weight.

(a) Design the beam based upon ultimate strength. Neglect shear and deflection.

(b) Assuming that the 20-kip load is all live load, check to determine if the beam section selected in (a) is adequate if the maximum live-load deflection is limited to $L/360$.

(Answers given in Appendix G.)

2. A continuous beam over three spans supports concentrated loads at 8-ft intervals as shown in Fig. 11.14. There is no uniform load —neglect the beam weight. One size beam section of A36 Steel is to be used throughout. Neglect shear and deflection. Design the beam assuming complete lateral support.

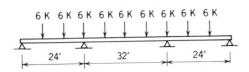

Figure 11.14

3. The continuous beam shown in Fig. 11.15 is to be made up of three separate sections, spliced at points marked S. Use A36 Steel. Design the beam maintaining a constant nominal depth. Neglect shear, deflection, and the beam weight. Assume complete lateral support. (Answer given in Appendix G.)

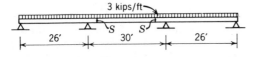

Figure 11.15

4. The continuous beam shown in Fig. 11.16 is to be made up of three separate sections, spliced together at points marked S. Design the beam using A36 Steel. Assume complete lateral support, and neglect shear, deflection, and beam weight.

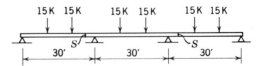

Figure 11.16

5. A three-span continuous beam is shown in Fig. 11.17. One size beam section is used throughout, and there is no uniform load. What exterior span length (L) should be used to achieve the most economical solution? (Answer given in Appendix G.)

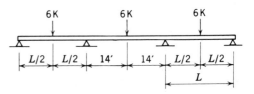

Figure 11.17

11.8
Frames

The 1989 AISC ASD Specification allows ultimate-strength design by the plastic method for braced and unbraced frames so long as a load factor of 1.7 is used for live and dead load, or 1.3 is used if the live and dead load act in conjunction with wind or earthquake forces. Special care, however, must be taken to ensure against local and overall buckling.

The analysis of frames by the plastic method is much too involved for comprehensive treatment in this text. Therefore,

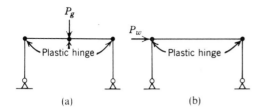

Figure 11.18/Load-mechanism relationship.

only basic principles will be discussed, and the majority of these in reference to simple rectangular frames. In addition to the references listed at the end of Art. 11.1, there are a number of texts devoted exclusively to this topic.[3]

The location of plastic hinges in frames to form a mechanism varies with type of loading, as well as with shape and other physical properties of the frame. It is important

[3]B. G. Neal, *The Plastic Methods of Structural Analysis*, Wiley, 1956; Fritz Engineering Laboratories, *Lecture Notes, Plastic Design of Multi-Story Frames*, Lehigh University, 1965; *Plastic Design of Braced Multi-story Steel Frames*, American Iron and Steel Institute and AISC.

to emphasize that each mechanism, for each condition of loading, requires a separate analysis, and the superimposing of moment diagrams is *not* valid.

This principle is illustrated in Fig. 11.18. The gravity load P_g requires three plastic hinges, located as shown in Fig. 11.18a, to create mechanism for this pinned base frame. However, the lateral load P_w requires only two plastic hinges for mechanism in the same frame (Fig. 11.18b).

With elastic design, it was possible to analyze several combinations of independent, variable loading conditions in order to obtain critical conditions for which the structure must be designed. As noted above, however, the superimposing of moment diagrams for various loading conditions does not apply in plastic design. Each possible loading combination will have a failure mechanism of its own; therefore, each combination must be considered separately.

Figure 11.19 illustrates five separate frames, and the location of the minimum number of plastic hinges needed for mechanism with the given loading.

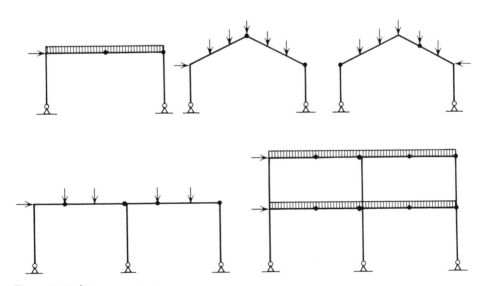

Figure 11.19/Frame mechanisms.

There are various analytical procedures which may be applied to frames. The two most widely used are the *mechanism method* and the *equilibrium method*. The mechanism method is recommended for frames having many redundant reactions. The method involves an energy theory wherein each plastic hinge is assumed to have a virtual rotation such that total internal work can be equated to external work. The external work is represented by the displacement of the supported loads. The mechanism method will be described first in reference to a simple rectangular frame. A discussion of the equilibrium method, including an illustrative example, will follow.

Mechanism Method / This method utilizes the principle of virtual work. The frame under analysis is assumed to have just reached a mechanism state, and each plastic hinge, developing a plastic moment M_p, goes through a rotation θ so small that it can be treated as virtual. The internal work at each plastic hinge can then be represented as the product of M_p and θ.

The sum of all the work at each plastic hinge, required for mechanism, represents the total internal work W_i. The external work is represented by the sum of the products of loads and their respective displacements (Δ). And, finally, the law of conservation of energy permits equating external work to internal work.

Consider the loaded frame shown in Fig. 11.20a. Three separate analyses are required: one for the beam-type mechanism requiring three plastic hinges and caused by the vertical load only (Fig. 11.20b), one for the panel-type mechanism requiring two plastic hinges and caused by the lateral load only (Fig. 11.20c), and one for the combined mechanism requiring two plastic hinges and caused by the combined vertical and lateral loads (Fig. 11.20d). The main task is to discover which mechanism requires the *least* load, thereby establishing the limiting capacity of the frame, i.e., its ultimate strength. To further simplify the task, assume $h = L/3$ and $P_w = P_g/4$; also, assume beam and columns to be all of the same size.

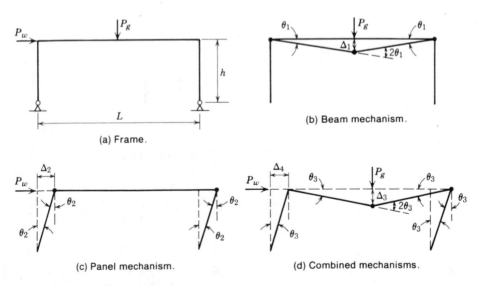

(a) Frame.

(b) Beam mechanism.

(c) Panel mechanism.

(d) Combined mechanisms.

Figure 11.20/Mechanism method.

Beam Mechanism / From geometry and the knowledge that for small angles $\tan \theta = \theta$, it can be observed that $\Delta_1 = \theta_1 L/2$. The load P_g moves through the distance Δ_1; therefore, the total external work is

$$W_e = P_g \Delta_1 = \frac{P_g \theta_1 L}{2}$$

The total internal work is the sum of the product of M_p and the angle for each plastic hinge. Working from left to right, the internal work is

$$W_i = M_p \theta_1 + M_p 2\theta_1 + M_p \theta_1 = 4M_p \theta_1$$

Equating external work to internal work and simplifying,

$$W_e = W_i = \frac{P_g \theta_1 L}{2} = 4M_p \theta_1$$

and

$$P_g = \frac{8M_p}{L}$$

Panel Mechanism / The external work is

$$W_e = P_w \Delta_2$$

and

$$\Delta_2 = \theta_2 h$$

However, since $h = L/3$,

$$\Delta_2 = \frac{\theta_2 L}{3}$$

Therefore,

$$W_e = \frac{P_w \theta_2 L}{3}$$

The internal work is calculated from two plastic hinges:

$$W_i = M_p \theta_2 + M_p \theta_2 = 2M_p \theta_2$$

Therefore,

$$W_e = W_i = \frac{P_w \theta_2 L}{3} = 2M_p \theta_2$$

$$P_w = \frac{6M_p}{L}$$

Also, since $P_w = P_g/4$,

$$P_g = \frac{24M_p}{L}$$

Combined Mechanism / The external work is

$$W_e = P_g \Delta_3 + P_w \Delta_4 = P_g \theta_3 \frac{L}{2} + P_w \theta_3 h$$

$$= \frac{P_g \theta_3 L}{2} + \frac{P_w \theta_3 L}{3}$$

Replacing P_w with $P_g/4$,

$$W_e = \frac{P_g \theta_3 L}{2} + \frac{P_g \theta_3 L}{12} = \frac{7P_g \theta_3 L}{12}$$

The internal work is

$$W_i = M_p 2\theta_3 + M_p 2\theta_3 = 4M_p \theta_3$$

Therefore,

$$W_e = W_i = \frac{7P_g \theta_3 L}{12} = 4M_p \theta_3$$

and

$$P_g = \frac{48M_p}{7L}$$

Summarizing, compare the results of the three analyses:

$$P_g = \frac{8M_p}{L} \quad \text{(beam mechanism)}$$

$$P_g = \frac{24M_p}{L} \quad \text{(panel mechanism)}$$

$$P_g = \frac{48M_p}{7L} \quad \text{(combined mechanism)}$$

It is apparent that the load for the combined mechanism is smallest and consequently that it will be the limiting load. The ultimate-strength load can be calculated from

$$P_g = \frac{48M_p}{7L}$$

One further check is required—that of assuring that nowhere in the frame is M_p

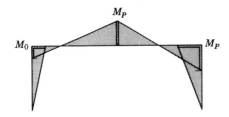

Figure 11.21/Moment diagram.

exceeded. Figure 11.21 illustrates the moment diagram for this check,[4] the moment at each section being determined by statics. This is an equilibrium check and is discussed in greater detail in what follows, and in conjunction with the discussion of the equilibrium method of analysis.

It should be apparent that the procedures just described do not become overly complicated, even when encountering more than three plastic hinges or variations in geometry. Consequently, the above method of analysis is preferred for frames that are more complicated than the simple rectangular frame.

Equilibrium Method / For frames with no more than three plastic hinges, the *equilibrium method* provides a simple solution quite similar to that used for continuous beams. Separate investigation is required for each possible mechanism caused by any variation in loading. Again, attention must be called to the fact that one cannot superimpose bending moment diagrams. Each solution is based upon statics and the fact that the maximum bending moment is limited to M_p—a plastic hinge. This permits a redistribution of other moments until a second hinge is formed, and so on, until a sufficient number of hinges have been formed to create a mechanism.

[4]When analyzing frames, it is often convenient to draw the moment diagram on the tension side of the frame. However, to maintain consistency in this text, this method will not be used.

Assume that the columns of the frame shown in Fig. 11.22a have one-half the ultimate bending strength of the beam, i.e.

$$M_{p1} = \tfrac{1}{2} M_{p2}$$

Three separate analyses are required (Fig. 11.22b, c and d). The beam-type mechanism requires three plastic hinges that are formed as shown (Fig. 11.22b). Then, from statics,

$$M_{p1} + M_{p2} = P_g \frac{L}{3}$$

The panel-type mechanism is shown in Fig. 11.22c. Two hinges are required, and

$$2M_{p1} = P_w h$$

The combined mechanism is shown in Fig. 11.22d, requiring only two hinges. The analysis is more complicated; however, from statics, one can conclude that

$$M_{p1} = \frac{PL}{3} - M_{p2} + \frac{P_w h}{2}$$

$$M_0 = \frac{PL}{3} - M_{p2} - \frac{P_w h}{2}$$

$$M_{p2} = \left(P_g - \frac{h}{L} P_w \right) \frac{L}{3} - M_0$$

A member subjected to a plastic moment M_p could also sustain a small axial load. However, as the axial load became larger, the plastic moment would have to be reduced. (See Footnote 1 references for a more detailed discussion). Chapter N of the AISC ASD Specification covers this aspect as well and provides formulas for various combinations of moment and axial load. This and other specification requirements will be discussed in conjunction with the example problem which follows.

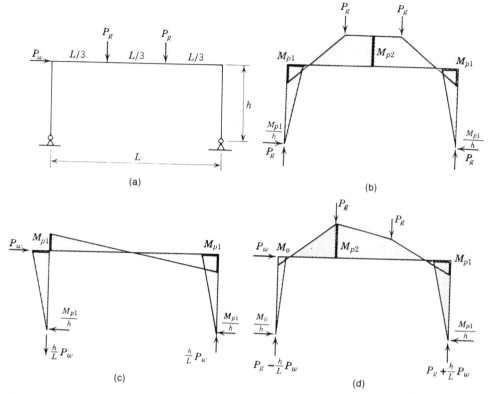

Figure 11.22/Equilibrium method.

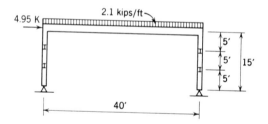

Figure 11.23/Example.

Example

Design the beam and columns constituting the steel frame shown in Fig. 11.23. A36 Steel is specified. The column bases are to be assumed pinned, and there are continuous connections at the knees. The sum of the gravity live, environmental, and dead

loads is 2.1 kips per ft. The lateral load from wind or earthquake is assumed concentrated at the knee (4.95 kips). Columns are braced about the weak axis at 5-ft intervals. Assume that the beam portion of the frame has adequate lateral support. Base the design upon the assumption that the beam has a Z-value approximately 4 times that of the columns.

The solution will be achieved in four parts. In the first (a), the trial size based upon bending only, resulting from gravity loads, will be established. In the second (b) and third (c), the trial size will be checked for bending only resulting from lateral loads. In the fourth (d), the trial sections will be examined for combined axial and bending loads and other specification requirements.

Solution (a)

Gravity-load bending only that could cause a beam-type mechanism.

(1) Increase gravity loads by applying the load factor:

$$w = LF(w_1) = 1.7(2.1) \approx 3.6 \text{ kips per ft}$$

(2) Determine the total simple-beam bending moment to be developed by beam and column:

$$M_t = \frac{3.6(40)^2}{8} = 720 \text{ ft-kips}$$

(3) Based on the assumed 4:1 ratio, determine the amount of positive moment to be developed by the beam, and the negative moment to be developed by the column at the knee:

$$M_{col} = \tfrac{1}{5}M_t = \tfrac{1}{5}(720) = 144 \text{ ft-kips}$$

$$M_{beam} = \tfrac{4}{5}M_t = \tfrac{4}{5}(720) = 576 \text{ ft-kips}$$

$$\text{Total} = 720 \text{ ft-kips}$$

(4) Make a sketch of the moment diagram based upon the above calculations (the shaded portion of Fig. 11.24).

(5) Determine the first trial size for the beam:

$$Z = \frac{M}{F_y} = \frac{576(12)}{36} = 192 \text{ in.}^3$$

Select a **W** 24×76 ($Z = 200$ in.^{3}). The beam is slightly overdesigned; therefore, calculate the plastic moment based upon its

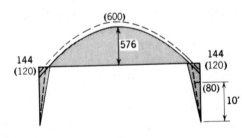

Figure 11.24/Moment diagram.

actual size:

$$M_{p(beam)} = \frac{200(36)}{12} = 600 \text{ ft-kips}$$

(6) Recalculate the required moment to be developed by the column, based upon adopting the **W** 24×76 for the beam:

$$M_{p(col)} = 720 - 600 = 120 \text{ ft-kips}$$

These bending moments are shown by the broken line in Fig. 11.24.

(7) Determine the first trial size for the column:

$$Z = \frac{M}{F_y} = \frac{120(12)}{36} = 40 \text{ in.}^3$$

A plastic section modulus of 40 in.^{3} would be required if there were no axial load. However, since there is an axial load, select a trial section having a somewhat larger Z-value. Select a **W** 12×35 ($Z = 51.2$ in.^{3}). The trial selections are in close agreement with the original requirements of a 4:1 ratio for the plastic section moduli, i.e.,

$$\frac{200}{51.2} \approx \frac{4}{1}$$

The check for detailed code requirements, as well as for combined bending and axial load, will be made after investigation of other type mechanisms.

Solution (b)

Bending from wind (or earthquake) action alone that could cause a panel mechanism. (This assumes that the dead load is negligible.)

(1) Increase lateral load by applying the load factor:

$$P_w = LF(P) = 1.3(4.95) = 6.44 \text{ kips}$$

(2) Determine the total wind moment:

$$M_w = P_w h = 6.44(15) = 96.6 \text{ ft-kips}$$

Figure 11.25/Wind only.

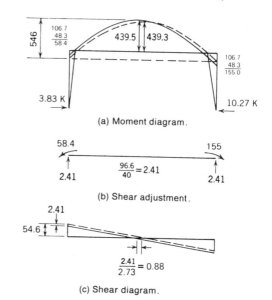

(a) Moment diagram.

(b) Shear adjustment.

$\frac{96.6}{40} = 2.41$

(c) Shear diagram.

$\frac{2.41}{2.73} = 0.88$

Figure 11.26

(3) Apportion the total wind moment to each knee (equal amounts in this case), and sketch the moment diagram (Fig. 11.25):

$$M_{knee} = \tfrac{1}{2}(96.6) = 48.3 \text{ ft-kips}$$

Since the plastic moment capability at the knee of the selected sections is 155 ft-kips, there can be no panel mechanism from the lateral load, and the trial sections are ruled safe.

Solution (c)

Bending from a combination of gravity loads and wind that could cause a combination mechanism.

(1) Increase wind and gravity loads by applying the load factor:

$$w = 1.3(2.1) = 2.73 \text{ kips per ft}$$
$$P_w = 1.3(4.95) = 6.44 \text{ kips}$$

(2) Calculate the total simple-beam bending moment to be developed by the beam (positive) and by the column at the knee (negative):

$$M_t = \frac{2.73(40)^2}{8} = 546 \text{ ft-kips}$$

This moment is shown as a broken line in Fig. 11.26a.

(3) Since the column is weaker than the beam at the knee where they are joined, it would reach its limiting plastic moment first at $M_p = 155$ ft-kips. It is observed that for the wind direction shown, the moments at the right knee caused by wind and by gravity loads are in the same direction.

Therefore, it is the combination of wind moment (48.3 ft-kips) and gravity-load moment that makes up the total plastic moment. Thus, the portion of the plastic moment available for gravity loads is $155 - 48.3 = 106.7$ ft-kips. The wind moments and gravity-load moments at the left knee are opposite in direction, so the combined effect is $106.7 - 48.3 = 58.4$ ft-kips. These combinations are shown in Fig. 11.26a.

(4) A mechanism at ultimate load requires a plastic hinge near the center of the beam, as well as one at the right knee. Consequently, it is necessary to determine if the maximum positive moment exceeds the plastic-moment capacity of the beam (600 ft-kips).

The positive moment, before inclusion of end moments for wind, was observed to be maximum at midspan and was $546 - 106.7 = 439.3$ ft-kips. When the knee moments (± 48.3 ft-kips) are accounted for, the location of the maximum positive moment shifts to the left and increases slightly (Fig. 11.26a). If the adjusted moment diagram is sketched to scale, it can be seen that this maximum positive moment is considerably less than that required for a plastic hinge (600 ft-kips). However, under other conditions, such a sketch might not be adequate to permit such a conclusion; calculations then would be required.

The location of the maximum positive moment and its value can be determined by referring to the shear diagram. Figure 11.26b shows the adjusted V values of the shearing forces (from the wind moments at the knees), and the position of the maximum positive moment shifts to the left of center (0.88 ft).

The area of shear diagram of the left portion is

$$A = \frac{52.19}{2}(19.12) = 498.9 \text{ ft-kips}$$

This area represents the difference in moment at the left knee and the maximum positive moment. Therefore,

$$M_{max} = 498.9 - 58.4 = 440.5 \text{ ft-kips}$$

Since $440.5 < 600$, a plastic hinge is not formed, and the first trial sections are accepted as safe.

Solution (d)

Check of combined bending and axial load and other specification restrictions.

(1) Refer to the most critical case—i.e., gravity load only in Solution (a)—and examine the trial section for combined loading (Section N3.1 of the 1989 AISC Specification). The following properties of the **W** 12×35 are listed for ready reference:

$$r_x = 5.25 \text{ in.}$$

$$r_y = 1.54 \text{ in.}$$

$$A = 10.3 \text{ in.}^2$$

$$Z = 51.2 \text{ in.}^3$$

The plastic axial load is

$$P_y = AF_y = 10.3(36) = 371 \text{ kips}$$

The plastic bending moment is

$$M_p = ZF_y = \frac{51.2(36)}{12} = 154 \text{ ft-kips}$$

The actual axial load is

$$P = \frac{wL}{2} = \frac{3.6(40)}{2} = 72 \text{ kips}$$

The actual bending moment (maximum) is

$$M = 117 \text{ ft-kips}$$

The slenderness ratio for the major axis (plane of bending) is

$$\frac{L_x}{r_x} = \frac{15(12)}{5.25} = 34.3 < C_c = 126.1 \qquad \text{OK}$$

and for the minor axis is

$$\frac{L_y}{r_y} = \frac{5(12)}{1.54} \approx 39 < 126.1 \qquad \text{OK}$$

The maximum strength of an axially loaded compression member is

$$P_{cr} = 1.7AF_a = 1.7(10.3)19.27 = 337 \text{ kips}$$

$$P = 72 \text{ kips} < 337 \text{ kips} \qquad \text{OK}$$

Also

$$f_a = \frac{P}{A} = \frac{72}{10.3} = 6.99 \text{ ksi} < 0.75F_y$$

$$0.75F_y = 27 \text{ ksi} \qquad \text{OK}$$

Since the columns are braced at 5-ft intervals in the weak direction,

$$M_m = \left[1.07 - \frac{(L_y/r_y)\sqrt{F_y}}{3160}\right]M_p$$

$$= \left[1.07 - \frac{39\sqrt{36}}{3160}\right]155 = 154$$

Checking the modified interaction formulas,

$$\frac{P}{P_{cr}} + \frac{C_m M}{\left(1 - \dfrac{P}{P_e}\right) M_m} \le 1.0$$

$$\frac{72}{337} + \frac{0.85(117)}{\left(1 - \dfrac{72}{\frac{23}{12}(10.3)98.18}\right)154}$$

$$= 0.89 < 1.0 \qquad \text{OK}$$

$$\frac{P}{P_y} + \frac{M}{1.18 M_p} \le 1.0$$

$$\frac{72}{371} + \frac{117}{1.18(154)} = 0.84 < 1.0 \qquad \text{OK}$$

(2) Check the column for maximum width-thickness ratios. For the flange,

$$\frac{b_f}{2t_f} = \frac{6.560}{2(0.52)} = 6.31 < 8.5 \qquad \text{OK}$$

For the web,

$$\frac{412}{\sqrt{F_y}} = \frac{412}{\sqrt{36}} = 68.7$$

$$\frac{d}{t} = \frac{10.50}{0.30} = 35 < 68.7 \qquad \text{OK}$$

(3) Check the column for lateral bracing. The critical section occurs between the top brace and the knee. The moment at the top brace is

$$M = \frac{10}{15}(117) = 78 \text{ ft-kips} \qquad (\text{Fig. 11.23})$$

The column is bent in a single curvature; therefore, M/M_p is positive, and

$$\frac{M}{M_p} = +\frac{78}{155} = +0.5$$

$$l_{cr} = r_y \left(\frac{1375}{F_y} + 25\right)$$

$$= 1.54 \left(\frac{1375}{36} + 25\right)$$

$$= 97 \text{ in.}$$

The actual unbraced distance of the column in the vicinity of the knee is 60 in. minus half the depth of the beam ($11.96 \approx$ 12 in.); therefore,

$$l_{cr} = 60 - 12 = 48 < 97 \text{ in.} \qquad \text{OK}$$

PROBLEMS

1. The unbraced steel frame shown in Fig. 11.27 has three equal vertical loads (P) applied at the quarter-points, and a lateral load (P) of the same magnitude applied at the knee. Assume that the ultimate strength of the beam and columns is the same. Find the value P in terms of M_p at ultimate strength. Use the mechanism method. (Answer given in Appendix G.)

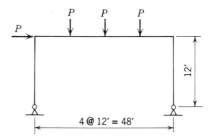

Figure 11.27

2. Assume that the steel for the unbraced frame shown in Fig. 11.27 is A36, and that there is adequate lateral support for columns and beam. Calculate the size of beam and column required, assuming that the beam has twice the strength of the column, and the loads (P) are 10.9 kips each. (The lateral load results from wind.) The column is braced about its weak axis at midheight. Use the equilibrium method. (Note that several solutions are possible, depending on which member size is selected first and on the final ultimate strength ratio of beam to column.)

12

COMPUTER-AIDED ANALYSIS AND DESIGN

12.1 Introduction

As was pointed out previously in this text, the performance of a structure—its safety and adequacy—is the responsibility of the designer, regardless of the aids he may use in discharging his responsibility. As with standards, guide specifications, and handbooks as guides to design, the computer and the programs available for its use are simply another, albeit powerful, aid to the designer. Properly used, the computer can add significantly to the designer's efficiency and effectiveness. It is for this reason that the electronic digital computer and its use in the analysis and design of structures are becoming commonplace.

Relative costs associated with computer usage continue to decline, computer services are more readily available, and designers are becoming more educated in their use. However, it must be stressed again that the computer and computer programs cannot substitute for a sound understanding of the properties of materials, and of the principles of structural analysis and design. As will be shown in this chapter, such understanding is essential if the designer is to be able to utilize the computer and to interpret and utilize its output.

Most of the information that is presented in this chapter would as readily apply to structures of materials other than steel; however, because this is a textbook devoted to the analysis and design of steel buildings, this discussion will quite naturally be focused on steel.

The computer is an electronic device; that is, its operations are performed by electronic rather than mechanical circuits. Because of this, the computer can carry out a large number of operations at incredible speeds. The electrical signals flow through the computer at speeds approaching that of light. Consider, for example, a computer instruction representing the addition of two numbers. It is not uncommon for computers to perform 100 million of these instructions per second.

The operations or processing capabilities of a computer include all of the usual mathematical procedures used in structural calculations. In addition, and most important, the computer has a memory-storage capability that enables the user to store a series of step-by-step instructions that can be recalled. These instructions must specifically and completely define the operations to be performed. Such a set of instructions is referred to as the *program*. Included in the program are instructions as to the manner in which data are to be received and the program is to operate on them, and the manner in which the results are to be displayed (printed). These data consists of the values of specific parameters relating to the structure (e.g., dimensions, loads, and physical properties such as area, modulus of elasticity, and moment of inertia).

Examples of such programs are shown in Appendixes E and F. The programs so illustrated are written in the FORTRAN language, which means that the instructions conform to a set of rules established for that language.

The two most common languages used for structural programs are FORTRAN and C. Most computers are equipped to accept programs written in one or both of these languages. Needless to say, a designer must be familiar with the language before he can create a new program written in that language. However, it is not always essential that the designer be experienced in the programming language to use an existing program, although that will be helpful in ensuring comprehension.

In Art. C.15, some of the basic FORTRAN instructions are described, including flowcharts.[1] Also, the two structural programs described in Appendixes E and F are written in FORTRAN. However, it is not necessary to understand FORTRAN in order to use these programs, and most structural programs are similar in that respect.

Programs could be prepared (and some are) that would use the structural theory and procedures used in other chapters of this book. However, it is generally conceded that programs of this kind are not particularly efficient and do not capitalize on the enormous capabilities of the computer. For example, the computer can manipulate a great deal of data and solve a large number of simultaneous equations very efficiently. One of the best ways to make use of these capabilities is to apply the principles of matrix algebra. (Appendix C provides an introduction to the principles of matrix algebra for those who are not familiar with the subject.)

Because of these considerations, two general procedures for structural analysis have been developed: the *flexibility method* and the *stiffness method*—both of which are based on well-established principles. Given that all parts of a structure are interrelated, the interrelationships are expressed as unknowns and then these unknowns are collectively expressed in a series of simultaneous equations. In the articles that follow, this approach will be discussed in some detail so as to illustrate the manner in which programs are prepared. Two independent programs with illustrative problems are shown in the appendixes, and in the last two articles of this chapter several of the existing programs that are available to designers are discussed.

[1] For further information on languages, the reader is referred to texts on the subject: for example, *A Guide to FORTRAN Programming*, Daniel S. McCracken, Wiley, New York, and *Programming in C*, by Miller and Quilici, Wiley, New York.

12.2
Structural Theory

The structural steel frame of any building will consist of many elements that could be assembled in a variety of geometric configurations. The connections between the elements may be either flexible or restrained. And, although the completed framework must be such that it is capable of enclosing three-dimensional space, doing so can be accomplished either by creating a three-dimensional framework directly, or by creating two or more independent two-dimensional framed components, usually at right angles to one another, which, when connected, will form a three-dimensional framework. The former kind is referred to as a space frame or space truss (Art. 12.6). The two-dimensional frames can be plane frames, plane trusses, or combinations of frame and truss elements.

Loads on two-dimensional frames are transferred in only two directions and in the same plane. Figure 12.1 illustrates a building structure consisting of two plane framed components. The plane truss receives surface gravity loads from the purlins and transfers these loads in the east-west direction. The plane frame receives loads from the intermediate trusses and transfers these loads in the north-south direction. Only plane structures will be treated here, even though many of the

principles that will be discussed have general applicability. This is consistent with other presentations in this book: that is, Chapter 9 contains information pertaining to plane trusses, and Chapter 10 contains information on plane frames.

In the introduction to this chapter, it was stated that much of the advantage to be gained from the use of the computer lies in its analytical capability. Analysis in this context is the determination of stress resultants (forces and moments) and displacements (deflections and rotations) at specific locations within the given structure. This is often referred to as structural theory. In Chapter 9, both mathematical and graphical methods were presented for determining stress resultants in plane, statically determinate trusses. In Chapter 10, the concept of statically indeterminate structures was presented, and methods were illustrated for determining stress resultants for continuous beams and frames. With introduction of the computer as an aid to analyses, it now will be necessary to introduce those new concepts that are pertinent to its use. However, this will be preceded by an overview of structural theory (Figure 12.2).

Building Structure / The starting point is a line drawing (sketch) of the proposed structure configuration (Fig. 12.1). Each part of the structure must be located, and the whole must satisfy the architectural requirements for the building. Furthermore, all materials and dimensions must be defined and specified.

Basic Theoretical Assumptions / Applicable theories are based on certain assumptions that either can be verified or are nearly verified. Among them are: the material of the structure is elastic and obeys Hooke's law (i.e., stress is proportional to strain); all loads are applied gradually, resulting in static rather than dynamic condi-

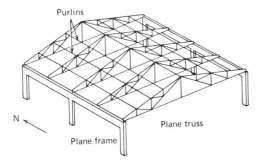

Figure 12.1/Plane truss and plane frame.

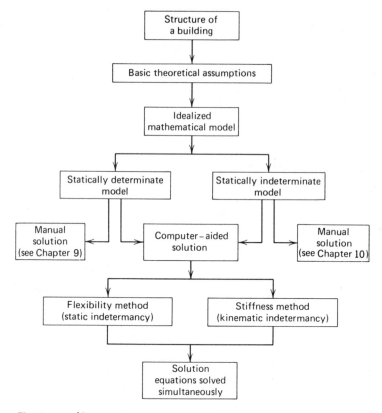

Figure 12.2/Overview of structural theory.

tions; a linear relationship is maintained between loads and resulting displacements, thereby making the laws of superposition applicable; members within the structure either are pinned to adjacent members or are continuous (built in to adjacent members); and points of structure support are (and must be) defined as pinned, fixed, elastic, or free in one or more directions.

Idealized Mathematical Structure / A most important step in this theoretical approach is the creation of a mathematical model, usually in the form of a figure sketch. In most problem-oriented texts, this step constitutes the beginning of the analysis. The model must simulate the real building structure, because it is this model that will be used to obtain the very accu-

rate solution that is possible. To the extent that the model fails to truly represent the real structure, the answers obtained will be inaccurate.

Establishing Determinacy / The configuration of the idealized mathematical model, together with the loading and basic assumptions applied, will place the structure in either the determinate or the indeterminate category. This was discussed in Chapter 10, where it was shown that the categorization depends on the type of connections used as well as the number, type, and location of supports external to the structure.

Manual Solution / Thus far in this book, reliance has been placed on manual solutions—that is, algebraic or graphical—which are readily accomplished using statically determinate models where the equations of statics can be applied. Solutions for statically indeterminate models are much more complex, however. This is particularly so when displacements (sidesway) are required, as is true with multistory and/or multibay structures. The equations of statics are then not sufficient, and the model requires additional equations that derive from the compatibility, or fit, of the structure.

Computer-Aided Solution / Computer-aided solution is applicable to either a determinate or an indeterminate model. It is particularly useful, however, when the model is complex and has a high degree of indeterminacy and/or when deflections are required in the solution. The procedure could be based on conventional theories used in the manual method as described in previous chapters, but usually it is based on one of the following two methods.

1. *Flexibility method*. This method takes axial forces, shear forces (reactions), bending moments, and the like as the basic unknowns of the system and considers the static indeterminacy of the model. (See Art. 12.3 and Art. 12.4.)

2. *Stiffness (or displacement) method*. This method takes displacements (translations and rotations) as the basic unknowns of the system and considers the kinematic indeterminacy of the structure. (This process is ideally suited for computer calculations and is described in more detail in subsequent articles.)

In general, either method (flexibility or stiffness) will generate a set of linear equations that must be solved simultaneously. Although this can be accomplished by any one of a number of procedures, the one most frequently used involves matrix algebra (Appendix C).

12.3
Basic Concepts

Steel, as an elastic material, obeys Hooke's law if the stresses to which it is subjected do not exceed the elastic limit. Under these conditions, stresses are proportional to strains, and a linear relationship exists between the applied loads and the resulting displacements of the idealized mathematical model. Consequently, the rules of superposition are valid and can be applied to set up the equilibrium equations.

An important feature of the rules of superposition is that the reaction components and displacements from separately applied loads, when added algebraically, produce the same reaction components and displacements that would result if the loads were applied simultaneously (Fig. 12.3). As will be shown in subsequent problems, the true usefulness of superposition is in its application to the more complex structures.

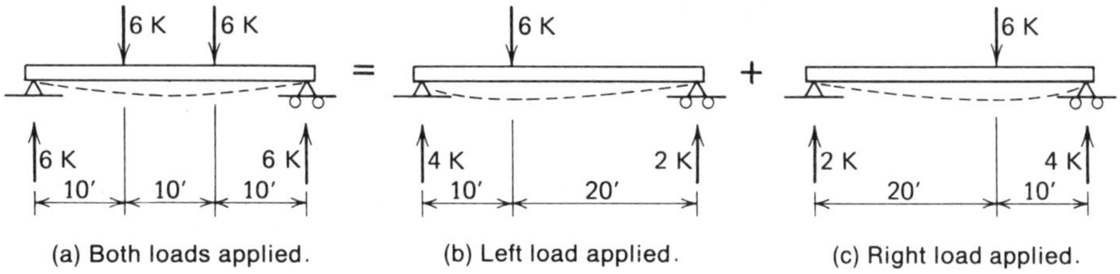

(a) Both loads applied.   (b) Left load applied.   (c) Right load applied.

Figure 12.3/Superposition.

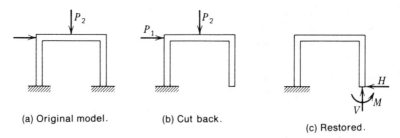

(a) Original model.　　　(b) Cut back.

(c) Restored.

Figure 12.4/Flexibility method.

The objective of the procedure is to prepare a set of independent, simultaneous equations in terms of specially designated unknowns. Of course, the number of unknowns must be equal to the number of equations. If the equations are few in number, they can easily be solved manually by using ordinary algebra or matrix algebra. However, if the equations are many, a computer can be programmed to solve them rapidly, using matrix algebra.

When system unknowns are expressed in terms of forces and moments resulting from static equilibrium, the process is known as the flexibility method. When the unknowns are expressed in terms of joint displacements and rotations, the process is known as the stiffness method or the displacement method.[2]

The flexibility method frequently requires understanding of the static indeterminacy of the structure (Art. 10.1). The number of redundant reaction components establishes the degree of structure indeterminacy. Equations for these redundant reaction components are added to the static-equilibrium equations to arrive at the total number of independent equations necessary for a complete solution.

Figure 12.4 shows a frame that is statically indeterminate to the third degree. The right reaction is arbitrarily selected as the one having redundant components. The

flexibility process requires that the structure be "cut back" to a determinate degree (Fig. 12.4b) and then restored to its initial form by reapplying the redundant components (usually one at a time). In this case, three additional independent equations could be set up. Described in Art. 10.2 was a method called "consistent deformation." Further observation shows that this method is the same as the flexibility method.

The stiffness method involves the kinematic indeterminacy of the structure (kinematics is the study of motion without regard to the forces that cause the motion). The degree of kinematic indeterminacy of a structure is established by the number of independent components of joint displacements (rotation, vertical translation, and horizontal translation) required to describe the response of the system to loads. In this connection, one neglects the very small change in the geometry of the structure due to the distortion of the members.

Figure 12.5a shows the same frame that was used to describe the flexibility method. Figure 12.5b shows the cut-back equivalent, which was obtained by fixing the two knee joints. Thus, any displacements (or motion) due to applied loads are prevented. This, at first, appears to require two restraint components at each knee joint (rotation and horizontal displacement). However, if the change in length of the horizontal member is neglected, and one knee joint is fixed against horizontal

[2]Note that the moment-distribution method used in Chapter 10 is a stiffness method.

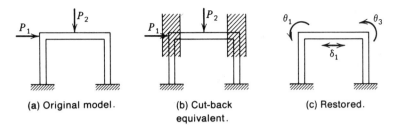

(a) Original model.   (b) Cut-back
                        equivalent.   (c) Restored.

Figure 12.5/Stiffness method.

translation, no horizontal restraint is required at the other knee. Therefore, two joint rotations and one joint translation are required to create kinematic determinacy.

Equations in terms of the two joint rotations and one joint translation are subsequently developed to restore equilibrium at the system nodes (joints) that was artificially created at the cut-back equivalent stage. This process of solution is very similar to the moment distribution process described in Chapter 10, where each joint was artificially fixed and then released, allowing it to rotate and balance out the moments.

12.4
Introduction to the Flexibility Method

The procedure for this method is described in reference to the two-span continuous beam shown in Fig. 12.6. To simplify the process, axial loads are not included; however, in any general problem-solving procedure, allowance should be made for their inclusion. Joints (or nodes) are numbered from right to left, and final reactions are as indicated in Fig. 12.6b.

The cut-back structure and its deflection are shown in Fig. 12.6c. Reactions R_1 and R_2 were designated as the redundant reactions and have been removed. Observe that each reaction component and deflec-

tion are labeled with two subscripts. The first subscript refers to the location of the reaction or deflection component, while the second refers to the cause or source of the component. In the situation shown in Fig. 12.6c, the zero subscript refers to the cut-back condition. In Fig. 12.6d the symbol δ_{21} denotes the deflection at point ② due to a unit load of point ①.

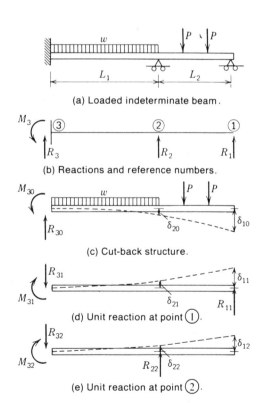

(a) Loaded indeterminate beam.

(b) Reactions and reference numbers.

(c) Cut-back structure.

(d) Unit reaction at point ①.

(e) Unit reaction at point ②.

Figure 12.6/Flexibility method.

Values for M_{30} and R_{30} can be easily calculated from statics; that is, $M_{30} = wL_1^2/2 + 2P(L_1 + L_2/2)$ and $R_{30} = wL_1 + 2P$. The deflections, δ_{10} and δ_{20}, are also calculated from static conditions. The computational effort involved depends on the loading and method used in determining deflections. This topic was introduced in Chapter 4.

The next step in the procedure accounts for the released redundant reactions. They are considered one at a time. Since the value of R_1 is unknown, a *unit* value R_{11} is assigned at that location and will be adjusted to the real value R_1 in the final equilibrium equation. Observe that R_{11} times R_1 is equal to R_1. Figure 12.6d shows the cut-back beam without the real loads, but with a unit load (R_{11}) applied at point ①. The deflection for this single load is shown by the dashed line. The quantity δ_{11} is the deflection at point ① due to a unit load applied at point ①, and δ_{21} is the deflection at point ② due to a unit load at point ①. These deflections can be calculated from statics by any of the methods described in Chapter 4. Also, M_{31} and R_{31} can be calculated as follows:

$$M_{31} = R_{11}(L_1 + L_2)$$

$$R_{31} = R_{11}$$

The above-stated process is repeated with a unit load R_{22} applied at point ②. Thus

$$M_{32} = R_{22}L_1$$

$$R_{32} = R_{22}$$

The final step in solving for the redundant reactions is accomplished by restoring the cut-back structure to its initial condition. Observe that reference point ① must have a zero deflection. Consequently, if R_{11} and R_{22} were the actual values $(R_1$ and $R_2)$ rather than unit values, the combined deflection values would add up to zero $(-\delta_{10} + \delta_{11} + \delta_{12} = 0)$. To account for the actual

reaction values, it is necessary to multiply the unit values of deflection by their corresponding actual reactions. This is done for both points ① and ② as follows:

$$-\delta_{10} + R_1\delta_{11} + R_2\delta_{12} = 0$$

$$-\delta_{20} + R_1\delta_{21} + R_2\delta_{22} = 0$$

These are observed to be linear simultaneous equations with only R_1 and R_2 as unknowns. The equations can be written in matrix form as follows:

$$\begin{bmatrix} \delta_{11} & \delta_{12} \\ \delta_{21} & \delta_{22} \end{bmatrix} \begin{bmatrix} R_1 \\ R_2 \end{bmatrix} = \begin{bmatrix} \delta_{10} \\ \delta_{20} \end{bmatrix}$$

The values of R_1 and R_2 can be determined by using matrix algebra (Appendix C).

Referring to Fig. 12.6 and superimposing (c), (d) and (e), the reaction components at ③ can be determined:

$$R_3 = R_{30} - R_1R_{31} - R_2R_{32}$$

$$M_3 = M_{30} - R_1M_{31} - R_2M_{32}$$

The matrix containing the elements δ_{11}, δ_{12}, δ_{21}, and δ_{22} is known as the flexibility matrix, and the elements themselves are known as the flexibility coefficients. The calculation of these elements can be accomplished by any one of a number of standard methods. However, computer programs prepared for solving complicated structures usually employ the method known as the *dummy unit load*. This method is sometimes discussed under the heading of virtual work.[3] Examples and assigned problems in this book are solved by using either the moment-area method (Art. 4.2) or the deflection formulas shown in Fig. 4.15 or in Part 2 of the AISC Manual.

[3]Refer to an appropriate textbook on structural theory.

Example

Calculate the reaction components for the continuous beam shown in Fig. 12.7a. Use the flexibility method and matrix algebra; $E = 29,000$ ksi and $I = 500$ in.^{4}.

Solution

(1) Label the reaction components (Fig. 12.7b). To eliminate round-off errors in the calculations, use a multiple of a basic length for actual lengths—in this case, 10 ft. Then $3l = 30$ ft, $2l = 20$ ft and $5l = 50$ ft. These are shown in Fig. 12.7b.

(2) Select R_1 and R_2 as the redundant reactions, and cut back the structure to a determinate form (Fig. 12.7c). Assume all forces and displacements to be in a positive direction (up, to the right, and coun-

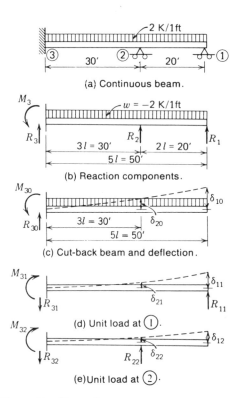

(a) Continuous beam.

(b) Reaction components.

(c) Cut-back beam and deflection.

(d) Unit load at ①.

(e) Unit load at ②.

Figure 12.7/Example.

terclockwise). Sketch the deflected shape, and label the deflection at the points where redundants were removed.

(3) Calculate the deflection in units of EI and l at points where redundant reactions were removed:

$$EI\delta_{10} = \frac{wL^4}{8} = \frac{-2k}{\text{ft}}\left[\frac{(5l)^4}{8}\right]$$

$$= -\frac{k}{\text{ft}}\left[\frac{625l^4}{4}\right]$$

The AISC Manual's equation for the deflection at any point x, due to a uniformly distributed load, is

$$EI\delta_x = \frac{wL^4}{24}\left[\left(\frac{x}{L}\right)^4 - 4\left(\frac{x}{L}\right)^3 + 6\left(\frac{x}{L}\right)^2\right]$$

Factoring out $(x/L)^2$ from the brackets,

$$EI\delta_x = \frac{wL^4}{24}\left(\frac{x}{L}\right)^2\left[\left(\frac{x}{L}\right)^2 - 4\left(\frac{x}{L}\right) + 6\right]$$

In this example problem, $x = 3l$, $L = 5l$, and $w = -2$. Substituting these terms in the deflection equation,

$$EI\delta_{20} = \frac{-2k(5l)^4}{\text{ft}\,24}\left(\frac{3l}{5l}\right)^2$$

$$\times\left[\left(\frac{3l}{5l}\right)^2 - 4\left(\frac{3l}{5l}\right) + 6\right]$$

$$= \frac{-k(5l)^4}{\text{ft}\,12}\left(\frac{9}{25}\right)\left(\frac{9}{25} - \frac{12}{5} + 6\right)$$

$$= \frac{-k\,625l^4}{\text{ft}\,12}\left(\frac{9}{25}\right)\left(\frac{9 - 60 + 150}{25}\right)$$

$$EI\delta_{20} = \frac{-k\,2971l^4}{\text{ft}\,4}$$

(4) Apply a unit load at the right reaction (Fig. 12.7d), and calculate the deflection

indicated:

$$EI\delta_{11} = \frac{R_{11}(L)^3}{3} = \frac{1(5l)^3}{3}$$

$$= \frac{125l^3}{3}$$

Using the moment-area method for calculating δ_{21},

$$EI\delta_{21} = \frac{1(2l)3l(3l)}{2} + \frac{1(3l)}{2}(3l)\left[\frac{2}{3}(3l)\right]$$

$$= 18l^3$$

(5) Apply a unit load at the center reaction (Fig. 12.7e), and calculate the deflections indicated:

$$EI\delta_{22} = \frac{R_{22}L^3}{3} = \frac{1(3l)^3}{3}$$

$$= 9l^3$$

Using the moment-area method for calculating δ_{12},

$$EI\delta_{12} = \frac{1(3l)}{2}3l\left[2l + \frac{2}{3}(3l)\right]$$

$$= 18l^3$$

(6) Restore the cut-back structure to its original condition by setting the sum of the deflections at points ① and ② to zero:

$$R_1\delta_{11} + R_2\delta_{12} + \delta_{10} = 0$$

$$R_1\delta_{21} + R_2\delta_{22} + \delta_{20} = 0$$

The factor EI occurs in each element and therefore cancels. Insert the calculated δ_{ij} formulas for each element:

$$R_1\frac{125l^3}{3} + R_2 18l^3 - \frac{625l^4}{4} = 0$$

$$R_1 18l^3 + R_2 9l^3 - \frac{297l^4}{4} = 0$$

Move the negative term to the right, and cancel l^3 from all elements. Substitute 10

for the remaining value of l:

$$\frac{125}{3}R_1 + 18R_2 = \frac{625(10)}{4}$$

$$18R_1 + 9R_2 = \frac{297(10)}{4}$$

In order to have whole numbers for the matrix, multiply the first equation by 6 and the second by $\frac{2}{9}$:

$$250R_1 + 108R_2 = 9375$$

$$4R_1 + 2R_2 = 165$$

(7) Rearrange the equations into matrix form, and solve for the redundant reactions (Appendix C):

$$\begin{bmatrix} 250 & 108 \\ 4 & 2 \end{bmatrix}\begin{bmatrix} R_1 \\ R_2 \end{bmatrix} = \begin{bmatrix} 9375 \\ 165 \end{bmatrix}$$

$$|A| = \begin{vmatrix} 250 & 108 \\ 4 & 2 \end{vmatrix} = 68$$

$$R_1 = \frac{\begin{vmatrix} 9375 & 108 \\ 165 & 2 \end{vmatrix}}{68} = 13.68 \text{ kips}$$

$$R_2 = \frac{\begin{vmatrix} 250 & 9375 \\ 4 & 165 \end{vmatrix}}{68} = 55.15 \text{ kips}$$

(8) Calculate the reaction components at the fixed end:

$$R_3 = 2(50) - 13.68 - 55.15$$

$$= 31.17 \text{ kips}$$

$$M_3 = \frac{2(50)^2}{2} - 13.68(50) - 55.15(30)$$

$$= 161.50 \text{ ft-kips}$$

12.5
Introduction to the Stiffness Method

The stiffness method contains certain features that make it more suitable than the flexibility method for general computer programs for complex structures. Therefore, it is introduced in a more general

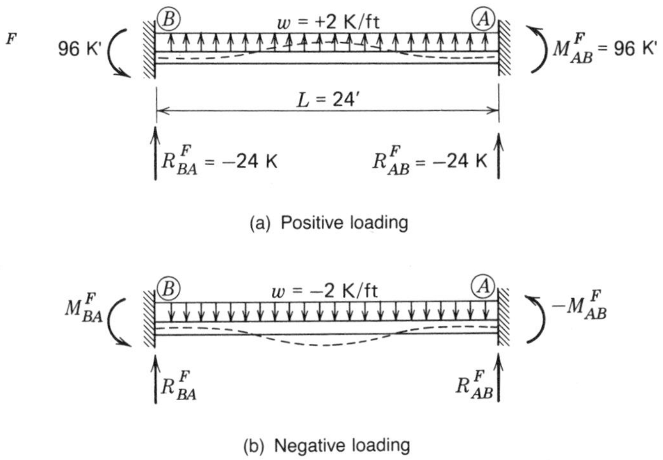

(a) Positive loading

(b) Negative loading

Figure 12.8/Sign convention.

manner and is developed more thoroughly in subsequent articles. The flexibility method and the stiffness method are similar in that each process generates a series of simultaneous equations that need to be solved for a set of unknown quantities. The basic difference in the two methods stems from the fact that the unknowns of the flexibility method are forces, while the unknowns of the stiffness method are displacements.

The general programs contained in Appendixes D and E are based on the stiffness (displacement) method. As is pointed out in Appendix C, the Cholesky square-root method requires that the matrices be positive definite. To assure that a structure stiffness matrix meets this criterion, a universal sign convention for all displacements and force components must be used. The positive direction is up, to the right, and counterclockwise. If unknown, the directions are assumed positive, and if subsequent calculations show them to be negative, they are in the negative direction.

It is customary to show all force components on the free-body diagram in the positive direction and to designate them as negative when appropriate. For example, see the free-body diagram for a fixed-end

beam shown in Fig. 12.8. Under a positive loading condition (Fig. 12.8a), it is obvious that both vertical reactions and the fixed-end moment at B are in the wrong direction and therefore require negative signs. The more common gravity loading is shown in Fig. 12.8b, where the load takes on a negative sign. Under these circumstances, only the fixed-end moment at A is negative.

When the redundant force components include bending moments, the process is very similar to the moment distribution procedure discussed in Chapter 10. A fixed-end moment is first calculated and subsequently adjusted by applying joint rotations. Because of the similarity of the two procedures, the following explanation employs the same symbols, notation, and descriptions used for moment distribution.

The method is explained in reference to the continuous beam shown in Fig. 12.9a. Axial loads have purposely been omitted in order to simplify the discussion; however, they will be introduced in later articles. The joints (or nodes) are labeled right to left A, B, and C. It is observed that the beam is kinematically indeterminate to the second degree. Both joints A and B could have some rotation and must, therefore,

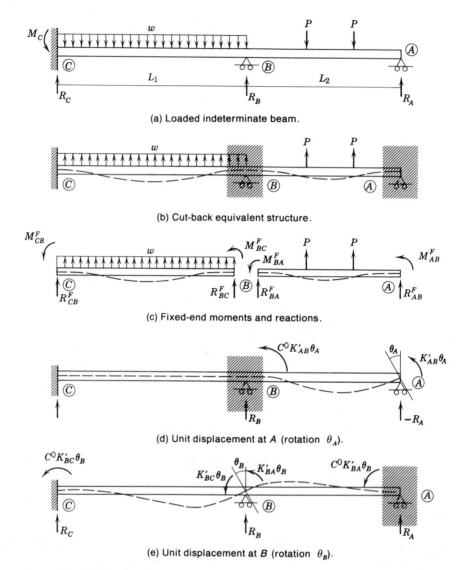

(a) Loaded indeterminate beam.

(b) Cut-back equivalent structure.

(c) Fixed-end moments and reactions.

(d) Unit displacement at A (rotation θ_A).

(e) Unit displacement at B (rotation θ_B).

Figure 12.9/Stiffness method.

be temporarily fixed and subsequently released, requiring adjustments in fixed-end force components.

Figure 12.9b shows the cut-back equivalent that is the kinematically determinate structure. The reaction force components for this fixed-end condition are shown in Fig. 12.9c. Observe that nodal points B and A are not in equilibrium—that is, M_{BA}^F is not necessarily equal to M_{BC}^F, and

M_{AB}^F is not equal to zero (the condition of the original structure). Consequently, adjustments to the fixed-end conditions are required.

These adjustments could readily be made if the actual amount of rotation at A and B were known. Since this is not the case, a unit rotation is independently applied at each location, and the influence of such action is determined at other locations.

The results are then summed and forced to meet equilibrium requirements by solving the equations for appropriate values of the unknown rotations θ_A and θ_B.

Figure 12.9d shows a displacement θ_A applied at joint A, all other joints being assumed as fixed. Article 10.6 defines a stiffness factor as the moment that generates a unit rotation and derives its value for a prismatic beam, namely

$$K'_{AB} = \frac{4EI}{L_2}$$

Observe that the actual moment at A for any rotation θ_A is equal to the product of the two, that is, $K'_{AB}\theta_A$.

Article 10.6 also describes the carry-over effect (a moment applied at one end of a fixed-end beam will carry over a moment to the other end). The sign will remain the same. Figure 12.9d shows the moment at B to be the product of the carry-over factor and the moment at A, namely, $C^0 K'_{AB}\theta_A$. It has been shown that the carry-over factor for prismatic beams is $\frac{1}{2}$.

Next, a rotation is applied at nodal point B. This requires the sum of two end moments, one for beam BA and one for beam BC. The carry-over effects to points A and C also are noted.

Finally, the laws of superposition permit the addition of the conditions shown in Fig. 12.9 (c), (d), and (e) to establish equilibrium. For joint A

$$M^F_{AB} + K'_{AB}\theta_A + C^0 K'_{BA}\theta_B = 0$$

For joint B

$$M^F_{BC} + M^F_{BA} + C^0 K'_{AB}\theta_A + K'_{BC}\theta_B + K'_{BA}\theta_B = 0$$

Using a C^0 value of $\frac{1}{2}$ for prismatic beams and rearranging the terms,

$$K'_{AB}\theta_A + \tfrac{1}{2}K'_{BA}\theta_B = -M^F_{AB}$$
$$\tfrac{1}{2}K'_{AB}\theta_A + (K'_{BA}+K'_{BC})\theta_B = -M^F_{BC}+M^F_{BA}$$

The above two equations can be expressed in matrix form as follows:

$$\underbrace{\begin{bmatrix} K'_{AB} & \tfrac{1}{2}K'_{BA} \\ \tfrac{1}{2}K'_{AB} & K'_{BA} + K'_{BC} \end{bmatrix}}_{\text{Stiffness matrix}} \underbrace{\begin{bmatrix} \theta_A \\ \theta_B \end{bmatrix}}_{\text{Displacement vector}}$$

$$= \underbrace{\begin{bmatrix} -M^F_{AB} \\ -(M^F_{BC} + M^F_{BA}) \end{bmatrix}}_{\text{Load vector}}$$

The unknown values of θ in the displacement vector are determined by matrix-algebra methods.

The elements in the stiffness matrix are known as stiffness coefficients and require further discussion. They are observed to have some similarities with the flexibility matrix described in Art. 12.4. Substituting the general symbol δ_{ij} for each of the elements in the matrix and noting the definition of each,

$K'_{AB} = \delta_{11}$, the moment at 1 due to a unit rotation at 1.

$\tfrac{1}{2}K'_{BA} = \delta_{12}$, the moment at 1 due to a unit rotation at 2.

$\tfrac{1}{2}K'_{AB} = \delta_{21}$, the moment at 2 due to a unit rotation at 1.

$K'_{BA} + K'_{BC} = \delta_{22}$, the moment at 2 due to a unit rotation at 2.

This pattern of elements identified by the two subscripts makes it fairly easy to prepare a stiffness matrix for a complex structure. Furthermore, Maxwell's law of reciprocal displacements makes the matrix symmetric with respect to its principal diagonal, i.e., the displacement at one point A in a structure due to a load applied at another point B is exactly the same as the displacement at point B if the same load is applied at point A. Expressed in equation form,

$$\delta_{ij} = \delta_{ji} \qquad \text{if} \quad i \neq j$$

The displacement vector will have one unknown value for each degree of indetermi-

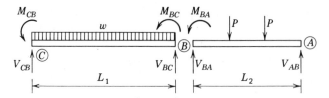

Figure 12.10/End moments and shears.

nacy, located at each node that requires an artificial fixed end. The load vector will have one value (ΣM^F) for each node that requires an artificial fixed end.

Once the θ value at each artificially restrained nodal point has been determined, the final reaction components can be calculated in terms of this θ value and conditions of static equilibrium.

Figure 12.10 represents the loaded continuous beam of Fig. 12.9a, shown as a free-body diagram for each beam. The final moment at C is the fixed-end moment at C plus the carry-over factor times the actual rotation at B, or

$$M_{CB} = M^F_{CB} + \tfrac{1}{2}K'_{BC}\theta_B$$

The final moment at B is equal to the fixed-end moment at BA plus the carry-over factor at A times the stiffness factor times the actual rotation at A plus the factor at B times the actual rotation at B, or

$$M_{BA} = M^F_{BA} + \tfrac{1}{2}K'_{AB}\theta_A + K'_{BA}\theta_B$$

The final moment at B also may be calculated by using the left segment, i.e.,

$$M_{BC} = M^F_{BC} + K'_{BC}\theta_B$$

The end shears are calculated from the loads and end moments as follows:

$$V_{AB} = -P - \frac{M_{BA}}{L_2}$$

$$V_{BA} = -P + \frac{M_{BA}}{L_2}$$

$$V_{BC} = -\frac{wL_1}{2} - \frac{M_{BC} + M_{CB}}{L_1}$$

$$V_{CB} = -\frac{wL_1}{2} + \frac{M_{BC} + M_{CB}}{L_1}$$

Example

Calculate the end moments and shears for the continuous beam shown in Fig. 12.7a. Use the stiffness method and matrix algebra. $E = 29,000$ ksi and $I = 500$ in.[4]. (This is the same problem used as the example for the flexibility method.)

Solution

(1) Sketch the structure, show the loads, and label the nodal points (Fig. 12.11).

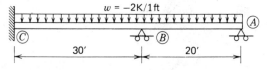

Figure 12.11/Example.

(2) Calculate the stiffness factor for the beams:

$$K'_{AB} = K'_{BA} = \frac{4EI}{L}$$

$$= \frac{4(29,000)500}{20(12)^2} = 20,139 \text{ ft-kips}$$

$$K'_{BC} = K'_{CB} = \frac{4(29,000)500}{30(12)^2}$$

$$= 13,426 \text{ ft-kips}$$

(3) Calculate the fixed-end moments:

$$M^F_{AB} = -M^F_{BA} = \frac{(-2)(20)^2}{12}$$

$$= -66.67 \text{ ft-kips}$$

$$M^F_{BC} = -M^F_{CB} = \frac{(-2)(30)^2}{12}$$

$$= -150.00 \text{ ft-kips}$$

(4) Determine the load vector:

$$-M^F_{AB} = -(-66.67)$$

$$= 66.67 \text{ ft-kips}$$

$$-(M^F_{BC} + M^F_{BA}) = -(66.67 - 150.00)$$

$$= 83.33 \text{ ft-kips}$$

(5) Determine the stiffness matrix (to simplify the calculations, a value of 10,000 is factored out of the stiffness values):

$$K'_{AB} = K'_{BA} = 10,000(2.01)$$

$$K'_{BC} = K'_{CB} = 10,000(1.34)$$

Using 2.01 and 1.34 for the stiffnesses means that the θ values will be 10,000 times larger than their real values. However, this effect will be canceled when θ is multiplied by K'_{ij} to get the balancing moments. This procedure should not be used when calculating deflections.

Referring to the subscript definition of the elements in the stiffness matrix, the follow-

ing values are determined:

$$\delta_{11} = K'_{AB} = 2.01$$

$$\delta_{12} = \delta_{21} = \tfrac{1}{2}K'_{BA} = 1.01$$

$$\delta_{22} = K'_{BA} + K'_{BC} = 2.01 + 1.34 = 3.35$$

(6) Set up the matrix, and solve for the rotations at points A and B:

$$\begin{bmatrix} 2.01 & 1.01 \\ 1.01 & 3.35 \end{bmatrix} \begin{bmatrix} \theta_A \\ \theta_B \end{bmatrix} = \begin{bmatrix} 66.67 \\ 83.33 \end{bmatrix}$$

$$|A| = \begin{vmatrix} 2.01 & 1.01 \\ 1.01 & 3.35 \end{vmatrix} = 5.71$$

$$\theta_A = \frac{\begin{bmatrix} 66.67 & 1.01 \\ 83.33 & 3.35 \end{bmatrix}}{5.71} = 24.36$$

$$\theta_B = \frac{\begin{bmatrix} 2.01 & 66.67 \\ 1.01 & 83.33 \end{bmatrix}}{5.71} = 17.53$$

(7) Calculate end moments for the beams:

$$M_{BA} = M^F_{BA} + \tfrac{1}{2}K'_{AB}\theta_A + K'_{BA}\theta_B$$

$$M_{BA} = 66.67 + \tfrac{1}{2}(2.01)(24.36)$$

$$+ (2.01)(17.53)$$

$$= 126.39 \text{ ft-kips}$$

$$M_{CB} = M^F_{CB} + \tfrac{1}{2}K'_{BC}\theta_B$$

$$= 150.00 + \tfrac{1}{2}(1.34)(17.53)$$

$$= 161.75 \text{ ft-kips}$$

(8) Determine the values of the end shears for the beams:

$$V_{AB} = -\left[\frac{-2(20)}{2}\right] - \frac{126.4}{20} = 13.68 \text{ kips}$$

$$V_{BA} = -\left[\frac{-2(20)}{2}\right] + \frac{126.4}{20} = 26.32 \text{ kips}$$

$$V_{BC} = -\left[\frac{-2(30)}{2}\right]$$

$$-\left[\frac{-126.4 + 161.7}{30}\right] = 28.82 \text{ kips}$$

$$V_{CB} = -\left[\frac{-2(30)}{2}\right]$$

$$+\left[\frac{-126.4 + 161.7}{30}\right] = 31.17 \text{ kips}$$

PROBLEMS

1. Using the flexibility method, solve the continuous beam shown in Fig. 12.12 for the redundant reactions at A and B. Calculate the bending moments at B and C.

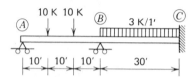

Figure 12.12/Problems 1 and 3.

2. Using the flexibility method, solve the continuous beam shown in Fig. 12.13 for the redundant reactions at B and C. Calculate the bending moments in the beam at points B and C. (Answers given in Appendix G.)

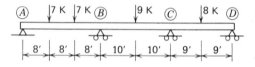

Figure 12.13/Problems 2 and 4.

3. Solve Problem 1 above by the stiffness method.

4. Solve Problem 2 above by the stiffness method.

12.6
General Structural Systems

Generally, all structures can be classified as being either *membrane* or *framed*. Typical membrane structures are shells and balloons (pneumatic), where the surface itself becomes the load-bearing element. Framed structures, on the other hand, consist of an assembly of individual elements, independent of the surfacing skin. These individual elements are shaped so that their length is long compared with

their transverse dimensions. Framed structures are typical of steel buildings.

There are five distinct types of framed structures, each type being categorized on the basis of certain aspects of its shape and degrees of freedom present in the system. These structures are shown in Fig. 12.14 and are referred to as a plane truss, plane frame, grid, space truss, or space frame. For purposes of clarity, all elements of the frames shown in Fig. 12.14 are indicated by a single straight line or as having a rectangular cross section; in reality, however, they may consist of double angles, tubular or I-shaped sections, or a cross section of any shape. The types of structure shown in Fig. 12.14 are independent of the cross-sectional shapes of the elements themselves.

Each type depends on the physical arrangement of the elements within a coordinate system and whether the elements are attached to one another by flexible (pinned) or moment-resisting connections. These two features (arrangement and connection) combine to define further the degree of freedom for the system.

A coordinate system is shown for each type in Fig. 12.14. The origin may be placed at any convenient location; however, the location where the arrows indicate positive directions of the coordinate axes is the usual one. The space truss and space frame require three-dimensional coordinates (x, y, and z). Each of the nodes (in this instance, connections) can be located and identified with x, y, and z dimensions. The long axis of the individual elements does not have to coincide with the axes of the coordinate system, since components of elements can be utilized instead. Such is the case for the diagonal web elements of the space truss (Fig. 12.14d). Only a two-dimensional coordinate system is required to describe the plane frame, plane truss, and grid. Observe that the plane

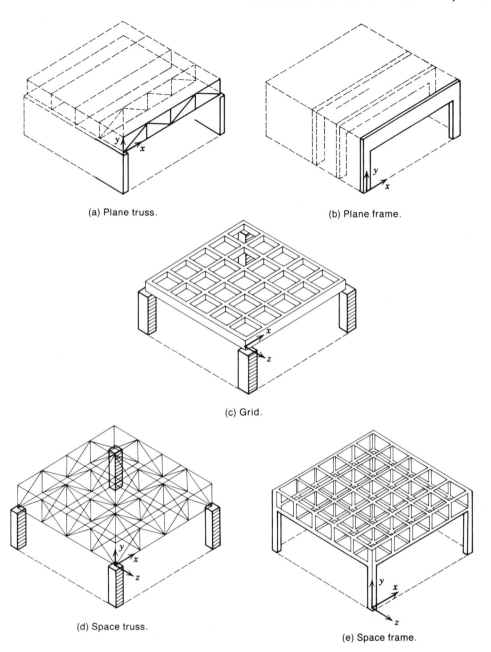

(a) Plane truss.

(b) Plane frame.

(c) Grid.

(d) Space truss.

(e) Space frame.

Figure 12.14/Types of framed structures.

truss and plane frame lie in the x-y plane, while the grid lies in the x-z plane.

The stiffness method of analysis is based on the displacement of the joints in the system (the unknowns in the matrix). The joint displacements represent the manner in which the framed structure may *freely* respond to applied loads. Consequently, the number of the component directions of the displacements, which also correspond to the unknown force components at the ends of each element, becomes the number of degrees of freedom at that joint. Also, the number of degrees of freedom for an entire structure is defined as the total number of unknown displacements at the joints in the structure.

In structural systems where the elements are attached to one another through moment-resisting connections, the moment component of the end reaction must be included in the joint displacement analysis; otherwise, only the direct axial and shear force displacements are present. Figure 12.15 shows the component reactions at the ends of elements (representing the degrees of freedom of the joint) for each type of framed structure shown in Fig. 12.14.

In order to be classified as a space frame (Fig. 12.14e), the elements must have rigid connections. Therefore, in addition to the three-dimensional axial forces F_x, F_y, and F_z, three end moments are present, each related to an axis of the coordinate system. M_z occurs in the x-y plane, and causes rotation about the z axis. M_y is in the x-z plane, causing rotation about the y axis, and M_x is in the y-z plane, causing rotation about the x axis. Since there are six reaction components, the space frame has six degrees of freedom at each joint.

The plane truss lies in the x-y plane and has elements pinned at the joint (Fig. 12.15a). Therefore, there are no moments, and only two degrees of freedom occur (F_x and F_y). Likewise, the plane frame lies in the x-y plane (Fig. 12.15b), but the members have moment-resisting connections, creating M_z. Consequently, the plane frame has three degrees of freedom at a joint (F_x, F_y, and M_z).

The grid lies in the y-z plane. It has moment-resisting connections, but is loaded and deformed in such a manner that there are no axial loads, thus creating three degrees of freedom (F_y, M_z, and M_x). The space truss (Fig. 12.15d), being three-dimensional and having pinned ends, has three degrees of freedom at a joint (F_x, F_y, and F_z).

Finally, adding up the numbers of degrees of freedom for all joints in a structural system will produce the number of internal degrees of freedom for the entire structure. It also should be mentioned that in some framed structures it is possible to have a combination of fixed and pinned joints, but their location in the geometric layout must be adequate for a stable structure.

The remaining articles in this chapter deal principally with the plane truss and the plane frame. Grids, space trusses, and space frames are referenced to packaged computer programs discussed in Art. 12.17.

To utilize the displacement method for the analysis of a framed structure, it is first necessary to determine the stiffness of a single member of the structure. The stiffness of a member is described by a stiffness matrix, the elements of which are known as stiffness coefficients. A typical member stiffness coefficient will be denoted by the symbol s_{ij} and is defined as the action (force or couple) induced on the member at line of action i due to a unit displacement (linear displacement or rotation) of the member at line of action j, the other displacements being prevented. A line of action may correspond to either a displacement or a rotation. Truss and

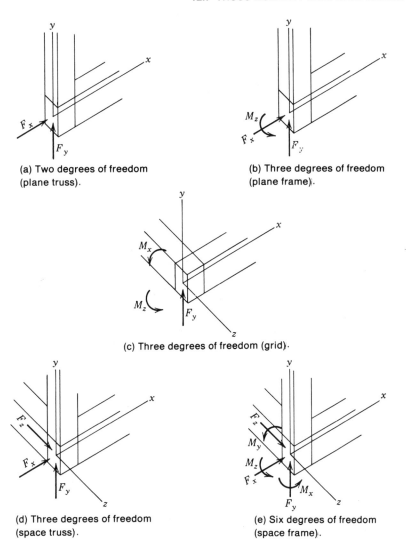

(a) Two degrees of freedom
(plane truss).

(b) Three degrees of freedom
(plane frame).

(c) Three degrees of freedom (grid).

(d) Three degrees of freedom
(space truss).

(e) Six degrees of freedom
(space frame).

Figure 12.15/Degrees of freedom at ends of members for framed structures.

frame member stiffness matrices will be developed in Arts. 12.7 and 12.8, respectively.

12.7
Truss-Element Stiffness Matrix

A typical element of a pin-jointed, plane truss structure is shown in Fig. 12.16. The

left end of the element is labeled A and the right end B; however, this choice of labeling is arbitrary and could be reversed if desired. Associated with the element is a set of coordinate axes $(x_e, y_e, \text{and } z_e)$, which are referred to as *element-oriented axes* or *element axes*. Each of the axes originates at the A end of the element. The positive x_e axis is directed along the element toward end B, the positive y_e axis

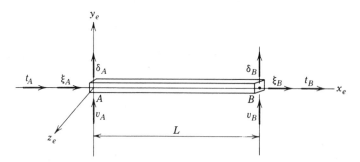

Figure 12.16/Typical truss element.

is directed upward, and the positive z_e axis is directed outward from the plane of the paper, forming a right-hand coordinate system.

Each element of a truss has associated with it a set of element axes. The positive directions of these axes are established once the A and B ends are designated. These element axes should not be confused with the structure axes discussed in Art. 12.6. Three examples of element axes are shown for the truss in Fig. 12.17.

Referring again to Fig. 12.16, the quantities t_A and v_A denote the end actions (thrust and shear forces) at end A of the element, and t_B and v_B denote the end actions at end B. The displacement of end

A in the x_e direction is denoted by ξ_A, and the displacement in the y_e direction is denoted by δ_A. Corresponding quantities at end B are denoted ξ_B and δ_B. All actions and displacements are considered positive when acting in the positive sense of the element axes.

It is assumed that the truss is loaded only at the joints, and since the truss is pin-jointed (making the element axially loaded only), the end shears v_A and v_B are, within the limitations of first-order theory, independent of the end displacements and always zero. The end thrusts t_A and t_B, however, are related to the end displacements ξ_A and ξ_B. To establish this relationship, the change in length due to axial force of the element must be considered. Let the total axial change in length of the element be denoted by λ, which is equal to the unit strain ϵ times the total length L, or $\lambda = \epsilon L$. By definition (Appendix A), the unit strain is equal to the unit stress divided by the modulus of elasticity (E) of the element. Also, the unit stress is equal to the axial load divided by the cross-sectional area of the element. Thus, expressed in equation form,

$$\lambda = \epsilon L = \frac{f_a L}{E} = \frac{PL}{AE}$$

Solving the equation for P yields

$$P = \frac{\lambda AE}{L}$$

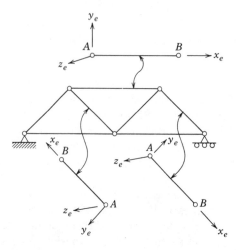

Figure 12.17/Examples of element axes.

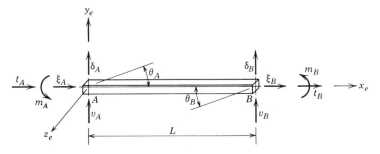

Figure 12.18/Typical frame element.

Since the axial change in length λ may be formulated in terms of end displacements ξ_A and ξ_B, the thrust at end A may be expressed as

$$t_A = (\xi_A - \xi_B) \frac{AE}{L}$$

The thrust t_B, obtained by considering the static equilibrium of the element, is

$$t_B = -(\xi_A - \xi_B) \frac{AE}{L}$$

The relationships between the end actions and the end displacements may be summarized as follows:

$$\xi_A \left(\frac{AE}{L} \right) + \xi_B \left(\frac{-AE}{L} \right) = t_A$$

$$\delta_A(0) = v_A$$

$$\xi_A \left(\frac{-AE}{L} \right) + \xi_B \left(\frac{AE}{L} \right) = t_B$$

$$\delta_B(0) = v_B$$

Expressed in matrix form, these become

$$\begin{bmatrix} \dfrac{EA}{L} & 0 & \dfrac{-EA}{L} & 0 \\ 0 & 0 & 0 & 0 \\ \dfrac{-EA}{L} & 0 & \dfrac{EA}{L} & 0 \\ 0 & 0 & 0 & 0 \end{bmatrix} \begin{bmatrix} \xi_A \\ \delta_A \\ \xi_B \\ \delta_B \end{bmatrix} = \begin{bmatrix} t_A \\ v_A \\ t_B \\ v_B \end{bmatrix}$$

or, in matrix notation,[4]

$$\mathbf{s}_e^{\,k} \mathbf{d}_e^{\,k} = \mathbf{a}_e^{\,k}$$

The superscript k in this equation refers to the kth element of the truss, while the subscript e refers to the element axes.

The above equation in matrix notation expresses the end actions t_A, v_A, t_B, and v_B in terms of the end displacements ξ_A, ξ_B, δ_A, and δ_B. The matrix $\mathbf{s}_e^{\,k}$ is the stiffness matrix of a typical element of a pin-jointed plane truss, and the elements of $\mathbf{s}_e^{\,k}$ are the stiffness coefficients which describe the stiffness of the element.

12.8
Frame-Element Stiffness Matrix

A frame-element stiffness matrix may be developed by using a procedure similar to that used to develop the truss-element stiffness matrix. A typical element of a frame is shown in Fig. 12.18. Observe that there are three degrees of freedom at each end and that a right-hand coordinate system, which is consistent with that used with the truss element in the previous article, is used. As before, these element axes must not be confused with the structure axes discussed in Art. 12.6.

[4]The boldface type in this and subsequent articles denotes a vector, a submatrix, or a matrix.

The quantities t_A, v_A, and m_A denote the end actions (thrust, shear, and moment) at end A of the element, and t_B, v_B, and m_B denote the corresponding end actions at end B. The displacement in the x_e direction of end A is denoted by ξ_A, and the displacement in the y_e direction is denoted by δ_A. Rotation about the z axis of end A is denoted by θ_A. Corresponding quantities at end B are denoted by ξ_B, δ_B, and θ_B, respectively. All quantities are shown in the positive sense.

The definition of a stiffness coefficient may be used as the basis for development of the elements of the stiffness matrix (Art. 12.6); that is, displacements may be imposed, in succession, at each of the six lines of action at the ends of the element, and corresponding end actions calculated. These end actions, when induced by a unit displacement, are stiffness coefficients and may be assembled into the element stiffness matrix.

The development is initiated by imposing a rotation θ_A at end A while maintaining zero displacements at all other lines of action. This condition is shown in Fig. 12.19. The bending moment at end A, caused by the rotation θ_A, is given by the expression

$$m_A = K'_{AB}\theta_A$$

in which K'_{AB} is the stiffness factor defined in Art. 10.6. As discussed in Art. 12.5, the carry-over factor C^0_{AB} provides a means of expressing the magnitude of the moment induced at end B as a fraction of the moment applied at end A:

$$m_B = K'_{AB}\theta_A C^0_{AB}$$

For prismatic elements, the value of the C^0_{AB} is $\frac{1}{2}$; however, for nonprismatic elements its value depends on the variation of the moment of inertia with respect to length.

Referring once again to Fig. 12.19, the application of the equations of statics re-

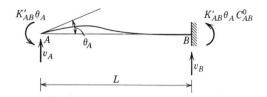

Figure 12.19/End actions due to displacement θ_A.

sults in end shears as follows:

$$v_A = +\frac{K'_{AB}\theta_A(1+C^0_{AB})}{L}$$

$$v_B = -\frac{K'_{AB}\theta_A(1+C^0_{AB})}{L}$$

Continuing in the same manner, a rotation θ_B is imposed at end B while maintaining zero displacements at all other lines of action. This condition is shown in Fig. 12.20. The expressions for end moments and shears induced by this rotation are

$$m_B = K'_{BA}\theta_B$$

$$m_A = K'_{BA}\theta_B C^0_{BA}$$

$$v_A = +\frac{K'_{BA}\theta_B(1+C^0_{BA})}{L}$$

$$v_B = -\frac{K'_{BA}\theta_B(1+C^0_{BA})}{L}$$

To develop expressions for end actions m and v due to a displacement δ_A in the y_e direction at end A, it is convenient to assume that the element is supported by pins at each end when the displacement is first imposed (Fig. 12.21a). This assumption implies, within the limitations of the

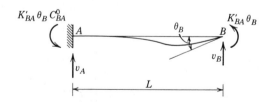

Figure 12.20/End actions due to displacement θ_B.

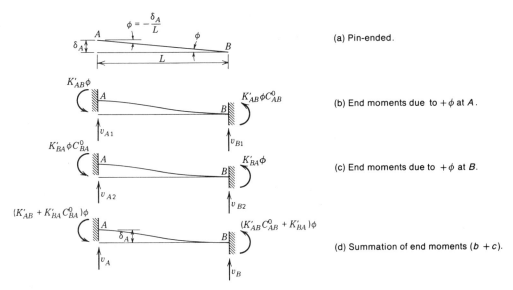

Figure 12.21/End actions due to displacement δ_A.

first-order small-deflection theory, that no end actions are induced as the result of the displacement. However, the displacement does produce rotations $\phi = -\delta_A/L$ at ends A and B. Since the definition of a stiffness coefficient requires that all displacements, except the displacement of interest, must be null, the rotations at A and B must be reduced to zero. This may be accomplished by assuming that ends A and B are now rotationally constrained and by rotating first one (Fig. 12.21b) and then the other (Fig. 12.21c) through an angle $\phi = +\delta_A/L$. These rotations cause end actions that are added together (Fig. 12.21d) to form the expressions for end actions induced by the displacement δ_A. In Fig. 12.21b, the end shears are

$$v_{A1} = \frac{K'_{AB}\phi(1 + C^0_{AB})}{L} = -v_{B1}$$

and the end shears in Fig. 12.21c are

$$v_{A2} = \frac{K'_{BA}\phi(1 + C^0_{BA})}{L} = -v_{B2}$$

Also, the expressions for the final end actions due to the displacement δ_A (Fig.

12.21d) are

$$m_A = \frac{(K'_{AB} + K'_{BA}C^0_{BA})\delta_A}{L}$$

$$m_B = \frac{(K'_{AB}C^0_{AB} + K'_{BA})\delta_A}{L}$$

$$v_A = \frac{(K'_{AB} + K'_{AB}C^0_{AB} + K'_{BA} + K'_{BA}C^0_{BA})\delta_A}{L^2}$$

$$v_B = -\frac{(K'_{AB} + K'_{AB}C^0_{AB} + K'_{BA} + K'_{BA}C^0_{BA})\delta_A}{L^2}$$

The expressions for end actions due to a displacement δ_B at end B (Fig. 12.22), obtained in a manner similar to that described above, are

$$m_A = -\frac{(K'_{AB} + K'_{BA}C^0_{BA})\delta_B}{L}$$

$$m_B = -\frac{(K'_{AB}C^0_{AB} + K'_{BA})\delta_B}{L}$$

$$v_A = -\frac{(K'_{AB} + K'_{AB}C^0_{AB} + K'_{BA} + K'_{BA}C^0_{BA})\delta_B}{L^2}$$

$$v_B = \frac{(K'_{AB} + K'_{AB}C^0_{AB} + K'_{BA} + K'_{BA}C^0_{BA})\delta_B}{L^2}$$

The lines of action for a plane-frame element are numbered as shown in Fig. 12.23;

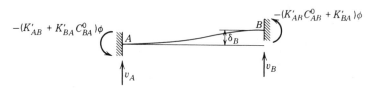

Figure 12.22/End actions due to displacement δ_B.

that is, t_A and ξ_A occur along line of action 1, v_A and δ_A occur along line of action 2, and m_A and θ_A occur along line of action 3. Corresponding quantities at end B occur along lines of action 4, 5, and 6, respectively. Using this notation, the end actions resulting from a unit rotation ($\theta_A = 1$) at end A may be identified as follows:

$$s_{23} = \frac{K'_{AB}(1 + C^0_{AB})}{L}$$

$$s_{33} = K'_{AB}$$

$$s_{53} = -s_{23}$$

$$s_{63} = K'_{AB}C^0_{AB}$$

These end actions are the stiffness coefficients, and the subscript notation is the same as that described for the fundamental stiffness method (Art. 12.5). In each of these equations, the subscripts ij of the parameter s denote the line of action i of the induced action and the line of action j of the imposed displacement. Thus, the imposed displacement ($\theta_A = 1$) occurs at line of action 3, and the induced actions v_A, m_A, v_B, and m_B occur at lines of action 2, 3, 5, and 6, respectively.

The stiffness coefficients related to unit displacements along lines of action 6, 2, and 5 are obtained in the same manner.

Figure 12.23/Notation for lines of action for a plane-frame element.

These coefficients are summarized as follows:

1. Stiffness coefficients related to a unit displacement along line of action 6 (i.e., $\theta_B = 1$):

$$s_{26} = \frac{K'_{BA}(1 + C^0_{BA})}{L}$$

$$s_{36} = K'_{BA}C^0_{BA}$$

$$s_{56} = -s_{26}$$

$$s_{66} = K'_{BA}$$

2. Stiffness coefficients related to a unit displacement along line of action 2 (i.e., $\delta_A = 1$):

$$s_{22} = \frac{K'_{AB}(1 + C^0_{AB}) + K'_{BA}(1 + C^0_{BA})}{L^2}$$

$$s_{32} = \frac{K'_{AB} + K'_{BA}C^0_{BA}}{L}$$

$$s_{52} = -s_{22}$$

$$s_{62} = \frac{K'_{AB}C^0_{AB} + K'_{BA}}{L}$$

3. Stiffness coefficients related to a unit displacement along line of action 5 (i.e., $\delta_B = 1$):

$$s_{25} = -\frac{K'_{AB}(1 + C^0_{AB}) + K'_{BA}(1 + C^0_{BA})}{L^2}$$

$$s_{35} = -\frac{K'_{AB} + K'_{BA}C^0_{BA}}{L}$$

$$s_{55} = -s_{25}$$

$$s_{65} = -\frac{K'_{AB}C^0_{AB} + K'_{BA}}{L}$$

The stiffness coefficients developed here are, within the limitations of first-order analysis, independent of the thrusts t_A and t_B and the axial displacements ξ_A and ξ_B. The effects of the thrusts and displacements themselves are identical to those developed for the plane-truss element in the previous article; therefore, the results of that development may be incorporated, without modification, directly into the present derivation. When $\xi_A = 1$,

$$ s_{11} = \frac{AE}{L} $$

$$ s_{41} = -\frac{AE}{L} $$

and when $\xi_B = 1$,

$$ s_{14} = -\frac{AE}{L} $$

$$ s_{44} = \frac{AE}{L} $$

The plane-frame element has six degrees of freedom. Each of the lines of action may be associated with a row and a column of a 6×6 matrix; hence, the stiffness coefficients may be assembled into an element stiffness matrix that relates end actions to end displacements. Thus,

$$
\begin{bmatrix}
s_{11} & & & s_{14} & & \\
& s_{22} & s_{23} & & s_{25} & s_{26} \\
& s_{32} & s_{33} & & s_{35} & s_{36} \\
s_{41} & & & s_{44} & & \\
& s_{52} & s_{53} & & s_{55} & s_{56} \\
& s_{62} & s_{63} & & s_{65} & s_{66}
\end{bmatrix}
\begin{bmatrix}
\xi_A \\ \delta_A \\ \theta_A \\ \xi_B \\ \delta_B \\ \theta_B
\end{bmatrix}
=
\begin{bmatrix}
t_A \\ v_A \\ m_A \\ t_B \\ v_B \\ m_B
\end{bmatrix}
$$

or in matrix notation

$$ \mathbf{s}_e^{\,k}\mathbf{d}_e^{\,k} = \mathbf{a}_e^{\,k} $$

The superscript k in this equation refers to the kth element of the frame, while the subscript e refers to the element axis. This equation expresses the end actions t_A, v_A, m_A, t_B, v_B, and m_B in terms of the end

displacements ξ_A, δ_A, θ_A, ξ_B, δ_B, and θ_B. The matrix $\mathbf{s}_e^{\,k}$ is the stiffness matrix of a typical element of a plane frame having continuity of members. The elements of $\mathbf{s}_e^{\,k}$ are the stiffness coefficients which describe the stiffness of the element.

The 6×6 matrix developed above, containing 20 independently derived coefficients and representing the stiffness matrix of a frame element, may be further simplified. A basic relationship known as Maxwell's law of reciprocal displacements states that the displacement at some point a produced by a unit load applied at some point b must be equal to the displacement at point b produced by the unit load applied at point a. Applying this law to the derived coefficients, it is clear that $s_{ij} = s_{ji}$ $(i \ne j)$ and that the stiffness matrix must be symmetric about the main diagonal (upper left to lower right).

The equality $s_{32} = s_{23}$, resulting from application of Maxwell's law, produces a relationship that may be utilized to simplify some of the expressions encountered in the current development, and is often useful when analyzing structures which contain nonprismatic members by the moment-distribution method. This relationship expresses the requirement that $K'_{AB}(C^0_{AB}) = K'_{BA}(C^0_{BA})$.

The application of Maxwell's law reduces the number of independent stiffness coefficients by seven, while the equalities previously identified ($s_{53} = -s_{23}$, $s_{56} = -s_{26}$, $s_{52} = -s_{22}$, and $s_{55} = -s_{25}$) reduce the number of independent quantities by four. In addition, it was shown that $s_{14} = -s_{11}$ and $s_{44} = s_{11}$, which reduces by two the remaining independent quantities; hence, there are only seven independent stiffness quantities remaining. They may be summarized as follows: If the axial-displacement stiffness quantity is identified as κ_a, and the remaining six independent quantities are identified by the symbols κ_i $(1 \le i \le 6)$, the element stiffness matrix may be expressed as

$$
\mathbf{s}_e^k =
\begin{bmatrix}
\kappa_a & & & -\kappa_a & & \\
 & \kappa_1 & \kappa_2 & & -\kappa_1 & \kappa_3 \\
 & \kappa_2 & \kappa_4 & & -\kappa_2 & \kappa_5 \\
-\kappa_a & & & \kappa_a & & \\
 & -\kappa_1 & -\kappa_2 & & \kappa_1 & -\kappa_3 \\
 & \kappa_3 & \kappa_5 & & -\kappa_3 & \kappa_6
\end{bmatrix}
$$

in which κ_a is the axial-displacement stiffness quantity, and

$$\kappa_1 = \frac{K'_{AB}\left(1+2C^0_{AB}\right)+K'_{BA}}{L^2}$$

$$\kappa_2 = \frac{K'_{AB}\left(1+C^0_{AB}\right)}{L}$$

$$\kappa_3 = \frac{K'_{AB}C^0_{AB}+K'_{BA}}{L}$$

$$\kappa_4 = K'_{AB}$$

$$\kappa_5 = K'_{AB}C^0_{AB}$$

$$\kappa_6 = K'_{BA}$$

Article 10.6 describes and develops the stiffness factor K' and the carry-over factor C^0 of a prismatic element, that is, $K' = 4EI/L$ and $C^0 = \frac{1}{2}$. Thus, for a prismatic element, $K'_{AB} = K'_{BA}$ and $C^0_{AB} = C^0_{BA}$. Substituting these quantities and terms into the derived expressions, κ_1 through κ_6 above, and noting that $\kappa_a = AE/L$, the element stiffness matrix is reduced to the form shown below:

It must be emphasized that this matrix is valid for *prismatic* frame elements only.

12.9
Rotation Matrix

The concept of a set of coordinate axes associated with each element of a structure (element-oriented axes or element axes) was introduced in Arts. 12.7 and 12.8. The orientation of each of these sets of axes depends on the orientation of the element with which they are associated. In a subsequent article (12.11), another set of axes is discussed, which is associated with the complete structure (*structure-oriented axes*, or *structure axes*) and whose orientation is fixed with respect to the structure. The intent here is to relate end action and displacement components, which are oriented with respect to element axes, to corresponding quantities that are oriented with respect to structure axes. In general, for plane trusses and plane frames, the element axis z_e coincides with the structure axis z_s, while the element axes x_e and y_e are frequently rotated through an angle α with respect to the structure axes x_s and y_s. This relationship is shown in Fig. 12.24.

If actions t_i and v_i or displacements ξ_i and δ_i $(i = A$ or $B)$ are represented by

$$
\mathbf{s}_e^k =
\begin{bmatrix}
\dfrac{AE}{L} & & & -\dfrac{AE}{L} & & \\[2mm]
 & \dfrac{12\,EI}{L^3} & \dfrac{6\,EI}{L^2} & & -\dfrac{12\,EI}{L^3} & \dfrac{6\,EI}{L^2} \\[2mm]
 & \dfrac{6\,EI}{L^2} & \dfrac{4\,EI}{L} & & -\dfrac{6\,EI}{L^2} & \dfrac{2\,EI}{L} \\[2mm]
-\dfrac{AE}{L} & & & \dfrac{AE}{L} & & \\[2mm]
 & -\dfrac{12\,EI}{L^3} & -\dfrac{6\,EI}{L^2} & & \dfrac{12\,EI}{L^3} & -\dfrac{6\,EI}{L^2} \\[2mm]
 & \dfrac{6\,EI}{L^2} & \dfrac{2\,EI}{L} & & -\dfrac{6\,EI}{L^2} & \dfrac{4\,EI}{L}
\end{bmatrix}
$$

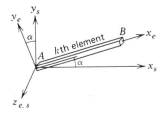

Figure 12.24/Relationship of a set of element axes to structure axes.

their resultant vectors, they may be associated with either the element or the structure axes. This relationship is shown in Fig. 12.25, where the symbol $\mathbf{a}_A^{\,k}$ represents the vector of end actions (Fig. 12.25a) and $\mathbf{d}_A$ represents the vector of end displacements (Fig. 12.25b) at end A. The symbols $\mathbf{a}_B^{\,k}$ and $\mathbf{d}_B$ represent corresponding vectors at end B. As before, the superscript k refers to the kth element. When referred to the element axes, the coordinates of $\mathbf{a}_i^{\,k}$ are t_i and v_i (representing end actions), while the coordinates of $\mathbf{d}_i$ are ξ_i and δ_i (representing end displacements). When referred to the structure axes, the corresponding coordinates of $\mathbf{a}_i^{\,k}$ will be T_i and V_i, and those of $\mathbf{d}_i$ will be Ξ_i and Δ_i; however, to distinguish between coordinates that pertain to the element axes and those that pertain to the structure axes, the vectors $\mathbf{a}_i^{\,k}$ and $\mathbf{d}_i$ will be denoted by the symbols $\mathbf{A}_i^{\,k}$ and $\mathbf{D}_i$, respectively, when the coordinates they represent refer to the structure axes. It must be remembered that the vector represented by $\mathbf{A}_i^{\,k}$ is the same vector that is represented by $\mathbf{a}_i^{\,k}$ and that

the vector represented by $\mathbf{D}_i$ is the same vector that is represented by $\mathbf{d}_i$. This is summarized as follows ($i = A$ or B):

Vector	Member-axis coordinates	Structure-axis coordinates
$\mathbf{a}_i^{\,k}$ or $\mathbf{A}_i^{\,k}$	$\begin{bmatrix} t_i \\ v_i \end{bmatrix}$	$\begin{bmatrix} T_i \\ V_i \end{bmatrix}$
$\mathbf{d}_i$ or $\mathbf{D}_i$	$\begin{bmatrix} \xi_i \\ \delta_i \end{bmatrix}$	$\begin{bmatrix} \Xi_i \\ \Delta_i \end{bmatrix}$

The purpose of this notation is to represent particular coordinates that refer to either the element axes or to the structure axes. This notation is convenient when carrying out several of the operations of the displacement method of analysis. These operations, which are developed in subsequent articles, frequently involve a change from coordinates that refer to one set of axes to coordinates that refer to the other set. Such a change may be efficiently accomplished by making use of a rotation matrix. Coordinates that refer to one set of axes, when multiplied by a rotation matrix, are changed into coordinates that refer to the other set of axes.

The development of the rotation matrices (usually a standard treatment in basic mathematics texts) is included here to provide a complete discussion of the stiffness method of analysis.

A set of structure axes and a set of element axes, rotated through an angle α

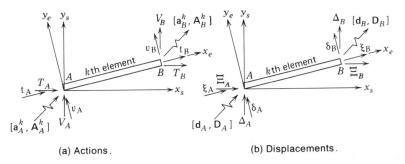

(a) Actions. (b) Displacements.

Figure 12.25/End actions and displacements represented as single vectors.

with respect to the structure axes, are shown in Fig. 12.26. The z axes for the element axes and the structure axes coincide and are perpendicular to the paper at the origin; therefore, they are not shown. A vector ρ is shown emanating from the origin and terminating at a point P which has the coordinates x'_s and y'_s relative to the structure axis and coordinates x'_e and y'_e relative to the element axes. Lines PAB (perpendicular to the structure axis x_s) and PDC (perpendicular to the element axis x_e) are drawn, creating the angle α at point P.

By considering geometry, the following relationships may be determined: line lengths $\overline{PA} = y'_e / \cos \alpha$, $\overline{AB} = x'_s \tan \alpha$, and $\overline{BC} = y'_s \tan \alpha$. Using these relationships, the coordinates relative to one axis can be expressed in terms of coordinates relative to the other axes. Thus

$$x'_e = \cos \alpha \left(x'_s + y'_s \tan \alpha \right)$$

from which

$$x'_e = x'_s \cos \alpha + y'_s \sin \alpha$$

Also

$$y'_s = \frac{y'_e}{\cos \alpha} + x'_s \tan \alpha$$

from which

$$y'_e = - x'_s \sin \alpha + y'_s \cos \alpha$$

These equations can be expressed in matrix form as follows:

$$\begin{bmatrix} \cos \alpha & \sin \alpha \\ -\sin \alpha & \cos \alpha \end{bmatrix} \begin{bmatrix} x'_s \\ y'_s \end{bmatrix} = \begin{bmatrix} x'_e \\ y'_e \end{bmatrix}$$

in which the matrix

$$\mathbf{R} = \begin{bmatrix} \cos \alpha & \sin \alpha \\ -\sin \alpha & \cos \alpha \end{bmatrix}$$

is the rotation matrix that transforms structure-axis coordinates to element-axis coordinates.

Conversely, the rotation matrix that transforms element-axis coordinates to structure-axis coordinates is the inverse of matrix $\mathbf{R}$:

$$\mathbf{R}^{-1} = \begin{bmatrix} \cos \alpha & -\sin \alpha \\ \sin \alpha & \cos \alpha \end{bmatrix}$$

Plane-truss elements have two action coordinates $\{t_i \, v_i\}$ and two displacement coordinates $\{\xi_i \, \delta_i\}$ at each end of an element; therefore, the rotation matrix $\mathbf{R}$ may be used without modification to transform coordinates that refer to the structure axes into those that refer to the element axes,

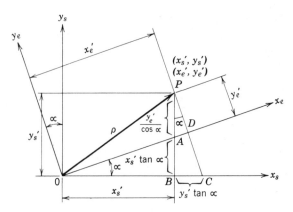

Figure 12.26/Relationship of element-axis coordinates to structure-axis coordinates.

that is,

$$\mathbf{a}_i^k = \begin{bmatrix} \cos\alpha & \sin\alpha \\ -\sin\alpha & \cos\alpha \end{bmatrix} \begin{bmatrix} T_i \\ V_i \end{bmatrix} = \begin{bmatrix} t_i \\ v_i \end{bmatrix}$$

$$(i = A \text{ or } B)$$

and

$$\mathbf{d}_i = \begin{bmatrix} \cos\alpha & \sin\alpha \\ -\sin\alpha & \cos\alpha \end{bmatrix} \begin{bmatrix} \Xi_i \\ \Delta_i \end{bmatrix} = \begin{bmatrix} \xi_i \\ \delta_i \end{bmatrix}$$

$$(i = A \text{ or } B)$$

Plane-frame elements, on the other hand, have three action coordinates $\{t_i\, v_i\, m_i\}$ and three displacement coordinates $\{\xi_i\, \delta_i\, \theta_i\}$ at each end of the element. Therefore, the rotation matrix $\mathbf{R}$ must be modified prior to its use. It is observed that the axial force and shear force coordinates $\{t_i\, v_i\}$, and the axial displacement and transverse displacement coordinates $\{\xi_i\, \delta_i\}$, transform in the same manner as the corresponding plane-truss action and displacement coordinates, while the moment and rotation coordinates undergo no change with respect to a change in coordinate axes. These requirements are satisfied if the rotation matrix is altered by adding an additional row and column, the elements of which are null except for the last element that has the value of unity. Thus

$$\mathbf{a}_i^k = \begin{bmatrix} \cos\alpha & \sin\alpha & 0 \\ -\sin\alpha & \cos\alpha & 0 \\ 0 & 0 & 1 \end{bmatrix} \begin{bmatrix} T_i \\ V_i \\ M_i \end{bmatrix} = \begin{bmatrix} t_i \\ v_i \\ m_i \end{bmatrix}$$

$$(i = A \text{ or } B)$$

or $\mathbf{R}\mathbf{A}_i^k = \mathbf{a}_i^k$, and

$$\mathbf{d}_i = \begin{bmatrix} \cos\alpha & \sin\alpha & 0 \\ -\sin\alpha & \cos\alpha & 0 \\ 0 & 0 & 1 \end{bmatrix} \begin{bmatrix} \Xi_i \\ \Delta_i \\ \Theta_i \end{bmatrix} = \begin{bmatrix} \xi_i \\ \delta_i \\ \theta_i \end{bmatrix}$$

$$(i = A \text{ or } B)$$

or $\mathbf{R}\mathbf{D}_i = \mathbf{d}_i$.

The transformation matrices for both the plane-truss elements and the plane-frame elements are denoted by the symbol $\mathbf{R}$; however, the actual matrices used are those given in the previous expressions.

12.10
Transformed Element Stiffness Matrix

The stiffness matrices for a typical element of a truss and for a frame were presented in previous articles. Both matrices were developed with reference to the element-oriented set of coordinate axes, and express the end actions on the element in terms of the end displacements. Also, a rotation matrix, which may be used to change end actions and displacements from sets which refer to element axes to sets which refer to structure axes, was developed in Art. 12.9. The objective here is to develop a means of transforming stiffness matrices that refer to element axes into matrices that refer to the structure-oriented set of coordinate axes. Such a form is referred to as a *transformed* element stiffness matrix. This transformed matrix is used to formulate conveniently the equilibrium equations for the structure.

The beam element stiffness matrix that was developed in Art. 12.8 refers to the element axes and is shown again here for ready reference:

$$\begin{bmatrix} s_{11} & & & s_{14} & & \\ & s_{22} & s_{23} & & s_{25} & s_{26} \\ & s_{32} & s_{33} & & s_{35} & s_{36} \\ \hline s_{41} & & & s_{44} & & \\ & s_{52} & s_{53} & & s_{55} & s_{56} \\ & s_{62} & s_{63} & & s_{65} & s_{66} \end{bmatrix} \begin{bmatrix} \xi_A \\ \delta_A \\ \theta_A \\ \xi_B \\ \delta_B \\ \theta_B \end{bmatrix} = \begin{bmatrix} t_A \\ v_A \\ m_A \\ t_B \\ v_B \\ m_B \end{bmatrix}$$

In order to develop a procedure for transforming element stiffness matrices, it is expedient to partition the matrices as shown by the dashed lines. Each of the four submatrices pertains to the actions at end A or B as a result of displacements at end A or B. For example, the upper right three-by-three submatrix pertains to actions induced along lines of action 1, 2, and 3 (all at end A) because of unit displacements 4, 5, and 6 (all at end B). These are identified by the subscripts to the elements. Similar descriptions can be made for the remaining submatrices and the equations rewritten in matrix form in the following manner:

$$\begin{bmatrix} s_{AA}{}^k & | & s_{AB}{}^k \\ \hline s_{BA}{}^k & | & s_{BB}{}^k \end{bmatrix} \begin{bmatrix} \mathbf{d}_A \\ \hline \mathbf{d}_B \end{bmatrix} = \begin{bmatrix} \mathbf{a}_A{}^k \\ \hline \mathbf{a}_B{}^k \end{bmatrix}$$

where:

$\mathbf{a}_A{}^k$ = vector of actions at end A (kth element).

$\mathbf{a}_B{}^k$ = vector of actions at end B (kth element).

$s_{AA}{}^k$ = submatrix that relates the actions at end A to the displacements at end A (kth element).

$s_{AB}{}^k$ = submatrix that relates the actions at end A to the displacements at end B (kth element).

$s_{BA}{}^k$ = submatrix that relates the actions at end B to the displacements at end A (kth element).

$s_{BB}{}^k$ = submatrix that relates the actions at end B to the displacements at end B (kth element).

$\mathbf{d}_A$ = vector of displacements at end A.

$\mathbf{d}_B$ = vector of displacements at end B.

The development is continued by expanding this equation (matrix multiplication as discussed in Appendix C). Thus

$$\mathbf{a}_A{}^k = s_{AA}{}^k \mathbf{d}_A + s_{AB}{}^k \mathbf{d}_B$$

$$\mathbf{a}_B{}^k = s_{BA}{}^k \mathbf{d}_A + s_{BB}{}^k \mathbf{d}_B$$

The action and displacement terms in these equations refer to the element axes. In Art. 12.9 it was shown that terms such as these can be changed to refer to the structure axes by multiplying by a rotation matrix $\mathbf{R}$. By using the notation previously introduced (e.g., letters in uppercase notation refer to the structure axes), these equations may be expressed in terms that refer to the structure axes:

$$\mathbf{RA}_A{}^k = s_{AA}{}^k (\mathbf{RD}_A) + s_{AB}{}^k (\mathbf{RD}_B)$$

$$\mathbf{RA}_B{}^k = s_{BA}{}^k (\mathbf{RD}_A) + s_{BB}{}^k (\mathbf{RD}_B)$$

To solve for $\mathbf{A}_i$ ($i = A$ or B) in each of the above, it becomes necessary to premultiply both sides of the equation by the inverse of $\mathbf{R}$ (Appendix C), i.e.,

$$\mathbf{A}_A{}^k = \left[\mathbf{R}^{-1} s_{AA}{}^k \mathbf{R} \right] \mathbf{D}_A + \left[\mathbf{R}^{-1} s_{AB}{}^k \mathbf{R} \right] \mathbf{D}_B$$

$$\mathbf{A}_B{}^k = \left[\mathbf{R}^{-1} s_{BA}{}^k \mathbf{R} \right] \mathbf{D}_A + \left[\mathbf{R}^{-1} s_{BB}{}^k \mathbf{R} \right] \mathbf{D}_B$$

or

$$\begin{bmatrix} \mathbf{A}_A{}^k \\ \hline \mathbf{A}_B{}^k \end{bmatrix} = \begin{bmatrix} \mathbf{R}^{-1} & | & 0 \\ \hline 0 & | & \mathbf{R}^{-1} \end{bmatrix} \begin{bmatrix} s_{AA}{}^k & | & s_{AB}{}^k \\ \hline s_{BA}{}^k & | & s_{BB}{}^k \end{bmatrix} \begin{bmatrix} \mathbf{R} & | & 0 \\ \hline 0 & | & \mathbf{R} \end{bmatrix} \begin{bmatrix} \mathbf{D}_A \\ \hline \mathbf{D}_B \end{bmatrix}$$

This equation also may be expressed in the form

$$\mathbf{A}_s{}^k = \mathbf{S}_s{}^k \mathbf{D}_s$$

in which $\mathbf{A}_s^k$ is the vector of end action components and $\mathbf{D}_s$ is the vector of end displacement components referred to the structure axes. The term $\mathbf{S}_s$ is the transformed element stiffness matrix and is expressed as

$$\mathbf{S}_s{}^k = \begin{bmatrix} \mathbf{R}^{-1} & 0 \\ 0 & \mathbf{R}^{-1} \end{bmatrix} \begin{bmatrix} s_{AA}{}^k & s_{AB}{}^k \\ s_{BA}{}^k & s_{BB}{}^k \end{bmatrix} \begin{bmatrix} \mathbf{R} & 0 \\ 0 & \mathbf{R} \end{bmatrix} = \begin{bmatrix} \mathbf{S}_{AA}{}^k & \mathbf{S}_{AB}{}^k \\ \mathbf{S}_{BA}{}^k & \mathbf{S}_{BB}{}^k \end{bmatrix}$$

It represents the end actions on a typical element in terms of the end displacements when both the action and displacement coordinates refer to the structure-oriented axes.

Taking the stiffness matrix for a pin-jointed truss element (as developed in Art. 12.7) and premultiplying by the inverse of the rotation matrix and postmultiplying by the rotation matrix i.e.,

$$
\begin{bmatrix}
\cos\alpha & -\sin\alpha & 0 & 0 \\
\sin\alpha & \cos\alpha & 0 & 0 \\
0 & 0 & \cos\alpha & -\sin\alpha \\
0 & 0 & \sin\alpha & \cos\alpha
\end{bmatrix}
\begin{bmatrix}
\dfrac{EA}{L} & 0 & -\dfrac{EA}{L} & 0 \\
0 & 0 & 0 & 0 \\
-\dfrac{EA}{L} & 0 & \dfrac{EA}{L} & 0 \\
0 & 0 & 0 & 0
\end{bmatrix}
\begin{bmatrix}
\cos\alpha & \sin\alpha & 0 & 0 \\
-\sin\alpha & \cos\alpha & 0 & 0 \\
0 & 0 & \cos\alpha & \sin\alpha \\
0 & 0 & -\sin\alpha & \cos\alpha
\end{bmatrix}
$$

will produce the transformed element stiffness matrix

$$
\mathbf{S}_s^k
\begin{bmatrix}
S_1 & S_2 & -S_1 & -S_2 \\
 & S_3 & -S_2 & -S_3 \\
 & & S_1 & S_2 \\
(\text{symmetric}) & & & S_3
\end{bmatrix}
$$

in which

$$
S_1 = \left(\frac{AE}{L}\right)\cos^2\alpha
$$

$$
S_2 = \left(\frac{AE}{L}\right)\sin\alpha\cos\alpha
$$

$$
S_3 = \left(\frac{AE}{L}\right)\sin^2\alpha
$$

In a like manner, the stiffness matrix for a plane frame element (as developed in Art. 12.8) can be premultiplied by the matrix $\mathbf{R}^{-1}$ and postmultiplied by the matrix $\mathbf{R}$ to produce the transformed stiffness matrix for a plane frame prismatic element. This matrix is

$$
\mathbf{S}_s^k =
\begin{bmatrix}
S_1 & S_2 & -S_3 & -S_1 & -S_2 & -S_3 \\
 & S_4 & S_5 & -S_2 & -S_4 & S_5 \\
 & & S_6 & S_3 & -S_5 & S_7 \\
 & & & S_1 & S_2 & S_3 \\
 & & & & S_4 & -S_5 \\
(\text{symmetric}) & & & & & S_6
\end{bmatrix}
$$

in which

$$
S_1 = \kappa_a \cos^2\alpha + \kappa_1 \sin^2\alpha
$$

$$
S_2 = (\kappa_a - \kappa_1)(\sin\alpha)(\cos\alpha)
$$

$$
S_3 = \kappa_2 \sin\alpha
$$

$$
S_4 = \kappa_a \sin^2\alpha + \kappa_1 \cos^2\alpha
$$

$$
S_5 = \kappa_2 \cos\alpha
$$

$$
S_6 = \kappa_4
$$

$$
S_7 = \kappa_5
$$

and

$$
\kappa_a = \frac{EA}{L}
$$

$$
\kappa_1 = \frac{12\,EI}{L^3}
$$

$$
\kappa_2 = \frac{6\,EI}{L^2}
$$

$$
\kappa_4 = \frac{4\,EI}{L}
$$

$$
\kappa_5 = \frac{2\,EI}{L}
$$

PROBLEMS

1. Form the element stiffness matrix and the transformed element stiffness matrix for a pin-jointed plane-truss element having a length of 6 ft and composed of two $3\times3\times\frac{1}{4}$-in. angles of A36 Steel. The element axes are rotated through an angle of $-36°$ with respect to the

structure axes. Sketch the two sets of axes and show the position of the element.

2. Form the element stiffness matrix and the transformed element stiffness matrix for a plane frame element having a length of 8 ft 4 in. The element is a **W** 10×39 of A36 Steel, oriented so that bending occurs about the major axis of the element. The element axes are rotated through an angle of 110° with respect to the structure axes. Sketch the two sets of axes, and show the position of the element.

3. Calculate the end actions on the plane-frame element in Problem 2 if the end displacements are

$$\xi_A = 0 \qquad\qquad \xi_B = -0.0032$$

$$\delta_A = 0 \qquad\qquad \delta_B = -0.2501$$

$$\theta_A = -0.0035 \qquad \theta_B = -0.0005$$

(Answers given in Appendix G.)

4. Convert the end actions of Problem 3 to structure-oriented axes. (Answers given in Appendix G.)

12.11
Analytical Description of a Structural System

For the purpose of analysis, a framed structure is divided into components that are referred to as "elements" and "joints." A pin-jointed plane-truss element is defined as any tension or compression member of a truss, and a plane-frame element is defined as any beam, column, or secondary member of a frame. Unless truss elements are joined in such a way that they form stable triangles, the structural model is unstable and a linear analysis is not possible; therefore, truss members may not be subdivided for the purpose of analysis (i.e., truss elements must correspond to the actual members of a truss). Frame elements, however, are not limited by this requirement of stability; therefore, frame members may be subdivided into as many elements as may be convenient for analy-

sis. A *joint* is defined as the junction of two or more elements.

Since all information that is processed must be digitized when the displacement method is employed, a system must be devised in which numbers are assigned specific meanings associated with the structure. These numbers frequently are interrelated and must be unambiguous. Elements are numbered serially $1, 2, 3, \ldots, M$, where M is equal to the total number of elements, and joints are numbered serially $1, 2, 3, \ldots, N$, where N is equal to the total number of joints. The manner in which three different framed structures are divided into analytical components is shown in Fig. 12.27. Joint numbers are distinguished by small circles enclosing the numbers, and element numbers are distinguished by small squares enclosing the numbers. It should be noted that joints have been arbitrarily placed along the horizontal member of the portal frame shown in Fig. 12.27f. This option could be used, for example, to account for concentrated loads as joint loads or to obtain intermediate displacements or stress resultants.

The geometry of a structure is described in terms of the coordinates of the joints relative to a set of coordinate axes which are referred to as *structure-oriented axes* or *structure axes* (Art. 12.9). The structure axes are denoted by the symbols x_s, y_s, and z_s. This set of axes is associated with the complete structure, and its position and orientation are fixed with respect to the structure. The location of the origin and the orientation of the axes are arbitrary and may be conveniently chosen so that all joint coordinates are positive and easily determined. Normally, a plane structure is assumed to lie in the x-y plane. Examples of structure axes are shown in Fig. 12.27.

An element is uniquely defined by specifying its number and the numbers of the

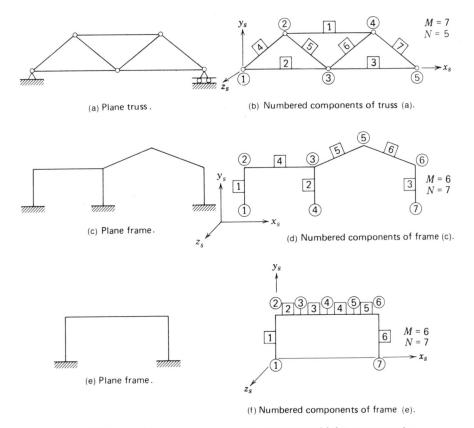

Figure 12.27/Division of framed structures into element and joint components.

joints to which it is connected. The structural characteristics of the elements are described in terms of moment of inertia (I), cross-sectional area (A), and modulus of elasticity (E). It also is necessary to adopt a system that uniquely identifies all lines of action at all joints and to link these to the joints where they occur. Two *global*, or *structure-oriented*, translational lines of action, selected to correspond with the x_s and y_s directions, are required at each joint to identify the forces $\{T, V_i\}$ and linear displacements $\{\Xi_i \Delta_i\}$ of the ends of the elements of the structure. In addition, if the structure is a plane frame, then a rotational line of action, selected to correspond to positive rotation about the z_s axis, is required at each joint to identify the moments $\{M_i\}$ and the angular dis-

placements $\{\Theta_i\}$ of the ends of the elements.[5]

To identify the global lines of action at every joint, it is necessary to number them in some systematic manner so that each line of action at every joint has a unique number. Several schemes, each having one or more desirable characteristics, have been proposed and may be used with success in conjunction with the displacement method of analysis. Of these, the scheme that is deemed most suitable for automatic computation is the one that links the identification number directly to the joint num-

[5]The angular displacements of the ends of truss elements usually are not determined; therefore, such determinations are not included in the procedure presented here.

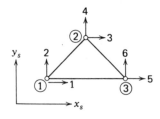

(a) Plane truss.

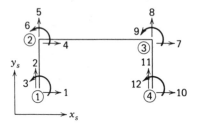

(b) Plane frame.

Figure 12.28/Examples of global lines of action.

ber. In this scheme, the translational lines of action at any joint j in the x_s and y_s directions are identified by the numbers

$$\beta j - (\beta - 1) \qquad \text{for } x_s$$

and

$$\beta j - (\beta - 2) \qquad \text{for } y_s$$

in which $\beta = 2$ for trusses and $\beta = 3$ for frames. The rotational line of action at any joint j of a plane frame is identified by the number $3j$. Examples of global lines of action are shown in Fig. 12.28. By using the numbering scheme described here, action and displacement components at any joint j can be identified as follows:

In the equations shown above (as before), the superscript k denotes the kth element number, and the subscript i denotes either the A end or the B end of the element, which may be conveniently identified by substituting the corresponding joint number for A or B. The expressions in the brackets [ ] designate the global line of action along which the action or displacement component occurs. Note that all elements meeting at a joint always have the same displacement components but, in general, have different action components.

12.12
Formulation of Structure Equilibrium Equations

Each joint of a structure must be in equilibrium under the action of applied loads ($W_{[\ ]}$) and the end actions ($A_{[\ ]}^k$) of the attached elements (Fig. 12.29). In general, the end actions include effects of loads placed directly on elements, but for purposes of this discussion it is assumed that end actions are the result of end displacements only. Loads directly on elements are brought into the formulation in Art. 12.15.

Referring to the free-body diagrams of the typical truss and frame joints shown in Fig. 12.29, the equilibrium requirement is expressed for the truss joint by the equation

$$\begin{bmatrix} W_{[2j-1]} \\ W_{[2j]} \end{bmatrix} - \Sigma \begin{bmatrix} A_{[2j-1]}^k \\ A_{[2j]}^k \end{bmatrix} = 0$$

and for the plane-frame joint by the equa-

Direction	Action Components		Displacement Components	
	Truss	Frame	Truss	Frame
x_s	$T_i = A_{[2j-1]}^k$	$T_i = A_{[3j-2]}^k$	$\Xi_i = D_{[2j-1]}$	$\Xi_i = D_{[3j-2]}$
y_s	$V_i = A_{[2j]}^k$	$V_i = A_{[3j-1]}^k$	$\Delta_i = D_{[2j]}$	$\Delta_i = D_{[3j-1]}$
Rotational	None	$M_i = A_{[3j]}^k$	None	$\Theta_i = D_{[3j]}$

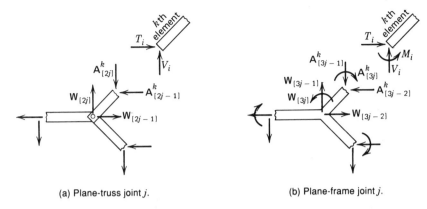

(a) Plane-truss joint j. (b) Plane-frame joint j.

Figure 12.29/Free-body diagrams of element end actions and joint loads.

tion

$$\begin{bmatrix} W_{[3j-2]} \\ W_{[3j-1]} \\ W_{[3j]} \end{bmatrix} - \Sigma \begin{bmatrix} A^k_{[3j-2]} \\ A^k_{[2j-1]} \\ A^k_{[3j]} \end{bmatrix} = 0$$

in which, for both equations, the indicated summation is taken over all the elements connected to the joint. Each of the above equations may be expressed in the alternate form

$$W_j - \Sigma A_j^{\ k} = 0$$

where

W_j = vector of directly applied joint loads.
A_j^k = vector of the kth-element end actions.

When the A end of the kth element of a structure occurs at joint j, and the B end occurs at joint i, the equation which relates end actions to end displacements for the kth element becomes

$$\begin{bmatrix} A_j^{\ k} \\ \hline A_i^{\ k} \end{bmatrix} = \begin{bmatrix} S_{jj}^{\ k} & S_{ji}^{\ k} \\ \hline S_{ij}^{\ k} & S_{ii}^{\ k} \end{bmatrix} \begin{bmatrix} D_j \\ \hline D_i \end{bmatrix}$$

This equation, when expanded (multiplied out), produces two equations for element

end actions:

$$A_j^{\ k} = S_{jj}^{\ k} D_j + S_{ji}^{\ k} D_i$$

and

$$A_i^{\ k} = S_{ij}^{\ k} D_j + S_{ii}^{\ k} D_i$$

The first of these pertains to joint j and may be substituted in the equilibrium equation. Thus

$$W_j - \Sigma \left(S_{jj}^{\ k} D_j + S_{ji}^{\ k} D_i \right) = 0$$

The second equation pertains to joint i and, of course, would be substituted into the equilibrium equation for that joint.

The formulation of these joint equilibrium equations for all joints of a structure produces a set of linear equations that expresses the relationship between joint loads and joint displacements. These equations may be expressed in the matrix form

$$W^c - S^c D^c = 0$$

or

$$S^c D^c = W^c$$

where S^c is the *complete* structure stiffness matrix, D^c is the column vector of all joint displacement components, and W^c is the column vector of all joint load components.

To illustrate the above procedure, the complete stiffness matrix for the seven-member truss shown in Fig. 12.30 will be formulated in a symbolic fashion:

1. Number the joints, enclose them in circles, and place them on the figure.

2. Number the members, enclose them in squares, and place them on the figure.

3. Calculate the identification number for lines of action at each joint as described in Art. 12.11.

(Steps 1, 2, and 3 are shown in Fig. 12.30.)

4. Prepare a summary table identifying the element number and its A end and B end:

Element	Ends	
	A	B
1	1	3
2	3	5
3	2	4
4	1	2
5	4	5
6	2	3
7	3	4

5. Write the end-action stiffness equation for each element of the truss:

Element 1: $\mathbf{A}_1^1 = \mathbf{S}_{11}{}^1\mathbf{D}_1 + \mathbf{S}_{13}{}^1\mathbf{D}_3$

$\mathbf{A}_3^1 = \mathbf{S}_{31}{}^1\mathbf{D}_1 + \mathbf{S}_{33}{}^1\mathbf{D}_3$

Element 2: $\mathbf{A}_3^2 = \mathbf{S}_{33}{}^2\mathbf{D}_3 + \mathbf{S}_{35}{}^2\mathbf{D}_5$

$\mathbf{A}_5^2 = \mathbf{S}_{53}{}^2\mathbf{D}_3 + \mathbf{S}_{55}{}^2\mathbf{D}_5$

Element 3: $\mathbf{A}_2^3 = \mathbf{S}_{22}{}^3\mathbf{D}_2 + \mathbf{S}_{24}{}^3\mathbf{D}_4$

$\mathbf{A}_4^3 = \mathbf{S}_{42}{}^3\mathbf{D}_2 + \mathbf{S}_{44}{}^3\mathbf{D}_4$

Element 4: $\mathbf{A}_1^4 = \mathbf{S}_{11}{}^4\mathbf{D}_1 + \mathbf{S}_{12}{}^4\mathbf{D}_2$

$\mathbf{A}_2^4 = \mathbf{S}_{21}{}^4\mathbf{D}_1 + \mathbf{S}_{22}{}^4\mathbf{D}_2$

Element 5: $\mathbf{A}_4^5 = \mathbf{S}_{44}{}^5\mathbf{D}_4 + \mathbf{S}_{45}{}^5\mathbf{D}_5$

$\mathbf{A}_5^5 = \mathbf{S}_{54}{}^5\mathbf{D}_4 + \mathbf{S}_{55}{}^5\mathbf{D}_5$

Element 6: $\mathbf{A}_2^6 = \mathbf{S}_{22}{}^6\mathbf{D}_2 + \mathbf{S}_{23}{}^6\mathbf{D}_3$

$\mathbf{A}_3^6 = \mathbf{S}_{32}{}^6\mathbf{D}_2 + \mathbf{S}_{33}{}^6\mathbf{D}_3$

Element 7: $\mathbf{A}_3^7 = \mathbf{S}_{33}{}^7\mathbf{D}_3 + \mathbf{S}_{34}{}^7\mathbf{D}_4$

$\mathbf{A}_4^7 = \mathbf{S}_{43}{}^7\mathbf{D}_3 + \mathbf{S}_{44}{}^7\mathbf{D}_4$

6. Substitute the above equations into the joint equilibrium equations, i.e.,

Joint 1: $\mathbf{W}_1 - \left[\left(\mathbf{S}_{11}{}^1\mathbf{D}_1 + \mathbf{S}_{13}{}^1\mathbf{D}_3 \right) \right.$
$\left. + \left(\mathbf{S}_{11}{}^4\mathbf{D}_1 + \mathbf{S}_{12}{}^4\mathbf{D}_2 \right) \right] = 0$

Joint 2: $\mathbf{W}_2 - \left[\left(\mathbf{S}_{22}{}^3\mathbf{D}_2 + \mathbf{S}_{24}{}^3\mathbf{D}_4 \right) \right.$
$\left. + \left(\mathbf{S}_{21}{}^4\mathbf{D}_1 + \mathbf{S}_{22}{}^4\mathbf{D}_2 \right) + \left(\mathbf{S}_{22}{}^6\mathbf{D}_2 + \mathbf{S}_{23}{}^6\mathbf{D}_3 \right) \right] = 0$

Joint 3: $\mathbf{W}_3 - \left[\left(\mathbf{S}_{31}{}^1\mathbf{D}_1 + \mathbf{S}_{33}{}^1\mathbf{D}_3 \right) \right.$
$+ \left(\mathbf{S}_{33}{}^2\mathbf{D}_3 + \mathbf{S}_{35}{}^2\mathbf{D}_5 \right) + \left(\mathbf{S}_{32}{}^6\mathbf{D}_2 + \mathbf{S}_{33}{}^6\mathbf{D}_3 \right)$
$\left. + \left(\mathbf{S}_{33}{}^7\mathbf{D}_3 + \mathbf{S}_{34}{}^7\mathbf{D}_4 \right) \right] = 0$

Joint 4: $\mathbf{W}_4 - \left[\left(\mathbf{S}_{42}{}^3\mathbf{D}_2 + \mathbf{S}_{44}{}^3\mathbf{D}_4 \right) \right.$
$\left. + \left(\mathbf{S}_{44}{}^5\mathbf{D}_4 + \mathbf{S}_{45}{}^5\mathbf{D}_5 \right) + \left(\mathbf{S}_{43}{}^7\mathbf{D}_3 + \mathbf{S}_{44}{}^7\mathbf{D}_4 \right) \right] = 0$

Joint 5: $\mathbf{W}_5 - \left[\left(\mathbf{S}_{53}{}^2\mathbf{D}_3 + \mathbf{S}_{55}{}^2\mathbf{D}_5 \right) \right.$
$\left. + \left(\mathbf{S}_{54}{}^5\mathbf{D}_4 + \mathbf{S}_{55}{}^5\mathbf{D}_5 \right) \right] = 0$

7. Finally, assemble the joint equilibrium equations in matrix form:

	Joint 1	Joint 2	Joint 3	Joint 4	Joint 5			
Joint 1	$\mathbf{S}_{11}{}^1 + \mathbf{S}_{11}{}^4$	$\mathbf{S}_{12}{}^4$	$\mathbf{S}_{13}{}^1$			$\mathbf{D}_1$		$\mathbf{W}_1$
Joint 2	$\mathbf{S}_{21}{}^4$	$\mathbf{S}_{22}{}^3 + \mathbf{S}_{22}{}^4 + \mathbf{S}_{22}{}^6$	$\mathbf{S}_{23}{}^6$	$\mathbf{S}_{24}{}^3$		$\mathbf{D}_2$		$\mathbf{W}_2$
Joint 3	$\mathbf{S}_{31}{}^1$	$\mathbf{S}_{32}{}^6$	$\mathbf{S}_{33}{}^1 + \mathbf{S}_{33}{}^2 + \mathbf{S}_{33}{}^6 + \mathbf{S}_{33}{}^7$	$\mathbf{S}_{34}{}^7$	$\mathbf{S}_{35}{}^2$	$\mathbf{D}_3$	$=$	$\mathbf{W}_3$
Joint 4		$\mathbf{S}_{42}{}^3$	$\mathbf{S}_{43}{}^7$	$\mathbf{S}_{44}{}^3 + \mathbf{S}_{44}{}^5 + \mathbf{S}_{44}{}^7$	$\mathbf{S}_{45}{}^5$	$\mathbf{D}_4$		$\mathbf{W}_4$
Joint 5			$\mathbf{S}_{53}{}^2$	$\mathbf{S}_{54}{}^5$	$\mathbf{S}_{55}{}^2 + \mathbf{S}_{55}{}^5$	$\mathbf{D}_5$		$\mathbf{W}_5$

Complete structure stiffness matrix
Vector of joint displacement components
Vector of joint load components

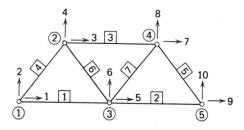

Figure 12.30/Symbolic truss example.

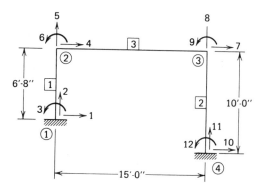

Figure 12.31/Numerical example.

It must be remembered that each $\mathbf{S}_{ij}^k$ represents a 2×2 submatrix and each $\mathbf{D}_i$ and $\mathbf{W}_i$ represents a vector having two elements; therefore, the equation shown above represents a set of 10 equations. The complete structure stiffness matrix, then, is a 10×10 matrix. Before this matrix can be used, the elements of each submatrix must be evaluated and the submatrices combined as indicated in the equation. It also should be noted that Maxwell's law applies to the complete structure stiffness matrix; therefore, it is symmetric.

Example

Formulate the complete stiffness matrix for the portal frame shown in Fig. 12.31. Use the following member properties for all members: $A = 8$ in.^{2}, $I = 200$ in.^{4}, and $E = 29,000$ ksi.

Solution

(1) Number (serially) the joints, elements, and end actions, and show them on the portal frame (Fig. 12.31).

(2) Make the table identifying elements and the A end and the B end:

| | Ends | |
Element	A	B
1	1	2
2	3	4
3	2	3

(3) Sketch each element, showing the element axes relative to the structure axes according to the numbering system established in steps (1) and (2) (Fig. 12.32).

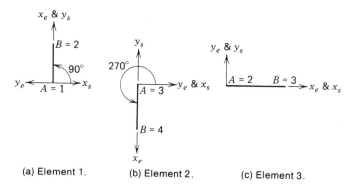

(a) Element 1.   (b) Element 2.   (c) Element 3.

Figure 12.32/Element orientations.

(4) Calculate the element stiffness coefficients relative to the element axis for each element:

Element 1:

$$\kappa_\alpha = \frac{AE}{L} = \frac{8(29)}{80} = 2.900 \text{ kips per in.}$$

$$\kappa_1 = \frac{12\,EI}{L^3} = \frac{12(29)200}{80^3} = 0.136 \text{ kips per in.}$$

$$\kappa_2 = \frac{6\,EI}{L^2} = \frac{6(29)200}{80^2} = 5.438 \text{ kips per in.}$$

$$\kappa_4 = \frac{4\,EI}{L} = \frac{4(29)200}{80} = 290.000 \text{ kips per in.}$$

$$\kappa_5 = \frac{2\,EI}{L} = \frac{2(29)200}{80} = 145.000 \text{ kips per in.}$$

Element 2 (calculated in a manner similar to that for Element 1):

$$\kappa_\alpha = 1.933, \qquad \kappa_1 = 0.040, \qquad \kappa_2 = 2.417$$

$$\kappa_4 = 193.333, \qquad \kappa_5 = 96.667$$

Element 3:

$$\kappa_\alpha = 1.289, \qquad \kappa_1 = 0.012, \qquad \kappa_2 = 1.074$$

$$\kappa_4 = 128.889, \qquad \kappa_5 = 64.445$$

(4) Assemble the coefficients (κ values) into the element stiffness matrix (relative to the element axes). Prepare the rotation matrix for the element. Premultiply the stiffness matrix by the inverse of the rotation matrix, and postmultiply that product by the rotation matrix. The end result of this multiplication is the transformed element stiffness matrix. In lieu of this matrix multiplication, the new coefficients can be calculated directly from the formulas developed in Art. 12.10. Subdivide this transformed element stiffness matrix (6×6) into four 3×3 submatrices, and identify these submatrices with respect to their end actions (the coefficient subscripts locate element positions in the complete structure stiffness matrix). Repeat this step for each of the three frame elements:

Frame Element 1:

Element stiffness matrix/1,000,000

$$\mathbf{s}_e^1 = \left[\begin{array}{ccc:ccc}
2.900 & 0.000 & 0.000 & -2.900 & 0.000 & 0.000 \\
0.000 & 0.136 & 5.438 & 0.000 & -0.136 & 5.438 \\
0.000 & 5.438 & 290.000 & 0.000 & -5.438 & 145.000 \\ \hdashline
-2.900 & 0.000 & 0.000 & 2.900 & 0.000 & 0.000 \\
0.000 & -0.136 & -5.438 & 0.000 & 0.136 & -5.438 \\
0.000 & 5.438 & 145.000 & 0.000 & -5.438 & 290.000
\end{array}\right]$$

Rotation matrix

$$\mathbf{R}_1 = \left[\begin{array}{ccc}
0.000 & 1.000 & 0.000 \\
-1.000 & 0.000 & 0.000 \\
0.000 & 0.000 & 1.000
\end{array}\right]$$

Transformed element stiffness matrix/1,000,000

$$\mathbf{S}_s^1 = \left[\begin{array}{ccc:ccc}
0.136 & 0.000 & -5.438 & -0.136 & 0.000 & -5.438 \\
0.000 & 2.900 & 0.000 & 0.000 & -2.900 & 0.000 \\
-5.438 & 0.000 & 290.000 & 5.438 & 0.000 & 145.000 \\ \hdashline
-0.136 & 0.000 & 5.438 & 0.136 & 0.000 & 5.438 \\
0.000 & -2.900 & 0.000 & 0.000 & 2.900 & 0.000 \\
-5.438 & 0.000 & 145.000 & 5.438 & 0.000 & 290.000
\end{array}\right] = \left[\begin{array}{c:c}
\mathbf{S}_{11}^1 & \mathbf{S}_{12}^1 \\ \hdashline
\mathbf{S}_{21}^1 & \mathbf{S}_{22}^1
\end{array}\right]$$

Frame Element 2:

Element stiffness matrix/1,000,000

$$
\mathbf{s}_e^2 =
\left[
\begin{array}{ccc|ccc}
1.933 & 0.000 & 0.000 & -1.933 & 0.000 & 0.000 \\
0.000 & 0.041 & 2.417 & 0.000 & -0.041 & 2.417 \\
0.000 & 2.417 & 193.333 & 0.000 & -2.417 & 96.667 \\
\hline
-1.933 & 0.000 & 0.000 & 1.933 & 0.000 & 0.000 \\
0.000 & -0.041 & -2.417 & 0.000 & 0.041 & -2.417 \\
0.000 & 2.417 & 96.667 & 0.000 & -2.417 & 193.333
\end{array}
\right]
$$

Rotation matrix

$$
\mathbf{R}_2 =
\begin{bmatrix}
0.000 & -1.000 & 0.000 \\
1.000 & 0.000 & 0.000 \\
0.000 & 0.000 & 1.000
\end{bmatrix}
$$

Transformed element stiffness matrix/1,000,000

$$
\mathbf{S}_s^2 =
\left[
\begin{array}{ccc|ccc}
0.041 & 0.000 & 2.417 & -0.041 & 0.000 & 2.417 \\
0.000 & 1.933 & 0.000 & 0.000 & -1.933 & 0.000 \\
2.417 & 0.000 & 193.333 & -2.417 & 0.000 & 96.667 \\
\hline
-0.041 & 0.000 & -2.417 & 0.041 & 0.000 & -2.417 \\
0.000 & -1.933 & 0.000 & 0.000 & 1.933 & 0.000 \\
2.417 & 0.000 & 96.667 & -2.417 & 0.000 & 193.333
\end{array}
\right]
=
\left[
\begin{array}{c|c}
\mathbf{S}_{33}^2 & \mathbf{S}_{34}^2 \\
\hline
\mathbf{S}_{43}^2 & \mathbf{S}_{44}^2
\end{array}
\right]
$$

Frame Element 3:

Element stiffness matrix/1,000,000

$$
\mathbf{s}_e^3 =
\left[
\begin{array}{ccc|ccc}
1.289 & 0.000 & 0.000 & -1.289 & 0.000 & 0.000 \\
0.000 & 0.012 & 1.074 & 0.000 & -0.012 & 1.074 \\
0.000 & 1.074 & 128.889 & 0.000 & -1.074 & 64.445 \\
\hline
-1.289 & 0.000 & 0.000 & 1.289 & 0.000 & 0.000 \\
0.000 & -0.012 & -1.074 & 0.000 & 0.012 & -1.074 \\
0.000 & 1.074 & 64.445 & 0.000 & -1.074 & 128.889
\end{array}
\right]
$$

Rotation matrix

$$
\mathbf{R}_3 =
\begin{bmatrix}
1.000 & 0.000 & 0.000 \\
0.000 & 1.000 & 0.000 \\
0.000 & 0.000 & 1.000
\end{bmatrix}
$$

Transformed element stiffness matrix/1,000,000

$$
\mathbf{S}_s^3 =
\left[
\begin{array}{ccc|ccc}
1.289 & 0.000 & 0.000 & -1.289 & 0.000 & 0.000 \\
0.000 & 0.012 & 1.074 & 0.000 & -0.012 & 1.074 \\
0.000 & 1.074 & 128.889 & 0.000 & -1.074 & 64.445 \\
\hline
-1.289 & 0.000 & 0.000 & 1.289 & 0.000 & 0.000 \\
0.000 & -0.012 & -1.074 & 0.000 & 0.012 & -1.074 \\
0.000 & 1.074 & 64.445 & 0.000 & -1.074 & 128.889
\end{array}
\right]
=
\left[
\begin{array}{c|c}
\mathbf{S}_{22}^3 & \mathbf{S}_{23}^3 \\
\hline
\mathbf{S}_{32}^3 & \mathbf{S}_{33}^3
\end{array}
\right]
$$

(5) Form the complete structure stiffness matrix. It may be helpful to first form this matrix symbolically, superimposing the 3×3 submatrices of the element matrices,

that is,

$$
\mathbf{S}^c = \begin{array}{c} \text{Joint} \\ \text{No. 1} \quad \text{No. 2} \quad \text{No. 3} \quad \text{No. 4} \end{array}
$$

$$
\mathbf{S}^c = \left[\begin{array}{c|c|c|c}
\mathbf{S}_{11}^1 & \mathbf{S}_{12}^1 & & \\
\hline
\mathbf{S}_{21}^1 & \mathbf{S}_{22}^1 + \mathbf{S}_{22}^3 & \mathbf{S}_{23}^3 & \\
\hline
 & \mathbf{S}_{32}^3 & \mathbf{S}_{33}^2 + \mathbf{S}_{33}^3 & \mathbf{S}_{34}^2 \\
\hline
 & & \mathbf{S}_{43}^2 & \mathbf{S}_{44}^2
\end{array}\right] \begin{array}{l} \text{Joint No. 1} \\ 2 \\ 3 \\ 4 \end{array}
$$

Complete structure stiffness matrix

0.136	0.000	−5.438	−0.136	0.000	−5.438						
0.000	2.900	0.000	0.000	−2.900	0.000						
−5.438	0.000	290.000	5.438	0.000	145.000						
−0.136	0.000	5.438	1.425	0.000	5.438	−1.289	0.000	0.000			
0.000	−2.900	0.000	0.000	2.912	1.074	0.000	−0.012	1.074			
−5.438	0.000	145.000	5.438	1.074	418.889	0.000	−1.074	64.445			
			−1.289	0.000	0.000	1.330	0.000	2.417	−0.041	0.000	2.417
			0.000	−0.012	−1.074	0.000	1.945	−1.074	0.000	−1.933	0.000
			0.000	1.074	64.445	2.417	−1.074	322.222	−2.417	0.000	96.667
						−0.041	0.000	−2.417	0.041	0.000	−2.417
						0.000	−1.933	0.000	0.000	1.933	0.000
						2.417	0.000	96.667	−2.417	0.000	193.333

12.13 Constraint Equations

The final equation developed in the previous article ($\mathbf{S}^c\mathbf{D}^c = \mathbf{W}^c$), relating the joint displacements and joint loads of an entire structure, does not have a meaningful solution as it stands (it implies that the structure is free in space). Obviously, the structure must be attached to a suitable foundation. This requirement also is evident when attempting a mathematical solution. To obtain a solution for the displacements, both sides of the equation must be multiplied by the inverse of the complete structure stiffness matrix ($[\mathbf{S}^c]^{-1}$), i.e.,

$$\mathbf{D}^c = [\mathbf{S}^c]^{-1}\mathbf{W}^c$$

Further study shows that a unique solution to this equation is not possible because the complete structure stiffness matrix ($\mathbf{S}^c$) is always singular (Art. C11); therefore, its inverse ($[\mathbf{S}^c]^{-1}$) does not exist.

A solution becomes possible, however, when the equation is modified by accounting for a sufficient number of constraints which prevent the structure from moving as a rigid body when external loads are applied. For a planar structure, a rigid-body movement is prevented if the structure is constrained along at least three lines of action in such a way that rigid-body translations in the direction of the x_s and y_s axes and rigid-body rotation about the z_s axis are prevented. The existence of these constraints is implied by the well-known equations that express the conditions for static equilibrium (Art. 2.4):

$$\sum F_x = 0, \qquad \sum F_y = 0, \qquad \sum M_z = 0$$

In general, the number of constraints that

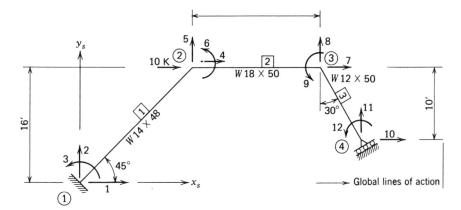

(a) Frame with identifying numbers.

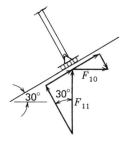

(b) Reaction constraints at joint 4.

Element	Ends A	B	A in^2	I_z in^4	E ksi/1000
1	1	2	14.1	485.0	29.0
2	2	3	14.7	800.0	29.0
3	3	4	14.7	394.0	29.0

(c) End numbers and element physical properties.

Figure 12.33/Portal frame with inclined columns.

may be imposed on a planar structure is not limited to three. Conditions of constraint may be specified along any of the lines of action, or they may be indirectly imposed by specified relationships among forces and specified relationships among displacements. Equations that express the various conditions of constraint are known as *constraint equations*. All such equations must be accounted for when solving the equation $\mathbf{S}^c \mathbf{D}^c = \mathbf{W}^c$.

If the displacements of a structure are defined in terms of m displacement components ($m = 2j$ for trusses and $m = 3j$ for frames), of which some or all are related by r constraint equations, then r of the components may be expressed in terms

of the remaining $m - r$ components. These $m - r$ components are referred to as *independent displacement components*.

To illustrate the concepts introduced here, the constraint equations for the plane frame shown in Fig. 12.33 are formulated and assembled into matrix form.

The support condition depicted at joint ① indicates that no movement is expected along lines of action 1, 2, and 3; however, to provide a more comprehensive example, it is assumed that because of poor foundation conditions, a settlement of 1 in. occurs along line of action 2 and a clockwise rotation of 0.004 radians occurs along line of action 3. Therefore, the conditions at joint ① may be expressed by the equations

$$D_1 = 0$$
$$D_2 = -1 \text{ in.}$$
$$D_3 = -0.004 \text{ radians}$$

Joint ④ is depicted as a *roller* that is constrained to translate along an inclined plane. Rotation of the joint about the z axis is unrestricted, and the moment at joint 4 is equal to 0. Also, because joint translation must occur along the plane, displacement components D_{10} and D_{11} must be related in such a way that their resultant always lies along the plane. This requirement is satisfied if the ratio of the two components is set equal to the tangent of the angle between the inclined plane and the x_s axis, that is,

$$\frac{D_{11}}{D_{10}} = \tan 30°$$

or

$$D_{11} = 0.577 D_{10}$$

The roller support condition at joint ④ also requires a zero reaction component parallel to the inclined plane. In other words, the inclined component from F_{11} must be canceled by the inclined component from F_{10} (Fig. 12.33b), and the shear at the B end of element ④ is zero. This requirement may be expressed in equation form as

$$F_{11} \sin 30° + F_{10} \cos 30° = v_B = 0$$

This end-action condition for element 3 can be expressed in terms of the element stiffness coefficients and displacement components. The element stiffness matrix that relates end actions to end displacements was developed in Art. 12.8. Removing from that matrix those coefficients pertaining to end shear (v_B),

$$v_B = s_{52}\delta_A + s_{53}\theta_A + s_{55}\delta_B + s_{56}\theta_B = 0$$

and substituting in the derived stiffness

coefficients,

$$v_B = \left[\frac{-12EI}{L^3}\right]\delta_3 + \left[\frac{-6EI}{L^2}\right]\theta_3$$
$$+ \left[\frac{12EI}{L^3}\right]\delta_4 + \left[\frac{-6EI}{L^2}\right]\theta_4 = 0$$

which simplifies to

$$\delta_3 - \delta_4 + \left[\frac{L}{2}\right]\theta_3 + \left[\frac{L}{2}\right]\theta_4 = 0$$

These relationships were derived with reference to the element axes and must be transformed to refer to the structure axes. The matrix for this transformation was developed in Art. 12.9, and removing from that matrix those terms pertaining to end shear results in the following equation:

$$\delta_i = \Xi_i(-\sin \alpha) + \Delta_i \cos \alpha$$

Substituting into this equation the appropriate values pertaining to joints ③ and ④ of the structure under consideration yields

$$\delta_3 = -D_7 \sin(-60°) + D_8 \cos(-60°)$$
$$\delta_4 = -D_{10} \sin(-60°) + D_{11} \cos(-60°)$$

Also

$$\left[\frac{L}{2}\right]\theta_3 = \left[\frac{120}{2\cos 30°}\right]D_9$$

$$\left[\frac{L}{2}\right]\theta_4 = \left[\frac{120}{2\cos 30°}\right]D_{12}$$

Consequently,

$$D_7 + 0.577D_8 - D_{10} - 0.577D_{11}$$
$$+ 80.000D_9 + 80.000D_{12} = 0$$

Substituting in the previously derived expression for D_{11} ($0.577D_{10}$) and solving for D_{10} in the above equation,

$$D_{10} = 0.750D_7 + 0.433D_8$$
$$+ 60.000D_9 + 60.000D_{12}$$

Also, from the relationship $D_{11} = 0.577 D_{10}$,

$$D_{11} = 0.433 D_7 + 0.250 D_8$$
$$+ 34.641 D_9 + 34.641 D_{12}$$

Thus, a total of five constraint equations, consisting of three equations stemming

from constraints at joint ① and two equations stemming from constraints at joint ④, are formulated for the frame shown in Fig. 12.33. For convenient reference they are summarized below and then are placed in a matrix form:

$$D_1 = 0$$

$$D_2 = -1 \text{ in.}$$

$$D_3 = -0.004 \text{ radians}$$

$$D_{10} = 0.750 D_7 + 0.433 D_8 + 60.000 D_9 + 60.000 D_{12}$$

$$D_{11} = 0.433 D_7 + 0.250 D_8 + 34.641 D_9 + 34.641 D_{12}$$

$$
\begin{bmatrix} D_1 \\ D_2 \\ D_3 \\ D_{10} \\ D_{11} \end{bmatrix}
=
\begin{bmatrix}
0 & 0 & 0 & 0 & 0 & 0 \\
1 & 0 & 0 & 0 & 0 & 0 \\
0 & 1 & 0 & 0 & 0 & 0 \\
0 & 0 & 0.750 & 0.433 & 60.000 & 60.000 \\
0 & 0 & 0.433 & 0.250 & 34.641 & 34.641
\end{bmatrix}
\begin{bmatrix} -1 \\ -0.004 \\ D_7 \\ D_8 \\ D_9 \\ D_{12} \end{bmatrix}
$$

The displacements of the entire frame shown in Fig. 12.33 are defined in terms of 12 displacement components (3 at each joint), some of which are related by the 5 constraint equations just developed; therefore, according to the relationship previously presented, there are 7 independent displacement components. The complete set of 12 displacement components may be expressed in matrix form in terms of these 7 components and the actual displacements at joint ①. Included in these equations are identities such as $D_4 = D_4$, $D_5 = D_5$, etc., for all of the independent components. The matrix that relates these components is referred to as the constraint matrix and is given by

$$
\begin{bmatrix} D_1 \\ D_2 \\ D_3 \\ D_4 \\ D_5 \\ D_6 \\ D_7 \\ D_8 \\ D_9 \\ D_{10} \\ D_{11} \\ D_{12} \end{bmatrix}
=
\begin{bmatrix}
0 & 0 & 0 & 0 & 0 & 0 & 0 & 0 & 0 \\
1 & 0 & 0 & 0 & 0 & 0 & 0 & 0 & 0 \\
0 & 1 & 0 & 0 & 0 & 0 & 0 & 0 & 0 \\
0 & 0 & 1 & 0 & 0 & 0 & 0 & 0 & 0 \\
0 & 0 & 0 & 1 & 0 & 0 & 0 & 0 & 0 \\
0 & 0 & 0 & 0 & 1 & 0 & 0 & 0 & 0 \\
0 & 0 & 0 & 0 & 0 & 1 & 0 & 0 & 0 \\
0 & 0 & 0 & 0 & 0 & 0 & 1 & 0 & 0 \\
0 & 0 & 0 & 0 & 0 & 0 & 0 & 1 & 0 \\
0 & 0 & 0 & 0 & 0 & 0.750 & 0.433 & 60.000 & 60.000 \\
0 & 0 & 0 & 0 & 0 & 0.433 & 0.250 & 34.641 & 34.641 \\
0 & 0 & 0 & 0 & 0 & 0 & 0 & 0 & 1
\end{bmatrix}
\begin{bmatrix} -1.00 \\ -0.004 \\ D_4 \\ D_5 \\ D_6 \\ D_7 \\ D_8 \\ D_9 \\ D_{12} \end{bmatrix}
$$

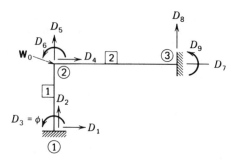

Figure 12.34/Simple knee frame with a load at the knee.

12.14
Solution Process

Once the constraint equations have been formulated and the complete set of displacement components have been expressed in terms of the independent displacement components (as discussed in the previous article), the solution of the equation as developed in Art. 12.12 ($\mathbf{S}^c\mathbf{D}^c = \mathbf{W}^c$) may be continued. To illustrate this process, let it be assumed that the displacements of the simple knee frame shown in Fig. 12.34 are defined in terms of nine displacement components. Then $\mathbf{S}^c$ (the complete structure stiffness matrix), $\mathbf{D}^c$ (the column vector of all joint displacement components), and $\mathbf{W}^c$ (the column vector of all joint load components) may be symbolically represented by the matrices shown below:

$$
\begin{bmatrix}
S_{11} & S_{12} & S_{13} & S_{14} & S_{15} & S_{16} & S_{17} & S_{18} & S_{19} \\
S_{21} & S_{22} & S_{23} & S_{24} & S_{25} & S_{26} & S_{27} & S_{28} & S_{29} \\
S_{31} & S_{32} & S_{33} & S_{34} & S_{35} & S_{36} & S_{37} & S_{38} & S_{39} \\
S_{41} & S_{42} & S_{43} & S_{44} & S_{45} & S_{46} & S_{47} & S_{48} & S_{49} \\
S_{51} & S_{52} & S_{53} & S_{54} & S_{55} & S_{56} & S_{57} & S_{58} & S_{59} \\
S_{61} & S_{62} & S_{63} & S_{64} & S_{65} & S_{66} & S_{67} & S_{68} & S_{69} \\
S_{71} & S_{72} & S_{73} & S_{74} & S_{75} & S_{76} & S_{77} & S_{78} & S_{79} \\
S_{81} & S_{82} & S_{83} & S_{84} & S_{85} & S_{86} & S_{87} & S_{88} & S_{89} \\
S_{91} & S_{92} & S_{93} & S_{94} & S_{95} & S_{96} & S_{97} & S_{98} & S_{99}
\end{bmatrix}
\begin{bmatrix}
D_1 \\ D_2 \\ D_3 \\ D_4 \\ D_5 \\ D_6 \\ D_7 \\ D_8 \\ D_9
\end{bmatrix}
=
\begin{bmatrix}
W_1 \\ W_2 \\ W_3 \\ W_4 \\ W_5 \\ W_6 \\ W_7 \\ W_8 \\ W_9
\end{bmatrix}
$$

Six constraint equations, expressing the support conditions at joints ① and ③, may be formulated for the structure. These equations are

$$D_1 = 0$$
$$D_2 = 0$$
$$D_3 = \phi$$
$$D_7 = 0$$
$$D_8 = 0$$
$$D_9 = 0$$

The complete set of displacement components, expressed in matrix form in terms of the independent components and the imposed displacement at joint ①, is

$$
\begin{bmatrix} D_1 \\ D_2 \\ D_3 \\ D_4 \\ D_5 \\ D_6 \\ D_7 \\ D_8 \\ D_9 \end{bmatrix} =
\begin{bmatrix}
C_{11} & C_{12} & C_{13} & C_{14} \\
C_{21} & C_{22} & C_{23} & C_{24} \\
C_{31} & C_{32} & C_{33} & C_{34} \\
C_{41} & C_{42} & C_{43} & C_{44} \\
C_{51} & C_{52} & C_{53} & C_{54} \\
C_{61} & C_{62} & C_{63} & C_{64} \\
C_{71} & C_{72} & C_{73} & C_{74} \\
C_{81} & C_{82} & C_{83} & C_{84} \\
C_{91} & C_{92} & C_{93} & C_{94}
\end{bmatrix}
\begin{bmatrix} \phi \\ D_4 \\ D_5 \\ D_6 \end{bmatrix}
$$

or

$$ \mathbf{D}^c = \mathbf{C}\mathbf{D}^i $$

in which $\mathbf{D}^i$ is the vector of *imposed displacements* and *independent displacement components*, and $\mathbf{C}$ is the matrix that relates $\mathbf{D}^c$ to $\mathbf{D}^i$. The matrix $\mathbf{C}$ will be referred to as the *constraint matrix*.

The matrix product $\mathbf{C}\mathbf{D}^i$ may be substituted into the equations $\mathbf{S}^c\mathbf{D}^c = \mathbf{W}^c$. Thus, $\mathbf{S}^c\mathbf{D}^c = \mathbf{S}^c\mathbf{C}\mathbf{D}^i = \mathbf{W}^c$:

$$
\begin{bmatrix}
S_{11} & S_{12} & S_{13} & S_{14} & S_{15} & S_{16} & S_{17} & S_{18} & S_{19} \\
S_{21} & S_{22} & S_{23} & S_{24} & S_{25} & S_{26} & S_{27} & S_{28} & S_{29} \\
S_{31} & S_{32} & S_{33} & S_{34} & S_{35} & S_{36} & S_{37} & S_{38} & S_{39} \\
S_{41} & S_{42} & S_{43} & S_{44} & S_{45} & S_{46} & S_{47} & S_{48} & S_{49} \\
S_{51} & S_{52} & S_{53} & S_{54} & S_{55} & S_{56} & S_{57} & S_{58} & S_{59} \\
S_{61} & S_{62} & S_{63} & S_{64} & S_{65} & S_{66} & S_{67} & S_{68} & S_{69} \\
S_{71} & S_{72} & S_{73} & S_{74} & S_{75} & S_{76} & S_{77} & S_{78} & S_{79} \\
S_{81} & S_{82} & S_{83} & S_{84} & S_{85} & S_{86} & S_{87} & S_{88} & S_{89} \\
S_{91} & S_{92} & S_{93} & S_{94} & S_{95} & S_{96} & S_{97} & S_{98} & S_{99}
\end{bmatrix}
\begin{bmatrix}
C_{11} & C_{12} & C_{13} & C_{14} \\
C_{21} & C_{22} & C_{23} & C_{24} \\
C_{31} & C_{32} & C_{33} & C_{34} \\
C_{41} & C_{42} & C_{43} & C_{44} \\
C_{51} & C_{52} & C_{53} & C_{54} \\
C_{61} & C_{62} & C_{63} & C_{64} \\
C_{71} & C_{72} & C_{73} & C_{74} \\
C_{81} & C_{82} & C_{83} & C_{84} \\
C_{91} & C_{92} & C_{93} & C_{94}
\end{bmatrix}
\begin{bmatrix} \phi \\ D_4 \\ D_5 \\ D_6 \end{bmatrix} =
\begin{bmatrix} W_1 \\ W_2 \\ W_3 \\ W_4 \\ W_5 \\ W_6 \\ W_7 \\ W_8 \\ W_9 \end{bmatrix}
$$

The solution process is continued by postmultiplying $\mathbf{S}^c$ by $\mathbf{C}$. This operation, the results of which are depicted below, produces a matrix which will be called the *product matrix*:

$$
\begin{bmatrix}
P_{11} & P_{12} & P_{13} & P_{14} \\
P_{21} & P_{22} & P_{23} & P_{24} \\
P_{31} & P_{32} & P_{33} & P_{34} \\
P_{41} & P_{42} & P_{43} & P_{44} \\
P_{51} & P_{52} & P_{53} & P_{54} \\
P_{61} & P_{62} & P_{63} & P_{64} \\
P_{71} & P_{72} & P_{73} & P_{74} \\
P_{81} & P_{82} & P_{83} & P_{84} \\
P_{91} & P_{92} & P_{93} & P_{94}
\end{bmatrix}
\begin{bmatrix} \phi \\ D_4 \\ D_5 \\ D_6 \end{bmatrix} =
\begin{bmatrix} W_1 \\ W_2 \\ W_3 \\ W_4 \\ W_5 \\ W_6 \\ W_7 \\ W_8 \\ W_9 \end{bmatrix}
$$

The next step consists of accounting for the known displacement ϕ by multiplying it by the elements of the first column of the product matrix, transferring the resulting terms $P_{i1}\phi$ to the right-hand side of the equation, and algebraically subtracting them from corresponding terms of the load vector. When this operation is completed, the load vector will be termed the *augmented load vector* and the product matrix will be known as the *condensed product matrix*. If several displacements had been imposed on the structure, each would have been multiplied by a corresponding column of the product matrix and the resulting terms transferred to the right-hand side of the equation. The results of this operation are

$$
\begin{bmatrix}
P_{12} & P_{13} & P_{14} \\
P_{22} & P_{23} & P_{24} \\
P_{32} & P_{33} & P_{34} \\
P_{42} & P_{43} & P_{44} \\
P_{52} & P_{53} & P_{54} \\
P_{62} & P_{63} & P_{64} \\
P_{72} & P_{73} & P_{74} \\
P_{82} & P_{83} & P_{84} \\
P_{92} & P_{93} & P_{94}
\end{bmatrix}
\begin{bmatrix}
D_4 \\
D_5 \\
D_6
\end{bmatrix}
=
\begin{bmatrix}
W_1 - P_{11}\phi \\
W_2 - P_{21}\phi \\
W_3 - P_{31}\phi \\
W_4 - P_{41}\phi \\
W_5 - P_{51}\phi \\
W_6 - P_{61}\phi \\
W_7 - P_{71}\phi \\
W_8 - P_{81}\phi \\
W_9 - P_{91}\phi
\end{bmatrix}
$$

At this point it is necessary to examine the nature of the joint load components that are accounted for in the complete load vector. These components stem from three sources: (1) loads applied directly on the joints; (2) loads applied directly on the elements; and (3) reactive forces acting along constrained lines of action. Joint loads produced by element loading are discussed in Art. 12.15 and are not germane to the present discussion; however, joint loads stemming from constrained lines of action influence the solution process because they constitute additional unknown quantities that must be determined by the process.

At every global line of action, *either* the displacement component *or* the load component may be specified; however, it is not possible to specify both. Thus, constraint equations also imply a relationship between force components that correspond to the constrained displacement components.

If the symbol W_i is used to denote only directly applied joint loads and the symbol F_i is used to denote only reactive forces acting along constrained lines of action, the total load acting along any constrained line of action may be expressed as

$$
W_i^* = W_i + F_i + \chi_i
$$

in which χ_i represents equivalent load

terms resulting from imposed displacement components. With this notation, the matrix equation above becomes

$$
\begin{bmatrix}
P_{12} & P_{13} & P_{14} \\
P_{22} & P_{23} & P_{24} \\
P_{32} & P_{33} & P_{34} \\
P_{42} & P_{43} & P_{44} \\
P_{52} & P_{53} & P_{54} \\
P_{62} & P_{63} & P_{64} \\
P_{72} & P_{73} & P_{74} \\
P_{82} & P_{83} & P_{84} \\
P_{92} & P_{93} & P_{94}
\end{bmatrix}
\begin{bmatrix}
D_4 \\
D_5 \\
D_6
\end{bmatrix}
=
\begin{bmatrix}
W_1^* \\
W_2^* \\
W_3^* \\
W_4 - P_{41}\phi \\
W_5 - P_{51}\phi \\
W_6 - P_{61}\phi \\
W_7^* \\
W_8^* \\
W_9^*
\end{bmatrix}
$$

These equations may be separated into two sets, with one set pertaining to the unknown displacement components and the other set pertaining to the unknown reaction components, i.e.,

$$
\begin{bmatrix}
P_{42} & P_{43} & P_{44} \\
P_{52} & P_{53} & P_{54} \\
P_{62} & P_{63} & P_{64}
\end{bmatrix}
\begin{bmatrix}
D_4 \\
D_5 \\
D_6
\end{bmatrix}
=
\begin{bmatrix}
W_4 - P_{41}\phi \\
W_5 - P_{51}\phi \\
W_6 - P_{61}\phi
\end{bmatrix}
$$

or

$$
\mathbf{SD} = \mathbf{W}
$$

in which $\mathbf{S}$ is the *structure stiffness matrix*, $\mathbf{D}$ is the vector of independent displacement components (*displacement vector*), and $\mathbf{W}$ is the vector of specified joint load components (*load vector*). Then

$$
\begin{bmatrix}
P_{12} & P_{13} & P_{14} \\
P_{22} & P_{23} & P_{24} \\
P_{32} & P_{33} & P_{34} \\
P_{72} & P_{73} & P_{74} \\
P_{82} & P_{83} & P_{84} \\
P_{92} & P_{93} & P_{94}
\end{bmatrix}
\begin{bmatrix}
D_4 \\
D_5 \\
D_6
\end{bmatrix}
=
\begin{bmatrix}
W_1^* \\
W_2^* \\
W_3^* \\
W_7^* \\
W_8^* \\
W_9^*
\end{bmatrix}
$$

or

$$
\mathbf{QD} = \mathbf{W}
$$

The first set may be solved for the unknown displacement components by using appropriate matrix-algebra methods as discussed in Appendix C. For example, using the inverse method, a solution may be indicated as follows:

$$\mathbf{D} = \mathbf{S}^{-1}\mathbf{W}$$

After solving for the unknown displacement components, the second set may be solved for the unknown reaction components:

$$
\begin{bmatrix} F_1 \\ F_2 \\ F_3 \\ F_7 \\ F_8 \\ F_9 \end{bmatrix}
=
\begin{bmatrix}
P_{12} & P_{13} & P_{14} \\
P_{22} & P_{23} & P_{24} \\
P_{32} & P_{33} & P_{34} \\
P_{72} & P_{73} & P_{74} \\
P_{82} & P_{83} & P_{84} \\
P_{92} & P_{93} & P_{94}
\end{bmatrix}
\begin{bmatrix} D_4 \\ D_5 \\ D_6 \end{bmatrix}
-
\begin{bmatrix}
W_1 + \chi_1 \\
W_2 + \chi_2 \\
W_3 + \chi_3 \\
W_7 + \chi_7 \\
W_8 + \chi_8 \\
W_9 + \chi_9
\end{bmatrix}
$$

Finally, element end actions may be evaluated for each element by postmultiplying the element stiffness matrix by the vector of element-oriented displacement components that occur at the ends of the element.

Example

The complete solution process will be illustrated by determining the displacement components, reactive components, and element end actions for the portal frame with inclined columns shown in Fig. 12.33. This portal frame was used as an example

problem in Art. 12.13 to illustrate the process of formulating constraint equations.

Solution

(1) Prepare the transformed element stiffness matrix for each element in the structure according to the method described in Art. 12.12; that is, calculate the element coefficients (κ-values) and assemble them into the stiffness matrix (s_e^k) relative to the element axes, sketch the element rotation and prepare the rotation matrix, and prepare the transformed element stiffness matrix ($\mathbf{S}_s^k$) and partition it into submatrices according to end actions and displacements:

Frame Element 1 (Fig. 12.35)

```
    L  =  271.529  INCHES

    A  =   14.100  INS**2

    I  =  485.000  INS**4

    E  =   29.000  KSI/1,000

ALPHA  =   45.000  DEGREES
```

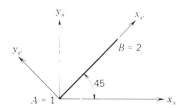

Figure 12.35/Element 1.

Element stiffness matrix/1,000,000

$$s_e^1 = \begin{bmatrix} 1.506 & 0.000 & 0.000 & -1.506 & 0.000 & 0.000 \\ 0.000 & 0.008 & 1.145 & 0.000 & -0.008 & 1.145 \\ 0.000 & 1.145 & 207.197 & 0.000 & -1.145 & 103.599 \\ \hline -1.506 & 0.000 & 0.000 & 1.506 & 0.000 & 0.000 \\ 0.000 & -0.008 & -1.145 & 0.000 & 0.008 & -1.145 \\ 0.000 & 1.145 & 103.599 & 0.000 & -1.145 & 207.197 \end{bmatrix}$$

Rotation matrix

$$R = \begin{bmatrix} 0.707 & 0.707 & 0.000 \\ -0.707 & 0.707 & 0.000 \\ 0.000 & 0.000 & 1.000 \end{bmatrix}$$

Transformed element stiffness matrix/1,000,000

$$S_s^1 = \begin{bmatrix} 0.757 & 0.749 & -0.809 & -0.757 & -0.749 & -0.809 \\ 0.749 & 0.757 & 0.809 & -0.749 & -0.757 & 0.809 \\ -0.809 & 0.809 & 207.197 & 0.809 & -0.809 & 103.599 \\ \hline -0.757 & -0.749 & 0.809 & 0.757 & 0.749 & 0.809 \\ -0.749 & -0.757 & -0.809 & 0.749 & 0.757 & -0.809 \\ -0.809 & 0.809 & 103.599 & 0.809 & -0.809 & 207.197 \end{bmatrix} = \begin{bmatrix} S_{11}^1 & S_{12}^1 \\ \hline S_{21}^1 & S_{22}^1 \end{bmatrix}$$

Frame Element 2 (Fig. 12.36):

```
    L =  216.000 INCHES
    A =   14.700 INS**2
    I =  800.000 INS**4
    E =   29.000 KSI/1,000
ALPHA =    0.000 DEGREES
```

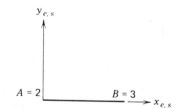

Figure 12.36/Element 2.

Element stiffness matrix/1,000,000

$$s_e^2 = \begin{bmatrix} 1.974 & 0.000 & 0.000 & -1.974 & 0.000 & 0.000 \\ 0.000 & 0.028 & 2.984 & 0.000 & -0.028 & 2.984 \\ 0.000 & 2.984 & 429.630 & 0.000 & -2.984 & 214.815 \\ \hline -1.974 & 0.000 & 0.000 & 1.974 & 0.000 & 0.000 \\ 0.000 & -0.028 & -2.984 & 0.000 & 0.028 & -2.984 \\ 0.000 & 2.984 & 214.815 & 0.000 & -2.984 & 429.630 \end{bmatrix}$$

Rotation matrix

$$R = \begin{bmatrix} 1.000 & 0.000 & 0.000 \\ 0.000 & 1.000 & 0.000 \\ 0.000 & 0.000 & 1.000 \end{bmatrix}$$

Transformed element stiffness matrix/1,000,000

$$S_s^2 = \left[\begin{array}{ccc|ccc} 1.974 & 0.000 & 0.000 & -1.974 & 0.000 & 0.000 \\ 0.000 & 0.028 & 2.984 & 0.000 & -0.028 & 2.984 \\ 0.000 & 2.984 & 429.630 & 0.000 & -2.984 & 214.815 \\ \hline -1.974 & 0.000 & 0.000 & 1.974 & 0.000 & 0.000 \\ 0.000 & -0.028 & -2.984 & 0.000 & 0.028 & -2.984 \\ 0.000 & 2.984 & 214.815 & 0.000 & -2.984 & 429.630 \end{array}\right] = \left[\begin{array}{c|c} S_{22}{}^2 & S_{23}{}^2 \\ \hline S_{32}{}^2 & S_{33}{}^2 \end{array}\right]$$

Frame Element 3 (Fig. 12.37):

$$\begin{aligned} L &= 138.567 \text{ INCHES} \\ A &= 14.700 \text{ INS}**2 \\ I &= 394.000 \text{ INS}**4 \\ E &= 29.000 \text{ KSI}/1,000 \\ ALPHA &= 300.000 \text{ DEGREES} \end{aligned}$$

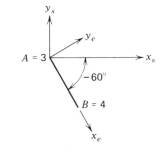

Figure 12.37/Element 3.

Element stiffness matrix/1,000,000

$$s_e^3 = \left[\begin{array}{ccc|ccc} 3.077 & 0.000 & 0.000 & -3.077 & 0.000 & 0.000 \\ 0.000 & 0.052 & 3.571 & 0.000 & -0.052 & 3.571 \\ 0.000 & 3.571 & 329.840 & 0.000 & -3.571 & 164.920 \\ \hline -3.077 & 0.000 & 0.000 & 3.077 & 0.000 & 0.000 \\ 0.000 & -0.052 & -3.571 & 0.000 & 0.052 & -3.571 \\ 0.000 & 3.571 & 164.920 & 0.000 & -3.571 & 329.840 \end{array}\right]$$

Rotation matrix

$$R = \left[\begin{array}{ccc} 0.500 & -0.866 & 0.000 \\ 0.866 & 0.500 & 0.000 \\ 0.000 & 0.000 & 1.000 \end{array}\right]$$

Transformed element stiffness matrix/1,000,000

$$S_s^3 = \left[\begin{array}{ccc|ccc} 0.808 & -1.310 & 3.092 & -0.808 & 1.310 & 3.092 \\ -1.310 & 2.320 & 1.785 & 1.310 & -2.320 & 1.785 \\ 3.092 & 1.785 & 329.840 & -3.092 & -1.785 & 164.920 \\ \hline -0.808 & 1.310 & -3.092 & 0.808 & -1.310 & -3.092 \\ 1.310 & -2.320 & -1.785 & -1.310 & 2.320 & -1.785 \\ 3.092 & 1.785 & 164.920 & -3.092 & -1.785 & 329.840 \end{array}\right] = \left[\begin{array}{c|c} S_{33}{}^3 & S_{34}{}^3 \\ \hline S_{43}{}^3 & S_{44}{}^3 \end{array}\right]$$

(2) Form the complete stiffness matrix by combining the submatrices for all the elements as indicated by the partitioning:

$$
S^c = \begin{bmatrix}
S_{11}^{1} & S_{12}^{1} & & & \\
S_{21}^{1} & S_{22}^{1}+S_{22}^{2} & S_{23}^{2} & & \\
& S_{32}^{2} & S_{33}^{2}+S_{33}^{3} & S_{34}^{3} & \\
& & S_{43}^{3} & S_{44}^{3}
\end{bmatrix}
$$

$$
S^c = 1{,}000{,}000 \begin{bmatrix}
0.757 & 0.749 & -0.809 & -0.757 & -0.749 & -0.809 & & & \\
0.749 & 0.757 & 0.809 & -0.749 & -0.757 & 0.809 & & & \\
-0.809 & 0.809 & 207.197 & 0.809 & -0.809 & 103.599 & & & \\
-0.757 & -0.749 & 0.809 & 2.731 & 0.749 & 0.809 & -1.974 & 0.000 & 0.000 \\
-0.749 & -0.757 & -0.809 & 0.749 & 0.785 & 2.174 & 0.000 & -0.028 & 2.984 \\
-0.809 & 0.809 & 103.599 & 0.809 & 2.174 & 636.827 & 0.000 & -2.984 & 214.815 \\
& & & -1.974 & 0.000 & 0.000 & 2.781 & -1.310 & 3.092 & -0.808 & 1.310 & 3.092 \\
& & & 0.000 & -0.028 & -2.984 & -1.310 & 2.348 & -1.198 & 1.310 & -2.320 & 1.785 \\
& & & 0.000 & 2.984 & 214.815 & 3.092 & -1.198 & 759.470 & -3.092 & -1.785 & 164.920 \\
& & & & & & -0.808 & 1.310 & -3.092 & 0.808 & -1.310 & -3.092 \\
& & & & & & 1.310 & -2.320 & -1.785 & -1.310 & 2.320 & -1.785 \\
& & & & & & 3.092 & 1.785 & 164.920 & -3.092 & -1.785 & 329.840
\end{bmatrix}
$$

(3) Formulate the constraint equations and prepare the constraint matrix (this step was shown as an illustrative problem in Art. 12.13):

$$
\begin{bmatrix} D_1 \\ D_2 \\ D_3 \\ D_4 \\ D_5 \\ D_6 \\ D_7 \\ D_8 \\ D_9 \\ D_{10} \\ D_{11} \\ D_{12} \end{bmatrix} =
\begin{bmatrix}
0.000 & 0.000 & 0.000 & 0.000 & 0.000 & 0.000 & 0.000 & 0.000 & 0.000 \\
1.000 & 0.000 & 0.000 & 0.000 & 0.000 & 0.000 & 0.000 & 0.000 & 0.000 \\
0.000 & 1.000 & 0.000 & 0.000 & 0.000 & 0.000 & 0.000 & 0.000 & 0.000 \\
0.000 & 0.000 & 1.000 & 0.000 & 0.000 & 0.000 & 0.000 & 0.000 & 0.000 \\
0.000 & 0.000 & 0.000 & 1.000 & 0.000 & 0.000 & 0.000 & 0.000 & 0.000 \\
0.000 & 0.000 & 0.000 & 0.000 & 1.000 & 0.000 & 0.000 & 0.000 & 0.000 \\
0.000 & 0.000 & 0.000 & 0.000 & 0.000 & 1.000 & 0.000 & 0.000 & 0.000 \\
0.000 & 0.000 & 0.000 & 0.000 & 0.000 & 0.000 & 1.000 & 0.000 & 0.000 \\
0.000 & 0.000 & 0.000 & 0.000 & 0.000 & 0.000 & 0.000 & 1.000 & 0.000 \\
0.000 & 0.000 & 0.000 & 0.000 & 0.000 & 0.750 & 0.433 & 60.000 & 60.000 \\
0.000 & 0.000 & 0.000 & 0.000 & 0.000 & 0.433 & 0.250 & 34.641 & 34.641 \\
0.000 & 0.000 & 0.000 & 0.000 & 0.000 & 0.000 & 0.000 & 0.000 & 1.000
\end{bmatrix}
\begin{bmatrix} -1.000 \\ -0.004 \\ D_4 \\ D_5 \\ D_6 \\ D_7 \\ D_8 \\ D_9 \\ D_{12} \end{bmatrix}
$$

(4) Prepare the condensed product matrix. This is done by postmultiplying the complete stiffness matrix (S^c) by the constraint matrix (C) to produce the product matrix (P). Then, the coefficients in the product matrix associated with the imposed displacements (the first two columns in this example) are removed from the matrix, leaving the condensed product matrix:

Condensed product matrix

$$
= 1{,}000{,}000 \begin{bmatrix}
-0.757 & -0.749 & -0.809 & 0.000 & 0.000 & 0.000 & 0.000 \\
-0.749 & -0.757 & 0.809 & 0.000 & 0.000 & 0.000 & 0.000 \\
0.809 & -0.809 & 103.599 & 0.000 & 0.000 & 0.000 & 0.000 \\
2.731 & 0.749 & 0.809 & -1.974 & 0.000 & 0.000 & 0.000 \\
0.749 & 0.785 & 2.174 & 0.000 & -0.028 & 2.984 & 0.000 \\
0.809 & 2.174 & 636.827 & 0.000 & -2.984 & 214.815 & 0.000 \\
-1.974 & 0.000 & 0.000 & 2.743 & -1.332 & 0.000 & 0.000 \\
0.000 & -0.028 & -2.984 & -1.332 & 2.335 & -2.984 & 0.000 \\
0.000 & 2.984 & 214.815 & 0.000 & -2.984 & 512.090 & -82.460 \\
0.000 & 0.000 & 0.000 & -0.769 & 1.332 & 0.000 & 0.000 \\
0.000 & 0.000 & 0.000 & 1.332 & -2.307 & 0.000 & 0.000 \\
0.000 & 0.000 & 0.000 & 0.000 & 0.000 & -82.460 & 82.460
\end{bmatrix}
$$

(5) Construct the augmented load vector. The first two columns of coefficients removed from the product matrix in step (4) are postmultiplied by the imposed displacements and added to the applied load as follows:

$$
1,000,000
\begin{bmatrix}
0.000 \\
0.000 \\
0.000 \\
0.010 \\
0.000 \\
0.000 \\
0.000 \\
0.000 \\
0.000 \\
0.000 \\
0.000 \\
0.000
\end{bmatrix}
+ 1,000,000
\begin{bmatrix}
-0.749 & 0.809 \\
-0.757 & -0.809 \\
-0.809 & -207.197 \\
0.749 & -0.809 \\
0.757 & 0.809 \\
-0.809 & -103.599 \\
0.000 & 0.000 \\
0.000 & 0.000 \\
0.000 & 0.000 \\
0.000 & 0.000 \\
0.000 & 0.000 \\
0.000 & 0.000
\end{bmatrix}
\begin{bmatrix}
-1.000 \\
-0.004
\end{bmatrix}
= 1,000,000
\begin{bmatrix}
0.746 \\
0.760 \\
1.638 \\
-0.736 \\
-0.760 \\
1.224 \\
0.000 \\
0.000 \\
0.000 \\
0.000 \\
0.000 \\
0.000
\end{bmatrix}
$$

| Applied loads | Columns 1 and 2 from product matrix | Augmented load vector |

(6) The condensed product matrix and augmented load vector are separated into two sets of equations. The first set pertains to the seven unknown joint displacements. As previously stated, this matrix is usually known simply as the stiffness matrix:

$$\mathbf{SD} = \mathbf{W}$$

$$
1,000,000
\begin{bmatrix}
2.731 & 0.749 & 0.809 & -1.974 & 0.000 & 0.000 & 0.000 \\
0.749 & 0.785 & 2.174 & 0.000 & -0.028 & 2.984 & 0.000 \\
0.809 & 2.174 & 636.827 & 0.000 & -2.984 & 214.815 & 0.000 \\
-1.974 & 0.000 & 0.000 & 2.743 & -1.332 & 0.000 & 0.000 \\
0.000 & -0.028 & -2.984 & -1.332 & 2.335 & -2.984 & 0.000 \\
0.000 & 2.984 & 214.815 & 0.000 & -2.984 & 512.090 & -82.460 \\
0.000 & 0.000 & 0.000 & 0.000 & 0.000 & -82.460 & 82.460
\end{bmatrix}
\begin{bmatrix}
D_4 \\
D_5 \\
D_6 \\
D_7 \\
D_8 \\
D_9 \\
D_{12}
\end{bmatrix}
= 1,000,000
\begin{bmatrix}
-0.736 \\
-0.760 \\
1.224 \\
0.000 \\
0.000 \\
0.000 \\
0.000
\end{bmatrix}
$$

| Stiffness matrix | Unknown displacements | Load vector |

The second set pertains to the reaction components and is used in step (8). This matrix is the **Q** matrix, which was previously defined:

$$
1,000,000
\begin{bmatrix}
-0.757 & -0.749 & -0.809 & 0.000 & 0.000 & 0.000 & 0.000 \\
-0.749 & -0.757 & 0.809 & 0.000 & 0.000 & 0.000 & 0.000 \\
0.809 & -0.809 & 103.599 & 0.000 & 0.000 & 0.000 & 0.000 \\
0.000 & 0.000 & 0.000 & -0.769 & 1.332 & 0.000 & 0.000 \\
0.000 & 0.000 & 0.000 & 1.332 & -2.307 & 0.000 & 0.000
\end{bmatrix}
\begin{bmatrix}
D_4 \\
D_5 \\
D_6 \\
D_7 \\
D_8 \\
D_9 \\
D_{12}
\end{bmatrix}
= 1,000,000
\begin{bmatrix}
0.746 \\
0.760 \\
1.638 \\
0.000 \\
0.000
\end{bmatrix}
+
\begin{bmatrix}
F_1 \\
F_2 \\
F_3 \\
F_{10} \\
F_{11}
\end{bmatrix}
$$

| Q matrix | Displacements | Load vector | Reactive forces |

(7) Solve for the unknown joint displacements, using the equations from the first set in step (6). Procedures for this are discussed in Appendix C. In this example, the solution is accomplished by partitioning the stiffness matrix and then generating the inverse of the

stiffness matrix. This inverse is then postmultiplied by the load vector:

$$\mathbf{D} = \mathbf{S}^{-1}\mathbf{W}$$

$$
\begin{bmatrix} D_4 \\ D_5 \\ D_6 \\ D_7 \\ D_8 \\ D_9 \\ D_{12} \end{bmatrix} =
\begin{bmatrix}
30.733 & -29.994 & 0.041 & 30.655 & 17.583 & 0.310 & 0.310 \\
-29.994 & 30.583 & -0.042 & -29.916 & -17.156 & -0.311 & -0.311 \\
0.041 & -0.042 & 0.002 & 0.042 & 0.025 & -0.001 & -0.001 \\
30.655 & -29.916 & 0.042 & 31.083 & 17.830 & 0.311 & 0.311 \\
17.583 & -17.156 & 0.025 & 17.830 & 10.661 & 0.181 & 0.181 \\
0.310 & -0.311 & -0.001 & 0.311 & 0.181 & 0.006 & 0.006 \\
0.310 & -0.311 & -0.001 & 0.311 & 0.181 & 0.006 & 0.018
\end{bmatrix}
\begin{bmatrix} -0.736 \\ -0.760 \\ 1.224 \\ 0.000 \\ 0.000 \\ 0.000 \\ 0.000 \end{bmatrix}
$$

Inverse of the stiffness matrix

(8) At this stage of the general solution process, all of the joint displacements are known. It is usually convenient to group all the joints' displacements together and place them in a tabular form as follows:

| | | *Deflections* | |
| | D(X) | D(Y) | ROTATION |
Joint	(IN.)	(IN.)	(RADS)
1	0.0000	−1.0000	−0.00400
2	0.2539	−1.2463	0.00394
3	0.2528	0.1436	0.00769
4	1.1747	0.6783	0.00769

(9) Solve for the reactive forces. This is done by rearranging the second set of equations prepared in Step (6) and substituting in the now known joint displacement values:

$$
\begin{bmatrix} F_1 \\ F_2 \\ F_3 \\ F_{10} \\ F_{11} \end{bmatrix} = 1{,}000{,}000
\begin{bmatrix}
-0.757 & -0.749 & -0.809 & 0.000 & 0.000 & 0.000 & 0.000 \\
-0.749 & -0.757 & 0.809 & 0.000 & 0.000 & 0.000 & 0.000 \\
0.809 & -0.809 & 103.599 & 0.000 & 0.000 & 0.000 & 0.000 \\
0.000 & 0.000 & 0.000 & -0.769 & 1.332 & 0.000 & 0.000 \\
0.000 & 0.000 & 0.000 & 1.332 & -2.307 & 0.000 & 0.000
\end{bmatrix}
\begin{bmatrix} 0.254 \\ -1.246 \\ 0.004 \\ 0.253 \\ 0.144 \\ 0.008 \\ 0.008 \end{bmatrix}
- 1{,}000{,}000
\begin{bmatrix} 0.746 \\ 0.760 \\ 1.638 \\ 0.000 \\ 0.000 \end{bmatrix}
$$

(10) Arrange the reactive forces in a tabular form as follows:

| | | *Reactions* | |
| | R(X) | R(Y) | COUPLE |
Joint	(kips)	(kips)	(kip-feet)
1	−7.845	−3.732	−1.347
4	−2.154	3.732	

(11) The last step in the solution process is to calculate and arrange in a tabular form all element end actions. They should be calculated with respect to the element-oriented axes; consequently, it is necessary to first transform the end displacement components pertaining to a given element into element-oriented axes. This is accomplished by postmultiplying the rotation matrix by the joint displacements calculated in step (7). The calculations

shown are for element 1. The displacements at the A end are

$$\begin{bmatrix} \xi_A \\ \delta_A \\ \theta_A \end{bmatrix} = \begin{bmatrix} 0.707 & 0.707 & 0.000 \\ -0.707 & 0.707 & 0.000 \\ 0.000 & 0.000 & 1.000 \end{bmatrix} \begin{bmatrix} 0 \\ -1 \\ -0.004 \end{bmatrix} = \begin{bmatrix} -0.707 \\ -0.707 \\ -0.004 \end{bmatrix}$$

The displacements at the B end are

$$\begin{bmatrix} \xi_B \\ \delta_B \\ \theta_B \end{bmatrix} = \begin{bmatrix} 0.707 & 0.707 & 0.000 \\ -0.707 & 0.707 & 0.000 \\ 0.000 & 0.000 & 1.000 \end{bmatrix} \begin{bmatrix} 0.254 \\ -1.246 \\ 0.004 \end{bmatrix} = \begin{bmatrix} -0.702 \\ -1.061 \\ 0.004 \end{bmatrix}$$

The end actions are obtained by postmultiplying each element stiffness matrix (s_e^k) by the end displacement vector, i.e., $a_e^k = s_e^k d_e$. This procedure was discussed in Art. 12.8. The calculations shown are for element 1.

A factor of 1,000,000 is associated with the element stiffness matrix shown below. When this factor is included in the calculation, the units of the end actions are pounds and inch-pounds. However, this factor may be omitted and units adjusted later when the final answers are placed in tabular form:

$$\begin{bmatrix} t_A \\ v_A \\ m_A \\ \hline t_B \\ v_B \\ m_B \end{bmatrix} = \begin{bmatrix} 1.506 & 0.000 & 0.000 & -1.506 & 0.000 & 0.000 \\ 0.000 & 0.008 & 1.145 & 0.000 & -0.008 & 1.145 \\ 0.000 & 1.145 & 207.197 & 0.000 & -1.145 & 103.599 \\ \hline -1.506 & 0.000 & 0.000 & 1.506 & 0.000 & 0.000 \\ 0.000 & -0.008 & -1.145 & 0.000 & 0.008 & -1.145 \\ 0.000 & 1.145 & 103.599 & 0.000 & -1.145 & 207.197 \end{bmatrix} \begin{bmatrix} -0.707 \\ -0.707 \\ -0.004 \\ \hline -0.702 \\ -1.061 \\ 0.004 \end{bmatrix} = \begin{bmatrix} -0.00819 \\ 0.00291 \\ -0.01616 \\ \hline 0.00819 \\ -0.00291 \\ 0.80602 \end{bmatrix}$$

The end actions for elements 2 and 3 are calculated in a similar manner.

Finally, the units are converted to kips (multiplying by 1000) and kip-feet (multiplying by 1000/12) and the answers are placed in a tabular form:

Member End Actions

Members	T(A) (kips)	V(A) (kips)	M(A) (kip-ft)	T(B) (kips)	V(B) (kips)	M(B) (kip-ft)
1	−8.186	2.909	−1.347	8.186	−2.909	67.170
2	2.155	−3.732	−67.170	−2.155	3.732	0.000
3	4.309	0.000	0.000	−4.311	0.000	0.000

12.15
Element Loads

The general solution process discussed in the previous articles deals with loads applied on the joints only. Frequently, loads are applied directly on the elements themselves and must be accounted for in two ways in the solution process. The loads are first expressed in terms of equivalent *fixed-end actions* at the ends of the elements. These end actions must be: (1) considered as additional loads at the joints; and (2) added to the element end actions that are caused by the joint displacement.

To account for the fixed-end actions as joint loads, it is necessary to take the *nega-*

Table 12.1
Fixed End Actions for Prismatic Frame Elements

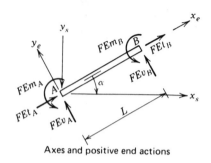

Axes and positive end actions

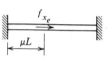

$$\text{FE}t_A = -(f_{x_e})(1-\mu)$$
$$\text{FE}v_A = 0$$
$$\text{FE}m_A = 0$$
$$\text{FE}t_B = -(f_{x_e})\mu$$
$$\text{FE}v_B = 0$$
$$\text{FE}m_B = 0$$

$$\text{FE}t_A = -(p_{x_e})L/2$$
$$\text{FE}v_A = 0$$
$$\text{FE}m_A = 0$$
$$\text{FE}t_B = -(p_{x_e})L/2$$
$$\text{FE}v_B = 0$$
$$\text{FE}m_B = 0$$

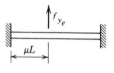

$$\text{FE}t_A = 0$$
$$\text{FE}v_A = -(f_{y_e})(1+2\mu)(1-\mu)^2$$
$$\text{FE}m_A = -(f_{y_e})L\mu(1-\mu)^2$$
$$\text{FE}t_B = 0$$
$$\text{FE}v_B = -(f_{y_e})\mu^2(3-2\mu)$$
$$\text{FE}m_B = (f_{y_e})L\mu^2(1-\mu)$$

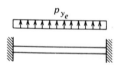

$$\text{FE}t_A = 0$$
$$\text{FE}v_A = -(p_{y_e})L/2$$
$$\text{FE}m_A = -(p_{y_e})L^2/12$$
$$\text{FE}t_B = 0$$
$$\text{FE}v_B = -(p_{y_e})L/2$$
$$\text{FE}m_B = (p_{y_e})L^2/12$$

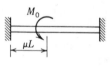

$$\text{FE}t_A = 0$$
$$\text{FE}v_A = (6M_0\mu/L)(1-\mu)$$
$$\text{FE}m_A = -M_0(1-4\mu+3\mu^2)$$
$$\text{FE}t_B = 0$$
$$\text{FE}v_B = -(6M_0\mu/L)(1-\mu)$$
$$\text{FE}m_B = M_0\mu(2-3\mu)$$

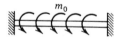

$$\text{FE}t_A = 0$$
$$\text{FE}v_A = m_0$$
$$\text{FE}m_A = 0$$
$$\text{FE}t_B = 0$$
$$\text{FE}v_B = -m_0$$
$$\text{FE}m_B = 0$$

Table 12.2
Load Transformation Formulas

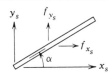

$$\begin{bmatrix} f_{x_e} \\ f_{y_e} \end{bmatrix} = \begin{bmatrix} \cos \alpha & \sin \alpha \\ -\sin \alpha & \cos \alpha \end{bmatrix} \begin{bmatrix} f_{x_s} \\ f_{y_s} \end{bmatrix}$$

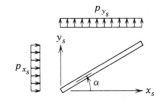

$$\begin{bmatrix} p_{x_e} \\ p_{y_e} \end{bmatrix} = \begin{bmatrix} \cos \alpha & \sin \alpha \\ -\sin \alpha & \cos \alpha \end{bmatrix} \begin{bmatrix} p_{x_s} \, \text{ABS}(\sin \alpha) \\ p_{y_s} \, \text{ABS}(\cos \alpha) \end{bmatrix}$$

in which ABS( ) denotes the absolute value of the argument within the parentheses.

For element dead loads
Gravity loads: $p_{x_e} = -A\gamma \sin \alpha$
$p_{y_e} = -A\gamma \cos \alpha$
in which A is the cross-sectional area of the element and γ is the weight per unit volume of the element material. The quantities A and γ must be expressed in consistent units.

tive of the actions, transform them to the structure-oriented axes, and then add them to the vector of *joint load components* which was presented in Arts. 12.12 and 12.14. The transformation is accomplished by postmultiplying the inverse of the rotation matrix ($\mathbf{R}^{-1}$) by the vector of fixed end actions (Art. 12.9).

By utilizing the expressions previously developed for end actions in terms of the member stiffness matrix and end displacements, the final values for end actions may be expressed in matrix form for truss elements[6] as

$$\begin{bmatrix} t_A \\ v_A \\ t_B \\ v_B \end{bmatrix} = [\mathbf{s}_e^k][\mathbf{d}_e] + \begin{bmatrix} \text{FE}t_A \\ \text{FE}v_A \\ \text{FE}t_B \\ \text{FE}v_B \end{bmatrix}$$

[6]Element loads are not accepted by program PLTRUSS, presented in Appendix D.

and for frame elements as

$$\begin{bmatrix} t_A \\ v_A \\ m_A \\ t_B \\ v_B \\ m_B \end{bmatrix} = [\mathbf{s}_e^k][\mathbf{d}_e] + \begin{bmatrix} \text{FE}t_A \\ \text{FE}v_A \\ \text{FE}m_A \\ \text{FE}t_B \\ \text{FE}v_B \\ \text{FE}m_B \end{bmatrix}$$

in which the symbols $\text{FE}t_A$, $\text{FE}v_A$, and $\text{FE}m_A$ denote the *fixed-end* thrust, shear, and bending moment, respectively, at end A, and $\text{FE}t_B$, $\text{FE}v_B$, and $\text{FE}m_B$ denote corresponding quantities at end B.

Fixed-end actions can be found by well-known, standard methods of analysis such as the area-moment method presented in Art. 4.2 and discussed in Art. 10.2. For convenient reference, however, formulas for fixed-end actions for prismatic frame elements for six conditions of loading are given in Table 12.1. By using the principle of superposition, these six conditions can be combined in a variety of ways to represent almost any practical loading condition.

Quite often, element loads are oriented with respect to the structure axes. These can be transformed into loads oriented with respect to the element axes by using the load transformation formulas given in Table 12.2. Once this has been accomplished, the fixed end actions can be calculated using the formulas in Table 12.1.

PROBLEMS

1. Prepare the stiffness matrices and solve for axial loads and joint displacements for the truss shown in Fig. 12.38. Assume top and bottom chords to be 4.5 in.^{2} and the diagonals to be 3.0 in.^{2}. (Answers given in Appendix G.)

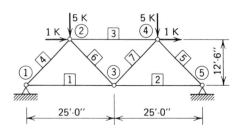

Figure 12.38

2. Prepare the stiffness matrices and solve for the reactions, end actions, and joint displacements for the frame shown in Fig. 12.39. Assume all elements to have an area of 8 in.^{2} and

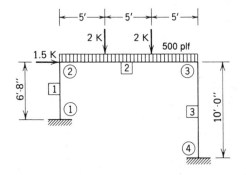

Figure 12.39

a moment of inertia of 200 in.^{4}. (Answers given in Appendix G.)

12.16
Utilization of Computers

The displacement (stiffness) method of analysis developed in the preceding articles is a very general method which, with appropriate modification, provides a unified and comprehensive approach to the solution of any type of *framed* structure. It is systematic and orderly, and does not require analytical decisions to define the unknowns for individual problems. Furthermore, the method involves only discrete operations, which are localized in their effects. Although it may be used effectively in conjunction with manual methods of calculation (similar to those used with moment distribution), the outstanding characteristic of the method is its eminent suitability for use in conjunction with a digital computer. To use the method on a digital computer, however, it must be expressed in *algorithmic* form. An *algorithm* is defined as a *formalized mathematical procedure whereby the desired results are obtained by following a prescribed, unambiguous sequence of operations*. When the prescribed sequence of operations is written in a form that may be implemented on a computer, it is known as a *program*.[7] To more fully understand and appreciate program formulation, a basic knowledge of some of the ways in which computers may be used is desirable; therefore, a brief discussion of the fundamental concepts of computer utilization follows.

Stated in simple terms, an electronic digital computer is a device that accepts a program and (usually) one or more sets of data to be processed.[8] The computer then

[7]Programs for the solution of plane trusses and frames are presented in Appendixes D and E, respectively.

[8]As used here, the word process is an inclusive term referring to all operations performed by the computer using the data.

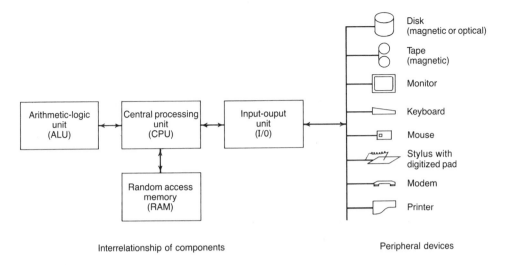

Interrelationship of components   Peripheral devices

Figure 12.40/Computer components.

automatically proceeds, under control of the program, to operate on the data by carrying out, in sequence, the operations defined by the program. Final results are obtained when the sequence of operations is completed for each set of data.

To more fully understand the components and functions of a computer, refer to the diagram shown in Fig. 12.40. There are four basic components, one of which (I/O) is attached to a variety of optional peripheral devices that a person uses to operate the computer.

In general, the CPU receives all the information from the I/O unit—program, data, instructions, etc.—and places much of this information in the memory unit. The CPU continues by following the directions of the program, selecting data and instructions from memory and passing it along to the ALU, where the actual numerical calculations take place. The CPU receives the results from the ALU and can then either return this information temporarily back to the memory or pass it on to the I/O unit, depending upon the instructions in the program.

The function of the memory unit is to store the program, data, and intermediate and final results. It accomplishes this task by the use of a semiconductor memory in the form of integrated circuits known as chips.

Information is communicated to a computer or received from a computer through the I/O unit, which, in turn can be attached to a variety of the peripheral devices as shown in Fig. 12.40. The most common of the devices are the disk drive, monitor, keyboard, and printer.

Typically, the instructions that the control unit obtains and interprets contain two items of information: the first specifies the operation that the arithmetic unit is to execute, while the second gives the memory locations containing the data on which the operation is to be performed. These instructions are encoded in unique sequences of binary digits, or *bits*, which are called *machine words*. The complete set of machine words that a computer can interpret is known as the *machine language* of the computer. To be processed, *all* program instructions must be expressed in machine language.

Coding a program in machine language is an incredibly tedious and time-consuming operation that usually requires the services of a very talented and skilled programmer —one who not only understands the pro-

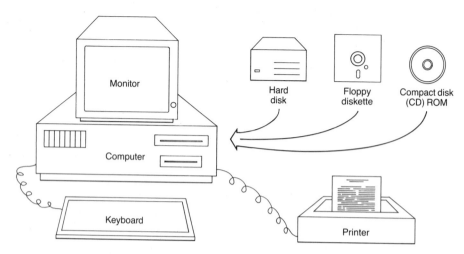

Figure 12.41/Typical PC assembly.

gram that is to be encoded but the internal operations of the computer as well. Fortunately, however, procedures have been developed that allow a programmer to express a program in a higher-level language that is subsequently translated by computer processing into appropriate machine language.

These *higher-level languages*, also known as *algorithmic languages*, are considerably easier and faster to use and do not require any knowledge of machine language. For example, two common engineering-type algorithmic languages are C and FORTRAN. Both of these require software known as *compilers* to convert the program to a machine language so that it becomes *executable* by the computer. Appendixes D and E contain structural programs written in FORTRAN. These programs have already been compiled for use with one of the Intel 80286, 80386, 80486, or 80586 processors, and linked with the necessary subroutines along with I/O instructions. The program, in this form, has been placed on the diskette contained in the envelope inside the back cover of this text. These processors have a specific kind of machine language often referred to as "IBM-compatible." Consequently, this diskette con-

tains the programs in executable form and is ready to be used in an appropriate computer. Instructions for its use are given in Appendixes D and E along with illustrative examples.

In common use today are personal computers, or PCs. These are computers that have components assembled by a variety of manufacturers, and many include the processors noted above, which are appropriate for use with the enclosed diskette.

Figure 12.41 represents a typical PC station. All disk drives can be externally or internally mounted. Usually the floppy drive is internal. Both the floppy and CD ROM disks are removable media that can be fed into the disk drive.

12.17
Computer Programs for Structural Analysis and Design

The two computer programs presented in this text, PLTRUSS and PLFRAME, are concise and simple to use. They can be used for rather large structures, and the associated expenses are virtually negligible. However, their use is limited to the analysis of two-dimensional trusses and frames

Table 12.3
Structural Analysis and Design Programs

Name of Program	Source
C-STRAAD	ECOM ASSOC. 8634 West Brown Deer Dr. Milwaukee, WI 53224
ETABS	Computers and Structures, Inc. 1995 University Ave. Berkeley, CA 94704
RISA-2D	RISA Technologies 17900 Sky Park Circle Suite 106 Irvine, CA 92714
SAP90	Computers and Structures, Inc. 1995 University Ave. Berkeley, CA 94704

consisting of prismatic elements. The analysis results in calculating and listing the deflection and axial force (PLTRUSS), or the angular rotation, shear, and bending moment (PLFRAME), at every joint. This calculation is viewed by many as the most difficult and time-consuming part of the analytical process in structural design, leaving only the sizing and detailing to be accomplished by manual methods.

A large number of computer programs for structural analysis and design are commercially available; some are priced between two and five thousand dollars. These programs are considerably more extensive than the two described herein. They not only make the analysis but also furnish the information for sizing and detailing. The analysis is usually based upon the same stiffness method that has been described in this chapter and used in the two programs PLTRUSS and PLFRAME.

Furthermore, many of these programs furnish a graphic image of the structure and operate in an interactive manner with respect to the input data. Some are even linked with drafting programs. In general, user's manuals and other forms of documentation, which explain how to use the programs effectively, also are available; therefore, the programs can be effectively utilized even by a person who is not familiar with all the underlying mathematical theory. A partial list of programs and where they may be obtained is presented in Table 12.3.

These programs may be used on almost any recently manufactured computer; however, there are some exceptions that warrant preliminary inquiry. A working understanding of the two programs PLTRUSS and PLFRAME is a good starting point for operating the more involved commercial programs. It must be remembered, however, that computer programs found in the public domain (including PLTRUSS and PLFRAME) are not, in general, certified; therefore, the responsibility for their proper use and for the validity of the results produced lies with the user and not with the authors or producers of the programs.

13

BUILDING DESIGN PROJECT

Design procedures and calculations are based on the 1989 AISC Specification.

13.1
Introduction

In Chapters 9 and 10, principal attention was given to one-story, single-span framing systems: systems with trusses spanning bearing walls or piers; braced frames; and rigid frames. It remains to apply the principles and design methods thus far developed to the design of a typical multistory, multibay, skeleton-frame system (Figs. 5.39 and 13.12). Therefore, attention in this final chapter will be focused on the design of a typical intermediate-height building of such construction.

Attention should be called to the fact that there is strong economic competition among steel, reinforced concrete, and composite steel and reinforced concrete framing for such intermediate-height buildings, i.e., buildings 3 to 14 stories in height. For taller buildings, however, the advantage of steel over reinforced concrete gains as the height increases. This is due principally to the relative strength-to-weight ratios between the two constructions and to the speed and ease of erection of the former.

As stated earlier, total building design begins with a careful consideration of function and site selection and, not infrequently, with a tight budget. The first two considerations set the internal and external environmental requirements, and the latter, the economic limits within which a solution must be sought.

It is the essential function of the designer to create an architecture which satisfies and aesthetically expresses these require-

529

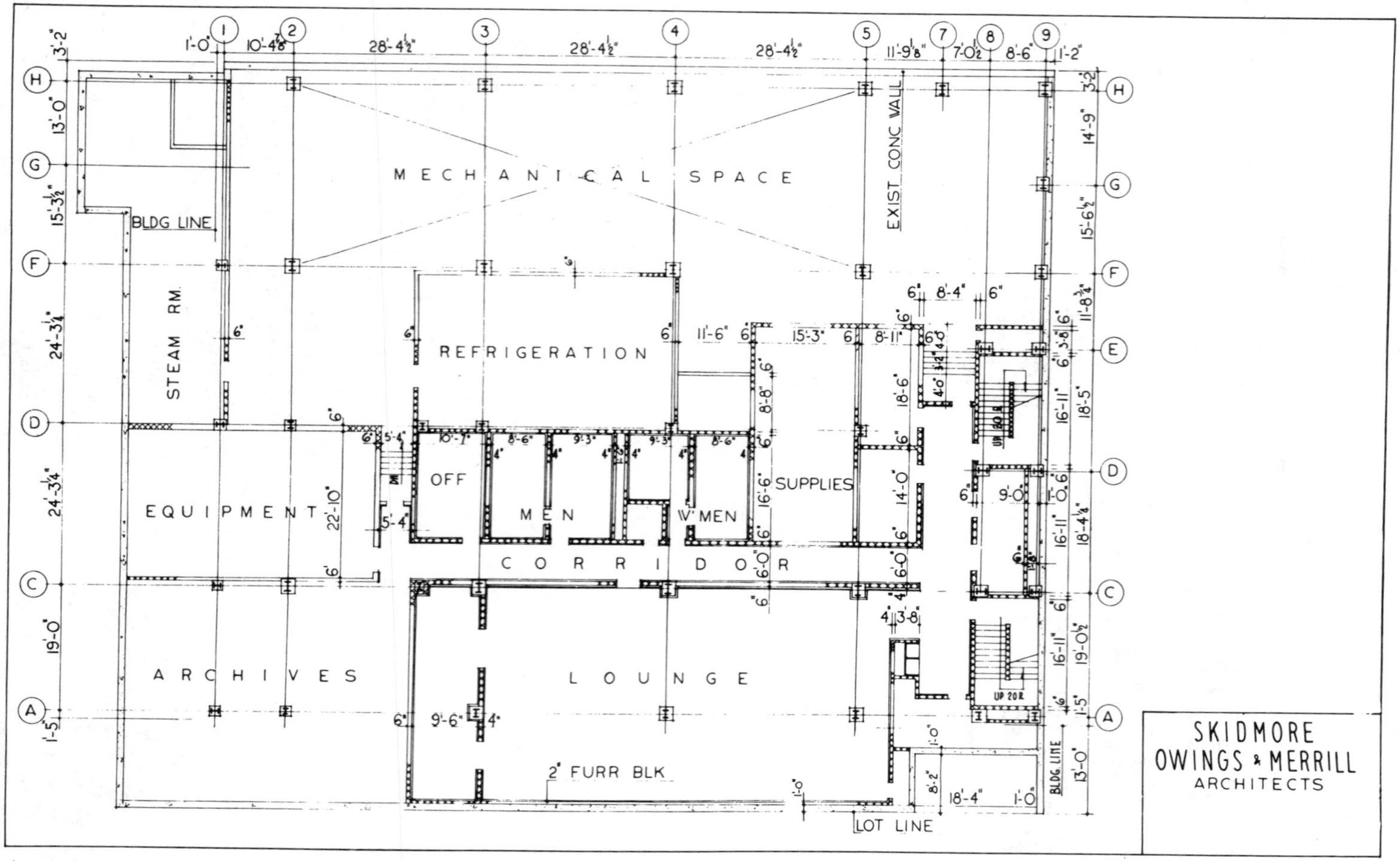

Figure 13.1/Foundation floor plan.

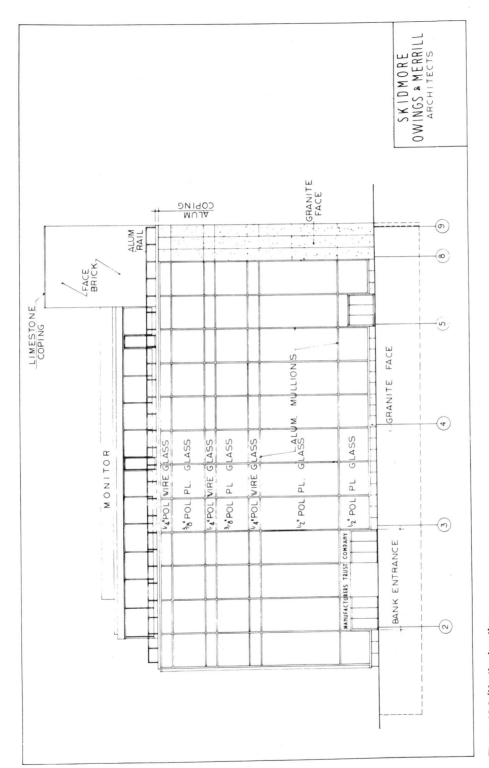

Figure 13.2/North elevation.

531

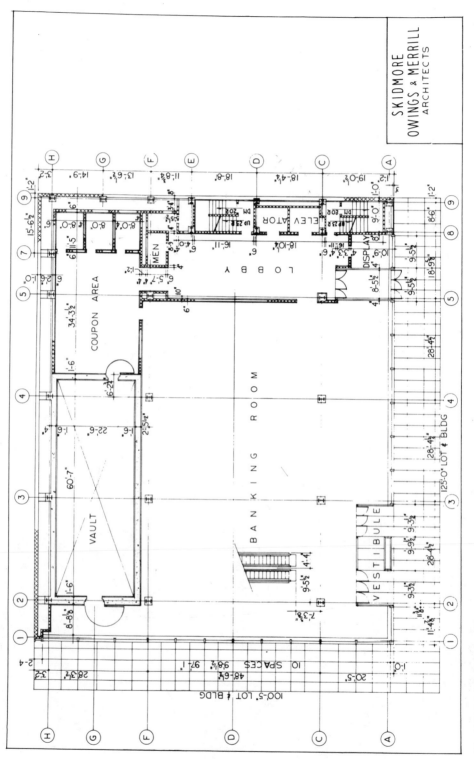

Figure 13.3 / First-floor plan.

532

ments. Rarely will the designer be seeking only the most economical solution, and rarely will he be able to achieve all that either he or the client wants. Limitations of time and materials, and limitations imposed by public regulations intended to protect public health, safety, and to some extent property—all these and more—will often affect his freedom of choice. Then too, increasingly the designer is faced with the need to provide a structure which has inherent adaptability to changing and even different functional requirements.

One of the most significant planning aids which has emerged as a partial answer to this bewildering array of requirements and limitations is the basic planning grid. That is, once a decision has been made to use a given structural material or combination of materials (e.g., steel, reinforced concrete, composite steel and reinforced concrete, or timber), a planning grid is established which reflects the economical span limits of that material or combination of materials and the structural capacity of accompanying flooring and vertical support members. This grid usually defines one or more patterns of rectangular or square spaces in the plan, with columns located at all or most points of grid intersect. All architectural planning is done within these imaginary grid lines. Obvious additional advantages are the ability to maintain dimensional control between floors and the ability to achieve economies by virtue of repetitive member sizes for any one floor and often for all but the roof and ground floor.

Some designers still feel that the use of the grid, and especially the added discipline of modular dimensioning, seriously restricts their freedom for creativity and individuality. For the majority of structures, however, the advantages would seem to far outweigh these presumed disadvantages. For example, Fig. 13.1 shows the foundation plan of a building which most

designers would agree is distinctive in character,[1] well designed (Fig. 13.2), and a pace-setter in its time. The definite grid pattern, defined by column spacings, is evident in the foundation plan. At the first-floor banking level (Fig. 13.3), greater floor area is achieved by longer-span members, and at the floors above there is a return to a more economical grid.

Obviously, the larger the building and the more repetitive the bay sizes and loadings, the greater is the degree of economy and efficiency attainable.

13.2 Design Problem

The remaining articles of this chapter will be devoted to the design of a small and purposely simplified commercial building. Particular emphasis will be placed on development of the framing system; the computation of loads; the determination of shears, bending moments, and direct loads; and the design of typical structural members.

Site drawings and architectural, structural, and mechanical/electrical drawings, together with the specifications, normally make up the contract documents for job bidding and construction. For the purposes to be served by this example, however, less than a full set of drawings and specifications will suffice. The architectural drawings of Figs. 13.4 through 13.10, for example, are adequate to show a building which might serve a small retail business or as an office building with first-floor commercial space. The building will be assumed to be located in a suburban area with ample land on three sides to provide

---

[1]Manufacturers Hanover Trust Company's Fifth Avenue Office in New York City, designed in 1953 by Skidmore, Owings, and Merrill, Architects, and Weiskopf and Pickworth, Structural Engineers.

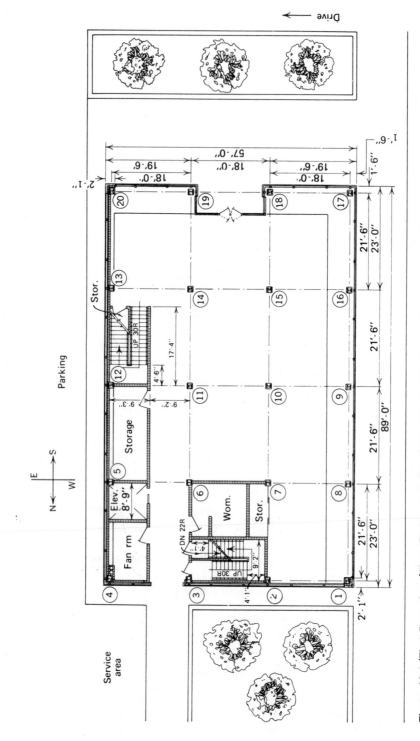

Figure 13.4./First-floor plan. (*Note*: Due to space available and need to show selected features, all plans and elevations are to approximate scale only and show limited detail. However, adequate architectural details are derivable.)

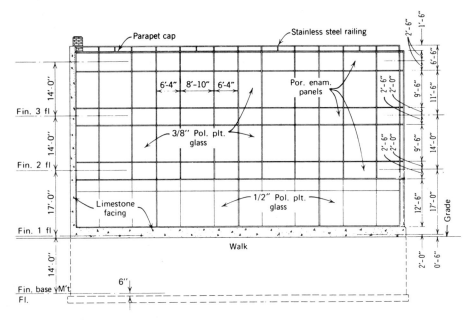

Figure 13.5/West elevation.

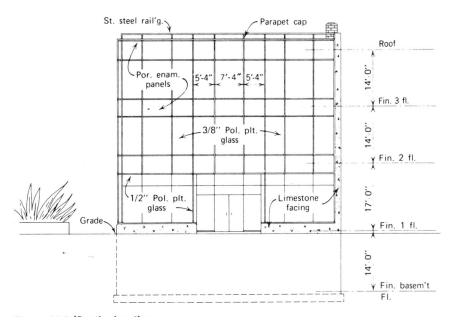

Figure 13.6/South elevation.

for access and egress. The west side faces the main thoroughfare; however, the public entrance will be located on the south side, which will be more accessible from the parking area located on the east side. The service entrance on the north side also is accessible through the parking area. With no immediately adjacent structures, there will be no need for masonry fire walls.

Plans / The basement plan is shown in Fig. 13.7. The freight and passenger elevator begins at this level, and the hydraulic equipment is located immediately adjacent to the elevator shaft. The principal mechanical (boiler) room is located at this level as well; however, the A.C. fan room is located immediately above at the first-floor level to permit air exhaust to the service area. Heavy storage and building

maintenance spaces are also located in the basement.

It will be noted that a basic 18-ft × 21-ft 6-in. planning grid has been established for this and all other floors. This grid will permit simplification of the frame and use of a considerable number of repetitive members. Further simplification is achieved by the above planning decisions, i.e., by placing the heaviest and least uniform loads in the basement. For example, by using a hydraulic elevator, there will be no need for hoisting-equipment loads at the top of the elevator shaft. By locating hydraulic equipment adjacent to the elevator shaft at the basement-floor level, the needed elevator-pit depth is minimized. Further, by locating building maintenance activities and heavy mechanical equipment in the basement—with the somewhat lighter A.C. fan equipment immediately

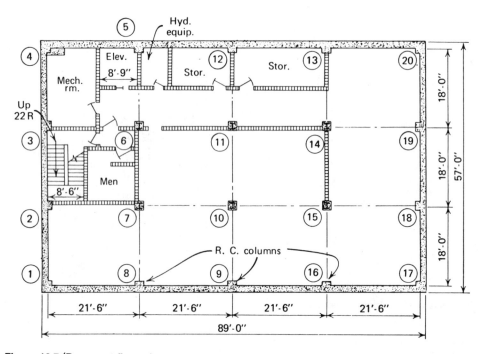

Figure 13.7/Basement floor plan.

above (at the first-floor level)—the need for a penthouse and cooling tower at the roof level has been eliminated. Finally, by using a reinforced-concrete *slab on ground* (i.e., on prepared fill), interior columns can be carried directly to independent footings beneath the slab, thus freeing them of these basement-floor loads. Admittedly, these simplifying decisions have been made here to permit a design problem to be carried through within the limits of a one-chapter presentation. However, they are not altogether unreasonable for the type of building under consideration.

Returning to the basement plan, it will be noted, too, that the fire stairway on the north side begins at this level. The front stairway does not.

The first-, second-, and third-floor plans (Figs. 13.4, 13.8 and 13.9) are quite similar. Special features at the first floor are the A.C. fan room previously mentioned, a

protected exit from the north-side stairwell to the service area, the main entrance, and elevated display platforms along the south and west walls. At the third floor, the principal difference is the absence of the front stairway.

From the plans thus far mentioned, it will be noted that ventilated interior toilet facilities for men and women are located on alternate floors.

The roof plan (Fig. 13.10) shows a drained surface surrounded by a parapet wall and interrupted only by a skylight and roof access above the rear stairwell.

Exterior Walls / The exterior foundation walls below grade are membrane-waterproofed reinforced concrete, supported on spread footings. Exterior columns in the basement story, formed partly within foundation walls, are also reinforced concrete (Fig. 13.7).

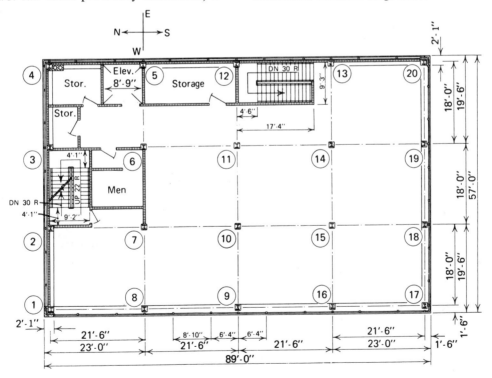

Figure 13.8/Second-floor plan.

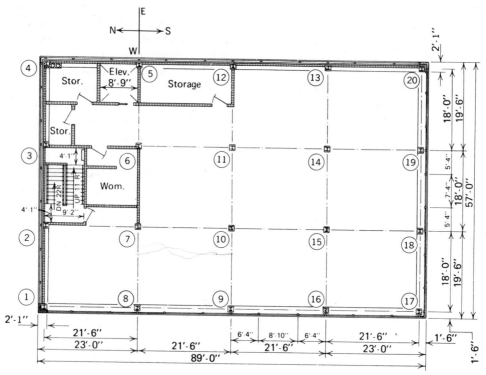

Figure 13.9/Third-floor plan.

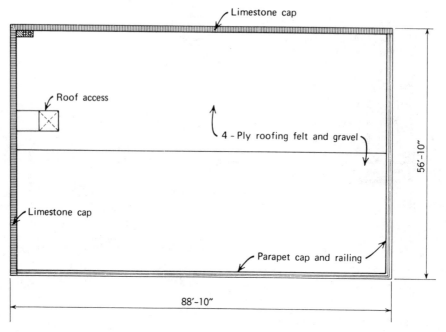

Figure 13.10/Roof plan.

First-floor framing impinging on exterior foundation walls is as follows:

1. Steel columns are carried below intersecting girders and cross-beams and are attached at their base, where they rest on the reinforced basement-story columns.

2. There are no spandrel beams; intermediate beams (north and south walls) rest on and are tied directly to reinforced-concrete foundation walls. (See Art. 13.7 for the definition of intermediate beams.)

3. Spandrel girders (east and west walls) rest on and are tied directly to reinforced-concrete walls.

4. The floor slab is turned down to join and become integral with the reinforced-concrete wall.

With all flexible steel-to-steel connections at the first-floor level, and with pin-connected interior column bases at the basement-floor level but with columns being made continuous from basement floor to second floor, the effect of the above-cited exterior-wall construction at and below the first floor is to create a rigid diaphragm at the first-floor level, freeing construction below this level of wind effects. Columns do not rise above the roof level.

The north and east exterior walls from foundation wall to parapet cap are windowless porcelain-enameled steel curtain walls, backed by 4-in. light aggregate concrete block and plastered in interior spaces (except in the elevator shaft).

At the west and south walls (Figs. 13.5 and 13.6), a 4-in. limestone veneer is used beneath the display windows. This veneer extends up the face of the building at the extreme east and north corners and returns an equal distance along the east and north sides. However, this veneer rests on the foundation walls and is only tied to columns for lateral support, i.e., the weight is transmitted directly to foundations. The

remainder of the south and west walls at the first-floor level is devoted to the entrance and plate-glass display windows.

The south and west walls at and above the second-floor level consist of a porcelain-enameled steel curtain wall with glazed portions at each floor. Backing, where it is used (Fig. 13.28), is 8-in. cinder block furred with 2-in. split furring tile and plastered.

The parapet wall at the roof level (Figs. 13.5, 13.6, and 13.10), extends 2 ft 6 in. above the top of the roof slab, and has a metal cap and railing, with built-up roofing turned up and flashed around the perimeter. The skylight is bounded by a low parapet, but, as previously noted, the structural frame does not extend above the roof level —a detail introduced here to simplify the frame for purposes of wind analysis.

The exact distance from the centerline of exterior columns to the outside face of walls will depend upon the spandrel detail; however, the distance will be taken as 18 in. for design purposes.

Fixed Partitions / Interior partitions shown in plan, enclosing the elevator shaft, stairways, and mechanical rooms, will be gypsum block, 6 in. thick and plastered ($\frac{3}{4}$ in.); all others will be considered movable and part of the live load.

Floors and Roof / The basement slab will be a 6-in. reinforced-stone-concrete slab on ground. The first, second, and third floors will consist of wire-mesh-reinforced 4-in. stone-concrete slabs supported on steel beams spaced not more than 8 ft apart—this spacing having been determined to be most economical for this type of slab (Fig. 13.11). There will be a 3-in. cinder fill and a 1-in. cement finish on top of each floor slab, and the undersides of all floor slabs will be unfinished. However, the second- and third-floor slabs will support a suspended metal-lath and gypsum-

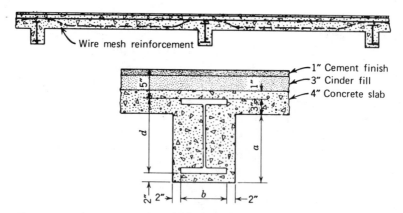

Figure 13.11/Typical beam fireproofing.

plaster ceiling with flush and recessed lighting fixtures.

The roof slab also will be 4 in. thick, topped with an average 3 in. of cinder fill over the entire roof area to permit development of a slight pitch to the finished surface for drainage, and covered with 4-ply roofing felt and gravel. This slab will also support a suspended ceiling, the same as beneath the second and third floors.

Fireproofing / It will be assumed that the code calls for fireproofing of all exposed structural steel. This will be accomplished by enclosing all framing members, horizontal and vertical, in a minimum of 2 in. of concrete (Fig. 13.11).

Unit Live Loads / On the basement slab, the live load will be 200 psf except in the boiler room, where it will be 300 psf. On the first floor, for a retail occupancy, the live load will be 100 psf, and on the second and third floors, 80 psf—allowing for office use as well. The roof will be assumed to take a 40-psf combined traffic and snow load. All stairs will be designed for a live load of 100 psf of horizontal projection.

The building will be assumed to be in a basic 20-psf wind-load area—the average height requiring a 20-psf horizontal wind

load for design. It is located in a zero earthquake zone.

References / The above live loads, and the weights of building materials used in computing dead loads, are taken from ASCE 7-88 and the Ninth Edition of the AISC Manual.[2] The structural design will be in accordance with the 1989 AISC Specification unless otherwise noted, using A36 Steel.

13.3
Floor Loads

The design load in pounds per square foot supported by the framing of any floor is equal to the sum of the live and dead loads on that floor. For example, the live load on the first floor is 100 psf. The weight of stone concrete is 144 lb per cu ft. Since the slab is 4 in. thick, its weight per square foot of floor area is $\frac{4}{12}(144) =$ 48 psf. The cinder fill and 1-in. cement finish weigh 15 and 12 psf, respectively.

[2]Some values in this and subsequent articles are at the low point of the range of weights of materials, and others are at the high or an intermediate point. In actual design, the values selected should be in concert with those for the materials specified for the job.

Therefore, the basic design load for the *first floor* is

Live load	100
Slab, fill, and finish (48 + 15 + 12)	75
Total load per sq ft	175

The basic design load for the *second and third floors* is

Live load	80
Slab, fill, and finish	75
Suspended ceiling— metal lath and gypsum plaster	10
Total load per sq ft	165

The design load for the *roof* is

Live load	40
4-in. concrete slab	48
3-in. cinder fill	15
Roofing, 4-ply felt and gravel	6
Suspended ceiling— metal lath and gypsum plaster	10
Total load per sq ft	119

The weight of steel stairs with concrete or terrazzo treads varies with the particular type employed, but an allowance of 50 psf of horizontal projection is ample. Using this value, the design load for the *stairs* is

Live load	100
Dead load	50
Total load per sq ft	150

13.4
Wall Loads

The weight of walls and fixed partitions per square foot of surface may also be computed and tabulated for reference. For example, the weight of 1 sq ft of the windowless exterior wall is determined as follows:

The weight of 4-in. light aggregate concrete block plastered on one side is 25 psf, and the weight of the curtain wall is 17.5 psf. Therefore, the weight of exterior windowless walls is 25 + 17.5 = 42.5 psf of surface.

Weights of this and other typical wall[3] and partition constructions are computed in a similar manner, and all are recorded below.

Exterior Walls

Curtain-wall alone		17.5 psf
Masonry back-up		
8-in. light aggregate concrete block		40
2-in. split-furring tile		8.5
Plaster		5
Total weight		53.5 psf
4-in. light aggregate concrete block		20 psf
Plaster		5
Total weight		25 psf
Glazing and frame		
$\frac{1}{2}$-in. plate		10 psf
$\frac{3}{8}$-in. plate		8 psf

Fixed Partitions

4-in. gypsum block	10 psf	10 psf
Plaster (two sides)	10	—
Plaster (one side)	—	5
Total weight	20 psf	15 psf
6-in. gypsum block		15 psf

13.5
Weight of Floor Framing

The area of the floor panel supported by any beam is equal to the span length multiplied by the sum of half the distances to

---

[3]A brick backing is used for the parapet wall above the roof level and is computed separately in Art. 13.23.

adjacent beams. Since panel dimensions are usually measured from center to center of supporting members, the product obtained by multiplying the design load in pounds per square foot by the panel area gives the total load brought to a beam. However, this load does not include the weight of the beam itself, nor that portion of the fireproofing which projects below the undersurface of the slab.

Since the weight of beams and girders, together with their fireproofing, is not known until after the members are designed, some system for estimating such weights must be adopted. One method is to include the weight of beams and fireproofing in the design load by assuming an equivalent load per square foot acting over the floor area. The value of the equivalent load depends upon the type of floor construction and the general arrangement of the framing. It is determined for any particular case by dividing the total weight of structural steel and fireproofing in a typical floor of a building of similar construction by the floor area. Considerable judgment must be exercised with this method if accurate results are to be obtained, since special conditions of framing often require beams of greater weight than the allowance provided by the equivalent load.

Because the weight of concrete-encased steel members is so great, the alternative method of actual calculation of such weights will be used. Figure 13.11 shows a typical rolled W-section supporting the conventional wire-fabric-reinforced slab. The cross-sectional area of the fireproofing is $a(b+4)$, where b is the flange width and a is the projection of the encased beam below the slab. Setting the top flange 1 in. below the surface of the structural slab,

$$a = d + 2 - 3 = (d - 1)$$

and

$$(d - 1)(b + 4) = FP$$
$$\text{(cross-sectional area)}.$$

Table 13.1
Weight of Concrete Fireproofing Per Lineal Foot for Selected S, M, and W Sections

Section		Beam depth d (in.)[a]	Flange width b (in.)[a]	Weight (plf)
W	8×13	8	4	56
S	8×18.4	8	4	56
W	8×18	$8\frac{1}{8}$	$5\frac{1}{4}$	66
W	8×24	$7\frac{7}{8}$	$6\frac{1}{2}$	72
W	10×15	10	4	72
W	10×45	$10\frac{1}{8}$	8	110
W	10×54	$10\frac{1}{8}$	10	128
W	10×68	$10\frac{3}{8}$	$10\frac{1}{8}$	132
M	12×11.8	12	$3\frac{3}{8}$	78
W	12×22	$12\frac{1}{4}$	4	90
W	12×26	$12\frac{1}{4}$	$6\frac{1}{2}$	118
W	12×40	12	8	132
W	12×53	12	10	154
W	12×79	$12\frac{3}{8}$	$12\frac{1}{8}$	183
M	14×18	14	4	104
W	14×22	$13\frac{3}{4}$	5	115
W	14×26	$13\frac{7}{8}$	5	117
W	14×30	$13\frac{7}{8}$	$6\frac{3}{4}$	138
W	14×43	$13\frac{5}{8}$	8	152
W	16×26	$15\frac{3}{4}$	$5\frac{1}{2}$	140
W	16×36	$15\frac{7}{8}$	7	164
W	16×40	16	7	165
W	16×50	$16\frac{1}{4}$	$7\frac{1}{8}$	170
W	18×46	18	6	171
W	18×50	18	$7\frac{1}{2}$	196
W	18×55	$18\frac{1}{8}$	$7\frac{1}{2}$	197
W	18×60	$18\frac{1}{4}$	$7\frac{1}{2}$	198
W	21×44	$20\frac{5}{8}$	$6\frac{1}{2}$	206
W	21×62	21	$8\frac{1}{4}$	245

[a]Taken from AISC Manual "Dimensions and Properties" tables.

Although stone concrete weighs about 150 psf, it is customary to assume a value of 144 for ease of calculation. Then, the weight of the fireproofing per lineal foot is

$$W_{FP} = \frac{(d-1)(b+4)}{144}(1)144$$

or

$$W_{FP} = (d-1)(b+4) \text{ plf}$$

Similar formulas can be developed for different conditions of encasement, and, of course, the areas and thus the weight can be computed for individual situations.

For the fireproofing specification for this building, i.e., beams surrounded by a minimum of 2 in. of concrete, and using the above method of computation, the per-lineal-foot weights of selected members which will be considered for use in this building have been computed and recorded in Table 13.1.

13.6
Choice of Sections

In general, wide flange sections and the miscellaneous light beams will be found to be more satisfactory than standard beams and channels; therefore, **W** and **M** shapes will be used for all basic framing needs. As noted in Table 13.1, the system of designating structural shapes on drawings is that used in previous chapters and recommended by AISC.

13.7
Framing System—General

Two common beam-to-girder framing systems are those employing third-point concentrations and those with center-point concentrations (Fig. 13.12). Girders are shown framed to columns, and intermediate beams to girders. Cross-beams are framed to column webs on grid lines. When the span of a girder is much over 16 ft, at least third-point concentration is desirable. As stated in Art. 13.5, the area of floor supported by any one beam is found by multiplying the span length by the sum of one-half the distance to adjacent beams. Span lengths are generally figured from center to center of supporting members, except where members frame to the flanges of large columns.

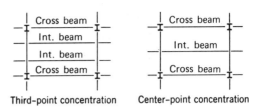

Figure 13.12/Floor framing systems.

The distance between beams, established for this design, is governed both by imposed loads and the inherent span limitations of the wire-mesh-reinforced 4-in. concrete slabs to be used. In this example problem, a maximum of 6 ft 2 in. is used for the initial assumption. (If composite steel and reinforced concrete were used, the spacing would be based on an economical balance between slab floor and size of supporting members.)

As mentioned under "Exterior Walls" in Art. 13.2, all connections from intermediate beam to girder, and from cross-beam and girder to column, at the first-floor level are flexible, i.e., will support only gravity loads and thus will neither create nor transfer end moments. This is justified by making the first-floor construction a rigid horizontal diaphragm. However, at the second-floor, third-floor, and roof levels, all connections from cross-beam to column and from girder to column will be moment-resisting. Intermediate-beam-to-girder connections, because they need resist no lateral (wind) load, will be flexible, as at the first-floor level. In general, flexible connections are more economical than moment-resisting connections.

One set of columns will run continuously from the basement floor (where they are assumed pinned) to a moment-resisting splice immediately above the second floor. This will create a continuous condition at the first- and second-floor levels (Fig. 13.13). The other set of columns will run continuously from the second floor, through the third floor, to the roof level,

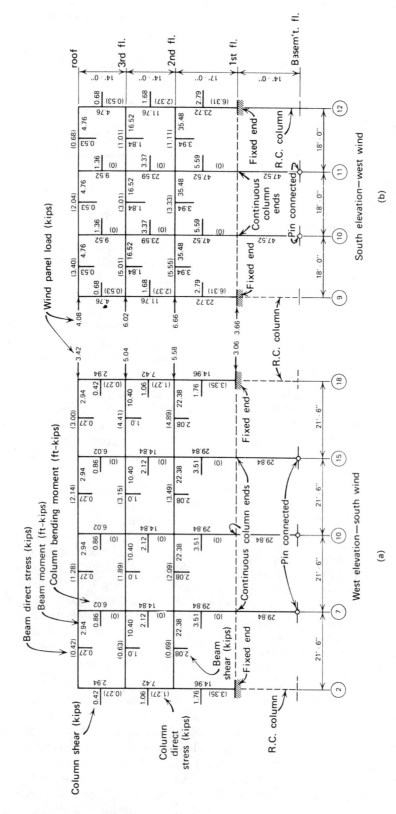

Figure 13.13/Wind bent analysis.

544

producing a continuous condition at the third and roof levels also. However, as previously noted, exterior columns will not extend below the foundation wall.

13.8
Structural Plan

The foundation plan is shown in Fig. 13.14, and the first-floor framing plan in Fig. 13.15. The most suitable arrangement of beams and columns is determined from a study of the architectural plan (Fig. 13.4). In practice, the exact dimensions are not recorded on the framing plans until most of the members have been designed, since clearances around the elevator well, stairways, etc., often necessitate slight changes in beam locations. In this problem, several dimensions have been recorded on the structural plans in order to facilitate the computations. It should be borne in mind

also that small dimensional changes may be made later without redesigning the members.

Beams and girders are not usually numbered on the framing plans, since numbers are subsequently assigned to members on the shop drawings prepared by the fabricator. However, separate *key* plans with beam numbers are sometimes used by the designer. In this problem, the beam under design is designated by its position in relation to a column. For example, the cross-beam spanning between columns ⑦ and ⑩ is denoted by —⑦→. The intermediate beam directly below when viewed from the lower edge of the drawing, is designated —⑦—. In all cases, the line with the arrow indicates the member under consideration.

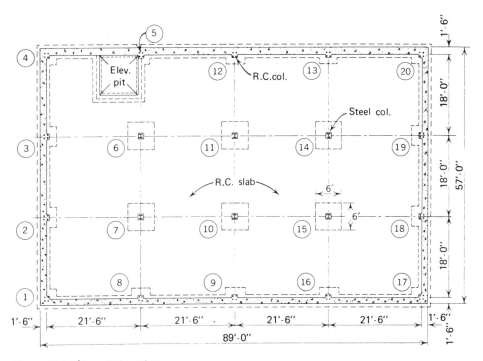

Figure 13.14/Foundation plan.

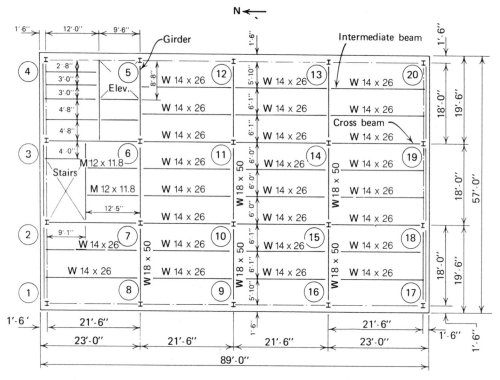

Figure 13.15/First-floor framing plan.

Unless otherwise noted, the tops of all floor beams and girders will be set 5 in. below the elevation of the finished floor, as shown in Fig. 13.11, the beams being coped as shown in Fig. 13.16 where they frame into the girders. The roof beams and girders frame, with their top flanges flush, 1 in. below the top surface of the concrete structural slab.

Design computations will be developed only for certain typical members in the text which follows. The correct sizes of

certain other beams and girders will be recorded on the plans, and many others will be left blank for use in individual problems.

13.9
Wind Design—General

As a prelude to design of typical members, it will be necessary to develop wind stresses to be added later to those resulting from gravity loads. One approximate method for this purpose is the *portal method*, which was introduced in Art. 9.40.

In this example building, floor loads of principal horizontal members are generally uniform, and horizontal loads on columns are applied in the plane of beams and girders. Interior columns carry a significantly greater gravity load than exterior

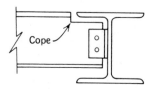

Figure 13.16/Beam detail.

columns. Therefore, the basic portal-method assumptions are reasonable.

13.10
Panel Wind Loads

Figure 13.13 (a) and (b) show line elevations of the short- and long-side framing. They will also serve as line sections through the building at the specific typical bents to be designed—designated by column number in plans, i.e., ②, ⑦, ⑩, ⑮, ⑱ and ⑨, ⑩, ⑪, ⑫.

Because all wind loads are brought to the columns at floor levels, the magnitude of the load is determined by first computing the panel area—one-half the distance between bents on either side of the column, times one-half the distance between floors above and below the floor in question (Figs. 13.5 and 13.6). Then multiply this area by the 20-psf wind load (note that although the beams are actually 5 in. below the floor levels, this slight discrepancy will be overlooked in the interest of simplicity). Thus, the following procedure is used:

1. *First-floor level and below.* There is no wind load below the first floor, because there is no exposure of below-ground construction to the wind. Also, the wind load represented by wind pressure over the lower half of the first-story height will be transmitted directly into the rigid diaphragm created by the first-floor construction and from this diaphragm to foundations, the result being no wind effect at or below the first-floor level.

2. *Second-floor level.* Columns ⑧, ⑨, and ⑯ on west side:

$$21.5\left(\frac{17}{2} + \frac{14}{2}\right) = 333 \text{ ft}^2$$

$$333(20) = 6660 \text{ lb}$$

Columns ① and ⑰ on west side:

$$\frac{22.5}{2}\left(\frac{17}{2} + \frac{14}{2}\right) = 174 \text{ ft}^2$$

$$174(20) = 3480 \text{ lb}$$

Columns ⑱ and ⑲ on south side:

$$18\left(\frac{17}{2} + \frac{14}{2}\right) = 279 \text{ ft}^2$$

$$279(20) = 5580 \text{ lb}$$

Columns ⑰ and ⑳ on south side:

$$\frac{19}{2}\left(\frac{17}{2} + \frac{14}{2}\right) = 147 \text{ ft}^2$$

$$147(20) = 2940 \text{ lb}$$

3. *Third-floor level.* Columns ⑧, ⑨, and ⑯ on west side:

$$21.5\left(\frac{14}{2} + \frac{14}{2}\right) = 301 \text{ ft}^2$$

$$301(20) = 6020 \text{ lb}$$

Columns ① and ⑰ on west side:

$$\frac{22.5}{2}\left(\frac{14}{2} + \frac{14}{2}\right) = 158 \text{ ft}^2$$

$$158(20) = 3160 \text{ lb}$$

Columns ⑱ and ⑲ on south side:

$$18\left(\frac{14}{2} + \frac{14}{2}\right) = 252 \text{ ft}^2$$

$$252(20) = 5040 \text{ lb}$$

Columns ⑰ and ⑳ on south side:

$$\frac{19}{2}\left(\frac{14}{2} + \frac{14}{2}\right) = 133 \text{ ft}^2$$

$$133(20) = 2660 \text{ lb}$$

4. *Roof.* Columns ⑧, ⑨, and ⑯ on west side:

$$21.5\left(\frac{14}{2}+2.5\right)=204 \text{ ft}^2$$

$$204(20)=4080 \text{ lb}$$

Columns ① and ⑰ on west side:

$$\frac{22.5}{2}\left(\frac{14}{2}+2.5\right)=107 \text{ ft}^2$$

$$107(20)=2140 \text{ lb}$$

Columns ⑱ and ⑲ on south side:

$$18\left(\frac{14}{2}+2.5\right)=171 \text{ ft}^2$$

$$171(20)=3420 \text{ lb}$$

Columns ⑰ and ⑳ on south side:

$$\frac{19}{2}\left(\frac{14}{2}+2.5\right)=90 \text{ ft}^2$$

$$90(20)=1800 \text{ lb}$$

These loads are shown applied in Fig. 13.13 (a) and (b) for the bents running through column ⑩, which will be used later to illustrate column design. (Similar calculations would be needed for each longitudinal and transverse bent.)

13.11
Column Shears

Referring again to Art. 9.40, it is assumed that interior columns will take twice the shear of exterior columns. Then:

1. *For wind from the south (west elevation).*

Shear in 3rd-story columns:
Total shear in 3rd story = 3.42 kips

Shear in exterior columns ② and ⑱
= 3.42/8 = 0.42 kip

Shear in interior columns ⑦, ⑩, and ⑮ = 3.42/4 = 0.86 kip

Shear in 2nd-story columns:
Total shear in 2nd story = 3.42 + 5.04
= 8.46 kips

Shear in exterior columns ② and ⑱
= 8.46/8 = 1.06 kips

Shear in interior columns ⑦, ⑩, and ⑮ = 8.46/4 = 2.12 kips

Shear in 1st-story columns:
Total shear in 1st story = 3.42 + 5.04
+ 5.58 = 14.04 kips

Shear in exterior columns ② and ⑱
= 14.04/8 = 1.76 kips

Shear in interior columns ⑦, ⑩, and ⑮ = 14.04/4 = 3.51 kips

2. *For wind from the west (south elevation).*

Shear in 3rd-story columns:
Total shear in 3rd story = 4.08 kips

Shear in exterior columns ⑨ and ⑫
= 4.08/6 = 0.68 kip

Shear in interior columns ⑩ and ⑪
= 4.08/3 = 1.36 kips

Shear in 2nd-story columns:
Total shear in 2nd story
= 4.08 + 6.02 = 10.10 kips

Shear in exterior columns ⑨ and ⑫
= 10.10/6 = 1.68 kips

Shear in interior columns ⑩ and ⑪
= 10.10/3 = 3.37 kips

Shear in 1st-story columns:
Total shear in 1st story
= 4.08 + 6.02 + 6.66 = 16.76 kips

Shear in exterior columns ⑨ and ⑫
= 16.76/6 = 2.79 kips

Shear in interior columns ⑩ and ⑪
= 16.76/3 = 5.59 kips

13.12
Beam and Girder Direct Stress

The direct compressive stress in floor cross-beams or girders—depending on whether it is the north-south or east-west bent that is under analysis—which distribute the load as shear to columns remote from the point of load at the exterior wall, is found by deducting that portion of the load taken by each column from the point of load to the most remote column on the far side of the building.

- *Roof — west elevation.*
Direct stress in beam 18-15
    = 3.42 − 0.42 = 3.00 kips
Direct stress in beam 15-10
    = 3.00 − 0.86 = 2.14 kips
Direct stress in beam 10-7
    = 2.14 − 0.86 = 1.28 kips
Direct stress in beam 7-2
    = 1.28 − 0.86 = 0.42 kip

- *Third floor — west elevation.*
Direct stress in beam 18-15
    = 5.04 − 0.63 = 4.41 kips
Direct stress in beam 15-10
    = 4.41 − 1.26 = 3.15 kips
Direct stress in beam 10-7
    = 3.15 − 1.26 = 1.89 kips
Direct stress in beam 7-2
    = 1.89 − 1.26 = 0.63 kip

(Note that only the panel load is used and not the cumulative shear from stories above.)

- *Second floor — west elevation.*
Direct stress in beam 18-15
    = 5.58 − 0.69 = 4.89 kips
Direct stress in beam 15-10
    = 4.89 − 1.40 = 3.49 kips
Direct stress in beam 10-7
    = 3.49 − 1.40 = 2.09 kips
Direct stress in beam 7-2
    = 2.09 − 1.40 = 0.69 kip

- *Roof — south elevation.*
Direct stress in girder 9-10
    = 4.08 − 0.68 = 3.40 kips
Direct stress in girder 10-11
    = 3.40 − 1.36 = 2.04 kips
Direct stress in girder 11-12
    = 2.04 − 1.36 = 0.68 kip

- *Third floor — south elevation.*
Direct stress in girder 9-10
    = 6.02 − 1.01 = 5.01 kips
Direct stress in girder 10-11
    = 5.01 − 2.00 = 3.01 kips
Direct stress in girder 11-12
    = 3.01 − 2.00 = 1.01 kips

- *Second floor — south elevation.*
Direct stress in girder 9-10
    = 6.66 − 1.11 = 5.55 kips
Direct stress in girder 10-11
    = 5.55 − 2.22 = 3.33 kips
Direct stress in girder 11-12
    = 3.33 − 2.22 = 1.11 kips

These stresses are shown in parentheses directly above each beam or girder at midspan.

13.13
Column Bending Moment

The bending moment at the top and bottom of each column (assuming point of inflection at column midheight) is equal to the column shear times one-half the story height. The bending moment in an interior column is twice that of exterior columns. For example, for the third-story columns of the south elevation:

Columns ⑨ and ⑫ :

$$M = 0.68(7)$$
$$= 4.76 \text{ ft-kips}$$

Columns ⑩ and ⑪ :

$$M = 1.36(7)$$
$$= 9.52 \text{ ft-kips}$$

The remainder of the moments are computed in a similar manner and recorded parallel to the columns at midheight.

13.14
Beam and Girder Bending Moments

The bending moments at the ends of floor cross-beams and girders are equal to the sums of the moments in the exterior columns immediately above and below the floor under consideration. The moments in cross-beams or girders of the same floor are equal and are independent of aisle width.

For example, for column ⑫, south elevation:

Roof girder moments
= 4.76 + 0 = 4.76 ft-kips
Third-floor girder moments
= 4.76 + 11.76 = 16.52 ft-kips
Second-floor girder moments
= 11.76 + 23.72 = 35.48 ft-kips

These moments are shown directly beneath each girder.

13.15
Beam and Girder Shears

The cross-beam and girder shears are computed from the fact that simple-beam moments result from equal end shears; therefore, the shear equals the moment divided by half the span; e.g., at the south elevation:

Shear in roof girders
= 4.76/9 = 0.53 kip
Shear in 3rd-floor girders
= 16.52/9 = 1.84 kips
Shear in 2nd-floor girders
= 35.48/9 = 3.94 kips

These values are shown vertically beneath each beam or girder center (assumed point of contraflexure). For the west elevation, use the moment divided by 2, i.e., 21.5/2 = 10.75.

13.16
Column Direct Stress

The column direct stress is computed directly from the beam or girder shear. Since the beam or girder shears of any floor are equal when the aisles are of equal width, there is no direct stress in the interior columns.

At the west elevation, the direct stress in column ⑱ is tension, and in column ②, compression. Likewise, at the south elevation, the direct stress in column ⑨ is tension, and in column ⑫, compression. These values are shown vertically in parentheses at each column; e.g., for column ⑨ at the south elevation:

Third-story column = 0.53 kip
Second-story column
= 0.53 + 1.84 = 2.37 kips
First-story column
= 2.37 + 3.94 = 6.31 kips

DESIGN OF MEMBERS
13.17
Typical Members—First Floor

Typical members are those which occur most frequently in the design. It is evident from a study of the first-floor framing plan of Fig. 13.15 that the conditions governing the design of beam —⑦→ apply to several other members as well. Therefore, this beam will be designed first. Attention is again called to the fact that first-floor framing will have flexible connections and will thus be designed for gravity loading only. This applies to girders, intermediate beams, and cross-beams, including beam

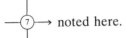 noted here.

As described in Art. 13.12 and shown in Fig. 13.15, there are no spandrel beams at the first-floor level. The panels on the west and east walls will be somewhat wider than 6 ft 0 in., but this slight difference will be neglected:

$$\text{Span} = 21.5 \text{ ft}$$

Beam —(7)→ floor area supported

$$= 6(21.5) = 129 \text{ ft}^2$$

Load (Art. 13.3) $= 175(129) = 22,600$ lb

The uniform load moment is

$$M = \frac{WL}{8} = \frac{22,600(21.5)12}{8}$$

$$\simeq 729,000 \text{ in.-lb}$$

The trial required section modulus is

$$S = \frac{M}{F_b} = \frac{729,000}{24,000} = 30.4 \text{ in.}^3$$

Referring to the AISC Manual "Allowable Stress Design Selection" table for shapes used as beams, and remembering that the weight of the beam and its fireproofing still must be factored in, select the **W** 12×26 with $S = 33.4$ in.^{3}.

Referring next to Table 13.1, the weight of fireproofing for the **W** 12×26 is 118 plf. The added weight of beam and fireproofing then is $(118 + 26)$ $21.5 \simeq 3100$ lb. The total load then is

$$3,100 + 22,600 \simeq 25,700 \text{ lb}$$

Revising the calculations for M and S_{req},

$$M = \frac{25,700(21.5)12}{8} = 829,000 \text{ in.-lb}$$

$$S_{\text{req}} = \frac{M}{F_b} = \frac{829,000}{24,000} = 34.5 \text{ in.}^3$$

$34.5 > 33.4$; therefore, select the next larger, lightest-weight section. This is the **W** 14×26 with $S = 35.3$ in.^{3}. The weight of fireproofing for this section is actually slightly less than for the **W** 12×26, and the beam weight is the same. Each end reaction is 12.85 kips, and the **W** 14×26 is compact for A36 Steel.

As explained in Chapter 5, the importance of live-load deflection in design is determined by its potential affect on integral or adjacent materials and on any aesthetic or psychological influence. The deflection of beams encased in concrete and made integral with a reinforced floor slab will be influenced by the stiffness of the total construction. Nevertheless, it is reasonable to assume that the live-load deflection (Δ) should be limited to $L/360$, or $21.5(12)/360 = 0.72$ in. The total live load is

$$100(129) = 12,900 \text{ lb}$$

and

$$I_{\text{req}} = \frac{5WL^3}{384E\Delta}$$

$$= \frac{5(12,900)21.5^3(1728)}{384(29,000,000)0.72}$$

$$= 138 \text{ in.}^4$$

For the **W** 14×26, $I = 245 > 138$ in.^{4}, which is satisfactory.

In this construction, it is apparent that the live load is not a sufficiently large percentage of the total load to create critical or even near-critical conditions.

The span of beam —(19)— and similar

beams will be a few inches longer than

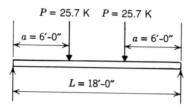

Figure 13.17/First-floor girder.

21.5 ft, due to the offset from the column centerline at the south wall to achieve embedment in the foundation wall. However, as this distance will be something less than 6 in., it will be neglected and a **W** 14×26 accepted as above. The end reactions are 12.85 kips, as before:

Girder —⑪— span = 18 ft

(center to center of columns)

The girder supports two typical beams at each third-point of the span. Since the beam reactions are each 12.85 kips, the load at each third-point is 25.7 kips. The load diagram is shown in Fig. 13.17. The weight of the girder and its fireproofing will be added later to that shown.

The maximum bending moment due to gravity loads occurs at the center and is

$$M = Pa = 25,700(6) = 154,200 \text{ ft-lb},$$
$$= 1,850,000 \text{ in.-lb}$$

The required section modulus is

$$S = \frac{M}{F_b} = \frac{1,850,000}{24,000} = 77.1 \text{ in.}^3$$

From the AISC "Allowable Stress Design Selection Table," select a **W** 21×44 as a trial section, with a section modulus of 81.6 in.³. This section is compact for A36 Steel and also is laterally restrained by having its top flange embedded in the floor slab. The total weight of the girder and fireproofing (Table 13.1) is

$$18(44+206) = 4500 \text{ lb}$$

The bending moment due to the girder weight and fireproofing is

$$M = \frac{WL}{8} = \frac{4500(18)12}{8} = 122,000 \text{ in.-lb}$$

The total moment is $122,000 + 1,850,000 = 1,972,000$ in.-lb.

The total required section modulus is

$$S = \frac{M}{F_b} = \frac{1,972,000}{24,000} = 82.2 > 81.6 \text{ in.}^3$$

It will be necessary to use the next larger most economical size, the **W** 18×50, which is also a compact section. The end reactions are

$$25.7 + \frac{\left[\dfrac{18(50+196)}{2}\right]}{1000} = 27.9 \text{ kips}$$

Checking the **W** 18×50 for deflection under live load with $\Delta_{allow} = L/360$,

$$P = \frac{21.5(6)100}{1000} = 12.9 \text{ kips}$$

$$\Delta_{allow} = \frac{L}{360} = \frac{18(12)}{360} = 0.60 \text{ in.}$$

$$I_{req} = \frac{23PL^3}{648E\Delta} \qquad \text{(Fig. 4.15)}$$

$$= \frac{23(12.9)18^3(1728)}{648(29,000)0.60}$$

$$= 265 < 800 \text{ in.}^4 \qquad \text{provided}$$

The first-floor beam —⑥—, neglecting the 4-in. partial partition, is designed as follows:

$$\text{Span} = 12.33 \text{ ft}$$

Floor area supported $= 6.08(12.33) = 75.0 \text{ ft}^2$

Load (Art. 13.3)

$$= 175(75) = 13,100 \text{ lb} = 13.1 \text{ kips}$$

The uniform load moment is

$$M = \frac{WL}{8} = \frac{13.1(12.33)12}{8}$$

$$\simeq 242 \text{ in.-kips}$$

Then

$$S_{\text{req}} = \frac{M}{F_b} = \frac{242}{24} = 10.1 \text{ in.}^3$$

From the AISC "Allowable Stress Design Selection Table" a **M** 12×10, which has $S = 10.3$ in.^{3} and is compact for A36 Steel, is selected as a trial section.

The added weight of beam and fireproofing is calculated as $10 + 80 = 90$ plf, and the added uniform load as $90(12.33) = 1110$ lb. Then

$$M = \frac{WL}{8} = \frac{(13.1 + 1.1)12.33(12)}{8}$$

$$= 263 \text{ in.-kips}$$

and

$$S_{\text{req}} = \frac{M}{F_b} = \frac{263}{24} = 10.96 \simeq 11 > 10.3 \text{ in.}^3$$

Select the **M** 12×11.8 with $S = 12$ in.^{3} and $I = 71.9$ in.^{4}. The **M** 12×11.8 with fireproofing is found to weigh $11.8 + 78$ plf (Table 13.1) = 90 plf, which is approximately the same as the **M** 12×10. Checking the live-load deflection,

$$\Delta_{\text{allow}} = \frac{L}{360} = \frac{12.33(12)}{360} = 0.41 \text{ in.}$$

$$I_{\text{req}} = \frac{5}{384} \left[\frac{7.5(12.33)^3 1728}{29,000(0.41)} \right]$$

$$= 26.6 < 71.9 \text{ in.}^4 \qquad \text{provided}$$

The end reactions are $14.2/2 = 7.1$ kips.

13.18
Typical Intermediate Beams—Second and Third Floors (Figs. 13.18 and 13.19)

Because intermediate beams are not in the plane of the wind bents, they will be connected to their respective supporting members (generally the girders) with flexible connections and thus will be expected to support gravity loads only. The total gravity load on the second and third floors (Art. 13.3) is 165 psf, which is only slightly less than on the first floor. The member sizes, therefore, probably will be the same or only slightly reduced. Nevertheless, it will be necessary to ascertain end reactions.

Span = 21.5 ft

Beam ⎯⑦⎯ floor area supported

$$= 6(21.5) = 129 \text{ ft}^2$$

$$\text{Load} = 165(129) = 21,300 \text{ lb}$$

$$= 21.3 \text{ kips}$$

The uniform-load moment is

$$M = \frac{WL}{8} = \frac{21.3(21.5)12}{8} = 687 \text{ in.-kips}$$

and

$$S_{\text{req}} = \frac{M}{F_b} = \frac{687}{24} = 28.6 \text{ in.}^3$$

From the AISC Manual "Allowable Stress Design Selection Table," and allowing for the weight of beam and fireproofing, select the **W** 12×26 as a trial section. It has $S = 33.4$ in.^{3} and $I = 204$ in.^{4}.

The added weight of beam and fireproofing (Table 13.1) is $26 + 118 = 144$ plf, and $144(21.5) = 3100$ lb = 3.1 kips. Then

$$M = \frac{(21.3 + 3.1)21.5(12)}{8} = 787 \text{ in.-kips}$$

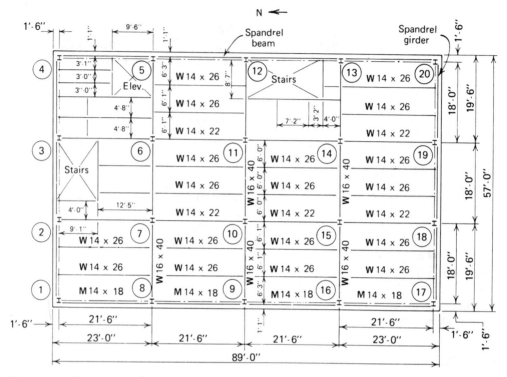

Figure 13.18/Second-floor framing plan.

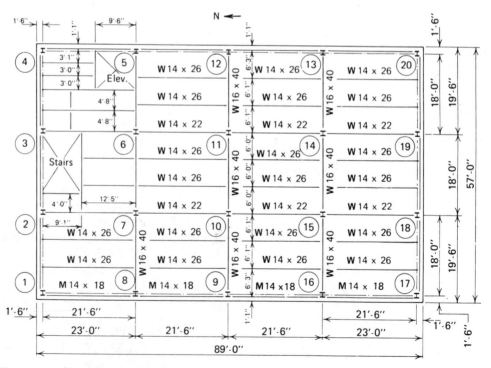

Figure 13.19/Third-floor framing plan.

and

$$S_{req} = \frac{M}{F_b} = \frac{787}{24} = 32.79 \approx 32.8 < 33.4 \text{ in.}^3$$

However, a 14-in.-depth member would be better; therefore, check the **W** 14×26, which is of the same weight. It has $S = 35.3$ in.^{3} and $I = 245$ in.^{4}. The weight of beam and fireproofing is $26 + 117 = 143$ plf; therefore

$$143(21.5) - 3075 \approx 3.1 \text{ kips}$$

$$M = 787 \text{ in.-kips}$$

and

$$S_{req} = 32.8 < 35.3 \text{ in.}^3$$

The reactions are $24.4/2 = 12.2$ kips.

Checking the live-load deflection,

$$\text{Live load} = 80(129) = 10{,}320 \text{ lb}$$

$$I_{req} = \frac{5}{384}\left[\frac{10.3(21.5)^3 1728}{29{,}000(0.72)}\right]$$

$$= 110 \text{ in.}^4$$

The **W** 14×26 provides $I = 245 > 110$ in.^{4}. Use the **W** 14×26 for intermediate beams at the second and third floors (as shown in Figs. 13.18 and 13.19).

13.19
Typical Intermediate
Beams—Roof (Fig. 13.20)

These members are designed in the same manner as the second- and third-floor intermediate beams except that the 119-psf

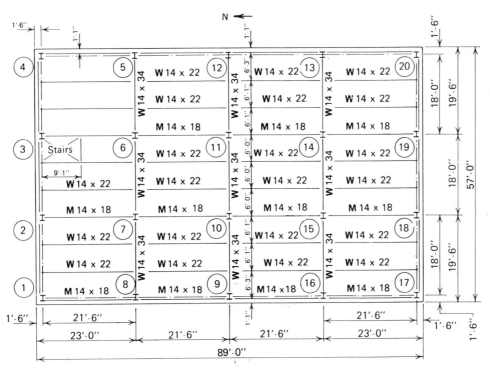

Figure 13.20/Roof framing plan.

design floor load is considerably less (Art. 13.3). A **W** 12×22 will be found to be adequate for beams such as ⑦ ; however, a **W** 14×22 is of better depth. The reactions are 9.17 kips.

13.20
Stair Framing

The design of beams around stairwells presents no special difficulty, but the work is likely to be tedious because of the irregular spacings and loads. Steel stairs are carried either on stringers which transmit the entire load to the beams at the ends of the well, or by hangers and struts supported by the side beams. The system to be used in any particular case is usually undetermined at the time the structural plans are being prepared. Therefore, it is necessary to design the stairwell framing so that the load may be supported by either the side or the end beams.

When there are two landings in any flight, as is the case with the rear stairway in this building, care must be taken to provide support for both landings.

13.21
Typical Cross-Beams and Girders—Second and Third Floors and Roof (Figs. 13.18, 13.19, and 13.20)

The gravity load on the second and third floors is 165 psf, and on the roof, 119 psf (Art. 13.3). However, as shown in Arts. 13.18 and 13.19, only intermediate beams such as ⑦ at these levels can be designed for gravity loads only, as were all members at the first-floor level (Art. 13.17).

The reason, of course, is that cross-beams such as ⑦ and girders such as ⑪ at the second- and third-floor and roof levels will be rigidly connected to columns and thus become an integral part of wind bents (or frames) through the building. As such, these members must be designed for lateral (wind) loading as well as gravity loading. Further, the rigid connections to columns which will be used to tie the frame together and develop lateral load resistance will also create restrained beam and girder ends for gravity loads. Therefore, at this point in the analysis and design, it will be necessary to treat the north-south and east-west three-story frames above the first-floor level (Fig. 13.13) as integral units and turn to indeterminate methods (Chapter 10).

As a prelude to frame analysis and member design, the following judgments will be made:

1. As previously noted, one column section will be used from basement floor to second-floor level—except at the exterior, where columns rest on foundation-wall columns immediately below the first-floor level. A second section will be used from second-floor level to roof. It will be assumed that ends of columns are restrained at floor levels, i.e., at the first-, second-, and third-floor levels and at the roof. Because only two-column sections are used, the splice will occur at the second floor—actually some 18 in. above the second-floor level. It will be assumed that no significant moments will be developed at any floor level due to gravity loads at another floor level.

2. No alternate-span live loading will be investigated. Seldom will such "pattern loading" govern the design when the ratio of live load to dead load is less than 3.

3. Because the frames are symmetrical—and because of assumptions 1 and 2 above—a direct moment distribution can be made with no sidesway correction for gravity loads (Chapter 10).

Three separate analyses will be required for each of the two typical frames selected for analysis here (Fig. 13.13): one each at the roof and the third- and second-floor levels.

Referring again to Chapter 10, it is seen that relative distribution factors (I/L values) first must be established for each member. To do this, member sizes must be approximated. This can be done in a number of ways to achieve the desired degree of accuracy. When the frame analyses are completed and members are designed, it will be necessary to check the latter against these assumptions. If the difference is significant, new assumptions should be made and the process repeated.

Example 1

Each roof cross-beam such as

must support a gravity load of 119(6)21.5 = 15,350 lb = 15.4 kips. Assume a weight of beam plus fireproofing equal to 140(21.5) = 3010 lb = 3.0 kips. The beam direct stress from wind (Fig. 13.13) is 3.0 kips. The total is 15.4 + 3.0 + 3.0 = 21.4 kips.[4] Referring to the AISC "Allowable

---

[4]It must be noted that gravity loads and direct stress (wind) are not actually additive (see Section H, AISC Specification). They are added for trial-section selection here, simply to allow for the possibility that gravity-load and wind-load moments (which are additive) might later otherwise be found to require a larger size and redesign. As will be seen in Art. 13.26, only the second-floor cross-beams are so affected after applying the allowable stress increase due to wind. In this one case, the trial section still proves adequate. One could apply the Section H provisions or simply ignore the wind direct stress if small, as is the case here, in selecting trial sections.

Uniform Load" tables for A36 Steel, **W** shapes, it is seen that a **W** 14×22 will support 21.5 kips on a 21.5-ft span, simply supported and uniformly loaded. Its weight plus fireproofing is $115 + 22 = 137 < 140$ plf, assumed. Use the **W** 14×22 as the trial section.

Example 2

Each second-floor girder such as ⑨ must support four intermediate beam reactions of 12.2 kips each. The concentrated load equivalent is $2.67P = 2.67(24.4) = 65.1$ kips. Assume a weight of girder plus fireproofing equal to $230(18) = 4.14$ kips. The girder direct stress (Fig. 13.13) is 3.33 kips. The total, then, is $65.1 + 4.1 + 3.1 = 72.3$ kips.[4] Referring again to the AISC "Allowable Uniform Load" tables, a **W** 18×46 will support 69 kips on an 18-ft span, only a little less than 72.3 kips. Its weight plus fireproofing is $171 + 45 = 217$ plf.

In summary, member sizes, and thus I/L values, are assumed as in Table 13.2. These values are recorded on the line drawing of the north-south frame through columns ②, ⑦, ⑩, ⑮, and ⑱ (Fig. 13.21a), and the east-west frame through columns ⑨, ⑩, ⑪, and ⑫ (Fig. 13.21b).

The analysis will be started with the east-west frame (Fig. 13.21b). The fixed-end moments are as follows:

1. Roof girders—**W** 14×30:

Weight of girder plus fireproofing (Table 13.1) $= 30 + 138 = 168 = 0.17$ k/ft

Intermediate beam reactions $= 9.17 + 9.17 = 18.3$ kips. Referring to Fig. 10.8,

$$M^F = \frac{wL^2}{12} + \frac{2PL}{9} = \frac{0.17(18)^2}{12} + \frac{2(18.3)18}{9}$$
$$= 77.8 \text{ ft-kips}$$

Table 13.2

Member	Section	I/L^a
Roof cross-beams	W 14×22	$199/21.5 = 9.2 \rightarrow 1.0$
3rd-floor cross-beams	W 14×22	$199/21.5 = 9.2 \rightarrow 1.0$
2nd-floor cross-beams	W 14×30	$291/21.5 = 13.5 \rightarrow 1.5$
Roof girders	W 14×30	$291/18 = 16.2 \rightarrow 1.0$
3rd-floor girders	W 16×36	$448/18 = 24.8 \rightarrow 1.5$
2nd-floor girders	W 18×46	$712/18 = 39.6 \rightarrow 2.4$
3rd- and 2nd-story exterior columns	W 10×45	$248/14 = 17.8 \rightarrow 1.1$ x-x $53.4/14 = 3.8 \rightarrow 0.4$ y-y
1st-story exterior columns	W 10×68	$394/17 = 23.2 \rightarrow 1.4$ x-x $134/17 = 7.9 \rightarrow 0.8$ y-y
3rd- and 2nd-story interior columns	W 10×54	$303/14 = 21.8 \rightarrow 1.4$ x-x $103/14 = 7.4 \rightarrow 0.8$ y-y
1st-story interior columns (and basement)	W 12×79	$662/17 = 39 \rightarrow 2.4$ x-x $216/17 = 12.7 \rightarrow 1.4$ y-y

[a]*Note.* If I/L for roof cross-beams are equaled to one, and 3rd- and 2nd-floor and column (y-y) values are proportioned accordingly, the values on the right are obtained, i.e., $x = 13.5/9.2 = 1.47 \approx 1.5$. In like manner, girder and column (x-x) values are adjusted by dividing by 16.1.

2. Third-floor girders—W 16×36:

Weight of girder plus fireproofing $= 36 + 164 = 200 = 0.2$ k/ft

Intermediate beam reactions $= 12.2 + 12.2 = 24.4$ kips. Thus

$$M^F = \frac{wL^2}{12} + \frac{2PL}{9} = \frac{0.2(18)^2}{12} + \frac{2(24.4)18}{9}$$
$$= 103 \text{ ft-kips}$$

3. Second-floor girders—W 18×46:
$$M^F = 104 \text{ ft-kips}$$

4. Roof cross-beams—W 14×22: Weight of beam plus fireproofing $= 22 + 115 = 137 \approx 0.14$ k/ft. Thus

Live load + dead load $= 119(6) = 714$
$$= 0.71 \text{ k/ft}$$

$$M^F = \frac{wL^2}{12} = \frac{(0.14+0.71)21.5^2}{12}$$
$$= 32.7 \text{ ft-kips}$$

5. Third-floor cross-beams—W 14×22:

$$M^F = 43.5 \text{ ft-kips}$$

6. Second-floor cross-beams—W 14×30:

$$M^F = 44.7 \text{ ft-kips}$$

The moment-distribution analysis follows the procedure presented in Chapter 10. Three cycles are used. No carry-over is made to column fixed ends, because no useful purpose would be served. Referring to Fig. 13.21 and reviewing the procedure for determining distribution factors (Example 1, Art. 10.5), the distribution factor (DF) for any given member is equal to its K' value divided by $\Sigma K'$ of all members framing into the joint. Beginning with the left-hand joint (south elevation, Figs. 13.21

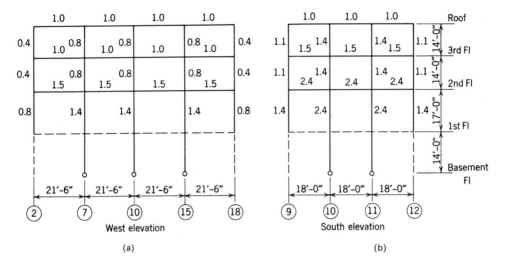

Figure 13.21/Relative I/L values.

and 13.22):

• For the roof exterior joint

$$\sum K' = 1.1 + 1.0 = 2.1$$

$$\mathrm{DF} = \frac{K'}{\sum K'} = \frac{1.1}{2.1} = 0.52$$

$$\mathrm{DF} = \frac{K'}{\sum K'} = \frac{1.0}{2.1} = 0.48$$

• For the first interior joint

$$\sum K' = 1.4 + 1.0 + 1.0 = 3.4$$

$$\mathrm{DF} = \frac{K'}{\sum K'} = \frac{1.0}{3.4} = 0.29$$

$$\mathrm{DF} = \frac{1.4}{3.4} = 0.41 \simeq 0.42$$

$$\mathrm{DF} = \frac{1.0}{3.4} = 0.29$$

Values of DF for other joints are calculated in a like manner.

Figures 13.22 through 13.27 show the moment distributions for the girders and cross-beams of the two frames of Fig. 13.21. Below each distribution diagram, free-body, shear, and moment diagrams are shown.

13.22
Design of Second- and Third-Floor and Roof Girders and Cross-Beams for Gravity Load Only

Tabulating computed maximum moments and referring to AISC charts for allowable moments in beams, a preliminary check may be made as to the adequacy of the original member-size assumptions. Because both girders and beams are embedded in the slab, they are laterally braced over the entire span in each case. The results are shown in Table 13.3.

The assumed sizes were sufficiently close to proceed. A final check of the adequacy of all assumptions as to member sizes will be made after design of columns.

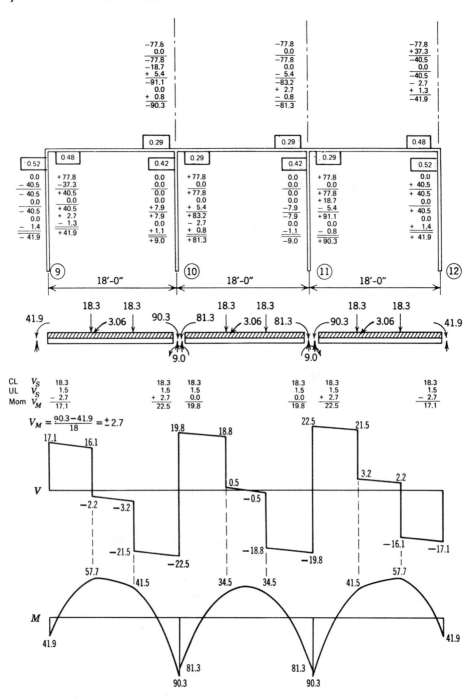

$$V_M = \frac{90.3 - 41.9}{18} = \pm 2.7$$

Figure 13.22/Roof girders.

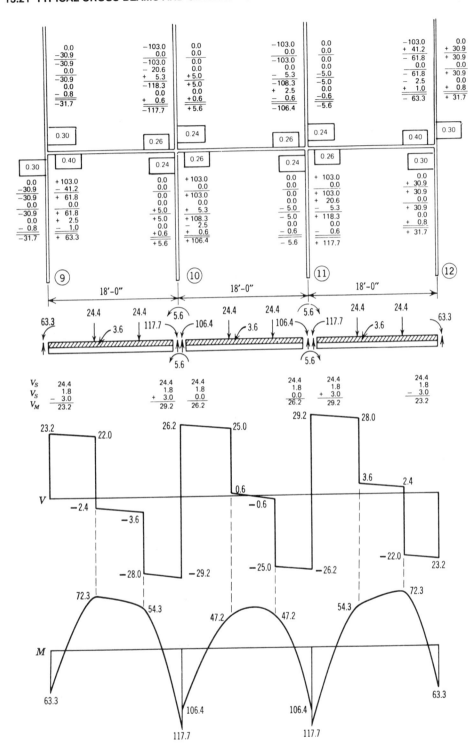

Figure 13.23/Third-floor girders.

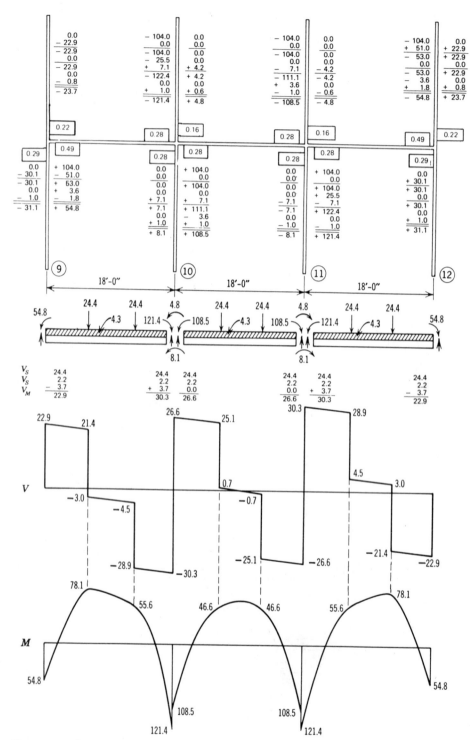

Figure 13.24/Second-floor girders.

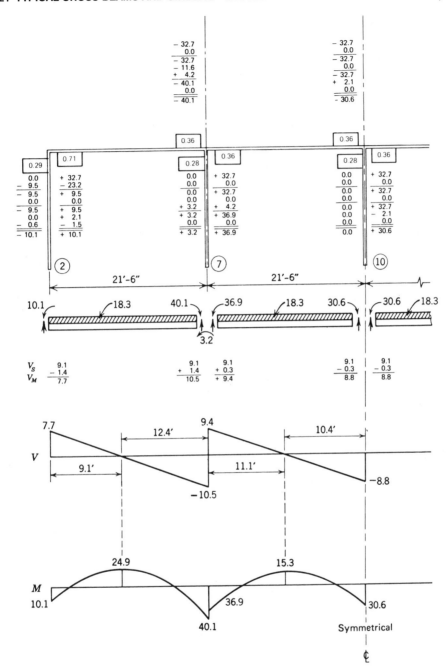

Figure 13.25/Roof cross-beams.

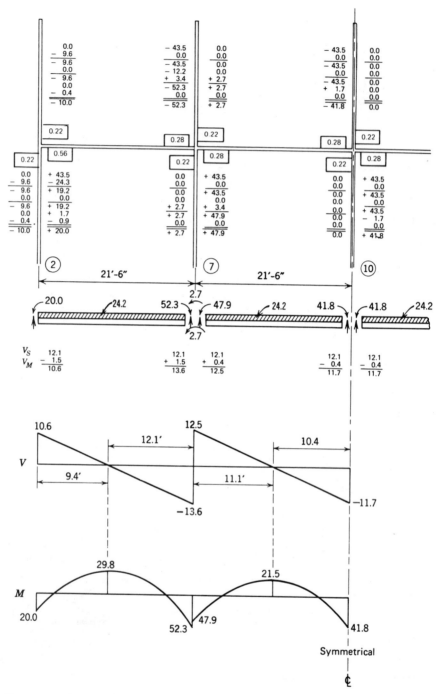

Figure 13.26/Third-floor cross-beams.

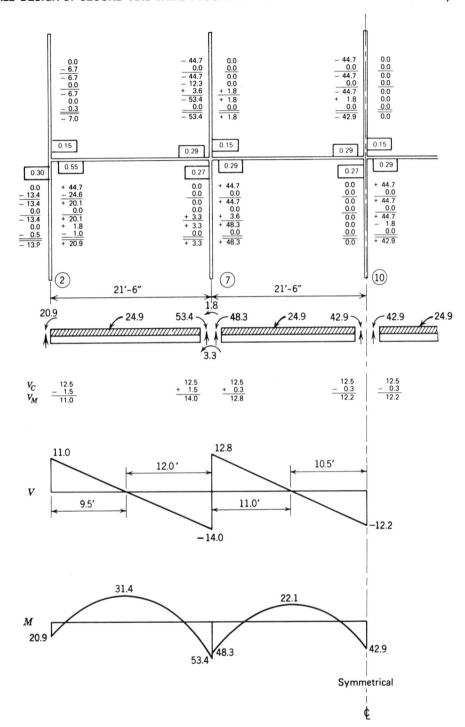

Figure 13.27/Second-floor cross-beams.

Table 13.3

	Maximum Moment (ft-kips)	Section		Assumed Section	
Roof girders	90.3	**W**	16×31		
			or	**W**	14×30
		W	14×34		
3rd-floor girders	117.7	**W**	16×40	**W**	16×36
2nd-floor girders	121.4	**W**	16×40	**W**	18×46
Roof cross-beams	40.1	**M**	14×18	**W**	14×22
3rd-floor cross-beams	52.3	**W**	14×22	**W**	14×22
2nd-floor cross-beams	53.4	**W**	14×22	**W**	14×30

13.23
Design of Second- and Third-Floor and Roof Spandrel Beams and Girders

Spandrel beams are designed like any other beam, once the load diagram has been drawn. As previously mentioned, spandrel beams at the first floor have been eliminated by joining slab and foundation wall at the east and west walls.

Typical Spandrel Beam —⑧→ —**Second Floor** / From Fig. 13.28, it will be seen that the curtain wall receives its primary support from zee angles at the column, and secondary support, 6 ft 4 in. inward from each column (Fig. 13.5), from zee angles bolted to the spandrel. Horizontally, the curtain-wall load will be divided among all four points of support, i.e., one at each column and one at 6 ft 4 in. from each column. (For safety, the column supports and horizontal curtain-wall members also will be designed to carry the full load if need be.)

Vertically, the curtain wall does not extend below the second floor, i.e., there is no curtain-wall load carried to the first floor. The weight of first-story display windows joined with the curtain wall (Fig. 13.28) will be carried directly to the first-floor slab and thus to foundation walls.

The 2-ft-4-in.-high cinder-block and tile back-up between columns (Fig. 13.28) rests directly over the spandrel. The 3-in. fill and 1-in. cement topping over the base floor slab do not extend to the spandrel. No reduction will be made, however, nor will a load addition be made for the drape track, falsework, mechanical equipment, or extra thickness of concrete fireproofing beneath the spandrel.

The loading diagram for this spandrel beam is shown in Fig. 13.29, and load calculations follow. Inch dimensions are reduced to decimals of a foot for convenience. Area A = floor load (uniform); B = weight of curtain wall below the top of the base slab and over the exterior of the back-up (concentrated); C = weight of the back-up (uniform); D = portion of the glazed curtain-wall above the back-up (concentrated).

The concentrated loads constitute such a small percentage of the total load that they will be converted to a concentrated load equivalent:

$$\text{C.L.E.} = \frac{8Pa}{L} = \frac{8(0.9)6.33}{21.5} = 2.12 \text{ kips}$$

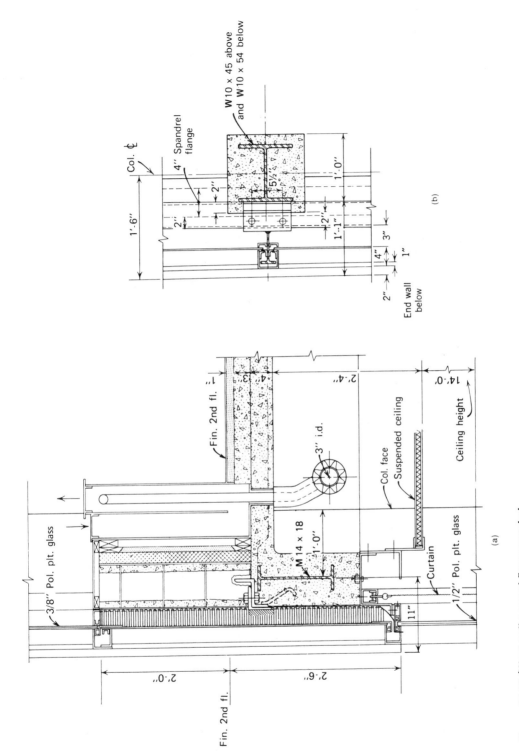

Figure 13.28/West-wall second-floor spandrel.

567

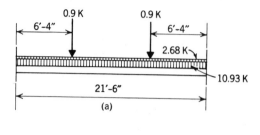

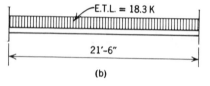

Figure 13.29/Second floor spandrel beam.

It is now necessary to assume a spandrel section to obtain the weight of the beam plus fireproofing—assume 130 plf, or 2.8 kips. Referring to Art. 13.9 and Fig. 13.13, it may be assumed that wind loads on the exterior west wall frame will be one-half those shown; therefore, assume a direct stress of 4.89/2 = 2.45 kips.[5] The total assumed load, then, is 13.6 + 2.12 + 2.80 + 2.45 = 21 kips. Referring to the AISC "Allowable Uniform Load" tables, a **W** 14 × 22 will support 21.5 kips on this span. Recalling that actual sizes likely will be less because of end restraint, accept an **M** 14 × 18 (see Table 13.4). The weight of the **M** 14 × 18 plus fireproofing (Table 13.1) is 18 + 104 = 122 plf = 2.6 kips, and the total gravity load is 13.6 + 2.12 + 2.6 = 18.3 kips.

It is reasonable to assume that relative I/L and distribution factors will be approximately the same as those for the interior frame of Fig. 13.13; therefore, a proportional maximum moment will be calculated (Art. 13.21). The maximum moment for the interior second-floor cross-beam was 53.4 ft-kips (Fig. 13.27), for a total

[5]*Ibid.*

uniform load of 24.9 kips. Then,

$$M = \frac{18.3}{24.9}(53.4) = 39.3 \text{ ft-kips}$$

Referring again to the AISC charts and tables for allowable moments in beams, it is seen that the **M** 14 × 18 is the most economical selection for gravity loads. The reactions will be 9.15 kips.

Typical Spandrel Beam ⟶─⑧─⟶—**Third Floor** / At the third floor, this beam will carry a somewhat larger concentrated load due to the glazed curtain wall beneath as well as above the floor level. This will increase each concentrated load (P) by

$$8(3.17 + 4.42)4.5 = 273 \simeq 270 \text{ lb}$$

and

$$\text{C.L.E.} = 2.12 + \frac{8(0.27)6.33}{21.5} = 2.74 \text{ kips}$$

The total gravity load—assuming the **M** 14 × 18 again to be adequate—is

$$13.6 + 2.74 + 2.6 = 18.9 \text{ kips}$$

Then, referring to Fig. 13.26,

$$M = \frac{18.9}{24.2}(52.3) = 40.8 \text{ ft-kips}$$

Referring to the AISC charts and tables for allowable moments in beams, the **M** 14 × 18 is again adequate. The reactions will be 9.45 kips.

Typical Spandrel Beam ⟶─⑧─⟶—**Roof** / These beams support a 2-ft 6-in. curtain-walled parapet, surmounted by a 1-ft 6-in. cap plate and railing (Fig. 13.5). The weights of the parapet and the curtain wall

Table 13.4
Loads on second floor spandrel

	Uniform (W)	Concentrated (P)
$A = 165(21.5)3.08^a$	10,930 lb	—
$B = 17.5(3.17+4.42)4.5$	—	598 lb
$C = 53.5(21.5)2.33$	2,680	—
$D = 8(3.17+4.42)5.0$	—	304
Totals	13,610	902
	$\approx 13,600$ lb	≈ 900 lb

^{a}*Note.* Although 6 ft-2 in. is used here, when columns are designed, 6 ft-3 in. is found to be correct. This latter dimension is shown on framing plans and does not require a different member size.

below the roof level are as follows:

B = curtain wall alone	17.5 psf
C = parapet—curtain-wall backing of 8-in. low-absorption clay brick	89.0 psf
D = Glazing ($\frac{3}{8}$-in. plate)	8.0 psf
E = Railing and cap plate	5.0 plf

The resulting loads on the spandrel beam are shown in Table 13.5. Then

$$\text{C.L.E.} = \frac{8(0.93)6.33}{21.5} = 2.2 \text{ kips}$$

The total gravity load, assuming the **M** 14×18 again, is $12.8 + 2.2 + 2.6 = 17.6$ kips. Then, referring to Fig. 13.25,

$$M = \frac{17.6}{18.3}(40.1) = 39.1 \text{ ft-kips}$$

From the AISC allowable-moment charts, it is seen that the **M** 14×18 is again adequate. The reactions are 8.8 kips.

Typical Spandrel Girder —⑱—**Second Floor and Above** / The design of these members will not be undertaken here; however, the point must be made that the concentrated loads of intermediate beams and curtain wall will not coincide, and because the former are so large in proportion to the uniform loads, a C.L.E. procedure should not be used. Instead the south wall frame above the second-floor level should be analyzed, as in Figs. 13.22 to 13.24.

13.24

Typical Interior Column ⑩— Gravity Loads

From Figs. 13.22 to 13.27, it will be noted that the bending moments due to continuity in this column are very small, i.e., zero on the y-y axis and ranging from 0 to 9.0

Table 13.5
Loads on roof spandrel

	Uniform (W)	Concentrated (P)
$A = 119(21.5)3.08$	7,880 lb	—
$B = 17.5(3.17+4.42)5.0$	—	660 lb
$C = 89(21.5)2.5$	4,780	—
$D = 8(3.17+4.42)4.5$	—	270
$E = 5(21.5)$	110	—
Totals	12,770	930 lb
	$\approx 12,800$ lb	

ft-kips on the *x-x* axis. Therefore, these moments will be neglected.

The direct loads will be taken directly from the shear diagrams. Referring to Art. 13.21, the following assumed column sizes may be tabulated and their weights, with a minimum of 2 in. of fireproofing, may be determined (Table 13.1):

3rd-story column	**W** 10×54	14 ft	2.5 kips
2nd-story column	**W** 10×54	14 ft	2.5 kips
1st-story column	**W** 12×79	17 ft	4.5 kips
Basement column	**W** 12×79	14 ft	3.7 kips

No live-load reduction is permitted for the roof; however, reductions can be taken for other floors (UBC, Chapter 23):

Third Floor

1. Area supported by column ⑩ = $21.5(18) = 387$ ft^2.

2. Live load on the third floor (Art. 13.3) = 80 psf.

3. Dead load on the third floor = $85(387) + 3690$ (weight of one **W** 16×40 girder) + 8966 (weight of one **W** 14×22 and two **W** 14×26 beams) = 45,551 lb.;

Unit dead load = $45{,}551/387 \approx 117$ psf

4. Then, from UBC:

(a) $387 > 150$ ft^2.
(b) $0.08(387) = 31$ per cent.
(c)

$$R = 23\left[1 + \frac{D}{L}\right]$$
$$= 23\left[1 + \frac{117}{80}\right] = 56.6 \text{ per cent}$$

(d) Maximum reduction = 56.6 per cent.
5. Use 31 per cent; therefore, the allowable unit reduction is $80(0.31) = 24.8$ psf, and the total reduction is $24.8(387)/1000 = 9.6$ kips.

Second Floor

1. Area supported by column ⑩ = $2(21.5)18 = 774$ ft^2.

2. Average live load from third and second floor = 80 psf.

3. Average dead load from third and second floor = 117 psf.

4. Then, from UBC:

(a) $774 > 150$ ft^2.
(b) $0.08(774) = 62$ per cent.
(c)

$$R = 23\left[1 + \frac{117}{80}\right] = 56.6 \text{ per cent}$$

(d) Maximum reduction = 56.6 per cent.
5. Use 56.6 per cent; therefore, the allowable unit reduction is $80(0.566) = 45.3$ psf, and the total reduction is $45.3(774)/1000 = 35.1$ kips.

First Floor

1. Area supported by column ⑩ = $3(21.5)18 = 1161$ ft^2.

2. Average live load from third, second, and first floors = $[2(80) + 100]/3 = 86.7$ psf.

3. Average dead load from third, second, and first floors = $2(45{,}551) + 75(387) + 4428$(weight of one **W** 18×50 girder) + 9224(weight of three **W** 14×26 beams) $\approx 133{,}800$ lb. Thus,

Unit load = $133{,}800/1161 \approx 115$ psf

4. Then, from UBC:

(a) $1161 > 150$ ft^2.
(b) $0.08(1161) = 93$ per cent.
(c)

$$R = 23\left[1 + \frac{115}{86.7}\right] = 53.5 \text{ per cent}$$

(d) Maximum reduction = 54 per cent.
5. Use 53.5 per cent; therefore, the allowable unit reduction is $86.7(0.535) = 46.4$ psf

and the total reduction is 46.4 psf(1161)/1000 = 53.9 kips.

Then the column gravity loads are as follows:

From roof	59.9 kips	(8.8 + 8.8 + 22.5 + 19.8)
Col. wt.	2.5	
	62.4 kips at bottom of 3rd-story column	
From 3rd fl.	78.8 kips	(11.7 + 11.7 + 29.2 + 26.2)
L.L. Reduc.	−9.6	at 31 per cent
	131.6 kips at top of 2nd-story column	
Col. wt.	2.5	
	134.1 kips at bottom of 2nd-story column	
From 2nd fl.	81.3 kips	(12.2 + 12.2 + 30.3 + 26.6)
L.L. Reduc.	−35.1	at 56.6 per cent
	180.3 kips at top of 1st-story column	
Col. wt.	4.5	
	184.8 kips at bottom of 1st-story column	
From 1st fl.	81.3 kips	(no continuity from beams or girders)
L.L. Reduc.	−53.9	at 53.5 per cent
	212.2 kips at top of basement column	
Col. wt.	3.7	
	215.9 kips at bottom of basement column	

Because this example structure has no diagonal bracing system or shear walls, i.e., masonry walls which are in the plane of one or several frames, there is no lateral

support which can be relied upon to prevent sidesway (lateral displacement of the top of columns). The curtain walls are not adequate for this purpose. Therefore, it is necessary to take particular care in selecting K values for column design to establish effective lengths (Chapter 6).

Referring to Fig. 6.4, it is seen that the following conditions apply to the basement section of interior column ⑩:

1. At the base of the basement column, a simple base plate with anchor bolts will be used. This theoretically is a pin-end condition, but in fact does provide some restraint. The column end would have to be designed as a true frictionless pin to be otherwise. Therefore, the base may be considered to be of the type "rotation free, translation fixed."

2. The top of the basement column is laterally braced by the first-floor diaphragm-like construction. This will restrain translation or sidesway. The connections (beam end, girder to column) are flexible, so they will not prevent rotation. However, the column is continuous through this diaphragm floor, and the first-story portion of the column will tend to prevent rotation. Therefore, there is partial restraint at the top.

3. The condition described above is seen to fall somewhere between classifications *b* and *d* of Fig. 6.4. Knowing that other factors will dictate the final size of the column, a conservative K value of 1.0 is selected.

4. The effective length, then, is

$$KL = 1.0(14) = 14 \text{ ft}$$

5. The total computed gravity load is 215.9 kips. Referring to the AISC "Allowable Axial Load" (safe-load) tables for columns, select a **W** 8×48, which will carry 215 kips.

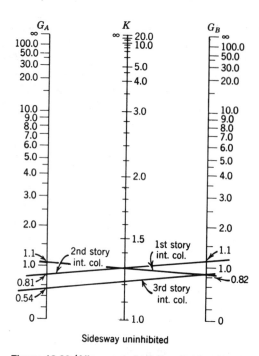

G_A K G_B

Sidesway uninhibited

Figure 13.30/Alignment chart for effective length of columns in continuous frames. Courtesy of the American Institute of Steel Construction. Original source: Column Research Council.

The first-story section of interior column

(10), extending from basement slab to 18 in. above the second floor, is the more critical, and establishing its effective length is more difficult. Referring to Fig. 13.30, it is seen that it is necessary to establish the values G_A and G_B for both axes.

1. The top of the first-story column will be free to displace (translate) laterally; however, some degree of resistance to joint rotation will be provided by the rigid beam-end girder-to-column connections. For the y-y axis, using the originally assumed **W** 10×54 section for the second- and third-story column ($I_y = 103$) and **W** 12×79 section for the basement and first-story column ($I_y = 216$), and the previously selected **W** 14×22 cross-beam

$(I_x = 199)$,

$$G_A = \frac{\sum \dfrac{I_c}{L_c}}{\sum \dfrac{I_g}{L_g}} = \frac{\dfrac{103}{14} + \dfrac{216}{17}}{2\left(\dfrac{199}{21.5}\right)} = 1.08$$

For the x-x axis, using the **W** 16×40 girder ($I_x = 518$), and $I_x = 662$ for the **W** 12×79 column,

$$G_A = \frac{\dfrac{306}{14} + \dfrac{662}{17}}{2\left(\dfrac{518}{18}\right)} = 1.06$$

2. The bottom of the first-story column can now be considered in more detail. In so doing, judgment must be applied. There is some degree of restraint provided by the basement portion of the continuous column and resulting from the fact that no lateral translation below the first-floor diaphragm can occur. This degree of restraint for the base of the first-floor column is determined by the relative stiffness of the first-floor column and the member offering the restraint, i.e., the basement column:

$$G_B = \frac{I_c/17}{I_c/17} = 0.82$$

3. Using the alignment chart of Fig. 13.30, and laying a straightedge between $G_A = 1.08$ and 1.06, and $G_B = 0.82$, it is seen that $K = 1.3$—the larger value. Therefore, the y-y axis is critical. The effective length, therefore, is

$$1.3(17) = 22 \text{ ft}$$

and $P = 184.7$ kips.

4. From the AISC column safe-load tables, select a **W** 12×53 with a capacity of 189 kips. Note that G_A was based upon a column size of **W** 12×79. A recalculation of G_A based upon the **W** 12×53 would give an even smaller value. Therefore, no further refinement is needed for gravity-load design. Since this column size is the larger of the two (basement and first floor), the **W** 12×53 will be temporarily accepted for the two-story basement-through-first-story column.

The second-through-third-story column must be designed in a similar manner.

1. At the roof, where the members are the **W** 10×54 assumed column ($I_x = 303$ and $I_y = 103$), **M** 14×18 cross-beams ($I_x = 148$), and **W** 14×34 girders ($I_x = 340$),

x-x axis:

$$G_A = \frac{\dfrac{303}{14}}{2\left(\dfrac{340}{18}\right)} = 0.58$$

y-y axis:

$$G_A = \frac{\dfrac{103}{14}}{2\left(\dfrac{148}{21.5}\right)} = 0.54$$

2. At the third floor, where the members are the **W** 10×54 assumed column, **W** 14×22 cross-beams ($I_x = 199$), and **W** 16×40 girder ($I_x = 518$),

x-x axis:

$$G_B = \frac{\dfrac{303}{14} + \dfrac{303}{14}}{2\left(\dfrac{518}{18}\right)} = 0.75$$

y-y axis:

$$G_B = \frac{\dfrac{103}{14} + \dfrac{103}{14}}{2\left(\dfrac{199}{21.5}\right)} = 0.79$$

3. At the second floor, where the assumed members are the **W** 10×54 column, **W** 12×79 column ($I_x = 662$ and $I_y = 216$), **W** 14×22 cross-beams ($I_x = 199$), and **W** 16×40 girders ($I_x = 518$), referring to the calculations for the first-story column,

x-x axis:

$$G_C = 1.06$$

y-y axis:

$$G_C = 1.10$$

4. Using the above values and the alignment chart of Fig. 13.30:

For the second-story column:

$$\text{Effective length} = 1.3(14) = 18.2 \text{ ft}$$
$$P = 134.1 \text{ kips}$$

Select a **W** 10×45, with a capacity of 156 kips. The assumed size was **W** 10×54.

For the third-story column:

$$\text{Effective length} = 1.2(14) = 16.8 \approx 17 \text{ ft}$$
$$P = 62.4 \text{ kips}$$

The **W** 10×45 is adequate here as well. Therefore, a **W** 10×45 will be temporarily accepted for the second- and third-story interior column ⑩, for gravity loads.

13.25

Typical Exterior Column ⑨— Gravity Loads

From Fig. 13.22, it is seen that a typical moment in exterior column ⑨ from the girder is significant (41.9 ft-kips), whereas the spandrel-induced moment in columns[6] such as ⑧, ⑨, and ⑯ (less than those for columns ⑦, ⑩, and ⑮, from Fig.

[6]With reference to the minor axis of the column.

13.25) is either zero or small and can thus be safely neglected.

Spandrel beams are connected to the outer flange of columns at their flanges (Fig. 13.28), bringing the reaction very near to the outer flange face. The rigid connection of girders to column flanges also brings the girder reactions very near to the column flange face. For all practical purposes, therefore, the vertical loads may be assumed to be symmetrically imposed on the columns (Fig. 13.31).

Loads and Moments—Roof / From Fig. 13.22, the girder reaction is 17.1 kips, and, as previously noted, the moment is 41.9 ft-kips.

Loads and Moments—Third Floor / The spandrel reactions are 9.45 kips each, the girder reaction is 23.2 kips, and the girder moment is 63.3 ft-kips (Fig. 13.23), divided evenly, 31.7 above and 31.7 below the floor level.

Loads and Moments—Second Floor / The spandrel reactions are 9.15 kips each, the girder reaction is 22.9 kips, and the girder moment is 54.8 ft-kips (Fig. 13.24), divided 23.7 above and 31.1 below the floor level.

The moments due to eccentricity of gravity loads (difference between girder and spandrel beam loads) will be distributed one-half each to the column above and column below. This is not entirely accurate at the second-floor level because of the difference in relative stiffnesses of columns, but will be accepted because of the negligible effect. Live-load reductions are calculated in the same manner as for the interior column (floor area = 193.5 ft^2), and are as follows:

Roof	0.0
3rd floor	2.4 kips at 15.5 per cent
2nd floor	9.6 kips at 31.0 per cent

The column weights, using previously as-

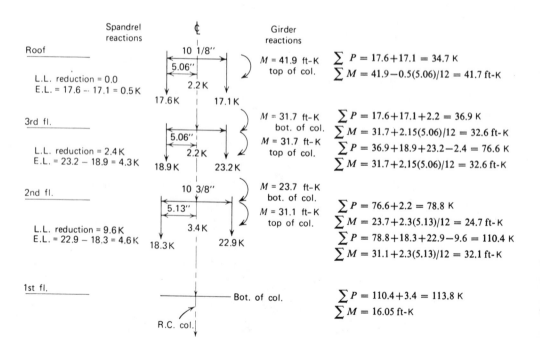

Figure 13.31/Exterior column ⑨.

sumed sizes, are

3rd-story
column   **W** 10×45   14 ft, 2.2 kips

2nd-story
column   **W** 10×45   14 ft, 2.2 kips

1st-story
column   **W** 10×68   17 ft, 3.4 kips

It should be noted that the exterior column does not extend below the first-floor level; it lands on the foundation-wall column.

These values, as well as P and M calculations are shown in Fig. 13.31. Bending will be considered for the x-x axis only, because of the obviously negligible moment on the y-y axis.

Essentially the same end conditions apply, i.e., the same as for the interior column. The primary difference is the nearly fixed-end condition at the base of the exterior column, making this condition of the type "rotation fixed, translation fixed."

It is now necessary to ascertain effective lengths for design. Referring to Art. 13.24 for the procedure, G values will be calculated only for the first assumed column size and will not be recalculated for any subsequent adjusted column size.

1. At the roof, where the assumed column is a **W** 10×45 ($I_x = 248$ and $I_y = 53.4$),

x-x axis:

$$G_A = \frac{\dfrac{248}{14}}{\dfrac{340}{18}} = 0.94$$

y-y axis:

$$G_A = \frac{\dfrac{53.4}{14}}{2\left(\dfrac{148}{21.5}\right)} = 0.28$$

2. At the third floor,

x-x axis:

$$G_B = \frac{\dfrac{248}{14} + \dfrac{248}{14}}{\dfrac{518}{18}} = 1.23$$

y-y axis:

$$G_B = \frac{\dfrac{53.4}{14} + \dfrac{53.4}{14}}{2\left(\dfrac{199}{21.5}\right)} = 0.41$$

3. At the second floor, where the assumed column below is a **W** 10×68 ($I_x = 394$ and $I_y = 134$),

x-x axis:

$$G_C = \frac{\dfrac{248}{14} + \dfrac{394}{17}}{\dfrac{518}{18}} = 1.42$$

y-y axis:

$$G_C = \frac{\dfrac{53.4}{14} + \dfrac{134}{17}}{2\left(\dfrac{199}{21.5}\right)} = 0.63$$

4. Consider the base of the exterior column, where the steel column terminates on top of the pilastered exterior foundation wall. Normally, when steel columns terminate at standard base plates that are not *specifically* designed to provide restraining conditions, there is only moderate inherent restraint and the AISC recommends a G value of 10. However, the proposed solution has significantly different inherent characteristics. The proposed detail is to extend the column through the first-floor beams (even though they are attached with flexible connections) and terminate them a few inches below the first floor with a standard base plate. This provides the necessary means of developing a nearly fully restrained end. The floor beam can develop a lateral force in one direction, and the anchor bolts a lateral force in the opposite direction, thus making the necessary couple for a restrained end. For the sake of simplicity, fully restrained ends will be assumed for the wind analysis only, and a slightly more conservative estimate will be used for the design condition; i.e., a G value of 1.0 will be used rather than the value of zero usually assigned to fully restrained ends.

Referring again to the alignment chart (Fig. 13.30), the K values in Table 13.6 are selected.

Design of First-Story Column ⑨, Gravity Loads / From Fig. 13.31, the equivalent load is

$$P + P' = P + MB_x$$

$$= 110.4 + [32.1(0.264)12]$$

$$= 110.4 + 101.6 = 212 \text{ kips}$$

The effective length is approximated as

$$KL = 1.27(17) = 21.5 \text{ ft}$$

From the AISC safe-load tables, the assumed **W** 10×68 will carry a concentric load of 252 kips at this length. The **W** 10×49 is not adequate. Try the

W 10×54, with carrying capacity = 205 kips:

$$A = 15.8 \text{ in.}^2, \qquad r_x = 4.37 \text{ in.}$$
$$S_x = 60.0 \text{ in.}^3, \qquad r_y = 2.56 \text{ in.}$$
$$S_y = 20.6 \text{ in.}^3, \qquad d/2 = 5.06 \text{ in.}$$

Make the stability check (Chapter 6) for the x-x axis.

1. The maximum unit axial and bending stresses are

$$f_a = \frac{P}{A} = \frac{110.4}{15.8} = 6.99 \text{ ksi}$$

$$f_b = \frac{M}{S_x} = \frac{32.1(12)}{60.0} = 6.42 \text{ ksi}$$

2. The ratio $K_y L_y / r_y$ is

$$\frac{1.27(17)12}{2.56} = 101$$

3. The allowable axial stress (AISC Table C-36) is

$$F_a = 12.85 \text{ ksi}$$

4. The ratio of axial stresses is

$$\frac{f_a}{F_a} = \frac{6.99}{12.85} = 0.54$$

Since $0.54 > 0.15$, the modified interaction formula must be satisfied.

5. Determine the amplification factor:

$$\frac{K_b L_b}{r_b} = \frac{K_x L_x}{r_x} = \frac{1.37(17)12}{4.37} = 64.0$$

Table 13.6

Story	Axis	G values		K
3rd	x-x	$G_A = 0.94$	$G_B = 1.24$	1.32
	y-y	$G_A = 0.28$	$G_B = 0.41$	1.11
2nd	x-x	$G_B = 1.23$	$G_C = 1.40$	1.40
	y-y	$G_B = 0.41$	$G_C = 0.62$	1.18
1st	x-x	$G_C = 1.42$	$G_D = 1.0$	1.37
	y-y	$G_C = 0.63$	$G_D = 1.0$	1.27

From AISC Table 8,

$$F'_e = 36.46 \text{ ksi}$$

Then

$$1 - \frac{f_a}{F'_e} = 1 - \frac{6.99}{36.46} = 0.81$$

6. The reduction factor, from H1 of the 1989 AISC Specification, is

$$C_m = 0.85$$

7. Testing the stress ratios,

$$\frac{f_a}{F_a} + \left[\frac{C_m}{1 - \dfrac{f_a}{F'_e}}\right]\frac{f_b}{F_b} = 0.54 + \left(\frac{0.85}{0.81}\right)\frac{6.42}{22.00}$$

$$= 0.54 + 0.31 = 0.85 < 1.0$$

8. Making an end check,

$$\frac{f_a}{0.6F_y} + \frac{f_b}{F_b} = \frac{6.99}{22.00} + \frac{6.42}{22.00} = 0.61 < 1.0$$

The **W** 10×54 (rather than the **W** 10×68 assumed) is accepted. It is possible that even a lighter section might be acceptable for these conditions; however, further refinement will not be made until after investigation for wind loads. (See also note at end of this article regarding changed values of K.)

Design of Second- and Third-Story Column ⑨, Gravity Loads / It is not readily apparent which is the critical design condition (Fig. 13.31). A preliminary check will be made for the more likely conditions, using B_x for the assumed **W** 10×45, i.e., $B_x = 0.271$.

- Top of third-story column,

$$P' = P + MB_x$$
$$= 34.7 + [41.7(0.271)12]$$
$$= 34.7 + 135.6 = 170.3 \text{ kips}$$
$$KL = 1.32(14) = 18.5 \text{ ft}$$

- Top of second-story column,

$$P' = 76.6 + [32.6(0.271)12]$$
$$= 76.6 + 106.0 = 182.6 \text{ kips}$$
$$KL = 1.40(14) = 19.6 \text{ ft}$$

- Bottom of second-story column,

$$P' = 78.8 + [24.7(0.271)12]$$
$$= 78.8 + 80.3 = 159.1 \text{ kips}$$

Design for the Top of the Second-Story Column / The equivalent load is 181.2 kips, and the effective length is 19.6 ft. At this length, the **W** 10×45 will carry 143 kips. It will be checked more accurately.

$$A = 13.3 \text{ in.}^2, \qquad r_x = 4.32 \text{ in.}$$
$$S_x = 49.1 \text{ in.}^3, \qquad r_y = 2.01 \text{ in.}$$
$$S_y = 13.3 \text{ in.}^3, \qquad d/2 = 5.06 \text{ in.}$$

1. $f_a = \dfrac{P}{A} = \dfrac{76.6}{13.3} = 5.76 \text{ ksi}$

$\quad f_b = \dfrac{M}{S_x} = \dfrac{32.6(12)}{49.1} = 7.97 \text{ ksi}$

2. $\dfrac{K_y L_y}{r_y} = \dfrac{1.18(14)12}{2.01} = 98.6$

3. $F_a = 13.15 \text{ ksi}$

4. $\dfrac{f_a}{F_a} = \dfrac{5.76}{13.15} = 0.44.$ Since $0.44 > 0.15$, the modified interaction formula must be satisfied.

5. $\dfrac{K_b L_b}{r_b} = \dfrac{K_x L_x}{r_x} = \dfrac{1.40(14)12}{4.32} = 54.44$

$\qquad F'_e = 50.40 \text{ ksi}$

$\qquad 1 - \dfrac{f_a}{F'_e} = 1 - \dfrac{5.76}{50.40} = 0.89$

6. $C_m = 0.85$

7. Testing the stress ratios,

$$0.44 + \left(\frac{0.85}{0.89}\right)\frac{7.97}{22.00} = 0.44 + 0.335$$

$$= 0.78 < 1.0$$

8. End check:

$$\frac{5.76}{22.00} + \frac{7.97}{22.00} = 0.62 < 1.0$$

Before the **W** 10×45 is accepted for gravity loads, check the **W** 10×39:

$$A = 11.5 \text{ in.}^2, \qquad r_x = 4.27 \text{ in.}$$
$$S_x = 42.1 \text{ in.}^3, \qquad r_y = 1.98 \text{ in.}$$
$$S_y = 11.3 \text{ in.}^3, \qquad d/2 = 4.97 \text{ in.}$$

1. $f_a = \dfrac{P}{A} = \dfrac{76.6}{11.5} = 6.67$ ksi

$$f_b = \frac{32.6(12)}{42.1} = 9.29 \text{ ksi}$$

2. $\dfrac{K_y L_y}{r_y} = \dfrac{1.18(14)12}{1.98} = 100$

3. $F_a = 12.98$ ksi

4. $\dfrac{f_a}{F_a} = \dfrac{6.67}{12.98} = 0.514 > 0.15$

5. $\dfrac{K_b L_b}{r_b} = \dfrac{K_x L_x}{r_x} = \dfrac{1.4(14)12}{4.27} = 55.1$

$$F_e' = 49.20 \text{ ksi}$$

$$1 - \frac{f_a}{F_e'} = 1 - \frac{6.67}{49.20} = 0.86$$

6. $C_m = 0.85$

7. $0.514 + \left(\dfrac{0.85}{0.86}\right)\dfrac{9.29}{22.00} = 0.514 + 0.417$

$$= 0.931 < 1.0$$

8. End check:

$$\frac{6.67}{22.00} + \frac{9.29}{22.00} = 0.725 < 1.0$$

Adopt the **W** 10×39 as satisfactory for gravity load.

Note. If new K values (for adjusted G values) were derived for the use of the **W** 10×54 and the **W** 10×39, it would be found that the stress-ratio check (step 7) would result in values of 0.826 and 0.908, respectively—i.e., even more conservative.

13.26
Wind Check for Typical Horizontal Members

The typical members which follow have thus far been designed for gravity loads (Art. 13.22). Needed pertinent information both for gravity and wind loads (Fig. 13.13) is recorded for ready reference. Note that the maximum moments occur at the supports and are additive for gravity load and wind (GL and W).

1. Roof girders—**W** 14×34:

    Span = 18 ft
    $M_{GL} = 90.3$ ft-kips (from Fig. 13.22)
    $M_W = 4.76$ (from Fig. 13.13)
    Total $= \overline{95.06} = 95.1$ ft-kips

2. Third-floor girders—**W** 16×40:

    Span = 18 ft
    $M_{GL} = 117.7$ ft-kips (from Fig. 13.23)
    $M_W = 16.52$ (from Fig. 13.13)
    Total $= \overline{134.22} = 134$ ft-kips

3. Second-floor girders—**W** 16×40:

    Span = 18 ft
    $M_{GL} = 121.4$ ft-kips (from Fig. 13.24)
    $M_W = 35.48$ (from Fig. 13.13)
    Total $= \overline{156.88} = 157$ ft-kips

4. Roof cross-beams—**W** 14×18:

    Span = 21.5 ft
    $M_{GL} = 40.1$ ft-kips (from Fig. 13.25)
    $M_W = 2.94$ (from Fig. 13.13)
    Total $= \overline{43.04} = 43.0$ ft-kips

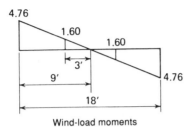

Wind-load moments

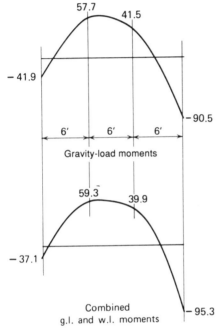

Gravity-load moments

Combined
g.l. and w.l. moments

Figure 13.32/Bending moments for roof girders.

5. Third-floor cross-beams—**W** 14×22:

Span = 21.5 ft
$M_{GL} = 52.3$ ft-kips (from Fig. 13.26)
$M_W = 10.4$ (from Fig. 13.13)
Total = $\overline{62.7}$ ft-kips

6. Second-floor cross-beams—**W** 14×22:

Span = 21.5 ft
$M_{GL} = 53.4$ ft-kips (from Fig. 13.27)
$M_W = 22.38$ (from Fig. 13.13)
Total = $\overline{75.78}$ = 75.8 ft-kips

The member end moments due to wind may be portrayed in a simple diagram (Fig. 13.32). The frame analysis by the portal method (Art. 13.9) was for a south wind on the short side and a west wind on the long side of the building. A reversal of wind would create an opposite end-movement effect; therefore, wind moments may be considered additive at points of support.

For the above members, then, the results are as summarized in Table 13.7. It is apparent that, when combining moments due to gravity load and wind, and then reducing that moment to three-fourths (which is the reverse of increasing the allowable stress by four-thirds), only the second-floor cross-beam is affected. Referring to the AISC charts for allowable moments in beams, a **W** 14×22 is still adequate for a moment of 58 ft-kips.

No member-size changes are required for wind.

Table 13.7

	M_{GL+W}	$\frac{3}{4}M_{GL+W}$	M_{GL}
Roof girders	95.1	71.3	< 90.5
Third-floor girders	134	101.0	<117.7
Second-floor girders	157	118.0	<121.4
Roof cross-beams	43.0	32.3	< 40.1
Third-floor cross-beams	62.7	47.0	< 52.3
Second-floor cross-beams	75.8	56.9	> 53.4

13.27
Wind Check for Typical Columns

Referring to Art. 13.24 and Figs. 13.13 and 13.15, the column design data below are recorded for ready reference. (Note that the moments for interior columns from gravity loads are considered to be negligible.)

---

1. Third-story interior column ⑩ —
W 10×45:

$$KL_{(x-x)} = 17 \text{ ft}$$
$$GL = 62.4 \text{ kips}$$
$$WL_D = 0.0$$
$$M_{W(x-x)} = 9.52 \text{ ft-kips}$$

$$KL_{(y-y)} = 17 \text{ ft}$$

$$M_{W(y-y)} = 6.02 \text{ ft-kips}$$

2. Second-story interior column ⑩ —
W 10×45:

$$KL_{(x-x)} = 18 \text{ ft}$$
$$GL = 134.1 \text{ kips}$$
$$WL_D = 0$$
$$M_{W(x-x)} = 23.59 \text{ ft-kips}$$

$$KL_{(y-y)} = 18.2 \text{ ft}$$

$$M_{W(y-y)} = 14.84 \text{ ft-kips}$$

3. First-story interior column ⑩ —
W 12×53:

$$KL_{(x-x)} = 22.0 \text{ ft}$$
$$GL = 184.8 \text{ kips}$$
$$WL_D = 0$$
$$M_{W(x-x)} = 47.52 \text{ ft-kips}$$

$$KL_{(y-y)} = 22 \text{ ft}$$

$$M_{W(y-y)} = 29.84 \text{ ft-kips}$$

4. Basement-story interior column ⑩ —
W 12×53:

$$KL_{(x-x)} = 14 \text{ ft}$$
$$GL = 215.9 \text{ kips}$$
$$WL_D = 0$$
$$M_{W(x-x)} = 47.52 \text{ ft-kips}$$

$$KL_{(y-y)} = 14 \text{ ft}$$

$$M_{W(y-y)} = 29.84 \text{ ft-kips}$$

5. Third-story exterior column ⑨ —
W 10×39:

$$KL_{(x-x)} = 18.5 \text{ ft}$$
$$P_{GL} = 34.7 \text{ kips (top of col.)}$$
$$P_{GL} = 36.9 \text{ kips (bot. of col.)}$$
$$WL_{D(x-x)} = 0.53 \text{ kips}$$
$$M_{W(x-x)} = 4.76 \text{ ft-kips}$$

$$KL_{(y-y)} = 15.5 \text{ ft}$$
$$M_{GL} = 41.7 \text{ ft-kips (top of col.)}$$
$$M_{GL} = 32.6 \text{ ft-kips (bot. of col.)}$$
$$WL_{D(y-y)} = 0$$
$$M_{W(y-y)} = 6.02/2 = 3.01 \text{ ft-kips}$$

6. Second-story exterior column ⑨ —
W 10×39:

$$KL_{(x\text{-}x)} = 19.6 \text{ ft}$$
$$P_{\text{GL}} = 76.6 \text{ kips (top of col.)}$$
$$P_{\text{GL}} = 78.8 \text{ kips (bot. of col.)}$$
$$\text{WL}_{\text{D}(x\text{-}x)} = 2.37 \text{ kips}$$
$$M_{\text{W}(x\text{-}x)} = 11.76 \text{ ft-kips}$$

$$KL_{(y\text{-}y)} = 16.5 \text{ ft}$$
$$M_{\text{GL}} = 32.6 \text{ ft-kips (top of col.)}$$
$$M_{\text{GL}} = 24.7 \text{ ft-kips (bot. of col.)}$$
$$\text{WL}_{\text{D}(y\text{-}y)} = 0$$
$$M_{\text{W}(y\text{-}y)} = 14.84/2 = 7.42 \text{ ft-kips}$$

7. First-story exterior column ⑨ —
W 10×54:

$$KL_{(x\text{-}x)} = 23.3 \text{ ft}$$
$$P_{\text{GL}} = 110.4 \text{ kips (top of col.)}$$
$$P_{\text{GL}} = 113.8 \text{ kips (bot. of col.)}$$
$$\text{WL}_{\text{D}(x\text{-}x)} = 6.31 \text{ kips}$$
$$M_{\text{W}(x\text{-}x)} = 23.7 \text{ ft-kips}$$

$$KL_{(y\text{-}y)} = 21.6 \text{ ft}$$
$$M_{\text{GL}} = 32.1 \text{ ft-kips (top of col.)}$$
$$M_{\text{GL}} = 15.05 \text{ ft-kips (bot. of col.)}$$
$$\text{WL}_{\text{D}(y\text{-}y)} = 0$$
$$M_{\text{W}(y\text{-}y)} = 29.84/2 = 14.92 \text{ ft-kips}$$

The sections chosen for gravity loads only will be checked for combined loads, allowing a one-third increase in stress for wind. These are the first- and second-story interior column ⑩, and the tops of the first- and second-story sections of exterior column ⑨. Because wind can come from any direction, moments will be considered to be additive.

Design of First-Story Column ⑩ —Gravity Loads Plus Wind / The adequacy of the previously designed **W** 12×53 (for gravity loads only) will be determined first.

Considering the first story only, a stability check will be made with reference to wind causing bending about the y-y axis:

1. $f_a = \dfrac{P}{A} = \dfrac{184.8}{15.6} = 11.85 \text{ ksi}$

$f_b = \dfrac{M}{S_y} = \dfrac{29.84(12)}{19.2} = 18.65 \text{ ksi}$

2. $\dfrac{K_y L_y}{r_y} = \dfrac{22(12)}{2.48} = 106.5 > \dfrac{K_x L_x}{r_x}$

3. $F_a = 12.14(1.33) = 16.13 \text{ ksi}$

4. $\dfrac{f_a}{F_a} = \dfrac{11.85}{16.13} = 0.735.$ Since $0.735 > 0.15$,

the modified interaction formula must be satisfied.

5. $\dfrac{K_b L_b}{r_b} = \dfrac{K_y L_y}{r_y} = 106.5$

$F'_e = 13.15(1.33) = 17.49 \text{ ksi}$

$1 - \dfrac{f_a}{F'_e} = 1 - \dfrac{11.85}{17.49} = 0.322$

6. $C_m = 0.85$

7. $\dfrac{f_a}{F_a} + \left[\dfrac{C_m}{1 - \dfrac{f_a}{F'_e}} \right] \dfrac{f_b}{F_b}$

$= 0.735 + \left(\dfrac{0.85}{0.322} \right) \dfrac{18.65}{(1.33)22}$

$= 0.735 + 1.682 = 2.417 > 1.0$

The **W** 12×53 is not adequate. Try the **W** 12×72:

$$A = 21.1 \text{ in.}^2, \qquad r_x = 5.31 \text{ in.}$$

$$S_x = 97.4 \text{ in.}^3, \qquad r_y = 3.04 \text{ in.}$$

$$S_y = 32.4 \text{ in.}^3$$

Try bending about the *y-y* axis:

1. $f_a = \dfrac{184.8}{21.1} = 8.76$ ksi

$f_b = \dfrac{29.84(12)}{32.4} = 11.05$ ksi

2. $\dfrac{K_y L_y}{r_y} = \dfrac{22(12)}{3.04} = 86.8 > \dfrac{K_x L_x}{r_x}.$

3. $F_a = 14.58(1.33) = 19.39$ ksi

4. $\dfrac{f_a}{F_a} = \dfrac{8.76}{19.39} = 0.452 > 0.15$

5. $\dfrac{K_b L_b}{r_b} = \dfrac{K_y L_y}{r_y} = 86.8$

6. $F'_e = 19.79(1.33) = 26.32$ ksi

$1 - \dfrac{f_a}{F'_e} = 1 - \dfrac{8.76}{26.32} = 0.667$

7. $C_m = 0.85$

8. $0.452 + \left(\dfrac{0.85}{0.667}\right)\dfrac{11.05}{22(1.33)}$

$= 0.452 + 0.481 = 0.933 < 1.0$

For bending about the *x-x* axis,

1. $f_a = 8.76$ ksi

$f_b = \dfrac{47.52(12)}{97.4} = 5.85$ ksi

2. $\dfrac{K_y L_y}{r_y} = 86.8 > \dfrac{K_x L_x}{r_x}$

3. $F_a = 19.39$ ksi

4. $\dfrac{f_a}{F_a} = 0.45 > 0.15$

5. $\dfrac{K_b L_b}{r_b} = \dfrac{K_x L_x}{r_x} = \dfrac{22(12)}{5.31} = 49.7$

6. $F'_e = 60.39(1.33) = 80.3$ ksi

$1 - \dfrac{f_a}{F'_e} = 1 - \dfrac{8.76}{80.3} = 0.891$

7. $C_m = 0.85$

8. $0.450 + \left(\dfrac{0.85}{0.891}\right)\dfrac{5.85}{22(1.33)}$

$= 0.450 + 0.191 = 0.641 < 1.0$

For the end check,

$$\dfrac{f_a}{0.6F_y} + \dfrac{f_b}{F_b} = \dfrac{8.76}{22(1.33)} + \dfrac{11.05}{22(1.33)}$$

$$= 0.677 < 1.0$$

The **W** 12×72 will be used instead of the **W** 12×53 that was selected for gravity loads only.

Second-Story Interior Column ⑩ — W 10×45 / Stability check for bending about the *y-y* axis:

1. $f_a = \dfrac{P}{A} = \dfrac{134.1}{13.3} = 10.08$ ksi

$f_b = \dfrac{M}{S_y} = \dfrac{14.84(12)}{13.3} = 13.4$ ksi

2. $\dfrac{K_y L_y}{r_y} = \dfrac{18.2(12)}{2.01} = 109$

3. $F_a = 11.81(1.33) = 15.71$ ksi

4. $\dfrac{f_a}{F_a} = \dfrac{10.08}{15.71} = 0.642 > 0.15$

5. $\dfrac{K_b L_b}{r_b} = \dfrac{K_y L_y}{r_y} = 109$

$F'_e = 12.57(1.33) = 16.72$ ksi

$1 - \dfrac{f_a}{F'_e} = 1 - \dfrac{10.08}{16.72} = 0.397$

6. $C_m = 0.85$

7. $0.642 + \left(\dfrac{0.85}{0.397}\right)\dfrac{13.4}{29.26}$

$= 0.642 + 2.14(0.458) = 1.63 > 1.0$

This section is undersized significantly. Try the **W** 10×49. Note that this is what is termed a square section (10×10 in.), whereas the **W** 10×45 is not. The introduction of a moment on the *y-y* axis renders all but square sections inadequate as a general rule:

$$A = 14.4 \text{ in.}^2, \qquad r_x = 4.35 \text{ in.}$$
$$S_x = 54.6 \text{ in.}^3, \qquad r_y = 2.54 \text{ in.}$$
$$S_y = 18.6 \text{ in.}^3$$

Stability check for bending about the y-y axis:

1. $f_a = \dfrac{134.1}{14.4} = 9.31 \text{ ksi}$

$f_b = \dfrac{14.84(12)}{18.7} = 9.52 \text{ ksi}$

2. $\dfrac{K_y L_y}{r_y} = \dfrac{18.2(12)}{2.54} = 86$

3. $F_a = 14.67(1.33) = 19.51 \text{ ksi}$

4. $\dfrac{f_a}{F_a} = \dfrac{9.31}{19.51} = 0.477 > 0.15$

5. $\dfrac{K_b L_b}{r_b} = \dfrac{K_y L_y}{r_y} = 86$

$F'_e = 20.16(1.33) = 26.81 \text{ ksi}$

$1 - \dfrac{9.31}{26.81} = 0.653$

6. $C_m = 0.85$

7. $0.477 + \left(\dfrac{0.85}{0.653}\right)\dfrac{9.52}{29.26}$

$= 0.477 + 0.425 = 0.901 < 1.0$

The end check and stability check for bending about the x-x axis would both be found satisfactory.

The **W** 10×49 will be used instead of **W** 10×45 that was selected for gravity loads only.

Design of First-Story Column ⑨ — Gravity Loads plus Wind / The adequacy of the previously designed **W** 10×54 (gravity loads only) will be checked. Checking wind moments about the x-x axis first at the top of the column, and referring to

item 7 at the beginning of this article,

$$P = 110.4 + 6.31 = 116.7 \text{ kips}$$
$$M_{(x\text{-}x)} = 32.10 + 23.76 = 55.86 \text{ ft-kips}$$

1. $f_a = \dfrac{P}{A} = \dfrac{116.7}{15.8} = 7.39 \text{ ksi}$

$f_b = \dfrac{M}{S_x} = \dfrac{55.86(12)}{60.0} = 11.17 \text{ ksi}$

2. $\dfrac{K_y L_y}{r_y} = \dfrac{21.6(12)}{2.56} = 101 > \dfrac{K_x L_x}{r_x}$

3. $F_a = 12.85(1.33) = 17.09 \text{ ksi}$

4. $\dfrac{f_a}{F_a} = \dfrac{7.39}{17.09} = 0.432 > 0.15$

5. $\dfrac{K_b L_b}{r_b} = \dfrac{K_x L_x}{r_x} = \dfrac{23.3(12)}{4.37} = 64.0$

$F'_e = 36.46(1.33) = 48.49 \text{ ksi}$

$1 - \dfrac{f_a}{F'_e} = 1 - \dfrac{7.39}{48.49} = 0.848$

6. $C_m = 0.85$

7. $0.432 + \left(\dfrac{0.85}{0.848}\right)\dfrac{11.17}{29.26}$

$= 0.432 + 0.383 = 0.815 < 1.0$

End check:

$$\dfrac{7.39}{22(1.33)} + \dfrac{11.17}{22(1.33)} = 0.634 < 1.0$$

Check the y-y axis. In this case, the column will be subjected to bending about both axes, i.e., the wind moment on the y-y axis (14.92 ft-kips) and the gravity-load girder moment on the x-x axis (32.1 ft-kips). In addition, there will be the direct gravity load of 110.4 kips. Referring to the general notes on columns in the AISC Manual, the interaction formula becomes

$$\dfrac{f_a}{F_a} + \dfrac{C_{mx} f_{bx}}{\left(1 - \dfrac{f_a}{F'_{ex}}\right) F_{bx}} + \dfrac{C_{my} f_{by}}{\left(1 - \dfrac{f_a}{F'_{ey}}\right) F_{by}} \le 1.0$$

1. $f_a = \dfrac{110.4}{15.8} = 6.99$ ksi

$f_{bx} = \dfrac{M_x}{S_x} = \dfrac{32.1(12)}{60.0} = 6.42$ ksi

$f_{by} = \dfrac{M_y}{S_y} = \dfrac{14.92(12)}{20.6} = 8.69$ ksi

2. $\dfrac{K_y L_y}{r_y} = 101 > \dfrac{K_x L_x}{r_x}$

3. $F_a = 17.09$ ksi

4. $\dfrac{f_a}{F_a} = \dfrac{6.99}{17.09} = 0.409 > 0.15$

5. $\dfrac{K_b L_b}{r_b} = \dfrac{K_y L_y}{r_y} = 101$

$F'_{ey} = 14.62(1.33) = 19.44$ ksi

$1 - \dfrac{f_a}{F'_{ey}} = 1 - \dfrac{6.99}{19.44} = 0.640$

$\dfrac{K_b L_b}{r_b} = \dfrac{K_x L_x}{r_x} = 63.7$

$F'_{ex} = 48.89$

$1 - \dfrac{f_a}{F'_{ex}} = 0.857$

6. $C_m = 0.85$

7. $0.409 + \left(\dfrac{0.85}{0.857}\right) \dfrac{6.42}{22(1.33)}$

$+ \left(\dfrac{0.85}{0.640}\right) \dfrac{8.69}{22(1.33)}$

$= 0.409 + 0.218 + 0.394 = 1.021$

This actually indicates the column is inadequate, but only to a very slight degree. The decision whether to accept the **W** 10×54 or to adopt the next larger section (**W** 10×60) must be based upon judgment, calling into play the accuracy and precision of all the assumptions and design procedures used. In this instance the **W** 10×54 will be considered safe.

End check:

$$\dfrac{6.99}{22(1.33)} + \dfrac{6.42}{22(1.33)} + \dfrac{8.69}{22(1.33)}$$

$$= 0.239 + 0.219 + 0.297$$

$$= 0.755 < 1.0$$

The **W** 10×54 adopted for gravity loads only is also satisfactory for gravity loads plus wind loads.

Second- and Third-Story Column

⑨—**Gravity Loads Plus Wind** / The adequacy of the previously designed **W** 10×45 (gravity loads only) will be checked. Checking wind moments first about the *x-x* axis at the top of the second-story column, and referring to item 6 at the beginning of this article,

$$P = 76.6 + 2.37 = 78.97; \qquad \text{use 79 kips}$$
$$M_x = 32.6 + 11.76 = 44.36 \text{ ft-kips}$$

1. $f_a = \dfrac{P}{A} = \dfrac{79}{13.3} = 5.94$ ksi

$f_b = \dfrac{M}{S_x} = \dfrac{44.36(12)}{49.1} = 10.84$ ksi

2. $\dfrac{K_y L_y}{r_y} = \dfrac{16.5(12)}{2.01} = 99 > \dfrac{K_x L_x}{r_x}$

3. $F_a = 13.10(1.33) = 17.42$ ksi

4. $\dfrac{f_a}{F_a} = \dfrac{5.94}{17.42} = 0.341 > 0.15$

5. $\dfrac{K_b L_b}{r_b} = \dfrac{K_x L_x}{r_x} = \dfrac{19.6(12)}{4.32} = 54.4$

$F'_e = 50.47(1.33) = 67.13$ ksi

$1 - \dfrac{f_a}{F'_e} = 1 - \dfrac{5.94}{67.13} = 0.912$

6. $C_m = 0.85$

7. $0.341 + \left(\dfrac{0.85}{0.913}\right) \dfrac{10.84}{22(1.33)}$

$$= 0.341 + 0.345 = 0.686 < 1.0$$

End check:

$$\frac{5.94}{22(1.33)} + \frac{10.84}{22(1.33)} = 0.573 < 1.0$$

Now check the **W** 10×45 for wind moments about the y-y axis. The critical section is still observed to occur at the top of the second-story column; and

$$P = 76.6 \text{ kips}$$
$$M_x = 32.6 \text{ ft-kips}$$
$$M_y = 7.42 \text{ ft-kips}$$

1. $f_a = \dfrac{76.6}{13.3} = 5.76$ ksi

$$f_{bx} = \frac{32.6(12)}{49.1} = 7.97 \text{ ksi}$$

$$f_{by} = \frac{7.42(12)}{13.3} = 6.69 \text{ ksi}$$

2. $\dfrac{K_y L_y}{r_y} = 99 > \dfrac{K_x L_x}{r_x}$

3. $F_a = 17.42$ ksi

4. $\dfrac{f_a}{F_a} = \dfrac{5.76}{17.42} = 0.331 > 0.15$

5. $\dfrac{K_b L_b}{r_b} = \dfrac{K_y L_y}{r_y} = 99$

$$F'_{ey} = 15.21(1.33) = 20.23 \text{ ksi}$$

$$1 - \frac{f_a}{F'_{ey}} = 1 - \frac{5.76}{20.23} = 0.715$$

$$\frac{K_b L_b}{r_b} = \frac{K_x L_x}{r_x} = 54.3$$

$$F'_{ex} = 66.79 \text{ ksi}$$

$$1 - \frac{f_a}{F'_{ex}} = 0.914$$

6. $C_m = 0.85$

7. $0.331 + \left(\dfrac{0.85}{0.914} \right) \dfrac{7.97}{22(1.33)}$

$$+ \left(\frac{0.85}{0.713} \right) \frac{6.69}{22(1.33)}$$

$$= 0.331 + 0.253 + 0.272$$

$$= 0.856 < 1.0$$

End check:

$$\frac{5.76}{29.26} + \frac{7.97}{29.26} + \frac{6.69}{29.26} = 0.698 < 1.0$$

The **W** 10×45 adopted for gravity loads only is also satisfactory for gravity load plus wind load.

13.28
Final Selections versus Assumptions

Final column sizes compared with original assumptions are shown in Table 13.8.

Portal Assumption / Agreement with the assumption that exterior columns have one-half the moment of inertia of interior columns is checked as follows:

	I_x	I_y
W 12×72	597	195
W 10×54	303	103
Ratio	1.97	1.89

These are close and are accepted.

Frame-Analysis Assumptions / It was necessary to estimate relative stiffness factors for the cross-beams, girders, and columns before the frame analysis could be undertaken (Art. 13.21). The values used in Art. 13.21 can be verified as shown in Table 13.9. The actual values are close enough to those originally assumed. No further refinement is necessary.

13.29
Framing Plans and Column Schedule

Figures 13.14, 13.15, and 13.18 through 13.20 show framing plans with all typical members recorded. Figure 13.33 shows a column schedule.

Table 13.8

Column	Final	Assumed
1st-story interior and basement	W 12×72	W 12×79
2nd- and 3rd-story interior	W 10×49	W 10×54
1st-story exterior	W 10×54	W 10×68
2nd- and 3rd-story exterior	W 10×45	W 10×45

Table 13.9

	Assumed	I/L	Actual	I/L
Roof cross-beams	W 14×22	1.0	M 14×18	1.0
3rd-floor cross-beams	W 14×22	1.0	W 14×22	1.3
2nd-floor cross-beams	W 14×30	1.5	W 14×22	1.3
Roof girders	W 14×30	1.0	W 14×34	1.0
3rd-floor girders	W 16×36	1.5	W 16×40	1.5
2nd-floor girders	W 18×46	2.4	W 16×40	2.1
1st-story interior columns	W 12×79	2.4_{x-x} 1.4_{y-y}	W 12×72	2.2_{x-x} 1.2_{y-y}
2nd- and 3rd-story interior columns	W 10×54	1.4_{x-x} 0.8_{y-y}	W 10×49	1.2_{x-x} 0.7_{y-y}
1st-story exterior columns	W 10×68	1.4_{x-x} 0.8_{y-y}	W 10×54	1.1_{x-x} 0.7_{y-y}
2nd- and 3rd-story exterior columns	W 10×45	1.1_{x-x} 0.4_{y-y}	W 10×45	1.1_{x-x} 0.4_{y-y}

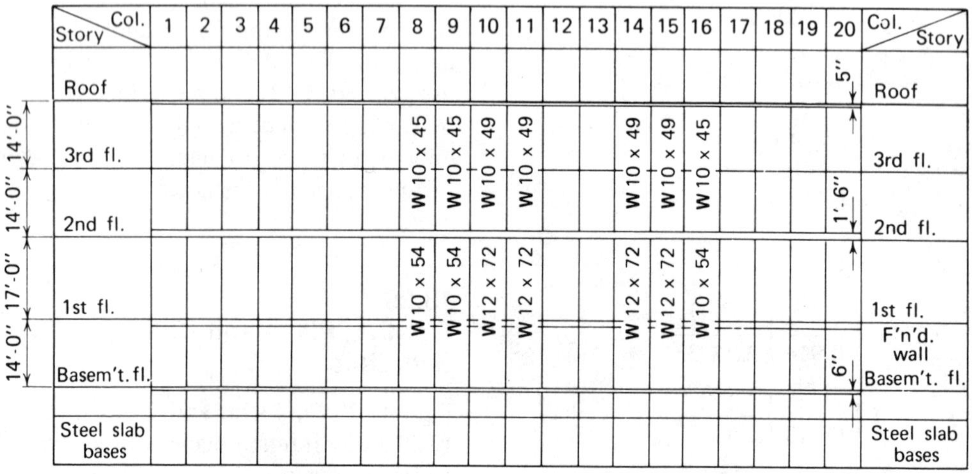

Figure 13.33/Column schedule.

13.30
Typical Connections

Only one connection of beam to column and one of girder to column will be designed. The joint selected is at column ⑩ at the second-floor level.

In general, all connections will be made using flexible framing angles with $\frac{3}{4}$-in. high-strength (H.S.) A325N bolts—bearing type with threads in the plane of shears and standard holes. These connections (for shear only) will be shop-fabricated and -welded, and the $\frac{3}{4}$-in. H.S. bolts will be field-placed. Where moment-resisting connections are necessary, i.e., for cross-beam and girder-to-column connections above the first floor, additional plates will be welded to flanges.

Referring to Fig. 13.34, it is seen that framing angles transferring shear to the columns will permit adequate erection clearance. Plates will be field-placed and -welded to develop end restraint, i.e., transfer end moments.

Connection from W 16 × 40 Girder to W 12 × 72 Column

Shear / 30.3 kips (Fig. 13.35a). For the shop-connected leg, allow 1-in. clearance between girder end and column flange (to allow for bottom-flange–plate weld).

A 4-in. connected leg will be used. Then, assuming three bolts will be required, the weld pattern, using $\frac{1}{4}$-in. fillet welds and E60XX electrodes, is as shown in Fig. 13.35a. $4 \times 4 \times \frac{5}{16}$ in. framing angles are adopted.

With the field-connected outstanding leg, shear is critical. The number of $\frac{3}{4}$-in. H.S. bolts required is

$$\frac{30.3}{9.3} = 3.26; \qquad \text{use 6}$$

Moment /

$$M_{GL} = 121.4(12) = 1460 \text{ in.-kips}$$

$$M_{GL+w} = \tfrac{3}{4}[1460 + 35.48(12)]$$

$$= 1420 < 1460 \text{ in.-kips}$$

The plate welded to the flanges must develop a force

$$F = \frac{1460}{16} = 91.3 \text{ kips}$$

If the plate is made 6 in. wide (allowing $\frac{1}{2}$ in. on either side for the fillet weld to the girder flange), the required thickness is

$$t = \frac{91.3}{22(6)}$$

$$= 0.69 \text{ in.;} \qquad \text{use a } \tfrac{3}{4}\text{-in. plate}$$

Using $\frac{3}{8}$-in. fillet welds and E60XX electrodes,

$$L = \frac{91.3}{2(4.7)} = 9.7 \text{ in.}$$

Use a 10-in. weld and a plate 12 in. long.

The weld at the column flange will be of a single-bevel butt type with back-up bars.

A diaphragm plate will be required between column flanges. The plate used to develop the cross-beam moment will be used to serve this function.

Connection from W 14 × 22 Cross-Beam to W 12 × 72 Column / The cross-beam will fit between column flanges.

Shear / 14 kips. For the shop-connected leg, standard $4 \times 3\frac{1}{2} \times \frac{5}{16}$-in. framing angles will be used. Assuming 4 H.S., A325N bolts will be adequate, the weld pattern is as shown in Fig. 13.35b. A $\frac{3}{16}$-in. fillet weld will be adequate. For the field-connected outstanding leg, using the $\frac{3}{4}$-in. H.S. bolts,

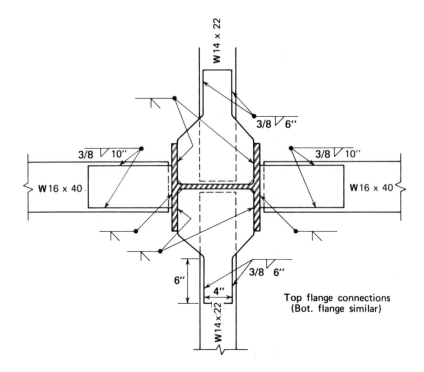

Top flange connections
(Bot. flange similar)

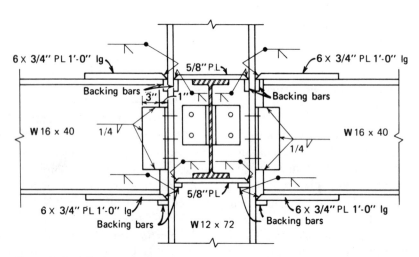

Figure 13.34/Wind connection.

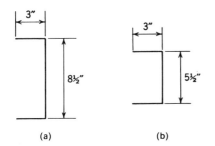

Figure 13.35/Framing angle weld.

the number required is

$$\frac{14}{9.3} = 1.51; \qquad \text{use 4}$$

Moment /

$$M_{GL} = 53.4(12) = 641 \text{ in.-kips}$$

$$M_{GL+W} = \tfrac{3}{4}[641 + 22.38(12)]$$

$$= 683 \text{ in.-kips}$$

Note that the maximum moment occurring at column ⑦ is used to standardize connections. Then the plate welded to the beam flanges must develop a force

$$F = \frac{683}{13.75} = 49.6 \text{ kips}$$

To permit this plate to be used as a diaphragm, it will be cut to fit between column flanges, welded to these flanges, and tapered to a 4-in. width for fillet welding to the cross-beam flange. Then

$$t = \frac{49.6}{22(4)}$$

$$= 0.56; \qquad \text{use a } \tfrac{5}{8}\text{-in. plate}$$

Using $\tfrac{3}{8}$-in. fillet welds at the cross-beam flange,

$$L = \frac{49.6}{2(4.7)} = 5.3; \qquad \text{use 6 in.}$$

Again, a single-bevel butt weld will be used between the plate and column flanges. With a full-penetration weld, back-up plates will be required. The resulting shear stress will be

$$f_v = \frac{49.6}{2(0.625)5} = 7.9 \text{ ksi}; \qquad \text{safe}$$

13.31
Conclusion

Although heavy loads were purposely selected so that deflection would not be controlling, deflection of critical members should always be checked. With this in mind, complete design of the structural frame of this building would require:

1. Checking girder and cross-beam size requirements around stairwells to be sure that symmetry is achieved.

2. Checking spandrels at the north and east walls; also for size agreement with the south and west walls.

3. Checking the remaining columns for size requirements—again, to achieve symmetry.

4. Design of remaining intermediate horizontal members.

5. Careful design of the connections at all floors, and particularly at exterior walls where spandrels are connected to the outer flanges of columns.

6. Design of column base plates: standard type at both the exterior walls and in the basement.

7. Design of reinforced concrete columns (pilasters) at the exterior walls.

8. Design of all column footings.

9. The $\tfrac{5}{8}$-in. plate also must be checked as a stiffener according to the 1989 AISC Specification.

A

STRESS AND DEFORMATION

The relation between the elastic limit, yield point, and ultimate strength of a material is clearly illustrated by means of a *stress-strain diagram*. Such a diagram is obtained from the data of a static tensile test, by plotting simultaneous values of the unit stress (total load on the specimen divided by original area of cross section) and the unit elongation (increase in gage length divided by the original gage length).

Figure A1 shows the stress-strain relationship for three different types of steel. The portion of the diagram for A36 Steel that is outlined by the dashed line is shown at a larger scale in Fig. A2. That part of the curve from the origin O to A is a straight line, since within this range the material behaves according to Hooke's law: stress is proportional to deformation. In the vicinity of point A, the curve deviates from a straight line, indicating that the proportional relationship no longer holds. Beyond point A, if the load had been released during the test, the specimen would not have recovered its original length, but would have taken a permanent set. The unit stress corresponding to point A on the curve is the *elastic limit* of the material.

At point B, a short distance beyond A, the curve becomes horizontal. This indicates that the specimen elongated without any increase in the load. The stress at B is called the *yield point*. Shortly beyond the yield point, the unit stress increased until the *ultimate strength* was reached at C. Failure actually begins at the ultimate strength as evidenced by "necking" of the specimen, although rupture does not occur until the *breaking stress* is reached at D.

It is evident from Fig. A2 that an exact determination of the elastic limit (point A) would be extremely difficult, whereas

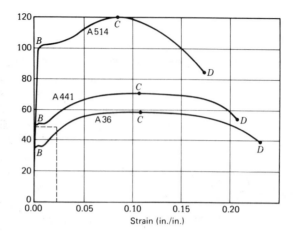

Figure A1/Stress-strain diagrams.

the yield point may be ascertained quite accurately. For this reason, it is customary to base allowable working stresses on the yield stress rather than the elastic limit. In addition to the stresses mentioned above, the stress-strain diagram also indicates the *modulus of elasticity* of the material tested. Modulus of elasticity is defined as the ratio of unit stress to unit deformation or,

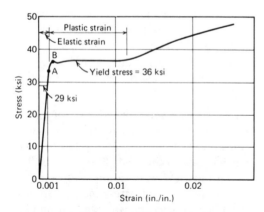

Figure A2/Partial stress-strain diagram for A36 Steel.

expressed mathematically,

$$E = \frac{f}{\epsilon}$$

If any point on the straight line portion (O to A) of the stress-strain diagram is selected, the value of E is found by dividing the ordinate (unit stress) by the abscissa (unit strain). For example, in the diagram shown in Fig. A2, an elongation of 0.001 in. corresponds to a unit stress of about 29,000 lb per sq in. Hence the modulus of elasticity is

$$E = \frac{f}{\epsilon} = \frac{29,000}{0.001} = 29,000,000 \text{ lb per sq in.}$$

It is important to observe that even though the yield stress is different for each type of steel, all steels have the same modulus of elasticity.

More complete discussions of this subject will be found in textbooks on mechanics of materials.

B

SUPPLEMENTARY BEAM DIAGRAMS

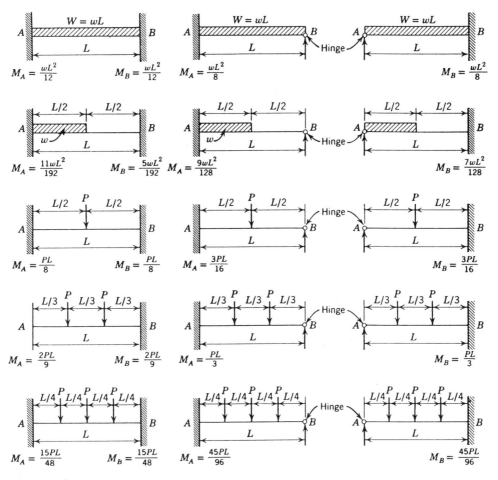

Figure B1/Fixed-end moments.

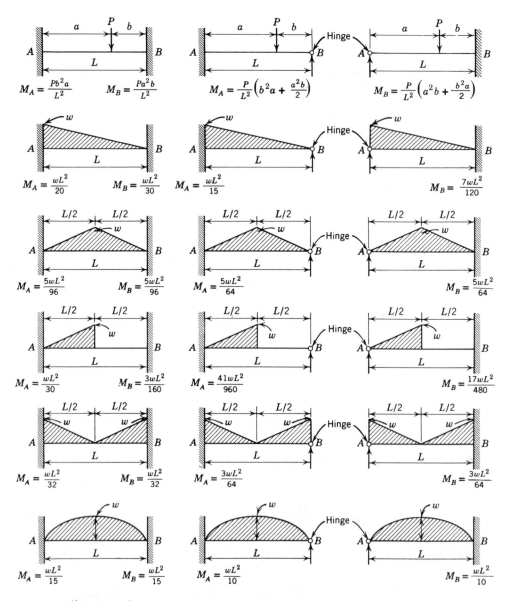

Figure B1/(Continued)

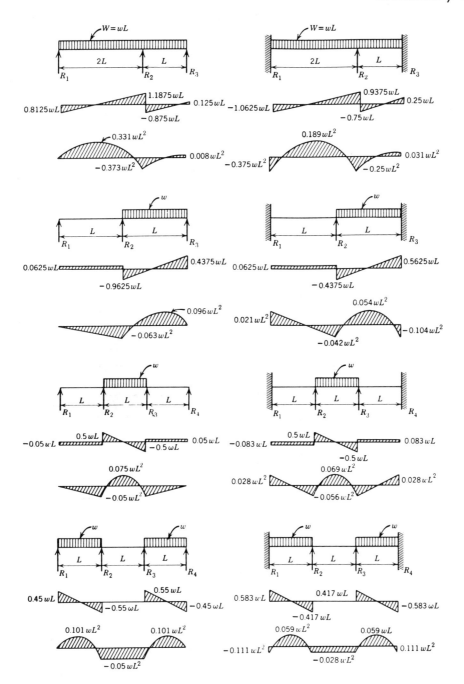

Figure B2/Continuous beams.

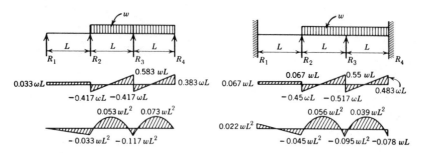

Figure B2/(Continued)

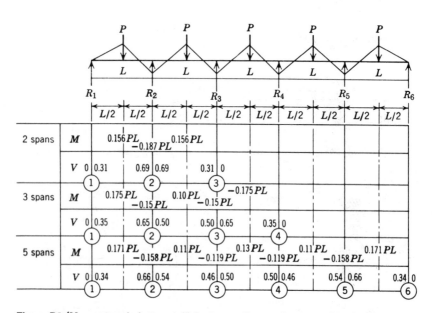

Figure B3/Moment and shear coefficients: continuous beams, midpoint concentrated loads.

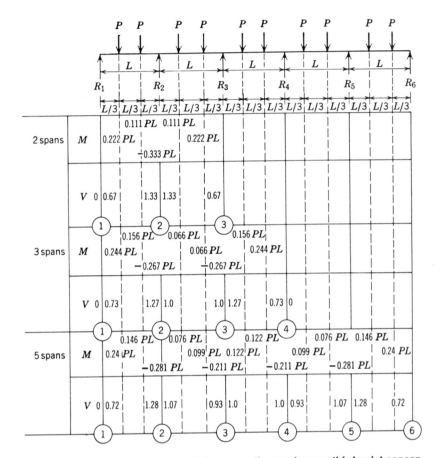

Figure B4/Moment and shear coefficients: continuous beams, third-point concentrated loads.

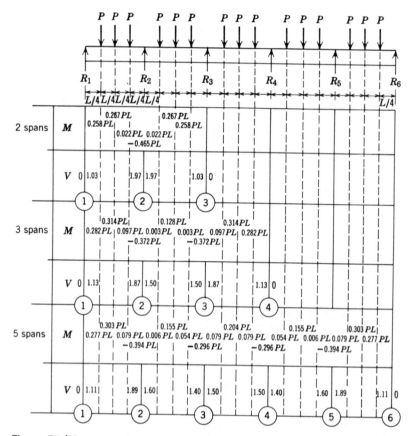

Figure B5/Moment and shear coefficients: continuous beams, quarter-point loads.

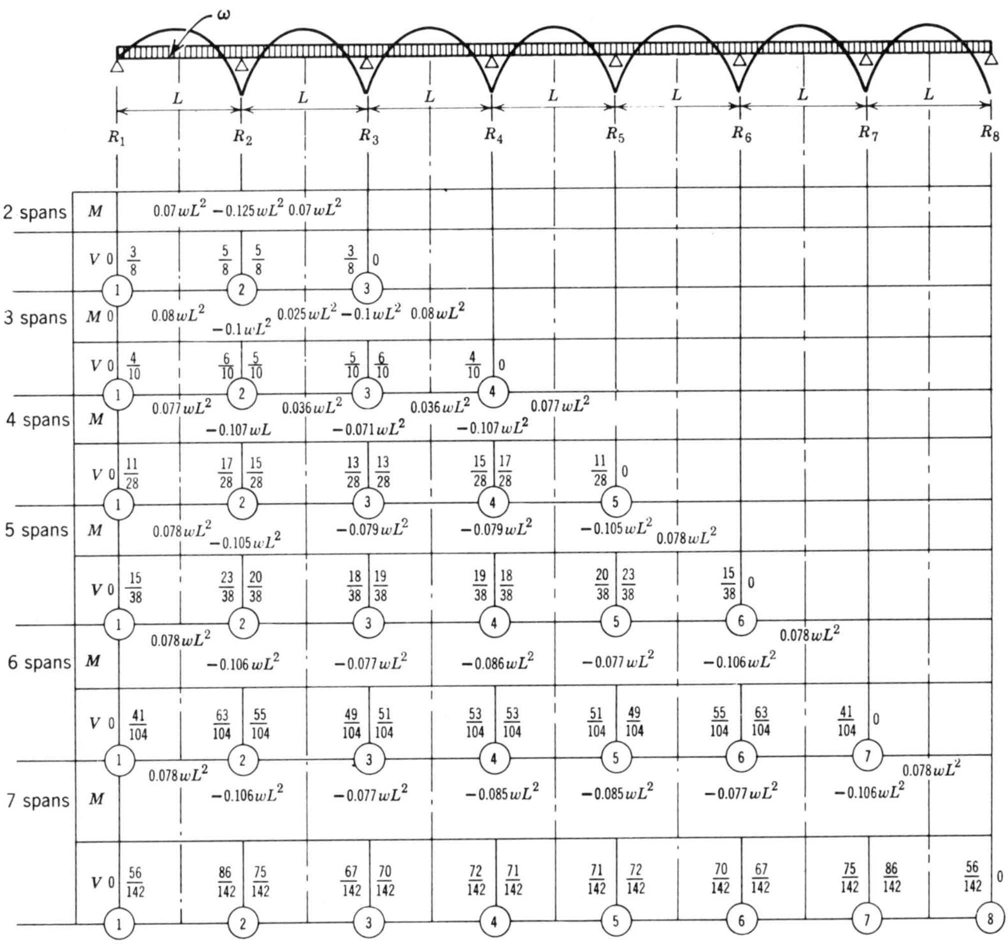

Figure B6/Moment and shear coefficients: continuous beams, uniform load.

C

FUNDAMENTALS OF MATRIX ALGEBRA

C.1 Introduction

Matrix algebra is a mathematical procedure that can be used to solve for the values of the unknown terms in a series of simultaneous linear algebraic equations. It is particularly applicable to problems that involve a large number of such equations and where a computer program can be written for carrying out the necessary steps for their solution. The purpose here is not to explain matrix algebra in its totality, but rather to present the fundamentals in sufficient detail to apply the procedure to the problems presented in Chapter 12.

C.2 Definition and Identification of Elements

Consider the following three simultaneous linear algebraic equations:

$$1x + 2y + 3z = 14$$
$$4x + 5y - 1z = 11$$
$$2x + 3y + 4z = 20$$

The term "linear" is used because the expressions in the equations are all of degree one, and the term "simultaneous" is used because of the requirement that the unknowns satisfy all equations simultaneously.

The elements of these equations may be grouped into three arrays, as follows:

$$\begin{bmatrix} 1 & 2 & 3 \\ 4 & 5 & -1 \\ 2 & 3 & 4 \end{bmatrix}$$

an array of the known *coefficients*;

$$\begin{bmatrix} x \\ y \\ z \end{bmatrix} \quad \text{or} \quad \begin{bmatrix} x & y & z \end{bmatrix}$$

an array of the unknown *variables*; and

$$\begin{bmatrix} 14 \\ 11 \\ 20 \end{bmatrix} \quad \text{or} \quad \{14 \quad 11 \quad 20\}$$

an array of the *constant* terms. The equations then may be expressed as follows:

$$\begin{bmatrix} 1 & 2 & 3 \\ 4 & 5 & -1 \\ 2 & 3 & 4 \end{bmatrix} \begin{bmatrix} x \\ y \\ z \end{bmatrix} = \begin{bmatrix} 14 \\ 11 \\ 20 \end{bmatrix}$$

These arrays are known as matrices and, as noted earlier, can be composed of a large number of elements. When these matrices are manipulated using the principles of matrix algebra—only some of which apply to ordinary algebra as well—the values of the unknown variables can be determined.

In order to readily identify the positions of elements in a matrix, the following procedure usually is adopted. The symbol a is used to denote any element of a matrix A, and a pair of subscripts to the symbol a are used to describe the location of that element within the matrix. The matrix consists of horizontal rows and vertical columns, and the first subscript identifies the row in which a is located and the second subscript identifies the column in which a is located; e.g.,

The subscripts i and j are general symbols that refer respectively to an element in the ith row and the jth column. The last row and column are designated n and m, respectively. The element itself may be almost any mathematical entity, such as a positive or negative number, unknown variable, trigonometric function, or algebraic expression. Square brackets (as shown above) are used to indicate a matrix whenever the elements of the array are shown; otherwise single capital letters, such as A and B, are used to represent the matrix.

The number of rows and columns indicate the order (or size) of a matrix. Thus, the rectangular matrix shown above is of order $n \times m$. A square matrix has the same number of rows and columns ($m = n$) and is said to be of order n. When a matrix is of order n, the elements $a_{11}, a_{22}, a_{33}, a_{ii}, \ldots, a_{nn}$ are said to constitute the principal diagonal (upper left corner to lower right corner). A matrix that has only one column or one row is known, respectively, as a column matrix or a row matrix.

The dimension of a matrix is defined as the number of subscripts required to uniquely locate an element in the matrix. Hence, row and column matrices are one-dimensional, while square and rectangular matrices are two-dimensional. Multidi-

	Col. 1	Col. 2	Col. 3	$\cdots$	Any Col. j	$\cdots$	Last Col. m
Row 1	a_{11}	a_{12}	a_{13}	$\cdots$	a_{1j}	$\cdots$	a_{1m}
Row 2	a_{21}	a_{22}	a_{23}	$\cdots$	a_{2j}	$\cdots$	a_{2m}
Row 3	a_{31}	a_{32}	a_{33}	$\cdots$	a_{3j}	$\cdots$	a_{3m}
$\vdots$	$\vdots$	$\vdots$	$\vdots$		$\vdots$		$\vdots$
Any row i	a_{i1}	a_{i2}	a_{i3}	$\cdots$	a_{ij}	$\cdots$	a_{im}
Last row n	a_{n1}	a_{n2}	a_{n3}	$\cdots$	a_{nj}	$\cdots$	a_{nm}

mensional matrices are often used in various types of mathematical analyses; however, in the analysis of structures, normally only one- and two-dimensional matrices are required.

C.3
Types of Matrices

Certain types of matrices appear frequently and have been given special names.

An identity matrix I, for example, is defined as a square matrix in which all the principal-diagonal elements (a_{ii}) are equal to unity while all the remaining elements are zero, e.g.,

$$I = \begin{bmatrix} 1 & 0 & 0 & 0 \\ 0 & 1 & 0 & 0 \\ 0 & 0 & 1 & 0 \\ 0 & 0 & 0 & 1 \end{bmatrix}$$

A lower triangular matrix L is defined as a square matrix in which $a_{ij} \neq 0$ for $i \geq j$ and $a_{ij} = 0$ for $i < j$, e.g.

$$L = \begin{bmatrix} a_{11} & 0 & 0 & 0 \\ a_{21} & a_{22} & 0 & 0 \\ a_{31} & a_{32} & a_{33} & 0 \\ a_{41} & a_{42} & a_{43} & a_{44} \end{bmatrix}$$

An upper triangular matrix U is defined as a square matrix in which $a_{ij} \neq 0$ for $i \leq j$ and $a_{ij} = 0$ for $i > j$, e.g.,

$$U = \begin{bmatrix} a_{11} & a_{12} & a_{13} & a_{14} \\ 0 & a_{22} & a_{23} & a_{24} \\ 0 & 0 & a_{33} & a_{34} \\ 0 & 0 & 0 & a_{44} \end{bmatrix}$$

A symmetric matrix is a square matrix whose elements are symmetric about the main diagonal, i.e., $a_{ij} = a_{ji}$ $(i \neq j)$.

C.4
Equality of Matrices

Two matrices A and B are defined to be equal only if $a_{ij} = b_{ij}$ for each pair of subscripts i and j; that is, two matrices are equal only if they have the same order and corresponding elements throughout. Therefore:

1. If A and B are any two matrices, either $A = B$ or $A \neq B$ (*determinative property*).

2. If A is any matrix, $A = A$ (*reflexive property*).

3. If $A = B$, then $B = A$ (*symmetric property*).

4. If $A = B$ and $B = C$, then $A = C$ (*transitive property*).

C.5
Addition and Subtraction of Matrices

The sum of two matrices A and B is defined as the matrix whose elements are obtained by taking the sum of corresponding elements in A and B:

$$S = A + B = \left[a_{ij} + b_{ij} \right]$$

For example,

$$\begin{bmatrix} 2 & -3 & 4 \\ 1 & 3 & 5 \end{bmatrix} + \begin{bmatrix} 1 & 2 & 3 \\ 1 & 4 & -3 \end{bmatrix} = \begin{bmatrix} 3 & -1 & 7 \\ 2 & 7 & 2 \end{bmatrix}$$

Two matrices must be of the same order to be conformable for addition.

It may be seen that the addition of matrices is both communicative and associative; i.e., if A, B, and C are conformable for addition, then

$$A + B = \left[a_{ij} + b_{ij} \right]$$
$$= \left[b_{ij} + a_{ij} \right]$$
$$= B + A \qquad \text{(commutative)}$$

and

$$A + (B + C) = \left[a_{ij} + (b_{ij} + c_{ij}) \right]$$
$$= \left[(a_{ij} + b_{ij}) + c_{ij} \right]$$
$$= (A + B) + C$$

(associative)

The negative N of a matrix $A = [a_{ij}]$ is defined as

$$N = -A = [-a_{ij}]$$

That is, the negative of a matrix A is formed by changing the sign of every element in A. Consequently, $A + (-A) = 0$, or simply $A - A = 0$, in which 0 represents a null matrix.

In general, the difference D of two matrices A and B is defined as

$$D = A + (-B)$$

That is, the difference D of two matrices A and B is defined as a matrix whose elements are obtained by taking the difference of the corresponding elements in A and B. Thus,

$$D = \left[a_{ij} - b_{ij} \right] = A - B$$

The same requirements for conformability apply to the subtraction of matrices as to the addition of matrices.

C.6
Scalar Multiplication of Matrices

A matrix $A = [a_{ij}]$ can be multiplied by a scalar quantity k by multiplying every element of the matrix by that quantity, i.e.,

$$kA = Ak = \left[ka_{ij} \right]$$

For example, the matrix

$$A = \begin{bmatrix} 1 & -3 & 4 \\ 2 & 3 & 1 \end{bmatrix}$$

multipied by the factor 2, is as follows:

$$2A = 2\begin{bmatrix} 1 & -3 & 4 \\ 2 & 3 & 1 \end{bmatrix} = \begin{bmatrix} 2 & -6 & 8 \\ 4 & 6 & 2 \end{bmatrix}$$

Furthermore, it may be seen that

$$(k + t)A = kA + tA$$
$$k(A + B) = kA + kB$$
$$k(tA) = (kt)A$$

in which k and t are scalar constants.

C.7
Multiplication of Matrices

The product P of two matrices A and B is defined as that matrix whose elements are obtained by taking the inner products of all rows in A and all columns in B. Each successive element in row 1 of matrix A is multiplied by the corresponding element in column 1 of matrix B, and, when the products are added together, the resultant value is the element in row 1, column 1 of the product matrix P. This procedure is repeated with the second row of A and the first column of B to form the second element in the first column of matrix B, and so on, until all the rows in matrix A are accounted for. The procedure is then repeated for each element in the next column in B and becomes the next column of elements in P. This process is best illustrated by the following example:

$$P = AB = \begin{bmatrix} 2 & 3 & 0 \\ 1 & 2 & 3 \\ 0 & 1 & 3 \end{bmatrix}\begin{bmatrix} 2 & 0 \\ 3 & -1 \\ -2 & 2 \end{bmatrix}$$

$$P = \begin{bmatrix} 2(2)+3(3)+0(-2) & 2(0)+3(-1)+0(2) \\ 1(2)+2(3)+3(-2) & 1(0)+2(-1)+3(2) \\ 0(2)+1(3)+3(-2) & 0(0)+1(-1)+3(2) \end{bmatrix} = \begin{bmatrix} 13 & -3 \\ 2 & 4 \\ -3 & 5 \end{bmatrix}$$

The multiplication procedure can be described by either of the following notations (the matrix A is of order $n \times r$ and B is of order $r \times m$):

$$P = AB$$
$$= [a_{i1}b_{1j} + a_{i2}b_{2j} + a_{i3}b_{3j} + \cdots a_{ir}b_{rj}]$$

or

$$P = AB = \left[\sum_{k=1}^{r} a_{ik}b_{kj} \right]$$

Matrix multiplication is readily accomplished through application of computer programs. (Art. C.15 describes a simple 17-step program for this purpose.)

It should be apparent that in order to be conformable for multiplication, the number of columns of A must be equal to the number of rows of B; otherwise there will not always be corresponding elements to multiply together. There is no restriction on the number of rows of A or the number of columns of B. If A is of order $n \times r$ and B is of order $r \times m$, the product (matrix P) will be of order $n \times m$. Thus,

$$[n \times r][r \times m] = [n \times m]$$

For example, consider the matrix A of order 5×4 and B of order 4×3; then the product P is as follows:

$$P = A_{5\times4}B_{4\times3}$$

and will be of order 5×3. Observe that the number of columns in A (4) must be the same as the number of rows in B.

Also, in the above expression, B is said to be premultiplied by A, and A to be postmultiplied by B. This distinction is necessary, since matrix multiplication is not ordinarily commutative; that is, except for special cases, the product AB is not equal to the product BA. Indeed, if A is of order $n \times r$ and B is of order $r \times m$ and $n \neq m$, the product AB can be obtained as described above, but the product BA is not defined.

The familiar rule of scalar algebra, that if a product is zero, then one of the factors must be zero, fails to hold for matrix multiplication, since it is possible for the product of two nonzero matrices to be a null (or zero) matrix. For example,

$$\begin{bmatrix} 2 & 4 \\ 3 & 6 \end{bmatrix}\begin{bmatrix} -2 & -8 \\ 1 & 4 \end{bmatrix} = \begin{bmatrix} 0 & 0 \\ 0 & 0 \end{bmatrix}$$

Furthermore, if $AB = AC$ or $BA = CA$, it cannot be generally concluded that $B = C$ even if $A \neq 0$, since it is possible to choose three matrices ($A \neq B \neq C$) such that $AB = AC$ or $BA = CA$. For example, consider

$$A = \begin{bmatrix} 1 & 0 \\ 2 & 0 \end{bmatrix}, \quad B = \begin{bmatrix} 2 & 3 \\ -4 & 1 \end{bmatrix},$$
$$C = \begin{bmatrix} 2 & 3 \\ 1 & 5 \end{bmatrix}$$

and the products

$$AB = AC = \begin{bmatrix} 2 & 3 \\ 4 & 6 \end{bmatrix}$$

C.8 Transpose of a Matrix

The transpose of a matrix A is denoted by the symbol A^t and is defined as the matrix that is formed by interchanging corresponding rows and columns of the matrix A. Therefore, if the original matrix A is

$$A = \begin{bmatrix} a_{11} & a_{12} & \cdots & a_{1n} \\ a_{21} & a_{22} & \cdots & a_{2n} \\ \vdots & \vdots & & \vdots \\ a_{m1} & a_{m2} & \cdots & a_{mn} \end{bmatrix}$$

the transpose of A is

$$A^t = \begin{bmatrix} a_{11} & a_{21} & \cdots & a_{m1} \\ a_{12} & a_{22} & \cdots & a_{m2} \\ \vdots & \vdots & & \vdots \\ a_{1n} & a_{2n} & \cdots & a_{mn} \end{bmatrix}$$

In certain matrix-algebra manipulation processes, the transpose becomes a useful tool. By the rules of addition and multiplication of matrices, several important relationships exist between a matrix and its transpose. If A^t and B^t are the transposes of matrices A and B, respectively, and if k is a scalar constant, these relationships must be summarized as follows:

1. $(A^t)^t = A$.
2. $(A + B)^t = A^t + B^t$.
3. $(kA)^t = kA^t$.
4. $(AB)^t = B^tA^t$.
5. If $C = AB$, then $C^t = B^tA^t$.

Example

For the transpose of the matrix

$$A = \begin{bmatrix} 1 & 2 & -1 & 3 \\ 0 & 2 & 1 & -2 \\ 1 & 0 & 3 & 2 \\ 2 & 1 & 3 & 4 \end{bmatrix}$$

one has

$$A^t = \begin{bmatrix} 1 & 0 & 1 & 2 \\ 2 & 2 & 0 & 1 \\ -1 & 1 & 3 & 3 \\ 3 & -2 & 2 & 4 \end{bmatrix}$$

Observe that the principal diagonal remains the same in both matrices.

C.9
Partitioning of Matrices

It is often desirable to subdivide large matrices into various rectangular groups of elements. These rectangular groups are known as submatrices, and the subdivided matrix is said to be partitioned. It is convenient to use dashed lines to indicate the partitioning. The matrix A which follows has been partitioned into four groups of elements, each of which may be considered to be a matrix and may be referred to as a submatrix:

$$A = \begin{bmatrix} a_{11} & a_{12} & a_{13} & a_{14} & a_{15} \\ a_{21} & a_{22} & a_{23} & a_{24} & a_{25} \\ a_{31} & a_{32} & a_{33} & a_{34} & a_{35} \\ \hline a_{41} & a_{42} & a_{43} & a_{44} & a_{45} \\ a_{51} & a_{52} & a_{53} & a_{54} & a_{55} \end{bmatrix}$$

If the submatrices are denoted

$$A_{11} = \begin{bmatrix} a_{11} & a_{12} & a_{13} \\ a_{21} & a_{22} & a_{23} \\ a_{31} & a_{32} & a_{33} \end{bmatrix}$$

$$A_{12} = \begin{bmatrix} a_{14} & a_{15} \\ a_{24} & a_{25} \\ a_{34} & a_{35} \end{bmatrix}$$

$$A_{21} = \begin{bmatrix} a_{41} & a_{42} & a_{43} \\ a_{51} & a_{52} & a_{53} \end{bmatrix}$$

$$A_{22} = \begin{bmatrix} a_{44} & a_{45} \\ a_{54} & a_{55} \end{bmatrix}$$

the matrix A may be written in the compact form

$$A = \begin{bmatrix} A_{11} & A_{12} \\ A_{21} & A_{22} \end{bmatrix}$$

Partitioned matrices can be added, subtracted, and multiplied, provided that the partitioning is performed in a manner such that the partitioned matrices are conformable for these operations. If two matrices are to be conformable for multiplication, the partitioning of the columns in the premultiplier must be the same as the partitioning of the rows in the postmultiplier. Consider the two matrices A and B

$$A = \begin{bmatrix} -1 & 2 & 3 \\ 0 & 1 & 2 \\ 1 & -1 & 3 \end{bmatrix}, \quad B = \begin{bmatrix} 0 & 2 \\ 1 & 3 \\ 0 & 1 \end{bmatrix}$$

The product P of these two matrices can be obtained by applying the procedures described in the previous article, and is

$$AB = \begin{bmatrix} 2 & 7 \\ 1 & 5 \\ -1 & 2 \end{bmatrix}$$

The same two matrices (A and B) will be

used to illustrate the procedure for partitioning and multiplying. The dashed lines define the partitioning of the original matrices:

$$A = \begin{bmatrix} -1 & 2 & \vdots & 3 \\ 0 & 1 & \vdots & 2 \\ \hdashline 1 & -1 & \vdots & 3 \end{bmatrix}, \qquad B = \begin{bmatrix} 0 & 2 \\ 1 & 3 \\ \hdashline 0 & 1 \end{bmatrix}$$

Observe that the two column-partition in A and the two-row partition in B make the subdivisions conformable for multiplication. The partitioning of rows in A and columns in B is immaterial and selected for convenience (two in A and one in B). The product now can be expressed in terms of these submatrices, or

$$P = \begin{bmatrix} A_{11} & A_{12} \\ A_{21} & A_{22} \end{bmatrix} \begin{bmatrix} B_{11} \\ B_{21} \end{bmatrix}$$

which is

$$P = \begin{bmatrix} A_{11}B_{11} + A_{12}B_{21} \\ A_{21}B_{11} + A_{22}B_{21} \end{bmatrix}$$

But

$$A_{11}B_{11} = \begin{bmatrix} -1 & 2 \\ 0 & 1 \end{bmatrix} \begin{bmatrix} 0 & 2 \\ 1 & 3 \end{bmatrix} = \begin{bmatrix} 2 & 4 \\ 1 & 3 \end{bmatrix}$$

$$A_{12}B_{21} = \begin{bmatrix} 3 \\ 2 \end{bmatrix} \begin{bmatrix} 0 & 1 \end{bmatrix} = \begin{bmatrix} 0 & 3 \\ 0 & 2 \end{bmatrix}$$

$$A_{21}B_{11} = \begin{bmatrix} 1 & -1 \end{bmatrix} \begin{bmatrix} 0 & 2 \\ 1 & 3 \end{bmatrix} = \begin{bmatrix} -1 & -1 \end{bmatrix}$$

$$A_{22}B_{21} = \begin{bmatrix} 3 \end{bmatrix} \begin{bmatrix} 0 & 1 \end{bmatrix} = \begin{bmatrix} 0 & 3 \end{bmatrix}$$

Therefore,

$$P = \begin{bmatrix} \begin{bmatrix} 2 & 4 \\ 1 & 3 \end{bmatrix} + \begin{bmatrix} 0 & 3 \\ 0 & 2 \end{bmatrix} \\ \begin{bmatrix} -1 & -1 \end{bmatrix} + \begin{bmatrix} 0 & 3 \end{bmatrix} \end{bmatrix} = \begin{bmatrix} 2 & 7 \\ 1 & 5 \\ -1 & 2 \end{bmatrix}$$

which is seen to be the same as the product obtained without the partitioning.

C.10
Determinants

Determinants are frequently used in matrix algebra and are particularly useful in solving linear equations. Therefore, portions of basic determinant theory will be presented here, together with several illustrative solutions.

Consider the following two linear simultaneous equations:

$$a_{11}x + a_{12}y = c_1$$
$$a_{21}x + a_{22}y = c_2$$

The values of x and y may be found by utilizing the usual algebraic process; namely, multiplying the first equation by a_{22} and the second by a_{12} and subtracting the second from the first, thereby eliminating the element containing the y term. The resulting equation can then be solved for x:

$$x = \frac{a_{22}c_1 - a_{12}c_2}{a_{11}a_{22} - a_{12}a_{21}}$$

A similar process yields the y term:

$$y = \frac{a_{12}c_2 - a_{21}c_1}{a_{11}a_{22} - a_{12}a_{21}}$$

In both cases, the denominator remains the same and is seen to contain all the coefficients of the original equations. This is the determinant of the a_{ij} matrix, and the symbol $|A|$ denotes its value, e.g.,

$$|A| = \begin{vmatrix} a_{11} & a_{12} \\ a_{21} & a_{22} \end{vmatrix}$$

In the particular case of a 2×2 matrix, the value of the determinant is obtained by taking the difference of the products of two diagonal elements in the matrix. The principal diagonal is from upper left to lower right. The product of the other diagonal is subtracted from the product of the principal diagonal. Therefore,

$$|A| = a_{11}a_{22} - a_{12}a_{21}$$

For example:

$$|A| = \begin{vmatrix} 1 & -2 \\ 3 & 2 \end{vmatrix}$$

$$|A| = (1)(2) - (-2)(3) = 2 + 6 = 8$$

If this procedure is applied to the three equations

$$a_{11}x + a_{12}y + a_{13}z = c_1$$
$$a_{21}x + a_{22}y + a_{23}z = c_2$$
$$a_{31}x + a_{32}y + a_{33}z = c_3$$

the resulting expression for x becomes

$$x = \frac{\begin{matrix} a_{12}a_{23}c_3 + a_{22}a_{33}c_1 + a_{32}a_{13}c_2 \\ - a_{12}a_{33}c_2 - a_{22}a_{13}c_3 - a_{32}a_{23}c_1 \end{matrix}}{\begin{matrix} a_{11}a_{22}a_{23} + a_{21}a_{32}a_{13} + a_{31}a_{12}a_{23} \\ - a_{11}a_{32}a_{23} - a_{21}a_{12}a_{33} - a_{31}a_{22}a_{13} \end{matrix}}$$

Again, the denominator is seen to contain all the coefficients of the original equations and is the determinant of the matrix of these coefficients. In this case it is a 3×3 determinant:

$$|A| = \begin{vmatrix} a_{11} & a_{12} & a_{13} \\ a_{21} & a_{22} & a_{23} \\ a_{31} & a_{32} & a_{33} \end{vmatrix}$$

The denominator can be rewritten as follows:

$$|A| = a_{11}(a_{22}a_{33} - a_{32}a_{23})$$
$$- a_{21}(a_{12}a_{33} - a_{32}a_{13})$$
$$+ a_{31}(a_{12}a_{23} - a_{22}a_{13})$$

Expressed in matrix form,

$$|A| = a_{11}\begin{vmatrix} a_{22} & a_{23} \\ a_{32} & a_{33} \end{vmatrix}$$
$$- a_{21}\begin{vmatrix} a_{12} & a_{13} \\ a_{32} & a_{33} \end{vmatrix} + a_{31}\begin{vmatrix} a_{12} & a_{13} \\ a_{22} & a_{23} \end{vmatrix}$$

This arrangement of elements reveals a unique pattern. The first column of elements (a_{11}, a_{21}, a_{31}) is said to have minors made up of the elements of the second and third column of the matrix but omitting the ith row. Therefore, the minor of a_{11} is the determinant

$$\begin{vmatrix} a_{22} & a_{23} \\ a_{32} & a_{33} \end{vmatrix}$$

Each element in the first column of the matrix has its minor (a 2×2 submatrix), the value of which can be determined by

derived principles of the 2×2 matrix; for example,

$$A = \begin{bmatrix} 1 & 2 & -3 \\ 2 & 0 & 1 \\ 3 & -1 & 2 \end{bmatrix}$$

$$|A| = 1\begin{vmatrix} 0 & 1 \\ -1 & 2 \end{vmatrix} - 2\begin{vmatrix} 2 & -3 \\ -1 & 2 \end{vmatrix}$$
$$+ 3\begin{vmatrix} 2 & -3 \\ 0 & 1 \end{vmatrix}$$

$$|A| = 1(0+1) - 2(4-3) + 3(2-0) = 5$$

Note that in forming the expression for the determinant of A, the signs of the minors alternate, i.e., $+, -, +, -, \cdots$. In general, minors exist for all elements of the determinants; however, this fact will not be used in this presentation.

One more observation is necessary at this point. Referring once more to the three simultaneous linear equations, it may be observed that the numerators of the expressions for the unknowns x, y, and z also may be expressed in terms of minors of a determinant. The determinant to be used in each case, however, is obtained by an appropriate combination of the coefficients of the unknowns and the known constants shown on the right-hand sides of the equations. For the three equations, these are

$$x = \frac{\begin{vmatrix} c_1 & a_{12} & a_{13} \\ c_2 & a_{22} & a_{23} \\ c_3 & a_{32} & a_{33} \end{vmatrix}}{|A|}$$

$$y = \frac{\begin{vmatrix} a_{11} & c_1 & a_{13} \\ a_{21} & c_2 & a_{23} \\ a_{31} & c_3 & a_{33} \end{vmatrix}}{|A|}$$

$$z = \frac{\begin{vmatrix} a_{11} & a_{12} & c_1 \\ a_{21} & a_{22} & c_2 \\ a_{31} & a_{32} & c_3 \end{vmatrix}}{|A|}$$

In each case, the numerator is a determinant whose elements are the original coefficients of the system with the exception that the coefficients of the unknown whose value is being determined is replaced by the corresponding constants of the system. Thus, this basic determinant theory can be applied directly to solve for the unknown values of a system. However, it is important to observe that the matrix must be square and the determinant of A nonzero. This scheme for solving sets of simultaneous equations is known as Cramer's rule. Further information may be found in almost any standard mathematical reference.

An example of determinant theory used in solving for unknowns in the equations introduced in Art. C.2 is as follows:

$$1x + 2y + 3z = 14$$
$$4x + 5y - 1z = 11$$
$$2x + 3y + 4z = 20$$

The determinant of the coefficient matrix is

$$|A| = \begin{vmatrix} 1 & 2 & 3 \\ 4 & 5 & -1 \\ 2 & 3 & 4 \end{vmatrix}$$
$$= 1\begin{vmatrix} 5 & -1 \\ 3 & 4 \end{vmatrix} - 4\begin{vmatrix} 2 & 3 \\ 3 & 4 \end{vmatrix} + 2\begin{vmatrix} 2 & 3 \\ 5 & -1 \end{vmatrix}$$
$$= 1(20+3) - 4(8-9) + 2(-2-15)$$
$$= -7$$

Solving for x, in terms of matrix theory,

$$x = \frac{\begin{vmatrix} 14 & 2 & 3 \\ 11 & 5 & -1 \\ 20 & 3 & 4 \end{vmatrix}}{|A|}$$

Then the determinant for the numerator is given by

$$14\begin{vmatrix} 5 & -1 \\ 3 & 4 \end{vmatrix} - 11\begin{vmatrix} 2 & 3 \\ 3 & 4 \end{vmatrix} + 20\begin{vmatrix} 2 & 3 \\ 5 & -1 \end{vmatrix}$$
$$= 14(20+3) - 11(8-9)$$
$$+ 20(-2-15) = -7$$

and the value of x is $-7/-7 = 1$. In a like manner, the values of y and z are calculated to be 2 and 3, respectively.

C.11
Inverse of a Matrix

The inverse matrix is defined only for square matrices whose determinants are not zero. A matrix that does not have an inverse is said to be singular. The product of a matrix and its inverse is the unit matrix. The matrix may be either premultiplied or postmultiplied by its inverse. If A denotes the matrix, A^{-1} is its inverse, and

$$AA^{-1} = A^{-1}A = I$$

C.12
Matrix Inversion by Gaussian Elimination

There are numerous methods available for determining the inverse of a nonsingular matrix; however, all of the exact methods are forms of Gaussian elimination and are equivalent to solving n simultaneous linear algebraic equations for n unknowns, n times. This procedure may be described in general terms for a large matrix as follows: consider A to be the given matrix and B to be the desired inverse of A. Bearing in mind that all elements a_{ij} are known values and that all elements b_{ij} are unknown values that must be determined, then

$$AB = I$$

or

$$\begin{bmatrix} a_{11} & a_{12} & \cdots & a_{1n} \\ a_{21} & a_{22} & \cdots & a_{2n} \\ \vdots & \vdots & & \vdots \\ a_{n1} & a_{n2} & \cdots & a_{nn} \end{bmatrix}$$
$$\times \begin{bmatrix} b_{11} & b_{12} & \cdots & b_{1n} \\ b_{21} & b_{22} & \cdots & b_{2n} \\ \vdots & \vdots & & \vdots \\ b_{n1} & b_{n2} & \cdots & b_{nn} \end{bmatrix}$$
$$= \begin{bmatrix} 1 & 0 & \cdots & 0 \\ 0 & 1 & \cdots & 0 \\ \vdots & \vdots & & \vdots \\ 0 & 0 & \cdots & 1 \end{bmatrix}$$

Applying the rule of matrix multiplication to A and the first column of B gives expressions for the elements of the first column of the unit matrix, i.e.,

$$a_{11}b_{11} + a_{12}b_{21} + \cdots + a_{1n}b_{n1} = 1$$

$$a_{21}b_{11} + a_{22}b_{21} + \cdots + a_{2n}b_{n1} = 0$$

$$\vdots$$

$$a_{n1}b_{11} + a_{n2}b_{21} + \cdots + a_{nn}b_{n1} = 0$$

These are a series of n equations with n unknowns. The unknowns of the system are the elements b_{i1} of the first column of the matrix B and may be determined by using the method of Gaussian elimination to solve the system. This procedure can be continued for all columns of the inverse matrix B. When the matrices are small, the process is simple and direct, but with larger matrices, the process becomes so cumbersome that use of computer programs is the only practical way to complete the necessary steps. An example of this procedure for a small 2×2 matrix is as follows: $A = [a_{ij}]$ is the coefficient matrix and $B = [b_{ij}]$ is its unknown inverse matrix. Then

$$\begin{bmatrix} a_{11} & a_{12} \\ a_{21} & a_{22} \end{bmatrix} \begin{bmatrix} b_{11} & b_{12} \\ b_{21} & b_{22} \end{bmatrix} = \begin{bmatrix} 1 & 0 \\ 0 & 1 \end{bmatrix}$$

Performing the indicated multiplication in the manner previously described, results in the following sets of equations:

$$\left. \begin{array}{l} a_{11}b_{11} + a_{12}b_{21} = 1 \\ a_{21}b_{11} + a_{22}b_{21} = 0 \end{array} \right\} \quad \text{set (a)}$$

$$\left. \begin{array}{l} a_{11}b_{12} + a_{12}b_{22} = 0 \\ a_{21}b_{12} + a_{22}b_{22} = 1 \end{array} \right\} \quad \text{set (b)}$$

The unknown elements for b_{11} and b_{21} of the inverse B are obtained by solving set (a), and the elements b_{12} and b_{22} are

obtained by solving set (b), i.e.,

$$b_{11} = \frac{\begin{vmatrix} 1 & a_{12} \\ 0 & a_{22} \end{vmatrix}}{|A|} = \frac{a_{22}}{|A|}$$

$$b_{21} = \frac{\begin{vmatrix} a_{11} & 1 \\ a_{21} & 0 \end{vmatrix}}{|A|} = \frac{-a_{21}}{|A|}$$

$$b_{12} = \frac{\begin{vmatrix} 0 & a_{12} \\ 1 & a_{22} \end{vmatrix}}{|A|} = \frac{-a_{12}}{|A|}$$

$$b_{22} = \frac{\begin{vmatrix} a_{11} & 0 \\ a_{12} & 1 \end{vmatrix}}{|A|} = \frac{a_{11}}{|A|}$$

Therefore the inverse matrix B is

$$B = \frac{1}{|A|} \begin{bmatrix} a_{22} & -a_{12} \\ -a_{21} & a_{11} \end{bmatrix}$$

where the determinant of the matrix A is

$$|A| = a_{11}a_{22} - a_{12}a_{21}$$

Example

Calculate the inverse of the matrix

$$A = \begin{bmatrix} 1 & 2 \\ 3 & -1 \end{bmatrix}$$

Solution

$$|A| = 1(-1) - 2(3) = -7$$

$$A^{-1} = \frac{1}{-7} \begin{bmatrix} -1 & -2 \\ -3 & 1 \end{bmatrix} = \begin{bmatrix} \frac{1}{7} & \frac{2}{7} \\ \frac{3}{7} & -\frac{1}{7} \end{bmatrix}$$

C.13
Matrix Inversion by Partitioning

The process of calculating the inverse of matrices larger than 2×2 becomes increasingly more difficult and cumbersome if the general procedures described in Art. C.12 are followed. An easier process involves partitioning of the original matrix, followed by the determination of the inverse of the submatrices, and substituting their values back into the original matrix. Consider a 5×5 matrix subdivided as shown below, for which the inverse matrix B is sought:

$$A = \begin{bmatrix} a_{11} & a_{12} & a_{13} & a_{14} & a_{15} \\ a_{21} & a_{22} & a_{23} & a_{24} & a_{25} \\ a_{31} & a_{32} & a_{33} & a_{34} & a_{35} \\ a_{41} & a_{42} & a_{43} & a_{44} & a_{45} \\ a_{51} & a_{52} & a_{53} & a_{54} & a_{55} \end{bmatrix}$$

The partitioning can be indicated by A_R, i.e.,

$$A_R = \begin{bmatrix} A_{11} & A_{12} \\ A_{21} & A_{22} \end{bmatrix}$$

where

$$A_{11} = \begin{bmatrix} a_{11} & a_{12} \\ a_{21} & a_{22} \end{bmatrix}$$

$$A_{12} = \begin{bmatrix} a_{13} & a_{14} & a_{15} \\ a_{23} & a_{24} & a_{25} \end{bmatrix}$$

$$A_{21} = \begin{bmatrix} a_{31} & a_{32} \\ a_{41} & a_{42} \\ a_{51} & a_{52} \end{bmatrix}$$

$$A_{22} = \begin{bmatrix} a_{33} & a_{34} & a_{35} \\ a_{43} & a_{44} & a_{45} \\ a_{53} & a_{54} & a_{55} \end{bmatrix}$$

Assume that the matrix B_R is the inverse of A_R partitioned in the same manner as A_R. Then, according to the definition of an inverse,

$$A_R B_R = I$$

$$= \begin{bmatrix} A_{11} & A_{12} \\ A_{21} & A_{22} \end{bmatrix} \begin{bmatrix} B_{11} & B_{12} \\ B_{21} & B_{22} \end{bmatrix}$$

$$= \begin{bmatrix} I & 0 \\ 0 & I \end{bmatrix}$$

Multiplying the elements in the brackets,

$$\left. \begin{aligned} A_{11}B_{11} + A_{12}B_{21} = I \quad (1) \\ A_{21}B_{11} + A_{22}B_{21} = 0 \quad (2) \end{aligned} \right\} \text{set 1}$$

$$\left. \begin{aligned} A_{11}B_{12} + A_{12}B_{22} = 0 \quad (3) \\ A_{21}B_{12} + A_{22}B_{22} = I \quad (4) \end{aligned} \right\} \text{set 2}$$

Equation (2) in set 1 can be solved for B_{21} in the following manner: Place $A_{21}B_{11}$ on the right side of the equation, and premultiply each side of the equation by the inverse of A_{22}, or

$$A_{22}^{-1}A_{22}B_{21} = A_{22}^{-1}(-A_{21}B_{11})$$

The product of A_{22} and A_{22}^{-1} is the identity matrix I, which when postmultiplied by B_{21} is simply B_{21}:

$$B_{21} = -A_{22}^{-1}A_{21}B_{11}$$

Substituting this matrix B_{21} into equation (1) of set 1,

$$A_{11}B_{11} + A_{12}(-A_{22}^{-1}A_{21}B_{11}) = I$$

$$(A_{11} - A_{12}A_{22}^{-1}A_{21})B_{11} = I$$

Solving for B_{11} by premultiplying both sides by the inverse of the matrix within the bracket,

$$B_{11} = (A_{11} - A_{12}A_{22}^{-1}A_{21})^{-1}$$

Also, solving for B_{12} in equation (3) of set B,

$$B_{12} = -A_{11}^{-1}A_{12}B_{22}$$

Substituting this matrix into equation (4) of set 2,

$$A_{21}(-A_{11}^{-1}A_{12}B_{22}) + A_{22}B_{22} = I$$

and

$$(-A_{21}A_{11}^{-1}A_{12} + A_{22})B_{22} = I$$

from which

$$B_{22} = (A_{22} - A_{21}A_{11}^{-1}A_{12})^{-1}$$

The derived expressions for the submatrices B_{11}, B_{12}, B_{21}, and B_{22} may be solved in order to get the inverse matrix B_R. However, a more tractable expression may be obtained for B_{11} and B_{21} by applying the permutability of an inverse as follows:

$$AA^{-1} = A^{-1}A = I$$

Thus

$$\begin{bmatrix} B_{11} & B_{12} \\ B_{21} & B_{22} \end{bmatrix} \begin{bmatrix} A_{11} & A_{12} \\ A_{21} & A_{22} \end{bmatrix} = \begin{bmatrix} I & 0 \\ 0 & I \end{bmatrix}$$

Performing the indicated multiplication,

$$B_{11}A_{11} + B_{12}A_{21} = I \qquad (5)$$

$$B_{21}A_{11} + B_{22}A_{21} = 0 \qquad (6)$$

Solving equations (5) and (6) for B_{11} and B_{21},

$$B_{11} = A_{11}^{-1} - B_{12}A_{21}A_{11}^{-1}$$

$$B_{21} = -B_{22}A_{21}A_{11}^{-1}$$

Therefore, to find the inverse A_R, it is necessary to first calculate the inverse A_{11}^{-1} and then simply perform the necessary subtraction and multiplication as indicated in the following summary:

$$B_{22} = \left(A_{22} - A_{21}A_{11}^{-1}A_{12} \right)^{-1}$$

$$B_{12} = -A_{11}^{-1}A_{12}B_{22}$$

$$B_{21} = -B_{22}A_{21}A_{11}^{-1}$$

$$B_{11} = A_{11}^{-1} - B_{12}A_{21}A_{11}^{-1}$$

Once the submatrices B_{11}, B_{12}, B_{21} and B_{22} have been calculated, they are assembled to form B_R, which has been defined as the inverse of A_R.

Example

Calculate the inverse of the following 4×4 matrix by partitioning:

$$A = \begin{bmatrix} 1 & -1 & 2 & 1 \\ 2 & -1 & 3 & 1 \\ \hline -1 & 0 & 3 & 2 \\ 2 & -1 & 1 & 3 \end{bmatrix} = A_R = \begin{bmatrix} A_{11} & A_{12} \\ A_{21} & A_{22} \end{bmatrix}$$

where

$$A_{11} = \begin{bmatrix} 1 & -1 \\ 2 & -1 \end{bmatrix}, \quad A_{12} = \begin{bmatrix} 2 & 1 \\ 3 & 1 \end{bmatrix}, \quad A_{21} = \begin{bmatrix} -1 & 0 \\ 2 & -1 \end{bmatrix}, \quad A_{22} = \begin{bmatrix} 3 & 2 \\ 1 & 3 \end{bmatrix}$$

then

$$|A_{11}| = -1 + 2 = 1$$

$$A_{11}^{-1} = \frac{1}{|A|} \begin{bmatrix} a_{22} & -a_{12} \\ -a_{21} & a_{11} \end{bmatrix} = \frac{1}{1} \begin{bmatrix} -1 & 1 \\ -2 & 1 \end{bmatrix}$$

$$\left(A_{21}A_{11}^{-1} \right) = \begin{bmatrix} -1 & 0 \\ 2 & -1 \end{bmatrix} \begin{bmatrix} -1 & 1 \\ -2 & 1 \end{bmatrix} = \begin{bmatrix} 1 & -1 \\ 0 & 1 \end{bmatrix}$$

$$\left(A_{21}A_{11}^{-1}\right)A_{12} = \begin{bmatrix} 1 & -1 \\ 0 & 1 \end{bmatrix}\begin{bmatrix} 2 & 1 \\ 3 & 1 \end{bmatrix} = \begin{bmatrix} -1 & 0 \\ 3 & 1 \end{bmatrix}$$

$$A_{22} - A_{21}A_{11}^{-1}A_{12} = \begin{bmatrix} 3 & 2 \\ 1 & 3 \end{bmatrix} - \begin{bmatrix} -1 & 0 \\ 3 & 1 \end{bmatrix} = \begin{bmatrix} 4 & 2 \\ -2 & 2 \end{bmatrix}$$

$$|A_{22} - A_{21}A_{11}^{-1}A_{12}| = 8 + 4 = 12$$

$$\left(A_{22} - A_{21}A_{11}A_{12}\right)^{-1} = \frac{1}{12}\begin{bmatrix} 2 & -2 \\ 2 & 4 \end{bmatrix} = \begin{bmatrix} \frac{1}{6} & -\frac{1}{6} \\ \frac{1}{6} & \frac{1}{3} \end{bmatrix} = B_{22}$$

$$A_{11}^{-1}A_{12} = \begin{bmatrix} -1 & 1 \\ -2 & 1 \end{bmatrix}\begin{bmatrix} 2 & 1 \\ 3 & 1 \end{bmatrix} = \begin{bmatrix} 1 & 0 \\ -1 & -1 \end{bmatrix}$$

$$\left(-A_{11}^{-1}A_{12}\right)B_{22} = \begin{bmatrix} -1 & 0 \\ 1 & 1 \end{bmatrix}\begin{bmatrix} \frac{1}{6} & -\frac{1}{6} \\ \frac{1}{6} & \frac{1}{3} \end{bmatrix} = \begin{bmatrix} -\frac{1}{6} & \frac{1}{6} \\ \frac{1}{3} & \frac{1}{6} \end{bmatrix} = B_{12}$$

$$-B_{22}\left(A_{21}A_{11}^{-1}\right) = \begin{bmatrix} -\frac{1}{6} & \frac{1}{6} \\ -\frac{1}{6} & -\frac{1}{3} \end{bmatrix}\begin{bmatrix} 1 & -1 \\ 0 & 1 \end{bmatrix} = \begin{bmatrix} -\frac{1}{6} & \frac{1}{3} \\ -\frac{1}{6} & -\frac{1}{6} \end{bmatrix} = B_{21}$$

$$B_{12}\left(A_{21}A_{11}^{-1}\right) = \begin{bmatrix} -\frac{1}{6} & \frac{1}{6} \\ \frac{1}{3} & \frac{1}{6} \end{bmatrix}\begin{bmatrix} 1 & -1 \\ 0 & 1 \end{bmatrix} = \begin{bmatrix} -\frac{1}{6} & \frac{1}{3} \\ \frac{1}{3} & -\frac{1}{6} \end{bmatrix}$$

$$A_{11}^{-1} - B_{12}A_{21}A_{11}^{-1} = \begin{bmatrix} -1 & 1 \\ -2 & 1 \end{bmatrix} - \begin{bmatrix} -\frac{1}{6} & \frac{1}{3} \\ \frac{1}{3} & -\frac{1}{6} \end{bmatrix} = \begin{bmatrix} -\frac{5}{6} & \frac{2}{3} \\ -\frac{7}{3} & \frac{7}{6} \end{bmatrix} = B_{11}$$

Next, having calculated the submatrices, B_{11}, B_{12}, B_{21} and B_{22}, form the matrix B which is the inverse of the matrix A:

$$B = \begin{bmatrix} -\frac{5}{6} & \frac{2}{3} & -\frac{1}{6} & \frac{1}{6} \\ -\frac{7}{3} & \frac{7}{6} & \frac{1}{3} & \frac{1}{6} \\ -\frac{1}{6} & \frac{1}{3} & \frac{1}{6} & -\frac{1}{6} \\ -\frac{1}{6} & -\frac{1}{6} & \frac{1}{6} & \frac{1}{3} \end{bmatrix}$$

Check the values by performing the indicated multiplication to prove that the product of A and B forms the identity matrix:

$$\begin{bmatrix} 1 & -1 & 2 & 1 \\ 2 & -1 & 3 & 1 \\ -1 & 0 & 3 & 2 \\ 2 & -1 & 1 & 3 \end{bmatrix}\begin{bmatrix} -\frac{5}{6} & \frac{2}{3} & -\frac{1}{6} & \frac{1}{6} \\ -\frac{7}{3} & \frac{7}{6} & \frac{1}{3} & \frac{1}{6} \\ -\frac{1}{6} & \frac{1}{3} & \frac{1}{6} & -\frac{1}{6} \\ -\frac{1}{6} & -\frac{1}{6} & \frac{1}{6} & \frac{1}{3} \end{bmatrix} = \begin{bmatrix} 1 & 0 & 0 & 0 \\ 0 & 1 & 0 & 0 \\ 0 & 0 & 1 & 0 \\ 0 & 0 & 0 & 1 \end{bmatrix}$$

C.14
Cholesky's Method

The need for an appropriate method of manipulating large arrays in order to solve a large number of simultaneous equations is readily apparent. Different methods have been developed, and all are used under various circumstances. Nearly all methods make use of the high-speed digital computer to operate on the large arrays.

It so happens that the matrices generated from structural problems are such that all the elements having nonzero values can be concentrated along the principal diagonal of the matrix. Furthermore, it has been found that the matrices are symmetrical from Maxwell's law of reciprocal deflections and that they typically display the property known as "positive-definiteness."

Consequently, the most efficient procedure for handling structural matrices is considered by many to be Cholesky's *square-root* method.

The first matrix that will be considered here consists of elements representing the coefficients of the simultaneous equations. The symbols A and a_{ij} are used to represent this matrix, which must be both symmetric and positive definite. Because of these unique properties of a structural matrix, it can be decomposed (or factored) into the product of a lower triangular matrix and an upper triangular matrix, each of which is the transpose of the other, i.e.,

$$A = U^t U$$

The symbol U represents the upper triangular matrix and U^t its transpose. This is also written as follows:

$$\begin{bmatrix} a_{11} & a_{12} & a_{13} & a_{14} & \cdots & a_{1n} \\ a_{21} & a_{22} & a_{23} & a_{24} & \cdots & a_{2n} \\ a_{31} & a_{32} & a_{33} & a_{34} & \cdots & a_{3n} \\ \vdots & \vdots & \vdots & \vdots & & \vdots \\ a_{n1} & a_{n2} & a_{n3} & a_{n4} & \cdots & a_{nn} \end{bmatrix}$$

$$= \begin{bmatrix} u_{11} & 0 & 0 & \cdots & 0 \\ u_{12} & u_{22} & 0 & \cdots & 0 \\ u_{13} & u_{23} & u_{33} & \cdots & 0 \\ \vdots & \vdots & \vdots & & \vdots \\ u_{1n} & u_{2n} & u_{3n} & \cdots & u_{nn} \end{bmatrix} \begin{bmatrix} u_{11} & u_{12} & u_{13} & u_{14} & \cdots & u_{1n} \\ 0 & u_{22} & u_{23} & u_{24} & \cdots & u_{2n} \\ 0 & 0 & u_{33} & u_{34} & \cdots & u_{3n} \\ \vdots & \vdots & \vdots & \vdots & & \vdots \\ 0 & 0 & 0 & 0 & \cdots & u_{nn} \end{bmatrix}$$

Performing the indicated multiplication of the matrices U^t and U and equating them to the elements in the matrix A produces the following:

$$a_{11} = u_{11}^2, \qquad \text{or} \quad u_{11} = \sqrt{a_{11}}$$

$$a_{12} = u_{11}u_{12}, \qquad \text{or} \quad u_{12} = \frac{a_{12}}{u_{11}}$$

$$a_{13} = u_{11}u_{13}, \qquad \text{or} \quad u_{13} = \frac{a_{13}}{u_{11}}$$

$$a_{22} = u_{12}^2 + u_{22}^2, \qquad \text{or} \quad u_{22} = \sqrt{a_{22} - u_{12}^2}$$

$$a_{23} = u_{12}u_{13} + u_{22}u_{23}, \qquad \text{or} \quad u_{23} = \frac{a_{23} - u_{12}u_{13}}{u_{22}}$$

$$a_{33} = u_{13}^2 + u_{23}^2 + u_{33}^2, \qquad \text{or} \quad u_{33} = \sqrt{a_{33} - \left(u_{13}^2 + u_{23}^2\right)}$$

$$a_{34} = u_{13}u_{14} + u_{23}u_{24} + u_{33}u_{34}, \qquad \text{or} \quad u_{34} = \frac{a_{34} - \left(u_{13}u_{14} + u_{23}u_{24}\right)}{u_{33}}$$

From the above relationships, two general formulas may be written for the upper triangular matrix U. For the principal diagonal

$$u_{ii} = \sqrt{a_{ii} - \sum_{k-1}^{k=i-1} u_{ki}^2},$$

where $\quad 1 < i = j$

For the other elements,

$$u_{ij} = \frac{a_{ij} - \sum_{k=1}^{k=i-1} u_{ki} u_{kj}}{u_{ii}},$$

where $\quad 1 < i < j$

Once the upper triangular matrix is formed, the lower triangular matrix is formed as its transpose.

Cholesky's method of solution is described with reference to the following three simultaneous equations:

$$a_{11}x_1 + a_{12}x_2 + a_{13}x_3 = b_1$$

$$a_{21}x_1 + a_{22}x_2 + a_{23}x_3 = b_2$$

$$a_{31}x_1 + a_{32}x_2 + a_{33}x_3 = b_3$$

where the a elements represent the coefficients (known), the b elements represent the constants (known), and the x elements represent the unknown variables. The above three equations can be expressed as

$$AX = B$$

First, the matrix A is factored into matrices U^t and U from the previously derived general formulas (decomposition). Hence, they become known and can be substituted for A in the above formula:

$$U^t U X = B$$

Next, a new vector Y is created as $UX = Y$. Consequently,

$$U^t Y = B$$

Using the three simultaneous equations noted above, this can be written in matrix form:

$$\begin{bmatrix} u_{11} & 0 & 0 \\ u_{12} & u_{22} & 0 \\ u_{13} & u_{23} & u_{33} \end{bmatrix} \begin{bmatrix} y_1 \\ y_2 \\ y_3 \end{bmatrix} = \begin{bmatrix} b_1 \\ b_2 \\ b_3 \end{bmatrix}$$

Since the u values and the b values are known, the y values can be determined by successive multiplication. This is known as the forward sweep, and general formulas for it are derived later in this article.

Next, return to the relationship for the new vector Y,

$$UX = Y$$

which can be written in matrix form for the three original simultaneous equations, as follows:

$$\begin{bmatrix} u_{11} & u_{12} & u_{13} \\ 0 & u_{22} & u_{23} \\ 0 & 0 & u_{33} \end{bmatrix} \begin{bmatrix} x_1 \\ x_2 \\ x_3 \end{bmatrix} = \begin{bmatrix} y_1 \\ y_2 \\ y_3 \end{bmatrix}$$

Since the u values and the y values are now known, the x values can be determined. This is known as the backward sweep, and general formulas for this solution also will be derived.

Returning first to the equations for the forward sweep ($U^t Y = B$), they are

$$\begin{bmatrix} u_{11} & 0 & 0 & \cdots & 0 \\ u_{12} & u_{22} & 0 & \cdots & 0 \\ u_{13} & u_{23} & u_{33} & \cdots & 0 \\ \vdots & \vdots & \vdots & & \vdots \\ u_{1n} & u_{2n} & u_{3n} & \cdots & u_{nn} \end{bmatrix}$$

$$\times \begin{bmatrix} y_1 \\ y_2 \\ y_3 \\ \vdots \\ y_n \end{bmatrix} = \begin{bmatrix} b_1 \\ b_2 \\ b_3 \\ \vdots \\ b_n \end{bmatrix}$$

Performing the indicated multiplication,

$$u_{11} y_1 = b_1$$

or

$$y_1 = \frac{b_1}{u_{11}}$$

then

$$u_{12}y_1 + u_{22}y_2 = b_2$$

or

$$y_2 = \frac{b_2 - u_{12}y_1}{u_{22}}$$

and finally

$$u_{13}y_1 + u_{23}y_2 + u_{33}y_3 = b_3$$

or

$$y_3 = \frac{b_3 - (u_{13}y_1 + u_{23}y_2)}{u_{33}}$$

The general equation for the y elements from the above relationships can be written as follows:

$$y_i = \frac{b_i - \sum\limits_{k=1}^{k=i-1} u_{ki}y_k}{u_{ii}} \qquad (i > 1)$$

Equations for the backward sweep ($UX = Y$) are

$$
\begin{bmatrix}
u_{11} & u_{12} & u_{13} & \cdots & u_{1,n-2} & u_{1,n-1} & u_{1n} \\
0 & u_{22} & u_{23} & \cdots & \cdot & \cdot & \cdot \\
0 & 0 & u_{33} & \cdots & \cdot & \cdot & \cdot \\
\vdots & \vdots & \vdots & \ddots & \cdot & \cdot & \cdot \\
0 & 0 & 0 & \cdots & u_{n-2,n-2} & u_{n-2,n-1} & u_{n-2,n} \\
0 & 0 & 0 & \cdots & 0 & u_{n-1,n-1} & u_{n-1,n} \\
0 & 0 & 0 & \cdots & 0 & 0 & n_{n,n}
\end{bmatrix}
\begin{bmatrix}
x_1 \\ x_2 \\ x_3 \\ \vdots \\ x_{n-2} \\ x_{n-1} \\ x_n
\end{bmatrix}
=
\begin{bmatrix}
y_1 \\ y_2 \\ y_3 \\ \vdots \\ y_{n-2} \\ y_{n-1} \\ y_n
\end{bmatrix}
$$

Performing the indicated multiplication, commencing at the bottom and working upward,

$$u_{nn}x_{nn} = y_n$$

or

$$x_n = \frac{y_n}{u_{nn}}$$

then

$$u_{n-1,n-1}x_{n-1} + u_{n-1,n}x_n = y_{n-1}$$

or

$$x_{n-1} = \frac{y_{n-1} - u_{n-1,n}x_n}{u_{n-1,n-1}}$$

and finally

$$u_{n-2,n-2}x_{n-2} + u_{n-2,n-1}x_{n-1} + u_{n-2,n}x_n = y_{n-2}$$

or

$$x_{n-2} = \frac{y_{n-2}(u_{n-2,n-1}x_{n-1} + u_{n-2,n}x_n)}{u_{n-2,n-2}}$$

The general equation for the x elements from the above relationships can be written as

$$x_i = \frac{y_i - \sum\limits_{k=i+1}^{k=n} u_{ik}x_k}{u_{ii}} \qquad (i < n)$$

Summary of Cholesky's Method

1. From the series of simultaneous equations, create the a_{ij} matrix.

2. Form the matrix U from the derived equations. ⎫
3. Form the transpose U^t. ⎬ Decompose

4. Calculate and form the matrix y_i from the derived general equation (forward sweep).

5. Calculate the x_i values from the derived general equations (backward sweep).

C.15
Fortran Programs
for Matrix Algebra

The three computer programs described and listed in this article are written in FORTRAN 77. The accompanying flowcharts correlate step by step with their respective programs. All programs are written as simple subroutines to a main program.

The programs named DECOMP and SOLVE must be included as subroutines when either of the two main programs, PLTRUSS (Appendix D) or PLFRAME (Appendix E), is used. They are included exactly as shown here and placed at the end of the main program as shown in the program listings in the appendixes. They are also included on the enclosed diskette with the executable version of PLTRUSS and PLFRAME.

1. MATMUL is an independent subroutine for matrix multiplication. It is not used, nor is it included in the listing or on the diskette. The purpose for including it in this appendix is to furnish an additional aid to the understanding of matrix multiplication described in Art. C.7 and the step-by-step procedure used in computer processing.

The program refers to two matrices, A and B. The matrix A is of order $M \times L$, while B is of order $L \times N$. When multiplied, a new matrix P is generated, and it is of order $M \times N$. The variable I is the *row counter* for A, while J is the *column counter* for matrix B. The variable K is the *inner-product counter*. The inner products are accumulated in SUM and assigned to the appropriate elements of the P matrix after K cycles from 1 to L.

The flowchart for MATMUL is shown in Fig. C1. The diamond-shaped step always refers to a question that must be answered yes or no. The return step means to return to the main program and terminate processing by the subroutine.

2. DECOMP is a subroutine for the decomposition of a positive definite, symmetric matrix using Cholesky's square-root method as described in Art. C.14. The stiffness matrix A is of order N. The variables I, J, and K are counters and are incremented by 1 before testing or making a computation. The upper triangle of the original matrix A is destroyed and replaced by the upper triangle of the decomposition of A. It should be noted that the reciprocals of the diagonal element of the

```
 1          SUBROUTINE MATMUL(A,B,P,M,L,N)
 2          DIMENSION A(120,120), B(120,120), P(120,120)
 3          I=0
 4       10 I=I+1
 5          IF (I.GT.M) RETURN
 6          J=0
 7       20 J=J+1
 8          IF (J.GT.N) GO TO 10
 9          K=0
10          SUM=0.0
11       30 K=K+1
12          IF (K.GT.L) GO TO 40
13          SUM=SUM+A(I,K)*B(K,J)
14          GO TO 30
15       40 P(I,J)=SUM
16          GO TO 20
17          END
```

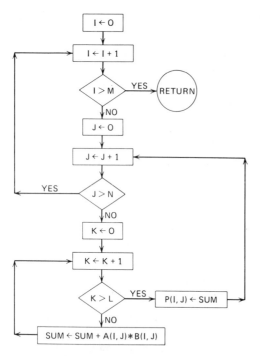

Figure C1/Flowchart for matrix multiplication.

decomposition are stored rather than the elements themselves. Consequently, operations in both DECOMP and SOLVE, involving division by a diagonal element, are converted to operations involving multiplication.

The flowchart for DECOMP is shown in Fig. C2. There are two returns from the subroutine. One occurs when I becomes greater than N and the processing is complete, while the other occurs only if the decomposition fails. The initial assumption is that the matrix is valid for this operation (true). A check is made whether SUM is equal to or less than zero. If so, the matrix is invalid (false) and returns to the main program. This return is a safeguard against an undefined arithmetic operation.

3. SOLVE is a subroutine that performs the operations for obtaining the vectors Y and X in the matrix equation described in Art. C.14, i.e.,

$$U'Y = B \qquad \text{(forward sweep)}$$

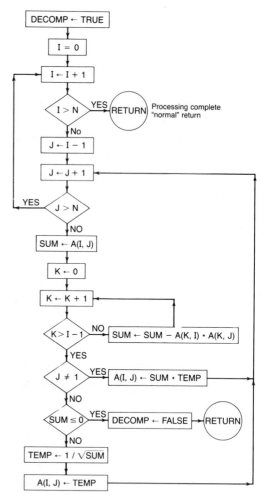

Figure C2/Flowchart for DECOMP.

and

$$UX = Y \qquad \text{(backward sweep)}$$

In an operation similar to that for DECOMP, the new elements overlie the old, thereby conserving computer storage space.

Figure C3 is the flowchart for SOLVE. The variable I is the row counter, and K, as before, is the inner-product counter. The top half (statements 3–13 inclusive) pertains to the forward sweep, and the bottom half (statements 14–24 inclusive), the backward sweep.

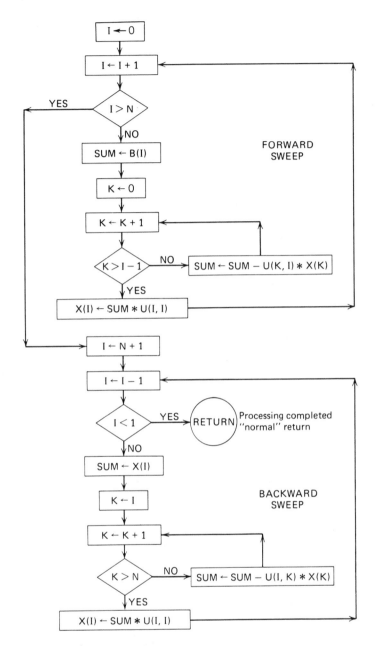

Figure C3/Flowchart for SOLVE.

```
              LOGICAL FUNCTION DECOMP(N,A)
              DIMENSION A(120,120)
              DECOMP=.TRUE.
              I=0
           10 I=I+1
              IF (I.GT.N) RETURN
              J=I-1
           20 J=J+1
              IF (J.GT.N) GO TO 10
              SUM=A(I,J)
              K=0
           30 K=K+1
              IF (K.GT.I-1) GO TO 40
              SUM=SUM-A(K,I)*A(K,J)
              GO TO 30
           40 IF (J.NE.I) GO TO 50
              IF (SUM.LE.0) GO TO 60
              TEMP=1.0/SQRT (SUM)
              A(I,J)=TEMP
              GO TO 20
           50 A(I,J)=SUM*TEMP
              GO TO 20
           60 DECOMP=.FALSE.
              RETURN
              END
```

```
 1            SUBROUTINE SOLVE(N,U,B,X)
 2            DIMENSION U(120,120),B(120),X(120)
 3            I=0
 4         10 I=I+1                    @'FORWARD' SWEEP
 5            IF (I.GT.N) GO TO 40
 6            SUM=B(I)
 7            K=0
 8         20 K=K+1
 9            IF (K.GT.I-1) GO TO 30
10            SUM=SUM-U(K,I)*X(K)
11            GO TO 20
12         30 X(I)=SUM*U(I,I)
13            GO TO 10
14         40 I=N+1
15         50 I=I-1                    @'BACKWARD8e ' SWEEP
16            IF (I.LT.1) RETURN       @PROCESSING COMPLETED
17            SUM=X(I)
18            K=I
19         60 K=K+1
20            IF (K.GT.N) GO TO 70
21            SUM=SUM-U(I,K)*X(K)
22            GO TO 60
23         70 X(I)=SUM*U(I,I)
24            GO TO 50
25            END
```

D

PLANE-
TRUSS
PROGRAM

D.1
Introduction

The name given to this computer program is PLTRUSS, and the language in which it currently is written is FORTRAN 77. The original version was written in 1975 with the intention of using a mainframe computer operating with input from punched cards in a batch process. The program since has undergone several revisions in an attempt to keep pace with the changing world of computer hardware and software. The version contained herein (on the diskette included with this text in a pocket on the inside of the back cover) and described in this appendix should run on most personal computers[1] using either MS or PC DOS (disk operating systems).

PLTRUSS is used to determine the axial forces in all members, the displacement of all joints, and the support reaction components of pin-jointed plane frame trusses of arbitrary stable configuration and any degree of indeterminancy. Only concentrated loads at the joints are permitted, along with the distributed dead load of the member.

The sizes of the arrays are declared such that the program is limited to trusses having not more than 100 members and 60 joints. For trusses exceeding these limits, a change would be required in the dimension statement made near the beginning of the program. Such a change, i.e., increasing the size of the arrays, also would require a memory capacity larger than is found in some personal computers.

The program is prepared so that any properly configured and stabilized truss can be analyzed for any number of independent loadings at the joints. For each loading

[1]See Article 12.16.

pattern for a given truss, the program determines the following:

1. Complete stiffness matrix.
2. Condensed stiffness matrix.
3. Q matrix.
4. Displacements of all joints.
5. Axial forces in all members.
6. Reactive forces at all truss supports.

However, the printed output resulting from using the included diskette in a computer contains only the fourth, fifth, and sixth items in the above list.

D.2
The Program

The main program for PLTRUSS and the subroutine programs DECOMP and SOLVE are listed below. Each line is a statement and is numbered consecutively from 1 to 493. Some numbered lines are intentionally left blank in order to improve the readability of the program. Each statement preceded by a letter C is a comment statement and is included only to aid in

understanding. All blank lines and comment statements could be omitted from the program with absolutely no affect on the results of the analysis.

This program should be self-explanatory to anyone with a fundamental knowledge of the FORTRAN language. The program proceeds in consecutive steps to:

1. Read in (accept) the data pertaining to the problem to be solved.
2. Prepare the stiffness matrix and all load vectors.
3. Call the subroutines to solve for the unknown displacements.

These subroutines are listed at the end of the main program, following the formatting instructions.

4. Make the necessary calculations for the axial loads.
5. Furnish instructions to the printing device for content and format of output information.

This program has been compiled as explained in Article 12.16 and placed on the included diskette in an executable form ready to use.

PLTRUSS Program

```
 1 C      STEEL BUILDINGS: ANALYSIS AND DESIGN
 2 C      CRAWLEY & DILLON
 3
 4 C      PLANE TRUSS ANALYSIS BY DISPLACEMENT METHOD
 5
 6        DIMENSION PID(3,18),
 7      1            JM(100), KM(100), AM(100), EM(100), GM(100),
 8      1            SM(100), CX(100), CY(100),
 9      1            X(60), Y(60), MCL(120),
10      1            S(120,120), Q(120,120), W(120), R(120), D(120)
11
12        REAL      K11, K12, K22
13
14        CHARACTER*1  XREST, YREST
15        CHARACTER*80 IFILE, OFILE
16
17        LOGICAL   DECOMP
18
19        INTEGER   FF
20        PARAMETER (FF=12)
21
22 C    * * * * * * * * * * *
23 C    * I / O   S E T U P *
24 C    * * * * * * * * * * *
25
26        WRITE(*,1000)
27        READ(*,1001) IFILE
28        OPEN
29      1            (UNIT=                1,
30      1            FILE=                 IFILE,
31      1            MODE=                 'READ',
32      1            ACCESS=               'SEQUENTIAL',
33      1            FORM=                 'FORMATTED',
34      1            ERR=                  901)
35
36        WRITE(*,1002)
37        READ(*,1001) OFILE
38        OPEN
39      1            (UNIT=                2,
40      1            FILE=                 OFILE,
41      1            MODE=                 'WRITE',
42      1            ACCESS=               'SEQUENTIAL',
43      1            FORM=                 'FORMATTED',
44      1            ERR=                  902)
45
46 C    * * * * * * * * * * * * * * * * * * * * * * * *
47 C    * P R O B L E M   I D E N T I F I C A T I O N *
48 C    * * * * * * * * * * * * * * * * * * * * * * * *
49
50    100 READ(1,500,ERR=903,END=904) ((PID(I,J), J=1,18), I=1,3)
51
52 C                PID(,): PROBLEM IDENTIFICATION (3 CARDS / 72 COLUMNS)
53
54 C    * * * * * * * * * * * * * * *
55 C    * T R U S S   G E O M E T R Y *
56 C    * * * * * * * * * * * * * * *
57
58 C      INPUT JOINT DATA
59
60        READ(1,510,ERR=903,END=904) NJ
```

```
61
62 C                      NJ:  NUMBER OF JOINTS
63
64        READ(1,510,ERR=903,END=904) (J, X(J), Y(J), INDEX=1,NJ)
65
66 C                       J:  JOINT NUMBER
67 C                    X(J):  X COORDINATE OF JOINT 'J' (FEET)
68 C                    Y(J):  Y COORDINATE OF JOINT 'J' (FEET)
69
70        NLA=2*NJ
71        DO 110 INDEX=1,NLA
72    110 MCL(INDEX)=0
73
74        READ(1,520,ERR=903,END=904) NRJ
75
76 C                     NRJ:  NUMBER OF RESTRAINED JOINTS
77
78        READ(1,520,ERR=903,END=904) (J, MCL(2*J-1), MCL(2*J), INDEX=1,NRJ)
79
80 C                       J:  RESTRAINED JOINT NUMBER
81 C                   MCL():  MATRIX CONDENSATION LIST ENTRY
82 C                            0: UNRESTRAINED LINE OF ACTION
83 C                            1:  RESTRAINED LINE OF ACTION
84
85 C     OUTPUT JOINT DATA
86
87        WRITE(2,600) CHAR(FF),((PID(I,J), J=1,18), I=1,3)
88        WRITE(2,610)
89        DO 120 J=1,NJ
90        XREST=' '
91        IF (MCL(2*J-1).GT.0) XREST='R'
92        YREST=' '
93        IF (MCL(2*J ).GT.0) YREST='R'
94        WRITE(2,620) J, X(J), Y(J), XREST, YREST
95        X(J)=12.0*X(J)
96        Y(J)=12.0*Y(J)
97    120 CONTINUE
98
99        NRL=0
100       DO 130 I=1,NLA
101       NRL=NRL+MCL(I)
102       DO 130 J=1,NLA
103   130 S(I,J)=0.0
104       NDF=NLA-NRL
105
106 C     INPUT MEMBER DATA
107
108       READ(1,530,ERR=903,END=904) NM
109
110 C                      NM:  NUMBER OF MEMBERS
111
112       WT=0.0
113       DO 140 INDEX=1,NM
114
115       READ(1,530,ERR=903,END=904) I, J, K, A, E, G
116
117 C                       I:  MEMBER NUMBER
118 C                       J:  'A' END MEMBER INCIDENCE
119 C                       K:  'B' END MEMBER INCIDENCE
120 C                       A:  CROSS-SECTIONAL AREA OF MEMBER (INS**2)
```

```
121 C                              E:   MODULUS OF ELASTICITY OF MEMBER (KSI/1000)
122 C                              G:   UNIT WEIGHT OF MEMBER (LBS/FT**3)
123
124         JM(I)=J
125         KM(I)=K
126         AM(I)=A
127         EM(I)=E
128         GM(I)=G
129
130         DELX=X(K)-X(J)
131         DELY=Y(K)-Y(J)
132         SPAN=SQRT(DELX**2+DELY**2)
133         SM(I)=SPAN/12.0
134         CX(I)=DELX/SPAN
135         CY(I)=DELY/SPAN
136         WT=WT+A*SPAN*G/3456000.0
137
138 C     CONTRIBUTION OF MEMBER TO TRUSS STIFFNESS MATRIX
139         J2=2*J
140         J1=J2-1
141         K2=2*K
142         K1=K2-1
143         TEMP=A*E/SPAN
144         K11=TEMP*CX(I)**2
145         K12=TEMP*CX(I)*CY(I)
146         K22=TEMP*CY(I)**2
147         S(J1,J1)=S(J1,J1)+K11
148         S(J1,J2)=S(J1,J2)+K12
149         S(J1,K1)=          -K11
150         S(J1,K2)=          -K12
151         S(J2,J1)=S(J2,J1)+K12
152         S(J2,J2)=S(J2,J2)+K22
153         S(J2,K1)=          -K12
154         S(J2,K2)=          -K22
155         S(K1,J1)=          -K11
156         S(K1,J2)=          -K12
157         S(K1,K1)=S(K1,K1)+K11
158         S(K1,K2)=S(K1,K2)+K12
159         S(K2,J1)=          -K12
160         S(K2,J2)=          -K22
161         S(K2,K1)=S(K2,K1)+K12
162    140 S(K2,K2)=S(K2,K2)+K22
163
164 C     OUTPUT MEMBER DATA
165
166         WRITE(2,600) CHAR(FF),((PID(I,J), J=1,18), I=1,3)
167         WRITE(2,630)
168         WRITE(2,640) (I, JM(I), KM(I), SM(I), AM(I), EM(I), I=1,NM)
169         WRITE(2,650) WT
170
171 C     CONDENSE TRUSS STIFFNESS MATRIX
172         I=0
173         M=0
174         DO 180 K=1,NLA
175         IF (MCL(K).NE.0) GO TO 160
176         I=I+1
177         IF (I.EQ.K) GO TO 180
178         DO 150 J=1,NLA
179    150 S(I,J)=S(K,J)
180         GO TO 180
```

```
181    160 M=M+1
182        DO 170 J=1,NLA
183    170 Q(M,J)=S(K,J)
184    180 CONTINUE
185        J=0
186        DO 200 K=1,NLA
187        IF (MCL(K).NE.0) GO TO 200
188        J=J+1
189        IF (J.EQ.K) GO TO 200
190        DO 190 I=1,NDF
191    190 S(I,J)=S(I,K)
192    200 CONTINUE
193
194 C       'DECOMPOSE' TRUSS STIFFNESS MATRIX
195    210 IF(.NOT.DECOMP(NDF,S)) GO TO 905
196
197 C     * * * * * * * * * * * * * * *
198 C     * S O L U T I O N   P H A S E *
199 C     * * * * * * * * * * * * * * *
200
201        READ(1,520,ERR=903,END=904) NLP
202
203 C                    NLP:  NUMBER OF LOADING PATTERNS TO BE ANALYZED
204
205        DO 390 LP=1,NLP
206
207        DO 220 INDEX=1,NLA
208        W(INDEX)=0.0
209    220 R(INDEX)=0.0
210
211        WRITE(2,600) CHAR(FF),((PID(I,J), J=1,18), I=1,3)
212        WRITE(2,660) LP, NLP
213
214        READ(1,520,ERR=903,END=904) NLJ, MWC
215
216 C                    NLJ:  NUMBER OF LOADED JOINTS IN LOADING PATTERN
217 C                    MWC:  MEMBER WEIGHT CODE
218 C                       0: EXCLUDE MEMBER WEIGHTS IN ANALYSIS
219 C                       1: INCLUDE MEMBER WEIGHTS IN ANALYSIS
220
221 C     GENERATE LOAD VECTOR
222        IF (NLJ.GT.0) GO TO 230
223        WRITE(2,662)
224        GO TO 240
225    230 DO 235 INDEX=1,NLJ
226        READ(1,510,ERR=903,END=904) J, WX, WY
227
228 C                     J:  LOADED JOINT NUMBER
229 C                    WX:  X DIRECTION JOINT LOAD (KIPS)
230 C                    WY:  Y DIRECTION JOINT LOAD (KIPS)
231
232        WRITE(2,664) J, WX, WY
233        W(2*J-1)=1000.0*WX
234    235 W(2*J  )=1000.0*WY
235
236    240 IF (MWC.GT.0) GO TO 245
237        WRITE(2,666)
238        IF (NLJ.GT.0) GO TO 260
239        GO TO 390
240 C      ASSIGN MEMBER WEIGHTS TO LOAD VECTOR
```

```
241     245 DO 250 I=1,NM
242         V=AM(I)*SM(I)*GM(I)/288.0
243         J2=2*JM(I)
244         K2=2*KM(I)
245         W(J2)=W(J2)-V
246     250 W(K2)=W(K2)-V
247         WRITE(2,668)
248
249 C       CONDENSE LOAD VECTOR
250     260 I=0
251         M=0
252         DO 270 K=1,NLA
253         L=MCL(K)
254         IF (L.NE.0) GO TO 265
255         I=I+1
256         W(I)=W(K)
257         GO TO 270
258     265 M=M+1
259         R(M)=-0.001*W(K)
260     270 CONTINUE
261
262 C       SOLVE EQUATIONS FOR UNKNOWN DEFLECTIONS
263         CALL SOLVE(NDF,S,W,D)
264
265 C       EXPAND DEFLECTION VECTOR TO INCLUDE KNOWN DEFLECTIONS
266         I=NLA
267         J=NDF+1
268     280 L=MCL(I)
269         IF (L.NE.0) GO TO 290
270         J=J-1
271         D(I)=0.000001*D(J)
272         GO TO 300
273     290 D(I)=0.0
274     300 I=I-1
275         IF (I.GT.0) GO TO 280
276
277 C       OUTPUT DEFLECTIONS
278         WRITE(2,600) CHAR(FF),((PID(I,J), J=1,18), I=1,3)
279         WRITE(2,670) LP, NLP
280         WRITE(2,672) (J, D(2*J-1), D(2*J), J=1,NJ)
281
282 C       CALCULATE AND OUTPUT MEMBER FORCES
283         WRITE(2,600) CHAR(FF),((PID(I,J), J=1,18), I=1,3)
284         WRITE(2,680) LP, NLP
285         DO 310 I=1,NM
286         J2=2*JM(I)
287         J1=J2-1
288         K2=2*KM(I)
289         K1=K2-1
290         DELTA=D(J1)-D(K1)
291         DELTB=D(J2)-D(K2)
292         TA=83.333333*(AM(I)*EM(I)/SM(I))*(CX(I)*DELTA+CY(I)*DELTB)
293         FORCE=-TA
294     310 WRITE(2,682) I, FORCE
295
296 C       CALCULATE REACTIVE FORCES
297         DO 330 I=1,NRL
298         SUM=0.0
299         DO 320 J=1,NLA
300     320 SUM=SUM+Q(I,J)*D(J)
```

```
301    330 R(I)=R(I)+1000.0*SUM
302
303 C       OUTPUT REACTIVE FORCES
304         WRITE(2,600) CHAR(FF),((PID(I,J), J=1,18), I=1,3)
305         WRITE(2,690) LP, NLP
306         I=0
307         DO 380 J=1,NJ
308         KX=MCL(2*J-1)
309         KY=MCL(2*J  )
310         IF (KX.EQ.0) GO TO 340
311         I=I+1
312         RX=R(I)
313    340 IF (KY.EQ.0) GO TO 350
314         I=I+1
315         RY=R(I)
316    350 IF ((KX.EQ.0).AND.(KY.EQ.0)) GO TO 380
317         IF ((KX.EQ.0).AND.(KY.EQ.1)) GO TO 370
318         IF ((KX.EQ.1).AND.(KY.EQ.0)) GO TO 360
319         WRITE(2,692) J, RX, RY
320         GO TO 380
321    360 WRITE(2,694) J, RX
322         GO TO 380
323    370 WRITE(2,696) J, RY
324    380 CONTINUE
325
326    390 CONTINUE
327
328 C    * * * * * * * * * * * * * * * * * * *
329 C    * E N D   S O L U T I O N   P H A S E *
330 C    * * * * * * * * * * * * * * * * * * * *
331
332         WRITE(*,1003)
333         GO TO 999
334
335 C    * * * * * * * * * * * * * * * *
336 C    * E R R O R   R E P O R T I N G *
337 C    * * * * * * * * * * * * * * * *
338
339    901 WRITE(*,1901)
340         GO TO 999
341
342    902 WRITE(*,1902)
343         GO TO 999
344
345    903 WRITE(*,1903)
346         GO TO 999
347
348    904 WRITE(*,1904)
349         GO TO 999
350
351    905 WRITE(*,1905)
352         GO TO 999
353
354    999 STOP
355
356 C       FORMAT SPECIFICATIONS
357    500 FORMAT(18A4)
358    510 FORMAT(I6,2F10.3)
359    520 FORMAT(3I6)
360    530 FORMAT(3I6,3F10.3)
```

```
361  600 FORMAT(A1,22X,'STEEL BUILDINGS:  ANALYSIS AND DESIGN'/
362    1          33X,'CRAWLEY & DILLON'//
363    1          19X,'PLANE TRUSS ANALYSIS BY DISPLACEMENT METHOD'////
364    1          4X,18A4/4X,18A4/4X,18A4///)
365  610 FORMAT(/29X,'* * * * * *  * * * * *'/
366    1          29X,'* J O I N T   D A T A *'/
367    1          29X,'* * * * * *  * * * * *'///
368    1          22X,'JOINT',5X,'COORDINATES',5X,'RESTRAINTS'/
369    1          33X,'X',7X,'Y',9X,'X  Y'/
370    1          31X,'(FEET)',2X,'(FEET)'//)
371  620 FORMAT(22X,I3,4X,2F8.3,6X,A1,2X,A1)
372  630 FORMAT(/28X,'* * * * * *  * * * * * *'/
373    1          28X,'* M E M B E R   D A T A *'/
374    1          28X,'* * * * * *  * * * * * *'///
375    1          9X,'MEMBER',5X,'END JOINTS',4X,'LENGTH',6X,
376    1            'AREA OF',6X,'MODULUS OF'/
377    1          22X,'A    B',15X,'CROSS-SECTION',3X,'ELASTICITY'/
378    1          34X,'(FEET)',6X,'(INS**2)',5X,'(KSI/1000)'//)
379  640 FORMAT(10X,I3,5X,2I5,2X,F10.3,3X,F10.3,4X,F10.3)
380  650 FORMAT(///
381    1          22X,'TOTAL WEIGHT OF TRUSS:',F8.3,1X,'TONS')
382  660 FORMAT(/20X,'LOADING PATTERN',I3,1X,'OF',I3,1X,
383    1            'LOADING PATTERNS'////
384    1          26X,'* * * * * * *  * * * * * *'/
385    1          26X,'* J O I N T   L O A D I N G *'/
386    1          26X,'* * * * * * *  * * * * * *'///
387    1          28X,'JOINT',8X,'LOADING'/
388    1          38X,'W(X)',5X,'W(Y)'/
389    1          37X,'(KIPS)',3X,'(KIPS)'//)
390  662 FORMAT(27X,'NO LOADED JOINTS IN PATTERN')
391  664 FORMAT(28X,I3,3X,2F9.3)
392  666 FORMAT(///
393    1          19X,'* * * * * * * * * * * * * * * * * * * *'/
394    1          19X,'* MEMBER WEIGHTS NOT INCLUDED IN ANALYSIS *'/
395    1          19X,'* * * * * * * * * * * * * * * * * * * *')
396  668 FORMAT(///
397    1          19X,'* * * * * * * * * * * * * * * * * * * *'/
398    1          19X,'* MEMBER WEIGHTS ARE INCLUDED IN ANALYSIS *'/
399    1          19X,'* * * * * * * * * * * * * * * * * * * *')
400  670 FORMAT(/20X,'LOADING PATTERN',I3,1X,'OF',I3,1X,
401    1            'LOADING PATTERNS'////
402    1          22X,'* * * * * * * *  * * * * * * * *'/
403    1          22X,'* J O I N T   D E F L E C T I O N S *'/
404    1          22X,'* * * * * * * *  * * * * * * * *'///
405    1          28X,'JOINT',7X,'DEFLECTIONS'/38X,'D(X)',7X,'D(Y)'/
406    1          36X,'(INCHES)    (INCHES)'//)
407  672 FORMAT(28X,I3,1X,2F11.4)
408  680 FORMAT(/20X,'LOADING PATTERN',I3,1X,'OF',I3,1X,
409    1            'LOADING PATTERNS'////
410    1          26X,'* * * * * *  * * * * * *'/
411    1          26X,'* M E M B E R   F O R C E S *'/
412    1          26X,'* * * * * *  * * * * * *'///
413    1          37X,'(KIPS)'//
414    1          32X,'TENSION      (+)'/32X,'COMPRESSION   (-)'///
415    1          32X,'MEMBER     FORCE'//)
416  682 FORMAT(33X,I3,F12.2)
417  690 FORMAT(/20X,'LOADING PATTERN',I3,1X,'OF',I3,1X,
418    1            'LOADING PATTERNS'////
419    1          25X,'* * * * * * *  * * * * * * *'/
420    1          25X,'* R E A C T I V E   F O R C E S *'/
```

```
421      1          25X,'* * * * * * * *  * * * * * * *'///
422      1          28X,'JOINT',8X,'REACTIONS'//
423      1          39X,'R(X)',5X,'R(Y)'/38X,'(KIPS)    (KIPS)'//)
424    692 FORMAT(28X,I3, 4X,2F9.3)
425    694 FORMAT(28X,I3, 4X, F9.3)
426    696 FORMAT(28X,I3,13X, F9.3)
427
428   1000 FORMAT(1H ,'Input file name = ',\)
429   1001 FORMAT(A80)
430   1002 FORMAT(1H ,'Output file name = ',\)
431   1003 FORMAT(1H ,'Analysis completed successfully')
432
433   1901 FORMAT(1H ,'Unable to open input file - analysis not performed')
434   1902 FORMAT(1H ,'Unable to open output file - analysis not performed')
435   1903 FORMAT(1H ,'Failure reading input file - analysis terminated')
436   1904 FORMAT(1H ,'Unexpected end of input file - analysis terminated')
437   1905 FORMAT(1H ,'Unstable configuration - analysis terminated')
438
439        END
440
441
442        LOGICAL FUNCTION DECOMP(N,A)
443        DIMENSION A(120,120)
444        DECOMP=.TRUE.
445        I=0
446     10 I=I+1
447        IF (I.GT.N) RETURN
448        J=I-1
449     20 J=J+1
450        IF (J.GT.N) GO TO 10
451        SUM=A(I,J)
452        K=0
453     30 K=K+1
454        IF(K.GT.I-1) GO TO 40
455        SUM=SUM-A(K,I)*A(K,J)
456        GO TO 30
457     40 IF (J.NE.I) GO TO 50
458        IF (SUM.LE.0) GO TO 60
459        TEMP=1.0/SQRT(SUM)
460        A(I,J)=TEMP
461        GO TO 20
462     50 A(I,J)=SUM*TEMP
463        GO TO 20
464     60 DECOMP=.FALSE.
465        RETURN
466        END
467
468
469        SUBROUTINE SOLVE(N,U,B,X)
470        DIMENSION U(120,120), B(120), X(120)
471        I=0
472     10 I=I+1
473        IF (I.GT.N) GO TO 40
474        SUM=B(I)
475        K=0
476     20 K=K+1
477        IF (K.GT.I-1) GO TO 30
478        SUM=SUM-U(K,I)*X(K)
479        GO TO 20
480     30 X(I)=SUM*U(I,I)
```

```
481          GO TO 10
482      40  I=N+1
483      50  I=I-1
484          IF (I.LT.1) RETURN
485          SUM=X(I)
486          K=I
487      60  K=K+1
488          IF (K.GT.N) GO TO 70
489          SUM=SUM-U(I,K)*X(K)
490          GO TO 60
491      70  X(I)=SUM*U(I,I)
492          GO TO 50
493          END
```

D.3
Computer Files

In a general sense all computer files can be divided into two main categories: program files and data files.

Program files contain instructions to the computer. The previous article (D.2) described a program file, and this file is on the inclosed diskette.

Data files are files created separately that contain information pertaining to the problem to be solved by the program file. The simplest type of data file is an *unformatted* file which contains only text and/or numbers—also referred to as an ASCII[2] file. This type must be used with the PLTRUSS program. Many data files that are created with other application programs (such as WordPerfect)[3] are *formatted* files. These files contain additional instructions such as for hard returns, centered text, boldface type, etc. Such files cannot be used directly with the PLTRUSS program.

Files created with DOS 2.0 or subsequent versions, following instructions from EDLIN or EDIT command mode, are unformated data files and can be used directly as the input information to the PLTRUSS program.

With data files created by the WordPerfect program, the hidden formatting instructions must be removed. This can be accomplished by using the Text In/Out feature of WordPerfect. After creating the formatted WordPerfect data file and using the Text Out function, naming and saving the file as a DOS file generates the unformatted (ASCII) version that can be used as the input data file for the PLTRUSS program. It is not necessary to save the formatted version, since the Text In function will automatically retrieve the ASCII version and convert it to the formatted version for subsequent adjustment and editing. The example problem in this appendix illustrates this unformatted input data file and is included with the structural programs on the diskette that is in the envelope on the back cover of this text. The name given to this file is on the diskette as TRUSEX.TX.

D.4
Data Input

Each truss, together with its loading(s), must be described within a named file in a

---

[2]ASCII stands for American Standard Code for Information Interchange.

[3]"WordPerfect" is the registered trademark of WordPerfect Corporation.

precise manner so that the FORTRAN program can interpret it for the analysis to be executed. Since this file will consist of only alphanumeric characters, they must occur not only in the correct sequence but also in a specific location within the data file.

The original version of this program, using FORTRAN V, had the alphanumeric data placed on punched cards containing 72 column locations across the card. Each card represented a single spaced line. This procedure can still be used. However, the current FORTRAN 77 language has a "short-field termination" feature that permits condensing the same data. Each single line can contain a string of characters, representing specific information, separated by commas, not spaces. The end of the string must be followed by a hard return (no space) to begin the next line. Also, no spaces should occur at the beginning of a line, and each line must contain data.

An exception to these rules is made for the first three lines, which are reserved for problem identification (see Table D1). The

Table D1
Truss Description Data

Sequence	Type of Data	Comments
1	Problem identification	Anything is allowed. Must consists of three lines. Lines may contain blank spaces.
2	Number of joints.	One integer at the beginning of the line. This sets up a counter for the next sequence of data.
3	Joint number: x coordinate. y coordinate.	One line per joint. Total number of lines equals the number of joints. Dimensions in feet carried out to the third decimal place.
4	Number of externally supported joints.	One integer at the beginning of the line. This sets up a counter for the next sequence of data.

5	Supported joint number:
	 Horizontal restraint?
	 Vertical restraint?
		One line for each joint that has external support:
		    Code 1 means yes.
		    Code 0 means no.

6	Number of members.	One integer at the beginning of the line. This sets up a counter for the next sequence of data.
7	Member number. Joint number at the "beginning" end of member. Joint number at the "ending" end of member. Cross-sectional area. Modulus of elasticity. Weight of member.	One line per member. Total number of lines equals total number of members. Identify beginning and end of member by joint numbers. Area in in.^{2} E in 1000 ksi. Weight in lb/ft^3. All dimensional quantities carried out to three decimal places.

Table D2
Truss Loading Data

Sequence	Type of Data	Comments
1	Number of loading patterns to be analyzed.	One integer at the beginning of the line. This is a counter. Lines in sequence 2 and 3 are to be repeated for each loading pattern.
2	Number of joints loaded in this loading pattern. Should weight of member be added?	One line only. The first integer sets up a counter. Code 1 means yes. Code 0 means no.
3	Joint number where load is applied: Load in X direction. Load in Y direction.	One line for each joint loaded. Total number of lines equals number of loaded joints. Load in kips carried out to three decimal places.

fourth line must begin with the technical data, i.e., the number of joints in the truss. Each joint in the truss is assigned a number to define its location (dimension) from the origin, which is usually established at the lower left-hand joint. Some of these joints must have external horizontal restraint (code 1) and/or vertical restraint (code 1). Code 0 is used for no restraint.

Members extend from joint to joint. Each member is assigned a number, and each has a beginning at a numbered joint and an ending at a numbered joint. Also, each member has a cross-sectional area, a modulus of elasticity, and a weight (lb per cu ft). This step 7 (Table D1) ends the truss description data.

Immediately following the truss description the loading data are placed (Table D2). The number of complete loading patterns to be analyzed sets up the first counter, and the number of joints loaded is an inner counter. Loads may have both vertical and horizontal components. Oth-

erwise a zero is required. Note that positive x is to the right and positive y is up.

D.5
Use of the Program

The data file must be created first, following the precise instructions described above. This can be accomplished by use of any one of several editing programs on a PC and must be unformatted as described in Art. D3. The following example shows a file that contains 31 lines of alphanumeric data, including the first three lines of problem identification. This file must have an assigned name according to the rules of the editing program being used. This file may be placed on a floppy diskette or on an internal or external hard disk.

The PLTRUSS program on the included diskette can be used directly, or it can be transferred to a hard disk. Room for the output data must be available either on a data diskette or on the hard disk. Once the program is loaded, all that is necessary is to input the word PLTRUSS. The program will prompt the user for both the name of the input file and the name to be assigned to the output data file. The analysis will then automatically take place with no further instructions.

The results of the analysis (output file) may then be printed using an attached printing device or displayed on the monitor for viewing. Preferably the output should be printed from the simple print command on DOS. Otherwise the printout may be uneven, to the extent that it is difficult to read.

Frequently, the initial results do not satisfy the user. This could be because the forces are too large or too small or the joint deflections are not satisfactory. Under these circumstances, the user can modify the input file and call for the analysis to be performed again. This cycle can be repeated as often as necessary to achieve the desired results.

The diskette contains instructions for an alternate procedure, prepared to help the beginner and to serve as a guide to the more experienced operator. The instructions on the disk label are to insert disk in drive A and, upon observing a DOS prompt on the monitor, enter A:. Then after observing the A:⟩ prompt, enter START. The Wiley logo will appear, followed by a copyright statement and instructions for continuing.

Continuing the process will lead to further instructions that offer the operator several options. If the truss input data file has been created and is ready for use, enter PLTRUSS and the procedure continues as previously described. If the input file has not been created and the operator desires further help, the option TRUSS GUIDE can be entered. The monitor will display additional information and instructions

helpful in creating the files. These instructions are most appropriate if either EDIT or the EDLIN procedures on DOS are used to create the files.

Example

Use the PLTRUSS program to analyze the truss shown in Fig. D1. Make two separate solutions: one for the vertical loads, and one for the horizontal loads. Assume the following cross-sectional areas: top chord 4.5 in.^{2}, bottom chord 3.5 in.^{2}, diagonals 3.0 in.^{2}, and verticals 2.0 in.^{2}. Include the weight of the members in the solution. The circled numbers represent joints, and the numbers within squares represent members. The origin for referencing dimensions is at the lower left-hand corner.

Solution

Prepare a data input file as shown in Fig. D2. Assign names to the input and output data files according to the rules of the PC being used, and run the PLTRUSS program. The results can be printed from the output file:

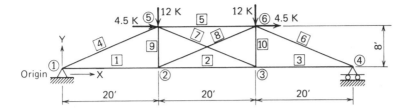

Figure D1/Example.

```
EXAMPLE
TRUSS EXAMPLE
April 1992

6
1,0.000,0.000
2,20.000,0.000
3,40.000,0.000
4,60.000,0.000
5,20.000,8.000
6,40.000,8.000
2
1,1,1
4,0,1
10
1,1,2,3.500,29.000,490.000
2,2,3,3.500,29.000,490.000
3,3,4,3.500,29.000,490.000
4,1,5,4.500,29.000,490.000
5,5,6,4.500,29.000,490.000
6,6,4,4.500,29.000,490.000
7,5,3,3.000,29.000,490.000
8,2,6,3.000,29.000,490.000
9,2,5,2.000,29.000,490.000
10,3,6,2.000,29.000,490.000
2
2,1
5,0.000,-12.000
6,0.000,-12.000
2,1
5,4.500,0.000
6,4.500,0.000
```

Figure D2/Data input file for example TRUSEX.TX on diskette.

STEEL BUILDINGS: ANALYSIS AND DESIGN
CRAWLEY & DILLON

PLANE TRUSS ANALYSIS BY DISPLACEMENT METHOD

EXAMPLE
TRUSS EXAMPLE
April 1992

```
* * * * * *  * * * * *
* J O I N T  D A T A *
* * * * * *  * * * * *
```

JOINT	COORDINATES		RESTRAINTS	
	X	Y	X	Y
	(FEET)	(FEET)		
1	.000	.000	R	R
2	20.000	.000		
3	40.000	.000		
4	60.000	.000		R
5	20.000	8.000		
6	40.000	8.000		

```
* * * * * *  * * * * * *
* M E M B E R  D A T A *
* * * * * *  * * * * * *
```

MEMBER	END JOINTS		LENGTH	AREA OF CROSS-SECTION	MODULUS OF ELASTICITY
	A	B	(FEET)	(INS**2)	(KSI/1000)
1	1	2	20.000	3.500	29.000
2	2	3	20.000	3.500	29.000
3	3	4	20.000	3.500	29.000
4	1	5	21.541	4.500	29.000
5	5	6	20.000	4.500	29.000
6	6	4	21.541	4.500	29.000
7	5	3	21.541	3.000	29.000
8	2	6	21.541	3.000	29.000
9	2	5	8.000	2.000	29.000
10	3	6	8.000	2.000	29.000

TOTAL WEIGHT OF TRUSS: 1.115 TONS

LOADING PATTERN 1 OF 2 LOADING PATTERNS

```
* * * * * *  * * * * * *
* J O I N T  L O A D I N G *
* * * * * *  * * * * * *
```

JOINT	LOADING	
	W(X)	W(Y)
	(KIPS)	(KIPS)
5	.000	-12.000
6	.000	-12.000

```
* * * * * * * * * * * * * * * * * * * * * *
* MEMBER WEIGHTS ARE INCLUDED IN ANALYSIS *
* * * * * * * * * * * * * * * * * * * * * *
```

```
* * * * * * * *  * * * * * * * *
* J O I N T  D E F L E C T I O N S *
* * * * * * * *  * * * * * * * *
```

JOINT	DEFLECTIONS	
	D(X)	D(Y)
	(INCHES)	(INCHES)
1	.0000	.0000
2	.0758	-.5411
3	.1482	-.5411
4	.2240	.0000
5	.1429	-.5414
6	.0811	-.5414

```
* * * * * *  * * * * * *
* M E M B E R  F O R C E S *
* * * * * *  * * * * * *
```

(KIPS)

TENSION   (+)
COMPRESSION (-)

MEMBER	FORCE
1	32.08
2	30.58
3	32.08
4	-34.55
5	-33.57
6	-34.55
7	1.61
8	1.61
9	-.22
10	-.22

LOADING PATTERN 1 OF 2 LOADING PATTERNS


```
* * * * * * *  * * * * * * * *
* R E A C T I V E   F O R C E S *
* * * * * * *  * * * * * * * *
```

```
      JOINT          REACTIONS

                  R(X)       R(Y)
                  (KIPS)     (KIPS)


        1          .000     13.115
        4                   13.115
```


LOADING PATTERN 2 OF 2 LOADING PATTERNS


```
* * * * * *  * * * * * * *
* J O I N T   L O A D I N G *
* * * * * *  * * * * * * *
```

```
      JOINT           LOADING
                   W(X)       W(Y)
                   (KIPS)     (KIPS)


        5          4.500       .000
        6          4.500       .000
```


```
* * * * * * * * * * * * * * * * * * * * * *
* MEMBER WEIGHTS ARE INCLUDED IN ANALYSIS *
* * * * * * * * * * * * * * * * * * * * * *
```


```
* * * * * * * * *  * * * * * * * *
* J O I N T   D E F L E C T I O N S *
* * * * * * * * *  * * * * * * * *
```

```
      JOINT           DEFLECTIONS
                   D(X)         D(Y)
                   (INCHES)     (INCHES)


        1          .0000         .0000
        2          .0191        -.0751
        3          .0322        -.0734
        4          .0442         .0000
        5          .0326        -.0762
        6          .0268        -.0725
```

LOADING PATTERN 2 OF 2 LOADING PATTERNS

```
* * * * * *  * * * * * *
* M E M B E R   F O R C E S *
* * * * * *  * * * * * *
```

(KIPS)

TENSION   (+)
COMPRESSION (−)

MEMBER	FORCE
1	8.08
2	5.52
3	5.08
4	.99
5	−3.13
6	−5.47
7	−.48
8	2.75
9	−.65
10	.55

```
* * * * * * *  * * * * * * *
* R E A C T I V E   F O R C E S *
* * * * * * *  * * * * * * *
```

JOINT	REACTIONS	
	R(X) (KIPS)	R(Y) (KIPS)
1	−9.000	−.085
4		2.315

E

PLANE-FRAME PROGRAM

E.1
Introduction

The name given to this computer program is PLFRAME, and the language in which it is currently written is FORTRAN 77. The original version was written in 1975 with the intention of using a mainframe computer operating with input from punched cards in a batch process. The program has since undergone several revisions in an attempt to keep pace with the changing world of computer hardware and software. The version contained herein (on the diskette included with this text in a pocket on the inside of the back cover) and described in this appendix should run on most personal computers[1] using either MS or PC DOS (disk operating systems).

PLFRAME is used to determine the shear, bending moment, and axial force at the ends of all members; the displacement of all joints; and the reaction components of rigid-jointed plane frames of arbitrary configuration and any degree of indeterminacy.

The sizes of the arrays are declared such that the program is limited to frames having not more than 80 members and 40 joints. For frames exceeding these limits a change would be required in the dimension statement made near the beginning of the program. Such a change, i.e., increasing the size of the arrays, also would require a memory capacity larger than is found in some personal computers.

The program is prepared so that any properly configured and stabilized frame of prismatic members can be analyzed for any number of independent loadings. Loads can be concentrated or uniformly distributed in addition to being moment-

[1]See Article 12.16.

641

loaded about the *z* axis. Concentrated loads or *z*-axis moment loads can occur at any point on a member or on a joint. Uniformly distributed loads and uniformly *z*-axis moment loads can occur on any or all members, provided they are continuous throughout the length of the member.

For each loading pattern for a given frame, the program determines the following:

1. Complete stiffness matrix.

2. Condensed stiffness matrix.

3. *Q* matrix.

4. Displacement of all joints.

5. Bending moment, shear, and axial forces at each end of each member.

6. Reactive forces at all frame supports.

However, the printed output contains only the fourth, fifth, and sixth items in the above list.

E.2
The Program

The main program for PLFRAME and the subroutine programs DECOMP and SOLVE are listed below. Each line is a statement and is numbered consecutively from 1 to 912. Some numbered lines are intentionally left blank in order to improve the readability of the program. Each line preceded by a letter C is a comment statement and is included only to aid in understanding. All blank lines and comment statements could be omitted from the program with absolutely no affect on the results of the analysis.

This program should be self-explanatory to anyone with a fundamental knowledge of the FORTRAN language. The program proceeds in consecutive steps to:

1. Read in (accept) the data pertaining to the problem to be solved.

2. Prepare the stiffness matrix and all load vectors.

3. Call the subroutines to solve for the unknown joint displacements and rotations.

4. Make the necessary calculations for bending moments, shears, and axial loads.

5. Furnish instructions to the printing device pertaining to content and format of output information.

This program has been compiled as explained in Article 12.16 and placed on the included diskette in an executable form ready for use.

PLFRAME Program

```
 1 C      STEEL BUILDINGS: ANALYSIS AND DESIGN
 2 C      CRAWLEY & DILLON
 3
 4 C      PLANE FRAME ANALYSIS BY DISPLACEMENT METHOD
 5
 6        DIMENSION PID(3,18),
 7      1           JM(80), KM(80), AM(80), ZM(80), EM(80), GM(80),
 8      1           SM(80), CX(80), CY(80),
 9      1           X(40), Y(40), MCL(120),
10      1           S(120,120), Q(120,120), W(120), R(120), D(120)
11
12        REAL      K11, K12, K13, K22, K23, K33, K36,
13      1           MO, MEA(80,6),
14      1           TA, VA, MA, TB, VB, MB
15
16        INTEGER   RAL(120)
17
18        CHARACTER*1  XREST, YREST, RREST, ORIENT
19        CHARACTER*80 IFILE, OFILE
```

```
20
21          LOGICAL    DECOMP
22
23          INTEGER    FF
24          PARAMETER (FF=12)
25
26  C     * * * * * * * * * *
27  C     * I / O   S E T U P *
28  C     * * * * * * * * * *
29
30          WRITE(*,1000)
31          READ(*,1001) IFILE
32          OPEN
33      1                (UNIT=                  1,
34      1                FILE=                   IFILE,
35      1                MODE=                   'READ',
36      1                ACCESS=                 'SEQUENTIAL',
37      1                FORM=                   'FORMATTED',
38      1                ERR=                    901)
39
40          WRITE(*,1002)
41          READ(*,1001) OFILE
42          OPEN
43      1                (UNIT=                  2,
44      1                FILE=                   OFILE,
45      1                MODE=                   'WRITE',
46      1                ACCESS=                 'SEQUENTIAL',
47      1                FORM=                   'FORMATTED',
48      1                ERR=                    902)
49
50  C     * * * * * * * * * * * * * * * * * * * * * * *
51  C     * P R O B L E M   I D E N T I F I C A T I O N *
52  C     * * * * * * * * * * * * * * * * * * * * * * *
53
54    100 READ(1,500,ERR=903,END=904) ((PID(I,J), J=1,18), I=1,3)
55
56  C              PID(,):  PROBLEM IDENTIFICATION (3 CARDS / 72 COLUMNS)
57
58  C     * * * * * * * * * * * * * * * *
59  C     * F R A M E   G E O M E T R Y *
60  C     * * * * * * * * * * * * * * * *
61
62  C     INPUT JOINT DATA
63
64          READ(1,502,ERR=903,END=904) NJ
65
66  C              NJ:  NUMBER OF JOINTS
67
68          READ(1,502,ERR=903,END=904) (J, X(J), Y(J), INDEX=1,NJ)
69
70  C                J:  JOINT NUMBER
71  C             X(J):  X COORDINATE OF JOINT 'J' (FEET)
72  C             Y(J):  Y COORDINATE OF JOINT 'J' (FEET)
73
74          NLA=3*NJ
75          DO 110 INDEX=1,NLA
76    110 MCL(INDEX)=0
77
78          READ(1,504,ERR=903,END=904) NRJ
79
80  C              NRJ:  NUMBER OF RESTRAINED JOINTS
81
82          READ(1,504,ERR=903,END=904)
83      1   (J, MCL(3*J-2), MCL(3*J-1), MCL(3*J), INDEX=1,NRJ)
84
85  C                J:  RESTRAINED JOINT NUMBER
86  C           MCL():  MATRIX CONDENSATION LIST ENTRY (X, Y, ROTATION)
87  C                   0: UNRESTRAINED LINE OF ACTION
```

```
88 C                              1:    RESTRAINED LINE OF ACTION
89
90 C       OUTPUT JOINT DATA
91
92         WRITE(2,600) CHAR(FF),((PID(I,J), J=1,18), I=1,3)
93         WRITE(2,602)
94         DO 120 J=1,NJ
95         XREST=' '
96         IF (MCL(3*J-2).GT.0) XREST='R'
97         YREST=' '
98         IF (MCL(3*J-1).GT.0) YREST='R'
99         RREST=' '
100        IF (MCL(3*J  ).GT.0) RREST='R'
101        WRITE(2,604) J, X(J), Y(J), XREST, YREST, RREST
102        X(J)=12.0*X(J)
103        Y(J)=12.0*Y(J)
104    120 CONTINUE
105
106        NRL=0
107        DO 130 I=1,NLA
108        NRL=NRL+MCL(I)
109        DO 130 J=1,NLA
110    130 S(I,J)=0.0
111        NDF=NLA-NRL
112
113 C      INPUT MEMBER DATA AND GENERATE 'COMPLETE' STIFFNESS MATRIX
114
115        READ(1,506,ERR=903,END=904) NM
116
117 C                  NM:   NUMBER OF MEMBERS
118
119        WT=0.0
120        DO 140 INDEX=1,NM
121
122        READ(1,506,ERR=903,END=904) I, J, K, A, Z, E, G
123
124 C                  I:   MEMBER NUMBER
125 C                  J:   'A' END MEMBER INCIDENCE
126 C                  K:   'B' END MEMBER INCIDENCE
127 C                  A:   CROSS-SECTIONAL AREA OF MEMBER (INS**2)
128 C                  Z:   SECOND MOMENT OF AREA OF MEMBER (INS**4)
129 C                  E:   MODULUS OF ELASTICITY OF MEMBER (KSI/1000)
130 C                  G:   UNIT WEIGHT OF MEMBER (LBS/FT**3)
131
132        JM(I)=J
133        KM(I)=K
134        AM(I)=A
135        ZM(I)=Z
136        EM(I)=E
137        GM(I)=G
138
139        DELX=X(K)-X(J)
140        DELY=Y(K)-Y(J)
141        SPAN=SQRT(DELX**2+DELY**2)
142        SM(I)=SPAN/12.0
143        CX(I)=DELX/SPAN
144        CY(I)=DELY/SPAN
145        WT=WT+A*SPAN*G/3456000.0
146
147 C      CONTRIBUTION OF MEMBER TO FRAME STIFFNESS MATRIX
148
149        K36=2.0*E*Z/SPAN
150        K33=2.0*K36
151        S23=3.0*K36/SPAN
152        S22=6.0*K36/SPAN**2
153        S11=   E*A/SPAN
154        K23=S23*CX(I)
155        K22=S11*CY(I)**2+S22*CX(I)**2
156        K13=S23*CY(I)
```

```
157          K12=(S11-S22)*CX(I)*CY(I)
158          K11=S11*CX(I)**2+S22*CY(I)**2
159
160          J3=3*J
161          J2=J3-1
162          J1=J3-2
163          K3=3*K
164          K2=K3-1
165          K1=K3-2
166
167          S(J1,J1)=S(J1,J1)+K11
168          S(J1,J2)=S(J1,J2)+K12
169          S(J1,J3)=S(J1,J3)-K13
170          S(J1,K1)=          -K11
171          S(J1,K2)=          -K12
172          S(J1,K3)=          -K13
173          S(J2,J1)=S(J2,J1)+K12
174          S(J2,J2)=S(J2,J2)+K22
175          S(J2,J3)=S(J2,J3)+K23
176          S(J2,K1)=          -K12
177          S(J2,K2)=          -K22
178          S(J2,K3)=           K23
179          S(J3,J1)=S(J3,J1)-K13
180          S(J3,J2)=S(J3,J2)+K23
181          S(J3,J3)=S(J3,J3)+K33
182          S(J3,K1)=           K13
183          S(J3,K2)=          -K23
184          S(J3,K3)=           K36
185          S(K1,J1)=          -K11
186          S(K1,J2)=          -K12
187          S(K1,J3)=           K13
188          S(K1,K1)=S(K1,K1)+K11
189          S(K1,K2)=S(K1,K2)+K12
190          S(K1,K3)=S(K1,K3)+K13
191          S(K2,J1)=          -K12
192          S(K2,J2)=          -K22
193          S(K2,J3)=          -K23
194          S(K2,K1)=S(K2,K1)+K12
195          S(K2,K2)=S(K2,K2)+K22
196          S(K2,K3)=S(K2,K3)-K23
197          S(K3,J1)=          -K13
198          S(K3,J2)=           K23
199          S(K3,J3)=           K36
200          S(K3,K1)=S(K3,K1)+K13
201          S(K3,K2)=S(K3,K2)-K23
202      140 S(K3,K3)=S(K3,K3)+K33
203
204 C       OUTPUT MEMBER DATA
205
206          WRITE(2,600) CHAR(FF),((PID(I,J), J=1,18), I=1,3)
207          WRITE(2,612)
208          WRITE(2,614) (I, JM(I), KM(I), SM(I), AM(I), ZM(I), EM(I), I=1,NM)
209          WRITE(2,616) WT
210
211 C       CONDENSE FRAME STIFFNESS MATRIX
212          I=0
213          M=0
214          DO 180 K=1,NLA
215          IF (MCL(K).NE.0) GO TO 160
216          I=I+1
217          IF (I.EQ.K) GO TO 180
218          DO 150 J=1,NLA
219      150 S(I,J)=S(K,J)
220          GO TO 180
221      160 M=M+1
222          DO 170 J=1,NLA
223      170 Q(M,J)=S(K,J)
224      180 CONTINUE
225          J=0
```

```
226          DO 200 K=1,NLA
227          IF (MCL(K).NE.0) GO TO 200
228          J=J+1
229          IF (J.EQ.K) GO TO 200
230          DO 190 I=1,NDF
231      190 S(I,J)=S(I,K)
232      200 CONTINUE
233
234 C        'DECOMPOSE' FRAME STIFFNESS MATRIX
235      210 IF(.NOT.DECOMP(NDF,S)) GO TO 905
236
237 C   * * * * * * * * * * * * * * * * *
238 C   * S O L U T I O N   P H A S E *
239 C   * * * * * * * * * * * * * * * * *
240
241          READ(1,508,ERR=903,END=904) NLP
242
243 C                    NLP:  NUMBER OF LOADING PATTERNS TO BE ANALYZED
244
245          DO 998 LP=1,NLP
246
247          DO 300 INDEX=1,NLA
248          W(INDEX)=0.0
249      300 R(INDEX)=0.0
250
251          WRITE(2,600) CHAR(FF),((PID(I,J), J=1,18), I=1,3)
252          WRITE(2,618) LP, NLP
253
254 C   * * * * * *   * * * * * *
255 C   * GENERATE LOAD VECTOR *
256 C   * * * * * *   * * * * * *
257
258 C   / / / / / / / /
259 C   / JOINT LOADS /
260 C   / / / / / / / /
261
262          READ(1,508,ERR=903,END=904) NLJ
263
264 C                    NLJ:  NUMBER OF LOADED JOINTS
265
266          IF (NLJ.GT.0) GO TO 310
267          WRITE(2,620)
268          GO TO 330
269
270      310 WRITE(2,622)
271          DO 320 INDEX=1,NLJ
272          READ(1,508,ERR=903,END=904) J, WX, WY, WM
273
274 C              J:   LOADED JOINT NUMBER
275 C             WX:   X DIRECTION JOINT LOAD (KIPS)
276 C             WY:   Y DIRECTION JOINT LOAD (KIPS)
277 C             WM:   Z-AXIS COUPLE (KIP-FEET)
278
279          WRITE(2,624)  J, WX, WY, WM
280
281          J3=3*J
282          J2=J3-1
283          J1=J3-2
284
285 C        ASSIGN LOADS IN POUNDS AND COUPLES IN INCH-POUNDS
286 C        TO LOAD AND REACTION VECTORS
287          W(J1)=W(J1)+ 1000.0*WX
288          W(J2)=W(J2)+ 1000.0*WY
289          W(J3)=W(J3)+12000.0*WM
290          IF (MCL(J1).NE.0) R(J1)=R(J1)- 1000.0*WX
291          IF (MCL(J2).NE.0) R(J2)=R(J2)- 1000.0*WY
292      320 IF (MCL(J3).NE.0) R(J3)=R(J3)-12000.0*WM
293
294 C   / / / /   / / / /
```

```
295 C     / MEMBER LOADS /
296 C    / / / /  / / / /
297
298   330 WRITE(2,626)
299       DO 340 I=1,NM
300       DO 340 J=1,6
301   340 MEA(I,J)=0.0
302
303 C     / MEMBER CONCENTRATED LOADS /
304
305       READ(1,510,ERR=903,END=904) NMC
306
307 C                 NMC:  NUMBER OF MEMBERS W/ CONCENTRATED LOADS
308
309       IF (NMC.GT.0) GO TO 350
310       WRITE(2,628)
311       GO TO 400
312
313   350 WRITE(2,630)
314       DO 390 INDEX=1,NMC
315       READ(1,510,ERR=903,END=904) M, NCL
316
317 C                   M:  LOADED MEMBER NUMBER
318 C                 NCL:  NUMBER OF CONCENTRATED LOAD POINTS
319
320       IF (NCL.EQ.0) GO TO 390
321       SPAN=12.0*SM(M)
322       COSA=CX(M)
323       SINA=CY(M)
324       TA=0.0
325       VA=0.0
326       MA=0.0
327       TB=0.0
328       VB=0.0
329       MB=0.0
330       DO 380 I=1,NCL
331       READ(1,512,ERR=903,END=904) K, A, FX, FY, MO
332
333 C                   K:  LOAD ORIENTATION
334 C                       0: MEMBER AXES
335 C                       1: STRUCTURE AXES
336 C                   A:  LOCATION OF LOADING (FRACTION OF SPAN)
337 C                  FX:  CONCENTRATED X DIRECTION LOAD (KIPS)
338 C                  FY:  CONCENTRATED Y DIRECTION LOAD (KIPS)
339 C                  MO:  CONCENTRATED Z-AXIS MOMENT (KIP-FEET)
340
341       IF (K.NE.0) GO TO 360
342
343 C     MEMBER AXES ORIENTATION
344       ORIENT='M'
345       WX=FX
346       WY=FY
347       WM=MO
348       GO TO 370
349
350 C     STRUCTURE AXES ORIENTATION
351   360 ORIENT='S'
352 C     TRANSFORM ORIENTATION TO MEMBER AXES
353       WX= FX*COSA+FY*SINA
354       WY=-FX*SINA+FY*COSA
355       WM=MO
356
357   370 IF (I.EQ.1) WRITE(2,632) M, A, ORIENT, FX, FY, MO
358       IF (I.GT.1) WRITE(2,634)    A, ORIENT, FX, FY, MO
359       WX= 1000.0*WX
360       WY= 1000.0*WY
361       WM=12000.0*WM
362
363 C     X(M) DIRECTION LOAD
```

```
364         TA=TA-WX*(1.0-A)
365         TB=TB-WX*A
366
367 C      Y(M) DIRECTION LOAD
368         TEMP=WY*SPAN*A*(1.0-A)
369         BMA=-TEMP*(1.0-A)
370         BMB= TEMP*A
371         MA=MA+BMA
372         MB=MB+BMB
373         TEMP=(BMA+BMB)/SPAN
374         VA=VA+TEMP-WY*(1.0-A)
375         VB=VB-TEMP-WY*A
376
377 C      Z-AXIS COUPLE
378         BMA=-WM*(1.0-A*(4.0-3.0*A))
379         BMB= WM*A*(2.0-3.0*A)
380         MA=MA+BMA
381         MB=MB+BMB
382         TEMP=(BMA+BMB+WM)/SPAN
383         VA=VA+TEMP
384     380 VB=VB-TEMP
385
386 C      ASSIGN END ACTIONS TO MEA(,) ARRAY AND W() AND R() VECTORS
387
388         MEA(M,1)=MEA(M,1)+TA
389         MEA(M,2)=MEA(M,2)+VA
390         MEA(M,3)=MEA(M,3)+MA
391         MEA(M,4)=MEA(M,4)+TB
392         MEA(M,5)=MEA(M,5)+VB
393         MEA(M,6)=MEA(M,6)+MB
394
395         J3=3*JM(M)
396         J2=J3-1
397         J1=J3-2
398         K3=3*KM(M)
399         K2=K3-1
400         K1=K3-2
401
402         W(J1)=W(J1)-TA*COSA+VA*SINA
403         W(J2)=W(J2)-TA*SINA-VA*COSA
404         W(J3)=W(J3)-MA
405         W(K1)=W(K1)-TB*COSA+VB*SINA
406         W(K2)=W(K2)-TB*SINA-VB*COSA
407         W(K3)=W(K3)-MB
408
409         IF (MCL(J1).NE.0) R(J1)=R(J1)+TA*COSA-VA*SINA
410         IF (MCL(J2).NE.0) R(J2)=R(J2)+TA*SINA+VA*COSA
411         IF (MCL(J3).NE.0) R(J3)=R(J3)+MA
412         IF (MCL(K1).NE.0) R(K1)=R(K1)+TB*COSA-VB*SINA
413         IF (MCL(K2).NE.0) R(K2)=R(K2)+TB*SINA+VB*COSA
414         IF (MCL(K3).NE.0) R(K3)=R(K3)+MB
415     390 CONTINUE
416
417 C   / MEMBER DISTRIBUTED LOADS /
418
419     400 READ(1,510,ERR=903,END=904) NMU
420
421 C                  NMU:  NUMBER OF MEMBERS W/ UNIFORMLY DISTRIBUTED LOADS
422
423         IF (NMU.GT.0) GO TO 410
424         WRITE(2,638)
425         GO TO 460
426
427     410 WRITE(2,640)
428         DO 450 INDEX=1,NMU
429         READ(1,510,ERR=903,END=904) M, NUL
430
431 C                  M:  LOADED MEMBER NUMBER
```

```
432 C                           NUL:  NUMBER OF DISTRIBUTED LOADS
433
434         IF (NUL.EQ.0) GO TO 450
435         SPAN=12.0*SM(M)
436         COSA=CX(M)
437         SINA=CY(M)
438         TA=0.0
439         VA=0.0
440         MA=0.0
441         TB=0.0
442         VB=0.0
443         MB=0.0
444         DO 440 I=1,NUL
445         READ(1,512,ERR=903,END=904) K, FX, FY, MO
446
447 C                        K:  LOAD ORIENTATION
448 C                            0: MEMBER AXES
449 C                            1: STRUCTURE AXES
450 C                       FX:  UNIFORMLY DISTRIBUTED X DIRECTION LOAD (KIPS/FT)
451 C                       FY:  UNIFORMLY DISTRIBUTED Y DIRECTION LOAD (KIPS/FT)
452 C                       MO:  UNIFORMLY DISTRIBUTED Z-AXIS MOMENT (K-FT/FOOT)
453
454         IF (K.NE.0) GO TO 420
455
456 C       MEMBER AXES ORIENTATION
457         ORIENT='M'
458         WX=FX
459         WY=FY
460         WM=MO
461         GO TO 430
462
463 C       STRUCTURE AXES ORIENTATION
464     420 ORIENT='S'
465 C       ADJUST FOR MEMBER INCLINATION
466         AFX=FX*ABS(SINA)
467         AFY=FY*ABS(COSA)
468         AMO=MO*ABS(COSA)
469 C       TRANSFORM ORIENTATION TO MEMBER AXES
470         WX= AFX*COSA+AFY*SINA
471         WY=-AFX*SINA+AFY*COSA
472         WM= AMO
473
474     430 IF (I.EQ.1) WRITE(2,642) M, ORIENT, FX, FY, MO
475         IF (I.GT.1) WRITE(2,644)    ORIENT, FX, FY, MO
476         WX=1000.0*WX/12.0
477         WY=1000.0*WY/12.0
478         WM=1000.0*WM
479
480 C       X(M) DIRECTION LOAD
481         TEMP=-WX*SPAN/2.0
482         TA=TA+TEMP
483         TB=TB+TEMP
484
485 C       Y(M) DIRECTION LOAD
486         TEMP=WY*SPAN/2.0
487         BMA=-TEMP*SPAN/6.0
488         BMB=-BMA
489         MA=MA+BMA
490         MB=MB+BMB
491         VA=VA-TEMP
492         VB=VB-TEMP
493
494 C       Z-AXIS COUPLE
495         VA=VA+WM
496     440 VB=VB-WM
497
498 C       ASSIGN END ACTIONS TO MEA(,) ARRAY AND W() AND R() VECTORS
```

```
499
500         MEA(M,1)=MEA(M,1)+TA
501         MEA(M,2)=MEA(M,2)+VA
502         MEA(M,3)=MEA(M,3)+MA
503         MEA(M,4)=MEA(M,4)+TB
504         MEA(M,5)=MEA(M,5)+VB
505         MEA(M,6)=MEA(M,6)+MB
506
507         J3=3*JM(M)
508         J2=J3-1
509         J1=J3-2
510         K3=3*KM(M)
511         K2=K3-1
512         K1=K3-2
513
514         W(J1)=W(J1)-TA*COSA+VA*SINA
515         W(J2)=W(J2)-TA*SINA-VA*COSA
516         W(J3)=W(J3)-MA
517         W(K1)=W(K1)-TB*COSA+VB*SINA
518         W(K2)=W(K2)-TB*SINA-VB*COSA
519         W(K3)=W(K3)-MB
520
521         IF (MCL(J1).NE.0) R(J1)=R(J1)+TA*COSA-VA*SINA
522         IF (MCL(J2).NE.0) R(J2)=R(J2)+TA*SINA+VA*COSA
523         IF (MCL(J3).NE.0) R(J3)=R(J3)+MA
524         IF (MCL(K1).NE.0) R(K1)=R(K1)+TB*COSA-VB*SINA
525         IF (MCL(K2).NE.0) R(K2)=R(K2)+TB*SINA+VB*COSA
526         IF (MCL(K3).NE.0) R(K3)=R(K3)+MB
527     450 CONTINUE
528
529 C   / MEMBER WEIGHT LOADS /
530
531     460 READ(1,510,ERR=903,END=904) MWC
532
533 C                   MWC:   MEMBER WEIGHT CODE
534 C                          0: EXCLUDE MEMBER WEIGHTS IN ANALYSIS
535 C                          1: INCLUDE MEMBER WEIGHTS IN ANALYSIS
536
537         IF (MWC.NE.0) GO TO 470
538         WRITE(2,648)
539         GO TO 700
540
541     470 WRITE(2,650)
542         DO 480 M=1,NM
543         SPAN=12.0*SM(M)
544         COSA=CX(M)
545         SINA=CY(M)
546         TEMP=0.00057870370*AM(M)*GM(M)
547         WX=-TEMP*SINA
548         WY=-TEMP*COSA
549         TEMP=WY*SPAN/2.0
550         MA=-TEMP*SPAN/6.0
551         MB=-MA
552         VA=-TEMP
553         VB=-TEMP
554         TA=-WX*SPAN/2.0
555         TB= TA
556
557 C       ASSIGN END ACTIONS TO MEA(,) ARRAY AND W() AND R() VECTORS
558
559         MEA(M,1)=MEA(M,1)+TA
560         MEA(M,2)=MEA(M,2)+VA
561         MEA(M,3)=MEA(M,3)+MA
562         MEA(M,4)=MEA(M,4)+TB
563         MEA(M,5)=MEA(M,5)+VB
564         MEA(M,6)=MEA(M,6)+MB
565
566         J3=3*JM(M)
567         J2=J3-1
```

```
568        J1=J3-2
569        K3=3*KM(M)
570        K2=K3-1
571        K1=K3-2
572
573        W(J1)=W(J1)-TA*COSA+VA*SINA
574        W(J2)=W(J2)-TA*SINA-VA*COSA
575        W(J3)=W(J3)-MA
576        W(K1)=W(K1)-TB*COSA+VB*SINA
577        W(K2)=W(K2)-TB*SINA-VB*COSA
578        W(K3)=W(K3)-MB
579
580        IF (MCL(J1).NE.0) R(J1)=R(J1)+TA*COSA-VA*SINA
581        IF (MCL(J2).NE.0) R(J2)=R(J2)+TA*SINA+VA*COSA
582        IF (MCL(J3).NE.0) R(J3)=R(J3)+MA
583        IF (MCL(K1).NE.0) R(K1)=R(K1)+TB*COSA-VB*SINA
584        IF (MCL(K2).NE.0) R(K2)=R(K2)+TB*SINA+VB*COSA
585        IF (MCL(K3).NE.0) R(K3)=R(K3)+MB
586    480 CONTINUE
587
588 C      CONDENSE LOAD VECTOR
589    700 I=0
590        M=0
591        DO 720 K=1,NLA
592        IF (MCL(K).NE.0) GO TO 710
593        I=I+1
594        W(I)=W(K)
595        GO TO 720
596    710 M=M+1
597        RAL(M)=K
598    720 CONTINUE
599
600 C      * * * * * * * * * * * * * * * * * * * * * * *
601 C      * SOLVE EQUATIONS FOR UNKNOWN DEFLECTIONS *
602 C      * * * * * * * * * * * * * * * * * * * * * * *
603
604        CALL SOLVE(NDF,S,W,D)
605
606 C      EXPAND DEFLECTION VECTOR TO INCLUDE KNOWN DEFLECTIONS
607        I=NLA
608        J=NDF+1
609    730 IF (MCL(I).NE.0) GO TO 740
610        J=J-1
611        D(I)=0.000001*D(J)
612        GO TO 750
613    740 D(I)=0.0
614    750 I=I-1
615        IF (I.GT.0) GO TO 730
616
617 C      OUTPUT DEFLECTIONS
618        WRITE(2,600) CHAR(FF),((PID(I,J), J=1,18), I=1,3)
619        WRITE(2,652) LP, NLP
620        WRITE(2,654) (J, D(3*J-2), D(3*J-1), D(3*J), J=1,NJ)
621
622 C      * * * * * * * * * * * * * * * * * * * * * * * *
623 C      * CALCULATE AND OUTPUT MEMBER END ACTIONS *
624 C      * * * * * * * * * * * * * * * * * * * * * * * *
625
626        WRITE(2,600) CHAR(FF),((PID(I,J), J=1,18), I=1,3)
627        WRITE(2,656) LP, NLP
628        DO 800 I=1,NM
629
630        SPAN=12.0*SM(I)
631        COSA=CX(I)
632        SINA=CY(I)
633
634        J3=3*JM(I)
635        J2=J3-1
636        J1=J3-2
```

```
637            K3=3*KM(I)
638            K2=K3-1
639            K1=K3-2
640
641            D1= D(J1)*COSA+D(J2)*SINA
642            D2=-D(J1)*SINA+D(J2)*COSA
643            D3= D(J3)
644            D4= D(K1)*COSA+D(K2)*SINA
645            D5=-D(K1)*SINA+D(K2)*COSA
646            D6= D(K3)
647
648            E=1000000.0*EM(I)
649            S36=2.0*E*ZM(I)/SPAN
650            S33=2.0*S36
651            S23=3.0*S36/SPAN
652            S22=6.0*S36/SPAN**2
653            S11=    E*AM(I)/SPAN
654
655            TA=MEA(I,1)+S11*(D1-D4)
656            VA=MEA(I,2)+S22*(D2-D5)+S23*(D3+D6)
657            MA=MEA(I,3)+S23*(D2-D5)+S33*D3+S36*D6
658            TB=MEA(I,4)-S11*(D1-D4)
659            VB=MEA(I,5)-S22*(D2-D5)-S23*(D3+D6)
660            MB=MEA(I,6)+S23*(D2-D5)+S36*D3+S33*D6
661
662            TA=TA/( 1000.0)
663            VA=VA/( 1000.0)
664            MA=MA/(12000.0)
665            TB=TB/( 1000.0)
666            VB=VB/( 1000.0)
667            MB=MB/(12000.0)
668        800 WRITE(2,658) I, TA, VA, MA, TB, VB, MB
669
670 C    * * * * * * * * * * * * * * * *
671 C    * CALCULATE REACTIVE FORCES *
672 C    * * * * * * * * * * * * * * * *
673
674            DO 910 I=1,NRL
675            SUM=0.0
676            DO 900 J=1,NLA
677        900 SUM=SUM+Q(I,J)*D(J)
678            K=RAL(I)
679        910 R(K)=R(K)/1000.0+1000.0*SUM
680
681 C        OUTPUT REACTIVE FORCES
682            WRITE(2,600) CHAR(FF),((PID(I,J), J=1,18), I=1,3)
683            WRITE(2,660) LP, NLP
684            DO 920 J=1,NJ
685            J3=3*J
686            J2=J3-1
687            J1=J3-2
688            K=MCL(J1)+MCL(J2)+MCL(J3)
689            IF (K.EQ.0) GO TO 920
690            IFMT=1
691            IF (MCL(J1).NE.0) IFMT=IFMT+1
692            IF (MCL(J2).NE.0) IFMT=IFMT+2
693            IF (MCL(J3).EQ.0) GO TO 915
694            IFMT=IFMT+4
695            R(J3)=R(J3)/12.0
696        915 IF (IFMT.EQ.1) WRITE(2,661) J
697            IF (IFMT.EQ.2) WRITE(2,662) J, R(J1)
698            IF (IFMT.EQ.3) WRITE(2,663) J,          R(J2)
699            IF (IFMT.EQ.4) WRITE(2,664) J, R(J1), R(J2)
700            IF (IFMT.EQ.5) WRITE(2,665) J,                    R(J3)
701            IF (IFMT.EQ.6) WRITE(2,666) J, R(J1),          R(J3)
702            IF (IFMT.EQ.7) WRITE(2,667) J,          R(J2), R(J3)
703            IF (IFMT.EQ.8) WRITE(2,668) J, R(J1), R(J2), R(J3)
704        920 CONTINUE
705
```

```
706    998 CONTINUE
707
708  C     * * * * * * * * * * * * * * * * * * *
709  C     * E N D   S O L U T I O N   P H A S E *
710  C     * * * * * * * * * * * * * * * * * * *
711
712        WRITE(*,1003)
713        GO TO 999
714
715  C     * * * * * * * * * * * * * * * * *
716  C     * E R R O R   R E P O R T I N G *
717  C     * * * * * * * * * * * * * * * * *
718
719    901 WRITE(*,1901)
720        GO TO 999
721
722    902 WRITE(*,1902)
723        GO TO 999
724
725    903 WRITE(*,1903)
726        GO TO 999
727
728    904 WRITE(*,1904)
729        GO TO 999
730
731    905 WRITE(*,1905)
732        GO TO 999
733
734    999 STOP
735
736  C     FORMAT SPECIFICATIONS
737
738  C     INPUT FORMATS
739    500 FORMAT(18A4)
740    502 FORMAT(I6,2F10.3)
741    504 FORMAT(4I6)
742    506 FORMAT(3I6,4F10.3)
743    508 FORMAT(I6,3F10.3)
744    510 FORMAT(2I6)
745    512 FORMAT(I6,4F10.3)
746
747  C     OUTPUT FORMATS
748    600 FORMAT(A1,22X,'STEEL BUILDINGS:   ANALYSIS AND DESIGN'/
749      1        33X,'CRAWLEY & DILLON'//
750      1        19X,'PLANE FRAME ANALYSIS BY DISPLACEMENT METHOD'////
751      1        4X,18A4/4X,18A4/4X,18A4///)
752    602 FORMAT(/29X,'* * * * * *   * * * * *'/
753      1        29X,'* J O I N T   D A T A *'/
754      1        29X,'* * * * * *   * * * * *'///
755      1        22X,'JOINT',5X,'COORDINATES',7X,'RESTRAINT'/
756      1        33X,'X',7X,'Y',8X,'CONDITION'/
757      1        31X,'(FEET)',2X,'(FEET)',6X,'X  Y  R'//)
758    604 FORMAT(22X,I3,4X,2F8.3,6X,A1,2X,A1,2X,A1)
759    612 FORMAT(/28X,'* * * * * *   * * * * * *'/
760      1        28X,'* M E M B E R   D A T A *'/
761      1        28X,'* * * * * *   * * * * * *'///
762      1        3X,'MEMBER',4X,'END JOINTS',6X,'LENGTH',5X,
763      1         'CROSS-SECTION PROPERTIES',4X,'MODULUS OF'/
764      1        43X,'AREA',9X,'I(ZZ)',7X,'ELASTICITY'/
765      1        15X,'A    B',8X,'(FEET)',6X,'(INS**2)',6X,'(INS**4)',5X
766      1         '(KSI/1000)'//)
767    614 FORMAT(4X,I3,4X,2I5,4X,F10.3,3X,F10.3,4X,F10.2,4X,F10.3)
768    616 FORMAT(///
769      1        22X,'TOTAL WEIGHT OF FRAME:',F8.3,1X,'TONS')
770    618 FORMAT(/20X,'LOADING PATTERN',I3,1X,'OF',I3,1X,
771      1            'LOADING PATTERNS'///)
772    620 FORMAT(/27X,'NO LOADED JOINTS IN PATTERN')
773    622 FORMAT(/26X,'* * * * * *   * * * * * *'/
774      1        26X,'* J O I N T   L O A D I N G *'/
```

```
775      1            26X,'* * * * * * *   * * * * * * *'///
776      1            19X,'JOINT',15X,'L O A D I N G'/
777      1            31X,'F(X)',8X,'F(Y)',7X,'COUPLE'/
778      1            30X,'(KIPS)',6X,'(KIPS)',4X,'(KIP-FEET)'//)
779    624 FORMAT(19X,I3,2X,3F12.3)
780    626 FORMAT(////
781      1            26X,'* * * * * * *   * * * * * * *'/
782      1            26X,'* M E M B E R   L O A D I N G *'/
783      1            26X,'* * * * * * * *   * * * * * * *'//)
784    628 FORMAT(/18X,'NO MEMBERS WITH CONCENTRATED LOADS IN PATTERN')
785    630 FORMAT(/1X,'MEMBER',20X,'C O N C E N T R A T E D',2X,
786      1                'L O A D I N G'//
787      1            10X,'LOCATION OF LOAD',3X,'ORIENTATION',4X,
788      1                'F(X)',10X,'F(Y)',10X,'M(O)'/
789      1            9X,'(FRACTION OF SPAN)',5X,'(AXES)',5X,
790      1                '(KIPS)',8X,'(KIPS)',6X,'(KIP-FEET)'/)
791    632 FORMAT(/1X,I4,1X,F14.3,14X,A1,3F14.3)
792    634 FORMAT(6X,F14.3,14X,A1,3F14.3)
793    638 FORMAT(////
794      1            18X,'NO MEMBERS WITH  DISTRIBUTED LOADS IN PATTERN')
795    640 FORMAT(///
796      1            8X,'MEMBER',5X,'U N I F O R M L Y',2X,
797      1                'D I S T R I B U T E D   L O A D I N G'//
798      1            21X,'ORIENTATION',4X,'W(X)',10X,'W(Y)',10X,'M(O)'/
799      1            24X,'(AXES)',3X,'(KIPS/FOOT)',3X,'(KIPS/FOOT)',3X,
800      1                '(K-FT/FOOT)'/)
801    642 FORMAT(/8X,I4,14X,A1,3F14.3)
802    644 FORMAT(26X,A1,3F14.3/)
803    648 FORMAT(////
804      1            19X,'* * * * * * * * * * * * * * * * * * * * * *'/
805      1            19X,'* MEMBER WEIGHTS NOT INCLUDED IN ANALYSIS *'/
806      1            19X,'* * * * * * * * * * * * * * * * * * * * * *')
807    650 FORMAT(////
808      1            19X,'* * * * * * * * * * * * * * * * * * * * * *'/
809      1            19X,'* MEMBER WEIGHTS ARE INCLUDED IN ANALYSIS *'/
810      1            19X,'* * * * * * * * * * * * * * * * * * * * * *')
811    652 FORMAT(/20X,'LOADING PATTERN',I3,1X,'OF',I3,1X,
812      1                'LOADING PATTERNS'////
813      1            22X,'* * * * * * * *   * * * * * * * *'/
814      1            22X,'* J O I N T   D E F L E C T I O N S *'/
815      1            22X,'* * * * * * * * *   * * * * * * * *'///
816      1            20X,'JOINT',8X,'D E F L E C T I O N S'/
817      1            30X,'D(X)',8X,'D(Y)',6X,'ROTATION'/
818      1            28X,'(INCHES)',4X,'(INCHES)',5X,'(RADS)'//)
819    654 FORMAT(20X,I3,2F12.4,F12.5)
820    656 FORMAT(/20X,'LOADING PATTERN',I3,1X,'OF',I3,1X,
821      1                'LOADING PATTERNS'////
822      1            22X,'* * * * * * * * * * * * * * * * * *'/
823      1            22X,'* M E M B E R   E N D   A C T I O N S *'/
824      1            22X,'* * * * * * * * * * * * * * * * * *'///
825      1            2X,'MEMBER',7X,'T(A)',8X,'V(A)',8X,'M(A)',8X,
826      1                'T(B)',8X,'V(B)',8X,'M(B)'/
827      1            14X,'(KIPS)',6X,'(KIPS)',6X,'(K-FT)',6X,
828      1                '(KIPS)',6X,'(KIPS)',6X,'(K-FT)'//)
829    658 FORMAT(2X,I4,2X,6F12.3)
830    660 FORMAT(/20X,'LOADING PATTERN',I3,1X,'OF',I3,1X,
831      1                'LOADING PATTERNS'////
832      1            25X,'* * * * * * * *   * * * * * * * *'/
833      1            25X,'* R E A C T I V E   F O R C E S *'/
834      1            25X,'* * * * * * * *   * * * * * * * *'///
835      1            19X,'JOINT',11X,'R E A C T I O N S'/
836      1            29X,'R(X)',8X,'R(Y)',8X,'COUPLE'/
837      1            28X,'(KIPS)',6X,'(KIPS)',5X,'(KIP-FEET)'//)
838    661 FORMAT(19X,I3)
839    662 FORMAT(19X,I3,F12.3)
840    663 FORMAT(19X,I3,12X,F12.3)
841    664 FORMAT(19X,I3,F12.3,F12.3)
842    665 FORMAT(19X,I3,12X,12X,F12.3)
843    666 FORMAT(19X,I3,F12.3,12X,F12.3)
```

```
844    667 FORMAT(19X,I3,12X,F12.3,F12.3)
845    668 FORMAT(19X,I3,F12.3,F12.3,F12.3)
846
847   1000 FORMAT(1H ,'Input file name = ',\)
848   1001 FORMAT(A80)
849   1002 FORMAT(1H ,'Output file name = ',\)
850   1003 FORMAT(1H ,'Analysis completed successfully')
851
852   1901 FORMAT(1H ,'Unable to open input file - analysis not performed')
853   1902 FORMAT(1H ,'Unable to open output file - analysis not performed')
854   1903 FORMAT(1H ,'Failure reading input file - analysis terminated')
855   1904 FORMAT(1H ,'Unexpected end of input file - analysis terminated')
856   1905 FORMAT(1H ,'Unstable configuration - analysis terminated')
857
858        END
859
860
861        LOGICAL FUNCTION DECOMP(N,A)
862        DIMENSION A(120,120)
863        DECOMP=.TRUE.
864        I=0
865     10 I=I+1
866        IF (I.GT.N) RETURN
867        J=I-1
868     20 J=J+1
869        IF (J.GT.N) GO TO 10
870        SUM=A(I,J)
871        K=0
872     30 K=K+1
873        IF(K.GT.I-1) GO TO 40
874        SUM=SUM-A(K,I)*A(K,J)
875        GO TO 30
876     40 IF (J.NE.I) GO TO 50
877        IF (SUM.LE.0) GO TO 60
878        TEMP=1.0/SQRT(SUM)
879        A(I,J)=TEMP
880        GO TO 20
881     50 A(I,J)=SUM*TEMP
882        GO TO 20
883     60 DECOMP=.FALSE.
884        RETURN
885        END
886
887
888        SUBROUTINE SOLVE(N,U,B,X)
889        DIMENSION U(120,120), B(120), X(120)
890        I=0
891     10 I=I+1
892        IF (I.GT.N) GO TO 40
893        SUM=B(I)
894        K=0
895     20 K=K+1
896        IF (K.GT.I-1) GO TO 30
897        SUM=SUM-U(K,I)*X(K)
898        GO TO 20
899     30 X(I)=SUM*U(I,I)
900        GO TO 10
901     40 I=N+1
902     50 I=I-1
903        IF (I.LT.1) RETURN
904        SUM=X(I)
905        K=I
906     60 K=K+1
907        IF (K.GT.N) GO TO 70
908        SUM=SUM-U(I,K)*X(K)
909        GO TO 60
910     70 X(I)=SUM*U(I,I)
911        GO TO 50
912        END
```

E.3
Computer Files

In a general sense all computer files can be divided into two main categories: program files and data files.

Program files contain instructions to the computer. The previous article (E.2) described a program file, and this file is on the enclosed diskette.

Data files are files created separately that contain information pertaining to the problem to be solved by the program file. The simplest type of data file is an *unformatted* file which contains only text and/or numbers—also referred to as an ASCII[2] file. This type must be used with the PLFRAME program. Many data files that are created with other application programs (such as WordPerfect[3]) are *formatted* files. These files contain additional instructions such as for hard returns, centered text, boldface type, etc. Such files cannot be used directly with the PLFRAME program.

Files created with DOS 2.0 or subsequent versions, following instructions from EDLIN or EDIT command modes, are unformatted data files that can be used directly as the input to the PLFRAME program.

Data files created by the WordPerfect program contain the hidden formatting instructions which must be removed. This can be accomplished by using the Text In/Out feature of WordPerfect. After creating the formatted WordPerfect data file and using the Text Out function, naming and saving the file generates the unformatted (ASCII) version that can be used as the input data file for the PLFRAME program. It is not necessary to save the formatted version, since the Text In function

will automatically retrieve the ASCII version and convert it to the formatted version for subsequent adjustment and editing. The example problem in this appendix illustrates this unformatted input data file and is included with the structural programs on the diskette that is in the envelope inside the back cover of this text. The name given to this file on the diskette is FRAMEX.TX.

E.4
Input Data

Each frame, together with its loading(s), must be described with a named file in a precise manner so that the FORTRAN program can interpret it for the analysis to be executed. Since this file will consist only of alphanumeric characters, they must occur not only in the correct sequence but also in a specific location within the data file.

The original version of this program, using FORTRAN V, had the alphanumeric data placed on punched cards containing 72 column locations across the card. Each card represented a single spaced line of data placed in specific column locations. This procedure can still be used. However, the current FORTRAN 77 language has a "short-field termination" feature that permits condensing the same data. Each single line can contain a string of characters representing specific information, separated by commas, not spaces. The end of the string must be followed by a hard return (no space) to begin the next line. Also, no spaces should occur at the beginning of the line, and each line must contain data.

An exception to these rules is made for the first three lines, which are reserved for problem identification (see Table E1). The fourth line must begin with the technical data, i.e., the number of joints in the frame. Each joint in the frame is assigned a num-

---

[2]ASCII stands for American Standard Code for Information Interchange.

[3]"WordPerfect" is the registered trademark of WordPerfect Corporation.

Table E1
Frame Description Data

Sequence	Type of Data	Comments
1	Problem identification.	Anything is allowed. Must consist of three lines. Lines may contain blank spaces.
2	Number of joints (or nodes).	One integer at the beginning of the line. This sets up a counter for the next sequence of data.
3	Joint number: *x* coordinate. *y* coordinate.	One line per joint. Total number of lines equals the number of joints. Dimensions in feet carried out to the third decimal place.
4	Number of externally supported joints.	One integer at the beginning of the line. This sets up a counter for the next sequence of data.
5	Supported joint number: Horizontal restraint? Vertical restraint? Moment restraint?	One line for each joint that has external support. Code 1 means yes. Code 0 means no.
6	Number of members.	One integer at the beginning of the line. This sets up a counter for the next sequence of data.
7	Member number. Joint number at the "beginning" end of the member. Joint number at the "ending" end of the member. Cross-sectional area. Moment of inertia. Modulus of elasticity. Weight of member.	One line per member. Total number of lines equals total number of members. Identify beginning and end of members by joint numbers. Area in in.^{2}. I in in.^{4}. E in 1000 ksi. Weight in lb/ft^3. (All units carried out to three decimal places.)

ber to identify its location (dimension) from the origin, which is usually established at the lower left-hand joint of the frame. Some of these joints must have external horizontal restraint (code 1), vertical restraint (code 1), and moment restraint (code 1). Code 0 is used when the restraint component is not present.

Members extend from joint to joint. Each member is assigned a number, and each has a beginnning at a numbered joint and an ending at a numbered joint. Also, each

member has a cross-sectional area, a moment of inertia, a modulus of elasticity, and a weight (lb per cu ft). This step 7 (Table E1) ends the frame description data.

The loading-data information is listed immediately following the frame description (see Table E2). The number of loading patterns to be analyzed is the first entry and becomes the first counter for the total number of runs to be made (sequence 1). Each loading pattern must contain some

Table E2
Frame Loading Data

Sequence	Type of Data	Comments
1	Number of loading patterns to be analyzed.	One integer at the beginning of the line. This is a counter for the number of times the following information is to be repeated.
2	Number of joints having concentrated loads in this loading pattern.	One integer at the beginning of the line. This is a counter for the number of times sequences 3 will be repeated. Must be 0 if no joints are loaded.

3	Joint number where a load is applied:
	  Load in x direction.
	  Load in y direction.
	  Moment load on z axis.
		One line for each loaded joint. Total number of lines equals total number of loaded joints. Concentrated load in kips, moment in kip-ft, each carried out to three decimal places.

4	Total number of members having concentrated loads in this loading pattern.	One integer at the beginning of the line. This is a counter for the number of times sequences 5 and 6 will be repeated. Must be 0 if no members have concentrated loads.
5	Assigned number of one of the members having concentrated loads. Number of points having concentrated loads.	Two integers at the beginning of the line, separated by a comma. The second integer is a counter for the number of times sequence 6 will be repeated.

6	Load orientation code
	Location of one loaded point:
	  Load in x direction.
	  Load in y direction.
	  Moment load on z axis.
		Code 0 means member axis. Code 1 means structure axis. Location in fraction of span. Total number of lines equals number of points having concentrated loads. Concentrated loads in kips, moment in kip-ft, all carried out to three decimal places.

7	Number of members having distributed loads in this loading pattern.	One integer at the beginning of the line. This is a counter for the number of times sequences 8 and 9 will be repeated. Must be 0 if no members have distributed loads.
8	Assigned number of a member having distributed loads. Number of distributed loads.	Two integers at the beginning of the line, separated by a comma. The second integer is a counter. This sequence must be repeated for each loaded member after the loads are described in the next sequence (9).

9	Load orientation code:
	  Uniform load in x direction.
	  Uniform load in y direction.
	  Uniform moment on z axis.
		Code 0 means member axis. Code 1 means structure axis. Total number of lines equals number of distributed loads on this member. Uniform load in kip/ft. Uniform moment in kip-ft/ft. (Each carried out to three decimal places.)

10	Should weight of member be included in the analysis?	A single integer at the beginning of the line. Code 0 means no. Code 1 means yes.

information for each location category, i.e., a zero must be used if no loading occurs in a location category. The categories are joints, point locations other than joints, and uniformly distributed locations.

Concentrated loads at the joints are described first (sequences 2, 3, Table E2). A counter is given (sequence 2) for the number of joints having concentrated loads. Next identify a joint number where concentrated loads or moment loads are applied, followed by x and y force components and the moment load about the z axis. Use zeros where appropriate. Forces to the right are x-positive, forces acting upward are y-positive, and z moments acting in a counterclockwise direction are positive. Actions in opposite directions to those just described must have a minus sign.

Following the joint loading data is the information pertaining to concentrated loads and moment loads at points (other than joints) on a member, i.e., sequences 4, 5, and 6 in Table E2. Counters are set up for the number of members having concentrated loads, followed by the data for each member having the load. For each numbered member, the number of point loads on that member is given (sequence 5). Finally one enters the load orientation (see Art. 12.9) and the location of the point in terms of fraction of span, expressed to three decimal places (sequence 6). This is followed by the direction ($+$ or $-$) and the magnitudes of the x component, y component, and z-moment component. This routine accounts for all the point loads on all the members.

The next line of data is the information pertaining to the uniformly distributed loads (sequences 7, 8, 9). First comes the total number of members having distributed loads, then the specific member number and the number of distributed loads (one load may have a member axis

and the other a structure axis), and finally the x, y load components and the z-axis moment load. The very last datum for each loading pattern (sequence 10) is the code for whether or not to include an additional load for member weight, which is automatically computed from the area and density of each member contained in sequence 7, Table E1.

E.5
Use of the Program

The data file must be created first, following precisely the instructions described above. This can be accomplished by use of any of several editing programs on a PC. The following example shows a file that contains 42 lines of alphanumeric data, including the first three lines of problem identification. This file must have an assigned name according to the rules of the editing program being used. The file may be placed on a floppy diskette or on an internal or external hard disk.

The PLFRAME program on the included diskette can be used directly, or it can be transferred to a hard disk. Room for the output data must be available either on a data diskette or on the hard disk. Once the program is loaded, all that is necessary is to input the word PLFRAME. The program will prompt the user for both the name of the input file and the name to be assigned to the output data file. The analysis then will automatically take place with no further instructions.

The results of the analysis (output file) may then be printed using an attached printing device or displayed on the monitor for viewing by following the instructions for operation of the PC being used.

Preferably the output should be printed from the simple print command on DOS. Otherwise the print-out may be uneven, to the extent that it is difficult to read.

Frequently the initial results do not satisfy the user. This may be because the forces (shear and axial) and/or the bending moment are too large or too small, or the joint deflections are not satisfactory. Under these circumstances, the user can modify the input file and call for the analysis to be performed again. This cycle can be repeated as often as necessary to achieve the desired results.

The diskette contains instructions for an alternate procedure, prepared to help the beginner and to serve as a guide to the more experienced operator. The instructions on the disk label are to insert disk in drive A and, upon observing a DOS prompt on the monitor, enter A:. Then after observing the A:⟩ prompt, enter START. The Wiley logo will appear, followed by a copyright statement and instructions for continuing.

Continuing the process will lead to further instructions that offer the operator several options. If the frame input data file has been created and is ready for use, enter PLFRAME and the procedure continues as previously described. If the input file has not been created and the operator desires further help, the option FRAME GUIDE can be entered. The monitor will display additional information and instructions helpful in creating the files. These instructions are most appropriate if either EDIT or the EDLIN procedures on DOS are used to create the files.

Example

Use the PLFRAME program to analyze the gable frame shown in Fig. E1. Make two separate solutions: one for the concentrated loads, and one for the distributed loads. This frame, with the concentrated loads, is the same as the one analyzed using moment distribution in Art. 12.14. The vertical distributed load represents the total dead load on the members, while the inclined distributed load (normal to the girder) represents wind uplift on one side. Assume both columns and girders are the same size (W 14 × 48). Since the weight of the member is included in the 800-plf dead load, do not include the weight of the members in the computer solution as an additional load. The circled numbers represent joints, and those within squares, members. The origin for referencing dimensions and loads is at the lower left-hand joint.

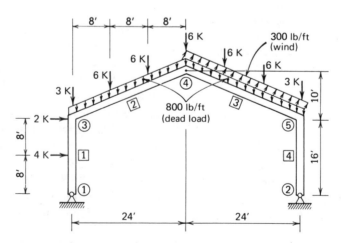

Figure E1/Example: rigid gable frame.

```
EXAMPLE
RIGID GABLE FRAME
APRIL 1992


5
1,0.000,0.000
2,48.000,0.000
3,0.000,16.000
4,24.000,26.000
5,48.000,16.000
2
1,1,1,0
2,1,1,0
4
1,1,3,14.100,485.000,29.000,490.000
2,3,4,14.100,485.000,29.000,490.000
3,4,5,14.100,485.000,29.000,490.000
4,5,2,14.100,485.000,29.000,490.000
2
3
3,2.000,-3.000,0.000
4,0.000,-6.000,0.000
5,0.000,-3.000,0.000
3
1,1
1,0.500,4.000,0.000,0.000
2,2
1,0.333,0.000,-6.000,0.000
1,0.667,0.000,-6.000,0.000
3,2
1,0.333,0.000,-6.000,0.000
1,0.667,0.000,-6.000,0.000
0
0
0
0
2
2,1
1,0.000,-0.800,0.000
3,2
1,0.000,-0.800,0.000
0,0.000,0.300,0.000
0
```

Figure E2/Data input file for example FRAMEX.TX
on diskette.

Solution

Prepare a data input file as shown in Fig. E2. This must be an unformatted (ASCII) file. Assign a name to this input file, such as FRAMEX.TX (this represents "frame example, text out"). Load the PLFRAME program and type in PLFRAME. The monitor will prompt for the input file. Type in FRAMEX.TX. The monitor will prompt for an output file name. Type in FRAMEX.OUT. The monitor will display Analysis completed successfully, FORTRAN STOP, PROGRAM TERMINATED, and the results are stored in a file named FRAMEX.OUT. The following results can be displayed on the monitor or printed from an attached printer. If a printer is not attached, the output file can be placed on a diskette to be taken to an available printer.

STEEL BUILDINGS: ANALYSIS AND DESIGN
CRAWLEY & DILLON

PLANE FRAME ANALYSIS BY DISPLACEMENT METHOD

EXAMPLE
RIGID GABLE FRAME
APRIL 1992


```
* * * * * *  * * * * *
* J O I N T   D A T A *
* * * * * *  * * * * *
```

JOINT	COORDINATES		RESTRAINT CONDITION		
	X (FEET)	Y (FEET)	X	Y	R
1	.000	.000	R	R	
2	48.000	.000	R	R	
3	.000	16.000			
4	24.000	26.000			
5	48.000	16.000			


```
* * * * * *  * * * * * *
* M E M B E R   D A T A *
* * * * * *  * * * * * *
```

MEMBER	END JOINTS		LENGTH (FEET)	CROSS-SECTION PROPERTIES AREA (INS**2)	I(ZZ) (INS**4)	MODULUS OF ELASTICITY (KSI/1000)
	A	B				
1	1	3	16.000	14.100	485.00	29.000
2	3	4	26.000	14.100	485.00	29.000
3	4	5	26.000	14.100	485.00	29.000
4	5	2	16.000	14.100	485.00	29.000

TOTAL WEIGHT OF FRAME: 2.015 TONS

LOADING PATTERN 1 OF 2 LOADING PATTERNS

```
* * * * * *  * * * * * *
* J O I N T  L O A D I N G *
* * * * * *  * * * * * *
```

| JOINT | LOADING | | |
	F(X) (KIPS)	F(Y) (KIPS)	COUPLE (KIP-FEET)
3	2.000	-3.000	.000
4	.000	-6.000	.000
5	.000	-3.000	.000

```
* * * * * * *  * * * * * *
* M E M B E R  L O A D I N G *
* * * * * * *  * * * * * *
```

CONCENTRATED LOADING

MEMBER	LOCATION OF LOAD (FRACTION OF SPAN)	ORIENTATION (AXES)	F(X) (KIPS)	F(Y) (KIPS)	M(O) (KIP-FEET)
1	.500	S	4.000	.000	.000
2	.333	S	.000	-6.000	.000
	.667	S	.000	-6.000	.000
3	.333	S	.000	-6.000	.000
	.667	S	.000	-6.000	.000

NO MEMBERS WITH DISTRIBUTED LOADS IN PATTERN

```
* * * * * * * * * * * * * * * * * * * * * *
* MEMBER WEIGHTS NOT INCLUDED IN ANALYSIS *
* * * * * * * * * * * * * * * * * * * * * *
```

```
* * * * * * * *  * * * * * * * *
* J O I N T  D E F L E C T I O N S *
* * * * * * * *  * * * * * * * *
```

| JOINT | DEFLECTIONS | | |
	D(X) (INCHES)	D(Y) (INCHES)	ROTATION (RADS)
1	.0000	.0000	-.00136
2	.0000	.0000	-.01088
3	.4532	-.0078	-.00502
4	.9447	-1.2076	.00142
5	1.4349	-.0091	-.00066

LOADING PATTERN 1 OF 2 LOADING PATTERNS

```
* * * * * * * * * * * * * * * * * * *
* M E M B E R   E N D   A C T I O N S *
* * * * * * * * * * * * * * * * * * *
```

MEMBER	T(A) (KIPS)	V(A) (KIPS)	M(A) (K-FT)	T(B) (KIPS)	V(B) (KIPS)	M(B) (K-FT)
1	16.667	-1.799	.000	-16.667	5.799	-60.789
2	12.456	9.616	60.789	-7.840	1.461	45.219
3	8.866	-1.000	-45.219	-13.481	12.077	-124.788
4	19.333	7.799	124.788	-19.333	-7.799	.000

```
* * * * * * * *  * * * * * * * *
* R E A C T I V E   F O R C E S *
* * * * * * * *  * * * * * * * *
```

JOINT	R E A C T I O N S		
	R(X) (KIPS)	R(Y) (KIPS)	COUPLE (KIP-FEET)
1	1.799	16.667	
2	-7.799	19.333	

LOADING PATTERN 2 OF 2 LOADING PATTERNS

NO LOADED JOINTS IN PATTERN

```
* * * * * * * *  * * * * * * *
* M E M B E R   L O A D I N G *
* * * * * * * *  * * * * * * *
```

NO MEMBERS WITH CONCENTRATED LOADS IN PATTERN

MEMBER	U N I F O R M L Y D I S T R I B U T E D L O A D I N G			
	ORIENTATION (AXES)	W(X) (KIPS/FOOT)	W(Y) (KIPS/FOOT)	M(O) (K-FT/FOOT)
2	S	.000	-.800	.000
3	S	.000	-.800	.000
	M	.000	.300	.000

```
* * * * * * * * * * * * * * * * * * * * * *
* MEMBER WEIGHTS NOT INCLUDED IN ANALYSIS *
* * * * * * * * * * * * * * * * * * * * * *
```

LOADING PATTERN 2 OF 2 LOADING PATTERNS

```
* * * * * * * *  * * * * * * * * *
* J O I N T   D E F L E C T I O N S *
* * * * * * * *  * * * * * * * * *
```

JOINT	D(X) (INCHES)	D(Y) (INCHES)	ROTATION (RADS)
	D E F L E C T I O N S		
1	.0000	.0000	-.00014
2	.0000	.0000	-.01030
3	.3746	-.0076	-.00557
4	.8774	-1.2267	.00219
5	1.3783	-.0071	-.00093

```
* * * * * * * * * * * * * * * * * *
* M E M B E R   E N D   A C T I O N S *
* * * * * * * * * * * * * * * * * *
```

MEMBER	T(A) (KIPS)	V(A) (KIPS)	M(A) (K-FT)	T(B) (KIPS)	V(B) (KIPS)	M(B) (K-FT)
1	16.088	-4.147	.000	-16.088	4.147	-66.351
2	10.015	13.255	66.351	-2.631	4.468	47.880
3	5.025	-1.278	-47.880	-12.410	11.201	-114.351
4	15.112	7.147	114.351	-15.112	-7.147	.000

```
* * * * * * *  * * * * * * * *
* R E A C T I V E   F O R C E S *
* * * * * * * *  * * * * * * * *
```

JOINT	R(X) (KIPS)	R(Y) (KIPS)	COUPLE (KIP-FEET)
	R E A C T I O N S		
1	4.147	16.088	
2	-7.147	15.112	

F

STEEL-
JOIST
LOAD
TABLES

Excerpted from Steel Joist Institute *Standard Specifications*, *Load Tables*, *and Weight Tables for Steel Joists and Joist Girders*, adopted by SJI 1990. All listed values are allowable loads in pounds per lineal foot. (See Art. 5.18.) Reprinted by permission of the Steel Joist Institute, 1205 48th Avenue North, Suite A, Myrtle Beach, SC 29577.

STANDARD LOAD TABLE [a]
OPEN WEB STEEL JOISTS, K-SERIES

Based on a Maximum Allowable Tensile Stress of 30,000 psi

Adopted by the Steel Joist Institute November 4, 1985; Revised to May 19, 1987.

The (top) figures in the following table give the TOTAL safe uniformly distributed load-carrying capacities, in pounds per linear foot, of K-Series Steel Joists. The weight of DEAD loads, including the joists, must be deducted to determine the LIVE load-carrying capacities of the joists. The load table may be used for parallel chord joists installed to a maximum slope of 1/2 inch per foot.

The (bottom) figures shown in this load table are the LIVE loads per linear foot of joist which will produce an approximate deflection of 1/360 of the span. LIVE loads which will produce a deflection of 1/240 of the span may be obtained by multiplying the (bottom) figures by 1.5. In no case shall the TOTAL load capacity of the joists be exceeded.

The approximate joist weights per linear foot shown in these tables do not include accessories.

The approximate moment of inertia of the joist, in inches[4] is: $I_J = 26.767 (W_{LL})(L^3)(10^{-6})$, where $W_{LL} = $ (bottom) figure in the Load Table; $L = $ (Span -0.33), in feet.

For the proper handling of concentrated and/or varying loads, see Section 5.5 in the Recommended (SJI) Code of Standard Practice.

Joist Designation	8K1	10K1	12K1	12K3	12K5	14K1	14K3	14K4	14K6	16K2	16K3	16K4	16K5	16K6	16K7	16K9
Depth (in.)	8	10	12	12	12	14	14	14	14	16	16	16	16	16	16	16
Approx. Wt. (lbs./ft.)	5.1	5.0	5.0	5.7	7.1	5.2	6.0	6.7	7.7	5.5	6.3	7.0	7.5	8.1	8.6	10.0
Span (ft.) → 8	550 550															
9	550 550															
10	550 480	550 550														
11	532 377	550 542														
12	444 288	550 455	550 550	550 550	550 550											
13	377 225	479 363	550 510	550 510	550 510											
14	324 179	412 289	500 425	550 463	550 463	550 550	550 550	550 550	550 550							
15	281 145	358 234	434 344	543 428	550 434	511 475	550 507	550 507	550 507							

668

Standard Load Table
Open Web Steel Joists, K-Series

Span	1	2	3	4	5	6	7	8	9	10	11	12	13	14	15	16
16	550/550	550/550	550/550	550/550	550/550	550/550	550/550	550/467	550/467	550/467	448/390	550/396	476/351	380/282	313/192	246/119
17	550/526	550/526	550/526	550/526	550/526	550/526	512/488	550/443	550/443	495/404	395/324	550/366	420/291	336/234	277/159	
18	550/490	550/490	550/490	550/490	550/490	508/456	456/409	550/408	530/397	441/339	352/272	507/317	374/245	299/197	246/134	
19	550/455	550/455	550/455	550/455	547/452	455/386	408/347	550/383	475/336	395/287	315/230	454/269	335/207	268/167	221/113	
20	550/426	550/426	550/426	550/426	493/386	410/330	368/297	525/347	428/287	356/246	284/197	409/230	302/177	241/142	199/97	
21	550/406	550/406	548/405	503/373	447/333	371/285	333/255	475/299	388/248	322/212	257/170	370/198	273/153	218/123		
22	550/385	550/385	498/351	458/323	406/289	337/247	303/222	432/259	353/215	293/184	234/147	337/172	249/132	199/106		
23	550/363	507/339	455/307	418/282	371/252	308/216	277/194	395/226	322/188	268/160	214/128	308/150	227/116	181/93		
24	550/346	465/298	418/269	384/248	340/221	283/189	254/170	362/199	295/165	245/141	196/113	282/132	208/101	166/81		
25	514/311	428/263	384/238	353/219	313/195	260/167	234/150	334/175	272/145	226/124	180/100					
26	474/276	395/233	355/211	326/194	289/173	240/148	216/133	308/156	251/129	209/110	166/88					
27	439/246	366/208	329/188	302/173	268/155	223/132	200/119	285/139	233/115	193/98	154/79					
28	408/220	340/186	306/168	281/155	249/138	207/118	186/106	265/124	216/103	180/88	143/70					
29	380/198	317/167	285/151	261/139	232/124	193/106	173/95									
30	355/178	296/151	266/137	244/126	216/112	180/96	161/86									
31	332/161	277/137	249/124	228/114	203/101	168/87	151/78									
32	311/147	259/124	233/112	214/103	190/92	158/79	142/71									

Standard Load Table/Open Web Steel Joists, K-Series
Based on a Maximum Allowable Tensile Stress of 30,000 psi

Joist Designation	18K3	18K4	18K5	18K6	18K7	18K9	18K10	20K3	20K4	20K5	20K6	20K7	20K9	20K10	22K4	22K5	22K6	22K7	22K9	22K10	22K11
Depth (in.)	18	18	18	18	18	18	18	20	20	20	20	20	20	20	22	22	22	22	22	22	22
Approx. Wt. (lbs./ft.)	6.6	7.2	7.7	8.5	9.0	10.2	11.7	6.7	7.6	8.2	8.9	9.3	10.8	12.2	8.0	8.8	9.2	9.7	11.3	12.6	13.8
Span (ft.)																					
18	550/550	550/550	550/550	550/550	550/550	550/550	550/550														
19	514/494	550/523	550/523	550/523	550/523	550/523	550/523														
20	463/423	550/490	550/490	550/490	550/490	550/490	550/490	517/517	550/550	550/550	550/550	550/550	550/550	550/550							
21	420/364	506/426	550/460	550/460	550/460	550/460	550/460	468/453	550/520	550/520	550/520	550/520	550/520	550/520							
22	382/316	460/370	518/414	550/438	550/438	550/438	550/438	426/393	514/461	550/490	550/490	550/490	550/490	550/490	550/548	550/548	550/548	550/548	550/548	550/548	550/548
23	349/276	420/323	473/362	516/393	550/418	550/418	550/418	389/344	469/402	529/451	550/468	550/468	550/468	550/468	518/491	550/518	550/518	550/518	550/518	550/518	550/518
24	320/242	385/284	434/318	473/345	526/382	550/396	550/396	357/302	430/353	485/396	528/430	550/448	550/448	550/448	475/431	536/483	550/495	550/495	550/495	550/495	550/495
25	294/214	355/250	400/281	435/305	485/337	550/377	550/377	329/266	396/312	446/350	486/380	541/421	550/426	550/426	438/381	493/427	537/464	550/474	550/474	550/474	550/474
26	272/190	328/222	369/249	402/271	448/299	538/354	550/361	304/236	366/277	412/310	449/337	500/373	550/405	550/405	404/338	455/379	496/411	550/454	550/454	550/454	550/454
27	252/169	303/198	342/222	372/241	415/267	498/315	550/347	281/211	339/247	382/277	416/301	463/333	550/389	550/389	374/301	422/337	459/367	512/406	550/432	550/432	550/432
28	234/151	282/177	318/199	346/216	385/239	463/282	548/331	261/189	315/221	355/248	386/269	430/298	517/353	550/375	348/270	392/302	427/328	475/364	550/413	550/413	550/413
29	218/136	263/159	296/179	322/194	359/215	431/254	511/298	243/170	293/199	330/223	360/242	401/268	482/317	550/359	324/242	365/272	398/295	443/327	532/387	550/399	550/399

Standard Load Table
Open Web Steel Joists, K-Series

	550/385	550/385	497/349	413/295	371/266	341/245	302/219	533/336	450/286	374/242	336/218	308/201	274/179	227/153	477/269	402/229	335/194	301/175	276/161	245/144	203/123
30	550/385	550/385	497/349	413/295	371/266	341/245	302/219	533/336	450/286	374/242	336/218	308/201	274/179	227/153	477/269	402/229	335/194	301/175	276/161	245/144	203/123
31	550/369	550/369	465/316	387/267	347/241	319/222	283/198	499/304	421/259	350/219	314/198	289/182	256/162	212/138	446/243	376/207	313/175	281/158	258/146	229/130	190/111
32	549/355	517/337	436/287	363/242	326/219	299/201	265/180	468/276	395/235	328/199	295/179	271/165	240/147	199/126	418/221	353/188	294/159	264/144	242/132	215/118	178/101
33	532/334	486/307	410/261	341/221	306/199	281/183	249/164	440/251	371/214	309/181	277/163	254/150	226/134	187/114	393/201	332/171	276/145	248/131	228/121	202/108	168/92
34	516/314	458/280	386/239	321/202	288/182	265/167	235/149	414/229	349/195	290/165	261/149	239/137	212/122	176/105	370/184	312/156	260/132	233/120	214/110	190/98	158/84
35	494/292	432/257	364/219	303/185	272/167	249/153	221/137	390/210	329/179	274/151	246/137	226/126	200/112	166/96	349/168	294/143	245/121	220/110	202/101	179/90	149/77
36	467/269	408/236	344/201	286/169	257/153	236/141	209/126	369/193	311/164	259/139	232/125	213/115	189/103	157/88	330/154	278/132	232/111	208/101	191/92	169/82	141/70
37	442/247	386/217	325/185	271/156	243/141	223/130	198/116	349/178	294/151	245/128	220/115	202/106	179/95	148/81							
38	419/228	366/200	308/170	256/144	230/130	211/119	187/107	331/164	279/139	232/118	208/106	191/98	170/87	141/74							
39	397/211	347/185	292/157	243/133	218/120	200/110	178/98	314/151	265/129	220/109	198/98	181/90	161/81	133/69							
40	377/195	330/171	278/146	231/123	207/111	190/102	169/91	298/140	251/119	209/101	188/91	172/84	153/75	127/64							
41	359/181	314/159	264/135	220/114	197/103	181/95	161/85														
42	342/168	299/148	252/126	209/106	188/96	173/83	153/79														
43	326/157	285/138	240/117	200/99	179/89	165/82	146/73														
44	311/146	272/128	229/109	191/92	171/83	157/76	139/68														

Standard Load Table/Open-Web Steel Joist, K-Series
Based on a Maximum Allowable Tensile Stress of 30,000 psi

Joist Designation	24K4	24K5	24K6	24K7	24K8	24K9	24K10	24K12	26K5	26K6	26K7	26K8	26K9	26K10	26K12	28K6	28K7	28K8	28K9	28K10	28K12
Depth (in.)	24	24	24	24	24	24	24	24	26	26	26	26	26	26	26	28	28	28	28	28	28
Approx. Wt. (lbs./ft.)	8.4	9.3	9.7	10.1	11.5	12.0	13.1	16.0	9.8	10.6	10.9	12.1	12.2	13.8	16.6	11.4	11.8	12.7	13.0	14.3	17.1
Span (ft.)																					
24	520	550	550	550	550	550	550	550													
	516	544	544	544	544	544	544	544													
25	479	540	550	550	550	550	550	550													
	456	511	520	520	520	520	520	520													
26	442	499	543	550	550	550	550	550	542	550	550	550	550	550	550						
	405	453	493	499	499	499	499	499	535	541	541	541	541	541	541						
27	410	462	503	550	550	550	550	550	502	547	550	550	550	550	550						
	361	404	439	479	479	479	479	479	477	519	522	522	522	522	522						
28	381	429	467	521	550	550	550	550	466	508	550	550	550	550	550	548	550	550	550	550	550
	323	362	393	436	456	456	456	456	427	464	501	501	501	501	501	541	543	543	543	543	543
29	354	400	435	485	536	550	550	550	434	473	527	550	550	550	550	511	550	550	550	550	550
	290	325	354	392	429	436	436	436	384	417	463	479	479	479	479	486	522	522	522	522	522
30	331	373	406	453	500	544	550	550	405	441	492	544	550	550	550	477	531	550	550	550	550
	262	293	319	353	387	419	422	422	346	377	417	457	459	459	459	439	486	500	500	500	500
31	310	349	380	424	468	510	550	550	379	413	460	509	550	550	550	446	497	550	550	550	550
	237	266	289	320	350	379	410	410	314	341	378	413	444	444	444	397	440	480	480	480	480
32	290	327	357	397	439	478	549	549	356	387	432	477	519	549	549	418	466	515	549	549	549
	215	241	262	290	318	344	393	393	285	309	343	375	407	431	431	361	400	438	463	463	463
33	273	308	335	373	413	449	532	532	334	364	406	448	488	532	532	393	438	484	527	532	532
	196	220	239	265	289	313	368	368	259	282	312	342	370	404	404	329	364	399	432	435	435

672

Span	1	2	3	4	5	6	7	8	9	10	11	12	13	14	15	16	17	18	19	20	21
34	257/179	290/201	315/218	351/242	388/264	423/286	502/337	516/344	315/237	343/257	382/285	422/312	459/338	516/378	516/378	370/300	412/333	456/364	496/395	516/410	516/410
35	242/164	273/184	297/200	331/221	366/242	399/262	473/308	501/324	297/217	323/236	360/261	398/286	433/310	501/356	501/356	349/275	389/305	430/333	468/361	501/389	501/389
36	229/150	258/169	281/183	313/203	346/222	377/241	447/283	487/306	280/199	305/216	340/240	376/263	409/284	486/334	487/334	330/252	367/280	406/306	442/332	487/366	487/366
37	216/138	244/155	266/169	296/187	327/205	356/222	423/260	474/290	265/183	289/199	322/221	356/242	387/262	460/308	474/315	312/232	348/257	384/282	418/305	474/344	474/344
38	205/128	231/143	252/156	281/172	310/189	338/204	401/240	461/275	251/169	274/184	305/204	337/223	367/241	436/284	461/299	296/214	329/237	364/260	396/282	461/325	461/325
39	195/118	219/132	239/144	266/159	294/174	320/189	380/222	449/261	238/156	260/170	289/188	320/206	348/223	413/262	449/283	280/198	313/219	346/240	376/260	447/306	449/308
40	185/109	208/122	227/133	253/148	280/161	304/175	361/206	438/247	227/145	247/157	275/174	304/191	331/207	393/243	438/269	266/183	297/203	328/222	357/241	424/284	438/291
41	176/101	198/114	216/124	241/137	266/150	290/162	344/191	427/235	215/134	235/146	262/162	289/177	315/192	374/225	427/256	253/170	283/189	312/206	340/224	404/263	427/277
42	168/94	189/106	206/115	229/127	253/139	276/151	327/177	417/224	205/125	224/136	249/150	275/164	300/178	356/210	417/244	241/158	269/175	297/192	324/208	384/245	417/264
43	160/88	180/98	196/107	219/118	242/130	263/140	312/165	406/213	196/116	213/126	238/140	263/153	286/166	339/195	407/232	230/147	257/163	284/179	309/194	367/228	407/252
44	153/82	172/92	187/100	209/110	231/121	251/131	298/154	387/199	187/108	204/118	227/131	251/143	273/155	324/182	398/222	220/137	245/152	271/167	295/181	350/212	398/240
45	146/76	164/86	179/93	199/103	220/113	240/122	285/144	370/185	179/101	194/110	217/122	240/133	261/145	310/170	389/212	210/128	234/142	259/156	282/169	334/198	389/229
46	139/71	157/80	171/87	191/97	211/106	230/114	272/135	354/174	171/95	186/103	207/114	229/125	250/135	296/159	380/203	201/120	224/133	248/146	270/158	320/186	380/219

Standard Load Table
Open-Web Steel Joist, K-Series

Joist Designation	24K4	24K5	24K6	24K7	24K8	24K9	24K10	24K12	26K5	26K6	26K7	26K8	26K9	26K10	26K12	28K6	28K7	28K8	28K9	28K10	28K12
Depth (In.)	24	24	24	24	24	24	24	24	26	26	26	26	26	26	26	28	28	28	28	28	28
Approx. Wt. (lbs./ft.)	8.4	9.3	9.7	10.1	11.5	12.0	13.1	16.0	9.8	10.6	10.9	12.1	12.2	13.8	16.6	11.4	11.8	12.7	13.0	14.3	17.1
Span (ft.) ↓																					
47	133 / 67	150 / 75	164 / 82	183 / 90	202 / 99	220 / 107	261 / 126	339 / 163	164 / 89	178 / 96	199 / 107	219 / 117	239 / 127	284 / 149	369 / 192	192 / 112	214 / 125	237 / 136	258 / 148	306 / 174	372 / 210
48	128 / 63	144 / 70	157 / 77	175 / 85	194 / 93	211 / 101	250 / 118	325 / 153	157 / 83	171 / 90	190 / 100	210 / 110	229 / 119	272 / 140	353 / 180	184 / 105	206 / 117	227 / 128	247 / 139	294 / 163	365 / 201
49									150 / 78	164 / 85	183 / 94	202 / 103	220 / 112	261 / 131	339 / 169	177 / 99	197 / 110	218 / 120	237 / 130	282 / 153	357 / 193
50									144 / 73	157 / 80	175 / 89	194 / 97	211 / 105	250 / 124	325 / 159	170 / 93	189 / 103	209 / 113	228 / 123	270 / 144	350 / 185
51									139 / 69	151 / 75	168 / 83	186 / 91	203 / 99	241 / 116	313 / 150	163 / 88	182 / 97	201 / 106	219 / 115	260 / 136	338 / 175
52									133 / 65	145 / 71	162 / 79	179 / 86	195 / 93	231 / 110	301 / 142	157 / 83	175 / 92	193 / 100	210 / 109	250 / 128	325 / 165
53																151 / 78	168 / 87	186 / 95	203 / 103	240 / 121	313 / 156
54																145 / 74	162 / 82	179 / 89	195 / 97	232 / 114	301 / 147
55																140 / 70	156 / 77	173 / 85	188 / 92	223 / 108	290 / 139
56																135 / 66	151 / 73	166 / 80	181 / 87	215 / 102	280 / 132

Standard Load Table/Open-Web Steel Joists, K-Series
Based on a Maximum Allowable Tensile Stress of 30,000 psi

Joist Designation	30K7	30K8	30K9	30K10	30K11	30K12
Depth (In.)	30	30	30	30	30	30
Approx. Wt. (lbs. / ft.)	12.3	13.2	13.4	15.0	16.4	17.6
Span (ft.) ↓						
30	550 / 543	550 / 543	550 / 543	550 / 543	550 / 543	550 / 543
31	534 / 508	550 / 520	550 / 520	550 / 520	550 / 520	550 / 520
32	501 / 461	549 / 500	549 / 500	549 / 500	549 / 500	549 / 500
33	471 / 420	520 / 460	532 / 468	532 / 468	532 / 468	532 / 468
34	443 / 384	490 / 420	516 / 441	516 / 441	516 / 441	516 / 441
35	418 / 351	462 / 384	501 / 415	501 / 415	501 / 415	501 / 415
36	395 / 323	436 / 353	475 / 383	487 / 392	487 / 392	487 / 392
37	373 / 297	413 / 325	449 / 352	474 / 374	474 / 374	474 / 374
38	354 / 274	391 / 300	426 / 325	461 / 353	461 / 353	461 / 353
39	336 / 253	371 / 277	404 / 300	449 / 333	449 / 333	449 / 333
40	319 / 234	353 / 256	384 / 278	438 / 315	438 / 315	438 / 315
41	303 / 217	335 / 238	365 / 258	427 / 300	427 / 300	427 / 300
42	289 / 202	320 / 221	348 / 240	413 / 282	417 / 284	417 / 284
43	276 / 188	305 / 206	332 / 223	394 / 263	407 / 270	407 / 270
44	263 / 176	291 / 192	317 / 208	376 / 245	398 / 258	398 / 258
45	251 / 164	278 / 179	303 / 195	359 / 229	389 / 246	389 / 246

Standard Load Table
Open-Web Steel Joists, K-Series

Joist Designation	30K7	30K8	30K9	30K10	30K11	30K12
Depth (In.)	30	30	30	30	30	30
Approx. Wt. (lbs. / ft.)	12.3	13.2	13.4	15.0	16.4	17.6
Span (ft.) ↓ 46	241 153	266 168	290 182	344 214	380 236	380 236
47	230 144	255 157	277 171	329 201	372 226	372 226
48	221 135	244 148	266 160	315 188	362 215	365 216
49	212 127	234 139	255 150	303 177	347 202	357 207
50	203 119	225 130	245 141	291 166	333 190	350 199
51	195 112	216 123	235 133	279 157	320 179	343 192
52	188 106	208 116	226 126	268 148	308 169	336 184
53	181 100	200 109	218 119	258 140	296 159	330 177
54	174 94	192 103	209 112	249 132	285 150	324 170
55	168 89	185 98	202 106	240 125	275 142	312 161
56	162 84	179 92	195 100	231 118	265 135	301 153
57	156 80	173 88	188 95	223 112	256 128	290 145
58	151 76	167 83	181 90	215 106	247 121	280 137
59	146 72	161 79	175 86	208 101	239 115	271 130
60	141 69	156 75	169 81	201 96	231 109	262 124

[a]The Steel Joist Institute publishes both specifications and load tables, each of which contains standards to be used in conjunction with the other.

STANDARD LOAD TABLE [a]
Based on a Maximum Allowable Tensile Stress of 30,000 psi

LONGSPAN STEEL JOISTS, LH- SERIES

Adopted by the Steel Joist Institute May 25, 1983
Revised to November 15, 1989

The (top) figures in the following table give the TOTAL safe uniformly-distributed load-carrying capacities, in pounds per linear foot, of LH-Series joists. The weight of DEAD loads, including the joists, must in all cases be deducted to determine the LIVE load-carrying capacities of the joists. The approximate DEAD load of the joists may be determined from the weights per linear foot shown in the tables.

The (bottom) figures in this load table are the LIVE loads per linear foot of joist which will produce an approximate deflection of 1/360 of the span. LIVE loads which will produce a deflection of 1/240 of the span may be obtained by multiplying the (bottom) figures by 1.5. In no case shall the TOTAL load capacity of the joists be exceeded.

This load table applies to joists with either parallel chords or standard pitched top chords. When top chords are pitched, the carrying capacities are determined by the nominal depth of the joists at center of the span. Standard top chord pitch is 1/8 inch per foot. If pitch exceeds this standard, the load table does not apply. This load table may be used for parallel chord joists installed to a maximum slope of 1/2 inch per foot.

When holes are required in top or bottom chords, the carrying capacities must be reduced in proportion to reduction of chord areas.

The top chords are considered as being stayed laterally by floor slab or roof deck

The approximate joist weights per linear foot shown in these tables do *not* include accessories.

Joist Desig-nation	Approx. Wt. in Lbs. per Linear Ft. (Joists only)	Depth in Inches	Safe[b] Load in Lbs. Between 21-24	Clear Span in Feet											
				25	26	27	28	29	30	31	32	33	34	35	36
18LH02	10	18	2000	468 313	442 284	418 259	391 234	367 212	345 193	324 175	306 160	289 147	273 135	259 124	245 114
18LH03	11	18	3300	521 348	493 317	467 289	438 262	409 236	382 213	359 194	337 177	317 161	299 148	283 136	267 124
18LH04	12	18	5500	604 403	571 367	535 329	500 296	469 266	440 242	413 219	388 200	365 182	344 167	325 153	308 141
18LH05	15	18	7500	684 454	648 414	614 378	581 345	543 311	508 282	476 256	448 233	421 212	397 195	375 179	355 164

Standard Load Table
Longspan Steel Joists, LH Series[b] (Continued)

Joist Desig-nation	Approx. Wt. in Lbs. per Linear Ft. (Joists only)	Depth in Inches	Safe[b] Load in Lbs. Between 21–24	25	26	27	28	29	30	31	32	33	34	35	36	37	38	39	40
18LH06	15	18	20700	809	749	696	648	605	566	531	499	470	443	418	396				
				526	469	419	377	340	307	280	254	232	212	195	180				
18LH07	17	18	21500	840	809	780	726	678	635	595	559	526	496	469	444				
				553	513	476	428	386	349	317	288	264	241	222	204				
18LH08	19	18	22400	876	843	812	784	758	717	680	641	604	571	540	512				
				577	534	496	462	427	387	351	320	292	267	246	226				
18LH09	21	18	24000	936	901	868	838	810	783	759	713	671	633	598	566				
				616	571	527	491	458	418	380	346	316	289	266	245				

Joist Desig-nation	Approx. Wt. in Lbs. per Linear Ft. (Joists only)	Depth in Inches	Safe[b] Load in Lbs. Between 22–24	25	26	27	28	29	30	31	32	33	34	35	36	37	38	39	40
20LH02	10	20	11300	442	437	431	410	388	365	344	325	307	291	275	262	249	237	225	215
				306	303	298	274	250	228	208	190	174	160	147	136	126	117	108	101
20LH03	11	20	13000	469	463	458	452	434	414	395	372	352	333	316	299	283	269	255	243
				337	333	317	302	280	258	238	218	200	184	169	156	143	133	123	114
20LH04	12	20	14700	574	566	558	528	496	467	440	416	393	372	353	335	318	303	289	275
				428	406	386	352	320	291	265	243	223	205	189	174	161	149	139	129
20LH05	14	20	15800	616	609	602	595	571	544	513	484	458	434	411	390	371	353	336	321
				459	437	416	395	366	337	308	281	258	238	219	202	187	173	161	150
20LH06	15	20	21100	822	791	763	723	679	635	596	560	527	497	469	444	421	399	379	361
				606	561	521	477	427	386	351	320	292	267	246	226	209	192	178	165
20LH07	17	20	22500	878	845	814	786	760	711	667	627	590	556	526	497	471	447	425	404
				647	599	556	518	484	438	398	362	331	303	278	256	236	218	202	187
20LH08	19	20	23300	908	873	842	813	785	760	722	687	654	621	588	558	530	503	479	457
				669	619	575	536	500	468	428	395	365	336	309	285	262	242	225	209
20LH09	21	20	25400	990	953	918	886	856	828	802	778	755	712	673	636	603	572	544	517
				729	675	626	581	542	507	475	437	399	366	336	309	285	264	244	227
20LH10	23	20	27400	1068	1028	991	956	924	894	865	839	814	791	748	707	670	636	604	575
				786	724	673	626	585	545	510	479	448	411	377	346	320	296	274	254

Clear Span in Feet

	Depth	28–32	33	34	35	36	37	38	39	40	41	42	43	44	45	46	47	48
24LH03	24	11500	342	339	336	323	307	293	279	267	255	244	234	224	215	207	199	191
			235	226	218	204	188	175	162	152	141	132	124	116	109	102	96	90
24LH04	24	14100	419	398	379	360	343	327	312	298	285	273	262	251	241	231	222	214
			288	265	246	227	210	195	182	169	158	148	138	130	122	114	107	101
24LH05	24	15100	449	446	440	419	399	380	363	347	331	317	304	291	280	269	258	248
			308	297	285	264	244	226	210	196	182	171	160	150	141	132	124	117
24LH06	24	20300	604	579	555	530	504	480	457	437	417	399	381	364	348	334	320	307
			411	382	356	331	306	284	263	245	228	211	197	184	172	161	152	142
24LH07	24	22300	665	638	613	588	565	541	516	491	468	446	426	407	389	373	357	343
			452	421	393	367	343	320	297	276	257	239	223	208	195	182	171	161
24LH08	24	23800	707	677	649	622	597	572	545	520	497	475	455	435	417	400	384	369
			480	447	416	388	362	338	314	292	272	254	238	222	208	196	184	173
24LH09	24	28000	832	808	785	764	731	696	663	632	602	574	548	524	501	480	460	441
			562	530	501	460	424	393	363	337	313	292	272	254	238	223	209	196
24LH10	24	29600	882	856	832	809	788	768	737	702	668	637	608	582	556	533	511	490
			596	559	528	500	474	439	406	378	351	326	304	285	266	249	234	220
24LH11	24	31200	927	900	875	851	829	807	787	768	734	701	671	642	616	590	567	544
			624	588	555	525	498	472	449	418	388	361	337	315	294	276	259	243

	Depth	33–40	41	42	43	44	45	46	47	48	49	50	51	52	53	54	55	56
28LH05	28	14000	337	323	310	297	286	275	265	255	245	237	228	220	213	206	199	193
			219	205	192	180	169	159	150	142	133	126	119	113	107	102	97	92
28LH06	28	18600	448	429	412	395	379	364	350	337	324	313	301	291	281	271	262	253
			289	270	253	238	223	209	197	186	175	166	156	148	140	133	126	120
28LH07	28	21000	505	484	464	445	427	410	394	379	365	352	339	327	316	305	295	285
			326	305	285	267	251	236	222	209	197	186	176	166	158	150	142	135
28LH08	28	22500	540	517	496	475	456	438	420	403	387	371	357	344	331	319	308	297
			348	325	305	285	268	252	236	222	209	196	185	175	165	156	148	140
28LH09	28	27700	667	639	612	586	563	540	519	499	481	463	446	430	415	401	387	374
			428	400	375	351	329	309	291	274	258	243	228	216	204	193	183	173
28LH10	28	30300	729	704	679	651	625	600	576	554	533	513	495	477	460	444	429	415
			466	439	414	388	364	342	322	303	285	269	255	241	228	215	204	193

Standard Load Table
Longspan Steel Joists, LH Series (Continued)

Joist Desig-nation	Approx. Wt. in Lbs. per Linear Ft. (Joists only)	Depth in Inches	Safe Load in Lbs. Between 33–40	Clear Span in Feet 41	42	43	44	45	46	47	48	49	50	51	52	53	54	55	56
28LH11	25	28	32500	780	762	736	711	682	655	629	605	582	561	540	521	502	485	468	453
				498	475	448	423	397	373	351	331	312	294	278	263	249	236	223	212
28LH12	27	28	35700	857	837	818	800	782	766	737	709	682	656	632	609	587	566	546	527
				545	520	496	476	454	435	408	383	361	340	321	303	285	270	256	243
28LH13	30	28	37200	895	874	854	835	816	799	782	766	751	722	694	668	643	620	598	577
				569	543	518	495	472	452	433	415	396	373	352	332	314	297	281	266

Joist Desig-nation	Approx. Wt. in Lbs. per Linear Ft. (Joists only)	Depth in Inches	Safe Load in Lbs. Between 38–48	Clear Span in Feet 49	50	51	52	53	54	55	56	57	58	59	60	61	62	63	64
32LH06	14	32	16700	338	326	315	304	294	284	275	266	257	249	242	234	227	220	214	208
				211	199	189	179	169	161	153	145	138	131	125	119	114	108	104	99
32LH07	16	32	18800	379	366	353	341	329	318	308	298	288	279	271	262	254	247	240	233
				235	223	211	200	189	179	170	162	154	146	140	133	127	121	116	111
32LH08	17	32	20400	411	397	383	369	357	345	333	322	312	302	293	284	275	267	259	252
				255	242	229	216	205	194	184	175	167	159	151	144	137	131	125	120
32LH09	21	32	25600	516	498	480	463	447	432	418	404	391	379	367	356	345	335	325	315
				319	302	285	270	256	243	230	219	208	198	189	180	172	164	157	149
32LH10	21	32	28300	571	550	531	512	495	478	462	445	430	416	402	389	376	364	353	342
				352	332	315	297	282	267	254	240	228	217	206	196	186	178	169	162
32LH11	24	32	31000	625	602	580	560	541	522	505	488	473	458	443	429	416	403	390	378
				385	363	343	325	308	292	277	263	251	239	227	216	206	196	187	179
32LH12	27	32	36400	734	712	688	664	641	619	598	578	559	541	524	508	492	477	463	449
				450	428	406	384	364	345	327	311	295	281	267	255	243	232	221	211
32LH13	30	32	40600	817	801	785	771	742	715	690	666	643	621	600	581	562	544	527	511
				500	480	461	444	420	397	376	354	336	319	304	288	275	262	249	238
32LH14	33	32	41800	843	826	810	795	780	766	738	713	688	665	643	622	602	583	564	547
				515	495	476	458	440	417	395	374	355	337	321	304	290	276	264	251
32LH15	35	32	43200	870	853	837	821	805	791	776	763	750	725	701	678	656	635	616	597
				532	511	492	473	454	438	422	407	393	374	355	338	322	306	292	279

Standard Load Table
Longspan Steel Joists, LH Series (Continued)

Designation	Wt	Depth	42-56	57	58	59	60	61	62	63	64	65	66	67	68	69	70	71	72
36LH07	16	36	16800	292	283	274	266	258	251	244	237	230	224	218	212	207	201	196	191
				177	168	160	153	146	140	134	128	122	117	112	107	103	99	95	91
36LH08	18	36	18500	321	311	302	293	284	276	268	260	253	246	239	233	227	221	215	209
				194	185	176	168	160	153	146	140	134	128	123	118	113	109	104	100
36LH09	21	36	23700	411	398	386	374	363	352	342	333	323	314	306	297	289	282	275	267
				247	235	224	214	204	195	186	179	171	163	157	150	144	138	133	127
36LH10	21	36	26300	454	440	426	413	401	389	378	367	357	347	338	328	320	311	303	295
				273	260	248	236	225	215	206	197	188	180	173	165	159	152	146	140
36LH11	23	36	28500	495	480	465	451	438	425	412	401	389	378	368	358	348	339	330	322
				297	283	269	257	246	234	224	214	205	196	188	180	173	166	159	153
36LH12	25	36	34100	593	575	557	540	523	508	493	478	464	450	437	424	412	400	389	378
				354	338	322	307	292	279	267	255	243	232	222	213	204	195	187	179
36LH13	30	36	40100	697	675	654	634	615	596	579	562	546	531	516	502	488	475	463	451
				415	395	376	359	342	327	312	298	285	273	262	251	240	231	222	213
36LH14	36	36	44200	768	755	729	706	683	661	641	621	602	584	567	551	535	520	505	492
				456	434	412	392	373	356	339	323	309	295	283	270	259	247	237	228
36LH15	36	36	46600	809	795	781	769	744	721	698	677	656	637	618	600	583	567	551	536
				480	464	448	434	413	394	375	358	342	327	312	299	286	274	263	252

Designation	Wt	Depth	47-64	65	66	67	68	69	70	71	72	73	74	75	76	77	78	79	80
40LH08	16	40	16600	254	247	241	234	228	222	217	211	206	201	196	192	187	183	178	174
				150	144	138	132	127	122	117	112	108	104	100	97	93	90	86	83
40LH09	21	40	21800	332	323	315	306	298	291	283	276	269	263	256	250	244	239	233	228
				196	188	180	173	166	160	153	147	141	136	131	126	122	118	113	109
40LH10	21	40	24000	367	357	347	338	329	321	313	305	297	290	283	276	269	262	255	249
				216	207	198	190	183	176	169	162	156	150	144	139	134	129	124	119
40LH11	22	40	26200	399	388	378	368	358	349	340	332	323	315	308	300	293	286	279	273
				234	224	215	207	198	190	183	176	169	163	157	151	145	140	135	130
40LH12	25	40	31900	486	472	459	447	435	424	413	402	392	382	373	364	355	346	338	330
				285	273	261	251	241	231	222	213	205	197	189	182	176	169	163	157
40LH13	30	40	37600	573	557	542	528	514	500	487	475	463	451	440	429	419	409	399	390
				334	320	307	295	283	271	260	250	241	231	223	214	207	199	192	185

681

Standard Load Table
Longspan Steel Joists, LH Series (Continued)

Joist Designation	Approx. Wt. in Lbs. per Linear Feet (Joists only)	Depth in Inches	Safe[b] Load in Lbs. Between 47–64	\multicolumn{16}{c}{Clear Span in Feet}
				65	66	67	68	69	70	71	72	73	74	75	76	77	78	79	80
40LH14	35	40	43000	656	638	620	603	587	571	556	542	528	515	502	490	478	466	455	444
				383	367	351	336	323	309	297	285	273	263	252	243	233	225	216	209
40LH15	36	40	48100	734	712	691	671	652	633	616	599	583	567	552	538	524	511	498	486
				427	408	390	373	357	342	328	315	302	290	279	268	258	248	239	230
40LH16	42	40	53000	808	796	784	772	761	751	730	710	691	673	655	638	622	606	591	576
				469	455	441	428	416	404	387	371	356	342	329	316	304	292	282	271

Joist Designation	Approx. Wt. in Lbs. per Linear Feet (Joists only)	Depth in Inches	Safe[b] Load in Lbs. Between 52–72	73	74	75	76	77	78	79	80	81	82	83	84	85	86	87	88
44LH09	19	44	20000	272	265	259	253	247	242	236	231	226	221	216	211	207	202	198	194
				158	152	146	141	136	131	127	122	118	114	110	106	103	99	96	93
44LH10	21	44	22100	300	293	286	279	272	266	260	254	249	243	238	233	228	223	218	214
				174	168	162	155	150	144	139	134	130	125	121	117	113	110	106	103
44LH11	22	44	23900	325	317	310	302	295	289	282	276	269	264	258	252	247	242	236	232
				188	181	175	168	162	157	151	146	140	136	131	127	123	119	115	111
44LH12	25	44	29600	402	393	383	374	365	356	347	339	331	323	315	308	300	293	287	280
				232	224	215	207	200	192	185	179	172	166	160	155	149	144	139	134
44LH13	30	44	35100	477	466	454	444	433	423	413	404	395	386	377	369	361	353	346	338
				275	265	254	246	236	228	220	212	205	198	191	185	179	173	167	161
44LH14	31	44	40400	549	534	520	506	493	481	469	457	446	436	425	415	406	396	387	379
				315	302	291	279	268	259	249	240	231	223	215	207	200	193	187	181
44LH15	36	44	47000	639	623	608	593	579	565	551	537	524	512	500	488	476	466	455	445
				366	352	339	326	314	303	292	281	271	261	252	243	234	227	219	211
44LH16	42	44	54200	737	719	701	684	668	652	637	622	608	594	580	568	555	543	531	520
				421	405	390	375	362	348	336	324	313	302	291	282	272	263	255	246

Standard Load Table
Longspan Steel Joists, LH-Series (*Continued*)

			52–72	73	74	75	76	77	78	79	80	81	82	83	84	85	86	87	88
44LH17	47	44	58200	790 / 450	780 / 438	769 / 426	759 / 415	750 / 405	732 / 390	715 / 376	699 / 363	683 / 351	667 / 338	652 / 327	638 / 316	624 / 305	610 / 295	597 / 285	584 / 276

			56–80	81	82	83	84	85	86	87	88	89	90	91	92	93	94	95	96
48LH10	48	21	20000	246 / 141	241 / 136	236 / 132	231 / 127	226 / 123	221 / 119	217 / 116	212 / 112	208 / 108	204 / 105	200 / 102	196 / 99	192 / 96	188 / 93	185 / 90	181 / 87
48LH11	48	22	21700	266 / 152	260 / 147	255 / 142	249 / 137	244 / 133	239 / 129	234 / 125	229 / 120	225 / 117	220 / 113	216 / 110	212 / 106	208 / 103	204 / 100	200 / 97	196 / 94
48LH12	48	25	27400	336 / 191	329 / 185	322 / 179	315 / 173	308 / 167	301 / 161	295 / 156	289 / 151	283 / 147	277 / 142	272 / 138	266 / 133	261 / 129	256 / 126	251 / 122	246 / 118
48LH13	48	29	32800	402 / 228	393 / 221	384 / 213	376 / 206	368 / 199	360 / 193	353 / 187	345 / 180	338 / 175	332 / 170	325 / 164	318 / 159	312 / 154	306 / 150	300 / 145	294 / 141
48LH14	48	32	38700	475 / 269	464 / 260	454 / 251	444 / 243	434 / 234	425 / 227	416 / 220	407 / 212	399 / 206	390 / 199	383 / 193	375 / 187	367 / 181	360 / 176	353 / 171	346 / 165
48LH15	48	36	44500	545 / 308	533 / 298	521 / 287	510 / 278	499 / 269	488 / 260	478 / 252	468 / 244	458 / 236	448 / 228	439 / 221	430 / 214	422 / 208	413 / 201	405 / 195	397 / 189
48LH16	48	42	51300	629 / 355	615 / 343	601 / 331	588 / 320	576 / 310	563 / 299	551 / 289	540 / 280	528 / 271	518 / 263	507 / 255	497 / 247	487 / 239	477 / 232	468 / 225	459 / 218
48LH17	48	47	57600	706 / 397	690 / 383	675 / 371	660 / 358	646 / 346	632 / 335	619 / 324	606 / 314	593 / 304	581 / 294	569 / 285	558 / 276	547 / 268	536 / 260	525 / 252	515 / 245

[a] The Steel Joist Institute publishes both specifications and load tables, each of which contains standards to be used in conjunction with the other.

[b] The safe uniform load for the clear spans shown in the shaded section is equal to (Safe Load) ÷ (Clear Span + 0.67). [The added 0.67 feet (8 inches) is required to obtain the proper length on which the Load Tables were developed.]

In no case shall the safe uniform load, for clear spans less than the minimum clear span shown in the shaded area, exceed the uniform load calculated for the minimum clear span listed in the shaded area.

To solve for *live* loads for clear spans shown in the shaded area (or lesser clear spans), multiply the live load of the shortest clear span shown in the Load Tables by (the shortest clear span shown in the Load Table + 0.67 feet)^2 and divide by (the actual clear span + 0.67 feet)^2. The live load shall *not* exceed the safe uniform load.

STANDARD LOAD TABLE[a]

Based on a Maximum Allowable Tensile Stress of 30,000 psi

FOR DEEP LONGSPAN STEEL JOISTS, DLH- SERIES

Adopted by the Steel Joist Institute May 25, 1983
Revised to November 15, 1989

The (top) figures in the following table give the TOTAL safe uniformly-distributed load-carrying capacities, in pounds per linear foot, of DLH-Series Joists. The weight of DEAD loads, including the joists, must in all cases be deducted to determine the LIVE load-carrying capacities of the joists. The approximate DEAD load of the joists may be determined from the weights per linear foot shown in the tables. All loads shown are for roof construction only.

The (bottom) figures in this table are the LIVE loads per linear foot of joist which will produce an approximate deflection of 1/360 of the span. LIVE loads which will produce a deflection of 1/240 of the span may be obtained by multiplying the (bottom) figures by 1.5. In no case shall the TOTAL load capacity of the joists be exceeded.

This load table applies to joists with either parallel chords or standard pitched top chords. When top chords are pitched, the carrying capacities are determined by the nominal depth of the joists at the center of the span. Standard top chord pitch is 1/8 inch per foot. If pitch exceeds this standard, the load table does not apply. This load table may be used for parallel chord joists installed to a maximum slope of 1/2 inch per foot.

When holes are required in top or bottom chords, the carrying capacities must be reduced in proportion to reduction of chord areas.

The top chords are considered as being stayed laterally by the roof deck.

The approximate joist weights per linear foot shown in these tables do *not* include accessories.

Joist Desig-nation	Approx. Wt. in Lbs. per Linear Ft. (Joists only)	Depth in Inches	Safe[a] Load in Lbs. in Between 61–88	89	90	91	92	93	94	95	96	97	98	99	100	101	102	103	104
												Clear Span in Feet							
52DLH10	25	52	26700	298	291	285	279	273	267	261	256	251	246	241	236	231	227	223	218
				171	165	159	154	150	145	140	136	132	128	124	120	116	114	110	107
52DLH11	26	52	29300	327	320	313	306	299	293	287	281	275	270	264	259	254	249	244	240
				187	181	174	169	164	158	153	149	144	140	135	132	128	124	120	117

684

Standard Load Table
Deep Longspan Steel Joists, DLH-Series (Continued)

Designation			61–88	89	90	91	92	93	94	95	96	97	98	99	100	101	102	103	104
52DLH12	29	52	32700	365	357	349	342	334	327	320	314	307	301	295	289	284	278	273	268
				204	197	191	185	179	173	168	163	158	153	149	144	140	135	132	128
52DLH13	34	52	39700	443	433	424	414	406	397	389	381	373	366	358	351	344	338	331	325
				247	239	231	224	216	209	203	197	191	185	180	174	170	164	159	155
52DLH14	39	52	45400	507	497	486	476	466	457	447	438	430	421	413	405	397	390	382	375
				276	266	258	249	242	234	227	220	213	207	201	194	189	184	178	173
52DLH15	42	52	51000	569	557	545	533	522	511	500	490	480	470	461	451	443	434	426	418
				311	301	291	282	272	264	256	247	240	233	226	219	213	207	201	195
52DLH16	45	52	55000	614	601	588	575	563	551	540	528	518	507	497	487	478	468	459	451
				346	335	324	314	304	294	285	276	267	260	252	245	237	230	224	217
52DLH17	52	52	63300	706	691	676	661	647	634	620	608	595	583	572	560	549	539	528	518
				395	381	369	357	346	335	324	315	304	296	286	279	270	263	255	247

Designation			66–96	97	98	99	100	101	102	103	104	105	106	107	108	109	110	111	112
56DLH11	26	56	28100	288	283	277	272	267	262	257	253	248	244	239	235	231	227	223	219
				169	163	158	153	149	145	140	136	133	129	125	122	118	115	113	110
56DLH12	30	56	32300	331	324	318	312	306	300	295	289	284	278	273	268	263	259	254	249
				184	178	173	168	163	158	153	150	145	141	137	133	130	126	123	119
56DLH13	34	56	39100	401	394	386	379	372	365	358	351	344	338	331	325	319	314	308	303
				223	216	209	204	197	191	186	181	175	171	166	161	157	152	149	145
56DLH14	39	56	44200	453	444	435	427	419	411	403	396	388	381	375	368	361	355	349	343
				249	242	234	228	221	214	209	202	196	190	186	181	175	171	167	162
56DLH15	42	56	50500	518	508	498	488	478	469	460	451	443	434	426	419	411	403	396	389
				281	272	264	256	248	242	234	228	221	215	209	204	198	192	188	182
56DLH16	46	56	54500	559	548	537	526	516	506	496	487	478	469	460	452	444	436	428	420
				313	304	294	285	277	269	262	254	247	240	233	227	221	214	209	204
56DLH17	51	56	62800	643	630	618	605	594	582	571	560	549	539	529	520	510	501	492	483
				356	345	335	325	316	306	298	289	281	273	266	258	251	245	238	231

Standard Load Table
Deep Longspan Steel Joists, DLH-Series (Continued)

Joist Desig- nation	Approx. Wt. in Lbs per Linear Ft. (Joists only)	Depth in Inches	Safe Load in Lbs Between 70–104	105	106	107	108	109	110	111	112	113	114	115	116	117	118	119	120
60DLH12	29	60	31100	295	289	284	279	274	270	265	261	256	252	248	244	240	236	232	228
				168	163	158	154	150	146	142	138	134	131	128	124	121	118	115	113
60DLH13	35	60	37800	358	351	345	339	333	327	322	316	311	306	301	296	291	286	282	277
				203	197	191	187	181	176	171	167	163	158	154	151	147	143	139	135
60DLH14	40	60	43000	398	391	383	376	370	363	356	350	344	338	332	327	321	316	310	305
				216	210	205	199	193	189	183	178	173	170	165	161	156	152	149	145
60DLH15	43	60	49300	467	458	450	442	434	427	419	412	405	398	392	385	379	373	367	361
				255	248	242	235	228	223	216	210	205	200	194	190	185	180	175	171
60DLH16	46	60	54200	513	504	494	485	476	468	460	451	444	436	428	421	414	407	400	393
				285	277	269	262	255	247	241	235	228	223	217	211	206	201	196	190
60DLH17	52	60	62300	590	579	569	558	548	538	529	519	510	501	493	484	476	468	460	453
				324	315	306	298	290	283	275	267	261	254	247	241	235	228	223	217
60DLH18	59	60	71900	681	668	656	644	632	621	610	599	589	578	568	559	549	540	531	522
				366	357	346	337	327	319	310	303	294	286	279	272	266	259	252	246

Joist Desig- nation	Approx. Wt. in Lbs per Linear Ft. (Joists only)	Depth in Inches	Safe Load in Lbs Between 75–112	113	114	115	116	117	118	119	120	121	122	123	124	125	126	127	128
64DLH12	31	64	30000	264	259	255	251	247	243	239	235	231	228	224	221	218	214	211	208
				153	150	146	142	138	135	132	129	125	122	119	116	114	111	109	106
64DLH13	34	64	36400	321	315	310	305	300	295	291	286	281	277	273	269	264	260	257	253
				186	181	176	171	168	163	159	155	152	148	144	141	137	134	131	128
64DLH14	40	64	41700	367	360	354	349	343	337	332	326	321	316	311	306	301	296	292	287
				199	193	189	184	179	174	171	166	162	158	154	151	147	143	140	136

Clear Span in Feet

Standard Load Table
Deep Longspan Steel Joists, DLH-Series (*Continued*)

		75–112	128	127	126	125	124	123	122	121	120	119	118	117	116	115	114	113
64DLH15	43	64	47800	331	336	341	347	352	358	363	369	375	381	387	394	400	407	414	421
				161	165	170	173	177	182	187	191	196	201	206	211	217	223	228	234
64DLH16	46	64	53800	370	376	382	388	394	401	407	414	421	428	435	443	450	458	466	474
				180	184	189	193	198	203	208	213	218	224	229	235	242	248	254	262
64DLH17	52	64	62000	426	432	439	446	454	461	468	476	484	492	501	509	518	527	536	546
				205	210	215	220	226	231	237	243	248	255	262	268	275	283	290	298
64DLH18	59	64	71600	491	499	507	515	523	532	540	549	559	568	578	587	598	608	619	630
				232	237	243	249	255	261	267	274	282	288	296	304	311	320	328	337

		80–120	136	135	134	133	132	131	130	129	128	127	126	125	124	123	122	121
68DLH13	37	68	35000	231	234	237	241	244	248	252	255	259	263	267	271	275	279	284	288
				121	124	127	130	133	135	138	142	145	149	152	155	159	164	168	171
68DLH14	40	68	40300	266	269	273	277	281	286	290	294	299	303	308	312	317	322	327	332
				130	133	135	138	141	145	148	152	155	159	163	167	171	175	179	184
68DLH15	40	68	45200	294	299	303	308	312	317	322	327	332	337	343	348	354	360	365	372
				145	148	152	155	158	162	166	170	174	178	182	187	191	196	201	206
68DLH16	49	68	53600	349	354	360	365	371	376	382	388	394	400	407	413	420	427	433	441
				171	174	178	182	186	190	195	199	204	209	214	219	225	230	236	242
68DLH17	55	68	60400	397	403	408	414	420	427	433	439	446	453	460	457	474	481	489	497
				194	198	203	208	212	217	222	228	232	238	244	249	256	262	268	275
68DLH18	61	68	69900	459	465	472	479	486	493	501	508	516	524	532	540	549	557	566	575
				219	225	230	234	240	246	251	257	263	269	276	283	289	297	304	311
68DLH19	67	68	80500	525	532	540	548	557	565	574	583	592	601	611	621	631	641	651	662
				248	254	260	266	272	278	285	291	298	305	313	320	328	336	344	353

Standard Load Table
Deep Longspan Steel Joists, DLH-Series (Continued)

Joist Desig-nation	Approx. Wt. in Lbs per Linear Ft. (Joists only)	Depth in Inches	Safe[b] Load in Lbs. Between 84–128	Clear Span in Feet															
				129	130	131	132	133	134	135	136	137	138	139	140	141	142	143	144
72DLH14	41	72	39300	303	298	294	290	285	281	277	274	270	266	262	259	255	252	248	245
				171	167	163	159	155	152	149	146	143	139	136	133	131	128	125	123
72DLH15	44	72	44900	347	342	336	331	326	322	317	312	308	303	299	295	291	286	282	279
				191	187	183	178	174	171	167	163	160	156	152	150	147	143	140	137
72DLH16	50	72	51900	401	395	390	384	378	373	368	363	358	353	348	343	338	334	329	325
				225	219	214	209	205	200	196	191	188	183	179	175	171	169	165	161
72DLH17	56	72	58400	451	445	438	432	426	420	414	408	402	397	391	386	381	376	371	366
				256	250	245	239	233	228	224	218	213	209	205	200	196	191	188	184
72DLH18	59	72	68400	528	520	512	505	497	490	483	479	470	463	457	450	444	438	432	426
				289	283	276	270	265	258	252	247	242	236	231	227	222	217	212	209
72DLH19	70	72	80200	619	609	600	591	582	573	565	557	549	541	533	526	518	511	504	497
				328	321	313	306	300	293	286	280	274	268	263	257	251	247	241	236

[a]The Steel Joist Institute publishes both specifications and load tables, each of which contains standards to be used in conjunction with the other.

[b]The safe uniform load for the clear spans shown in the shaded section is equal to (Safe Load) ÷ (Clear Span + 0.67). [The added 0.67 feet (8 inches) is required to obtain the proper length on which the Load Tables were developed.]

In no case shall the safe uniform load, for clear spans less than the minimum clear span shown in the shaded area, exceed the uniform load calculated for the minimum clear span listed in the shaded area.

To solve for *live* loads for clear spans shown in the shaded are (or lesser clear spans), multiply the live load of the shortest clear span shown in the Load Tables by (the shortest clear span shown in the Load Table + 0.67 feet)^2 and divide by (the actual clear span + 0.67 feet)^2. The live load shall *not* exceed the safe uniform load.

ANSWERS TO PROBLEMS

Chapter 2

1. $R_L = 7.0$ kips, $R_R = 6.5$ kips.

2. $R_L = 19.0$ kips, $R_R = 17.0$ kips.

3. $R_L = 5.5$ kips, $R_R = 20.3$ kips.

8. Shear at left end $= +7$ kips; at 6 ft from left end $= +7$ kips and -2 kips; at 14 ft from left end $= -2$ kips and -6.5 kips; at right end $= 6.5$ kips. Bending moment at left end $=$ zero; at 6 ft from left end $= +42$ ft-kips; at 14 ft from left end $= +26$ ft-kips; at right end $=$ zero.

9. Shear at left end $= +19$ kips; at 5 ft from left end $= +19$ kips and -1 kip; at 10 ft from left end $= -1$ kip, increasing linearly to -17 kips at right end. Bending moment at left end $=$ zero; at 5 ft from left end $= +95$ ft-kips; at 10 ft from left end $= +90$ ft-kips; at right end $=$ zero.

10. Shear at left end $= +5.5$ kips; at 5 ft from left end $= +5.5$ kips and -9.5 kips; at right reaction $= -9.5$ kips and $+10.8$ kips; at right end $= 6.0$ kips and zero. Bending moment at left end $=$ zero; at 5 ft from left end $= +27.5$ ft-kips; at right reaction $= -67.2$ ft-kips; at right end $=$ zero. Also, point of zero moment is at 7.9 ft from left end.

20. $R_L = 9.53$ kips, $R_R = 7.87$ kips; load equations:

$$L_{(0\text{-}4)} = -0.8, \qquad L_{(4\text{-}18)} = -0.8$$

shear equations:

$$V_{(0\text{-}4)} = -0.8x + 9.53, \qquad V_{(4\text{-}18)} = -0.8x + 6.53$$

moment equations:

$$M_{(0\text{-}4)} = -0.4x^2 + 9.53x$$
$$M_{(4\text{-}18)} = -0.4x^2 + 6.53x + 12.0$$

Maximum bending moment occurs at point where shear $=$ zero, or 8.15 ft from the left reaction. Shear at left end $= +9.53$ kips; at 4 ft from left end $= +6.33$ kips and $+3.33$ kips; at right reaction $= -7.87$ kips. Bending moment at left end $=$ zero; at 4 ft from left end $= +31.8$ ft-kips; at 8.15 ft from left end $= +38.6$ ft-kips (maximum value); at 14 ft from left end $= +25.4$ ft-kips; at right end $=$ zero.

22. Partial answer: maximum moment at 7.5 ft from left end. Maximum moment = 21 ft-kips.

Chapter 3

Following Art. 3.9:

1. $W = 46.9$ kips.

3. **W** 21×62, **S** 20×75.

4. **W** 10×12.

Following Art. 3.15:

1. $f_v = 6.47$ ksi, **W** = 102 kips.

2. (2) $f_v = 2.84$ ksi.

(3) $f_v = 4.16$ ksi for the **W** 21×62.
$f_v = 2.75$ ksi for the **S** 20×75.

(4) $f_v = 1.62$ ksi.

Chapter 4

1. (a) $El\Theta_{(0\text{-}10)} = 3.33x^2 - 555$.
$El\Theta_{(10\text{-}30)} = -1.67x^2 + 100x - 1055$.
$El\Delta_{(0\text{-}10)} = 1.11x^3 - 555x$.
$El\Delta_{(10\text{-}30)} = -0.56x^3 + 50x^2 - 1055x + 1660$.

(b) Maximum deflection occurs at 13.6 ft from left reaction.

(c) Maximum deflection = -0.85 in.

3. (a) $\Delta = 0.66$ in.

(b) $\Delta = 0.36$ in.

4. $\Delta = 1.15$ in.

6. **W** 10×68.

Chapter 5

Following Art. 5.4:

1. **W** 18×35.

2. **W** 21×44.

3. **W** 14×38.

4. **W** 14×22 (no change).

5. **W** 14×26.

6. **W** 16×31 (required $S = 34.4$ in.^{3}; I between supports = 282 in.^{4}; I for over-hang = 370 in.^{4}).

Following Art. 5.7:

1. (a) Beam is safe.

(b) $f_b = 22.3 > 22$ ksi. Beam is marginal but unsafe.

2. (a) $W = 47.7$ kips.

(b) $W = 10.5$ kips.

5. **W** 8×21.

7. **W** 16×50.

Following Art. 5.10:

2. $N = 2.39$ (use 3 in.).

3. **W** 16×45 or 21×44.

4. **W** 16×31.

6. **W** 14×34.

8. **W** 18×40.

Following Art. 5.19:

2. $W = 20.2$ kips. $\Delta = 0.74$ in.

6. $f_b = 24.1$ ksi, $F_b = 17.8$ ksi (not safe).

7. **PL** $8 \times 11 \times \frac{5}{8}$ in.

8. **W** 6×15.

Chapter 6

Following Art. 6.8:

1. $P = 137$ kips.

3. $P = 217$ kips.

5. $P = 317$ kips.

Following Art. 6.13:

1. **W** 8×40.

2. **W** 6×15.

4. 2 angles $3 \times 2\frac{1}{2} \times \frac{1}{4}$ in. with long legs back-to-back.

5. **W** 10×33.

8. $P = 306$ kips.

Following Art. 6.17:

1. Stability check = 0.836; End check = 0.895. Column is safe.

3. $P = 32$ kips.

5. Stability check = 0.850;
End check = 0.69.
Column is safe.

7. W 12×45.

8. W 10×49.

Following Art. 6.19:

4. Plate $1\frac{1}{4}$ in. thick.

Chapter 7

Following Art. 7.12:

1. $P = 53.6$ kips, end distance = $1\frac{3}{4}$ in.

3. $P = 48$ kips, end distance = $1\frac{1}{8}$ in.

6. Thickness of gusset plate = $\frac{3}{8}$ in. Use three A325**X** bolts in one line. End distance = $1\frac{1}{4}$ in. (angles), $2\frac{1}{4}$ in. (gusset).

Following Art. 7.16:

1. Framing angles $4 \times 3\frac{1}{2} \times \frac{1}{4}$ in. at $5\frac{1}{2}$ in. long. Two bolts in connected leg, 4 bolts in outstanding legs.

2. Use $\frac{3}{4}$ in. bolts.

5. Use $6 \times 4 \times \frac{7}{8}$ in. seat angle on each side with four $\frac{3}{4}$-in. bolts through girder web.

Following Art. 7.20:

1. Total stress = 10.53 kips on extreme right bolt. Connection is safe.

3. Design force = 8.76 kips. Use $\frac{5}{8}$ in. bolts.

Following Art. 7.26:

1. Plate thickness = 0.424 in., use $\frac{1}{2}$ in. Length of overlap = 4.4 in., use $4\frac{1}{2}$ in.

3. $L = 3$ in.

Following Art. 7.29:

1. $R = 26.2$ kips.

2. $D = 0.312$ in., use $\frac{5}{16}$-in. weld.

Following Art. 7.32:

2. (a) $R = 55.1$ kips.
   (b) $R = 56.2$ kips.

3. $R = 35.4$ kips.

4. $R = 42.6$ kips (connected leg controls).

Chapter 8

Following Art. 8.7:

1. (a) ASCE 7-88:
External forces plus internal suction.
Roof pressure $p = 13.2$ psf (outward).
Windward eaves $w = 381$ lb per ft.
Leeward eaves $w = 147$ lb per ft.
External forces plus internal pressure.
Roof pressure $p = 21.6$ psf (outward).
Windward eaves $w = 234$ lb per ft.
Leeward eaves $w = 294$ lb per ft.

   (b) UBC:
Roof pressure $p = 14$ psf (outward).
Wind eaves $w = 255$ lb per ft.
Leeward eaves $w = 179$ lb per ft.

Following Art. 8.11:

1. NEHRP, $V = 75$ kips.
UBC, $V = 90$ kips.

2. Diaphragm load: transverse, $F_Q = 53.6$ kips; longitudinal, $F_Q = 34.7$ kips. Bottom of shear walls: end wall, $V = 20$ kips (each segment); side wall, $V = 9.1$ kips (middle segment). $V = 19.1$ kips (each end segment).

Chapter 9

Following Art. 9.36:

1. Snow load: int. = 2.66 kips; ext. = 1.33 kips.

3. Top chord = 26.6 and 21.5 kips compression. Bottom chord = 24.6 and 14.8 kips tension. Web verticals = 4 kips compression. Web diagonals to ridge = 8 kips tension. Web diagonals near heel = 5.2 kips compression.

6. 2C 8×13.75.

Following Art. 9.40:

1. (a) Member A-1 = 6.7 kips (comp.),
      4-5 = 4.48 kips (comp.).
      Maximum shear in columns = 2 kips.
      Maximum moment in columns = 20 ft-kips.

(c) Member A-7 = 7.0 kips (ten.), 4-5 = 5.37kips (comp.).
Maximum shear in columns = 3 kips.
Maximum moment in columns = 24 ft-kips.

Chapter 10

Following Art. 10.3:

1. End reactions = 13.5 kips. Middle reaction = 45 kips.

3. $M = 152$ ft-kips.

Following Art. 10.5:

1. $M_A = 75.9$ ft-kips, $M_B = 48$ ft-kips.

4. $R_A = 10.5$ kips, $R_B = 41.5$ kips, $R_C = 23.0$ kips. Negative moments: $M_B = 67.5$ ft-kips, $M_C = 30$ ft-kips. Positive moments: M at 5.25 ft from left support = 28 ft-kips, M at 11 ft from middle support = 51 ft-kips.

Following Art. 10.9:

1. (2) $M_A = 156$ ft-kips, $M_B = 101$ ft-kips.
   (3) $M = 126$ ft-kips.
   (4) $M = 67.7$ ft-kips.

3. $M_B = 28.6$ ft-kips, $M_C = 28.6$ ft-kips.

Following Art. 10.10:

1. Member AB: shear = 4 kips, moment = 24 ft-kips, axial load = 15 kips; Member BC: shear = 15 kips, moment = 120 ft-kips, axial load = 4 kips.

3. Member AB and BC: shear = 5.1 kips, moment = 40 ft-kips, axial load = 21.8 kips.

Following Art. 10.15:

1. $V_A = 5.5$ kips, $V_C = 6.5$ kips, $H_A = 2$ kips, $H_C = 6$ kips.
$M_A = 3.3$ ft-kips, $M_B = 23.4$ ft-kips, $M_C = 33.3$ ft-kips.

3. $V_A = 13.7$ kips, $V_D = 16.3$ kips, $H_A = 1.3$ kips, $H_D = 7.3$ kips. $M_B = 51.5$ ft-kips, $M_C = 87.5$ ft-kips, $M = 70.2$ ft-kips at midspan.

Chapter 11

Following Art. 11.2:

1. (a) web $w/t = 20 < 68.7$ OK. Flange $w/t = 28 < 32$ OK.
   (b) $M_b = 95$ ft-kips.
   (c) $M_y = 142.5$ ft-kips.
   (d) $M_p = 185$ ft-kips.
   (e) $l_{cr} = 20$ ft.

2. Major axis: $M_p = 192$ ft-kips, $u = 1.13$. Minor axis: $M_p = 31.5$ ft-kips.

Following Art. 11.5:

1. W 16×26.
2. W 12×22.

Following Art. 11.6:

1. W 16×31.
2. W 16×31.

Following Art. 11.7:

1. (a) W 21×44.
   (b) $L/360 = 1.0$ in.; Maximum Δ occurs at 17.2 ft from interior support and is 0.43 in., satisfactory.

3. Use W 18×50 for all spans. However, other combinations could also work.

5. $L = 21.1$ ft.

Following Art. 11.8:

1. $P = M_p/15$.

Chapter 12

Following Art. 12.5:

2. $R_A = 5.28$ kips, $R_B = 14.31$ kips. $R_C = 8.50$ kips, $R_D = 2.91$ kips. $M_B = -41.28$ ft-kips, $M_C = -19.62$ ft-kips.

Following Art. 12.10:

3. Element stiffness matrix ÷ 1000 (kips and
in.):

$$
\begin{bmatrix}
3.335 & 0.000 & 0.000 & -3.335 & 0.000 & 0.000 \\
0.000 & 0.072 & 3.636 & 0.000 & -0.072 & 3.636 \\
0.000 & 3.636 & 242.440 & 0.000 & -3.636 & 121.220 \\
-3.335 & 0.000 & 0.000 & 3.335 & 0.000 & 0.000 \\
0.000 & -0.072 & -3.636 & 0.000 & 0.072 & -3.636 \\
0.000 & 3.636 & 121.220 & 0.000 & -3.636 & 242.440
\end{bmatrix}
$$

Transformed element stiffness matrix ÷ 1000
(kips and in.):

$$
\begin{bmatrix}
0.454 & -1.094 & -3.418 & -0.454 & 1.094 & -3.418 \\
 & 2.953 & -1.244 & 1.094 & -2.953 & -1.244 \\
 & & 242.440 & 3.418 & 1.244 & 121.220 \\
 & & & 0.454 & -1.094 & 3.418 \\
 & & & & 2.953 & 1.244 \\
\text{Symmetric} & & & & & 242.440
\end{bmatrix}
$$

End actions are as follows: $t_A = 10.67$ kips, $v_A = 3.66$ kips, $m_A = 0.37$ in-kips; $t_B = -10.67$ kips,
$v_B = -3.66$ kips, $m_B = 365.77$ in-kips.

4. End actions converted to structure axes are as follows: $T_A = -7.09$ kips, $V_A = 8.77$ kips, $M_A = 0.37$
in-kips; $T_B = 7.09$ kips, $V_B = -8.77$ kips, $M_B = 365.77$ in-kips.

Following Art. 12.15:

**1. Complete stiffness matrix per 1,000,000 (lb
and in.):**

$$
\begin{bmatrix}
0.64006 & 0.20506 & -0.20506 & -0.20506 & -0.43500 & 0.00000 & 0.00000 & 0.00000 & 0.00000 & 0.00000 \\
0.20506 & 0.20506 & -0.20506 & -0.20506 & 0.00000 & 0.00000 & 0.00000 & 0.00000 & 0.00000 & 0.00000 \\
-0.20506 & -0.20506 & 0.84512 & 0.00000 & -0.20506 & 0.20506 & -0.43500 & 0.00000 & 0.00000 & 0.00000 \\
-0.20506 & -0.20506 & 0.00000 & 0.41012 & 0.20506 & -0.20506 & 0.00000 & 0.00000 & 0.00000 & 0.00000 \\
-0.43500 & 0.00000 & -0.20506 & 0.20506 & 1.28012 & 0.00000 & -0.20506 & -0.20506 & -0.43500 & 0.00000 \\
0.00000 & 0.00000 & 0.20506 & -0.20506 & 0.00000 & 0.41012 & -0.20506 & -0.20506 & 0.00000 & 0.00000 \\
0.00000 & 0.00000 & -0.43500 & 0.00000 & -0.20506 & -0.20506 & 0.84512 & 0.00000 & -0.20506 & 0.20506 \\
0.00000 & 0.00000 & 0.00000 & 0.00000 & -0.20506 & -0.20506 & 0.00000 & 0.41012 & 0.20506 & -0.20506 \\
0.00000 & 0.00000 & 0.00000 & 0.00000 & -0.43500 & 0.00000 & -0.20506 & 0.20506 & 0.64006 & -0.20506 \\
0.00000 & 0.00000 & 0.00000 & 0.00000 & 0.00000 & 0.00000 & 0.20506 & -0.20506 & -0.20506 & 0.20506
\end{bmatrix}
$$

		Member forces (kips)[a]	
Load vector (lb)	Displacement vector (in.)	Member	Force
R_1	0.0000	1	0.50
R_2	0.0000	2	-0.50
1000	0.0088	3	-5.00
-5000	0.0307	4	-6.36
0	0.0011	5	-7.78
0	-0.0359	6	-0.71
1000	-0.0027	7	0.71
-5000	-0.0296		
R_9	0.0000		
R_{10}	0.0000		

[a]Tension +, compression −.

2. Complete stiffness matrix per 1,000,000 (lb and in.):

$$
\begin{bmatrix}
0.136 & 0.000 & -5.437 & -0.136 & 0.000 & -5.437 & 0.000 & 0.000 & 0.000 & 0.000 & 0.000 & 0.000 \\
0.000 & 2.900 & 0.000 & 0.000 & -2.900 & 0.000 & 0.000 & 0.000 & 0.000 & 0.000 & 0.000 & 0.000 \\
-5.437 & 0.000 & 289.985 & 5.437 & 0.000 & 144.993 & 0.000 & 0.000 & 0.000 & 0.000 & 0.000 & 0.000 \\
-0.136 & 0.000 & 5.437 & 1.425 & 0.000 & 5.437 & -1.289 & 0.000 & 0.000 & 0.000 & 0.000 & 0.000 \\
0.000 & -2.900 & 0.000 & 0.000 & 2.912 & 1.074 & 0.000 & -0.012 & 1.074 & 0.000 & 0.000 & 0.000 \\
-5.437 & 0.000 & 144.993 & 5.437 & 1.074 & 418.874 & 0.000 & -1.074 & 64.444 & 0.000 & 0.000 & 0.000 \\
0.000 & 0.000 & 0.000 & -1.289 & 0.000 & 0.000 & 1.329 & 0.000 & 2.417 & -0.040 & 0.000 & 2.417 \\
0.000 & 0.000 & 0.000 & 0.000 & -0.012 & -1.074 & 0.000 & 1.945 & -1.074 & 0.000 & -1.933 & 0.000 \\
0.000 & 0.000 & 0.000 & 0.000 & 1.074 & 64.444 & 2.417 & -1.074 & 322.222 & -2.417 & 0.000 & 96.667 \\
0.000 & 0.000 & 0.000 & 0.000 & 0.000 & 0.000 & -0.040 & 0.000 & -2.417 & 0.040 & 0.000 & -2.417 \\
0.000 & 0.000 & 0.000 & 0.000 & 0.000 & 0.000 & 0.000 & -1.933 & 0.000 & 0.000 & 1.933 & 0.000 \\
0.000 & 0.000 & 0.000 & 0.000 & 0.000 & 0.000 & 2.417 & 0.000 & 96.667 & -2.417 & 0.000 & 193.333
\end{bmatrix}
$$

Load vector (lb. and in.):   Displacement vector (in. and rad.):

$$
\begin{bmatrix}
R_1 \\
R_2 \\
R_3 \\
1500 \\
-5375 \\
-192500 \\
0 \\
-5375 \\
192500 \\
R_{10} \\
R_{11} \\
R_{12}
\end{bmatrix}
\qquad
\begin{bmatrix}
0.0000 \\
0.0000 \\
0.0000 \\
0.0303 \\
-0.0018 \\
-0.00094 \\
0.0283 \\
0.0032 \\
0.00057 \\
0.0000 \\
0.0000 \\
0.0000
\end{bmatrix}
$$

Member end actions

Member	T(A) (kips)	V(A) (kips)	M(A) (kip-ft)	T(B) (kips)	V(B) (kips)	M(B) (kip-ft)
1	5.364	-1.016	2.312	-5.364	1.016	-9.084
2	2.516	5.364	9.084	-2.516	6.136	-14.871
3	6.136	2.516	14.871	-6.136	-2.516	10.286

Index